$x^2 + 9x + 18$

Algebra in Action

A Course in Groups, Rings, and Fields

Pure and Applied
UNDERGRADUATE TEXTS · 27

The Sally
SERIES

Algebra in Action

A Course in Groups, Rings, and Fields

Shahriar Shahriari

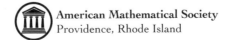

American Mathematical Society
Providence, Rhode Island

2010 *Mathematics Subject Classification.* Primary 00-01; Secondary 20-01, 13-01, 12-01.

The image on the cover is by the author.

For additional information and updates on this book, visit
www.ams.org/bookpages/amstext-27

Library of Congress Cataloging-in-Publication Data

Names: Shahriari, Shahriar.
Title: Algebra in action : a course in groups, rings, and fields / Shahriar Shahriari.
Description: Providence, Rhode Island : American Mathematical Society, [2017] | Series: Pure
  and applied undergraduate texts ; volume 27 | Includes bibliographical references and index.
Identifiers: LCCN 2016036490 | ISBN 9781470428495 (alk. paper)
Subjects: LCSH: Algebra, Abstract–Textbooks. | Algebra–Textbooks. | AMS: General – Instruc-
  tional exposition (textbooks, tutorial papers, etc.). msc | Group theory and generalizations
  – Instructional exposition (textbooks, tutorial papers, etc.). msc | Commutative algebra –
  Instructional exposition (textbooks, tutorial papers, etc.). msc | Field theory and polynomials
  – Instructional exposition (textbooks, tutorial papers, etc.). msc
Classification: LCC QA162 .S465 2017 | DDC 512/.02–dc23 LC record available at https://lccn.loc.
  gov/2016036490

To my partner Nanaz and to our sons Kiavash and Neema

Contents

Preface

For many years, abstract algebra has been one of my favorite classes to teach at Pomona College, and this text has grown out of that experience. My students, by and large, have been eager second- or third-year undergraduates who have had no prior experience with the material. When they start out, they are more or less comfortable with proofs, have had a solid linear algebra class, and are familiar with the arithmetic (but not the theory) of complex numbers. The goal is to give the students a rigorous and motivated introduction to groups, rings, and fields, and to go deep enough into each subject to see the power of abstract thinking and to be convinced that the subject is full of unexpected results. There is more than enough material here for a one-year course, but appropriate selections can be made for a one-semester course as well. While the text is unmistakably for undergraduates and assumes no prior familiarity with the subject, it hopes to nudge students toward thinking like mathematicians by putting a premium on building intuition and by expecting the students to be actively involved in the learning of the material. It has been my experience that after going through this material, the students are amply prepared for graduate level courses in algebra.

Algebraic structures abound in contemporary mathematics, and abstract algebra provides the language for studying them. Consider the following problems:

- You have 47 colors and you want to color a cube by assigning a (not necessarily distinct) color to each face. How many "different" colorings are possible? If you can get from one coloring to another by rotating the cube, then count the two colorings as the same. (Page 158)

- Can you find the solutions to $x^5 - 10x + 5 = 0$ precisely? (Example 28.18)

- Can you describe all the integer solutions to $x^2 - 3y^2 = z^2$? Or to $y^3 = x^2 + 4$? (Problem 28.4.10 and Proposition 20.18)

- Each of 100 briefcases contains the name of one contestant. Each of the 100 contestants gets to privately examine the contents of 50 of the 100 briefcases. The contestant will be successful if she/he finds the briefcase that has her/his own name in it. There is no communication between the contestants, but they can agree on a common strategy before the contest. Is there a strategy that with a probability of more than 30% assures the success of *every* one of the contestants? (Problem 3.3.5)

- Can you double a cube? More precisely, given one edge of a cube, can you construct—using a straightedge and compass—another line segment such that a cube with this new line segment as its side will have a volume twice as much as the original cube? (Corollary 23.18)

- Does there exist a real number α such that rational linear combinations of 1, α, α^2, ... include every one of $\sqrt[3]{47}$, $\sqrt[5]{17}$, and $18 - 2\sqrt[7]{19}$? (Question 25.48)

- Let X be a set with 47 elements. Choose 169 one-to-one and onto functions from X to X such that if you compose any two of the functions, you get another function in the set. I predict the following:
  - (a) If f and g are two of your functions, then $f(g(x)) = g(f(x))$ for all $x \in X$.
  - (b) The identity function is the function i such that $i(x) = x$ for all $x \in X$. If f is any of your functions and if you start composing f with itself, then after 169 iterations you will always get i.

Can you prove either prediction? (Problem 6.3.1)

Each of these questions will be answered in this text, but what is somewhat surprising is that abstract algebra provides a common framework for answering them. While it is possible to attack many (but not all) of these problems individually and without recourse to deep theory, an abstract axiomatic development of the properties of algebraic structures will give us the tools and the language necessary to think about them conceptually. The result is a far-reaching, powerful, and—dare I say—beautiful theory.

Historically, different strands have come together to create the common language of algebraic structures that is at the core of modern abstract algebra. One thread is the attempt to solve algebraic equations. In fact, the word "algebra" is from the Arabic "al-jabr" and translates to "completion" or "restoration", referring to moving a negative quantity to the other side of an equation where it becomes positive. In the medieval Islamic world, where algebra started to become a discipline separate from geometry, the central problem was that of solving of equations. The quadratic equation allows us to solve any equation of degree 2, and similar— but more complicated—formulas for cubic and quartic equations were found in sixteenth century Italy. The quest for solving the quintic resulted in deeper studies of permutations and eventually the advent of group theory and the Galois theory of fields. From this beginning, group theory has evolved into mathematicians' preferred language for the study of symmetries in whatever context. A second thread was the investigation of Diophantine equations—that is, finding integer solutions to equations with many variables such as $x^n + y^n = z^n$. This, together with other problems in number theory, led to the desirability of doing arithmetic and number theory with collections of numbers other than integers. Commutative ring theory is what resulted. Having arisen from old historical roots, the methods and techniques of abstract algebra permeate all of modern mathematics.

This text introduces groups, rings, and fields to a student who is seeing these concepts for the first time and yet wants to gain a somewhat sophisticated taste of the material. The choice of material and the mix of results and problems reflects this pedagogical aim. As such the book is not comprehensive, and the proofs are not necessarily the sleekest proofs available. While the text tells an astounding story—starting from very meager beginnings and building a sophisticated edifice— the main task will be for you the reader to engage the material directly. The large selection of problems will facilitate your endeavor. Since you are assumed to be new to abstract algebra, the writing is somewhat conversational and verbose toward the beginning and becomes more terse as the text progresses. An attempt is made to give you a taste of how mathematicians think about the subject, and so, in addition to the usual definitions and theorems, the text tries to help build your intuition for the material. The proofs of some of the theorems are relegated to the problems. This is because proofs are important, and I want you to figure out some of them for yourself. Since the reader is learning this material for the first time, sometimes topics are repeated. A topic may make its first appearance in an exercise followed by a fuller treatment later. Sometimes a more specialized result is presented before the more general result. To facilitate self-study, many problems have hints, some have short answers, and over 100 problems are solved completely. The hints, short answers, and solutions are all at the back of the book. You are urged to start working on a problem without looking at the back. Only when you are truly frustrated—a necessary part of the learning process—should you look to see if there is a hint, an answer, or a solution. If you have done a problem, but there is a solution at the back, then I urge you to read the solution anyway since it may provide a bit of additional insight. The problems that are important to the development of the subject have boldfaced numbers, while those with a complete solution in the back have italicized numbers.

Groups, rings, and fields have much in common, and an important part of the modern treatment of abstract algebra is an emphasis on the similarities of these and other algebraic structures. However—in the last analysis—to get deep and powerful results, you have to go beyond the commonalities. Groups are mathematicians' way of thinking about and working with symmetry; commutative rings came about when a need arose to do arithmetic in more general settings than the integers; and field theory originated in the pursuit of solving polynomial equations in one variable. While one of the stories of this book is that all of these things are related to each other, another part of the narrative develops the distinct personalities of each of groups, rings, and fields. As a student of the subject, you need to develop different and separate intuition for each of the structures. It is possible, in an introductory class, to give a survey of these subjects, focusing on their common aspects, and not go too deeply into any of them. My approach has been to go far enough in each topic to showcase some aspect of the deeper theory while constantly bringing out the commonalities.

To the Instructor. The group theory portion of the text has three somewhat unusual features. If you use this text, it will be very hard to avoid the first feature (and you may be better off with a different text if you don't buy into this approach), but the other two features are quite optional.

First, group actions are introduced very early. In the mathematical world outside group theory, groups appear and show their properties when they act on other objects. Group actions not only get the students to look at groups as "groups of symmetries", they bring much rhyme and reason to the study of group theory itself. Many important subgroups are stabilizers of actions, the orbits of an action provide a systematic way of partitioning interesting sets, and much of introductory group theory can be organized—as you can see from the table of contents—around various actions of groups on groups. In addition, in Galois theory, studying the action of the Galois groups on the roots of an equation becomes central. For an introductory class, the early introduction of actions may seem as overburdening the students with another level of abstract constructions. This may be true, but what is gained in perspective and intuition is well worth the price. In fact, I believe that actions actually make group theory easier.

Second, Hasse diagrams of posets—lattice diagrams in most cases of interest here—and homomorphism diagrams (instead of exact sequences) are introduced, and students are encouraged to use them to visualize what is going on and to help in arguing proofs. It has been my experience that if you gain facility with these diagrams—which are ubiquitous in notebooks of professional mathematicians—many statements and many proofs turn from abstract and mysterious to straightforward arguments.

Third, normal subgroups, quotient groups, and homomorphisms are introduced somewhat late. Homomorphisms are defined early, but their serious treatment waits until Chapter 11. As I will explain, I have reasons for doing so, but, if you prefer, you can easily change course. One could go to normal subgroups and Chapter 10 right after Chapter 6. Normal subgroups and homomorphisms are very important and their study is at the core of group theory. However, my experience has been that they are also difficult concepts for the first time learner, and much is to be gained if the student develops a variety of intuitions about groups before tackling these concepts. When I teach abstract algebra at Pomona College, I follow the order of this book. Hence, the students will see alternating groups and Sylow theorems early. These give the students a feel for finite group theory and allow the construction of many examples. In addition, the students will have worked with orbits of actions extensively. By the time they are asked to consider quotient groups, the construction will almost seem natural.

Because of these features, the instructor has to be careful not to get bogged down in the first few chapters. There is much material here but you can move briskly. In fact, the writing is meant to be read by the students, and this should help the instructor move through the introductory material more quickly. In my own teaching, I often have students read a section and do some of the more computational problems before I discuss the topic in class, and, in fact, I leave the development of some topics entirely to the students. Reading a mathematics text is an important skill, and my hope is that the many remarks and expository discussions will be helpful in this regard.

To give you a sense of what I do with the material, in the first semester of the course I cover group theory (skipping Sections 5.3, 7.3, 11.6, and 12.4 and Chapters 13 and 14) and ring theory (through Chapter 18 proving ED $\Rightarrow$ PID $\Rightarrow$ UFD, but skipping Sections 16.4, 17.2, and 18.5). In the second semester I start with Chapter 19 on polynomial rings and go through Galois theory (skipping Section 19.6 and and Chapter 20, and the discussion of algebraic closures in Section 24.2).

I have kept the main part of the text to material that I want the students to read—any section or chapter that can be skipped in a first reading is marked by an asterisk—but there are many extra problems and mini-projects that can be used to explore topics not covered in the text. Using these and the references provided, the students should be able to design many independent mini-projects. The website for this book (`www.ams.org/bookpages/amstext-27`) maintained by the publisher has a detailed syllabus for a year-long course based on this text and other supplementary material.

The three parts—groups, rings, and fields—presented here constitute the first volume of an eventual two-volume text. The second volume will cover modules over a PID, an introduction to algebraic geometry via Gröbner bases, and representation theory.

Acknowledgments. I have collected the material for this book over the course of many years, and, as a result I am indebted to many mathematicians and many books. However, I learned algebra primarily from Marty Isaacs at the University of Wisconsin–Madison. I went to Madison with no particular interest in algebra, but Marty's graduate course in algebra (which years later became Isaacs [**Isa94**]) was a revelation. All of a sudden, not only could I follow the individual steps in the arguments, but the questions, the techniques, and the whole enterprise made sense. A small part of the attraction was Marty's emphasis on group actions and on lattice diagrams. Marty's indirect and direct influence can be seen on every page of this book. If I have been able to transmit even a small part of the excitement that I felt when taking his class, then I will claim the book a success. In addition to Marty's classes and books, over the years I have relied on the many wonderful texts on abstract algebra. Some of my favorites are Herstein [**Her75**], Hadlock [**Had78**], Hartley and Hawkes [**HH76**], Stewart [**Ste15**], Dummit and Foote [**DF04**], Bhattacharya, Jain, and Nagpaul [**BJN94**], and Goodman [**Goo98**]. I also want to thank my many students. Not only have they constantly alerted me to typos and mistakes, but their enthusiasm, engagement, and positive feedback convinced me to write the text. It is a cliché to say that the book would not have been possible without the support of my family. But it is true. The book is dedicated to my partner Nanaz and our sons Kiavash and Neema who heard the excuse "I am writing a book" way too often. Finally, I acknowledge my late father Parviz Shahriari, who, as my high school algebra teacher in 10th grade, got me interested in mathematics and whose many books, such as "Raveshhaye Jabr" [**Sha70**], made high school algebra actually fun.

Part 1

(Mostly Finite) Group Theory

Chapter 1

Four Basic Examples

...where, prior to defining groups, four already familiar examples of groups are explored, the common underlying structure is revealed, and, as a bonus, some possibly familiar prerequisites on symmetries, functions, integers, and matrices are reviewed.

Abstract algebra is the systematic study of *algebraic structures*, and the first algebraic structure that we will study is something called a *group*. A group is a set of elements together with an operation on the elements, and it has the added requirement that the operation must follow a few (innocuous looking) rules. As such, a group is both abstract—we will not specify what the elements of the set are, they could be anything—and general. Being so general, at least theoretically, should make the results applicable in a wide range of situations. Two questions remain. The first is whether this is an object worth studying. Before we embark on studying groups, we need to know that groups appear and play a role in our mathematical lives. This would help convince us that we want to know groups better. The second question goes in the opposite direction. If our definition is so general—which is what allows for its ubiquitous appearance—then how can we possibly say and prove anything intelligent about it? The first question—where can we find groups in mathematics—will be partially addressed in this chapter. We will start with four seemingly different examples and see that we can approach all of them through a common abstract lens. These examples will be the basis of our definition of a group. Each example will have its own context, but we shall see that much can be captured by the idea of a set together with an operation that follows certain rules. The second question—can we really prove anything profound about such a general object—is in some sense the subject of the whole first part of the book. We hope to convince (and surprise) you that the answer is an emphatic *YES*.

Each of the four sections in this chapter introduce a mathematical object that you may have seen before. In each case, we identify the set to be studied, and we see a natural "operation" on the elements. We will also abstract a set of rules that this operation follows. Noticing that the rules—at least the rules that we have chosen to highlight—are identical in the four cases, we will then, in the next chapter, abstract the notion of a group and begin a rigorous study of its properties. All throughout the study of group theory, we will come back to the examples from this chapter to investigate our ideas and illustrate our results.

There is one other ulterior motive for the examples in this chapter. They will allow us to standardize our notation and "be on the same page" on some background material (e.g., symmetries, bijective functions, elementary properties of integers, invertible matrices).

Group theory is the study of symmetries, and we will start by looking at symmetries of geometrical objects.

1.1. Symmetries of a Square

Suppose we remove a square from a plane, move it in some way, then put the square back into the space it originally occupied. We want to describe in some reasonable fashion all the ways this can be done. To keep track of what we have done to the square, we will number its corners. As an example, in Figure 1.1 we rotated the square 90 degrees counterclockwise.

Figure 1.1. The 90 degree counterclockwise rotation of a square

We could also rotate the square 450 degrees (counterclockwise). However, both the 90 degree rotation and the 450 degree rotation—as well as a clockwise 270 degree rotation—give the same final result, i.e., if we start with squares in the same initial position, they will end up in the same final position after these moves. Thus, we think of these—the 90 and 450 degree counterclockwise rotations as well as the 270 degree clockwise rotation—as the same. They are not really different.

Each of the possible moves is called a (*rigid*) *symmetry* of the square.

How many of these moves are there? There are four possible locations for the corner labeled 1, and two different possibilities for everything else (once we have decided on the location of one of the corners). Thus, there are $2 \times 4 = 8$ symmetries of the square. In Figure 1.2, we have depicted these symmetries visually, and we have given the following names to the eight symmetries of the square:

$$R_0, R_{90}, R_{180}, R_{270}, H, V, D, D'.$$

A reflection about the horizontal axis—the axis that goes through the middle of opposite sides—is denoted by H. A reflection about the vertical axis (through the middle of opposite sides) is denoted by V. Two other reflections, about the two diagonal axes (through opposite vertices) are denoted by D and D'. The first one, D, is the reflection about the main (matrix) diagonal, and D' is the reflection about the diagonal axis through the bottom left and the top right corners. Finally, the symbol R_θ denotes a counterclockwise rotation of θ degrees, with the center

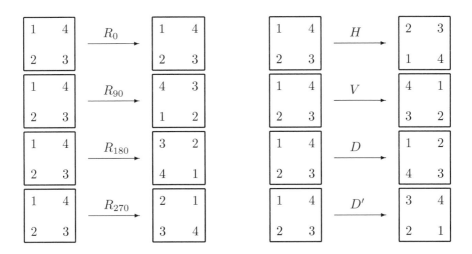

Figure 1.2. The symmetries of a square

of rotation at the center of the square. Note that, in our list, we have included $R_0 = R_{360}$. This is the same as doing nothing. You take the square out of its place and put it back exactly the same way.

Combining Symmetries. We can combine two symmetries of a square by performing them one after the other. For example, doing R_{90} and then V is the same as D' (see Figure 1.3). We think of this operation as a form of "multiplication", and we write it as

$$V R_{90} = D'.$$

Remark 1.1. Note that when we write $V R_{90}$, we mean do R_{90} first, and then do V. This is consistent with the more common notation for function composition. When we write fg, usually we mean do g first and then f, i.e., $fg(x) = f(g(x))$. Actually, for the purposes of abstract algebra, it would make as much (and it could be argued more) sense to define fg to mean to first do f and then g. For the sake

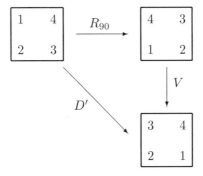

Figure 1.3. Combining symmetries

of not confusing the reader, we will not do so here, but you should be aware that in many advanced algebra books this alternate notation is used.

When we focus not on just one symmetry of the square but on the complete set of the symmetries together with the operation of combining them, we have our first example of a group. For completeness, we will state this as a definition.

Definition 1.2 (The dihedral group). The set of symmetries of a square, together with the "multiplication" defined above, is called the *dihedral group of order* 8 and is denoted by D_8.

In other words, as a set $D_8 = \{R_0, R_{90}, R_{180}, R_{270}, H, V, D, D'\}$. But D_8 is more than a set. It is a set with a *nice* operation.

Remark 1.3. As we shall see, the operation on D_8 has some nice properties. But it does not have all the properties of the usual multiplication of numbers. For example (see Figure 1.4) $R_{90}H = D'$, but $HR_{90} = D$, and so $R_{90}H \neq HR_{90}$.

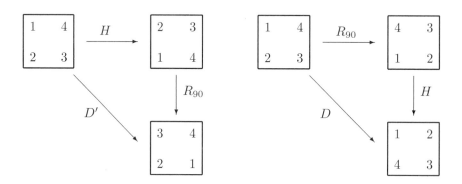

Figure 1.4. The multiplication in D_8 is not commutative.

Now that we have a multiplication for our set, we can create a multiplication table for D_8. We have started such a table in Figure 1.5. You may want to find some more of the entries before proceeding.

Historically, the introduction of symbols and using algorithms for manipulating the symbols was an important step in the development of algebra. Following that model, we also can simplify our calculations of multiplications of the symmetries.

We will denote R_0 by e and R_{90} by a. We do not need a new symbol for R_{180} since $R_{180} = R_{90}R_{90} = aa$, and we can denote this by a^2. Likewise, $R_{270} = aaa = a^3$. But then what is a^4? Here we see a difference with the algebra of numbers. For $a = R_{90}$, we have $a^4 = R_{360} = R_0 = e$. Higher powers of a will continue to give us a, a^2, a^3, or e. Hence we will never get any of the reflections by just using a and e. So, we let b denote H. Then we have $b^2 = e$. So far, we have introduced three symbols, e, a, and b. We found other elements by finding powers of these. We can also multiply the various elements. Multiplying by e has no effect, since $ex = xe = x$ for all $x \in D_8$. But, multiplying a and b does give us some of the

	R_0	R_{90}	R_{180}	R_{270}	H	V	D	D'
R_0	R_0							
R_{90}	R_{90}	R_{180}			D'		H	
R_{180}	R_{180}	R_{270}			V			
R_{270}								
H		D						
V								
D								
D'								

Figure 1.5. The (to-be-completed) multiplication table for D_8. Note that the entry in row labeled H and column labeled R_{90} is D since $HR_{90} = D$, and this means that R_{90} followed by H is D.

other elements of D_8. You can check that $ab = D'$, $a^2b = V$, and $a^3b = D$, and as a result we now have all the eight elements of D_8:

$$D_8 = \{e, a, a^2, a^3, b, ab, a^2b, a^3b\}.$$

But we also had seen that $ba = D$, and thus in working with the symmetries of a square, we also have $ba = a^3b$. This equation—as well as equations such as $a^4 = e$ and $b^2 = e$—is called a *relation* in D_8. There are, of course, many other relations, but, as will become clear later on, the rest of the relations are consequences of the ones we already have. For example, $a^5 = a$. However, we could derive this as $a^5 = a^4a = ea = a$. Since a group is the combination of a set and an operation, it will be more complete to include the necessary relations. Hence we write

$$D_8 = \{e, a, a^2, a^3, b, ab, a^2b, a^3b \mid a^4 = e, b^2 = e, ba = a^3b\}.$$

We should really include the relation $ex = x = xe$ for all $x \in D_8$, but we treat the element e as special. It is called the *identity element* of the group, and it has the property—which will be assumed when we write a set of elements and their relations—that $ex = x = xe$ for all $x \in D_8$.

We can condense the notation further by writing

$$D_8 = \langle a, b \mid a^4 = b^2 = e, ba = a^3b \rangle.$$

This is read as D_8 is *generated* by a and b with the given relations. This means that D_8 is the collection of all words in a and b—a word in a and b is a finite string of a's and b's such as $a^3ba^2b^4a$—subject to the given relations. It is always implicitly assumed that $ex = xe = x$ for all $x \in D_8$.[1]

With our new notation at hand, we can find products in D_8 algorithmically and without resorting to geometric drawings. We present two examples.

Example 1.4. If we do a series of symmetries of the square one after the other, by necessity—since D_8 is a list of *all* the symmetries of a square—the combined effect is that of one of the eight symmetries in D_8. For example, what is H followed by R_{90} followed by H and followed by R_{90}? Using our notation in terms of generators

---

[1]In this text, whenever we consider D_8, the symbols a and b stand for R_{90} and H, respectively. It makes more sense to remind the reader of this meaning of a and b every time, but we use D_8 as an example so often that fixing a and b as R_{90} and H will be handy.

and relations, we calculate (the reader should check the result of this algebraic manipulation by drawing diagrams)

$$R_{90}HR_{90}H = abab = a(ba)b = a(a^3b)b = (aa^3)(bb) = a^4b^2 = ee = e = R_0.$$

Example 1.5. What is R_{180} followed by V?

$$
\begin{aligned}
VR_{180} &= (a^2b)a^2 = a^2(ba)a = a^2(a^3b)a = (a^2a^3)(ba) \\
&= a^5(a^3b) = (a^5a^3)b = a^8b = b = H.
\end{aligned}
$$

In doing the calculations in Examples 1.4 and 1.5, in addition to the relations in D_8 and the special property of the identity element (that is, $ex = xe = x$ for all $x \in D_8$), we used the associative property (that is, $(xy)z = x(yz)$ for all $x, y, z \in D_8$). Informally, if x, y, and z are symmetries of a square, then both $(xy)z$ and $x(yz)$ are the same as doing z, then y, and finally x. Hence, they are equal.

A final property of the elements of D_8 is that of invertibility. The effect of every element of D_8 on the square can be "undone". In other words, you can follow each of the symmetries with another symmetry so that the combined effect is that of e, doing nothing. To repeat, if x is an element of D_8, then there is another element y of D_8 such that $xy = yx = e$. We usually denote such a y as x^{-1} and call it the *inverse* of x. In other words, an element y—while keeping its own perfectly acceptable name y—can also be x^{-1} for some other element x.

Example 1.6. In D_8, we have $a^3 = a^{-1}$, and $b = b^{-1}$. This means that, based on the context, we may refer to R_{270} as a^3 or as a^{-1}. Both of these designations are referring to the same element.

Summarizing and thinking a bit more abstractly, we list the four crucial—but quite mundane looking—properties of the multiplication in D_8.

Properties of Multiplication in D_8

(a) **Closure.** If $x, y \in D_8$, then $xy \in D_8$.

(b) **Associativity.** If $x, y, z \in D_8$, then $(xy)z = x(yz)$.

(c) **Identity.** One of the elements of D_8, that is e, is special. It has the property that $xe = ex = x$, $\forall x \in D_8$.[2]

(d) **Inverses.** For every element $x \in D_8$, there exists another element $y \in D_8$ such that $xy = yx = e$.

We have listed only a select number of properties of D_8. For our purposes, these happen to be the crucial ones. We will see, in the following sections, a number of other examples where we have a set and an operation on the set with the same exact properties as above. We will then define a *group* as a set with an operation that satisfies the above rules. The examples will have convinced us that the abstract study of a group might be productive.

Note to the Reader: In the Problems sections throughout this book, italicized numbers have a complete solution in Appendix C, bold-faced numbers indicate important results that will be used later, and numbers that are both italicized and

---

[2]When convenient, we use the standard notation $\forall$ to mean "for all", and $\exists$ to mean "there exists".

bold-faced have both characteristics. In addition some problems have hints (see Appendix A) and others have short answers (see Appendix B).

Problems

1.1.1. Complete the multiplication table for D_8. Find at least one interesting pattern in the table.

1.1.2. Consider the multiplication table for D_8. Does any element of D_8 appear more than once in any given row? Can you prove your assertion without resorting back to the table?

1.1.3. List the symmetries of an isosceles triangle.

1.1.4. (a) List the symmetries of a rectangle.
    (b) Write the multiplication table for the symmetries of a rectangle.

Center and Centralizer

Definition 1.7 (Center of D_8). We say x and y *commute* if $xy = yx$. The *center* of D_8, denoted $\mathbf{Z}(D_8)$, is a set consisting of all elements in D_8 that commute with every element of D_8.

Definition 1.8 (The centralizer of an element). If $x \in D_8$, the *centralizer* of x, denoted by $\mathbf{C}_{D_8}(x)$, is the set of all elements in D_8 that commute with x.

1.1.5. (a) Find the center of D_8.
    (b) Find $\mathbf{C}_{D_8}(R_{90})$ and $\mathbf{C}_{D_8}(H)$.

1.1.6. Let D_6 denote the set of symmetries of an equilateral triangle. Find the multiplication table for D_6. What is the center of D_6?

1.2. 1-1 and Onto Functions

Functions appear everywhere in mathematics, and 1-1 and onto functions are especially important. Our second example of a group will be the set of all 1-1 and onto functions from a set to itself.

We are assuming that the reader is familiar with the basic concepts of a map—in this book "map" is synonymous with "function" and "mapping"—an image of a map, a 1-1 map, an onto map, and function composition. For completeness, we give the necessary definitions below, but we will go through this introductory material rather quickly.

Definition 1.9 (Maps, 1-1 and onto). Let X and Y be sets. A *map* (or a *mapping* or a *function*) f from X to Y, denoted by $f : X \rightarrow Y$, is a rule that assigns to each element of X precisely one element of Y. The sets X and Y are respectively called the *domain* and the *codomain* (or *target set*) of f.

A mapping $f : X \to Y$ is *1-1* (or *injective*) if, for every $x_1, x_2 \in X$,

$$f(x_1) = f(x_2) \quad \Rightarrow \quad x_1 = x_2.$$

A mapping $f : X \to Y$ is *onto* (or *surjective*) if, for every $y \in Y$, there exists an $x \in X$ with $f(x) = y$. A map that is both 1-1 and onto is called a *bijection* or a *1-1 correspondence*.

Example 1.10. If the sets X and Y are both the set of positive real numbers $\mathbb{R}^{>0}$, then the map $f : X \to Y$ defined by $f(x) = x^2$ is both 1-1 and onto. However, if X and Y were the set of all real numbers $\mathbb{R}$, then the map $f : X \to Y$ given by the same formula is neither 1-1 nor onto. (Why?)

Remark 1.11. To give a complete definition of a map f, you need to specify both the domain X and the codomain Y as well as the rule for assigning each element of X to an element of Y. So, for example, unless the context makes the domain and the codomain clear, it is not enough to say "Consider the function $f(x) = x^2$." Also, in a properly defined function, such as $g : \mathbb{R} \to \mathbb{R}$ defined by $g(x) = x^2$, the name of the function is g and *not* $g(x)$. The expression $g(x)$ denotes the value of the function g at x (where x is an element of the already specified domain), and the equation $g(x) = x^2$ or the notation $x \mapsto x^2$ gives the rule that defines the function.

Definition 1.12 (Function composition). Let X, Y, and Z be sets, and assume $f : X \to Y$, $g : Y \to Z$ are maps. Then a new map $gf : X \to Z$ is defined[3] by

$$gf(x) = g(f(x)) \quad \text{for} \quad x \in X.$$

The function gf is called the *composition* of f and g.

We depict function composition with the diagram in Figure 1.6. Such a figure is an example of a commutative diagram.

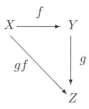

Figure 1.6. Function composition as a commutative diagram

Definition 1.13 (Commutative diagrams). Given three sets X, Y, and Z and maps $f : X \to Y$, $g : Y \to Z$, and $h : X \to Z$, we say that the triangle in Figure 1.7 *commutes* or that it is a *commutative diagram* if $h = gf$. Likewise, the square in Figure 1.7 is a commutative diagram if $gf = kh$.

More generally, a diagram of sets and maps is a commutative diagram if following two paths through the diagram, from any of the sets to any of the other sets, gives the same result (by function composition).

---

[3]You may be used to the notation $g \circ f$ for gf.

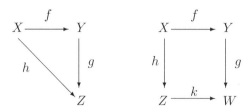

Figure 1.7. The triangle and the square are commutative diagrams if and only if $h = gf$ and $gf = kh$, respectively.

Example 1.14. Let $\mathbb{R}$ and $\mathbb{R}^2$ denote the set of real numbers and the set of points in the plane, respectively. Let $f\colon \mathbb{R}^2 \to \mathbb{R}$ and $\lambda\colon \mathbb{R} \to \mathbb{R}^2$ be defined by

$$f(x,y) = \frac{2xy}{x^2 + y^2}, \text{ and } \lambda(t) = (\sin(t), \cos(t)).$$

If $g\colon \mathbb{R} \to \mathbb{R}$ is defined by $g(t) = \sin(2t)$, then the diagram in Figure 1.8 commutes. This is because

$$(f\lambda)(t) = f(\lambda(t)) = f(\sin(t), \cos(t)) = \frac{2\sin(t)\cos(t)}{\sin^2(t) + \cos^2(t)} = \sin(2t) = g(t).$$

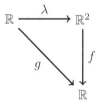

Figure 1.8. A commutative diagram of maps

Algebra is the study of structures, and, as such, instead of studying one object with certain properties, we often focus on the collection of all objects with the given properties. Looking at the whole may reveal connections and allow constructions that would not have been apparent if we considered only one object at a time. While the study of a single bijective map may have value, we turn to the collection of *all* bijective maps on a given set.

Definition 1.15 (Perm(Ω))**.** For any (possibly infinite) set Ω, define

$$\text{Perm}(\Omega) = \{\, f\colon \Omega \to \Omega \mid f \text{ is 1-1 and onto}\}.$$

Thus Perm(Ω) (where Perm stands for permutations) is the set of all bijective maps on Ω. (Other authors use Sym(Ω) or S_Ω for what we have called Perm(Ω).) Our immediate aim is to see that the set Perm(Ω) has a structure similar to that of the symmetries of a square. First, we have to have a "multiplication" for this set. The next theorem says that function composition provides us with an operation on Perm(Ω), and, hence, we can think of function composition as the "multiplication" for Perm(Ω).

Theorem 1.16. *Let X, Y, and Z be sets, and assume $f\colon X \to Y$, and $g\colon Y \to Z$.*

(a) *If f and g are 1-1, then so is gf.*

(b) *If f and g are onto, then so is gf.*

In particular, for a set Ω if $f, g \in \mathrm{Perm}(\Omega)$, then $gf \in \mathrm{Perm}(\Omega)$.

Proof. First note that $gf : X \to Z$ is a map from X to Z.

(a) Assume that $gf(x_1) = gf(x_2)$. This means that $g(f(x_1)) = g(f(x_2))$ which implies that $f(x_1) = f(x_2)$ since g is a 1-1 function. However, f is also a 1-1 function, and so $x_1 = x_2$, proving that gf is also 1-1.

(b) Let $z \in Z$ be an arbitrary element. The map g is onto, and thus there exists an element $y \in Y$ such that $g(y) = z$. Since f is onto, we have an element $x \in X$ such that $f(x) = y$. Now $gf(x) = g(f(x)) = g(y) = z$, and thus gf is onto. $\square$

Now that function composition is a multiplication for $\mathrm{Perm}(\Omega)$, we can, just as in the case of the symmetries of a square, use exponents to denote repeated multiplication. Hence, if Ω is a set and $f \in \mathrm{Perm}(\Omega)$, we use f^2 for ff and f^3 for fff, and so on.

Among the symmetries of a square, we had the special element $R_0 = e$, the identity element. The "identity function" is the corresponding element of $\mathrm{Perm}(\Omega)$.

Definition 1.17 (The identity map). Let Ω be an arbitrary set. The mapping $1_\Omega : \Omega \to \Omega$ defined by $1_\Omega(x) = x$ is called the *identity map* on Ω.

Just as in D_8, clearly, $f1_\Omega = f = 1_\Omega f$ for every set Ω and $f \in \mathrm{Perm}(\Omega)$. We now turn to inverses.

Remark 1.18. As you may have noticed, we choose much of the notation in analogy with usual multiplication. The notation 1_Ω has two components, 1 and Ω. The 1 is meant to remind you of the number 1 which is the identity for multiplication of numbers. The Ω is the domain of the function and, in some sense, it is where the function lives.

Definition 1.19 (Inverse of a map). Let X and Y be sets, and let $f\colon X \to Y$ and $g\colon Y \to X$ be mappings.

We say g is the *inverse* of f if $fg = 1_Y$, and $gf = 1_X$. Such a g—if it exists—is denoted by f^{-1}.

Theorem 1.20. *Let X and Y be sets. Then a map $f\colon X \to Y$ has an inverse if and only if f is 1-1 and onto.*

Proof. ($\Rightarrow$) Assume that f has an inverse g. Thus $g : Y \to X$, and $fg = 1_Y$ and $gf = 1_X$.

Claim 1: f is 1-1.

Proof of Claim 1: Assume that $f(x_1) = f(x_2)$. Apply g to both sides and use the fact that $gf = 1_X$ to get

$$x_1 = gf(x_1) = gf(x_2) = x_2.$$

CLAIM 2: f is onto.

PROOF OF CLAIM 2: Let y be an arbitrary element of Y. We need to find an element of X that is mapped onto y. Let $x = g(y)$. Now using the fact that $fg = 1_Y$, we get

$$f(x) = f(g(y)) = fg(y) = 1_Y(y) = y.$$

($\Leftarrow$) Assume that f is 1-1 and onto. Define $g : Y \to X$ as follows: Let $y \in Y$. Since f is 1-1 and onto, there exists a unique element x in X such that $f(x) = y$. Let $g(y) = x$. This rule makes g a map since x is unique. The fact that g is the inverse of f follows directly from the definition of g and the definition of an inverse. (Can you write down the proof? You are asked to do so in Problem 1.2.9.) $\square$

Theorem 1.21. *Let X and Y be sets, and let $f : X \longrightarrow Y$ be a 1-1, onto map. Then $f^{-1} : Y \to X$ is 1-1 and onto also. In particular, for a set Ω, every $f \in \mathrm{Perm}(\Omega)$ has an inverse in $\mathrm{Perm}(\Omega)$.*

Proof. This follows from Theorem 1.20. Problem 1.2.10 asks you to write down a proof. $\square$

To summarize, for the study of 1-1, onto functions on a set Ω, we have defined a new set $\mathrm{Perm}(\Omega)$ together with an operation (function composition). This operation again has the same four properties (closure, associativity, the existence of identity, and the existence of inverses) as the symmetries of geometrical objects.

Proposition 1.22. *Let Ω be any non-empty set. Then the set $\mathrm{Perm}(\Omega)$ together with the operation of function composition has the following properties:*

(a) ***Closure.*** *If σ and τ are in $\mathrm{Perm}(\Omega)$, then so is $\sigma\tau$.*

(b) ***Associativity.*** *If σ, τ, and μ are elements of $\mathrm{Perm}(\Omega)$, then $\sigma(\tau\mu) = (\sigma\tau)\mu$.*

(c) ***Identity.*** *The set $\mathrm{Perm}(\Omega)$ contains the identity map 1_Ω which has the property that $\sigma 1_\Omega = \sigma = 1_\Omega \sigma$ for every $\sigma \in \mathrm{Perm}(\Omega)$.*

(d) ***Inverses.*** *For every $\sigma \in \mathrm{Perm}(\Omega)$ there exists an element $\sigma^{-1} \in \mathrm{Perm}(\Omega)$ with the property that $\sigma\sigma^{-1} = 1_\Omega = \sigma^{-1}\sigma$.*

Cardinality. In this part of the book, we are mostly interested in finite groups— groups that have only a finite number of elements. However, much of the general theory will be developed for all groups, and often we will make an effort to present proofs that can be generalized to infinite settings. For our purposes, we do not need a deep understanding of issues of size—known as cardinality—for infinite sets. However, the following definitions will be useful.

Definition 1.23 ($|A|$)**.** If A is any finite set, then $|A|$ will denote the *number of elements* in A. The (somewhat vague) notation $|A| = \infty$ will be used to indicate that A is an infinite set.

For any two (not necessarily finite) sets A and B, we write $|A| = |B|$, and say that A and B have the same *cardinality*, if there exists a bijection $f : A \to B$.

Note that for two *finite* sets A and B, to say that A and B have the same cardinality is the same as saying that they have the same size.

Example 1.24. The set $A = \{x, y, z, w\}$ has the same cardinality as the set $[4] = \{1, 2, 3, 4\}$. In fact, sets that have the same cardinality as $[4]$ are exactly the sets with four elements.

The set $\mathbb{Z}^{>0}$ of positive integers has the same cardinality as the set $2\mathbb{Z}^{>0}$ of even positive integers. (You are asked to give a proof in Problem 1.2.3.)

We will not pursue these ideas here—it will take us off track—but if you have not already seen them, there are many interesting issues to be explored. For example, a set is called *countable* if it is finite or has the same cardinality as the set of positive integers $\mathbb{Z}^{>0}$. You can show—you may want to give these a try—that the set of rational numbers $\mathbb{Q}$ is countable, while the set of real numbers $\mathbb{R}$ is not.

The Symmetric Group. The special case when the set Ω has a finite number of elements is particularly noteworthy and will be pursued in this section (and again in Chapter 3). Since we are focusing on the properties of the 1-1 and onto functions on a set, the actual properties of the elements of the set are immaterial. Hence any set of size $|\Omega|$ can replace Ω.

Definition 1.25 (S_n)**.** Let n be a positive integer, and let $[n] = \{1, 2, \ldots, n\}$. The collection of 1-1 and onto functions from $[n]$ to $[n]$, that is, $\mathrm{Perm}([n])$, is called the *symmetric group of degree n*, and is denoted by S_n. Hence

$$S_n = \mathrm{Perm}([n]) = \{f : [n] \to [n] \mid f \text{ is 1-1 and onto}\}.$$

Elements of S_n are called *permutations* of $[n]$.

Example 1.26. Let $n = 3$. An example of $f \in S_3$ might be the function that is defined as follows.

$$f : \begin{array}{ccc} 1 & \longmapsto & 2 \\ 2 & \longmapsto & 3 \\ 3 & \longmapsto & 1 \end{array}$$

There are a number of notations that can be used for identifying elements of S_n. For example, we could depict the function f by the diagram below.

Another way to denote f is by the notation

$$f = \begin{pmatrix} 1 & 2 & 3 \\ 2 & 3 & 1 \end{pmatrix}.$$

In this notation, the top row has the elements of the domain of the function, and right below each element we write its image. Without any loss of generality, the top row can always be written as 1, 2, 3 in that order. This notation makes it clear that f is merely a permutation of $\{1, 2, 3\}$. In fact, we do not really need the top row—it is the same for all elements of S_3—so we can say that f *is* the permutation 231. Using this permutation notation, we would have

$$S_3 = \{123, 132, 213, 231, 312, 321\}.$$

For the purposes of abstract algebra, however, a different notation called *cycle notation* is preferred.

Cycle Notation for Elements of S_n**.** Thinking of the function f above as $1 \longmapsto 2 \longmapsto 3 \longmapsto 1$, we ask where does f send 1? The answer is 2. Hence we start by writing down (1 2. Now we ask where does f send 2? The answer is 3, and we continue: (1 2 3. Now 3 is sent back to 1. This is indicated by closing the parenthesis: (1 2 3), and writing $f = (1\ 2\ 3)$. Thus f behaves as a cycle: 1 goes to 2, 2 goes to 3, and 3 goes back to 1.

As an example, consider

$$\begin{pmatrix} 1 & 2 & 3 & 4 & 5 \\ 2 & 1 & 5 & 3 & 4 \end{pmatrix} \in S_5.$$

The cycle notation for this element is

$$\begin{pmatrix} 1 & 2 \end{pmatrix}\begin{pmatrix} 3 & 5 & 4 \end{pmatrix}.$$

This means 1 goes to 2, and 2 goes to 1, whereas 3 goes to 5, 5 goes to 4, and 4 goes to 3.

As another example, using cycle notation, instead of $\begin{pmatrix} 1 & 2 & 3 & 4 & 5 \\ 4 & 5 & 2 & 1 & 3 \end{pmatrix}$, we would write $\begin{pmatrix} 1 & 4 \end{pmatrix}\begin{pmatrix} 2 & 5 & 3 \end{pmatrix}$.

Usually we omit fixed-points; that is, instead of writing $\begin{pmatrix} 1 & 2 & 3 \end{pmatrix}\begin{pmatrix} 4 \end{pmatrix}$, we often write $\begin{pmatrix} 1 & 2 & 3 \end{pmatrix}$. Hence, you have to tell from the context whether $\begin{pmatrix} 1 & 2 & 3 \end{pmatrix}$ is meant as an element of S_3, or S_4, or S_{47}.

We record the above discussion in the following definitions:

Definition 1.27 (Cycles, cycle decomposition, and cycle type)**.** Let n be a positive integer, and let S_n be the symmetric group of degree n. The element of S_n that sends a_1 to a_2, a_2 to a_3, ..., and a_m to a_1, and fixes every other member of $[n]$ is denoted by $(a_1\ a_2\ a_3 \cdots a_m)$. This element is called a *cycle of length m* or an *m-cycle*.

Every element of S_n can be written as a product of disjoint cycles (i.e., cycles that have no numbers in common). This is called the *cycle decomposition* of the element.

If $\sigma \in S_n$ is the product of disjoint cycles of lengths $n_1, n_2, \ldots, n_r$ with $n_1 \leq n_2 \leq \cdots \leq n_r$ (including its 1-cycles), then the sequence of integers $n_1, n_2, \ldots, n_r$ is called the *cycle type* of σ.

Note that the cycle $(a_1\ a_2\ \cdots\ a_m)$ can also be written as $(a_2\ a_3\ \cdots\ a_m\ a_1)$, and, in fact, we can start a cycle with any of its elements.

The assertions in the above definition (e.g., every element of S_n can be written as a product of disjoint cycles) are quite reasonable as soon as you look at a few examples. We will give a proof in Lemma 3.2.

Example 1.28. Consider

$$\sigma = \begin{pmatrix} 1 & 2 & 3 & 4 & 5 & 6 & 7 & 8 & 9 \\ 2 & 6 & 4 & 3 & 7 & 1 & 8 & 9 & 5 \end{pmatrix} \in S_9.$$

The cycle decomposition of σ is $(1\ 2\ 6)(3\ 4)(5\ 7\ 8\ 9)$ and its cycle type is $2, 3, 4$.

As another example, the cycle type of $(1\ 2\ 3)(4)(5\ 7)(6) \in S_7$ is $1, 1, 2, 3$ (sometimes abbreviated as $1^2, 2, 3$).

Just as in the more general case of $\mathrm{Perm}(\Omega)$, given two elements σ and τ of S_n, $\sigma\tau$ denotes the usual function composition. In other words, to find the effect of $\sigma\tau$ we first apply τ and then apply σ.

Example 1.29. We have $(2\ 4\ 3)(3\ 2\ 1) = (1\ 2)(3\ 4)$. You may think of this as follows.

$$
\begin{array}{ccccc}
1 & \longmapsto & 3 & \longmapsto & 2 \\
2 & \longmapsto & 1 & \longmapsto & 1 \\
3 & \longmapsto & 2 & \longmapsto & 4 \\
4 & \longmapsto & 4 & \longmapsto & 3 \\
& (3\ 2\ 1) & & (2\ 4\ 3) &
\end{array}
$$

When faced with finding the product $(2\ 4\ 3)(3\ 2\ 1)$, usually, instead of drawing the above diagram, we start with 1 and ask where it goes. We first look at the rightmost cycle and, in that cycle, 1 goes to 3. We then go to one cycle to the left, and ask what that does to 3. In this case the answer is 2. Hence we write $(1\ 2$. We then say 2 goes to 1 (in the rightmost cycle), and 1 is fixed (in one cycle to the left). Hence 2 goes to 1, and we can close the parenthesis: $(1\ 2)$. We then repeat the same procedure with 3 and 4.

Example 1.30. Going back to S_3, here is a list of its elements (for this first example, we give both notations for each element).

$$
f = \begin{pmatrix} 1 & 2 & 3 \\ 2 & 3 & 1 \end{pmatrix} = \begin{pmatrix} 1 & 2 & 3 \end{pmatrix}
$$

$$
g = \begin{pmatrix} 1 & 2 & 3 \\ 2 & 1 & 3 \end{pmatrix} = \begin{pmatrix} 1 & 2 \end{pmatrix}
$$

$$
ff = f^2 = \begin{pmatrix} 1 & 2 & 3 \\ 3 & 1 & 2 \end{pmatrix} = \begin{pmatrix} 1 & 3 & 2 \end{pmatrix}
$$

$$
f^3 = g^2 = \begin{pmatrix} 1 & 2 & 3 \\ 1 & 2 & 3 \end{pmatrix} = 1_{[3]} = e
$$

$$
fg = gf^2 = \begin{pmatrix} 1 & 2 & 3 \\ 2 & 3 & 1 \end{pmatrix}\begin{pmatrix} 1 & 2 & 3 \\ 2 & 1 & 3 \end{pmatrix} = \begin{pmatrix} 1 & 2 & 3 \\ 3 & 2 & 1 \end{pmatrix} = \begin{pmatrix} 1 & 3 \end{pmatrix}
$$

$$
f^2 g = gf = \begin{pmatrix} 1 & 2 & 3 \\ 1 & 3 & 2 \end{pmatrix} = \begin{pmatrix} 2 & 3 \end{pmatrix}
$$

Thus, using the cycle notation, we have (the identity function can be denoted by $1_{[3]}$, e, or just 1)

$$
S_3 = \{1, \begin{pmatrix} 1 & 2 & 3 \end{pmatrix}, \begin{pmatrix} 1 & 3 & 2 \end{pmatrix}, \begin{pmatrix} 1 & 2 \end{pmatrix}, \begin{pmatrix} 1 & 3 \end{pmatrix}, \begin{pmatrix} 2 & 3 \end{pmatrix}\}.
$$

On the other hand, to show the similarity with symmetries, we can use the generators $f = (1\ 2\ 3)$ and $g = (1\ 2)$ to write

$$
S_3 = \{1_{[3]} = e, f, f^2, g, fg, f^2 g \mid f^3 = g^2 = e, gf = f^2 g\}.
$$

Just as for D_8, we take for granted the special role of e, and the fact that $ex = xe = e$ for all $x \in S_3$. When we write S_3 in this format, we can see that S_3 looks identical to the symmetries of an equilateral triangle (find these).

As with D_8, we can use generators and relations to further condense the notation:

$$S_3 = \langle f, g \mid f^3 = g^2 = e, gf = f^2 g \rangle.$$

Recall that, for a positive integer n, the notation $n!$ is read n *factorial* and is defined by $n! = n(n-1) \cdots 2 \times 1$.

Lemma 1.31. $|S_n| = n!$

Proof. In how many ways can we permute n elements? The first element can go to any of n places, the second to any of the $n-1$ remaining places, and the nth has only one choice. Hence, the total number of choices is $n(n-1)(n-2) \cdots 2 \cdot 1 = n!$. □

Example 1.32. As a harbinger of things to come, we note that two examples we have explored—that is, D_8 and S_4—are related.

Every symmetry of the square—that is, every element of D_8—can be thought of as a function on the set $\{1, 2, 3, 4\}$ of corners of the square. In other words, every element of D_8 gives a permutation of $\{1, 2, 3, 4\}$. If we just look at the collection of these permutations, we get a set that behaves exactly like D_8 and at the same time is a subset of S_4. If we want to use fancier language, we say that we can think of D_8 as a subset of S_4 by identifying every element of D_8 with the permutation it induces on the corners of a square, thus:

$$e = \begin{pmatrix} 1 & 2 & 3 & 4 \\ 1 & 2 & 3 & 4 \end{pmatrix} = 1_{[4]}$$

$$a = \begin{pmatrix} 1 & 2 & 3 & 4 \end{pmatrix}$$

$$a^2 = \begin{pmatrix} 1 & 2 & 3 & 4 \end{pmatrix} \begin{pmatrix} 1 & 2 & 3 & 4 \end{pmatrix} = \begin{pmatrix} 1 & 3 \end{pmatrix} \begin{pmatrix} 2 & 4 \end{pmatrix}$$

$$b = \begin{pmatrix} 1 & 2 \end{pmatrix} \begin{pmatrix} 3 & 4 \end{pmatrix}$$

and so on. Thus, we can write

$$D_8 = \{ e = 1_{[4]}, a = \begin{pmatrix} 1 & 2 & 3 & 4 \end{pmatrix}, a^2 = \begin{pmatrix} 1 & 3 \end{pmatrix} \begin{pmatrix} 2 & 4 \end{pmatrix}, a^3 = \begin{pmatrix} 1 & 4 & 3 & 2 \end{pmatrix},$$
$$b = \begin{pmatrix} 1 & 2 \end{pmatrix} \begin{pmatrix} 3 & 4 \end{pmatrix}, ab = \begin{pmatrix} 1 & 3 \end{pmatrix}, a^2 b = \begin{pmatrix} 1 & 4 \end{pmatrix} \begin{pmatrix} 2 & 3 \end{pmatrix}, a^3 b = \begin{pmatrix} 2 & 4 \end{pmatrix} \}.$$

From this point of view, D_8 is a subset of S_4. While you can think of this as another notation for expressing D_8, the fact is that we already had a definition for D_8. To keep things straight—or you may think it to be a bit pedantic—we sometimes use a different, and somewhat more precise, language. Formally, we have two different groups:

$$D_8 = \{ e, a, a^2, a^3, b, ab, a^2 b, a^3 b \mid a^4 = b^2 = e, ba = a^3 b \},$$
$$G = \{ 1_{[4]}, \begin{pmatrix} 1 & 2 & 3 & 4 \end{pmatrix}, \begin{pmatrix} 1 & 3 \end{pmatrix} \begin{pmatrix} 2 & 4 \end{pmatrix}, \begin{pmatrix} 1 & 4 & 3 & 2 \end{pmatrix}, \begin{pmatrix} 1 & 2 \end{pmatrix} \begin{pmatrix} 3 & 4 \end{pmatrix},$$
$$\begin{pmatrix} 1 & 3 \end{pmatrix}, \begin{pmatrix} 1 & 4 \end{pmatrix} \begin{pmatrix} 2 & 3 \end{pmatrix}, \begin{pmatrix} 2 & 4 \end{pmatrix} \}.$$

The second group G is a subset of S_4 while the first one—at least formally—is not. However, it is clear that D_8 and G behave in exactly the same way, and we want to say that they are the same. As we shall see later (see Section 2.4) we say that D_8 and G are *isomorphic* groups. In the study of group theory, isomorphic groups are considered the same. Using this language, an alternative to saying that D_8 is a subset of S_4, is to say that S_4 contains an isomorphic copy of D_8.

Problems

1.2.1. Let $\Omega = \mathbb{Z}$ be the set of integers. Define $f : \Omega \to \Omega$ by $f(x) = x + 5$. Is $f \in \mathrm{Perm}(\Omega)$? If so, what is its inverse? If n is a positive integer, then what is $f^n(x)$? What if instead of $\mathbb{Z}$, we had $\Omega = \mathbb{Z}^{\geq 0}$, the set of non-negative integers?

1.2.2. Let $\Omega = \mathbb{Z}$ be the set of integers. Define $f : \Omega \to \Omega$ by

$$f(x) = \begin{cases} x + 1 & \text{if } x \text{ is even,} \\ x - 1 & \text{if } x \text{ is odd.} \end{cases}$$

Is $f \in \mathrm{Perm}(\Omega)$? If so, what is its inverse? What is f^2? What about f^3?

1.2.3. Show that the set $\mathbb{Z}^{>0}$ of positive integers has the same cardinality as the set $2\mathbb{Z}^{>0}$ of even positive integers.

1.2.4. Let $\sigma = (1\ 3\ 5)(2\ 4)$ and $\tau = (1\ 5)(2\ 3)$ be elements of S_5. Find σ^2, $\sigma\tau$, $\tau\sigma$, and $\tau\sigma^2$.

1.2.5. Construct a complete multiplication table for S_3. What is the center (see Definition 1.7) of S_3? If $f = (1\ 2\ 3)$, what is $\mathbf{C}_{S_3}(f)$, the centralizer of f in S_3?

1.2.6. Let $f = (1\ 2\ 3) \in S_3$. Find the maps in the following sequence

$$1_{[3]}, f, f^2, f^3, f^4, f^5, \ldots.$$

Do you see a pattern?

1.2.7. Let S, T, and R be sets. Assume that $f : S \to T$, and $g : T \to R$ are maps. Assume that gf is onto. Does f have to be onto? Does g have to be onto? In both cases either prove that the map is onto or give a counterexample.

1.2.8. Let R, S, and T be sets. Let $f : R \to S$, and $g : S \to T$ be maps. Assume that we know that gf is 1-1.
   (a) Must f be 1-1? Either prove that it is or find a counterexample.
   (b) Must g be 1-1? Either prove that it is or find a counterexample.

1.2.9. In the proof of the second direction of Theorem 1.20, it was assumed that f is 1-1 and onto, and, based on that assumption, a function g was defined. It was then claimed that g is the inverse of f. Write down a proof of this claim.

1.2.10. Write down a proof of Theorem 1.21.

1.2.11. In an algebra book you read the following definition: The function $g : Y \to X$ is the inverse of the function $f : X \to Y$ if the two diagrams in Figure 1.9 commute.

   Is this any different from our definition of inverses? Can you draw one diagram—with four nodes and five arrows—that is commutative if and only if g is the inverse of f?

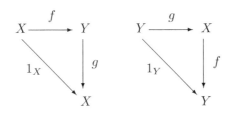

Figure 1.9. The two diagrams commute if and only if f and g are inverses.

Left Inverse, Right Inverse

Definition 1.33 (Left inverse, right inverse)**.** Let X and Y be sets, and let $f : X \to Y$ be a map. A map $g : Y \to X$ is called the *left inverse* of f, if gf is the identity map on X. Likewise, g is called a *right inverse* of f if fg is the identity map on Y. Finally, the map g is the *inverse* of f if it is both a right and a left inverse of f.

1.2.12. Give an example of a map f that has a left inverse, but not an inverse.

1.2.13. Give the definition of the left inverse using a commuting diagram.

1.2.14. Let X be a non-empty set, and let $f : X \to Y$ be a map.
Which one of the following statements imply which other ones? For every true implication give a proof, and for all others give a counterexample.
(a) f is 1-1.
(b) f is onto.
(c) f has a left inverse.

1.2.15. (a) Give an example of a map f that has a right inverse, but not an inverse.
(b) Show that f has a right inverse if and only if f is onto.

1.2.16. Let $\mathbb{Z}^{\geq 0}$ be the set of non-negative integers. Define $f : \mathbb{Z}^{\geq 0} \to \mathbb{Z}^{\geq 0}$ by $f(n) = 3n + 1$. How many left inverses does f have? What about right inverses?

1.2.17. If f, g are mappings of S into S and fg is a *constant* function (this means that there exists $\alpha \in S$ such that $f(g(s)) = \alpha$ for all $s \in S$), then
(a) What can you say about f if g is onto?
(b) What can you say about g if f is onto?

1.2.18. Does the exponential function $E : \mathbb{R} \to \mathbb{R}$, defined by $E(x) = e^x$, have an inverse? If the answer is yes, find the inverse. If the answer is no, how can we modify the domain and/or codomain so that the function has an inverse?

1.2.19. Repeat the previous problem for the sine function $S : \mathbb{R} \to \mathbb{R}$ defined by $S(x) = \sin(x)$.

1.2.20. Let S be a set with a finite number of elements, and let $f : S \to S$ be a map.
   (a) If f is onto, can f not be 1-1?
   (b) If f is 1-1, can f not be onto?
   (c) Do your conclusions remain valid even if S has an infinite number of elements?

1.2.21. As usual, let $(0, 1) = \{x \in \mathbb{R} \mid 0 < x < 1\}$. Can you find a 1-1 and onto map $f : (0, 1) \to \mathbb{R}$?

1.2.22. As usual, let $[0, 1) = (0, 1) \cup \{0\}$ and $[0, 1] = (0, 1) \cup \{0, 1\}$. Can you find a 1-1, onto function $f : [0, 1) \to [0, 1]$?

1.3. Integers mod n and Elementary Properties of Integers

Definition 1.34. The following notation will be fixed for the rest of the book:

   $\mathbb{Z}$ is the set of *integers*: $\ldots, -2, -1, 0, 1, 2, \ldots$.

   $\mathbb{Z}^{\geq 0}$ (or $\mathbb{N}$) is the set of *natural numbers*: $0, 1, 2, \ldots$.[4]

   $\mathbb{Z}^{>0}$ is the set of *positive integers*: $1, 2, 3, \ldots$.

   $\mathbb{Q}$ is the set of *rational numbers*.

   $\mathbb{R}$ is the set of *real numbers*.

   $\mathbb{C}$ is the set of *complex numbers*.

Let n be an integer greater than 1, and consider the set $\{0, 1, \ldots, n-2, n-1\}$ with n elements. We want to define an operation akin to the usual addition on this set. Clearly, the usual addition of numbers will not work, since, for example, $(n-1) + (n-1) = 2n - 2$ will not be in the set anymore. In other words, this set is not closed under the usual addition. So, we introduce a different operation. For two elements a and b in this set, we define $a + b$ to be the *remainder* of the ordinary sum of a and b when divided by n. This new operation will be called *addition mod n* (or modulo n).

For example, if $n = 8$, then addition mod 8 gives $7 + 5 = 4$.

Likewise, we define ab to be the remainder of the ordinary product of a and b when divided by n, and we call this new operation *multiplication* mod n (or *modulo n*). We record this definition:

Definition 1.35 ($\mathbb{Z}/n\mathbb{Z}$). Let n be an integer greater than 1. The set $\{0, 1, \ldots, n-1\}$ with the operations of addition and multiplication mod n will be called the set of *integers* mod n (or *integers modulo n*) and is denoted by $\mathbb{Z}/n\mathbb{Z}$ (We read this as $\mathbb{Z}$ mod n or as $\mathbb{Z}$ mod $n\mathbb{Z}$.)

$\mathbb{Z}_n$, $\mathbb{Z}/n$, and $\mathbb{Z}/(n)$ are other commonly used notations for $\mathbb{Z}/n\mathbb{Z}$.

Hence, for example, $\mathbb{Z}/7\mathbb{Z} = \{0, 1, \ldots, 6\}$, and the result of $a + b$ is the remainder of $a + b$ when divided by 7. Thus $6 + 4 = 3$, and $3 + 4 = 0$.

---

[4] Some authors start the natural numbers with 1.

Remark 1.36. Given an addition and a multiplication, we already also have subtraction and division (although some restrictions may apply). For example, -3 means the element that if added to 3 gives zero, while $1/3$ is the element that when multiplied by 3 gives 1. Thus, in $\mathbb{Z}/5\mathbb{Z}$, $-3 = 2$ and $1/3 = 2$. Now $2 - 3 = 2 + (-3)$ and $4/3 = 4 \cdot 1/3$, and hence we can subtract and divide. Given the definitions, for addition and multiplications modulo n, it is clear that we can always subtract (if $a \in \mathbb{Z}/n\mathbb{Z}$, then $-a = n - a$). However, it is not clear—and not always true—that we can divide by all non-zero elements. We will address this issue soon.

Example 1.37. In $\mathbb{Z}/6\mathbb{Z}$: $3+4 = 1$, $3+5 = 2$, $3 \times 4 = 0$, $-3 = 3$, $-2 = 4$, $1/5 = 5$, and so on.

Remark 1.38. The use of the $+$ (and $\cdot$) sign is somewhat ambiguous. When we use $+$ sometimes we mean ordinary addition, while at other times we mean addition mod n. The context will make the meaning clear.

Our third basic example of a group is $(\mathbb{Z}/n\mathbb{Z}, +)$, where $+$ is addition mod n. We note that, just like D_8 and S_n, this operation on $\mathbb{Z}/n\mathbb{Z}$—even though we are writing it as "addition" and not "multiplication"—has the same four properties (closure, associativity, existence of identity, existence of inverses).

Zero is the identity, and it is clear that addition mod n is closed and that there are additive inverses for each element. It is not so obvious that this operation is associative. This needs a proof, and the proof depends on the rudiments of basic number theory. We shall return to this shortly.

We got a group (i.e., a set with an operation that satisfies the four properties) when we considered $\mathbb{Z}/n\mathbb{Z}$ with *addition* as the operation. What if we use multiplication as the operation? Then the operation will continue to be closed and associative—this needs a proof—and there is an (multiplicative) identity, namely 1. But inverses need more care. Recall that an inverse of an element x is another element y such that xy and yx are equal to the identity. Here since the operation is commutative, given any x, we want y such that $xy = 1$ (where the multiplication is mod n).

Consider

$$\mathbb{Z}/5\mathbb{Z} = \{0, 1, 2, 3, 4\}.$$

Note that 2 and 3 are mutual multiplicative inverses (that is $2 \cdot 3 = 1$), 1 and 4 are their own inverse, but 0 has no inverse. Thus $(\mathbb{Z}/5\mathbb{Z}, \cdot)$ is not a group.

On the other hand, $(\mathbb{Z}/5\mathbb{Z} - \{0\}, \cdot)$ *does have* our favorite four properties.

What about $(\mathbb{Z}/6\mathbb{Z} - \{0\}, \cdot)$?

We ask the readers to answer this question on their own. After a few problems, we continue this section with a brief introduction to the basic properties of integers. You may be familiar with these, in which case, you should skim through them. We will use these facts to prove associativity for both addition and multiplication mod n, and to see if we can create a group (that is a set and an operation with the four basic properties of closure, associativity, identity, and inverses) from integers mod n and multiplication mod n. In the next chapter, the same basic properties will be useful in developing basic facts about cyclic groups.

Problems

1.3.1. (a) Find $-\frac{3}{4} - 4$ in $\mathbb{Z}/7\mathbb{Z}$.

   (b) In $\mathbb{Z}/12\mathbb{Z}$ does every non-zero element have a multiplicative inverse (i.e., for $a \in \mathbb{Z}/12\mathbb{Z}$, can we find b such that $ab = 1$)?

   (c) In $\mathbb{Z}/7\mathbb{Z}$ does every non-zero element have a multiplicative inverse?

   (d) We want to know for which integers $n > 1$ every non-zero element of $\mathbb{Z}/n\mathbb{Z}$ has a multiplicative inverse. Look at some examples and make a conjecture. You do not have to prove your conjecture.

1.3.2. Consider the addition operation on $\mathbb{Z}/7\mathbb{Z}$. Start with the element $a = 3$ and find $2a = a + a$, $3a = a + a + a$, and so on until at least $20a$. Do you notice a pattern? Now change a to 4 and repeat what you did. Make a general conjecture based on the patterns that you found. Repeat what you did for $\mathbb{Z}/6\mathbb{Z}$. Is there any difference?

1.3.3. In $\mathbb{Z}/n\mathbb{Z}$, choose an element a and consider

$$a, 2a, 3a, \ldots, na.$$

For which elements $a \in \mathbb{Z}/n\mathbb{Z}$ does the above list give exactly every element of $\mathbb{Z}/n\mathbb{Z}$? Make a conjecture.

1.3.4. Let $\mathbb{Z}/9\mathbb{Z} = \{0, 1, \ldots, 8\}$ with addition and multiplication mod 9. Define $f : \mathbb{Z}/9\mathbb{Z} \to \mathbb{Z}/9\mathbb{Z}$ by $f(a) = 5a$. Does f have an inverse? Is f 1-1? Is f onto? Answer the same questions for the map $g : \mathbb{Z}/9\mathbb{Z} \to \mathbb{Z}/9\mathbb{Z}$ defined by $g(a) = 3a$.

1.3.5. (a) Which elements of $\mathbb{Z}/12\mathbb{Z}$ should we throw out so that the remaining elements form a group with the operation being multiplication mod 12?

   (b) By looking at some more examples, make a conjecture generalizing your answer to the previous part.

1.3.6. Let $G = \{3, 9, 15, 21\}$, and let the operation on G be multiplication mod 24. Is the operation closed? Is there an identity element? Does every element have an inverse?

Elementary Properties of Integers. Number theory is concerned with the properties of integers and arithmetic operations on them. In other words, the properties of $\mathbb{Z}$ with the operations of $+$ and $\cdot$. The best abstract context for studying such a set with *two* related operations is that of rings. Thus, later in this book when studying rings, we will develop many properties of the integers rigorously and from first principles. We will see then that the fact that the integers have the division algorithm (Theorem 1.47) will be crucial and from it many properties (including unique factorization into primes) follow.

   Here we will only consider the basics that we need for understanding integers mod n (and cyclic groups in the next chapter). We will be sketchy in our proofs since the proofs are not too hard and will be done in a much more general context later.

An important method of proof in discrete mathematics is induction, and we assume that the reader is familiar with it. In fact, the principle of induction is equivalent to the so-called "well-ordering" axiom that can be just as useful in many proofs. We accept this axiom as one of the basic properties of the integers.

Axiom 1.39 (Well-ordering principle). Let S be a non-empty collection of non-negative integers, then S has a smallest member. In other words, there is an $m \in S$ such that $m \leq s$ for all $s \in S$.

To be complete, we now give a series of—hopefully familiar—definitions.

Definition 1.40 (Divisibility). Given two integers a and b, if b is *divisible* by a (i.e., if when dividing b by a we have no remainder), then we write

$$a \mid b.$$

This means that $b = ak$ for some integer k. If a does not divide b, we write $a \nmid b$. If b is divisible by a, we also say that a *divides* b, a is a *divisor* of b, or b is a *multiple* of a.

For a positive integer n, and any two integers a and b, we write $a = b \bmod n$ if a and b have the same remainder when divided by n. In other words, $a = b \bmod n$ if and only if $n \mid b - a$.

Note that some of the notation may be confusing to you. When we divide b by a and we want the quotient, we write $\frac{b}{a}$ or b/a. Now if this division does not have a remainder, then we say a divides b and write $a \mid b$.

Definition 1.41 (Prime numbers). An integer $a > 1$ is prime if its only positive integer divisors are 1 and a. (For now, we only consider positive integers as primes. Later, in ring theory, we will enlarge the definition.)

Definition 1.42 (Greatest common divisor). Let $a, b \in \mathbb{Z}$. The *greatest common divisor* of a and b, denoted by $\gcd(a, b)$, is an integer d such that:

(a) $d > 0$, and

(b) $d \mid a$, $d \mid b$, and

(c) if $c \in \mathbb{Z}$ with $c \mid a$ and $c \mid b$, then $c \mid d$.

Example 1.43. The greatest common divisor of 8 and 10 is 2.

Note that the definition of the greatest common divisor leaves some questions unanswered. Do greatest common divisors always exist? Must the greatest common divisor of a and b actually be the largest among the divisors of both a and b? Answering these is straightforward (give it a try, and see Problem 1.3.8). We will come back to a thorough treatment of greatest common divisors in a more general setting in Section 18.5.

Definition 1.44 (Relatively prime). Two integers a and b are *relatively prime* if $\gcd(a, b) = 1$.

Definition 1.45 (Euler's ϕ function). Let n be a positive integer. Define $\phi(n)$ to be the number of positive integers less than or equal to n that are relatively prime to n. In other words,

$$\phi(n) = |\{a \in \mathbb{Z} \mid 1 \leq a \leq n \text{ and } \gcd(a, n) = 1\}|.$$

The function $\phi(n)$ is called the *Euler ϕ-function* or the *Euler totient function*.

Example 1.46. Since $\{1, 3, 7, 9\}$ are all the positive integers less than 10 that are relatively prime to 10, we have $\phi(10) = 4$. Likewise, $\phi(8) = 4$, since $\{1, 3, 5, 7\}$ are the positive integers less than 8 that are relatively prime to 8.

Two main theorems—called the *fundamental theorem of arithmetic* and the *division algorithm*—are central to finding the properties of integers.

Theorem 1.47 (The division algorithm). *Given $n \in \mathbb{Z}$ with $n \geq 1$ and $m \in \mathbb{Z}$, there exists unique integers q and r such that*

$$m = qn + r \qquad and \qquad 0 \leq r < n.$$

Proof. Noticing that eventually we want to find the smallest non-negative r with $r = m - qn$, we consider the set

$$M = \{m - kn \mid k \in \mathbb{Z}\}.$$

First notice that since k is allowed to be positive or negative, regardless of the values of m and n, the set M has some non-negative integers. Let r be the smallest non-negative integer in M—we are using the well-ordering principle of non-negative integers. So far, by construction, we have that $r \geq 0$ and $r = m - qn$ for some integer q.

We now argue that $r < n$, since if r was not less than n, then M would have contained a non-negative integer smaller than r, namely $r - n$.

So, as required, we have two integers q and r such that $m = qn + r$ and $0 \leq r < n$. It remains to show that q and r are unique.

To show uniqueness, assume that another pair q_1 and r_1 have the same properties, i.e., $m = q_1 n + r_1$ and $0 \leq r_1 < n$. We have to prove that $r_1 = r$ and $q_1 = q$. Since $r_1 = m - q_1 n$, we have r_1 is a non-negative integer in M. Hence $r \leq r_1$, since r was the smallest non-negative integer in M. We conclude that $0 \leq r_1 - r < n$.

We now have $r_1 - r = (m - q_1 n) - (m - qn) = n(q - q_1)$. Since $r_1 - r \geq 0$, we have to have $q - q_1 \geq 0$. If the latter is actually bigger than zero, then $r_1 - r \geq n$, which is a contradiction. Hence, $q - q_1 = 0$, which also means $r_1 - r = 0$. So $q = q_1$ and $r = r_1$, and the proof is complete. $\qquad\square$

Theorem 1.48 (The fundamental theorem of arithmetic). *Let $n > 1$ be an integer. Then n can be factored uniquely into a product of primes.*

In other words, there is a finite list of distinct primes $p_1 < p_2 < \cdots < p_r$ and a corresponding list of positive integers $\alpha_1, \ldots, \alpha_r$ such that

$$n = p_1^{\alpha_1} p_2^{\alpha_2} \cdots p_r^{\alpha_r}.$$

The uniqueness of the factorization means that whenever we factor n into powers of distinct primes, ordering the primes from the smallest to the largest, then we get exactly the above factorization.

Proof. See Chapter 18 and especially Corollary 18.52. $\qquad\square$

Using the division algorithm and the fundamental theorem of arithmetic, we can prove many basic and useful facts about integers. When we do ring theory, we will see that the fundamental theorem of arithmetic follows from the division algorithm. This fact can guide our exposition in two—contradictory—ways. We could say that we should not need to use the fundamental theorem since it is a consequence of the division algorithm—and indeed we do not have to. On the other hand, while a large class of commutative rings (sets with two operations satisfying certain rules) that satisfy the division algorithm also have unique factorization, the converse is not true. Thus, for the sake of later generalizing the arguments to more general settings, we should use unique factorization instead of the division algorithm whenever we can. In this section, we are only interested in the properties of ordinary integers, and we will mainly use the division algorithm—which is the stronger result. However, when doing the problems, the reader is welcomed to use either result. We relegate a careful discussion of the finer points to the chapters on ring theory.

Theorem 1.49. *Let $a, b \in \mathbb{Z}$, and assume that at least one of them is non-zero. Then $\gcd(a, b)$ exists, and, in fact, $\gcd(a, b)$ is the smallest positive integer that can be written as an integer linear combination of a and b. In other words, there exist $m_0, n_0 \in \mathbb{Z}$ such that*

$$\gcd(a, b) = m_0 a + n_0 b,$$

and no integer linear combination of a and b is both positive and smaller than $\gcd(a, b)$.

Proof. We let

$$M = \{ma + nb \mid m, n \in \mathbb{Z}\}.$$

Since at least one of a and b are not zero, M must have some non-zero numbers in it. Also, $ma + nb \in M$ implies that $-(ma + nb) = (-m)a + (-n)b$ is also in M. So M must have some positive elements.

Using the well-ordering principle, we now know that among the positive numbers in M there is a smallest one. Let d be the smallest positive integer in M.

CLAIM: $d = \gcd(a, b)$.

PROOF OF CLAIM: Since d is an element of M, we have that $d = m_0 a + n_0 b$ for some integers m_0 and n_0.

Note that if another integer c divides both a and b, then c divides $m_0 a + n_0 b = d$. Hence it only remains to show that $d \mid a$ and $d \mid b$. Since a and b themselves are in M, the proof will be complete if we show the more general fact that d divides every element of M.

Hence let $x = ma + nb$ be an arbitrary element of M, and we want to show that $d \mid x$. By the division algorithm, $x = dq + r$ with $0 \le r < d$. Replacing x by $ma + nb$, replacing d by $m_0 a + n_0 b$, and solving for r, we get

$$r = x - dq = (m - qm_0)a + (n - qn_0)b \in M.$$

Recall that d was the smallest positive integer in M. Now, r is also in M, and $0 \le r < d$. Hence we must have $r = 0$. It follows that $x = dq$ and $d \mid x$. $\qquad\square$

Consequences for $\mathbb{Z}/n\mathbb{Z}$**.** Recall that we defined the elements of $\mathbb{Z}/n\mathbb{Z}$ to be $\{0, 1, 2, \ldots, n-1\}$, and we defined two operations $+$ and $\cdot$ for it. These were defined as the remainders of the usual sum and product when divided by n. If we focus on the addition, then we know it is clear that $(\mathbb{Z}/n\mathbb{Z}, +)$ is closed under addition, has an identity (namely 0), and every element has an inverse (namely, the inverse of a is $n - a$). Addition is also associative. But this needs a proof and the proof uses the division algorithm.

Theorem 1.50 (Associativity in $(\mathbb{Z}/n\mathbb{Z}, +)$)**.** *Let a, b and $c \in \mathbb{Z}/n\mathbb{Z}$. Then*

$$(a + b) + c = a + (b + c) \quad \text{in } \mathbb{Z}/n\mathbb{Z}.$$

Proof. By the division algorithm, we have $a + b = nq_1 + r_1$ with $0 \le r_1 < n$, and $r_1 + c = nq_2 + r_2$ with $0 \le r_2 < n$. By definition we have $(a + b) + c = r_2$ in $\mathbb{Z}/n\mathbb{Z}$.

Similarly, $b + c = nq_3 + r_3$ with $0 \le r_3 < n$, and $a + r_3 = nq_4 + r_4$ with $0 \le r_4 < n$. Thus again in $\mathbb{Z}/n\mathbb{Z}$ we have $a + (b + c) = r_4$. We have to show that $r_2 = r_4$.

Now in $\mathbb{Z}$ we have $a + b + c = nq_1 + r_1 + c = n(q_1 + q_2) + r_2$, and also $a + b + c = a + nq_3 + r_3 = n(q_3 + q_4) + r_4$. We also know that $0 \le r_2, r_4 < n$. Now the division algorithm says that when you divide $a + b + c$ by n both the quotient and the remainder are *unique*. Hence $r_2 = r_4$ (and also $q_1 + q_2 = q_3 + q_4$). $\qquad\square$

Turning to multiplication, we know that $(\mathbb{Z}/n\mathbb{Z}, \cdot)$—where $\cdot$ is multiplication mod n—does not have the four properties of a group, because not every element is invertible. However, if we throw out all the non-invertible elements, then we do get a group (i.e., a set with an operation that is closed, associative, has an identity, and has inverses). This is actually a special case of a more general construction. In any ring (see Definition 15.4) the set of invertible elements form a group with respect to multiplication (see Proposition 15.11).

Hence we make the following important definition.

Definition 1.51 (Units of $\mathbb{Z}/n\mathbb{Z}$)**.** Let n be a positive integer. Define

$$(\mathbb{Z}/n\mathbb{Z})^{\times} = \{\, a \in \mathbb{Z}/n\mathbb{Z} \mid \exists b \in \mathbb{Z}/n\mathbb{Z} \text{ such that } ab = 1 \}.$$

We call $(\mathbb{Z}/n\mathbb{Z})^{\times}$ the *group of units of* $\mathbb{Z}/n\mathbb{Z}$. Many authors use the notation U_n for $(\mathbb{Z}/n\mathbb{Z})^{\times}$.

Example 1.52. $(\mathbb{Z}/10\mathbb{Z})^{\times} = \{1, 3, 7, 9\}$.

Example 1.53. $(\mathbb{Z}/8\mathbb{Z})^{\times} = \{1, 3, 5, 7\}$.

We claimed—by calling $(\mathbb{Z}/n\mathbb{Z})^{\times}$ a *group*—that the set $(\mathbb{Z}/n\mathbb{Z})^{\times}$ together with multiplication mod n is another example of a group. However, we still need to justify that this set together with multiplication mod n does indeed satisfy the four properties of a group. Since by definition all elements of this set are invertible, and the set contains 1, we only need closure and associativity. Assuming associativity, closure is easy. If $a, b \in (\mathbb{Z}/n\mathbb{Z})$ and are both invertible, then their product ab will also be invertible. To see this, assume that c is the inverse of a and d is the inverse of b. Then dc will be the inverse of ab since $(dc)(ab) = d(ca)b = db = 1$.

Hence we only need to show associativity of the product.

Theorem 1.54 (Associativity in $(\mathbb{Z}/n\mathbb{Z}, \cdot)$). *Let a, b and c* $\in \mathbb{Z}/n\mathbb{Z}$*, and let* $\cdot$ *denote multiplication* mod *n. Then*

$$(a \cdot b) \cdot c = a \cdot (b \cdot c).$$

Proof. By the definition of $a \cdot b$ (recall that $a \cdot b$ is the remainder of ab when divided by n), we know that $ab = k_1 n + a \cdot b$ and $bc = k_2 n + b \cdot c$. We now have

$$\begin{aligned} abc - k_1 nc &= (a \cdot b)c = k_3 n + (a \cdot b) \cdot c, \\ abc - k_2 na &= a(b \cdot c) = k_4 n + a \cdot (b \cdot c). \end{aligned}$$

Thus $abc = q_1 n + (a \cdot b) \cdot c$ and $abc = q_2 n + a \cdot (b \cdot c)$ where q_1 and q_2 are integers. So both $(a \cdot b) \cdot c$ and $a \cdot (b \cdot c)$ are the remainders of abc when divided by n. Now the *uniqueness* in the division algorithm says that $(a \cdot b) \cdot c = a \cdot (b \cdot c)$. $\qquad\square$

But which elements of $\mathbb{Z}/n\mathbb{Z}$ are the invertible ones? The next theorem gives a useful characterization.

Theorem 1.55. *Let n be a positive integer. Then*

$$(\mathbb{Z}/n\mathbb{Z})^{\times} = \{\, a \in \mathbb{Z}/n\mathbb{Z} \mid \gcd(a, n) = 1 \,\}.$$

Proof. We need to show that the set of invertible elements in $\mathbb{Z}/n\mathbb{Z}$ is precisely the set of elements that are relatively prime to n.

First assume that $a \in \mathbb{Z}/n\mathbb{Z}$ is invertible, and let $d = \gcd(a, n)$. We will show that $d = 1$. Since a is invertible, let $b \in \mathbb{Z}/n\mathbb{Z}$ be its inverse. Hence, the product of a and b in $\mathbb{Z}/n\mathbb{Z}$ is 1, which means that the remainder of ab when divided by n is 1, and so $ab = nq + 1$. Rearranging, we get $1 = ab - nq$. But d divides both a and n and hence it must divide $ab - nq$. This means that $d \mid 1$ which implies $d = 1$.

Conversely assume $a \in \mathbb{Z}/n\mathbb{Z}$ and $\gcd(a, n) = 1$. We have to show that a is invertible. By Theorem 1.49, we know that, for some integers u and v, we have $au + nv = 1$. Let r be remainder of u when divided by n. Hence $0 \leq r < n$ and $u = qn + r$. We have $a(qn + r) + nv = 1$ which gives

$$ar = (-aq - v)n + 1.$$

Since, in the division algorithm, the quotient and the remainder are unique, we conclude that the remainder of ar when divided by n is 1. This means that r is the (multiplicative) inverse of a in $\mathbb{Z}/n\mathbb{Z}$ and the proof is complete. $\qquad\square$

Corollary 1.56. *Let n be a positive integer, and let* $\phi(n)$ *denote the Euler* ϕ *function. Then*

$$|(\mathbb{Z}/n\mathbb{Z})^{\times}| = \phi(n).$$

Corollary 1.57. *Let p be a prime number. Then*

$$(\mathbb{Z}/p\mathbb{Z})^{\times} = \{1, 2, \ldots, p - 1\}.$$

In particular, in $\mathbb{Z}/p\mathbb{Z}$ *every non-zero element has a multiplicative inverse.*

Remark 1.58. Let p be a prime number, and consider $(\mathbb{Z}/p\mathbb{Z}, +, \cdot)$ the set of integers mod p with the *two* operations of addition and multiplication mod p. Then in this set, we can add, subtract, multiply, and divide by non-zero elements (and get another element of the set), and both operations behave nicely. Other examples of the same phenomenon include the set of rational numbers $\mathbb{Q}$, the set of real numbers $\mathbb{R}$, and the set of complex numbers $\mathbb{C}$. These are examples of what we will call fields. They will be studied in some detail later in the text (see Definition 15.14 and Part 3), but we will come back to them shortly in Section 1.4. The field $(\mathbb{Z}/p\mathbb{Z}, +, \cdot)$ is an example of a *finite* field, and—in addition to $\mathbb{Z}/p\mathbb{Z}$—we will also use the notation $\mathbb{F}_p$ for it.

1.3.7. What is $\gcd(-8, 0)$?

1.3.8. Does every pair of integers possess a greatest common divisor? Prove your assertion using Definition 1.42. Is the greatest common divisor of two integers always the largest integer among the common divisors of the two integers? Again use Definition 1.42 to justify your assertions.

1.3.9. Find the multiplication table for $(\mathbb{Z}/8\mathbb{Z})^\times$ explicitly.

1.3.10. The groups $(\mathbb{Z}/5\mathbb{Z})^\times$ and $(\mathbb{Z}/4\mathbb{Z}, +)$ both have four elements. Are they related? In the multiplication table for each group, can you replace the names of the elements with e, a, b, and c appropriately so that the two tables look identical?

1.3.11. Let i, r, and k be positive integers, and let $s = \gcd(i, r)$. Assume that $r \mid ik$. Prove that $\frac{r}{s} \mid k$.

1.3.12. Let n, m, and k all be positive integers. Assume that

$$n \mid mk - 1.$$

Prove that $\gcd(n, m) = 1$.

1.3.13. Let a, b, and c be integers.
(a) Prove that if $\gcd(a, b) = 1$ and $a \mid bc$, then $a \mid c$.
(b) Prove that if $\gcd(a, b) = 1$ and $\gcd(a, c) = 1$, then $\gcd(a, bc) = 1$.

1.3.14. Let n be an arbitrary integer greater than 1, and let k and x be integers relatively prime to n. Let k' and x', respectively, be the multiplicative inverses of k and x mod n—in other words, kk' and xx' each have remainder 1 when divided by n.
(a) Show that kn divides $k^2 k' x - kx$.
(b) Show that kn divides $k^2 (k')^2 xx' - kk'$.

1.3.15. Let p be a prime, and $G = (\mathbb{Z}/p\mathbb{Z})^\times$.
(a) Show that $p - 1$ is its own inverse in G.
(b) Show that 1 and $p - 1$ are the only elements of G that are their own inverses.
(c) (Wilson's theorem.) Show that $(p - 1)! \equiv (p - 1) \bmod p \equiv -1 \bmod p$.

1.3.16. Let a and b be positive integers. Let $a = p_1^{\alpha_1} p_2^{\alpha_2} \cdots p_k^{\alpha_k}$ and $b = p_1^{\beta_1} p_2^{\beta_2} \cdots p_k^{\beta_k}$ where $\alpha_i, \beta_i \geq 0$ and, for $1 \leq i \leq k$, p_i are distinct primes. (Note that to get the same set of primes in both factorizations, we are allowing a zero exponent.) Show that $\gcd(a, b) = p_1^{\gamma_1} p_2^{\gamma_2} \cdots p_k^{\gamma_k}$, where $\gamma_i = \min(\alpha_i, \beta_i)$.

In particular, a and b are relatively prime if and only if they do not have any common prime divisors.

Least Common Multiple

Definition 1.59 (Least common multiple). Let $a, b \in \mathbb{Z}$. Then $\ell = \operatorname{lcm}(a, b)$ the *least common multiple* of a and b, is an integer such that:

(a) $\ell > 0$, and

(b) $a \mid \ell$, $b \mid \ell$, and

(c) If $d \in \mathbb{Z}$ with $a \mid d$ and $b \mid d$, then $\ell \mid d$.

1.3.17. Let a and b be positive integers. Let $a = p_1^{\alpha_1} p_2^{\alpha_2} \cdots p_k^{\alpha_k}$ and $b = p_1^{\beta_1} p_2^{\beta_2} \cdots p_k^{\beta_k}$ where $\alpha_i, \beta_i \geq 0$ and, for $1 \leq i \leq k$, p_i are distinct primes. (Note that to get the same set of primes in both factorizations, we are allowing a zero exponent.) Show that $\operatorname{lcm}(a, b) = p_1^{\delta_1} p_2^{\delta_2} \cdots p_k^{\delta_k}$, where $\delta_i = \max(\alpha_i, \beta_i)$.

1.3.18. Let a and b be positive integers. What can you say about the product of $\gcd(a, b)$ and $\operatorname{lcm}(a, b)$? By looking at some examples, make a conjecture. Can you prove your conjecture?

1.3.19. Let a and b be positive integers. If $\gcd(a, b) = 1$, then what can you say about $\operatorname{lcm}(a, b)$? Prove your assertion.

1.4. Invertible Matrices

For our final example, we let F stand for one of $\mathbb{Q}$ (rational numbers), $\mathbb{R}$ (real numbers), $\mathbb{C}$ (complex numbers), or $\mathbb{Z}/p\mathbb{Z}$ (integers mod p) with p a prime, and we call each of these sets of numbers a *field*. In each of these cases, we can do addition, subtraction, multiplication, and division by non-zero numbers. In fact, the definition of a field is more general (see Definition 1.65 on page 32, Problems 1.4.10–1.4.12, and Section 15.2), and we can expand the above list considerably. However, for our illustrative purposes, it is enough for F to be one of $\mathbb{Q}$, $\mathbb{R}$, $\mathbb{C}$, or $\mathbb{Z}/p\mathbb{Z}$ for p a prime.

Definition 1.60 (The general linear group). Let n be a positive integer, and let F be $\mathbb{Q}$, $\mathbb{R}$, $\mathbb{C}$, or $\mathbb{Z}/p\mathbb{Z}$ for p a prime. Then the *general linear group* $\mathrm{GL}(n, F)$ is defined by

$$\mathrm{GL}(n, F) = \{ n \times n \text{ invertible matrices with entries in } F \}.$$

We write $\mathrm{GL}(n, p)$ instead of $\mathrm{GL}(n, \mathbb{Z}/p\mathbb{Z})$.

Example 1.61. The group $\mathrm{GL}(n, \mathbb{R})$ consists of invertible $n \times n$ matrices with real entries, and we can list all the elements of the group $\mathrm{GL}(2, 2)$:

$$\mathrm{GL}(2, 2) = \left\{ \begin{pmatrix} 1 & 0 \\ 0 & 1 \end{pmatrix}, \begin{pmatrix} 0 & 1 \\ 1 & 0 \end{pmatrix}, \begin{pmatrix} 1 & 0 \\ 1 & 1 \end{pmatrix}, \begin{pmatrix} 0 & 1 \\ 1 & 1 \end{pmatrix}, \begin{pmatrix} 1 & 1 \\ 1 & 0 \end{pmatrix}, \begin{pmatrix} 1 & 1 \\ 0 & 1 \end{pmatrix} \right\}.$$

Remark 1.62. If the scalars of a mathematical object X are from a particular set F (in this section F is usually one of the fields $\mathbb{Q}$, $\mathbb{R}$, $\mathbb{C}$, or $\mathbb{Z}/p\mathbb{Z}$), we often say that "X is *over* F". So, for example, instead of saying that M is a 3×3 matrix with entries in $\mathbb{C}$, we could say that M is a 3×3 matrix over the complexes. If $p(x)$ is a polynomial with coefficients in $\mathbb{Z}/3\mathbb{Z}$, we could say that $p(x)$ is a polynomial over $\mathbb{Z}/3\mathbb{Z}$. As a final example, if we say that V is vector space over $\mathbb{Q}$, we mean that V is a vector space and the scalars are from the field of rationals.

We note that $\mathrm{GL}(n, F)$ with the operation of matrix multiplication has the by-now-familiar four properties of a group. Closure is easy to prove (if A and B are invertible matrices, then $B^{-1}A^{-1}$ is the inverse of AB, and hence the latter is invertible). Matrix multiplication is associative (to prove this, we either use a somewhat tedious calculation or resort to the relationship between matrices and linear transformations). The identity matrix is always in $\mathrm{GL}(n, F)$, and we have deliberately only included the invertible matrices.

Much of the matrix theory that you know works for matrices with entries in a field. These will be developed rigorously in the sequel, but for now, and for the sake of having more examples, we accept that we can find determinants of matrices with entries in a field routinely and that the determinant function continues to have its usual properties (e.g., $\det(AB) = \det(A)\det(B)$). This allows us to define a subset of $\mathrm{GL}(n, F)$ that itself—together with the operation of matrix multiplication—has the four properties of closure, associativity, the existence of identity, and the existence of inverses.

Definition 1.63 (The special linear group). Let n be a positive integer, and let F be $\mathbb{Q}$, $\mathbb{R}$, $\mathbb{C}$, or $\mathbb{Z}/p\mathbb{Z}$ for p a prime. Let $\mathrm{GL}(n, F)$ be the general linear group of invertible $n \times n$ matrices with entries in F. Define

$$\mathrm{SL}(n, F) = \{A \in \mathrm{GL}(n, F) \mid \det A = 1\}.$$

Then $\mathrm{SL}(n, F)$ is called the *special linear group*. Again, instead of $\mathrm{SL}(n, \mathbb{Z}/p\mathbb{Z})$, we write $\mathrm{SL}(n, p)$.

Clearly, matrix multiplication in $\mathrm{SL}(n, F)$ continues to be associative. The identity matrix has determinant 1, and so it is in $\mathrm{SL}(n, F)$. Also, since $\det AB = \det A \det B$, we have closure. If the determinant of a matrix is d, then the determinant of its inverse is d^{-1}, and so the inverses of elements in $\mathrm{SL}(n, F)$ are in $\mathrm{SL}(n, F)$ as well. Thus we, again, have a set and an operation that satisfies all the four basic properties.

Both $\mathrm{GL}(n, p)$ and $\mathrm{SL}(n, p)$ are finite groups (i.e., have a finite number of elements). We record here their sizes. In the problems you are guided to a proof of this result.

Theorem 1.64. *Let n be a positive integer, and let p be a prime. Then*

$$|\mathrm{GL}(n, p)| = (p^n - 1)(p^n - p)(p^n - p^2) \cdots (p^n - p^{n-1}),$$

$$|\mathrm{SL}(n, p)| = \frac{1}{p - 1} |\mathrm{GL}(n, p)|.$$

Problems

1.4.1. What is the center of $GL(2,2)$ (see Definition 1.7)?

1.4.2. Let $a = \begin{bmatrix} 2 & 2 \\ 2 & 1 \end{bmatrix} \in GL(2,3)$. What is the inverse of a? Is $a \in SL(2,3)$?

1.4.3. Recall that $\mathbb{R}^n$ is the set of all n-tuples with entries in $\mathbb{R}$. In analogy, for a prime p, we define $(\mathbb{Z}/p\mathbb{Z})^n$ to be the set of all n-tuples with entries in $\mathbb{Z}/p\mathbb{Z}$. In other words,

$$(\mathbb{Z}/p\mathbb{Z})^n = \{(a_1, a_2, \ldots, a_n) \mid a_1, \ldots, a_n \in \mathbb{Z}/p\mathbb{Z}\}.$$

Just as in $\mathbb{R}^n$, we can add elements of $(\mathbb{Z}/p\mathbb{Z})^n$, and we can multiply them by scalars. For $\mathbb{R}^n$ the scalars are real numbers, and the scalars for $(\mathbb{Z}/p\mathbb{Z})^n$ are elements of $\mathbb{Z}/p\mathbb{Z}$. (In fact, both $\mathbb{R}^n$ and $(\mathbb{Z}/p\mathbb{Z})^n$ are examples of vector spaces.)
   (a) How many elements does $(\mathbb{Z}/p\mathbb{Z})^n$ have?
   (b) Let x be a fixed non-zero n-tuple in $(\mathbb{Z}/p\mathbb{Z})^n$. (In other words, $x \in (\mathbb{Z}/p\mathbb{Z})^n$ and not all of the entries of x are zero.) How many elements of $(\mathbb{Z}/p\mathbb{Z})^n$ are a scalar multiple of x?

1.4.4. How many elements does $GL(2,3)$ have? Justify your answer without an appeal to Theorem 1.64. Can you extend your argument to $GL(2,p)$ where p is an arbitrary prime?

1.4.5. **The size of** $GL(n,p)$**.** For this problem, accept as given that an $n \times n$ matrix is invertible if and only if its rows are linearly independent. Let A be a mystery invertible $n \times n$ matrix with entries in $\mathbb{Z}/p\mathbb{Z}$. In other words, $A \in GL(n,p)$.
   (a) How many possible choices are there for the first row of A?
   (b) If you know the first row of A, then how many choices are there for the second row of A?
   (c) If you know the first two rows of A, then how many choices are there for the third row of A?
   (d) Prove the statement about $|GL(n,p)|$ in Theorem 1.64.

1.4.6. List the elements of $SL(2,2)$. What are the possible values for a determinant of a matrix over $\mathbb{Z}/2\mathbb{Z}$? What can you say about the relationship between $GL(n,2)$ and $SL(n,2)$?

1.4.7. Fix a positive integer n. What are the possible values for a determinant of an invertible $n \times n$ matrix over $\mathbb{Z}/3\mathbb{Z}$? Do each of these values occur the same number of times (i.e., are the number of matrices that have determinant equal to α the same as the number of matrices with determinant equal to β)? Can you prove your assertion?

1.4.8. How many elements does $SL(2,3)$ have? Justify your answer without an appeal to Theorem 1.64.

1.4.9. **The size of** $SL(n,p)$**.** For this problem, accept as given that $\det(AB) = \det(A)\det(B)$ for $n \times n$ matrices A and B, and that a matrix (with entries in a field) is invertible if and only if its determinant is non-zero. Let M be an arbitrary matrix in $\mathrm{GL}(n,p)$.

(a) How many possible choices are there for $\det(M)$?

(b) Let k be an arbitrary non-zero element of $\mathbb{Z}/p\mathbb{Z}$. Give an explicit $n \times n$ matrix A with $\det(A) = k$.

(c) Let α and β be non-zero elements of $\mathbb{Z}/p\mathbb{Z}$. Let $\mathcal{S}_1$ be the set of elements of $\mathrm{GL}(n,p)$ with determinant equal to α. Likewise, let $\mathcal{S}_2 = \{M \in \mathrm{GL}(n,p) \mid \det(M) = \beta\}$. Let A be an $n \times n$ matrix with determinant equal to β/α. Prove that the map $f : \mathcal{S}_1 \to \mathcal{S}_2$ defined by $f(N) = AN$ is an actual 1-1, onto function. Conclude that $|\mathcal{S}_1| = |\mathcal{S}_2|$.

(d) Prove the statement about $|\mathrm{SL}(n,p)|$ in Theorem 1.64.

Fields. In this section, we have assumed F to be one of $\mathbb{Q}$, $\mathbb{R}$, $\mathbb{C}$, or $\mathbb{Z}/p\mathbb{Z}$ where p is a prime. In fact, we can let F be any field. We will study fields in much more detail later on (see Section 15.2), but, as a prelude, we give the definition here.

Informally, a field is a set with two operations (we call one addition and the other multiplication) such that both operations are commutative and that we can do addition, subtraction, multiplication, and division by non-zero elements.

Definition 1.65 (Field). Let F be a set with two binary operations $+$ and $\cdot$. Assume that

(a) $(F, +)$ is closed, associative, commutative, has an identity (called zero), and every element has an (additive) inverse.

(b) $(F - \{0\}, \cdot)$ is closed, associative, commutative, has an identity (called one), and every (non-zero) element has an (multiplicative) inverse.

(c) We have the distributive law: $a \cdot (b + c) = a \cdot b + a \cdot c$.

Then $(F, +, \cdot)$ is called a *field*.

1.4.10. In Problem 1.3.1—see also Corollary 1.57—you were asked if every non-zero element of $\mathbb{Z}/12\mathbb{Z}$ has a multiplicative inverse. Is this the same as asking if $\mathbb{Z}/12\mathbb{Z}$ is a field? For which n is $\mathbb{Z}/n\mathbb{Z}$ a field? Why?

1.4.11. Define
$$\mathbb{Q}[\sqrt{3}] = \{a + b\sqrt{3} \mid a, b \in \mathbb{Q}\}.$$

Is $\mathbb{Q}[\sqrt{3}]$ a field? (The two operations are the usual addition and multiplication of numbers.)

1.4.12. Define $\mathbb{R}[x]$ to be the set of polynomials in the variable x and with coefficients in $\mathbb{R}$, the real numbers. Addition and multiplication is the usual addition and multiplication of polynomials. Is $(\mathbb{R}[x], +)$ a group? Is $(\mathbb{R}[x], +, \cdot)$ a field?

1.5. More Problems and Projects

Problems

1.5.1. **The Infinitude of Primes.** In this problem, you are asked to reconstruct a novel proof of the infinitude of primes.[5]

Let A be a subset of the integers. We say that A is *periodic* if there exists an integer m—called a *period* of A—such that, for all integers x, we have $x \in A$ if and only if $x + m \in A$.

(a) Prove that if A and B are periodic subsets of the integers, then so is $A \cup B$. What is a period for $A \cup B$? Conclude that finite unions of periodic sets are periodic.

(b) Prove that if A is a periodic subset of the integers, then so is the complement of A.

(c) Can a finite set be periodic?

(d) For a prime p, define
$$S_p = \{np \mid n \in \mathbb{Z}\}.$$

Show that S_p is periodic.

(e) Show that, other than $+1$ and -1, every integer is in at least one S_p for some prime p.

(f) Assume that the number of primes is finite. Then, using the previous parts, argue that the complement of the union of all the sets S_p must be periodic, and that this complement is the non-periodic set $\{1, -1\}$. Conclude that the number of primes must be infinite.

(g) To make sure that you understood this proof, give an example of an infinite number of periodic sets whose union is not periodic. Prove your assertions.

---

[5] Adapted from Cass and Wildenberg [**CW03**].

Elementary Matrices and the General Linear Group. You are probably familiar with elementary row operations and elementary matrices from linear algebra. We give the relevant definitions for completeness.

Definition 1.66 (Elementary row operations). Let F be a field, and let A be an $m \times n$ matrix with entries from F. The elements of F are called scalars, and the following operations on the rows of A are called *elementary row operations (over F)*:

(a) Switch two rows of A.

(b) Multiply a row of A with a non-zero scalar.

(c) Add to one row a scalar multiple of another row.

Definition 1.67 (Row equivalence). Let A and B be two $m \times n$ matrices with entries from a field F. If you can get B from A through a sequence of elementary row operations, then we say A is *row equivalent* to B.

Definition 1.68 (Elementary matrices). Let m be a positive integer, and let I_m be the $m \times m$ identity matrix. Assume that the matrix E is obtained from I_m using one elementary row operation over a field F. Then E is called an *elementary matrix (over F)*.

The following two results are standard results from elementary linear algebra and will be assumed:

Proposition 1.69. *Let F be a field, and assume all matrices have entries in F. Let A and B be $m \times n$ matrices, and let X be an $n \times 1$ matrix of unknowns.*

(a) *If A and B are row equivalent, then the set of solutions to the system $AX = 0$ is identical to the set of solutions to the system $BX = 0$.*

(b) *If E is an elementary matrix obtained from I_m by an elementary row operation, then EA is the matrix obtained from A by the same elementary row operation.*

(c) *An elementary matrix is invertible, and its inverse is also an elementary matrix.*

One significance of elementary matrices and elementary row operations is the following:

Theorem 1.70. *Let F be a field. Let A be an $n \times n$ matrix over F. Then the following are equivalent:*

(a) *The matrix A is invertible.*

(b) *The only solution to the system $AX = 0$ is $X = 0$.*

(c) *The matrix A is row equivalent to the identity matrix I_n.*

(d) *The matrix A is a product of elementary matrices.*

Now $\mathrm{GL}(n, F)$ consists of invertible $n \times n$ matrices over F. By Theorem 1.70, elementary matrices are in $\mathrm{GL}(n, F)$, and every element of $\mathrm{GL}(n, F)$ is a product of elementary matrices. Hence, we conclude:

Proposition 1.71. $\mathrm{GL}(n, F)$ *is generated by the set of $n \times n$ elementary matrices over F.*

1.5.2. Sketch a proof of Proposition 1.69.

1.5.3. Give a proof—possibly by consulting a linear algebra text—of Theorem 1.70.

1.5.4. Find the elementary matrices in $GL(2,2)$. Do you need all of them to generate $GL(2,2)$? How many suffice?

1.5.5. List the elementary matrices in $GL(2,3)$, and from among them find a (small) set of generators for $GL(2,3)$.

1.5.6. **The center of** $GL(n,F)$. Let F be a field (the rationals, the reals, the complexes, or $\mathbb{Z}/p\mathbb{Z}$). Show that $\mathbf{Z}(GL(n,F))$, the center of the general linear group, is $\{\lambda I_n \mid \lambda \in F\}$, the set of matrices that are constant multiples of the identity matrix. You may find the following steps helpful:
Step 1: For $1 \leq i,j \leq n$, let $E_{i,j}$ be the $n \times n$ matrix that has a one in the (i,j) entry and zeros elsewhere. Show that $B_{i,j} = I_n + E_{i,j}$ is an elementary matrix and hence an element of $GL(n,F)$.
Step 2: Assume that $A \in \mathbf{Z}(GL(n,F))$. Show that $AB_{i,j} = B_{i,j}A$ implies that $AE_{i,j} = E_{i,j}A$.
Step 3: If $A \in \mathbf{Z}(GL(n,F))$, by comparing $AE_{i,j}$ with $E_{i,j}A$, conclude that $A = \lambda I_n$ for some $\lambda \in F$.

Mathematical Induction and the Well-Ordering Principle. In studying the properties of integers as well as other mathematical statements that refer to an infinite sequence of integers, a very powerful tool is mathematical induction.

Axiom 1.72 (Principle of mathematical induction). Given an infinite sequence of statements,

$$P(1), P(2), \ldots, P(n), \ldots,$$

in order to prove that all of them are true, it is enough to show that $P(1)$ is true, and that, for $k \geq 1$, the truth of $P(k)$ implies the truth of $P(k+1)$.

Can we "prove" that mathematical induction works? This depends on the axioms that you accept to begin your work. Mathematical induction itself or one of its equivalents need to be taken as an axiom of set theory. If you choose an equivalent formulation, then you can "prove" the principle of mathematical induction. If you choose the principle of induction as an axiom, then, of course, you cannot prove it. An axiom that is equivalent to the principle of induction is the well-ordering principle of Axiom 1.39. Recall that the well-ordering principle says that a non-empty set of non-negative integers has a smallest member. Here you are asked to show that the two principles are equivalent.

1.5.7. Prove that the principle of mathematical induction is true if and only if the well-ordering principle is true.

Figure 1.10. Cardboard pieces available for tiling a chessboard.

1.5.8. Let n be a positive integer, and assume that we have a $2^n \times 2^n$ chessboard and an unlimited supply of cardboard pieces, each made of three squares as shown in Figure 1.10. The squares on the cardboard pieces are identical with the squares on the chessboard.

  (a) The integers $2^{2n} - 1$, 2^{2n}, and $2^{2n} + 1$ are three consecutive integers. Which one(s) is divisible by 3?

  (b) Can you tile the chessboard using an appropriate number of cardboard pieces?

  (c) Cut out one of the squares in the chessboard. Can you now tile the chessboard using an appropriate number of cardboard pieces? Does it matter which square is cut out?

Chapter 2

Groups: The Basics

...where a group is defined, and a basic language for studying groups is developed by exploring cyclic groups, orders, isomorphisms, direct products, and subgroups.

2.1. Definitions and Examples

We saw four examples—symmetries of a square, the set of 1-1, onto maps on a set, integers modulo n, and invertible matrices over a field—where we had a set of elements and an operation. The operation had four basic properties: closure, associativity, existence of identity, and existence of inverses. The examples were somewhat varied in that the elements of these sets and the operations were not really that similar. Based on these examples, we abstract out the notion of a group. A *group* will be a set together with an operation that follows the familiar four rules. This is an abstract object in that we do not know what the elements of the set are. The real question is whether we can say anything deep or profound given the little that we have to work with. By the end of the book we hope to have answered this question satisfactorily.

To define the notion of a group, it is helpful—although not really necessary—to first define Cartesian products and binary operations.

Definition 2.1 (Direct product). If X and Y are two sets, then

$$X \times Y = \{(x, y) \mid x \in X, y \in Y\}$$

is the *direct product* (or *Cartesian product*) of X and Y.

Thus the direct product of X and Y is a new set consisting of ordered pairs of elements, the first element of each pair is from X and the second element is from Y.

Example 2.2. If $X = \{\square, \clubsuit, \heartsuit\}$ and $Y = \{\aleph, \flat\}$, then

$$X \times Y = \{(\square, \aleph), (\square, \flat), (\clubsuit, \aleph), (\clubsuit, \flat), (\heartsuit, \aleph), (\heartsuit, \flat)\}.$$

Example 2.3. Let $\mathbb{R}$ denote the set of real numbers. Then

$$\mathbb{R} \times \mathbb{R} \times \mathbb{R} = \mathbb{R}^3 = \{(a, b, c) \mid a, b, c \in \mathbb{R}\}.$$

In other words, the plane $\mathbb{R}^2$ and the three dimensional space $\mathbb{R}^3$—familiar to us from calculus and linear algebra—are examples of direct products.

Example 2.4. We have

$$\mathbb{Z}/2\mathbb{Z} \times \mathbb{Z}/3\mathbb{Z} = \{(0, 0), (0, 1), (0, 2), (1, 0), (1, 1), (1, 2)\}.$$

Definition 2.5 (Binary operations). For a set S, a map $b \colon S \times S \to S$ is called a *binary operation* on S.

In other words, a binary operation is a rule that we can apply to any (ordered) pair of elements of S and get a new element of S. If $*$ is a binary operation, we write $a * b$ instead of $*(a, b)$.

Definition 2.6 (Groups). Let G be a non-empty set. Let $*$ be a binary operation on G. Then G and the operation $*$ form a *group* if the following hold:

(a) If a and b are arbitrary elements of G, then so is $a * b$ *(closure of the product)*.

(b) For all elements a, b, and c of G, we have $a * (b * c) = (a * b) * c$ *(associativity of the product)*.

(c) Among the elements of G, there is an element, denoted by e, with the property that $a * e = e * a = a$ for every $a \in G$ *(existence of identity)*.

(d) For every element $a \in G$, there exists an element, denoted by a^{-1}, such that $a * a^{-1} = e = a^{-1} * a$ *(existence of inverses)*.

Note that to define a group, we need to specify a set G *and* an operation $*$. Thus it would have been more precise to say that $(G, *)$ is a group if .... We will name the operation only if there is some danger of confusion. In fact, most of the time we will suppress $*$ altogether. In other words, we write ab instead of $a * b$. We often—but not always—use e for the identity element. If there are a number of groups and we want to specifically talk about the identity of the group G, we denote it by e_G.

Also note that the above definition does have at least one redundancy. A binary operation is closed by definition, and hence, since we had specified that $*$ is a binary operation, we did not have to stipulate that $*$ be closed.

Group theory is the study of groups and, as such, is an abstract undertaking. We would like to know what else follows, given that a set with a given operation is a group. In other words usually we are not interested in what the elements of the group are, and what other properties they have. Of course, you might say that if we only know that we are starting with a set with an operation that follows only the above innocent looking axioms, then there could not be much that we can say that has any content. However, amazing things are true and can be proved about groups. For example, even though the axioms do not say anything about commutativity—and, we have already seen examples of non-commutative groups, namely D_8 and S_3—we can (and will) prove that any group with $2,209$ elements has to be commutative!

We have already seen a number of examples of groups, but we list them and others here as well. When we are developing the theory of groups, we will be quite careful with proofs. For every claim that we make, we will either give a proof or ask the reader to complete the argument. However, when presenting examples, we may be more sketchy. Sometimes, providing a complete proof may take us too far off track. Often, as we develop more theory, we will be able to justify our examples more easily.

Example 2.7. Symmetries of any geometric object form a group.

Example 2.8. As a specific example of a geometric object, let P_n be the regular polygon in a plane with n sides. The symmetries of P_n are denoted by D_{2n} which is called the dihedral group of order $2n$.[1]

The elements of D_{2n} are

(a) rotations by $0, 2\pi/n, \ldots, 2(n-1)\pi/n$ radians, and

(b) reflections: if n is even, axes go through opposite vertices or opposite sides; if n is odd, axes go from a vertex to the midpoint of the opposite side.

If we let ρ denote a rotation by $2\pi/n$ radians and we let δ denote any one of the reflections, then

$$D_{2n} = \{e, \rho, \rho^2, \ldots, \rho^{n-1}, \delta, \rho\delta, \rho^2\delta, \ldots, \rho^{n-1}\delta \mid \rho^n = \delta^2 = e, \delta\rho = \rho^{n-1}\delta\}.$$

Example 2.9. If Ω is any set, then $\mathrm{Perm}(\Omega)$, the set of 1-1, onto functions on Ω, with the operation of function composition, is a group.

Example 2.10. The set of integers $\mathbb{Z}$ with the operation of addition is a group. The identity element is 0, and the inverse of an integer a is $-a$.

Example 2.11. The set of integers mod n with the operation of addition mod n, $(\mathbb{Z}/n\mathbb{Z}, +)$, is a group.

Example 2.12. With the operation of multiplication mod n, $((\mathbb{Z}/n\mathbb{Z})^{\times}, \cdot)$, the set of elements of $\mathbb{Z}/n\mathbb{Z}$ with a multiplicative inverse—the so-called units of $\mathbb{Z}/n\mathbb{Z}$—is a group.

Example 2.13. Each of $(\mathbb{Q}, +)$, $(\mathbb{R}, +)$, and $(\mathbb{C}, +)$ are groups. However, each of these sets—in addition to addition—have a multiplication. Zero is the only non-invertible element (under multiplication), and each of these sets without the zero and with multiplication as the operation is also a group. If F is any one of $\mathbb{Q}$, $\mathbb{R}$, or $\mathbb{C}$—in fact, F can be any field—then we denote by $F^{\times}$ or $(F^{\times}, \cdot)$ the group of non-zero elements of F together with multiplication. So $\mathbb{Q}^{\times}$, $\mathbb{R}^{\times}$, and $\mathbb{C}^{\times}$ are all examples of groups.

The non-zero elements of a field are also called its *units*. In Section 15.2, we define an algebraic structure—called a ring with identity—more general than a field. The units in a ring with identity will also form a group. (See Proposition 15.11.)

---

[1]Note that we denote by D_{2n} the symmetries of an n-sided regular polygon. We use the subscripts $2n$ since this is the number of elements in D_{2n}. Some authors refer to this same group as D_n preferring to focus on the number of sides of the polygon. Often group theorists use D_{2n}, and topologists use D_n.

Example 2.14. If n is a positive integer and F is any of $\mathbb{Q}$, $\mathbb{R}$, $\mathbb{C}$, or $\mathbb{Z}/p\mathbb{Z}$ where p is a prime, then $(\mathrm{GL}(n, F), *)$, the set of invertible $n \times n$ matrices with entries in F, is a group. Here $*$ is matrix multiplication, and the identity element is the $n \times n$ identity matrix, I_n.

Example 2.15. If n and F are as in the previous example, then the set of $n \times n$ matrices with entries in F and with determinant 1, $(\mathrm{SL}(n, F), *)$, is a group. Here also $*$ is matrix multiplication, and e is I_n.

Example 2.16. Let $C^0(-\infty, \infty) = \{f : \mathbb{R} \to \mathbb{R} \mid f \text{ is continuous}\}$ be the set of $\mathbb{R}$-valued continuous functions on $\mathbb{R}$. If f and g are functions in $C^0(-\infty, \infty)$, then we define a new function $f + g$ with domain and codomain $\mathbb{R}$, using the rule $(f + g)(x) = f(x) + g(x)$. This operation is called pointwise addition of functions, and $(C^0(-\infty, \infty), +)$ is a group.

Before we begin investigating groups, we make two definitions and give two questions that will guide our work.

Definition 2.17 (Order of a group). Let G be a group. G is a *finite group* if it has a finite number of elements; otherwise, G is an *infinite group*. The number of elements of G, denoted by $|G|$, is called the *order of G*.

Definition 2.18 (Commuting elements and abelian groups). Let $(G, *)$ be a group, and let a and b be elements of G. We say that a and b *commute* if $a * b = b * a$. We say that G is *abelian* if every pair of elements commute.

We will be concentrating on the study of finite groups, and, to appreciate the structural information that we will be able to deduce, we will have two guiding questions in mind.

Question 2.19 (Guiding questions).

- How do we prove that a group is abelian? Is there ever an alternative to checking that $ab = ba$ for every pair of elements $a, b \in G$?

- For a particular n, how many "different" groups of order n do we have?

As we prove new results, we keep coming back to see if we can make any progress on the two questions above.

Problems

2.1.1. Let I_n be the $n \times n$ identity matrix. Is

$$\{rI_n \mid r > 0, r \in \mathbb{R}\}$$

a group under matrix multiplication?

2.1.2. Let θ be a real number and define

$$R_\theta = \begin{bmatrix} \cos(\theta) & -\sin(\theta) \\ \sin(\theta) & \cos(\theta) \end{bmatrix}.$$

(a) R_θ is called a rotation matrix. Can you explain why?

(b) Show

$$R_\theta R_\mu = R_?,$$

$$R_\theta^{-1} = R_?.$$

(c) Let

$$G = \{R_\theta \mid \theta \in \mathbb{R}\}.$$

Show that G is a group under matrix multiplication.

2.1.3. Let $\mathbb{Z}$ denote the set of integers, and let

$$G = \{ \begin{bmatrix} 1 & a & 0 \\ 0 & 1 & 0 \\ 0 & 0 & 1 \end{bmatrix} \mid a \in \mathbb{Z}\}.$$

Prove that G together with the usual matrix multiplication forms a group.

2.1.4. **GL$(n, \mathbb{Z})$ and SL$(n, \mathbb{Z})$.** Let $\mathrm{M}_{n \times n}(\mathbb{Z})$ be the set of $n \times n$ matrices with integer entries.
(a) Does the set of invertible matrices in $\mathrm{M}_{n \times n}(\mathbb{Z})$ form a group?
The set of invertible matrices in $\mathrm{M}_{n \times n}(\mathbb{Z})$ whose matrix inverse is also a matrix in $\mathrm{M}_{n \times n}(\mathbb{Z})$ is denoted by GL$(n, \mathbb{Z})$. In addition, the set of matrices in $\mathrm{M}_{n \times n}(\mathbb{Z})$ that have determinant 1 is denoted by SL$(n, \mathbb{Z})$.
(b) Let $A \in \mathrm{M}_{n \times n}(\mathbb{Z})$. Show that $A \in$ GL$(n, \mathbb{Z})$ if and only if $\det(A) = \pm 1$.
(c) Is GL$(n, \mathbb{Z})$ a group?
(d) Is SL$(n, \mathbb{Z})$ a group?

2.1.5. Let $\mathrm{M}_{2 \times 2}(\mathbb{Z}/6\mathbb{Z})$ be the set of 2×2 matrices with entries in $\mathbb{Z}/6\mathbb{Z}$.
(a) Can you find a matrix in $\mathrm{M}_{2 \times 2}(\mathbb{Z}/6\mathbb{Z})$ whose determinant is non-zero and yet is not invertible?
(b) Does the set of invertible matrices in $\mathrm{M}_{2 \times 2}(\mathbb{Z}/6\mathbb{Z})$ form a group?

2.1.6. Let n be a positive integer. For which n is S_n abelian? Prove your assertion.

2.1.7. In Definition 1.9, we defined a function (or a map) as a *rule* that assigns to each element of X precisely one element of Y. You may object that we had not properly defined what we mean by a "rule". We can give an alternate definition of a function using sets and direct products:

Let X and Y be sets. A *function* (or *map* or *mapping*) f from X to Y is a subset of $X \times Y$ with the property that for each $x \in X$ there is exactly one element of the form $(x, y) \in X \times Y$ in f. If f is a function and if $(x, y) \in f$, then we write $f(x) = y$.

(a) Are you convinced that the definition above can be substituted for Definition 1.9?
(b) Give a set theoretic definition—with no mention of functions—of a binary operation.

2.1.8. Let G be a group of functions from a set S to itself with multiplication defined as function composition. Find an example with $|G| \geq 2$ such that G does not contain any 1-1 functions.

Power Set and Symmetric Difference

Definition 2.20. Let X be any set, and let 2^X be the *power set* of X, i.e., 2^X is the set of all subsets of X. For A and B subsets of X, $A - B$ denotes the set consisting of all elements of A that are not in B. The *symmetric difference* of A and B is denoted by $\triangle$, and is defined by

$$A \triangle B = (A - B) \cup (B - A).$$

2.1.9. Show that $(2^X, \triangle)$ is an abelian group. Make sure that you prove associativity, that you identify the identity, and that you find every element's inverse. (You can use Venn diagrams when appropriate.)

2.2. Cancellation Properties

In some sense all the properties of a group can be found in its multiplication table. This table has all the information about a group and hence any theorem about a group is, in the last analysis, a statement about the multiplication table of the group. This point of view, as we shall see, is not too productive. However, the first things that we want to prove about groups are, in fact, patterns that are clear if you look at some group multiplication tables. The first thing that you will notice in a group table is that there are no repeat entries in any row or any columns. How do we express this fact mathematically? Consider the row of the element a in a group G. The elements in this row are of the form ax for some $x \in G$. The fact that there are no repeats in this row can be expressed by saying that $ax \neq ay$ for $x \neq y$. Another way of saying this is that if $ax = ay$, then $x = y$. In other words, we can cancel like elements on the left. Similarly, no repeats in the columns translates to canceling like terms on the right of any equation. We shall now prove these:

Lemma 2.21 (Cancellation properties)**.** *Let G be a group, and let $a, b, c \in G$. Then*

(a) $ab = ac \Rightarrow b = c$.

(b) $ba = ca \Rightarrow b = c$.

Proof. (a)

$$
\begin{aligned}
ab &= ac \\
\Rightarrow \quad a^{-1}(ab) &= a^{-1}(ac) \\
\Rightarrow \quad (a^{-1}a)b &= (a^{-1}a)c \\
\Rightarrow \quad b &= c.
\end{aligned}
$$

(b) Similar to the previous part. □

Lemma 2.22. *Let G be any group. Then*

(a) *The identity element e is* unique. *That is, no other element x has the property that $xg = g = gx$ for all $g \in G$.*

(b) *Every element $a \in G$ has a* unique *inverse.*

(c) *For every $a \in G$, we have $(a^{-1})^{-1} = a$.*

(d) *If a and b are elements of G, then $(ab)^{-1} = b^{-1}a^{-1}$.*

Proof. (a) Assume that e and e' are both identities for the group G. Then $ee' = e$ since e' is an identity element, and also $ee' = e'$ since e is an identity element. So
$$e = ee' = e'.$$

(b) Assume that b, c are inverses of a. Then
$$ba = e = ca.$$

Since $ba = ca$, cancel a and get $b = c$.

(c) We have that $a^{-1}a = aa^{-1} = e$. This means that a is the inverse of a^{-1}. Thus $(a^{-1})^{-1} = a$.

(d) The claim is that the inverse of ab is $b^{-1}a^{-1}$. We check this by multiplying the two to see if we get the identity. We have $(ab)(b^{-1}a^{-1}) = e$ as well as $(b^{-1}a^{-1})(ab) = e$, and, hence, ab and $b^{-1}a^{-1}$ are inverses of each other. We write this as $(ab)^{-1} = b^{-1}a^{-1}$, and the proof is complete. □

We now show that there was some redundancy in the definition of a group. We will use this to get a useful criterion for showing that a set with an operation is indeed a group.

Definition 2.23. Assume that $*$ is an associative binary operation on a set G. Then $(G, *)$—or G for short—is called a *semigroup*.[2]

Theorem 2.24. *Let G be a non-empty semigroup. Assume that G has a left identity and that every element of G has a left inverse. That is, there exists an element $e \in G$ such that, for every $a \in G$, $ea = a$, and, for every $a \in G$, there exists an element, denoted by a^{-1}, such that $a^{-1}a = e$.*

Then G is a group.

Proof. First, note that in the proof of Lemma 2.21, to show that we had left cancellation, we only used left identity and left inverses. Thus we can conclude that we have left cancellation in G.

CLAIM 1: e is also a right identity.

PROOF OF CLAIM 1: Let $a \in G$. We need $ae = a$. We know a has a left inverse a^{-1} in G, and so
$$a^{-1}a = e = ee = a^{-1}ae,$$
which implies, by left cancellation, that
$$a = ae.$$

CLAIM 2: Let $a \in G$. Then a^{-1} is also a right inverse.

---

[2]Some authors insist that a semigroup be a non-empty set. Also a semigroup with an identity element—which then has to be non-empty—is often called a *monoid*. A group then is a monoid where every element is invertible.

PROOF OF CLAIM 2:

$$a^{-1}e = a^{-1} = ea^{-1} = (a^{-1}a)a^{-1} = a^{-1}(aa^{-1}),$$

so

$$a^{-1}e = a^{-1}(aa^{-1}).$$

Thus, by left cancellation, we have

$$e = aa^{-1}. \qquad \square$$

We now show that to guarantee that a finite semigroup is a group, we only need the cancellation laws.

Theorem 2.25. *Let G be a non-empty semigroup with a finite number of elements. Assume that $ax = ay$ implies $x = y$, and $ua = wa$ implies $u = w$ for all $a, x, y, w, u \in G$.*

Then G is a group.

Proof. Let $G = \{a_1, a_2, \ldots, a_n\}$. We need to find an identity and then prove that every element has an inverse.

Let b be a fixed element of G. How many distinct elements does the set $\{ba_1, ba_2, \ldots, ba_n\}$ have? If $ba_i = ba_j$, then by left cancellation we have $a_i = a_j$, and so this set has n elements. All of these elements are in G, and G has only n elements. Thus

$$G = \{ba_1, ba_2, \ldots, ba_n\}.$$

So, in particular, $b = ba_k$ for some $1 \le k \le n$.

First, we note that $bb = (ba_k)b = b(a_kb)$. Canceling the b's on the left, we get $b = a_kb$. In other words, the element a_k is the left identity for the element b.

Second, we show that a_k is indeed a left identity for all of G. Let $c \in G$. Since $G = \{ba_1, ba_2, \ldots, ba_n\}$, every element of G—and, in particular, c—is also of this form. Hence, $c = ba_j$ for some $1 \le j \le n$. Now, we have $a_kc = a_kba_j = ba_j = c$, and so a_k is the left identity for G.

Third, we show that every element of G has a left inverse. Let $c \in G$. We want c^{-1}. Consider $\{a_1c, a_2c, \ldots, a_nc\}$. Again this set has n distinct elements by right cancellation, and hence it is the same as G. Thus $a_\ell c = a_k$ for some $1 \le \ell \le n$. So a_ℓ is a left inverse for c.

Now G is a group by Theorem 2.24 since we have a left identity and every element has a left inverse. $\qquad \square$

Remark 2.26. It is amusing—and of interest to logicians—to know that it is possible to define groups using just a single axiom. In 1952, G. Higman and B. H. Neumann [**HN52**] (see also Kunen [**Kun92**] and McCune and Sands [**MS96**]) showed that a non-empty set G with a binary operation "/" is a group (the operation of the group—as well as the identity and inverses—can then be defined in terms of "/"), if and only if for all $x, y, z \in G$, we have

$$(x/((((x/x)/y)/z)/(((x/x)/x)/z))) = y.$$

Problems

2.2.1. Let G be a group. Prove that $(ab)^{-1} = a^{-1}b^{-1}$ for all a and b in G if and only if G is abelian.

2.2.2. Let G be a group. Show that, for all $a, b \in G$, we have $(ab)^2 = a^2b^2$ if and only if G is abelian.

2.2.3. If G is a group in which $a^2 = e$ for all $a \in G$, show that G is abelian.

2.2.4. (a) If G is a finite group of even order, show that there must be an element $a \neq e$, such that $a^{-1} = a$.
(b) Give an example to show that the conclusion of part (a) above does not hold for groups of odd order.

2.2.5. Find the multiplication table of all groups of order 3. Can a group of order 3 be non-abelian?
Note: We consider two groups the "same" if after relabeling (the elements), their multiplication tables become identical. (See Section 2.4 for a more formal definition.)

2.2.6. Let $\{M_i \mid i = 1, \ldots, k\}$ be a set of $n \times n$ matrices (over a field) that form a group under matrix multiplication. Let $M = M_1 + \cdots + M_k$.
(a) What can you say about $M_i M$?
(b) What can you say about M^2?
(c) Give an example.

2.3. Cyclic Groups and the Order of an Element

Every group has one operation which we are calling multiplication. Just as in integers, we can use repeated multiplication to define exponentiation.

Definition 2.27 (Exponentiation). Let n be a positive integer. Let G be a group, and let $a \in G$. We define $a^n = \underbrace{a \cdot a \cdots a}_{n \text{ times}}$, $a^0 = e$, and $a^{-n} = (a^{-1})^n$.

Example 2.28. The matrix $a = \begin{bmatrix} 0 & 1 \\ 1 & 1 \end{bmatrix}$ is an element of the group $\mathrm{GL}(2,3)$. Remembering that the scalar operations are in $\mathbb{Z}/3\mathbb{Z}$, we can find various powers of a. As examples:

$$a^2 = \begin{bmatrix} 0 & 1 \\ 1 & 1 \end{bmatrix}\begin{bmatrix} 0 & 1 \\ 1 & 1 \end{bmatrix} = \begin{bmatrix} 1 & 1 \\ 1 & 2 \end{bmatrix},$$

$$a^3 = \begin{bmatrix} 0 & 1 \\ 1 & 1 \end{bmatrix}\begin{bmatrix} 0 & 1 \\ 1 & 1 \end{bmatrix}\begin{bmatrix} 0 & 1 \\ 1 & 1 \end{bmatrix} = \begin{bmatrix} 1 & 2 \\ 2 & 0 \end{bmatrix},$$

$$a^{-1} = \begin{bmatrix} 2 & 1 \\ 1 & 0 \end{bmatrix},$$

$$a^{-2} = (a^{-1})^2 = \begin{bmatrix} 2 & 1 \\ 1 & 0 \end{bmatrix}\begin{bmatrix} 2 & 1 \\ 1 & 0 \end{bmatrix} = \begin{bmatrix} 2 & 2 \\ 2 & 1 \end{bmatrix}.$$

Lemma 2.29. *Let G be a group, and let $a \in G$. Let $m, n \in \mathbb{Z}$. Then*

(a) $a^m a^n = a^{m+n}$, *and*

(b) $(a^n)^{-1} = a^{-n}$, *and*

(c) $(a^m)^n = a^{mn}$.

Proof. We leave the proofs to the reader, except, as an example, we give a proof of the second assertion. We want to prove that a^{-n} is the inverse of a^n.

This is clear if $n = 0$. If n is a positive integer, then

$$a^n \cdot a^{-n} = \underbrace{a \cdots (a\,a^{-1}) \cdots a^{-1}}_{\substack{n \text{ times} \quad n \text{ times}}} = e,$$

and similarly $a^{-n} a^n = e$, proving that a^n is the inverse of a^{-n}.

If n is a negative integer, then let $k = -n > 0$. Then we need to show, for $k > 0$, that a^k is the inverse of a^{-k}. We just proved this, and so, regardless of the sign of n, we have $a^{-n} = (a^n)^{-1}$.   □

Inside every group, there are many smaller groups. The next theorem gives the simplest way of constructing a group inside an already existing one.

Theorem 2.30. *Let $(G, \cdot)$ be a group, and let $a \in G$. Let*

$$H = \{\, a^k \mid k \in \mathbb{Z} \,\} = \{e, a, a^{-1}, a^2, a^{-2}, \ldots\}.$$

Then $(H, \cdot)$ is an abelian group.

Proof. The set H consists of all integer (positive, negative, and zero) powers of the element a. Hence elements such as a^{47} as well as a^{-47}, a^3, and a^0 are in H. To prove that H is an abelian group, we need to show that the four conditions for a group hold, and that the operation is commutative.

(a) We have closure by Lemma 2.29(a) since the product of any two elements of H continues to be in H: $a^m a^n = a^{m+n}$.

(b) Associativity of the product is inherited from G. In other words, the operation was already associative when applied to the larger set of elements of G. Hence it continues to be associative when restricted to only elements of H.

(c) The identity element of G is $a^0 = e$ and belongs to H.

(d) By Lemma 2.29(b), the inverse of a^n is a^{-n} which also belongs to H.

(e) If m and n are integers, then $m + n = n + m$. This results in H being abelian since

$$a^m a^n = a^{m+n} = a^{n+m} = a^n a^m.$$ □

Example 2.31. Let $G = (\mathbb{Q} - \{0\}, \cdot)$ be the group of non-zero rational numbers under multiplication. Let $a = 1/2 \in G$. Then the abelian group promised by Theorem 2.30 is

$$H = \{1, 1/2, 2, 1/4, 4, 1/8, 8, \ldots\} = \{2^n \mid n \in \mathbb{Z}\}.$$

The group H constructed in Theorem 2.30 is generated by one element and has a particularly basic structure. We give a name to groups of this kind:

Definition 2.32 (Cyclic groups). If G is a group and there exists $a \in G$ such that $G = \{a^k \mid k \in \mathbb{Z}\}$, then G is called a *cyclic group* and a is called a *generator* of G. We write $G = \langle a \rangle$.

Example 2.33. The group D_8 is not a cyclic group, since it cannot be generated by only one element.

However, inside D_8 we do have many cyclic groups. For example, $\langle R_{90} \rangle = \{R_0, R_{90}, R_{180}, R_{270}\} = \{e, a, a^2, a^3\}$.

Example 2.34. The group $(\mathbb{Z}/5\mathbb{Z}, +) = \{0, 1, 2, 3, 4\}$ is cyclic. This is because $\langle 1 \rangle = \{1, 1+1 = 2, 1+1+1 = 3, 1+1+1+1 = 4, 1+1+1+1+1 = 0\} = \{0, 1, 2, 3, 4\}$. In fact, 1, 2, 3, and 4 are all generators.

Example 2.35. The group $(\mathbb{Z}/5\mathbb{Z})^\times = \{1, 2, 3, 4\}$ (where the operation is multiplication modulo 5) is also cyclic, since

$$\langle 2 \rangle = \{2, 2^2 = 4, 2^3 = 3, 2^4 = 1\}.$$

However, for this group, 4 is not a generator, since $\langle 4 \rangle = \{4, 4^2 = 1\}$.

Remark 2.36. If $(G, \star)$ is a group, we can, of course, use any symbol instead of the group operation $\star$. However, the two most common notations are multiplication and addition. When using the multiplication notation we write ab for $a * b$, and when using the additive notation we write $a + b$ for $a * b$. At first, this can cause some confusion. Since, for example, in the additive notation $2a = a + a$, while in the multiplicative notation we have $aa = a^2$. The important thing to remember is that we only use the additive notation for abelian groups; in other words, $a + b = b + a$, while ab may not be the same as ba. To be applicable more generally, when stating theorems, we tend to use the multiplicative notation. However, if the group under discussion is abelian, we could switch to the additive notation. Figure 2.1 gives a list of some corresponding statements in both notations.

multiplicative notation	additive notation
ab	$a + b$
$e = 1$	$e = 0$
$a^0 = 1$	$0a = 0$
a^{-1}	$-a$
$a(bc) = (ab)c$	$a + (b + c) = (a + b) + c$
a^n	na
$a^{-n} = (a^n)^{-1}$	$(-n)a = n(-a)$

Figure 2.1. The multiplicative versus the additive notation. The latter is used only for abelian groups.

A priori, if we want to list the elements of a cyclic group and if a is the generator for the cyclic group, then we have to list all the positive powers $a, a^2, \ldots$, all the negative powers $a^{-1}, a^{-2}, \ldots$, and the identity element a^0 (in the additive notation these would be $a, 2a, \ldots, -a, -2a, \ldots$, and $0a$). However, the next theorem shows that, in the case of *finite* cyclic groups, the task is easier. When the group is finite, we only start with a and write down its *positive* powers in order. At some point, we are guaranteed to get e, and, at exactly that point, we have a complete list of the elements of the group.

Theorem 2.37. *Let* $G = \langle a \rangle$ *be a cyclic group of finite order. Then there is a positive integer* k *such that* $a^k = e$, *and if* m *is the smallest positive integer with* $a^m = e$, *then* $|G| = m$ *and* $G = \{a, a^2, \ldots, a^{m-1}, a^m = e\}$.

Proof. By definition $G = \langle a \rangle = \{e, a, a^{-1}, a^2, a^{-2}, \ldots\}$. The point of this theorem is that, in the case when we know that G is finite, we do not need to use all the powers of a, and only a small subset (as specified above) will suffice.

CLAIM: There is a positive integer k such that $a^k = e$.

PROOF OF CLAIM: We know that G is a finite group, but $\{e, a, a^2, \ldots\}$ is an infinite list of symbols. Hence there must be repeats in this list. Assume $a^t = a^s$ for some positive integers t and s with $t > s$.

Then $a^t a^{-s} = e$, which implies that $a^{t-s} = e$. So as promised by the claim, there is a positive integer k—namely $t - s$—such that $a^k = e$.

Let m be the smallest positive integer with $a^m = e$.

Consider $\{a, a^2, \ldots, a^{m-1}, a^m = e\}$. There are no repeats in this list, since if $a^i = a^j$ for $i > j$, then $a^{i-j} = e$, and $i - j < m$, contradicting the minimality of m.

We now have to show that every element of G appears in the short list

$$\{a, a^2, \ldots, a^{m-1}, a^m = e\}.$$

A typical element of G is a^k with $k \in \mathbb{Z}$. A priori, if $k < 0$ or if $k > m$, then it is not clear that a^k appears in our short list. To prove that it indeed does, using the division algorithm, write $k = qm + r$, where $0 \leq r < m$. Now

$$a^k = a^{qm+r} = (a^m)^q a^r = a^r.$$

Thus a^k is the same as a^r, and the latter is in our short list. Hence, $\{a, a^2, \ldots, a^m = e\}$ contains every element of the group G, and the proof is complete. $\qquad\square$

Example 2.38. Let $G = D_8$. This is a finite group, and hence Theorem 2.37 applies. What is the cyclic group generated by a? The theorem says that we have to start with a and find consecutive powers until we reach e. We get a, a^2, a^3, and $a^4 = e$. Hence, we know that $\langle a \rangle = \{a, a^2, a^3, e\}$. Because of Theorem 2.37, we know that this is a group, and, for example, the inverse of each of these elements is guaranteed to be one of them also. Similarly, what is the cyclic group generated by ab? Again, we find consecutive powers of ab until we get to e. We have $\langle ab \rangle = \{ab, e\}$.

Definition 2.39 (Order of an element). Let G be a group, and let $a \in G$. The *order* of a, denoted by $o(a)$, is defined to be $|\langle a \rangle|$.

Equivalently, $o(a)$ is the smallest positive integer d such that $a^d = e$. If no such d exists, then we say that $o(a) = \infty$. Other notation used for the order of a include $\mathrm{ord}(a)$ and $\mathrm{ord}_G(a)$.

Example 2.40. Let $G = (\mathbb{Z}, +)$. Then $o(1)$—as well as the order of every other non-zero element—is ∞.

Example 2.41. Let $G = D_8$. We record the orders of the elements.

D_8							
elements	e	a	a^2	a^3	b	ab	a^2b	a^3b
orders	1	4	2	4	2	2	2	2

Example 2.42. For the group $G = (\mathbb{Z}/6\mathbb{Z}, +)$, we have the following.

$\mathbb{Z}/6\mathbb{Z}$						
elements	0	1	2	3	4	5
orders	1	6	3	2	3	6

For example, to see that $o(5) = 6$, using the additive notation—which means that a^3 becomes $3a$—we calculate:

$$\langle 5 \rangle = \{5, 5 + 5 = 4, 5 + 5 + 5 = 3, 5 + 5 + 5 + 5 = 2,$$
$$5 + 5 + 5 + 5 + 5 = 1, 5 + 5 + 5 + 5 + 5 + 5 = 0\}.$$

We conclude that $o(5) = 6$ and that 5 generates G.

Remark 2.43. Note that we have used the word "order" in two different ways: the order of a group, and the order of an element. These are related but not quite the same—and this can cause confusion when you are starting out. Let G be a group, and let $x \in G$. We have defined the order of x to be the order of $\langle x \rangle$. In other words, inside the group G we construct the cyclic group $\langle x \rangle$, and then the size of this cyclic group is the order of the element $x \in G$. Given what we proved in Theorem 2.37, this is much less complicated than it may sound. Consider the following sequence of elements of G: $x, x^2, \ldots, x^n, \ldots$. If none of these elements is e, then x has *infinite order*; otherwise, let d be the smallest positive integer such that $x^d = e$. Then the *order* of x—as well as the order of $\langle x \rangle$—is d, and we write $o(x) = d$.

Example 2.44. Let $G = \mathrm{GL}(2,3)$, and let $a = \begin{bmatrix} 0 & 1 \\ 1 & 1 \end{bmatrix} \in G$. How do we find the group generated by a? We start with a and find consecutive powers until we reach identity:

$$\langle a \rangle = \{a = \begin{bmatrix} 0 & 1 \\ 1 & 1 \end{bmatrix}, a^2 = \begin{bmatrix} 1 & 1 \\ 1 & 2 \end{bmatrix}, a^3 = \begin{bmatrix} 1 & 2 \\ 2 & 0 \end{bmatrix}, a^4 = \begin{bmatrix} 2 & 0 \\ 0 & 2 \end{bmatrix}, a^5 = \begin{bmatrix} 0 & 2 \\ 2 & 2 \end{bmatrix},$$
$$a^6 = \begin{bmatrix} 2 & 2 \\ 2 & 1 \end{bmatrix}, a^7 = \begin{bmatrix} 2 & 1 \\ 1 & 0 \end{bmatrix}, a^8 = \begin{bmatrix} 1 & 0 \\ 0 & 1 \end{bmatrix} = e\}.$$

We conclude that the order of $\begin{bmatrix} 0 & 1 \\ 1 & 1 \end{bmatrix}$ in $\mathrm{GL}(2,3)$ is 8.

The following proposition will be used often. The proof is outlined here, and you are asked to complete it in Problem **2.3.8**.

Proposition 2.45. *Let G be any group, and let $x \in G$. Assume $o(x) = m$ and $x^s = e$, where m and s are positive integers. Then m divides s.*

Proof Outline. We know that m—being the order of x—is the smallest positive integer with $x^m = e$. We also know $x^s = e$—which means that $s \geq m$—and we want to show that $m \mid s$. Often, when we want to prove that one integer m divides another integer s, we use the division algorithm to write $s = qm + r$ with $0 \leq r < m$. We then proceed to show that r must be zero. We ask you, the reader, to provide this last step (Problem **2.3.8**). ☐

At this point, it will be useful to establish a number of facts about orders of elements and generators of cyclic groups. We list some of the relevant questions here, but you are asked to investigate and discover the answers, as well as give the relevant proofs, in the problems.

Question 2.46 (On order of elements and generators of cyclic groups). Let G be a group, and let $x, y \in G$.

(a) (Problem 2.3.11) How are $o(x)$ and $o(x^{-1})$ related?

(b) (Problem 2.3.12) How are $o(x)$ and $o(x^2)$ related?

(c) (Problem **2.3.13**) How are $o(x)$ and $o(yxy^{-1})$ related?

(d) (Problems 2.3.15 and **2.3.16**) What can you say about $o(xy)$ if $xy = yx$ and if $o(x)$ and $o(y)$ are distinct primes?

Assume $G = \langle g \rangle$ is a cyclic group of order n.

(e) (Problem 2.3.10) If $x \in G$, then what can you say about x^n? How are $o(x)$ and n related?

(f) (Problem **2.3.19**) For which positive integers m is g^m a generator for G?

(g) (Problem 2.3.20) How many generators does G have?

Finally, Problem **2.3.18** asks you to prove that if x is an element of a finite group G and $i > 0$, then

$$o(x^i) = \frac{o(x)}{\gcd(i, o(x))}.$$

Problems

2.3.1. Find the order of each of the elements of the following groups: $\mathbb{Z}/12\mathbb{Z}$, S_3, and $GL(2, 2)$.

2.3.2. Find the orders of $(1\ 5\ 2)(3\ 4)$ and $(1\ 5)(2\ 4)$ in the group S_5.

2.3.3. What is the order of $\begin{bmatrix} -1 & -1 \\ 0 & -1 \end{bmatrix}$ in the group $SL(2, 3)$?

2.3.4. Find the order of each of the elements of the group $((\mathbb{Z}/8\mathbb{Z})^\times, \cdot)$. Is this group cyclic? Do the same for the group $((\mathbb{Z}/10\mathbb{Z})^\times, \cdot)$.

2.3.5. Find the order of each of the elements of the group in Problem 1.3.6. (See also Problem 2.7.11.) Is this group cyclic?

2.3.6. Find all the generators of the following cyclic groups:

$$(\mathbb{Z}/6\mathbb{Z}, +), ((\mathbb{Z}/5\mathbb{Z})^\times, \cdot), (2\mathbb{Z}, +), ((\mathbb{Z}/11\mathbb{Z})^\times, \cdot).$$

On Orders of Elements in Groups. In Problems 2.3.7–**2.3.18**, you are asked to prove a number of useful facts—the most important are boldfaced—about orders of elements of groups.

2.3.7. Show that a finite group of even order has to have at least one element of order 2.

2.3.8. **Proof of Proposition 2.45.** Complete the proof of Proposition 2.45: If x is an element of order m in a group G and if, for a positive integer s, we have $x^s = e$, then m divides s.

2.3.9. Let ℓ be an integer greater than 1, and let G be a finite group with no element of order ℓ. Can there exist $a \in G$ with $\ell \mid o(a)$? Prove your assertion.

2.3.10. Assume that G is a cyclic group of order n. Let $b \in G$. What can you say about b^n? How is $o(b)$ related to n?

2.3.11. Let G be a group, and let $x \in G$. How are $o(x)$ and $o(x^{-1})$ related? Prove your assertion.

2.3.12. Let G be a group, and let $x \in G$. How are $o(x)$ and $o(x^2)$ related? Does you answer depend on whether $o(x)$ is odd or even? Prove your assertions.

2.3.13. Let G be a group, and let $x, y \in G$. Show that $o(yxy^{-1}) = o(x)$.

2.3.14. Let G be a group, and let $a, b \in G$. Show that $o(ab) = o(ba)$.

2.3.15. Give an example of a group G and elements $x, y \in G$ with $o(xy) < \min(o(x), o(y))$. Give an example of a group G and elements $x, y \in G$ with $o(xy) > o(x)o(y)$.

2.3.16. Let G be a group, and let $x, y \in G$. Assume that $xy = yx$, $o(x) = p$, and $o(y) = q$, where p and q are distinct prime numbers. What can you say about $o(xy)$?

2.3.17. Let G be a group, let $a \in G$, and let ℓ be an integer greater than 1. Assume that $o(a)$ is finite and relatively prime to ℓ. Show that there exists $x \in G$ with $x^\ell = a$.

2.3.18. Let G be a finite group. Let $x \in G$, and let $i > 0$. Then prove that

$$o(x^i) = \frac{o(x)}{\gcd(i, o(x))}.$$

On Generators of Cyclic Groups. If a finite cyclic group is generated by an element g, then which powers of g are also generators? How many generators does a finite cyclic group have? You are asked to answer these two questions in Problems *2.3.19*–2.3.20.

2.3.19. (a) Let $G = \{e, a, \ldots, a^9 \mid a^{10} = e\}$ be the cyclic group of order 10. For which m, is $G = \langle a^m \rangle$?

(b) Let $G = \langle g \rangle$ be a cyclic group of order n. For which m is g^m a generator of G? Why?

2.3.20. Let G be a cyclic group of order n. How many generators does G have?

Shuffling a Deck of Cards. Number the cards in a deck $1, \ldots, k$. After a shuffle, the cards in the deck will be in a different order. Thus we can view any shuffle as a permutation of $1, \ldots, k$ or as a 1-1, onto function from $[k] = \{1, \ldots, k\}$ to itself. Thus any fixed shuffle of a deck of k cards is an element of S_k. For example, the (somewhat silly) shuffle where we take the bottom card and put it right underneath the top card corresponds to the element $(1)(2\ 3\ 4\ \ldots\ k)$, i.e., the top card stays fixed and card 2 goes in place of 3, 3 goes in the place of 4, $\ldots$, and card k goes in place of 2.

The Perfect Riffle Shuffle. For the purpose of this shuffle, assume that the number of cards is even and denote it by $2n$. Also number the cards, $1, \ldots, 2n$. After cutting the cards in half, one stack will consist of cards 1 through n and the other stack will consist of cards $n+1$ through $2n$. Doing a perfect riffle shuffle will result in the following order:

$$1, n+1, 2, n+2, 3, n+3, \ldots, n-1, 2n-1, n, 2n.$$

2.3.21. Consider a fixed shuffle of a deck of cards. Does repeating this fixed shuffle some finite (positive) number of times bring the deck eventually back to its original order? Why?

2.3.22. We have a deck consisting of ten cards. What element of S_{10} corresponds to the perfect riffle shuffle of this deck? What is the order of this element?

2.3.23. We have a deck consisting of $2n$ cards. For the following values of n find the smallest number of consecutive perfect riffle shuffles that will bring the deck back to its original order:

(a) $n = 5$,
(b) $n = 7$,
(c) $n = 8$,
(d) $n = 26$.

Is there anything surprising?

2.3.24. Let G be a finite group with no element of order 3. Further, assume that $(ab)^3 = a^3 b^3$ for all $a, b \in G$.

(a) Let $a \in G$. Show that there exists $x \in G$ with $x^3 = a$.

(b) Let $a, b \in G$. Show that $ab^2 = b^2 a$.

(c) Let $a, b \in G$. Show that $a^2 b^2 = b^2 a^2$.

(d) Show that G is abelian.[3]

2.3.25. Let n be an integer greater than 1. Look at examples of $((\mathbb{Z}/n\mathbb{Z})^\times, \cdot)$. For which n is $(\mathbb{Z}/n\mathbb{Z})^\times$ cyclic? Make a conjecture. You do not have to prove your conjecture.

2.4. Isomorphisms

When are two groups the same? Of course, if you take the elements of a given group and color them red, then the new red group will still have the same group theoretic properties as the original group. So, even though these two groups look different, as far as group theory is concerned, they are considered the same. In fact, the same group often appears in many different forms and in different contexts. It is useful and important to realize that the underlying groups in these different situations are actually the same group. In this section we define when two groups are the same (or isomorphic). At this stage, we will give the definition and discuss some of the issues near the surface. Later—see Chapter 11—we will come back to this important notion and take a more serious look.

Definition 2.47 (Isomorphic groups). Let $(G, *)$ and $(H, \cdot)$ be groups. Then G and H are *isomorphic*, if there exists a map $\phi \colon G \to H$, called an *isomorphism*, such that

(a) the map ϕ is 1-1,

(b) the map ϕ is onto, and

(c) for all $a, b \in G$, we have $\phi(a * b) = \phi(a) \cdot \phi(b)$.

If G and H are isomorphic, we write $G \cong H$.

The map ϕ is the map that allows translation between the two groups. The definition says that there is a way of assigning to each element of the group G an element of the group H, such that the two multiplication tables look the same. In other words, assume that $\phi : G \to H$ is an isomorphism of groups, and order the elements of G arbitrarily as g_1, g_2, …. Now order the elements of H as $\phi(g_1)$, $\phi(g_2)$, …. Since ϕ is 1-1 and onto, we have listed every element of H exactly once. Now using these orders, write the multiplication table for G and for H. The (i, j) entry in the table for G will be $g_i g_j$, while the (i, j) entry in the multiplication table of H will be $\phi(g_i) \cdot \phi(g_j) = \phi(g_i * g_j)$. So we can get the multiplication table of H by starting with the multiplication table of G and replacing every $g \in G$ with $\phi(g) \in H$. Since the multiplication tables of two isomorphic groups are the same, if you have a question about a group that can be answered using the multiplication table, then the answer will be the same for any other group that is isomorphic to it.

Note that talking about a "multiplication table" of a group certainly makes sense for finite groups and even for countable groups (see page 14). However, for infinite uncountable groups, we cannot write down a multiplication table. For such

---

[3]This problem—without the intermediate steps—is from Herstein [**Her75**, Problem 24, Section 2.5].

+	0	1	2	3
0	0	1	2	3
1	1	2	3	0
2	2	3	0	1
3	3	0	1	2

$\cdot$	e	a	a^2	a^3
e	e	a	a^2	a^3
a	a	a^2	a^3	e
a^2	a^2	a^3	e	a
a^3	a^3	e	a	a^2

Figure 2.2. The multiplication tables for $(\mathbb{Z}/4\mathbb{Z}, +)$ and $\langle a \mid a^4 = e \rangle$

groups, the intuition of the above argument remains and can be made precise using the isomorphism ϕ and without recourse to multiplication tables.

Example 2.48. The two groups $(\mathbb{Z}/4\mathbb{Z}, +) = \{0, 1, 2, 3\}$ and $\langle a \rangle = \{e, a, a^2, a^3\} \subset D_8$ are isomorphic, and one isomorphism is given by the map $\phi : \mathbb{Z}/4\mathbb{Z} \to \langle a \rangle$ defined by $\phi(k) = a^k$. We can show formally that this map is indeed an isomorphism, but in this case, we can simply look at the whole map:

$$\phi : \begin{cases} 0 \mapsto e \\ 1 \mapsto a \\ 2 \mapsto a^2 \\ 3 \mapsto a^3 \end{cases}$$

It is clear that the map is 1-1 and onto, and also $\phi(i+j) = a^{i+j} = a^i a^j = \phi(i)\phi(j)$. As promised, the map ϕ gives a relabeling that shows that the two multiplication tables are actually the same. See Figure 2.2.

Example 2.49. Let $\mathbb{Z}$ denote the set of integers, and let

$$G = \left\{ \begin{bmatrix} 1 & a & 0 \\ 0 & 1 & 0 \\ 0 & 0 & 1 \end{bmatrix} \mid a \in \mathbb{Z} \right\}.$$

First, we show that G is a group when the operation is the usual matrix multiplication (this was actually Problem 2.1.3). The set is closed under the operation since

$$\begin{bmatrix} 1 & a & 0 \\ 0 & 1 & 0 \\ 0 & 0 & 1 \end{bmatrix} \begin{bmatrix} 1 & b & 0 \\ 0 & 1 & 0 \\ 0 & 0 & 1 \end{bmatrix} = \begin{bmatrix} 1 & a+b & 0 \\ 0 & 1 & 0 \\ 0 & 0 & 1 \end{bmatrix},$$

and $a + b$ is an integer. The identity matrix is the identity of the group and the inverse of $\begin{bmatrix} 1 & a & 0 \\ 0 & 1 & 0 \\ 0 & 0 & 1 \end{bmatrix}$ is $\begin{bmatrix} 1 & -a & 0 \\ 0 & 1 & 0 \\ 0 & 0 & 1 \end{bmatrix}$, which, since $-a$ is an integer, is also in G. What about associativity? Since matrix multiplication is already associative, the operation in G is also associative and there is no need for a separate proof.

We now show that the group G is isomorphic to $(\mathbb{Z}, +)$, the group of integers under addition. We define $\phi : G \longrightarrow \mathbb{Z}$, by

$$\phi \left(\begin{bmatrix} 1 & a & 0 \\ 0 & 1 & 0 \\ 0 & 0 & 1 \end{bmatrix} \right) = a.$$

It is straightforward to show that ϕ is 1-1 and onto. We also have

$$\phi\left(\begin{bmatrix} 1 & a & 0 \\ 0 & 1 & 0 \\ 0 & 0 & 1 \end{bmatrix}\begin{bmatrix} 1 & b & 0 \\ 0 & 1 & 0 \\ 0 & 0 & 1 \end{bmatrix}\right) = \phi\left(\begin{bmatrix} 1 & a+b & 0 \\ 0 & 1 & 0 \\ 0 & 0 & 1 \end{bmatrix}\right)$$

$$= a+b$$

$$= \phi\left(\begin{bmatrix} 1 & a & 0 \\ 0 & 1 & 0 \\ 0 & 0 & 1 \end{bmatrix}\right) + \phi\left(\begin{bmatrix} 1 & b & 0 \\ 0 & 1 & 0 \\ 0 & 0 & 1 \end{bmatrix}\right),$$

completing the proof that ϕ is an isomorphism.

We next show that two finite cyclic groups of the same order are isomorphic. This means that, as far as group theoretical properties are concerned, there is at most one cyclic group for each order. Since, for each positive integer n, $(\mathbb{Z}/n\mathbb{Z}, +)$ is a cyclic group of order n, it follows that for each order, there is exactly one cyclic group. In other words, there is one cyclic group of order 5 and one cyclic group of order 216 and so on. This also justifies why we can think (and write) $\mathbb{Z}/n\mathbb{Z}$ whenever we are considering a cyclic group of order n.

Theorem 2.50. *Let G and H be finite cyclic groups. If $|G| = |H|$, then $G \cong H$.*

Proof. Assume that the order of the two groups is m and that the elements a and x are the generators for G and H, respectively. In other words, $G = \langle a \rangle = \{a, a^2, \ldots, a^m = e\}$, and $H = \langle x \rangle = \{x, x^2, \ldots, x^m = e\}$.

Let the map $\phi\colon G \to H$ be defined by $\phi(a^i) = x^i$, for $1 \leq i \leq m$.

Now ϕ is clearly both 1-1 and onto. Also

$$\phi(a^i a^j) = \phi(a^{i+j}) = x^{i+j} = x^i x^j = \phi(a^i)\phi(a^j).$$

Hence ϕ is an isomorphism, and $G \cong H$. $\qquad \square$

The definition of an isomorphism had three conditions. For a map $\phi : G \to H$ to be an isomorphism, you need ϕ to be 1-1 and onto, and you need $\phi(xy) = \phi(x)\phi(y)$ for all $x, y \in G$. If a map ϕ is 1-1 and onto—to say that ϕ is a *one-to-one correspondence* or a *bijection* means the same thing—then for every element of G there is exactly one element of H and vice versa. This is certainly necessary for an isomorphism, but it is not enough, since these conditions do not say anything about the multiplication table. Hence, the last condition is crucial, leading us to the following definition:

Definition 2.51 (Homomorphism). Let $(G, *)$ and $(H, \cdot)$ be groups. A map $\phi\colon G \to H$ is called a *homomorphism* if

$$\phi(a * b) = \phi(a) \cdot \phi(b).$$

In later chapters—see Chapter 11—we shall see the importance of studying homomorphisms. (Homomorphisms are the analog, for groups, of linear transformations for vector spaces.) For now we know that a 1-1, onto homomorphism is an isomorphism, and we prove a straightforward lemma:

Lemma 2.52. *Let G and H be groups, and assume that $\phi : G \to H$ is a homomorphism. Then $\phi(e_G) = e_H$.*

Proof. A priori we do not know what $\phi(e_G)$ is, but we do know that it is an element of H. Hence let $h = \phi(e_G)$. We want to show that $h = e_H$. We have

$$he_H = h = \phi(e_G) = \phi(e_G e_G) = \phi(e_G)\phi(e_G) = hh.$$

Now from the equation $he_H = hh$—and the cancellation property—we get that $e_H = h$, and the proof is complete. $\qquad\qquad\qquad\qquad\qquad\qquad\qquad\qquad\qquad\qquad\Box$

We asserted that two groups that are isomorphic have the same group theoretic properties. Of course, for every group theoretic property, we need to prove this assertion. But if two finite groups are isomorphic, then we can arrange their elements in a way that the two multiplication tables look identical. This means that the two groups share all properties that are determined by the multiplication table. For example, if two groups are isomorphic and one is abelian, then so must be the other. Also the list of orders of group elements of two isomorphic groups are the same. As you gain experience, you will know that you can easily turn this intuitive argument—e.g., if two groups are isomorphic and one is abelian, then so must be the other since it has the same multiplication table—into a rigorous proof. We give here two such examples. The proofs are straightforward, do not directly mention the multiplication table—and hence can be applied to infinite groups as well—and are typical of proofs that show that isomorphic groups have similar properties.

Theorem 2.53. *Let G and H be groups, and assume that $\phi : G \to H$ is an isomorphism.*

(a) *The group G is abelian if and only if H is abelian.*

(b) *If $x \in G$ and x or $\phi(x)$ have finite order, then $o(x) = o(\phi(x))$.*

Proof. (a) ($\Rightarrow$) Assuming G is abelian, we show that H is abelian. Let $a, b \in H$. Since ϕ is onto, we have two elements x and y in G such that $\phi(x) = a$ and $\phi(y) = b$. Now

$$ab = \phi(x)\phi(y) = \phi(xy) = \phi(yx) = \phi(y)\phi(x) = ba.$$

($\Leftarrow$) For the converse, assuming H is abelian, we show that G is abelian. Let $x, y \in G$. Then we have

$$\phi(xy) = \phi(x)\phi(y) = \phi(y)\phi(x) = \phi(yx).$$

Since $\phi(xy) = \phi(yx)$ and ϕ is 1-1, we conclude that $xy = yx$.

(b) Let $x \in G$. Then for any positive integer m, the following statements are equivalent:

$$x^m = e_G \quad \Leftrightarrow \quad \phi(x^m) = \phi(e_G) = e_H \quad \Leftrightarrow \quad \phi(x)^m = e_H.$$

Now assume that $o(x) = k$, and we want to prove $o(\phi(x)) = k$. We know—by what we just proved—that $\phi(x)^k = e_H$. We are not quite done proving that the order of $\phi(x)$ is k since we have to show that k is the smallest power of $\phi(x)$ that gives identity. But if ℓ was a smaller power such that $\phi(x)^\ell = e_H$, then—again by the chain of if-and-only-if statements that we proved—we would have $x^\ell = e_G$. But this would contradict the fact that $o(x) = k$. Hence k is the smallest power such that $\phi(x)^k = e_H$, and so $o(\phi(x)) = k$. The proof in the other direction is similar. $\square$

In the above proofs we did not use every property of isomorphisms for every part of the proof. As a good exercise, for example, assume H is abelian and $\phi : G \to H$ is a map, then what properties of ϕ guarantee that G is also abelian?

Problems

2.4.1. The groups $\mathbb{Z}/6\mathbb{Z}$, S_3, $\mathrm{GL}(2,2)$, and D_6 (the symmetries of an equilateral triangle) are all groups of order 6. Which ones are isomorphic? If two of the groups are isomorphic, give the relabeling explicitly. If two of the groups are not isomorphic, then give a reason.

2.4.2. Are the groups $(\mathbb{Z}/6\mathbb{Z}, +)$ and $(\mathbb{Z}/7\mathbb{Z})^\times$ isomorphic? If so, give an explicit relabeling of the elements that shows the isomorphism.

2.4.3. Are the groups $(\mathbb{Z}/12\mathbb{Z}, +)$ and $(\mathbb{Z}/13\mathbb{Z})^\times$ isomorphic?

2.4.4. Let $H = \{2^n \mid n \in \mathbb{Z}\}$, and let $\cdot$ denote ordinary multiplication. Show that $(H, \cdot)$ is isomorphic to $(\mathbb{Z}, +)$.

2.4.5. Let G be an infinite cyclic group. Prove that $G \cong (\mathbb{Z}, +)$.

2.4.6. Recall that $G = \mathrm{GL}(2, \mathbb{R})$ is the group of all invertible 2×2 matrices with entries in the reals $\mathbb{R}$.
   (a) We want to see if inside G there is a group isomorphic to D_8. In other words, can we find a group isomorphic to D_8 consisting of 2×2 invertible matrices with real entries?
   Let S be the square in $\mathbb{R}^2$ with vertices at $(\pm 1, \pm 1)$. Think of elements of D_8 as linear transformations from $\mathbb{R}^2$ to $\mathbb{R}^2$. For example a is the linear transformation that rotates every vector $90°$. Find a 2×2 matrix for each of these linear transformations. Do they form a group isomorphic to D_8?
   (b) Can you find a group isomorphic to D_8 consisting of elements of $\mathrm{GL}(2, \mathbb{C})$?
   (c) Can you find a group isomorphic to D_8 consisting of elements of $\mathrm{GL}(2, 3)$?

2.4.7. Let $G = \mathrm{GL}(2, \mathbb{C})$ be the group of invertible 2×2 matrices with entries in the complex numbers $\mathbb{C}$.

Let $H = \langle \begin{pmatrix} 0 & i \\ i & 0 \end{pmatrix}, \begin{pmatrix} 0 & 1 \\ -1 & 0 \end{pmatrix} \rangle$ be a group consisting of some of the elements of G. What is the order of H? Is H isomorphic to D_8?

Quaternion Group of Order 8

Definition 2.54 (Q_8)**.** Let $Q_8 = \{\pm 1, \pm i, \pm j, \pm k\}$ be a set with eight elements. Define a product on Q_8 as follows: 1 is the identity, -1 gets multiplied as usual, and the rest are multiplied similar to the cross product of vectors, i.e.,

$$ij = k, \ jk = i, \ ki = j, \ ji = -k, \ kj = -i, \ ik = -j, \ i^2 = j^2 = k^2 = -1.$$

The set Q_8 with this operation is a group and is called the *quaternion group* of order 8.

2.4.8. Is Q_8 isomorphic to D_8 or to the group in Problem *2.4.7*?

2.4.9. Find the multiplication table of all groups of order 4.

2.4.10. Can you find a group of invertible 2×2 matrices with integer entries that is isomorphic to D_{12}?

2.4.11. Can you find a group of invertible 2×2 integer matrices isomorphic to Q_8?

Elementary Properties of Homomorphisms. The purpose of this section was to define isomorphic groups so that two seemingly different groups can be identified as the "same". As a byproduct, we defined the more general notion of a group homomorphism (Definition 2.51). Group homomorphisms will take center stage in Chapter 11, where their importance in the study of groups becomes clear. Here and in Section 2.6—as a prelude—we ask you to prove a few elementary properties.

2.4.12. Let G and H be groups, and let $\phi : G \to H$ be a group homomorphism. For $x \in G$, prove that $\phi(x^{-1}) = \phi(x)^{-1}$.

2.4.13. Give an example of an abelian group G, a non-abelian group H, and a homomorphism $\phi : G \to H$.

2.4.14. Let G be abelian, let H be a group, and let $\phi : G \to H$ be an onto homomorphism. Does H have to be abelian?

2.4.15. Let G be cyclic, let H be a group, and let $\phi : G \to H$ be an onto homomorphism. Does H have to be cyclic? What if ϕ was not known to be onto?

2.4.16. Give an example of two groups G and H, an element $x \in G$, and a homomorphism $\phi : G \to H$ such that $o(x)$ does not equal $o(\phi(x))$.

2.4.17. Let G and H be groups, and let $\theta : G \longrightarrow H$ be a homomorphism. Let $g \in G$ with $o(g) < \infty$. What can you say about the relation of $o(g)$ and $o(\theta(g))$?

2.5. Direct Products (New Groups from Old Groups)

In this section, given two groups G and H, we construct, a new group denoted by $G \times H$.

Definition 2.55 (Direct product of groups). Let $(G, \circ)$ and $(H, \cdot)$ be groups. Recall that

$$G \times H = \{(g, h) \mid g \in G, h \in H\}.$$

Now, define the operation $*$ on $G \times H$ as

$$(g_1, h_1) * (g_2, h_2) = (g_1 \circ g_2, h_1 \cdot h_2).$$

$(G \times H, *)$ is called the *direct product* of G and H.

Lemma 2.56. *Let G and H be groups, then $G \times H$ is a group.*

Proof. The proof is straightforward. If (g_1, h_1) and (g_2, h_2) are elements of $G \times H$, then we know that $g_1, g_2 \in G$ and $h_1, h_2 \in H$. That means that $g_1 g_2 \in G$ since G is a group and the operation in G is closed. Likewise, $h_1 h_2 \in H$. So we conclude that $(g_1 g_2, h_1 h_2) \in G \times H$. The proof of associativity of the product also follows from the same property in G and H. If e_G is the identity for the group G and e_H is the identity for the group H, then (e_G, e_H) is the identity for the group $G \times H$. Finally, the inverse of $(g, h) \in G \times H$ is (g^{-1}, h^{-1}). □

Example 2.57. Consider $\mathbb{Z}/2\mathbb{Z} \times \mathbb{Z}/2\mathbb{Z} = \{(0,0), (0,1), (1,0), (1,1)\}$.

We use the additive notation for this abelian group. Thus, for example, $(1,1) + (1,0) = (0,1)$. In this group every element has order 2. Recall that $\mathbb{Z}/4\mathbb{Z}$ is also a group of order 4. However, $\mathbb{Z}/4\mathbb{Z}$ is cyclic, and thus it has at least one element of order 4 (in fact, it has two elements of order 4). For this reason, $\mathbb{Z}/2\mathbb{Z} \times \mathbb{Z}/2\mathbb{Z}$ is not isomorphic to $\mathbb{Z}/4\mathbb{Z}$. The group $\mathbb{Z}/2\mathbb{Z} \times \mathbb{Z}/2\mathbb{Z}$ is called the *Klein 4-group*.

Example 2.58. Consider the group $\mathbb{Z}/2\mathbb{Z} \times \mathbb{Z}/3\mathbb{Z}$.

This group has six elements:

$$\mathbb{Z}/2\mathbb{Z} \times \mathbb{Z}/3\mathbb{Z} = \{(0,0), (1,0), (0,1), (1,1), (0,2), (1,2)\}.$$

The identity element $(0,0)$ as usual has order 1, while $(1,0)$ has order 2. There are two elements of order 3: $(0,1)$, and $(0,2)$. Finally the elements $(1,1)$ and $(1,2)$ have order 6. This means that this group is cyclic and isomorphic to $\mathbb{Z}/6\mathbb{Z}$. If we let $a = (1,1)$, then—writing additively—$2a = (0,2)$, $3a = (1,0)$, $4a = (0,1)$, $5a = (1,2)$, and $6a = (0,0)$.

The above two examples suggest the following natural and interrelated questions which you are asked to explore in the problems.

Question 2.59 (Problems **2.5.7** and 2.5.8)**.** Let m and n be positive integers, and let $G = \mathbb{Z}/m\mathbb{Z} \times \mathbb{Z}/n\mathbb{Z}$.

(a) What is the order of $(1,1)$ in G?

(b) When is $G = \langle (1,1) \rangle$?

(c) For which choices of m and n is G cyclic?

Groups of Order 1, 2, 3, and 4. For every positive integer n we know at least one group of order n. This is the cyclic group of order n. Since we also know that two cyclic groups of the same order are isomorphic, we conclude that for every order there is exactly one cyclic group of that order. To be concrete, when thinking of a cyclic group of order n, we can think of $\mathbb{Z}/n\mathbb{Z}$ or $\{e, a, a^2, \ldots, a^{n-1}\}$ where

$a^n = e$. It is easy to see that $\mathbb{Z}/1\mathbb{Z}$ is the only group of order 1 and $\mathbb{Z}/2\mathbb{Z}$ is the only group of order 2. With more work (see Problem 2.2.5), we see that $\mathbb{Z}/3\mathbb{Z}$ is the only group of order 3 and (see Problem **2.4.9**) that there are exactly (up to isomorphism) two groups of order 4: the cyclic group of order 4, $\mathbb{Z}/4\mathbb{Z}$, and the Klein 4-group, $\mathbb{Z}/2\mathbb{Z} \times \mathbb{Z}/2\mathbb{Z}$.

Problems

2.5.1. For a set X, recall Definition 2.20 (page 42) of 2^X, the power set of X, and $\triangle$, the symmetric difference of two sets. Also recall from Problem 2.1.9 that $(2^X, \triangle)$ is an abelian group.
   (a) Let $X = \{1, 2\}$. Find a familiar group that is isomorphic to $(2^X, \triangle)$.
   (b) Do the same when $X = \{1, 2, 3\}$.

2.5.2. Give two distinct groups of order 25, and provide a reason for why they are not isomorphic.

2.5.3. Does there exist a non-cyclic group of order 99? If the answer is yes, then find two non-isomorphic groups of order 99.

2.5.4. Find a familiar group that is isomorphic to $((\mathbb{Z}/8\mathbb{Z})^\times, \cdot)$.

2.5.5. Find a familiar group that is isomorphic to each of

$$((\mathbb{Z}/15\mathbb{Z})^\times, \cdot), \ ((\mathbb{Z}/16\mathbb{Z})^\times, \cdot), \ ((\mathbb{Z}/20\mathbb{Z})^\times, \cdot).$$

2.5.6. Is $\mathbb{Z}/8\mathbb{Z} \times \mathbb{Z}/2\mathbb{Z}$ isomorphic to $\mathbb{Z}/4\mathbb{Z} \times \mathbb{Z}/4\mathbb{Z}$? Why or why not?

2.5.7. (a) Let m and n be integers greater than 1. What is the order of the element $(1, 1)$ in $\mathbb{Z}/m\mathbb{Z} \times \mathbb{Z}/n\mathbb{Z}$? Make a conjecture.
   (b) Under what conditions would $(1, 1)$ be a generator for $\mathbb{Z}/m\mathbb{Z} \times \mathbb{Z}/n\mathbb{Z}$?

2.5.8. Let m and n be positive integers with $\gcd(m, n) = 1$. Prove that $\mathbb{Z}/m\mathbb{Z} \times \mathbb{Z}/n\mathbb{Z} \cong \mathbb{Z}/mn\mathbb{Z}$.

2.5.9. Let H and K be finite groups, and let $h \in H$ and $k \in K$. What is $o((h, k))$ in $H \times K$?

2.5.10. Let m and n be positive integers with $\gcd(m, n) = 1$, and let ϕ denote Euler's ϕ-function (Definition 1.45). Consider the group $\mathbb{Z}/n\mathbb{Z} \times \mathbb{Z}/m\mathbb{Z}$.
   (a) Show that (a, b) is a generator for the group $\mathbb{Z}/n\mathbb{Z} \times \mathbb{Z}/m\mathbb{Z}$ if and only if a and b are, respectively, generators for $\mathbb{Z}/n\mathbb{Z}$ and $\mathbb{Z}/m\mathbb{Z}$.
   (b) Show that the number of generators of the group $\mathbb{Z}/n\mathbb{Z} \times \mathbb{Z}/m\mathbb{Z}$ is $\phi(n)\phi(m)$.
   (c) Prove that for relatively prime integers m and n,
$$\phi(nm) = \phi(n)\phi(m).$$

2.5.11. Assume that $G \times H$ is an abelian group. Can we conclude that G and H are abelian?

2.5.12. Find a familiar group that is isomorphic to the group in Problem 1.3.6. (Also see Problems 2.3.5 and 2.7.11.)

2.5.13. Let H and K be sets, and let A be a subset of $H \times K$. Is A equal to $X \times Y$ for some $X \subseteq H$ and $Y \subseteq K$?

2.5.14. Let H and K be groups. Define $\phi : H \times K \to H$ by $\phi(h, k) = h$ for all $(h, k) \in H \times K$. Is ϕ a homomorphism? When is ϕ an isomorphism?

2.5.15. Let H and K be groups, and define $\theta : H \to H \times K$ by $\theta(h) = (h, e)$ for all $h \in H$. Is θ a homomorphism? Can it be an isomorphism?

2.5.16. A poem from a brass plate in St. Mawgan Church, England, commemorating Hanniball Basset who died in 1709, reads:[4]

Shall wee all dye
Wee shall dye all
All dye shall wee
Dye all wee shall

Let $G = \{\text{Shall}, \text{wee}, \text{all}, \text{dye}\}$. Is G a group with the above poem as its multiplication table (you have to add the column and row headings)? If so, find a familiar group that is isomorphic to G.

2.6. Subgroups

Many times, inside a given group, there are many other groups. These substructures tend to be important in understanding the group in question. This happens in two ways: First, understanding the substructures in a group tends to be a good way to find the properties of the larger group. Second, and this will become clear later, we try to change most questions about a group to ones about specific substructures. In other words, particular substructures of a group will give us information about particular properties of the group. The previous statements are quite vague, but will be developed fully as we proceed. First, we will define subgroups, that is, the smaller groups within a larger group.

Definition 2.60 (Subgroups). Let $(G, *)$ be a group, and let H be a non-empty subset of G. The subset H is a *subgroup* of G if $(H, *)$ is a group.

For any group G, $\{e\}$ and G are automatically subgroups. They are called the *trivial subgroups* of G.

If H is a subgroup of G, we write $H \leq G$.[5] If H is a subgroup of G and, in addition, $H \neq G$, we say that H is a *proper subgroup* of G, and we write $H < G$.

Example 2.61. When we write $\{1, -1\} \leq (\mathbb{Q} - \{0\}, \cdot)$, we mean that, in the group of non-zero rationals with multiplication, the set consisting of the two elements 1 and -1 is a subgroup. This is because these two elements do form a group themselves—with the same multiplication operation as in $\mathbb{Q} - \{0\}$. In fact, this group is isomorphic to $\mathbb{Z}/2\mathbb{Z}$, as can be seen by comparing their multiplication

---

[4]From Andersen [**And**]. This is an example—one of the oldest in Western Europe—of a Latin square of order 4. A Latin square of order n is an $n \times n$ array whose entries are from a set of n symbols and such that each symbol appears exactly once in each row and in each column. A group multiplication table is an example of a Latin square. The oldest known Latin squares are from around the year 1000 CE and are in Arabic. See Andersen [**And**] for a history of Latin squares.

[5]While many texts use the notation $H \leq G$ to denote that H is a subgroup of G, not all do. Certainly, $\leq$ has other meanings as well and will denote a subgroup only in the context of group theory.

$$
\begin{array}{c|cc}
\cdot & 1 & -1 \\
\hline
1 & 1 & -1 \\
-1 & -1 & 1
\end{array}
\qquad\qquad
\begin{array}{c|cc}
+ & 0 & 1 \\
\hline
0 & 0 & 1 \\
1 & 1 & 0
\end{array}
$$

Figure 2.3. The groups $(\{1, -1\}, \cdot)$ and $(\mathbb{Z}/2\mathbb{Z}, +)$ are isomorphic.

tables—see Figure 2.3—or by recalling that, up to isomorphism, there is only one group of order 2.

Example 2.62. Let $G = \mathbb{Z}/2\mathbb{Z} \times \mathbb{Z}/3\mathbb{Z} = \{(0,0), (0,1), (0,2), (1,0), (1,1), (1,2)\}$. Let $H = \{0\} \times \mathbb{Z}/3\mathbb{Z} = \{(0,0), (0,1), (0,2)\}$. It is clear that H is a group isomorphic to $\mathbb{Z}/3\mathbb{Z}$, since every element of H has an immaterial 0 in the first coordinate. Hence $H \leq G$.

Lemma 2.63. *Let G be a group, and let $H \leq G$. Then the identity of H is the same as the identity of G.*

Proof. Both H and G must have an identity element. We call these elements e_H and e_G, respectively. Both of these elements are elements of G, and we know that $he_H = e_H h = h$ for every $h \in H$, while $ge_G = e_G g = g$ for every $g \in G$. Hence,

$$e_H e_H = e_H = e_H e_G.$$

Now, G is a group, and so, in the equation $e_H e_H = e_H e_G$, we cancel e_H on the left and get $e_H = e_G$. $\qquad\square$

The next lemma—whose proof we will leave to the reader—says that in deciding whether a subset is a subgroup, we only need to check three things. The subset should be non-empty, closed under the operation, and closed under taking inverses.

Lemma 2.64. *Let $(G, *)$ be a group, and let H be a non-empty subset of G. Then H is a subgroup of G if and only if*

(a) *the set H is closed with respect to $*$, i.e., $h_1 * h_2 \in H$ whenever $h_1, h_2 \in H$, and*

(b) *the set H is closed with respect to taking inverses, i.e., if $h \in H$, then $h^{-1} \in H$.*

Example 2.65. Let $G = (\mathbb{Z}, +) = \langle 1 \rangle$, and let $H = 2\mathbb{Z} = \langle 2 \rangle$ be the set of even integers. Since the sum of two even numbers and the negative of an even number are even, we know that H is closed with respect to addition and taking inverses. Hence, by Lemma 2.64, H is a subgroup of G. It would not have been difficult to directly verify that H is a subgroup, but using the lemma allowed us not to worry about associativity or the existence of the identity.

Now we turn our attention to constructing subgroups. We start with formalizing what we mean by the *smallest* (or *the minimal*) set having a given property.

Definition 2.66. Let X be a set, and let P be a property that subsets of X may or may not have. We say that the set Y is *the smallest subset of X that has property P* (or *the unique minimal subset of X with respect to property P*), if $Y \subseteq X$, Y has property P, and if $Z \subseteq X$ and Z has property P, then $Y \subseteq Z$.

Such a Y may or may not exist. If such a Y does exist, then it is equal to the intersection of all subsets of X with property P.

Hence—at least in this book—when we talk about the smallest subset with a given property, we are not directly referring to size. The smallest subset with a given property is the unique minimal subset with respect to that property: it has the given property and is contained in *every* subset with the property.

Lemma 2.67. *Let G be a group, with $a \in G$. Then $\langle a \rangle = \{a^k \mid k \in \mathbb{Z}\}$ is the smallest subgroup of G containing a.*

Proof. We know that $\langle a \rangle$ is a subgroup of G, and $a \in \langle a \rangle$. Furthermore, if $a \in H \leq G$, then all integer powers of a must be in H as well (since H is a subgroup). Hence $\langle a \rangle \subseteq H$.

Thus $\langle a \rangle$ is a subgroup of G that contains a, and it is contained in any subgroup of G that contains a. This is exactly what it means—see Definition 2.66—to say that $\langle a \rangle$ is the smallest subgroup of G containing a. $\square$

We can generalize the idea of the subgroup generated by one element as follows:

Definition 2.68 (Subgroup generated by X). Let G be a group, and let X be a subset of G. Then the *subgroup generated by X*, written $\langle X \rangle$, is the smallest subgroup of G containing X. Equivalently, it is the intersection of all subgroups of G that contain X as a subset.

Example 2.69. As usual, let $D_8 = \langle a, b \mid a^4 = b^2 = e, ba = a^3 b \rangle$. At this point, we do not yet have tools to easily classify all the subgroups of D_8. Instead we have to use something akin to brute force, and generate as many subgroups as we can. Later, our theorems will tell us much about what we can expect.

Every group, D_8 included, has two trivial subgroups: $\{e\}$ and the whole group D_8. Each element of the group also generates a subgroup. In the case of D_8, we have five elements of order 2—the four reflections and the 180 degree rotation—and each of these generate a subgroup isomorphic to $\mathbb{Z}/2\mathbb{Z}$. These are $\langle a^2 \rangle$, $\langle b \rangle$, $\langle ab \rangle$, $\langle a^2 b \rangle$, and $\langle a^3 b \rangle$. We also have two elements of order 4, and these generate the same subgroup of order 4, namely $\langle a \rangle = \langle a^3 \rangle$. This group is a cyclic group of order 4 and, hence, is isomorphic to $\mathbb{Z}/4\mathbb{Z}$. There are no more cyclic subgroups.

Are there any other subgroups? We can try subgroups generated by two elements. Some combinations—such as $\langle a, b \rangle$—give the whole group. But two combinations give a proper subgroup: $\langle a^2, b \rangle = \{e, a^2, b, a^2 b\}$ and $\langle a^2, ab \rangle = \{e, a^2, ab, a^3 b\}$. Each of these is a non-cyclic group of order 4, and hence (you would know this if you have done Problem *2.4.9*) is isomorphic to the Klein 4-group, $\mathbb{Z}/2\mathbb{Z} \times \mathbb{Z}/2\mathbb{Z}$. Alternatively, you would directly see that the multiplication table for each of these two groups is identical—after a proper renaming—to the multiplication table of $\mathbb{Z}/2\mathbb{Z} \times \mathbb{Z}/2\mathbb{Z}$.

There are no other subgroups, and so D_8 has ten subgroups.

Example 2.70. Let $G = D_8 \times D_8$, and let
$$H = \langle a \rangle \times \langle b \rangle = \{(e, e), (e, b), (a, e), (a, b), (a^2, e), (a^2, b), (a^3, e), (a^3, b)\}.$$
Then H is a subgroup of G, and a set of generators for H is $\{(a, e), (e, b)\}$.

Is H isomorphic to another group that we know? Can you find other sets of generators?

Remark 2.71. Right at the beginning of the study of groups, it seems as though most proofs in group theory involve playing around with elements, and thus the subject resembles high school algebra. However, group theory is the study of structures, and we try to get away from considering elements as much as possible. We are not as much interested in how one element gets multiplied with another, as how all elements with a given property are situated in the group, and how they interact with collections of other elements with other properties. Because of this viewpoint, subgroups are very important. As we develop more terminology and get past the very elementary properties of groups, we will concentrate on analyzing the subgroup structure of a group. This will prove to be very fruitful.

As an example, to see how close a group is to being abelian, we will not examine each element and see whether it commutes with others or not. Instead we will look at the subgroup consisting of all elements that commute with everything. The bigger this subgroup, the closer the group to being abelian. In this way the theorems about subgroups seem to give us extra information about commutativity. All of this will become clear later. For now, we will limit ourselves to centralizers and the center. These subgroups tell us about commutativity, and will be used quite often.

Recall that two elements g and h of a group G *commute* if $gh = hg$.

Definition 2.72 (Centralizer and center). Let G be a group, and let X be a subset of G. Then $\mathbf{C}_G(X)$, the *centralizer of X in G*, is the collection of all elements of G that commute with *every* element of X.

In particular, if we fix $g \in G$, then $\mathbf{C}_G(\{g\})$ is the centralizer of g in G, is denoted by $\mathbf{C}_G(g)$, and consists of the elements of G that commute with g. Also, $\mathbf{C}_G(G)$ is called the *center* of G, is denoted by $\mathbf{Z}(G)$, and consists of all the elements of the group that commute with *every* element of the group.

Proposition 2.73. *Let G be a group, and let X be a subset of G. Then $\mathbf{C}_G(X)$ is a subgroup of G.*

Proof. First note that the identity element commutes with every element of G, hence $e \in \mathbf{C}_G(X)$, and $\mathbf{C}_G(X)$ is non-empty. In light of Lemma 2.64, we have to show that $\mathbf{C}_G(X)$ is closed under multiplication and under taking inverses.

Let z_1 and z_2 be elements of $\mathbf{C}_G(X)$; we need to show that $z_1 z_2 \in \mathbf{C}_G(X)$. We do this by showing that $z_1 z_2$ commutes with every element $x \in X$:

$$(z_1 z_2)x = z_1(z_2 x) = z_1(x z_2) = (z_1 x)z_2 = (x z_1)z_2 = x(z_1 z_2).$$

To complete the proof, we have to show that the inverse of every element in $\mathbf{C}_G(X)$ is also in $\mathbf{C}_G(X)$. Let $z \in \mathbf{C}_G(X)$. To show that $z^{-1} \in \mathbf{C}_G(X)$, we need to show that $z^{-1}x = xz^{-1}$ for every $x \in X$. For any $x \in X$, we know that $zx = xz$. Multiplying on the left and on the right by z^{-1}, we get $xz^{-1} = z^{-1}x$. This completes the proof. $\square$

We will now prove that, in the case of finite subsets, the task of determining whether a subset is a subgroup can be made even simpler—we do not even have to check for the existence of inverses.

Lemma 2.74. *Let $(G, *)$ be a group, and let H be a non-empty, finite subset of G. Assume H is closed under $*$. Then H is a subgroup of G.*

Proof. Since H is a non-empty subset of a group G, by Lemma 2.64, to determine whether H is a subgroup, we only have to check that H is closed under taking products and inverses. Hence, to prove this lemma, we need to show that if H is closed under taking products, it will automatically be closed under taking inverses. So, let $a \in H$. We need to show that $a^{-1} \in H$.

Assume $H = \{a_1, a_2, \ldots, a_n\}$. Consider the set $\{a_1 a, a_2 a, \ldots, a_n a\}$. Every element of this set is in H, since H is closed under multiplication. Furthermore, no two elements of this set are equal (since $a_i a = a_j a$ implies that $a_i = a_j$). Thus this set has n elements from H, and H had exactly n elements. So we conclude that

$$H = \{a_1 a, a_2 a, \ldots, a_n a\}.$$

Thus, in this list, one of the elements must be a. So, for some k with $1 \leq k \leq n$, we have $a_k a = a$. By right cancellation, $a_k = e$, and so we know that $e = a_k \in H$.

Now, since the list above includes all elements of H, it must also include e (since we just proved that $e \in H$). So there exists j, with $1 \leq j \leq n$, such that $a_j a = a_k = e$. Multiplying on the right by a^{-1} (which is guaranteed to exist in the group G), we get that $a_j = a^{-1}$. So $a^{-1} = a_j \in H$, and the proof is complete. $\square$

Finally, we can prove that a subgroup of a cyclic group is always cyclic. This severely restricts the possibilities for subgroups of cyclic groups.

Theorem 2.75. *Let $G = \langle a \rangle$ be a cyclic group, and let $\{e\} \neq H \leq G$. Let k be the smallest positive integer such that $a^k \in H$. Then $H = \langle a^k \rangle$.*

In particular, every subgroup of a cyclic group is cyclic.

Proof. The element a is the generator for G, and hence all elements of G—including all elements of H—are powers of a. We are assuming that H is not the identity group and so H contains some power of a other than a^0. Now, if $a^\ell \in H$, then its inverse $a^{-\ell}$ must also be in H, and—as long as $a^\ell \neq e$—either ℓ or $-\ell$ are positive. We conclude that there *are* positive integers m with $a^m \in H$. Now k is the smallest such positive integer, and we want to prove that $H = \langle a^k \rangle$.

We need to prove that $H \subseteq \langle a^k \rangle$ and $\langle a^k \rangle \subseteq H$. The latter is straightforward since $a^k \in H$, and so every power of a^k will have to be in H as well. So assume $x \in H$; we have to show $x \in \langle a^k \rangle$.

We have that $x \in H \subseteq G = \langle a \rangle$, and so $x = a^t$ for some $t \in \mathbb{Z}$. We need to show that $a^t \in \langle a^k \rangle$, and so we should show that $a^t = (a^k)^m$ for some integer m.

Using the division algorithm, we know that $t = mk + r$, where $0 \leq r < k$. Our aim is to show that $r = 0$.

Now $a^t = a^{mk+r} = a^{mk} a^r$. Solving for a^r, we get

$$a^r = a^{t-mk} = a^t (a^k)^{-m}.$$

Both a^t and a^k—hence also $(a^k)^{-m}$—are elements of H. This means that the product $a^t (a^k)^{-m}$ is in H as well. We conclude that $a^r \in H$.

But k was the smallest positive integer with $a^k \in H$, and yet $a^r \in H$ and $r < k$. So the only choice is that $r = 0$, and this completes the proof. $\quad\square$

A number of important facts and concepts about subgroups are left for you to tackle in the problems. We list some of them here:

Question 2.76 (Problems 2.6.21 and **2.6.22**). Let G be a cyclic group of order n. Let $m \leq n$ be a positive integer. How many subgroups of order m does G have?

Proposition 2.77 (Problem **2.6.24**). *If the only subgroups of a finite group G are the two trivial subgroups $\{e\}$ and G and if $|G| > 1$, then G is a cyclic group and its order is a prime number.*

Proposition 2.78 (New subgroups from old). *Let H and K be subgroups of a group G, and let x be an arbitrary element of G. There are several ways to construct (possibly) new subgroups of G:*

(a) *(Problem **2.6.4**) The intersection of H and K, $H \cap K$, is a subgroup of G.*

(b) *(Problem **2.6.27**) The set $xHx^{-1} = \{xhx^{-1} \mid h \in H\}$—called a* conjugate *of H—is a subgroup of G.*

(c) *(Problems 2.6.31–**2.6.33**) The set $HK = \{hk \mid h \in H, k \in K\}$—called the* product *of H and K—is* not *always a subgroup. In fact, HK is a subgroup of G if and only if $HK = KH$.*

Problems

2.6.1. List the orders of elements and the orders of subgroups of $\mathbb{Z}/2\mathbb{Z} \times \mathbb{Z}/2\mathbb{Z} \times \mathbb{Z}/2\mathbb{Z}$.

2.6.2. Let $G = (\mathbb{Z}/12\mathbb{Z}, +)$. Find all subgroups of G.

2.6.3. Find all subgroups of $(\mathbb{Z}/18\mathbb{Z}, +)$.

2.6.4. Intersection of two subgroups. Let G be a group, and let H and K be subgroups of G. Show that $H \cap K$ is a subgroup of G.

2.6.5. Let G be a group, and let x be an element of G that is *not* in the center. As usual, let $\mathbf{Z}(G)$ and $\mathbf{C}_G(x)$, respectively, denote the center of the group and the centralizer of x in G. Show that

$$\mathbf{Z}(G) < \mathbf{C}_G(x) < G.$$

In other words, $\mathbf{Z}(G)$ is a *proper* subgroup of $\mathbf{C}_G(x)$ which in turn is a *proper* subgroup of G.

2.6.6. Give a proof of Lemma 2.64.

2.6.7. Let X be a set, let Y be a subset of X, and let P be a property that subsets of X may or may not have. Referring back to Definition 2.66, prove that if Y is the smallest subset of X with property P, then Y is the intersection of all subsets of X that have property P. Is the converse necessarily true?

2.6.8. Let G be a group, and let X be a set. Let I be the intersection of all subgroups of G that contain X. Show that I is the smallest subgroup of G that contains X. Conclude that $I = \langle X \rangle$.

2.6.9. Let G be a group, and assume that a and b are two elements of order 2 in G. If $ab = ba$, then what can you say about $\langle a, b \rangle$?

2.6.10. Can D_8 be generated by two reflections?

2.6.11. If possible, explicitly find a subgroup of S_4 that is isomorphic to $\mathbb{Z}/2\mathbb{Z} \times \mathbb{Z}/2\mathbb{Z}$.

2.6.12. Let $G = \langle x, y \mid x^7 = y^3 = e, yxy^{-1} = e \rangle$. What is $|G|$? Find a familiar group that is isomorphic to G.

2.6.13. Let $G = \langle x, y \mid x^7 = y^3 = e, yxy^{-1} = x \rangle$. What is $|G|$? Find a familiar group that is isomorphic to G.

2.6.14. Let $G = \left\{ \begin{pmatrix} a & b \\ 0 & d \end{pmatrix} \in \mathrm{GL}(2, \mathbb{R}) \right\}$. Is G a subgroup of $\mathrm{GL}(2, \mathbb{R})$?

2.6.15. Let $\mathrm{SL}(2, \mathbb{Z})$ be the group of 2×2 (invertible) matrices with determinant 1 over the integers. (See Problem 2.1.4.) Define

$$\Gamma_0(47) = \left\{ \begin{bmatrix} a & b \\ c & d \end{bmatrix} \in \mathrm{SL}(2, \mathbb{Z}) \mid c \text{ is divisible by } 47 \right\}.$$

Is $\Gamma_0(47)$ a subgroup of $\mathrm{SL}(2, \mathbb{Z})$?

2.6.16. Let F be one of $\mathbb{Q}$, $\mathbb{R}$, $\mathbb{C}$, or $\mathbb{Z}/p\mathbb{Z}$ where p is a prime (more generally F can be any field). Let $G \leq \mathrm{GL}(n, F)$ be a group of matrices (with matrix multiplication). Let $H = \{A^t \mid A \in G\}$ be the set of the transposes of the matrices in G. Prove that H is a subgroup of $\mathrm{GL}(n, F)$ as well.

2.6.17. Let $G = \mathrm{SL}(2, p)$ be the group of 2×2 matrices with determinant 1 and with entries in $\mathbb{Z}/p\mathbb{Z}$. Let

$$H = \left\{ \begin{bmatrix} \lambda & \mu \\ 0 & \lambda^{-1} \end{bmatrix} \mid \lambda, \mu \in \mathbb{Z}/p\mathbb{Z}, \lambda \neq 0 \right\}.$$

(a) Show that H is a subgroup of G. (Note that λ^{-1} is the multiplicative inverse of λ in $\mathbb{Z}/p\mathbb{Z}$.)

(b) In the case of $p = 3$, find a familiar group that is isomorphic to H.

2.6.18. **Subgroups and homomorphisms.** Let G and H be groups, and let $\theta : G \longrightarrow H$ be a homomorphism. Let K be a subgroup of G. Is $\theta(K)$ necessarily a subgroup of H?

2.6.19. **Kernels of homomorphisms.** Let G and H be groups, and let $\theta : G \longrightarrow H$ be a homomorphism. The set $\{x \in G \mid \theta(x) = e\}$ is called the *kernel* of θ and is denoted by $\ker(\theta)$. Show that $\ker(\theta)$ is a subgroup of G.

2.6.20. **Inverse images of subgroups.** Let G and H be groups, and let $\theta : G \longrightarrow H$ be a homomorphism. Let K be a subgroup of H. The set of elements of G that are mapped into K are denoted by $\theta^{-1}(K)$. In other words,

$$\theta^{-1}(K) = \{g \in G \mid \theta(g) \in K\}.$$

Is $\theta^{-1}(K)$ necessarily a subgroup of G?

Subgroups of cyclic groups. We proved in Theorem 2.75 that every subgroup of a cyclic group is cyclic. In Problem **2.6.22**—which uses Problem 2.6.21—you are asked to find the *number* of subgroups of each order for a finite cyclic group. Problem 2.6.23 then uses this result to give an identity for the Euler totient function.

2.6.21. Let $G = \langle g \rangle$ be a cyclic group of order n, and let $H \leq G$. Let k be the smallest positive integer such that $g^k \in H$. Show that k divides n.

2.6.22. Let G be a cyclic group of order n. Let $m \leq n$ be a positive integer. How many subgroups of order m does G have? Prove your assertion.

2.6.23. Let n be an integer bigger than 1, and let $\phi(n)$ be the Euler totient function. Use Problems **2.6.22** and 2.3.20 to show

$$\sum_{d \mid n} \phi(d) = n.$$

2.6.24. **Groups with only trivial subgroups.** The finite group G has more than one element and no non-trivial subgroups. Prove that G is cyclic of order p, where p is a prime number.

2.6.25. Let G be a finite group with $H \leq G$ and $a \in G - H$. Let s be the smallest positive integer such that $a^s \in H$, and let n be some other positive integer with $a^n \in H$. Prove that $s \mid n$.

2.6.26. Suppose that G is a finite group with the property that the order of every non-identity element is a prime number. Suppose that $\mathbf{Z}(G)$ is not $\{e\}$. Prove that every non-identity element of G has the same order. Would the conclusion have to remain valid if $\mathbf{Z}(G) = \{e\}$?

2.6.27. **Conjugate subgroups.** Let G be a group, let $H \leq G$, and let $x \in G$. We use the notation xHx^{-1} to denote the set of elements $\{xhx^{-1} \mid h \in H\}$. ($xHx^{-1}$ is called a *conjugate* of H.)
(a) Prove that xHx^{-1} is a subgroup of G.
(b) If H is finite, then how are $|H|$ and $|xHx^{-1}|$ related?
(c) Prove that H is isomorphic to xHx^{-1}.

2.6.28. Let A be a subgroup of $H \times K$ where H and K are groups. Is A of the form $X \times Y$ where X and Y are subgroups of H and K?

2.6.29. H and K are groups. $G = H \times K$. What is the smallest subgroup of G that contains both $H \times \{e\}$ and $\{e\} \times K$?

2.6.30. H and K are groups. Let $G = H \times K$. Assume H is abelian. What can we say about $\mathbf{C}_G(H \times \{e\})$?

Product of Two Subgroups. We often want to construct new subgroups of a group from subgroups that we already have. A construction that is very important but *does not always work* is the "product" of two subgroups, and it is given here.

Definition 2.79 (*AB*). If A and B are subgroups of a group G, define $AB = \{ ab \mid a \in A, b \in B \}$.

Hence, to find the elements of AB, we have to multiply every element of A with every element of B. The result will be a *subset* of the original group. It will be important to know when AB is an actual subgroup—as opposed to a subset. In Problem **2.6.33**, you are asked to show that AB is a subgroup if and only if $AB = BA$. We will be coming back to this important construct quite often.

2.6.31. (a) Let $G = D_8$, $A = \langle a^2 \rangle$, and $B = \langle b \rangle$. Find AB.
(b) Let $G = \mathbb{Z}/2\mathbb{Z} \times \mathbb{Z}/4\mathbb{Z}$, $A = \langle (1,0) \rangle$, and $B = \langle (0,2) \rangle$. Find AB.

2.6.32. (a) If G is abelian and A and B are subgroups of G, prove that AB is a subgroup of G.
(b) Give an example of a group G and two subgroups A and B of G, such that AB is not a subgroup of G.

2.6.33. **When is the product of two subgroups a subgroup?** Assume that H and K are subgroups of a group G. Show that HK is a group if and only if $HK = KH$.

2.6.34. **Union of subgroups.** Suppose $G = H \cup K$ where H and K are subgroups. Show that $G = H$ or $G = K$.

2.6.35. Suppose $G = \bigcup_\alpha H_\alpha$ where the H_α are proper subgroups of G. Assume $xy = yx$ whenever $x \in H_\alpha$ and $y \in H_\beta$ with $\alpha \neq \beta$. Prove that G is abelian.

2.7. More Problems and Projects

Problems

2.7.1. **The discrete Heisenberg group.** Let $G = \mathrm{SL}(3, \mathbb{R})$ be the group of 3×3 matrices with determinant 1 with entries in the reals. Let

$$x = \begin{bmatrix} 1 & 1 & 0 \\ 0 & 1 & 0 \\ 0 & 0 & 1 \end{bmatrix}, \ y = \begin{bmatrix} 1 & 0 & 0 \\ 0 & 1 & 1 \\ 0 & 0 & 1 \end{bmatrix} \in G.$$

(a) Find x^{-1}, x^n, y^{-1}, and y^m, where n and m are positive integers.
(b) Let $z = xyx^{-1}y^{-1}$. Find z^ℓ, where ℓ is a positive integer.
(c) The discrete Heisenberg group is the subgroup of G generated by x and y. It is denoted by $\mathbf{H}_3(\mathbb{Z})$. In other words,

$$\mathbf{H}_3(\mathbb{Z}) = \langle x, y \rangle.$$

Verify that z is in the center of $\mathbf{H}_3(\mathbb{Z})$.

(d) Prove that

$$\mathbf{H}_3(\mathbb{Z}) = \{ \begin{bmatrix} 1 & a & b \\ 0 & 1 & c \\ 0 & 0 & 1 \end{bmatrix} \mid a, b, c \in \mathbb{Z} \}.$$

2.7.2. **Heisenberg group over a field.** Let F be one of $\mathbb{Q}$, $\mathbb{R}$, $\mathbb{C}$, or $\mathbb{Z}/p\mathbb{Z}$ for p a prime. (In fact, F can be any field.) In analogy with Problem 2.7.1, define the *Heisenberg group* over F to be

$$\mathbf{H}_3(F) = \left\{ \begin{bmatrix} 1 & a & b \\ 0 & 1 & c \\ 0 & 0 & 1 \end{bmatrix} \mid a, b, c \in F \right\}.$$

(a) Prove that $\mathbf{H}_3(F)$ is a group under matrix multiplication.
(b) Find a familiar group that is isomorphic to $\mathbf{H}_3(\mathbb{Z}/2\mathbb{Z})$.

Affine Groups

Definition 2.80. Let $\mathbb{R}$ be the set of all real numbers. Pick a and b to be two of your favorite real numbers, and define the map $T_{a,b} : \mathbb{R} \longrightarrow \mathbb{R}$ by

$$T_{a,b}(r) = ar + b.$$

Now define

$$\mathrm{Aff}(1, \mathbb{R}) = \{ T_{a,b} \mid a, b \in \mathbb{R}, \ a \neq 0 \}.$$

The operation on $\mathrm{Aff}(1, \mathbb{R})$ is function composition.

More generally, let n be a positive integer, and let F be one of $\mathbb{Q}$, $\mathbb{R}$, $\mathbb{C}$, or $\mathbb{Z}/p\mathbb{Z}$ where p is a prime (in fact, F can be any field). Writing the elements of F^n as column vectors, we define

$$\mathrm{Aff}(n, F) = \{ T : F^n \to F^n \mid T(x) = Ax + b, \text{ where } A \in \mathrm{GL}(n, F), \text{ and } b \in F^n \}.$$

Then $\mathrm{Aff}(n, F)$ with function composition is called the *affine group* of degree n over F.

2.7.3. Let $G = \mathrm{Aff}(1, \mathbb{R})$ be the affine group of degree 1 over the reals.
(a) Verify that G with the operation of function composition is a group.
(b) What is the identity element of G?
(c) For real numbers a and b with $a \neq 0$, let $T_{a,b}$ be the element of G defined above. What is the inverse of $T_{a,b}$?
(d) What is the center of G?

2.7.4. Find the number of elements of $\mathrm{Aff}(2, \mathbb{Z}/2\mathbb{Z})$.

2.7.5. Let n be a positive integer, and let F be a field. Verify that the affine group $\mathrm{Aff}(n, F)$ is a group.

2.7.6. Find familiar groups that are isomorphic to $\mathrm{Aff}(1, \mathbb{Z}/2\mathbb{Z})$ and $\mathrm{Aff}(1, \mathbb{Z}/3\mathbb{Z})$.

2.7.7. We define several subsets of $\mathrm{Aff}(1, \mathbb{R})$:

$$
\begin{aligned}
N &= \{T_{1,b} \in \mathrm{Aff}(1, \mathbb{R}) \mid b \in \mathbb{R}\}, \\
A &= \{T_{a,0} \in \mathrm{Aff}(1, \mathbb{R}) \mid a \in \mathbb{R}, a \neq 0\}, \\
H &= \{T_{1,b} \in \mathrm{Aff}(1, \mathbb{R}) \mid b \in \mathbb{Z}\}.
\end{aligned}
$$

(a) Show that N and A are subgroups of $\mathrm{Aff}(1, \mathbb{R})$.
(b) Is H a subgroup of N?
(c) What can you say about $N \cap A$?
(d) What can you say about NA?

2.7.8. **The group of rational points on the unit circle.**[6] A point in the complex plane with both coordinates rational is called a *rational point*. The set of rational points on the unit circle will be denoted by $C(\mathbb{Q})$:

$$
C(\mathbb{Q}) = \{a + bi \in \mathbb{C} \mid a, b \in \mathbb{Q}, \text{ and } a^2 + b^2 = 1\}.
$$

(a) Show that, with the usual multiplication of complex numbers, $C(\mathbb{Q})$ is a group.
(b) Is $i \in C(\mathbb{Q})$? What is $o(i)$?
(c) Let $x = \frac{3}{5} + \frac{4}{5}i$. Is $x \in C(\mathbb{Q})$? Find x^2, x^3, and x^{-1}.
(d) A triple (a, b, c), where a, b, and c are integers, $c \neq 0$, and $a^2 + b^2 = c^2$ is called a *Pythagorean triple*. Given an element of $C(\mathbb{Q})$, can you construct a Pythagorean triple? Given a Pythagorean triple, can you construct an element of $C(\mathbb{Q})$? Under this correspondence, when do two different Pythagorean triples give the same element of $C(\mathbb{Q})$?
(e) In $C(\mathbb{Q})$, what is the order of $3/5 + 4/5i$? Make a conjecture.

Approximate Subgroups

Definition 2.81. Let G be a group, and let A be a finite subset of G. Assume that $e \in A$, and if $a \in A$, then $a^{-1} \in A$. Let k be an integer greater than 0. We say that A is a k-approximate group if there exists $X \subseteq G$ with $|X| \leq k$ and such that $AA \subseteq XA$.[7]

2.7.9. Let G be a group, and let A be a finite subset of G. Assume that $e \in A$, and if $a \in A$, then $a^{-1} \in A$. Let $k, k' \geq 1$ be parameters.

(a) Prove that $A \subseteq G$ is a 1-approximate group if and only if A is a subgroup of G.
(b) Prove that if $A \subseteq G$ is a k-approximate group, then it is a k'-approximate group if $k' \geq k$.
(c) Find all 2-approximate groups inside D_8.

---

[6]Adapted from Tan [**Tan96**] where, using elementary ring theory, the structure of the group of rational points on the unit circle is investigated.

[7]For a superb exposition—aimed at professional mathematicians—of approximate algebraic structures, their applications in additive number theory, and their connections to other areas of mathematics, see Ben Green [**Gre**].

2.7.10. **The number of centralizers of a finite group.** Let G be a finite
group. A subgroup of G may be the centralizer of an element of G. Count
the number of such subgroups of G, and call the answer $\#\mathrm{Cent}(G)$.[8]
   (a) Show that G is abelian if and only if $\#\mathrm{Cent}(G) = 1$.
   (b) Show that any non-abelian group G is the union of its proper cen-
       tralizers.
   (c) Show that $\#\mathrm{Cent}(G) \neq 2$.
   (d) Show that $\#\mathrm{Cent}(G) \neq 3$.
   (e) What is $\#\mathrm{Cent}(D_8)$?

An Unlikely Group. In Problem 1.3.6, you were asked to show that $\{3, 9, 15, 21\}$
forms a group if the operation is multiplication mod 24. In the next problem, we
generalize this example.[9] (Also see Problems 2.3.5, and 2.5.12.)

2.7.11. Let n be an integer greater than 1, and let k be a positive integer relatively
prime to n. Let

$$G = k(\mathbb{Z}/n\mathbb{Z})^{\times} = \{kx \in \mathbb{Z} \mid x \in \mathbb{Z} \text{ with } 1 \leq x < n, \text{ and } \gcd(x, n) = 1\}.$$

In what follows, the binary operation on the set G will be multiplication
mod kn. (Note that the multiplication is mod kn and *not* mod n.)
   (a) Find the elements of G for $n = 8$ and $k = 7$. Using multiplication
       mod 56, construct the multiplication table for G and verify that it
       is a group. What is the identity element?
   (b) Let k' be the inverse of k in $((\mathbb{Z}/n\mathbb{Z})^{\times}, \cdot)$. In other words, $1 \leq k' < n$
       and remainder of kk' when divided by n is 1. Show that kk' is the
       identity of G (where the operation is multiplication mod kn).
   (c) Let $y \in G$. Then $y = kx$ with $x \in (\mathbb{Z}/n\mathbb{Z})^{\times}$. Let x' be the multi-
       plicative inverse of x mod n. Show that $kk'x'k'$ is the inverse of y in
       G.
   (d) Prove that G with multiplication mod kn is a group.
   (e) Is the group in Problem 1.3.6 an example of G? If the answer is yes,
       then what is n and what is k?

2.7.12. Use Theorem 2.25 to give an alternate proof that G defined in Problem
2.7.11 is a group.

2.7.13. **An implication axiom for abelian groups.** McCune and Sands
[**MS96**] proved the following:
   Let G be a non-empty set with a binary operation $+$ and a unary
operation $'$ such that for all $x, y, z, u \in G$,

$$x + y = z + u \implies (y' + z) + u = x.$$

---

[8]This problem is adapted from Belcastro and Sherman [**BS94**] where it is proved that $\#\mathrm{Cent}(G) = 4$ if and only if $G/\mathbf{Z}(G) \cong \mathbb{Z}/2\mathbb{Z} \times \mathbb{Z}/2\mathbb{Z}$, and $\#\mathrm{Cent}(G) = 5$ if and only if $G/\mathbf{Z}(G)$ is isomorphic to $\mathbb{Z}/3\mathbb{Z} \times \mathbb{Z}/3\mathbb{Z}$ or S_3. You will learn the meaning of $G/\mathbf{Z}(G)$ in Chapter 10.
[9]Adapted from Berger [**Ber05**]. For a variation, see Green [**Gre00**, Theorem 1 and 2].

Then $(G, +)$ is an abelian group (with $x' = -x$ for all $x \in G$).

Find a copy of their paper, decipher the proof—their proof is a translation, into a human-readable form, of a proof found by a computer program—and write an exposition of it.

Periods of Periodic Maps on $\mathbb{R}$

Remark 2.82. Problems 2.7.14–2.7.17 are for those who have had a course in mathematical analysis.

Definition 2.83 (Periodic maps). Let $\mathbb{R}$ denote the set of real numbers. Let D be a non-empty subset of $\mathbb{R}$, and let p be a real number. A function $f \colon D \to \mathbb{R}$ has *period p on D* if:

(a) For any real number x, x is in D if and only if $x + p$ is in D.

(b) For all $x \in D$, $f(x + p) = f(x)$.

We say that f is *periodic on D* if f has some positive period on D.

2.7.14. Let G be a subgroup of $\mathbb{R}$ under addition. Let G^+ denote the set of positive elements in G.

Suppose that $G \neq \{0\}$, and let $r_0 = \inf(G^+) = $ greatest lower bound of G^+. Show that G is cyclic if and only if $r_0 > 0$. Furthermore, if G is cyclic, then $G = \langle r_0 \rangle$.

2.7.15. Let G be a subgroup of $\mathbb{R}$ under addition. Prove that if G is not cyclic, then G is a dense subset of $\mathbb{R}$.

2.7.16. Let f be a periodic function on a non-empty set $D \subseteq \mathbb{R}$, and let P be the set of periods of f on D. Prove that P is an additive subgroup of $\mathbb{R}$.

2.7.17. Let f be a non-constant periodic function on a non-empty set $D \subseteq \mathbb{R}$. Suppose that f is continuous at a point a in D. Prove that f has a least positive period p_0 on D and the set of periods of f on D consists of all integral multiples of p_0.

Chapter 3

The Alternating Groups

...where basic facts about permutations and the symmetric group are proved, and the alternating groups which are subgroups of the symmetric groups are defined.

So far, we have been considering symmetric groups as a source of examples for groups. The symmetric groups play an important part in group theory for two reasons. The first reason is that permutations are important in many branches of mathematics and, hence, the symmetric group appears in many places. The second reason is that the structure of symmetric groups is very rich. In fact, according to Cayley's theorem—which shall be discussed later—every finite group is isomorphic to a subgroup of some symmetric group. In this short chapter, we shall introduce an important subgroup—called the *alternating group*—of the symmetric group.

3.1. Permutations, Cycles, and Transpositions

Recall that $[n] = \{1, \ldots, n\}$, and $S_n = \mathrm{Perm}([n])$ is the group of 1-1, onto maps from $[n]$ to $[n]$. The operation for this group is function composition. We have used this group as an example, and in our calculations we have already been using the cycle notation for representing the elements of S_n. We now want to go back and develop the basic ideas more rigorously.

Definition 3.1 (Cycles and transpositions). Let $i_1, i_2, \ldots, i_m$ be distinct elements of $[n]$, and let the map $\sigma : [n] \to [n]$ be defined by

$$\sigma : \begin{cases} i_1 \mapsto i_2 \\ i_2 \mapsto i_3 \\ \vdots \\ i_{m-1} \mapsto i_m \\ i_m \mapsto i_1 \\ k \mapsto k \qquad \text{for each } k \in [n] \setminus \{i_1, \ldots, i_m\}. \end{cases}$$

Then σ is an element of S_n, it is denoted by $(i_1\ i_2\ \cdots\ i_m)$ and is called an *m-cycle* or a cycle of length m.

Two cycles $(i_1\ i_2\ \cdots\ i_m)$ and $(j_1\ j_2\ \cdots\ j_\ell)$ are *disjoint*, if $\{i_1, i_2, \ldots, i_m\} \cap \{j_1, j_2, \ldots, j_\ell\} = \emptyset$.

A 2-cycle is called a *transposition*.

So, for example, $(3\ 5)$ is a transposition in S_5 (and in fact it is in every S_n for $n \geq 5$). This elements swaps 3 with 5 and keeps 1, 2, and 4, fixed.

Lemma 3.2. *Let n be a positive integer, and let S_n be the symmetric group of degree n.*

(a) *If σ and τ are disjoint cycles in S_n, then $\sigma\tau = \tau\sigma$.*

(b) *Every element of S_n can be written as a cycle or a product of disjoint cycles.*

(c) *Every element of S_n can be written as a product of transpositions.*

Proof. (a) Let $k \in [n] = \{1, \ldots, n\}$. If both σ and τ fix k, then $\sigma(\tau(k)) = k = \tau(\sigma(k))$. Otherwise, since σ and τ are disjoint, exactly one of the two moves k. Without loss of generality assume $\tau(k) = k$ and σ moves k. Then $\sigma(k)$ is also moved by σ and fixed by τ, and so $\tau(\sigma(k)) = \sigma(k) = \sigma(\tau(k))$. As as result $\sigma\tau = \tau\sigma$.

(b) A permutation $\sigma \in S_n$ is a 1-1, onto map from $[n] = \{1, \ldots, n\}$ to $[n]$. We begin with $1 \in [n]$ and r epeatedly apply σ:

$$1, \sigma(1), \sigma^2(1), \ldots.$$

Every one of these elements are members of $[n]$, a finite set, and hence eventually there will be repeats. Now assume $\sigma^i(1) = \sigma^j(1)$ for some $i > j$. This is the same as $\sigma(\sigma^{i-1}(1)) = \sigma(\sigma^{j-1}(1))$. But σ is 1-1, and so it follows that $\sigma^{i-1}(1) = \sigma^{j-1}(1)$. Repeating this argument—or, more formally, using induction—we get that $\sigma^{i-j}(1) = 1$. Now let k be the smallest positive integer for which $\sigma^k(1) = 1$—having shown that $\sigma^{i-j}(1) = 1$, we know that such a k exists—then there are no repeats in

$$1, \sigma(1), \sigma^2(1), \ldots, \sigma^{k-1}(1).$$

If this list contains every element of $[n]$, then σ is the cycle

$$(1\ \sigma(1)\ \sigma^2(1)\ \cdots\ \sigma^{k-1}(1)).$$

Otherwise, choose an element of $[n]$ not in the above list, and again repeatedly apply σ to get a second cycle disjoint from the first one. Continuing in this fashion, we get that σ is a product of disjoint cycles. Except for rearranging the order of the cycles, this representation of σ is unique.

(c) Any cycle can be written as a product of transpositions:

$$(k_1\ k_2\ \cdots\ k_m) = (k_1\ k_m)(k_1\ k_{m-1}) \cdots (k_1\ k_2).$$

Now, since every element of S_n is a product of cycles, we conclude that every element of S_n is also a product of transpositions. $\square$

Example 3.3. $(1\ 2\ 3\ 4)(6\ 7\ 8) = (1\ 4)(1\ 3)(1\ 2)(6\ 8)(6\ 7)$.

The last part of the lemma says that one can achieve any permutations of n objects by repeatedly switching two at a time. For example, if you have objects A, B, and C in the order A, B, C and want to put them in the order C, A, B, then you can achieve this by first switching the first and the second object—resulting in the order B, A, C—and then switching the first and third object (resulting in the order C, A, B).

Recall from Definition 1.27 that a *cycle decomposition* of an element of $\sigma \in S_n$ is a product of disjoint cycles that equal σ. The sequence of non-decreasing integers that gives the length of the cycles in a cycle decomposition of σ is called the *cycle type* of σ.

In Problem 3.1.3, you are asked to write down an argument for the following:

Proposition 3.4. *Let n be a positive integer, and let $\sigma \in S_n$. Then the order of σ is the least common multiple of the lengths of the cycles in a cycle decomposition of σ.*

Problems

3.1.1. Let $\sigma = (a_1 \ a_2 \ \cdots \ a_m) \in S_n$. Find σ^{-1}.

3.1.2. Let $\sigma = (1\ 3\ 5)(2\ 4)$ and $\tau = (1\ 5)(2\ 3)$ be elements of S_5. Find the orders of σ, τ, σ^2, $\sigma\tau$, $\tau\sigma$, $\tau\sigma^2$.

3.1.3. **Order of elements in S_n.** Prove Proposition 3.4. In other words, prove that the order of an element $\sigma \in S_n$ is the least common multiple of the lengths of the cycles in its cycle decomposition.

3.1.4. What is the smallest positive integer n for which S_n has an element of order 15? What about an element of order 11?

3.1.5. Does S_7 have a subgroup isomorphic to $\mathbb{Z}/12\mathbb{Z}$? Either prove that it does not or exhibit such a subgroup.

3.1.6. Let x and y be two 3-cycles. Can xy be a 3-cycle? A 5-cycle? An element of order 2? In each case, either give an example, or prove that it is impossible.

3.1.7. What is the smallest positive integer m such that $g^m = e$ for all $g \in S_9$?

3.1.8. Let $\sigma = (1\ 2\ 3\ \cdots\ 12) \in S_{12}$. For $1 \le i \le 12$, find σ^i. For which i is σ^i a 12-cycle? What is σ^{4747}?

3.1.9. **Conjugate elements.** Let $\sigma, \tau \in S_n$. Define $\delta = \tau\sigma\tau^{-1}$. ($\sigma$ and δ are called *conjugate elements*; see Chapter 6.)
   (a) Show that if $\sigma(i) = j$, then $\delta(\tau(i)) = \tau(j)$.
   (b) Explain that the previous part says that if you apply τ to each of the *entries* in the cycle notation of σ, then you get the cycle notation for δ. In other words, if σ has cycle decomposition

$$(a_1 \ a_2 \ \cdots \ a_{k_1})(b_1 \ b_2 \ \cdots \ b_{k_2}) \cdots .$$

Then δ has cycle decomposition

$$(\tau(a_1)\ \tau(a_2)\ \cdots\ \tau(a_{k_1}))(\tau(b_1)\ \tau(b_2)\ \cdots\ \tau(b_{k_2}))\cdots.$$

(c) Illustrate the previous part by letting $\sigma = (1\ 4\ 3\ 8)(2\ 6\ 5)$, $\tau = (1\ 6\ 3)(7\ 5\ 2)$, and quickly writing down the cycle decomposition for $\tau\sigma\tau^{-1}$. Check your answer by actually finding the product $\tau\sigma\tau^{-1}$.

3.1.10. Let $\sigma = (3\ 4)(1\ 5\ 2\ 7)$ and $\delta = (6\ 7)(1\ 4\ 3\ 5)$ be elements of S_7. Are σ and δ conjugate? In other words, does there exists $\tau \in S_7$ such that $\delta = \tau\sigma\tau^{-1}$?

3.1.11. How many elements of order 5 does S_{12} have?

3.1.12. Let $\sigma \in S_n$ and assume that σ is the product of r disjoint cycles. (In calculating r, we do count the 1-cycles.) Show that σ can be written as a product of $n - r$ transpositions.

3.1.13. A *simple transposition* is a transposition of the form $(i\ i+1)$. Show that every element of S_n can be written as a product of simple transpositions.

3.1.14. Let $n > 1$ be an integer, and let $\tau = (1\ 2)$ and $\sigma = (1\ 2\ \ldots\ n)$ be elements of S_n.
(a) What is $\sigma\tau\sigma^{-1}$? What about $\sigma^2\tau\sigma^{-2}$?
(b) Show that $\langle \tau, \sigma \rangle$ contains every simple transposition (see Problem 3.1.13).
(c) Do τ and σ generate S_n?
(d) Let a and b be integers with $1 \le a < b \le n$. Let σ' be any n-cycle that sends a to b, and let $\tau' = (a\ b)$. Show that $\langle \sigma', \tau' \rangle = S_n$.

3.1.15. Let p be a prime number, and let a and b be integers with $1 \le a < b \le p$. Let $\tau = (a\ b)$ and $\sigma = (1\ 2\ \cdots\ p)$ be elements of S_p.
(a) Let $\sigma' = \sigma^{b-a}$. Is σ' a p-cycle? Where does σ' send a?
(b) Show that τ and σ generate S_p.
(c) In S_p, let δ be any transposition and let ρ be any p-cycle. Show that $S_p = \langle \delta, \rho \rangle$.
(d) Do $(1\ 3)$ and $(1\ 2\ 3\ 4)$ generate S_4?

3.2. Even and Odd Permutations and A_n

Finite permutations—that is, elements of S_n—come in one of two varieties: even or odd permutations.

Definition 3.5 (Even and odd permutations). An element of S_n—that is, a permutation of a finite number of elements—is called an *even permutation*, if it can be written as a product of an even number of transpositions. Likewise, an element of S_n that can be written as a product of an odd number of transpositions is called an *odd permutation*.

Note that the *factorization* into a product of transpositions is not unique. For example, $(1\ 2\ 3) = (1\ 3)(1\ 2) = (4\ 5)(1\ 3)(1\ 2)(4\ 5)$. Since we *can* write $(1\ 2\ 3)$ as a product of an even number of transpositions, then, by the definition, $(1\ 2\ 3)$ is an even permutation. However, could $(1\ 2\ 3)$ also be an odd permutation? In other words, maybe we could also find a way of writing this permutation as the product of an odd number of transpositions. This is *not* the case. In fact, this is an important result—which is not obvious and needs a proof—and we will prove it later in this section. So until we prove this result, we have to allow for the possibility that a permutation may be both even and odd. Even so, the next lemma follows immediately from the definition:

Lemma 3.6. *Let $\sigma, \tau \in S_n$, then $\sigma\tau$ is even if σ and τ are both even or both odd. If one of σ and τ are even and the other odd, then $\sigma\tau$ is odd.*

We are ready to define the alternating groups:

Definition 3.7 (The alternating group A_n)**.** Let n be a positive integer, and let A_n be the set of all even permutations in S_n. A_n is called the *alternating group of degree n*.

Corollary 3.8. *A_n is a subgroup of S_n.*

Proof. The product of two even permutations is even and, hence, A_n is closed under the operation of S_n. Since A_n is also non-empty and finite, it must be a subgroup by Lemma 2.74. □

How big is A_n? The following theorem will help to answer this:

Theorem 3.9. *Let n be an integer greater than one. In S_n, the number of even permutations equals the number of odd permutations.*

Proof. Recall that A_n is the set of even permutations in S_n. We will denote the set of odd permutations in S_n by B_n.

We will show that $|A_n| = |B_n|$ by giving a bijection between the two sets.

Let $\delta = (1\ 2) \in B_n$. Define $\phi\colon A_n \to B_n$ by $\sigma \mapsto \delta\sigma$.

Note that if $\sigma \in A_n$, then σ can be written as a product of an even number of transpositions. So $\delta\sigma$ can be written as a product of an odd number of transpositions, namely δ and the ones used to represent σ. Hence $\delta\sigma \in B_n$, and ϕ is indeed a map from A_n to B_n.

We first show that ϕ is 1-1. Assume $\phi(\sigma_1) = \phi(\sigma_2)$, then $\delta\sigma_1 = \delta\sigma_2$. By cancellation, we get $\sigma_1 = \sigma_2$, and hence ϕ is 1-1.

We now show that ϕ is onto. Let τ be an arbitrary element of B_n. Then τ can be written as a product of an odd number of transpositions. This means that $\delta\tau$ can be written as a product of an even number of transpositions, and so $\delta\tau \in A_n$. In addition, $\phi(\delta\tau) = \delta\delta\tau = \tau$, and hence τ is in the image of ϕ. This proves that ϕ is onto.

So, ϕ is a bijection from A_n to B_n, and so $|A_n| = |B_n|$. □

Every permutation is either odd or even, and we have shown that the number of even permutations is the same as the number of odd permutations. Does this mean that the number of even permutations—and hence the order of the group A_n—is half of the size of S_n? This fact, while being true, does not quite follow yet. To be able to make this claim, we have to prove that no permutation is both even and odd. (In the language of the proof of Theorem 3.9, we have to show that $A_n \cap B_n = \emptyset$.) We will prove this now.

Theorem 3.10. *No permutation is both even and odd.*

Proof. Assume, to the contrary, that there exists $\sigma \in S_n$ such that $\sigma = \tau_1 \tau_2 \cdots \tau_m$ with m odd, as well as $\sigma = \sigma_1 \sigma_2 \cdots \sigma_k$ with k even, and where τ_1, τ_2, ..., τ_m as well as σ_1, ..., σ_k are all transpositions. We seek a contradiction.

Note that $\sigma_i^{-1} = \sigma_i$. It follows that $e = \sigma\sigma^{-1} = \tau_1 \tau_2 \cdots \tau_m \sigma_k^{-1} \sigma_{k-1}^{-1} \cdots \sigma_1^{-1} = \tau_1 \tau_2 \cdots \tau_m \sigma_k \sigma_{k-1} \cdots \sigma_1$. Thus, the identity e can be written as the product of an odd number of transpositions.

Let ℓ be the smallest odd integer such that e is the product of ℓ transpositions, and let $e = \gamma_1 \gamma_2 \cdots \gamma_\ell$, where γ_1, γ_2, ..., γ_ℓ are transpositions.

Now $\ell \neq 1$, since no single transposition is the identity. Thus $\ell > 1$ (in fact, here is the place where we use the fact that ℓ is odd, since otherwise ℓ could have been zero).

Assume that $\gamma_\ell = (a\ b)$, with a and b distinct elements of $[n]$. What are the possibilities for the transposition $\gamma_{\ell-1}$? There are three possibilities: in $\gamma_{\ell-1}$, both a and b, just one of them, or neither of them, could be transposed.

CASE 1: $\gamma_{\ell-1}$ transposes both a and b. In other words, $\gamma_{\ell-1} = (a\ b)$.

In this case, we would have $\gamma_{\ell-1}\gamma_\ell = (a\ b)(a\ b)$. But then we could cancel these two transpositions and have a shorter product of an odd number of transpositions be equal to e, and this is not possible since ℓ was chosen to be minimal.

CASE 2: $\gamma_{\ell-1}$ transposes only one of a and b. Hence $\gamma_{\ell-1} = (a\ d)$ or $\gamma_{\ell-1} = (b\ d)$ with a, b, and d distinct elements of $[n]$.

In this case, we either have

$$\gamma_{\ell-1}\gamma_\ell = (a\ d)(a\ b) = (a\ b\ d) = (a\ b)(b\ d)$$

or

$$\gamma_{\ell-1}\gamma_\ell = (b\ d)(a\ b) = (a\ d\ b) = (a\ d)(b\ d).$$

CASE 3: $\gamma_{\ell-1}$ transposes neither a nor b, in which case we have $\gamma_{\ell-1} = (c\ d)$ with a, b, c, and d all distinct elements of $[n]$.

Here, we have

$$\gamma_{\ell-1}\gamma_\ell = (c\ d)(a\ b) = (a\ b)(c\ d).$$

So, the first case is impossible and, in the other two cases, we can move the transposition that moves a to the left. (Note that in every one of $(a\ b)(b\ d)$, $(a\ d)(b\ d)$ and $(a\ b)(c\ d)$, the last transposition does not move a.) We now repeat this same argument for $\gamma_{\ell-1}$ and $\gamma_{\ell-2}$, and further on, until the only transposition that moves a is γ_1.

But this is impossible, since if a is only moved by γ_1 and $e = \gamma_1 \gamma_2 \cdots \gamma_\ell$, we will have to conclude that the identity moved a. The contradiction proves the theorem. □

Corollary 3.11. *Let n be an integer greater than one. Then*

$$|A_n| = \frac{1}{2}|S_n| = \frac{n!}{2}.$$

Proof. Follows directly from Theorems 3.9 and 3.10. □

The alternating group A_n is an important subgroup of the symmetric group S_n. In fact, whenever we have a subgroup of S_n for a particular n, we have a collection of permutations, of n objects, that form a group. Such groups are called permutation groups:

Definition 3.12 (Permutation groups). Let Ω be any set, and let $H \leq \operatorname{Perm}(\Omega)$. Then the group H is called a *permutation group*. If Ω is a finite set of size n, then H is called a *permutation group of degree n*.

Example 3.13. Let
$$G = \{1_{[4]}, (1\ 2\ 3\ 4), (1\ 3)(2\ 4), (1\ 4\ 3\ 2), (1\ 2)(3\ 4), (1\ 3), (1\ 4)(2\ 3), (2\ 4)\}.$$
As we saw in Example 1.32, the group G is a subgroup of S_4 isomorphic to D_8. Hence G is a permutation group of degree 4. When we write $D_8 = \langle a, b \mid a^4 = b^2 = e, ba = a^3b \rangle$, then we are not thinking of D_8 as a permutation group. But, as we have seen, there is a permutation group, namely G above, isomorphic to D_8. When thinking of D_8 as a permutation group, we could have in mind the effect of the elements of D_8 on the corners of the square (and how they are permuted). This proves to be quite useful and will be taken up in the next chapter when we turn to group actions. In fact, identifying an abstract group as a group of permutations helps us in carrying out calculations—we know how to multiply permutations, and we do not need, for example, to look at generators and relations—as well as in discerning the properties of the group.

Example 3.14. Consider the symmetric group S_4. Let $H = \{\sigma \in S_4 \mid \sigma(4) = 4\}$. By definition, H consists of those permutations in S_4 that leave 4 fixed. Hence elements of H are exactly the permutations of $[3] = \{1, 2, 3\}$. This means that H is a permutation group isomorphic to S_3. Thus S_4 has a subgroup H that is isomorphic to S_3. In fact, S_4 has several such subgroups (for example, we can take all those permutations that leave 2 fixed). Similarly, if $m < n$ are positive integers, then S_n has many subgroups isomorphic to S_m. We usually think of this as
$$S_3 \leq S_4 \leq S_5 \leq \cdots.$$
Similarly, we also have
$$A_3 \leq A_4 \leq A_5 \leq \cdots.$$

Problems

3.2.1. Consider the following element of S_9 written in two-line format:
$$\begin{pmatrix} 1 & 2 & 3 & 4 & 5 & 6 & 7 & 8 & 9 \\ 3 & 1 & 2 & & & 7 & 8 & 9 & 6 \end{pmatrix}.$$

Assume you know that this permutation is an even permutation. Can you retrieve the lost images of 4 and 5?

3.2.2. Let x and y be two 3-cycles. Can xy be a 4-cycle? Either give an example, or prove that it is impossible.

3.2.3. Define $\phi : S_n \to \mathbb{Z}/2\mathbb{Z}$ by

$$\phi(x) = \begin{cases} 0 & \text{if } x \text{ is an even permutation,} \\ 1 & \text{if } x \text{ is an odd permutation.} \end{cases}$$

Show that ϕ is a group homomorphism.

3.2.4. How many elements of order 2 does A_5 have?

3.2.5. The alternating group A_6 has how many elements of order 3?

3.2.6. Let $G = \langle (1\ 2)(3\ 4), (1\ 2\ 3) \rangle$ be a subgroup of S_4.
(a) What are the elements of G?
(b) How is G related to A_4?

3.2.7. Is A_4 isomorphic to D_{12}? Why?

3.2.8. Is A_4 isomorphic to $S_3 \times \mathbb{Z}/2\mathbb{Z}$? Why?

3.2.9. Do all the 7-cycles in S_7 generate S_7? Prove your assertion and, if possible, generalize.

3.2.10. Let $n \geq 3$. Prove that A_n can be generated by all the 3-cycles in S_n.

3.2.11. For $n \geq 3$, does the set $\{(1\ 2\ 3), (1\ 2\ 4), \ldots, (1\ 2\ n)\} \subset S_n$ generate A_n?

3.2.12. Even permutations are those elements of S_n that can be written as a product of $2k$ transpositions for some positive integer k. By analogy, let the set $P_n(3)$ consist of those elements of S_n that can be written as a product of $3k$ transpositions for some positive integer k.
(a) Prove that $P_n(3)$ is a subgroup of S_n.
(b) What are the elements of $P_3(3)$?
(c) Is there a theorem for elements of $P_n(3)$ corresponding to Theorem 3.10? For example, can an element of S_n be an element of $P_n(3)$ and also be written as a product of $3k + 1$ transpositions, for some positive integer k?
(d) Is $P_n(3)$ an interesting subgroup of S_n? Prove your assertion.

3.3. More Problems and Projects

Problems

3.3.1. Mackiw [**Mac95**] gives a proof—using linear algebra and the Gram–Schmidt process—of the following result:
> A permutation in S_n cannot be written as the product of fewer than $n-r$ transpositions, where r is the number of disjoint cycles in the permutation.

Find a copy of the paper, decipher the proof, and write an exposition of it.

A Permutation Puzzle. (This was one of the problems highlighted in the Preface.) In one round of "Does everyone in your group want to be a millionaire?", a group of 100 contestants play as a team. In one room there are 100 briefcases lined up on a table that are numbered one through 100. In a random draw before the contest, each of the names of the contestants has been placed in one of the briefcases. Hence, when the contestants begin, each of the briefcases has the name of one of the contestants in it. The contestants go into the room one at a time and each is allowed to examine 50 of the 100 briefcases. Each individual contestant will be successful if they find the briefcase with their own name. However, to win anything, *every one* of the 100 contestants has to be successful. After a contestant has left the room, and, before the next one comes in, the room and the briefcases are returned to their original setup. In addition, after leaving the room, a contestant cannot discuss their finding with any of the other contestants. However, the 100 contestants can come up with a common strategy prior to starting the game. If every one of the 100 contestants can go in the room and, after examining 50 briefcases, finds the one with their own name in it, then each of the contestants gets one million dollars![1]

What should the team's strategy be? Could a strategy exists that assures success more than 1% of the time? How about more than 30% of the time?

I suggest that you try to think of a good strategy before reading on.

In the problems (see Problem 3.3.5 which depends on Problems 3.3.3 and 3.3.4) you will investigate the following strategy and prove the rather unbelievable fact that this strategy will succeed more than 30% of the time!

The Strategy. The contestants are numbered 1 through 100. When the contestant numbered i goes into the room, she will choose briefcase numbered i. If she was super-lucky and that briefcase had her name, then she is done; otherwise, she will choose the briefcase whose number is the number of the contestant whose name was in the briefcase that she opened (namely, in the briefcase numbered i). She will continue in this manner until she has opened the allowed 50 briefcases. Thus each time she opens the briefcase whose number—or rather the name of the contestant with that number—was in the previous briefcase. If somewhere along the way she finds her own name, then she stops, declares success, and leaves.

3.3.2. In the permutation puzzle described above, what are the chances of (group) success if each contestant chooses 50 briefcases randomly and independently of the others?

3.3.3. How many elements of S_{100} have exactly one cycle of size 60? If you randomly pick an element of S_{100}, what is the probability that the element will have exactly one cycle of size 60?

---

[1] The puzzle and its solution are adopted from Winkler [**Win06**], where it is reported that the puzzle originated with the Danish computer scientist Peter Bro Miltersen. The puzzle is also featured in Winkler [**Win07**, p. 12].

3.3.4. (a) Estimate $\frac{1}{51} + \frac{1}{52} + \cdots + \frac{1}{100}$ either by using integrals or by using some software program.

(b) If you randomly pick an element of S_{100}, what is the probability that it will have some cycle of size more than 50?

3.3.5. To analyze the permutation puzzle described here, consider the placement of names in the briefcases as a permutation of $1, \ldots, 100$. In other words, the initial placement of the names in the briefcases is an element of S_{100}. Prove that the suggested strategy will succeed if this element of S_{100} has no cycle of size more than 50. Using your answer to Problem 3.3.4, prove that if the contestants agree on the suggested strategy, then the probability of their success (i.e., *all* of them finding the right briefcase) is more than 30%.

Chapter 4

Group Actions

... where we introduce the fundamental notion of the action of a group on a set and we define Cayley graphs, stabilizers, and orbits of actions, as well as the regular and the conjugation actions.

4.1. Definition and Examples

So far, we have treated groups either as a collection of elements with a particularly well-behaved operation, or as a structure where much information can be gleaned from its substructures. Hence, when confronted with a group, we have looked internally. However, as the concept of group actions will show, there is much to be learned about a group, when it interacts with the outside world. Often, the properties and the structure of a group become evident when we see a group act. In fact, we have already seen this. We learned much about the group D_8 from the effects of its elements on the corners of a square, and the symmetric group was actually defined as the permutations of the set $\{1, \ldots, n\}$.

In this chapter, we formally introduce *group actions* which will become the unifying concept throughout our investigation of groups. Many seemingly different ideas will become examples of the use of this single concept. Using group actions is also one of the most important ways that group theory is used in other areas of mathematics. Graph theorists are interested in groups acting on graphs, topologists are interested in groups acting on surfaces and knots, and group theorists are interested in groups acting on groups.

The basic idea is very simple. We have a group acting on a set whenever every element of the group permutes the elements of the set in a way that the group multiplication is respected. We have already seen many examples of this. For example the group $\mathrm{Perm}(\Omega)$, consisting of 1-1 and onto functions on the set Ω, acts on Ω; that is every element of $\mathrm{Perm}(\Omega)$, being a 1-1 and onto function, permutes the elements of Ω. As another example, symmetry groups act on geometric objects.

For example D_8, the group of symmetries of a square, acts on the corners of the square. This is the idea that we want to make precise.

Definition 4.1 (Group action). Let G be a group, and let Ω be a set.

Then G *acts on* Ω if for every element $g \in G$ and every element $\alpha \in \Omega$, there exists an element $g \cdot \alpha \in \Omega$—think of this as $g(\alpha)$—such that

(a) for all $\alpha \in \Omega$, we have $e \cdot \alpha = \alpha$; and

(b) for all $g, h \in G$ and $\alpha \in \Omega$, we have $h \cdot (g \cdot \alpha) = (hg) \cdot \alpha$.

If the group G acts on the set Ω, then Ω is called a G-set.

Example 4.2. The natural action of S_n. Our first example is the natural action of the symmetric group S_n on the set $[n] = \{1, 2, \ldots, n\}$. We define $\sigma \cdot i = \sigma(i)$, for $\sigma \in S_n$ and $i \in [n]$. This is straightforward enough that you should be able— maybe with a bit more experience—to verify the conditions by just looking at them. However, for this first example, we will give the details.

Each element $\sigma \in S_n$ is already a 1-1 and onto map on the set $[n]$, and hence, for $\sigma \in S_n$ and $i \in [n]$, the expression $\sigma(i)$ makes sense and is an element of $[n]$. Hence, as promised in the definition of the action, for each element σ of the group and each element i of the set, we have defined an element $\sigma \cdot i = \sigma(i)$ of the set.

Clearly, $1_{[n]}$ acts trivially. In other words, $1_{[n]} \cdot i = 1_{[n]}(i) = i$, and hence $e \cdot \alpha = \alpha$ for all $\alpha \in [n]$. Also the group multiplication is a function composition— the product $\sigma\tau$ means do τ and then σ—and hence

$$(\sigma\tau) \cdot i = (\sigma\tau)(i) = \sigma(\tau(i)) = \sigma \cdot (\tau \cdot i).$$

The natural action of S_n on $[n]$ can often be used to define other actions, often on sets indexed by $[n]$ or by subsets of $[n]$. We give one example here. For other examples see Problems 4.1.4 and 4.3.9.

Example 4.3. An extension of the natural action of S_n. Let $G = S_4$, and let $\Omega = \{\{1, 2\}, \{1, 3\}, \{2, 3\}, \{1, 4\}, \{2, 4\}, \{3, 4\}\}$ be the set of subsets of size 2 of $[4] = \{1, 2, 3, 4\}$. In the previous example, we saw that S_4 has a natural action on $[4]$. This action also extends to Ω. In other words, define an action of G on Ω by

$$\sigma \cdot \{a, b\} = \{\sigma(a), \sigma(b)\}, \quad \text{for} \quad \sigma \in G, \ \{a, b\} \in \Omega.$$

It is straightforward to show that this is indeed an action.

Example 4.4. Action of D_8 on the diagonals of the square. The group D_8 acts on the diagonals of a square. Thinking of elements of D_8 as rotations and reflections, each element—when applied to the square as a symmetry—does something to the diagonals. Denoting the diagonals by the set $\Omega = \{d_1, d_2\}$, we see that the effect of $a = R_{90}$ on the square is to switch the place of d_1 and d_2. Hence, we define $a \cdot d_1 = d_2$ and $a \cdot d_2 = d_1$. The element $b = H$ does the same, while the diagonal reflection ab fixes both d_1 and d_2. Hence, we define the action of D_8 on Ω by what each of the symmetries does to the diagonals. Convince yourself that the two conditions for an action are satisfied.

One way to think of an action is to imagine that the set Ω is in front of you. When you throw one element of the group G at Ω, then the elements of Ω move around. The conditions of the action say that throwing e results in no movement, and instead of applying σ and then τ, you can take a shortcut and throw $\tau\sigma$ at the set. Now the effect of each element of G on Ω is actually a *permutation* of Ω. (Two different group elements could give the same permutation.) The fact that we get an actual permutation—that is, a bijection—follows directly from the definition of an action as we see in the proof given below.

Lemma 4.5. *Let the group G act on the set Ω. For $g \in G$ define the map $\sigma_g :$ $\Omega \to \Omega$ by $\sigma_g(\alpha) = g \cdot \alpha$. Then $\sigma_g \in \operatorname{Perm}(\Omega)$.*

In other words, the action of every element of the group gives a 1-1 and onto map on Ω. Thus in the case when $|\Omega| = n < \infty$, the action of every element of G gives an element of S_n.

Proof. First, σ_g is a map from Ω to Ω since for each $\alpha \in \Omega$, we have $g \cdot \alpha \in \Omega$. To show σ_g is 1-1, assume $\sigma_g(\alpha) = \sigma_g(\beta)$. This means $g \cdot \alpha = g \cdot \beta$ is an element of Ω. Letting g^{-1} act on this element, we get $g^{-1} \cdot (g \cdot \alpha) = g^{-1} \cdot (g \cdot \beta)$. Now this means that $g^{-1}g \cdot \alpha = g^{-1}g \cdot \beta$ which implies $e \cdot \alpha = e \cdot \beta$. It now follows that $\alpha = \beta$.

To show σ_g is onto, let $\beta \in \Omega$ and let $\alpha = g^{-1} \cdot \beta$. Now $g \cdot \alpha = g \cdot (g^{-1} \cdot \beta) = gg^{-1} \cdot \beta = e \cdot \beta = \beta$. Hence $\sigma_g(\alpha) = \beta$, and the proof is complete. $\square$

Example 4.6. Action of D_8 on the vertices of a square. The group D_8 acts on the vertices of a square. Denote the set of vertices by $\{1, 2, 3, 4\}$ as in the square on the left side of Figure 1.1. We can now write down—as we already have seen in Example 1.32—the permutations that each element of D_8 produces:

$$D_8 = \begin{cases} e \longleftrightarrow 1 \\ a \longleftrightarrow \begin{pmatrix} 1 & 2 & 3 & 4 \end{pmatrix} \\ a^2 \longleftrightarrow \begin{pmatrix} 1 & 3 \end{pmatrix}\begin{pmatrix} 2 & 4 \end{pmatrix} \\ a^3 \longleftrightarrow \begin{pmatrix} 1 & 4 & 3 & 2 \end{pmatrix} \\ b \longleftrightarrow \begin{pmatrix} 1 & 2 \end{pmatrix}\begin{pmatrix} 3 & 4 \end{pmatrix} \\ ab \longleftrightarrow \begin{pmatrix} 1 & 3 \end{pmatrix} \\ a^2b \longleftrightarrow \begin{pmatrix} 1 & 4 \end{pmatrix}\begin{pmatrix} 2 & 3 \end{pmatrix} \\ a^3b \longleftrightarrow \begin{pmatrix} 2 & 4 \end{pmatrix} \end{cases}$$

We had remarked before that the above set of permutations form a subgroup of S_4 that is isomorphic to D_8. At this point, to see this, we have to do a number of calculations. We have to show that these permutations are a subgroup of S_4, and then show that the map above is an isomorphism. The map is clearly 1-1 and onto, and so we have to show that the image of a product is the product of the images. Later on (see Theorem 11.28 and Example 11.32) we shall see that much follows from general facts with no computation whatsoever!

Example 4.7. We saw in Example 4.4 that D_8 acts on the diagonals of the square. In this action, a, b, a^3, and a^2b give the permutation $(1\ 2)$ (since they switch the two diagonals), while e, a^2, ab, and a^3b give the identity permutation.

Example 4.8. Action of $\mathrm{GL}(2, \mathbb{R})$ **on** $\mathbb{R}^2$**.** Let $G = \mathrm{GL}(2, \mathbb{R})$ be the group of invertible 2×2 matrices with real entries. Let $\Omega = \mathbb{R}^2$. Writing elements of Ω as column vectors, for $A \in G$ and $x \in \Omega$, define

$$A \cdot x = Ax,$$

where Ax is the usual matrix product of A and x. Then

$$I_2 \cdot x = x \quad \text{and} \quad A \cdot (B \cdot x) = A \cdot (Bx) = A(Bx) = (AB)x = AB \cdot x.$$

Hence, $\cdot$ defines an action. Every element of G does give a permutation of $\Omega = \mathbb{R}^2$ (even though this permutation is not an element of an S_n since Ω is an infinite set). For example, the effect of the action of $A = \begin{bmatrix} 0 & 1 \\ 1 & 0 \end{bmatrix}$ on $\mathbb{R}^2$ is a reflection of the points in the plane across the line $y = x$.

Groups Acting on Groups. In all of the examples so far, a group has acted on an external set Ω and, as we shall see in the coming chapters, this interaction with outside objects is revealing. We can learn much about the group from such actions. However, there are a number of important actions that can be defined internally. For these actions, the set Ω is either the group itself or a set constructed from the group in some way.

There are two fundamental ways for constructing actions internally and from within the group. In one family of actions—the so-called regular action, the translation action, and their relatives (see below)—the action is basically the group multiplication on the left. In the second family, you act by conjugation (see below). We shall see a number of different versions of these two types of actions and, in the process, learn much about groups. Various actions by left multiplication will play important roles in Chapters 5 and 7, while conjugation will be centrally featured in Chapters 6 and 10.

Acting by Left Multiplication. Using the group multiplication itself, we can construct a number of actions. Here, we will define the regular action and a generalization, which we call the translation action. In future chapters other variants of this action will be used as well.

Definition 4.9 (Regular action). Let G be any group, and let the set Ω be G also. Then define an action of G on Ω by $g \cdot x = gx$. Checking that this is indeed an action is straightforward:

$$e \cdot x = ex = x \quad \text{and} \quad g \cdot (h \cdot x) = g \cdot (hx) = ghx = gh \cdot x.$$

This action is called the *regular action*.

Definition 4.10 (Translation action). Let G be any group, and let $H \leq G$. Then define an action of H on $\Omega = G$ by

$$h \cdot g = hg.$$

It is straightforward to check—and, in fact, the demonstration is identical for the one for the regular action—that this is an action. We call this action the *translation action* (or the action of H on G by *left multiplication*), and, clearly, the regular action is the special case of this action when $H = G$.

Example 4.11. Let $G = D_8 = \langle a, b \mid a^4 = b^2 = e, ba = a^3b \rangle$, and $H = \langle b \rangle = \{e, b\}$. Then H acts on G by the translation action. We know that the action of e has no effect on G, but how does b act on G?

We have $b \cdot e = b$, $b \cdot a = ba = a^3b$, $b \cdot a^2 = ba^2 = a^2b$, $b \cdot a^3 = ab$, $b \cdot b = e$, $b \cdot ab = a^3$, $b \cdot a^2b = a^2$, and $b \cdot a^3b = a$. You may have noticed that we have just reproduced the row corresponding to b of the multiplication table of D_8.

Remark 4.12. If G is any group and $H \leq G$, then we have seen that H acts on $G = \Omega$ by $h \cdot g = hg$.

However, if we had defined $h \cdot g = gh$, then, in general, we would *not* have an action. This is because if h_1 and h_2 are two elements of H that do not commute, then

$$h_1 \cdot (h_2 \cdot g) = gh_2h_1 \neq gh_1h_2 = (h_1h_2) \cdot g.$$

On the other hand, if you really did need the multiplication on the right for an action, you can define $h \cdot g = gh^{-1}$. This gives an action (see Problem 4.1.2) quite similar to the translation action.

Acting by Conjugation. If g is an element of a group, then pre- and post-multiplying, respectively, by g and g^{-1} is called conjugation. We introduce two conjugation actions here. Both—as well as some close relatives—will prove very useful.

Definition 4.13 (Conjugation action). Let G be any group, and let $\Omega = G$. For $g \in G$ and $x \in \Omega$ define

$$g \cdot x = gxg^{-1}.$$

This defines an action of G on Ω called the *conjugation action*. The element gxg^{-1} is called a *conjugate* of x and is sometimes denoted by ${}^g x$.

This definition does indeed define an action since we have

$$e \cdot x = exe = x$$

and

$$g \cdot (h \cdot x) = g \cdot (hxh^{-1}) = ghxh^{-1}g^{-1} = (gh)x(gh)^{-1} = (gh) \cdot x.$$

If we change the set and let Ω be the set of subgroups of G, then we get another conjugation action. If $g \in G$ and $H \leq G$, we define $gHg^{-1} = \{ghg^{-1} \mid h \in H\}$. In other words, we pre- and post-multiply every element of H by g and g^{-1}, respectively, to get a set of elements, denoted by gHg^{-1}. It is a fact—and you were asked to prove it in Problem **2.6.27**—that gHg^{-1} is also a subgroup of G. The set gHg^{-1} is called the *conjugate* of H and is, sometimes, denoted by ${}^g H$.

Definition 4.14 (Another conjugation action). Let G be any group, and let Ω be the set of subgroups of G. Define an action of G on Ω by

$$g \cdot H = gHg^{-1}, \text{ for } g \in G, \text{ and } H \leq G.$$

This definition does indeed define an action (the proof is identical to the one for the conjugation action defined above) and is also called the *conjugation action (on subgroups of G)*.

In the rest of this chapter we will develop a few elementary tools for analyzing actions.

Problems

4.1.1. Let $G = \mathrm{GL}(n, \mathbb{R})$, and let $\Omega = \mathbf{M}_{n \times n}(\mathbb{R})$ be the set of all $n \times n$ matrices with real entries. Define an action of G on Ω by

$$P \cdot A = PAP^{-1} \text{ for } P \in G, A \in \Omega.$$

Prove that $\cdot$ is indeed an action. Would we have an action if $\cdot$ was defined as $P \cdot A = P^{-1}AP$? What about $P \cdot A = PA$?

4.1.2. Let G be a group, and let $H \leq G$. We define an action of H on $\Omega = G$ by $h \cdot g = gh^{-1}$. Is this indeed an action?

4.1.3. Let a group G act on a set Ω. Assume that $H \leq G$. Explain how—using the action of G—we can say that H also acts on Ω.

4.1.4. Let G be a subgroup of S_n. Hence every element of G is a permutation of $[n] = \{1, \ldots, n\}$. Let e_i be the element of $\mathbb{R}^n$ with a 1 in the ith coordinate and 0's in all other coordinates. The set $B = \{e_1, \ldots, e_n\}$ is the standard basis for $\mathbb{R}^n$. Define an action of G on B by

$$\sigma \cdot e_i = e_{\sigma(i)}.$$

Extend this action to an action of G on $\mathbb{R}^n$ as follows: If $v \in \mathbb{R}^n$, then, for some scalars α_1, ..., α_n, we have $v = \alpha_1 e_1 + \cdots + \alpha_n e_n$. For $\sigma \in G$, we define

$$\sigma \cdot v = \alpha_1 e_{\sigma(1)} + \alpha_2 e_{\sigma(2)} + \cdots + \alpha_n e_{\sigma(n)}.$$

(a) Let $n = 3$, let $G = S_3$, and let $v = (\sqrt{2}, -8, 4) \in \mathbb{R}^3$. Find $\sigma \cdot v$ and $\tau \cdot v$, where $\sigma = (1\ 2\ 3)$ and $\tau = (2\ 3)$.

(b) Show that the above definition does indeed give an action of G on $\mathbb{R}^n$.

(c) Can you generalize the above action to an action of any subgroup of S_n on any n-dimensional vector space with a designated basis?

4.1.5. Let G be a group, and let Ω be the set of all subgroups of G.

(a) Show that conjugation (see Definition 4.14) does define an action of G on Ω.

(b) Let $G = S_3$, $a = (1\ 2)$, $b = (1\ 3)$, and $c = (2\ 3)$. Also let $H = \langle a \rangle$. We know that G acts on its subgroups by conjugation. Find $a \cdot H$, $b \cdot H$, and $c \cdot H$.

4.1.6. Let the group G act on the set Ω. Let $L(\Omega) = \{\sigma : \Omega \to \mathbb{C}\}$ be the set of functions from Ω to the complex numbers. (Since we can add functions and multiply them by scalars, the set $L(\Omega)$ is actually a vector space over $\mathbb{C}$.) For $g \in G$ and $\sigma \in L(\Omega)$, define a function $g \cdot \sigma : \Omega \to \mathbb{C}$ by

$$(g \cdot \sigma)(x) = \sigma(g^{-1} \cdot x).$$

Show that this defines an action of G on $L(\Omega)$. Would we have got an action if we had defined $(g \cdot \sigma)(x) = \sigma(g \cdot x)$?

4.1.7. Let $G = D_8$ act on $\Omega = D_8$ by conjugation. Make an 8×8 table, where the rows are indexed by the elements of G and the columns are indexed by the elements of Ω. The entry in row g and column x is the result of the action of g on x (that is $g \cdot x$.) Complete the table.

(a) In each row do you get every element of Ω? Does any element of Ω ever appear in any row more than once? Is this a coincidence?

(b) Does an element of Ω ever appear in any column more than once? Is it possible for a column to contain only one element of Ω? When does this happen?

(c) Can two rows be identical?

(d) Is it possible to tell the order of the elements of the group from this table?

4.2. The Cayley Graph of a Group Action*

Let G be a group, and let G act on a set Ω. In this section we will explore a useful graphical way of looking at this action. To be able to have this graphical representation, we have to settle on a generating set for our group. So, let S be a set of generators for G, that is, $G = \langle S \rangle$. Since every element of G is a product of elements of S (and their inverses), for any $g \in G$ and $\alpha \in \Omega$, we can find $g \cdot \alpha$, if we know $s \cdot \alpha$ for all $s \in S$. Note that if $s \cdot \alpha = \beta$, then $s^{-1} \cdot \beta = \alpha$, and so if we know what s does, we also know what s^{-1} does.

Definition 4.15 (Cayley digraph of an action). Let G act on a set Ω, and let S be a set of generators for G. Define a directed graph (digraph), called the *Cayley digraph of the action*, as follows:

The vertices of the graph are the element of the set Ω, we choose a color for each $s \in S$. For $\alpha, \beta \in \Omega$, we put an s-colored directed edge (arc) from α to β if $s \cdot \alpha = \beta$.

To avoid clutter in our graphs, we adopt two conventions: First, we do not draw loops. In other words, if $s \cdot \alpha = \alpha$, then we do *not* draw an s-colored arc from α to α. Second, if there is an s-colored arc from α to β and another s-colored arc from β to α, then we replace these two directed edges with an undirected s-colored edge between α and β.[1]

Example 4.16. Let D_8 act on the $\{1, 2, 3, 4\}$, the vertices of a square. If we choose the $S = \{a, b\}$ as our generating set, then Figure 4.1 shows the Cayley digraph of this action. From this figure we can, for example, see that $(ab) \cdot 1 = a \cdot (b \cdot 1) = a \cdot 2 = 3$, while $(ba) \cdot 1 = b \cdot 2 = 1$.

Example 4.17. Let $G = D_8$ act on $\Omega = D_8$ by the regular action. Recall that this means that for $g \in G$ and $x \in \Omega$, we have $g \cdot x = gx$. If we choose the generating set to be $S = \{a, b\}$, then the Cayley digraph of this action is given in Figure 4.2.

---

[1]Some—in fact, many—authors insist that for every element $g \in S$, we should also include g^{-1} in S. In other words, S should be closed under taking inverses. If you adopt this convention, then if there is a g-colored arc from α to β, then there is a g^{-1}-colored arc from β to α. These two arcs are then replaced with one undirected g-colored edge between α and β, resulting in an undirected graph.

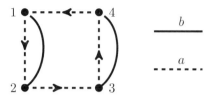

Figure 4.1. The Cayley digraph of the action of D_8 on $\{1,2,3,4\}$ with $S = \{a,b\}$

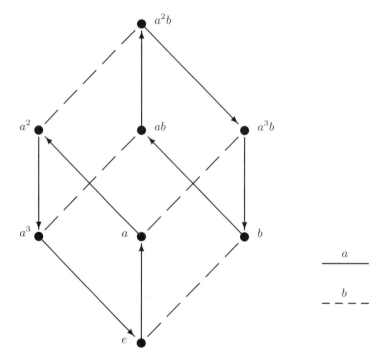

Figure 4.2. The Cayley digraph of the regular action of D_8 on D_8

Problems

4.2.1. Let G be a group that acts on a set Ω, and let S be a set of generators for G. Let $s \in S$. If in S we replace s with s^{-1}, will we still have a generating set? What will the effect be on the Cayley digraph of this action?

4.2.2. Figure 4.1 gives the Cayley digraph for the action of D_8 on the vertices of a square with respect to the generating set $S = \{a,b\}$. Change the generating set to $S' = \{a, ab\}$, and draw the corresponding Cayley digraph.

4.2.3. Figure 4.2 gives the Cayley digraph of the regular action of D_8 on D_8 with respect to the generating set $S = \{a,b\}$. Change the generating set to $S' = \{a, ab\}$, and draw the corresponding Cayley digraph.

4.2.4. Let $G = D_8 = \langle a, b \mid a^4 = b^2 = e, ba = a^3b \rangle$ act by conjugation on $\Omega = D_8$. Let $S = \{a, b\}$ be a set of generators for D_8. Draw the Cayley digraph of the action for this set of generators.

4.2.5. Let $G = D_8 = \langle a, b \mid a^4 = b^2 = e, ba = a^3b \rangle$, and let $\Omega = \{H \mid H \leq G\}$ be the set of subgroups of G. Let G act on Ω by conjugation. (Recall from Definition 4.14 that for $g \in G$ and $H \leq G$, for this action, we have $g \cdot H = gHg^{-1}$.) Let $S = \{a, b\}$ be a set of generators for D_8. Draw the Cayley digraph of the action for this set of generators.

4.2.6. Let $\Omega = \{\{1, 2\}, \{1, 3\}, \{2, 3\}, \{1, 4\}, \{2, 4\}, \{3, 4\}\}$ be the set of subsets of size 2 of $\{1, 2, 3, 4\}$. The action of S_4 on $\{1, 2, 3, 4\}$ extends naturally to an action of S_4 on Ω. In other words, if $g \in S_4$ and $\{a, b\} \in \Omega$, then $g \cdot \{a, b\} = \{g \cdot a, g \cdot b\}$. (See Example 4.3.)
(a) Show that $S = \{(1\ 2), (1\ 2\ 3\ 4)\}$ is a set of generators for S_4.
(b) Draw the Cayley digraph of the action for this set of generators.

4.3. Stabilizers

When G acts on a set Ω, then every element of the group permutes (moves around) the elements of the set. For every element in the set, its stabilizer is the collection of the group elements that leave it fixed in the group action. What becomes important is that this collection is not just a collection of elements of the group but a subgroup of G. Thus, among other things, this provides a way for constructing subgroups: have your group act on some set, and then find stabilizers.

Definition 4.18 (Stabilizers). Let G be a group, and let Ω be a set. Assume that G acts on Ω. Let $\alpha \in \Omega$. Define

$$\mathrm{Stab}_G(\alpha) = \{h \in G \mid h \cdot \alpha = \alpha\}.$$

In other words, $\mathrm{Stab}_G(\alpha)$ is the collection of elements of G that fix α. The set $\mathrm{Stab}_G(\alpha)$ is called the *stabilizer* of α in G. Some authors use the notation G_α for $\mathrm{Stab}_G(\alpha)$, and sometimes $\mathrm{Stab}_G(\alpha)$ is called the *isotropy* group of α.

Lemma 4.19. *Let the group G act on the set Ω, and let $\alpha \in \Omega$. Then $\mathrm{Stab}_G(\alpha)$ is a subgroup of G.*

Proof. Assume $g, h \in \mathrm{Stab}_G(\alpha)$. This means that $g \cdot \alpha = \alpha$ and $h \cdot \alpha = \alpha$. We need to show that $gh \in \mathrm{Stab}_G(\alpha)$. We have

$$gh \cdot \alpha = g \cdot (h \cdot \alpha)$$
$$= g \cdot \alpha$$
$$= \alpha.$$

We conclude that $gh \in \mathrm{Stab}_G(\alpha)$.

We also need to show that the inverse of an element in $\mathrm{Stab}_G(\alpha)$ is also in $\mathrm{Stab}_G(\alpha)$. Let $g \in \mathrm{Stab}_G(\alpha)$.

$$g \cdot \alpha = \alpha$$
$$\Rightarrow g^{-1} \cdot (g \cdot \alpha) = g^{-1} \cdot \alpha$$
$$\Rightarrow (g^{-1}g) \cdot \alpha = g^{-1} \cdot \alpha$$
$$\Rightarrow e \cdot \alpha = g^{-1} \cdot \alpha$$
$$\Rightarrow \alpha = g^{-1} \cdot \alpha.$$

This means that $g^{-1} \in \mathrm{Stab}_G(\alpha)$. $\qquad\square$

Remark 4.20. Note that because of the above lemma we can use actions to find subgroups. That is, as soon as we have an action, we look at the stabilizers of the set elements which are subgroups of the original group.

Example 4.21. If we number the corners of the square 1 through 4 counterclockwise starting with the top left corner (see Figure 1.2), then the group D_8 acts on $\Omega = \{1, 2, 3, 4\}$. In this action, we have $\mathrm{Stab}_{D_8}(1) = \{R_0, D\} = \{e, a^3b\}$, where, as usual, $a = R_{90}$ and $b = H$.

Example 4.22. The group D_8 also acts on $\{d_1, d_2\}$, the diagonals of the square (see Example 4.4). Let d_1 be the main (matrix) diagonal of the square, then the identity, the 180-degree rotation, and both of the diagonal reflections fix d_1. Hence $\mathrm{Stab}_{D_8}(d_1) = \{R_0, R_{180}, D, D'\} = \{e, a^2, ab, a^3b\}$.

Stabilizers for the Regular and the Conjugation Actions. We have seen (see page 88) that given any group, we have a number of natural actions built from the group multiplication. The regular action and the conjugation action (and their relatives) play an important role in our later analysis of the structure of groups. Here, we want to see what is the significance of stabilizers for these actions. In any particular action if the stabilizers turn out to be interesting subgroups, then we may give them additional names.

First the regular action. Let G be any group. G acts on G by regular action; that is, $g \cdot x = gx$. Let $x \in \Omega$, then what is $\mathrm{Stab}_G(x)$? This subgroup consists of all elements of G that do not move x. Hence, $g \in \mathrm{Stab}_G(x)$ if and only if $g \cdot x = x$. But $g \cdot x = x$ if and only if $gx = x$, and the latter happens if and only if $g = e$. Thus for the regular action, the stabilizers of the elements of G are uninteresting. They are just the trivial subgroup. (This does not have to be the case for some of the relatives of the regular action, as we shall see in the course of the text.)

Turning to the conjugation action, we recall that a group G acts on $\Omega = G$ by $g \cdot x = gxg^{-1}$, and this is called the conjugation action (see Definition 4.13). What is $\mathrm{Stab}_G(x)$ for $x \in G$? We have

$$\mathrm{Stab}_G(x) = \{g \in G \mid g \cdot x = x\}$$
$$= \{g \in G \mid gxg^{-1} = x\}$$
$$= \{g \in G \mid gx = xg\}.$$

So the elements of the stabilizer of $x \in G$ in the conjugation action are precisely the elements of G that commute with x. We already had a name for this subgroup—the centralizer of x in G, $\mathbf{C}_G(x)$. We record this fact:

Lemma 4.23. *Let G be a group, and let $x \in G$. In the action of G on G by conjugation, the stabilizer of x in G is the centralizer of x in G.*

Another very important stabilizer comes about in one of the relatives of the conjugation action. This is when the group G acts on the set Ω of all subgroups of G by conjugation (see Definition 4.14). In this case, the stabilizer of a subgroup is called the *normalizer* of the subgroup. Normalizers will be considered in the problems as well as in Chapters 7 and (especially) 10.

Problems

4.3.1. In a square S, let h_1 and h_2 be two line segments joining (and being perpendicular to) the middle of opposing edges. As usual let a denote 90-degree counterclockwise rotation, and let b denote a horizontal reflection. The group $D_8 = \langle a, b \rangle$ then acts on $\{h_1, h_2\}$. For example, if you rotate the group 90 degrees, then h_1 and h_2 are switched. Hence $a \cdot h_1 = h_2$ and $a \cdot h_2 = h_1$. Find $\mathrm{Stab}_{D_8}(h_1)$.

4.3.2. Let $G = \mathrm{GL}(n, \mathbb{R})$, and let $\Omega = \mathbf{M}_{n \times n}(\mathbb{R})$ be the set of all $n \times n$ matrices with real entries. In Problem 4.1.1, we defined an action of G on Ω by

$$P \cdot A = PAP^{-1} \text{ for } P \in G, A \in \Omega.$$

Let $n = 2$, and let $A = \begin{bmatrix} 1 & 1 \\ 0 & 1 \end{bmatrix}$. Describe $\mathrm{Stab}_G(A)$.

A Special Stabilizer: the Normalizer. In addition to the usual conjugation action of a group on itself, we also have defined a conjugation action of a group on the set of all of its subgroups (see Definition 4.14, and also Problems **2.6.27** and **4.1.5**). The set Ω in this action is the set of the subgroups of the group G, and the action of G on Ω is defined by $g \cdot H = gHg^{-1}$. Given this action, for each subgroup H, we can find the stabilizer. This stabilizer—which itself will be a subgroup since all stabilizers are subgroups—has a special name: the normalizer of H in G. The normalizer will play an important role in Chapter 10.

Definition 4.24 (Normalizer). Let G be a group, and let Ω be the set of subgroups of G. Let G act on Ω by conjugation (Definition 4.14), and let $H \in \Omega$. For this action, the stabilizer of H in G is called the *normalizer* of H in G and is denoted by $\mathbf{N}_G(H)$.

By definition, the normalizer of a subgroup H consists of those elements $g \in G$ such that $gHg^{-1} = H$. In other words,

$$\mathbf{N}_G(H) = \{g \in G \mid gHg^{-1} = H\}.$$

4.3.3. Let $G = D_8$, $H = \langle b \rangle$, and $K = \langle a^2 \rangle$. Find $\mathbf{N}_G(H)$ and $\mathbf{N}_G(K)$.

4.3.4. Let G be a group, and let $H \leq G$. Show that
$$\mathbf{N}_G(H) = \{x \in G \mid x^{-1}Hx = H\}.$$

4.3.5. Let G be a group, let $H \leq G$, and let $N = \mathbf{N}_G(H)$. Prove that $H \leq N$ and
$$\mathbf{N}_N(H) = N.$$

4.3.6. Let S_4 act on the subsets of size 2 of $\{1, 2, 3, 4\}$ (see Problem 4.2.6). What is the stabilizer of $\{1, 2\}$?

4.3.7. The group S_n acts on subsets of size k of $[n] = \{1, \ldots, n\}$ (see Problem 4.2.6 and 4.3.6). Let $\alpha = \{1, 2, \ldots, k\}$. What is $|\text{Stab}_{S_n}(\alpha)|$?

4.3.8. Let $G = S_4$, and let $V = \mathbb{R}^4$. In Problem 4.1.4, we defined an action of G on V.

   (a) For this action, what is the stabilizer of $(3, \sqrt{2}, 3, \sqrt{2})$? Find a familiar group that is isomorphic to this stabilizer.

   (b) For $g \in G$, let $W(g) = \{v \in V \mid g \cdot v = v\}$. Prove that $W(g)$ is a subspace of V. Find a basis for $W(g)$ when $g = (1\ 3) \in S_4$.

4.3.9. The group S_4 acts on the set $[4] = \{1, 2, 3, 4\}$. It also acts on the set $\{x_1, x_2, x_3, x_4\}$ by acting on the subscripts of the variables. Hence if $\sigma = (1\ 2\ 3) \in S_4$, then $\sigma \cdot x_3 = x_{\sigma \cdot 3} = x_1$ while $\sigma \cdot x_4 = x_{\sigma \cdot 4} = x_4$. Now let $\mathbb{R}[x_1, x_2, x_3, x_4]$ denote the set of polynomials in four variables—namely x_1, x_2, x_3, and x_4—and with real coefficients. The action of the group S_4 can be extended to an action of S_4 on $\mathbb{R}[x_1, x_2, x_3, x_4]$. So, for example, for $\sigma = (1\ 2\ 3)$, we have
$$\sigma \cdot (x_1^4 - 3x_2^2 + x_4) = x_2^4 - 3x_3^2 + x_4.$$
Let $p(x) = x_1^2 - x_2^2 - x_3^2 + x_4^2$, and find $\text{Stab}_{S_4}(p(x))$.

4.4. Orbits

Equivalence Relations. We begin this section with a review of equivalence relations. Skip to the subsection on orbits, if you are already familiar with equivalence relations and equivalence classes.

Definition 4.25 (Relations). Let X be a non-empty set. A *relation* R on the set X is a subset of $X \times X$. If $(a, b) \in R$, then we write $a \sim b$ or aRb.

In other words, when we have a relation on X, then some of the elements are related to some of the others, and the easiest way to say this is to say that we have a subset of $X \times X$.

Definition 4.26. Let R be a relation on a set X.

- If, for every $a \in X$, we have $a \sim a$, then we say that R is *reflexive*.

- If, for all $a, b \in X$, whenever $a \sim b$, then $b \sim a$, we say that R is *symmetric*.

- If, for all $a, b, c \in X$, the relations $a \sim b$ and $b \sim c$ implies $a \sim c$, then we say that R is *transitive*.

Definition 4.27 (Equivalence relations). An *equivalence relation* on a non-empty set X is a relation that has the reflexive, symmetric, and transitive properties.

Example 4.28. Let X be the set of points in a plane. For $x, y \in X$ we define $x \sim y$ if x and y are the same distance from the origin. Then $\sim$ is an equivalence relation.

Example 4.29. Let $\mathbb{Z}$ be the set of integers. Define $a \sim b$ if $a \leq b$. Then $\sim$ is not an equivalence relation since it does not have the symmetric property.

Example 4.30. Let $\mathbb{Z}$ be the set of integers. Define $a \sim b$ if $n \mid b - a$. In this case we write $a \equiv b \pmod{n}$. In other words, $a \sim b$ if a and b have the same remainder when divided by n. This relation is an equivalence relation.

Example 4.31. Let $\Omega = \mathbf{M}_{n \times n}(\mathbb{R})$ be the set of $n \times n$ matrices with real entries. For $A, B \in \Omega$, define $A \sim B$ if there exists an invertible matrix P such that $B = PAP^{-1}$. This relation is an equivalence relation (in linear algebra, we say that A is *similar* to B). Note that $A = I_n A I_n$ shows that $\sim$ is reflexive. If $A = PBP^{-1}$ and $B = QCQ^{-1}$, then $B = (P^{-1})A(P^{-1})^{-1}$ and $A = (PQ)C(PQ)^{-1}$ confirming that $\sim$ is symmetric and transitive.

Equivalence Classes. When we are given an equivalence relation on a set, we look at the subsets of the set consisting of elements related to each other. These subsets—which we will call equivalence classes—will partition our original set. We shall see that a judicious partition of a set can be of much use in studying the properties or in counting the number of elements of a set.

Definition 4.32 (Equivalence classes). Let X be a set, and let R be a relation on X. Let $a \in X$. We define the *class* of a by $\mathrm{cl}(a) = \{x \in X \mid aRx\}$. If the relation is an equivalence relation, then the class of a is called the *equivalence class* of a.

If R is an equivalence relation on X, then the set of equivalence classes partition X:

Lemma 4.33. *Let X be a set, and let $\sim$ be an equivalence relation on X. Then the equivalence classes partition X.*

Proof. Since the relation is reflexive, every element is in some equivalence class. We need to show that two distinct classes have an empty intersection. Assume that $\mathrm{cl}(x)$ and $\mathrm{cl}(y)$ are two distinct equivalence classes. Thus, without loss of generality, there exists an element $z \in \mathrm{cl}(x)$ such that $z \notin \mathrm{cl}(y)$. Assume that $w \in \mathrm{cl}(x) \cap \mathrm{cl}(y)$. Then $w \in \mathrm{cl}(y)$ implies that $y \sim w$. However, $w \in \mathrm{cl}(x)$ implies that $w \sim x$. So, by transitivity, $y \sim x$. We also know that $x \sim z$. Thus $y \sim z$, which means that $z \in \mathrm{cl}(y)$, and this is a contradiction. $\square$

Orbits. Equivalence relations are ubiquitous in mathematics. However, in group theory, much of what we need to accomplish will be in terms of equivalence classes that arise out of group actions. In fact, every group action gives us a very useful equivalence relation on the underlying set. We will define this relation next.

Lemma 4.34. *Let the group G act on the set Ω. Let $\alpha, \beta \in \Omega$. Define $\alpha \sim \beta$ if there exists $g \in G$ with $g \cdot \alpha = \beta$.*

Then $\sim$ is an equivalence relation on Ω.

Proof. We need to show that the relation is reflexive, symmetric, and transitive. Since, for $\alpha \in \Omega$, we have $e \cdot \alpha = \alpha$, we are assured that $\alpha \sim \alpha$. Hence $\sim$ is reflexive.

To show that $\sim$ is symmetric, assume that, for $\alpha, \beta \in \Omega$, we have $\alpha \sim \beta$. This means that there is a $g \in G$ with $g \cdot \alpha = \beta$. Since G is a group, $g^{-1} \in G$. Let g^{-1} act on $g \cdot \alpha = \beta$, and let $g^{-1} \cdot (g \cdot \alpha) = g^{-1} \cdot \beta$. Hence $g^{-1} \cdot \beta = \alpha$, and so $\beta \sim \alpha$.

Finally assume $\alpha \sim \beta$ and $\beta \sim \gamma$ for $\alpha, \beta, \gamma \in \Omega$. By the definition of $\sim$, this means that there are elements $g, h \in G$ with $g \cdot \alpha = \beta$, and $h \cdot \beta = \gamma$. Now

$$hg \cdot \alpha = h \cdot (g \cdot \alpha) = h \cdot \beta = \gamma,$$

and, hence, $\alpha \sim \gamma$. Being reflexive, symmetric, and transitive, the relation is an equivalence relation. $\square$

As soon as we have a group G act on a set Ω, we have the equivalence relation $\sim$ defined in the previous lemma—basically two elements of the set are related if you can get from one to the other using the group action—and, hence, we can consider the equivalence classes of this relation. These equivalence classes will play an important role in our theory and are called the orbits of the action.

Definition 4.35 (Orbits). Let G be a group, and let Ω be a set. Assume that G acts on Ω, and let $\alpha \in \Omega$. The *orbit* of α in this action is denoted by $\mathcal{O}_\Omega(\alpha)$ or $G\alpha$, and is defined as

$$\mathcal{O}_\Omega(\alpha) = G\alpha = \{\beta \in \Omega \mid \exists g \in G \text{ with } g \cdot \alpha = \beta\}.$$

The set of orbits of the action of G on Ω is denoted by Ω/G.

In other words, the orbit of α is the collection of elements of Ω that α is sent to under the action of G. In addition to $G\alpha$ and $\mathcal{O}_\Omega(\alpha)$, some authors use $\mathcal{O}_\alpha$, $\mathcal{O}(\alpha)$ or $^G\alpha$ to denote the orbit of α.

Corollary 4.36. *Let the group G act on the set Ω. Then the orbits of the action partition Ω.*

Proof. The orbits are the equivalence classes of the equivalence relation defined in Lemma 4.34, and hence they partition the set. $\square$

For some actions there is only one orbit. In other words, starting with any element of the set, you can get to any other element of the set (by the action of a group element). Such an action is called transitive:

Definition 4.37 (Transitive actions). Let G act on Ω. If the action has exactly one orbit, then we say the action is *transitive*.

Example 4.38. The dihedral group of order 8, D_8, is the group of symmetries of a square, and it acts on the set consisting of the corners of a square. Starting with any corner, you can get to any other corner by the action of the group. Hence, there is exactly one orbit consisting of all four corners of the square, and the action is transitive.

Example 4.39. Let G be the subgroup of S_8 generated by $(1\ 2\ 3)$ and $(5\ 6)$. A straightforward calculation shows

$$G = \{e, (1\ 2\ 3), (5\ 6), (1\ 2\ 3)(5\ 6), (1\ 3\ 2), (1\ 3\ 2)(5\ 6)\} = \langle (1\ 2\ 3)(5\ 6) \rangle \simeq \mathbb{Z}/6\mathbb{Z}.$$

Now S_8 acts on $[8] = \{1, 2, 3, 4, 5, 6, 7, 8\}$ in the natural way, and G being a subgroup of S_8 consists of permutations of $[8]$ and, hence, also acts on $[8]$. What are the stabilizers and the orbits? Recall that the stabilizer of a set element is the collection of all the group elements that do not move it. For example, which group elements do not move and fix 1? The identity and $(5\ 6)$. Hence, $\mathrm{Stab}_G(1) = \{e, (5\ 6)\}$. All the stabilizers are found as easily, and we have

$$\mathrm{Stab}_G(1) = \mathrm{Stab}_G(2) = \mathrm{Stab}_G(3) = \{e, (5\ 6)\},$$
$$\mathrm{Stab}_G(5) = \mathrm{Stab}_G(6) = \{e, (1\ 2\ 3), (1\ 3\ 2)\},$$
$$\mathrm{Stab}_G(4) = \mathrm{Stab}_G(7) = \mathrm{Stab}_G(8) = G.$$

Now, the orbit of a set element x is the collection of all *set* elements that you can get by the action of the elements of the group on x. For example, what is the orbit of 1? We have $e \cdot 1 = 1$, and $(1\ 2\ 3) \cdot 1 = 2$, and $(1\ 3\ 2) \cdot 1 = 3$. Hence 1, 2, and 3 are in the orbit of 1. The other group elements do not send 1 anywhere new, and so the orbit of 1 under this action, $\mathcal{O}_{[8]}(1)$, is the set $\{1, 2, 3\}$. The other orbits are found similarly:

$$\mathcal{O}_{[8]}(1) = \mathcal{O}_{[8]}(2) = \mathcal{O}_{[8]}(3) = \{1, 2, 3\}$$
$$\mathcal{O}_{[8]}(4) = \{4\}$$
$$\mathcal{O}_{[8]}(5) = \mathcal{O}_{[8]}(6) = \{5, 6\}$$
$$\mathcal{O}_{[8]}(7) = \{7\}$$
$$\mathcal{O}_{[8]}(8) = \{8\}.$$

Example 4.40. Let $G = \mathrm{GL}(n, \mathbb{R})$, and let $\Omega = \mathbf{M}_{n \times n}(\mathbb{R})$ be the set of all $n \times n$ matrices with real entries. In Problem 4.1.1 (see also Problem 4.3.2), we defined an action of G on Ω by

$$P \cdot A = PAP^{-1} \text{ for } P \in G, A \in \Omega.$$

Now fix $A \in \Omega$. What is the orbit of A? The orbit of A consists of those $n \times n$ matrices that we can get by having an element of G act on A. In other words, an $n \times n$ matrix B is in the orbit of A if $B = PAP^{-1}$, for some invertible $n \times n$ matrix P. In linear algebra, such matrices A and B are called *similar*. Thus the orbit of A, in this action, consists of all matrices that are similar to A.

This action of G on Ω was a conjugation action. We can also define an action of $G = \mathrm{GL}(n, \mathbb{R})$ on $\Omega = \mathbf{M}_{n \times n}(\mathbb{R})$ by left multiplication:

$$P \cdot A = PA \text{ for } P \in G, A \in \Omega.$$

Again fix $A \in \Omega$. What is the orbit of A in this action? By definition, the orbit of A consists of all matrices in Ω of the form PA, where P is an invertible matrix. A matrix—such as P—is invertible if and only if it is a product of elementary matrices (see Definition 1.68 and Theorem 1.70), and multiplying by an elementary matrix on the left is the same as performing an elementary row operation (Proposition 1.69). Hence, the matrix PA is what we get from A by a sequence of elementary row operations. We conclude that the orbit of A is the collection of all matrices that are *row equivalent* to A (see Definition 1.67).

Generators, Cayley Digraphs, and Orbits. Assume that you have a set S of generators for a group G. This means that every element of G is a (possibly repeated) product of elements of S and their inverses. Now, if G acts on a set Ω, you know the whole action if you know the effect of the elements of S. This is because $gh \cdot x = g \cdot (h \cdot x)$, and hence you know how every element acts if you know how the generators act.

The Cayley digraph of an action with respect to a set S of generators, conveniently displays the information about how the generators act. Hence, it is easy to read off orbits from a Cayley digraph of an action. Starting with any element of the set, you just see what other elements are connected to it (via forward or backward arcs).

Example 4.41. Consider the action in Example 4.39 again. The Cayley digraph of this action with respect to $S = \{(1\ 2\ 3), (5\ 6)\}$ is given in Figure 4.3. We can directly see the five orbits. The set $\{1, 2, 3\}$ is one orbit, as is $\{5, 6\}$. On the other hand, the singletons $\{4\}$, $\{7\}$, and $\{8\}$ are each an orbit themselves.

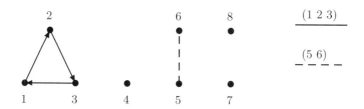

Figure 4.3. The Cayley digraph of the action of $G = \langle (1\ 2\ 3), (5\ 6) \rangle$ on $[8]$. The orbits are the connected components.

Orbits for the Regular and the Conjugation Actions

The Regular Action. Let G act on G by the regular action (i.e., $g \cdot x = gx$). Now fix $x \in G$, and let $g \in G$ be arbitrary. Regardless of your choices for x and g, you can act on x and get to g. This is because if you let $h = gx^{-1}$, then $h \cdot x = g$. Hence $\mathcal{O}_G(x) = G$. This means that, for this action, there is only one orbit. Hence the action is transitive. We recall that, for this action, $\mathrm{Stab}_G(x)$ is the set $\{g \in G \mid g \cdot x = x\} = \{e\}$.

The Conjugation Action. Recall—see Definition 4.13—that for any group G we defined an action on $\Omega = G$ called the conjugation action. For $g \in G$ and $x \in \Omega$, we defined

$$g \cdot x = gxg^{-1}.$$

Recall that the element gxg^{-1} is called a conjugate of x.

In this action the orbits are called the *conjugacy classes* of G. Thus each conjugacy class of G consists of elements of G that are conjugate to each other. As we shall see, conjugate elements share many properties. For example, in Problem

2.3.13, you were asked to show that two conjugate elements have the same order. Given the importance of conjugacy classes, we record their definition without mentioning actions:

Definition 4.42 (Conjugacy classes). Let G be a group, and let $x \in G$. Then the *conjugacy class* of x in G is denoted by $\mathrm{cl}_G(x)$, is a subset of G, and is defined by

$$\mathrm{cl}_G(x) = \{gxg^{-1} \mid g \in G\}.$$

Example 4.43. What are the conjugacy classes of S_3? The question can be rephrased as "What are the orbits of the conjugation action of S_3 on S_3?" Alternately, we can refer to Definition 4.42 directly without invoking actions. For each element of $x \in S_3$, we have to find the collection of all the elements of the group of the form gxg^{-1}.

Now if g and x commute, then $gxg^{-1} = x$ and there is nothing to compute. In other words, in the conjugation action, the elements of $\mathbf{C}_G(x)$ do not move x. In particular, the identity element does not get moved by the action of any of the elements. Hence, $\{e\}$ is one conjugacy class (that is, an orbit of the conjugation action).

To find the orbit of $x = (1\ 2\ 3)$, we calculate gxg^{-1} for all $g \in G$. However, we know that e, x, and x^2 commute with x, and hence, in this action, they do not move x. We have

$$(1\ 2)x(1\ 2)^{-1} = (1\ 3\ 2), \ (1\ 3)x(1\ 3)^{-1} = (1\ 3\ 2), \ (2\ 3)x(2\ 3)^{-1} = (1\ 3\ 2).$$

Hence, the conjugacy class of $(1\ 2\ 3)$ in S_3, $\mathrm{cl}_{S_3}(1\ 2\ 3)$, is the set $\{(1\ 2\ 3),(1\ 3\ 2)\}$.

We can continue in this manner or, to simplify matters, we can draw a Cayley digraph of the action with respect to a set of generators. Let $S = \{(1\ 2\ 3),(1\ 2)\}$ be the set of generators for S_3. The Cayley digraph of the conjugation action of S_3 with respect to S_3 is given in Figure 4.4. We see that S_3 has three conjugacy classes and they are

$$\{e\}, \ \{(1\ 2\ 3),(1\ 3\ 2)\}, \ \{(1\ 2),(2\ 3),(1\ 3)\}.$$

We should note that in the particular case of the group S_n, conjugating elements is easy (see Problem *3.1.9*). This results in a simple description of conjugacy classes of the symmetric group. (See Proposition 4.45.)

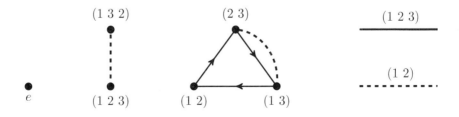

Figure 4.4. The Cayley digraph of the action of S_3 on S_3 by conjugation

Question 4.44. Can you make a conjecture about the relationship between $|\mathcal{O}_\Omega(\alpha)|$ and $|\mathrm{Stab}_G(\alpha)|$?

To make a conjecture, look at Examples 4.39 and 4.43. Whatever your conjecture is, you may want to see what it says in the extreme case when the action is transitive.

Problems

4.4.1. For each of the three properties of an equivalence relation, construct an example of a relation R that does not satisfy the given property, but does satisfy the other two. For each of your examples find the "equivalence" classes. Do they partition the original set?

4.4.2. Let X be a set. Assume that P_1, ..., P_n are subsets of X that partition X. Show that there is an equivalence relation $\sim$ on X such that P_1, ..., P_n are the equivalence classes of $\sim$. Does it matter that the number of subsets partitioning X is finite?

4.4.3. A relation R on the set P is called *circular* if aRb and bRc always imply cRa. Can you find a relation that is reflexive and circular but is not an equivalence relation? What about a relation that is an equivalence relation but not circular? Either find examples or prove that you cannot.

4.4.4. If R is a symmetric and transitive relation on a set P and aRb, then bRa by symmetry. Then transitivity forces aRa. Does this argument show that every symmetric and transitive relation is automatically reflexive?

4.4.5. As usual, let D_{10} denote the group of symmetries of a regular pentagon. Let g be one of the reflections, and let $H = \{e, g\}$ be a subgroup of D_{10}. Both D_{10} and H act on the corners of a regular pentagon. For each action, determine the number and size of the orbits.

4.4.6. Give a one sentence proof that conjugacy classes of a group partition the group. Then find all the conjugacy classes of D_8.

4.4.7. Let a group G act on itself by the conjugation action. What is the significance of conjugacy classes consisting of one element?

4.4.8. Let $G = \mathrm{SL}(2, 3)$ be the group of 2×2 matrices with determinant 1 over $\mathbb{Z}/3\mathbb{Z}$. Let $H = \{\begin{bmatrix} \lambda & \mu \\ 0 & \lambda^{-1} \end{bmatrix} \mid \lambda, \mu \in \mathbb{Z}/3\mathbb{Z}, \lambda \neq 0\}$. In Problem 2.6.17, you showed that H is a subgroup of G. Let

$$\Omega = (\mathbb{Z}/3\mathbb{Z})^2 = \{\begin{bmatrix} \alpha \\ \beta \end{bmatrix} \mid \alpha, \beta \in \mathbb{Z}/3\mathbb{Z}\},$$

be the set of 2×1 column vectors over $\mathbb{Z}/3\mathbb{Z}$. Both G and H act on Ω by matrix multiplication. Find the orbits of the action of H on Ω.

4.4.9. We have a group $G = \{x_1, x_2, \ldots, x_8\}$ with eight elements and a set $\Omega = \{1, 2, 3, 4, 5, 6, 7, 8\}$ also with eight elements. We know that G acts on Ω. As usual, the action of every group element can be described by a permutation of elements of Ω. We are told that these permutations are as follows:

$$
\begin{aligned}
x_1 &\longmapsto 1, \\
x_2 &\longmapsto 1, \\
x_3 &\longmapsto (2\ 4)(6\ 8), \\
x_4 &\longmapsto (2\ 4)(6\ 8), \\
x_5 &\longmapsto (5\ 7)(6\ 8), \\
x_6 &\longmapsto (5\ 7)(6\ 8), \\
x_7 &\longmapsto (2\ 4)(5\ 7), \\
x_8 &\longmapsto (2\ 4)(5\ 7).
\end{aligned}
$$

(a) What is $\mathrm{Stab}_G(2)$?

(b) What is $\mathcal{O}_\Omega(2)$?

(c) Find two subgroups of order 4 in G. Explain how you got them.

(d) Find a subgroup of order 2 in G, and explain how you found it.

(e) Assume that $S = \{x_3, x_5\}$ is a set of generators for G. Draw the Cayley digraph of the action of G on Ω with respect to S.

4.4.10. Let $G = \{a, b, c, d, e, f\}$, and let $\Omega = \{x, y, z, u, v, w\}$. We know that G is a group and Ω is a set. We also know that G acts on Ω. The following table tells us how every element of G acts on elements of Ω.

	x	y	z	u	v	w
a	w	y	z	v	u	x
b	x	u	z	v	y	w
c	w	v	z	u	y	x
d	x	y	z	u	v	w
e	x	v	z	y	u	w
f	w	u	z	y	v	x

So for example, $c \cdot x = w$ and $c \cdot u = u$.

(a) What is $(bc) \cdot y$?

(b) Can you find a subgroup of G with one element? What about a subgroup of G with two elements? What about a subgroup of G with three elements? If the answer is yes, give the elements of the subgroup, and in any case give adequate explanation for your answers.

(c) Can you find an orbit with three elements?

(d) Let H be the stabilizer of w in G. If we multiply $c \in G$ by every element of H—that is, find ch for all $h \in H$—we get what is called a *left coset* of H, and it is denoted by cH. What are the elements of the left coset cH?

4.4.11. Find two non-isomorphic groups with an action for each and a set of generators for each such that the two Cayley digraphs for these actions are the same (except for the labels).

4.4.12. Assume that the group G acts on the set Ω and that $\mathcal{O}$ is one of the orbits. Let $H \leq G$. Show that the same action can be restricted to give an action of the group H on the set $\mathcal{O}$. How many orbits does the action of G on $\mathcal{O}$ have?

4.4.13. Let $G = S_3$ and $V = \mathbb{R}^3$. In Problem 4.1.4 we defined an action of G on V. (Also see Problem 4.3.8.) What is the orbit of $(3, \sqrt{2}, 3)$ in this action? What about $(4, 4, 4)$? What are the possible orbit sizes for this action?

4.4.14. Let V be a vector space over $\mathbb{R}$. Let G be the group $(\mathbb{R} - \{0\}, \cdot)$. For $r \in G$ and $v \in V$, define $r \cdot v = rv$ where rv is the scalar product in the vector space. Is this an action of G on V? If so, what are the orbits and the stabilizers?

Conjugacy Classes in S_n. For the particular case of the symmetric group, S_n, conjugating elements, finding which elements are conjugate, and describing conjugacy classes are all relatively easy, thanks to the results in Problems **3.1.9** and **4.4.15**. For the record, we recap these results here:

Proposition 4.45. (Conjugacy in S_n.)

(a) *Let $\sigma, \tau \in S_n$, and let $\delta = \tau \sigma \tau^{-1}$. Then to find the cycle decomposition of δ, you need to apply τ to each of the entries in the cycle decomposition of σ. In other words, if σ has cycle decomposition*

$$(a_1\ a_2\ \cdots\ a_{k_1})(b_1\ b_2\ \cdots\ b_{k_2}) \cdots,$$

then δ has cycle decomposition

$$(\tau(a_1)\ \tau(a_2)\ \cdots\ \tau(a_{k_1}))(\tau(b_1)\ \tau(b_2)\ \cdots\ \tau(b_{k_2})) \cdots.$$

(b) *Two elements of S_n are conjugate in S_n if and only if they have the same cycle type.*

4.4.15. **Proof of Proposition 4.45b.** Using Problem **3.1.9**, prove that two elements of S_n are conjugate in S_n if and only if they have the same cycle type.

4.4.16. Find the number of conjugacy classes of S_4 and the number of elements in each of these classes.

4.4.17. Let G be a group, and assume that G acts on a set Ω. Let i be a positive integer, and let Λ_i be the set of i-tuples of *distinct* elements of Ω. In other words,

$$\Lambda_i = \{(a_1, \ldots, a_i) \mid a_1, \ldots, a_i \in \Omega,\ \text{and}\ a_k \neq a_j\ \text{for}\ 1 \leq k < j \leq i\}.$$

(a) Show that we can extend the action of G on Ω to an action of G on Λ_i by

$$g \cdot (a_1, a_2, \ldots, a_i) = (g \cdot a_1, g \cdot a_2, \ldots, g \cdot a_i), \quad \text{for}\ g \in G,\ (a_1, \ldots, a_i) \in \Lambda_i.$$

(b) Let $G = D_8$, and let Ω be the corners of a square. As usual, G acts on Ω. Let i be a positive integer, and, as in the previous part, extend this action to an action of G on Λ_i. Let F_i denote the number of orbits of the action of G on Λ_i. Find F_1, F_2, F_3, and F_4.

4.5. More Problems and Projects

Problems

4.5.1. A simple card trick[2]

The magician has a spectator choose a card, memorize it, and return it to the top of the deck. She then allows the spectator to cut the cards—split the deck into two by taking a set of cards from the top, and then switch the two parts—as many times as he would like. The magician spreads the cards face up and announces the chosen card.

Analyze and explain the above trick using the following steps:

Step 1: Consider a deck of 52 cards $C_1, C_2, \ldots, C_{52}$, and let H be the subgroup of S_{52} generated by $(1\ 2\ \cdots\ 51\ 52)$. If $\sigma \in H$, then define $\sigma \cdot C_i = C_{\sigma(i)}$. Show that this gives an action of H on the deck of cards.

Step 2: Let $\tau = (1\ 2\ \cdots\ 51\ 52)$, and put the deck in the order $C_1, C_2, \ldots, C_{52}$. Now, if we apply τ to the deck of cards, then what happens to the order of the cards? What if we apply τ^2? Show that any "cutting of the cards" can be achieved by the action of an element of H.

Step 3: Let $\Omega = \{\{C_1, C_2\}, \{C_2, C_3\}, \ldots, \{C_{52}, C_1\}\}$ be the set of consecutive pairs of cards in the original deck. Show that the action of H on the deck of cards results in an action of H on Ω. Conclude that cutting the deck does not change the set of consecutive pairs of cards.

Step 4: Explain the card trick.

4.5.2. Group of matrices with column sums 1. Let F denote one of $\mathbb{Q}$, $\mathbb{R}$, $\mathbb{C}$, or $\mathbb{Z}/p\mathbb{Z}$ for a prime p. (As usual, in fact, F can be any field.) We denote the set of n-dimensional column vectors with entries from F by F^n. The general linear group $G = \mathrm{GL}(n, F)$ acts on $\Omega = F^n$ by $M \cdot x = Mx$.

(a) What is the $\mathrm{Stab}_G(\begin{bmatrix} 1 \\ 1 \\ \vdots \\ 1 \end{bmatrix})$? What is the $\mathrm{Stab}_G(\begin{bmatrix} 1 \\ 0 \\ \vdots \\ 0 \end{bmatrix})$?

(b) We call a matrix whose column sums is 1, a *generalized stochastic matrix*. (A *stochastic matrix* is a generalized stochastic matrix

---

[2]Adapted from Ensley [**Ens99**], where this as well as another more intricate card trick is analyzed using group actions.

whose entries are non-negative.) Using the previous part and Problem 2.6.16, prove that the set of invertible generalized stochastic matrices forms a group under matrix multiplication.

(c) We denote the group of $n \times n$ invertible generalized stochastic matrices with entries in F by $S(n, F)$. Find a familiar group that is isomorphic to $S(2, \mathbb{Z}/2\mathbb{Z})$. Do the same for $S(2, \mathbb{Z}/3\mathbb{Z})$.[3]

4.5.3. **Action of $\mathrm{SL}(2, \mathbb{Z})$ on the complex upper half-plane.** Let $\mathrm{SL}(2, \mathbb{Z})$ be the group of 2×2 invertible matrices with determinant 1 over the integers $\mathbb{Z}$. (See Problems 2.1.4 and 2.6.15.) Let $\mathcal{H} = \{x + yi \mid y > 0\}$ denote the complex upper half-plane. For $A = \begin{bmatrix} a & b \\ c & d \end{bmatrix} \in \mathrm{SL}(2, \mathbb{Z})$ and $z \in \mathcal{H}$, define

$$A \cdot z = \frac{az + b}{cz + d}.$$

(a) For $A \in \mathrm{SL}(2, \mathbb{Z})$ and $z \in \mathcal{H}$, show that $A \cdot z \in \mathcal{H}$.
(b) Show that the above defines an action of $\mathrm{SL}(2, \mathbb{Z})$ on $\mathcal{H}$.
(c) The point $i = 0 + 1i$ is in $\mathcal{H}$. Find the stabilizer, in $\mathrm{SL}(2, \mathbb{Z})$, of i, and find a familiar group that is isomorphic to this stabilizer.
(d) Find at least four points in the orbit of i.
(e) What is the stabilizer of $1 + i$ under this action?

4.5.4. **Action of $\mathrm{SL}(2, \mathbb{Z})$ on quadratic forms.** Let $\mathrm{SL}(2, \mathbb{Z})$ be the group of 2×2 invertible matrices with determinant 1 over the integers $\mathbb{Z}$. An *integral binary quadratic form* is a homogeneous (no linear or constant terms) quadratic polynomial, $f(x, y) = ax^2 + bxy + cx^2$, in two variables and with integer coefficients.

(a) Let $f(x, y) = ax^2 + bxy + cx^2$ be an integral binary quadratic form, and let $M_f = \begin{bmatrix} a & b/2 \\ b/2 & c \end{bmatrix}$. Show that

$$f(x, y) = \begin{bmatrix} x & y \end{bmatrix} M_f \begin{bmatrix} x \\ y \end{bmatrix}.$$

(b) Let $A = \begin{bmatrix} p & q \\ r & s \end{bmatrix}$ be a 2×2 matrix with integer entries. Show that

$$f(px + ry, qx + sy) = \begin{bmatrix} x & y \end{bmatrix} A M_f A^t \begin{bmatrix} x \\ y \end{bmatrix}.$$

(c) Let $A = \begin{bmatrix} p & q \\ r & s \end{bmatrix} \in \mathrm{SL}(2, \mathbb{Z})$. For f, an integral binary quadratic form, define

$$A \cdot f(x, y) = f(px + ry, qx + sy).$$

Show that this defines an action of $\mathrm{SL}(2, \mathbb{Z})$ on the set of integral binary quadratic forms.

---

[3] Adapted from Poole [**Poo95**]. Also see Theorem 11.50 and the ensuing problems.

(d) The *discriminant* of the quadratic form $ax^2 + bxy + cy^2$ is defined to be $b^2 - 4ac$. Show that two quadratic forms in the same orbit of the above action have the same discriminant.

Remark: It can be shown that two quadratic forms in the same orbit assume the same values on $\mathbb{Z}^2$. In other words, if we plug in various integers for x and y, then the two quadratic forms give the same set of integers as values.[4]

---

[4]For a unique perspective on quadratic forms see Conway [**Con97**].

Chapter 5

A Subgroup Acts on the Group: Cosets and Lagrange's Theorem

...where we investigate one specific action—the action of a subgroup on the group by left multiplication—define cosets, prove Lagrange's theorem which restricts the sizes of a subgroup of a group, and see applications, including Euler's theorem, on congruences and the classification of groups of order p and $2p$.

5.1. Translation Action and Cosets

In the last chapter, we saw that, given an action of a group on a set, we have stabilizers and orbits. Stabilizers are automatically subgroups of G, and the set of orbits partition the set. These two little facts will be quite useful. In this chapter, we will investigate the consequences for a specific action, that is the translation action (one of the variants of the regular action).

Let $H \leq G$, for some group G. Recall that H acts on G by left multiplication, i.e., $h \cdot g = hg$, for $h \in H$ and $g \in G$. What are the stabilizers and the orbits?

Let $x \in G$, then the stabilizer in H of $x = \mathrm{Stab}_H(x) = \{h \in H \mid h \cdot x = x\} = \{e\}$. Hence the stabilizers for this action are not interesting.

Again, let $x \in G$. The orbit of x under this action is $\mathcal{O}_G(x) = \{y \in G \mid y = h \cdot x \text{ for some } h \in H\}$. This is exactly the set of elements that you get by multiplying the various elements of H by x. In other words, $\mathcal{O}_G(x) = \{hx \mid h \in H\}$. We denote this set by Hx.

We know that the orbits partition the group G, and the orbits of this specific action are so important that we give them a special name.

Definition 5.1 (Cosets). Let $H \leq G$, and let $x \in G$. Then $Hx = \{hx \mid h \in H\}$ is called a *right coset* of H in G.

By analogy, the set $xH = \{xh \mid h \in H\}$ is called a *left coset* of H in G.

Corollary 5.2. *Let G be any group, and let $H \leq G$. Then the right cosets of H in G partition G.*

Proof. Right cosets are orbits of the translation action, and orbits always partition the set.   □

Remark 5.3. Left cosets of a subgroup also partition G since they are also orbits of an action of H on G, this time by $h \cdot x = xh^{-1}$. (The reader should check this.)

Example 5.4. Let $G = S_3 = \{1, (1\ 2\ 3), (1\ 3\ 2), (1\ 2), (1\ 3), (2\ 3)\}$, and let $H = \{1, (1\ 2)\}$. The *right cosets* of H in G are

$$He = H = \{1, (1\ 2)\} = H(1\ 2),$$
$$H(1\ 3) = \{(1\ 3), (1\ 3\ 2)\} = H(1\ 3\ 2),$$
$$H(2\ 3) = \{(2\ 3), (1\ 2\ 3)\} = H(1\ 2\ 3).$$

Note that, for example, $H(1\ 3)$ and $H(1\ 3\ 2)$ are the *same* coset. Cosets are sets, and when two sets have the same elements, they are the same set.

We reiterate that the right cosets are the orbits of a specific action—that is, the action of H on G by left multiplication—and, hence, we can visualize them by drawing a Cayley digraph with respect to a set of generators. In this example, the group that acts is H and $S = \{(1\ 2)\}$ is a generating set for it. The Cayley digraph of this action with respect to S is given in Figure 5.1. As expected, there are three orbits each of size 2.

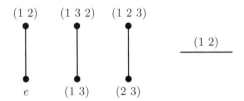

Figure 5.1. The Cayley digraph of the action of $H = \langle (1\ 2) \rangle$ on S_3 by left multiplication. The orbits are the right cosets of $H = \langle (1\ 2) \rangle$ in S_3.

Example 5.5. Let $G = D_8 = \langle a, b \mid a^4 = b^2 = e, \ ba = a^3 b \rangle$, and let $H = \langle a \rangle$. What are the right cosets of H in G?

We can answer this question directly and without going back to the action of H on G by left multiplication. The right cosets of H in G are the sets of the form Hx where $x \in G$. With a little calculation, we have

$$He = H = \{e, a, a^2, a^3\} = Ha = Ha^2 = Ha^3,$$
$$Hb = \{b, ab, a^2 b, a^3 b\} = Hab = Ha^2 b = Ha^3 b.$$

Hence, H has two right cosets in G, and each have four elements.

We could, of course, visualize these orbits by drawing the Cayley digraph of the action with respect to the generating set $S = \{a\}$. This Cayley digraph is given in Figure 5.2. (Compare this with Figure 4.2. Does the similarity makes sense to you?)

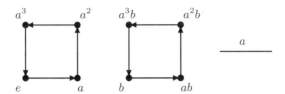

Figure 5.2. The Cayley digraph of the action of $H = \langle a \rangle$ on $G = D_8$ by left multiplication. The orbits are the right cosets of H in G.

It is important to note that $Hx = Hy$ does *not* mean that $x = y$. All it means is that x and y are in the same orbit in the action of H on G by left multiplication. The following lemma will be used often.

Lemma 5.6. *Let G be a group, let $H \leq G$, and let $x, y \in G$. Then the following are equivalent:*

(a) $Hx = Hy$,

(b) $y \in Hx$,

(c) $y = hx$ *for some* $h \in H$,

(d) $yx^{-1} \in H$.

In particular, $Hg = H$ if and only if $g \in H$.

Proof. Recall that Hx and Hy are orbits of the translation action. Moreover, Hx is the orbit of x, while Hy is the orbit of y. As such $x \in Hx$ and $y \in Hy$. The orbits partition G, and so to say $Hx = Hy$ is the same as saying that x and y are in the same orbit. Thus $Hx = Hy$ if and only if $y \in Hx$. Now note that $y \in Hx$ if and only if $y = hx$ for some $h \in H$. Since the expression $y = hx$ is equivalent to $yx^{-1} = h$, all the equivalences are now established.

In particular $Hg = H = He$ if and only if $g \in He = H$.   □

Definition 5.7 (Index). Let G be a group, and let $H \leq G$. The set of right cosets of H in G is denoted by G/H. We denote $|G/H|$ by $|G : H|$ and call it the *index* of H in G. In other words, if G/H is a finite set, then $|G : H|$ is the number of right cosets of H in G, and if the number of right cosets is infinite, then we write $|G : H| = \infty$.

Remark 5.8. The notation G/H for the set of cosets of H in G is standard and is a special case of the notation Ω/G for the set of orbits of the action of G on Ω (see Definition 4.35). The cosets *are* the orbits of the action of H on G (by left multiplication).

If $H \leq G$, then one of the right cosets of H in G is the subgroup H itself. To begin our study of right cosets, we show that every right coset has the same cardinality as the subgroup H.

Lemma 5.9. *Let G be a group, and let $H \leq G$. For an arbitrary element $g \in G$, we have*

(a) $|Hg| = |H|$,

(b) $|gH| = |H|$.

Remark 5.10. Note that if H is a finite set, then $|H|$ is just the number of elements of H, and $|H| = |Hg|$ just means that H and Hg have the same number of elements. On the other hand, if H is infinite, then $|H| = |Hg|$ means that these two sets have the same cardinality, which is another way of saying that there is a bijection from one to the other. (See Definition 1.23.)

Proof. To show that two sets have the same cardinality, we establish a bijection between them. Let $g \in G$, and define $\phi \colon H \to Hg$ by $\phi(h) = hg$.

CLAIM: ϕ is 1-1 and onto.

PROOF OF CLAIM: To show that the map is 1-1, assume $\phi(h_1) = \phi(h_2)$. This means $h_1 g = h_2 g$, and by canceling the g's, we get $h_1 = h_2$.

To show that the map is onto, let $x \in Hg$. This means that $x = hg$ for some $h \in H$. But this means that $\phi(h) = x$, and the map is onto.

The proof for left cosets is similar.   □

Remark 5.11. It may occur to you that we could have made the above proof a bit more transparent if we let $H = \{h_1, \ldots, h_n\}$. Then, by definition, $Hg = \{h_1 g, \ldots, h_n g\}$. We can then argue that $|Hg|$ is n since there are no repeats in the list—this follows just like the proof above from cancellation. Thus $|H| = n = |Hg|$.

This new presentation of the proof is certainly correct for finite subgroups H. However, our original version does not depend on H being finite, and proves that H and Hg have the same cardinality even if H is infinite.

We close the section by recording a number of straightforward observations about the index of a subgroup.

Lemma 5.12. *Let H be a subgroup of a group G. Then*

(a) *If $H = \{e\}$, then $|G : H| = |G|$.*

(b) *If $H = G$, then $|G : H| = 1$.*

(c) *"Left" index is the same as index. In other words, the set of left cosets of H in G has the same cardinality as the set of right cosets of H in G. In particular, if the index is finite, the number of left cosets of H in G is the same as the number of right cosets of H in G.*

Proof. The first two claims follow directly from the definition. For the last claim, we find a bijection between the set of left and right cosets of G.

Define

$$\phi : \{\text{the set of left cosets of } H \text{ in } G\} \to \{\text{the set of right cosets of } H \text{ in } G\}$$

by $\phi(xH) = Hx^{-1}$.

We first have to make sure that this map is well defined. This is because left (and right) cosets have aliases. Each left coset can have more than one name, and we have to make sure that our map sends a coset to the same place regardless of which name we use for the coset. Until we have done this, we cannot legitimately use the function ϕ as a function.

Thus assume $xH = yH$. We need to show that $Hx^{-1} = Hy^{-1}$. Now $xH = yH$ implies that $y = xh$ for some $h \in H$. We have

$$Hy^{-1} = H(xh)^{-1} = H(h^{-1}x^{-1}) = (Hh^{-1})x^{-1} = Hx^{-1}.$$

Now we show that the map is 1-1. Assume $\phi(xH) = \phi(yH)$. This implies that

$$Hx^{-1} = Hy^{-1} \Rightarrow y^{-1} = hx^{-1} \Rightarrow y = xh^{-1} \Rightarrow yH = xh^{-1}H = xH.$$

To show that ϕ is onto, let Hw be an arbitrary right coset of H in G. Then $\phi(w^{-1}H) = Hw$.

The proof is complete since we have shown ϕ to be a bijection. $\qquad\square$

Problems

5.1.1. Let $G = \mathbb{Z}/4\mathbb{Z} \times \mathbb{Z}/3\mathbb{Z}$ be the direct product of $(\mathbb{Z}/4\mathbb{Z}, +)$ and $(\mathbb{Z}/3\mathbb{Z}, +)$, and let $H = \langle (2, 0) \rangle$ be a subgroup of G. Find the right cosets of H in G.

5.1.2. Let $G = (\mathbb{Z}, +)$ be the group of integers, and let $H = (5\mathbb{Z}, +)$ be the subgroup of G consisting of all multiples of 5. Describe the right cosets of H in G.

5.1.3. Let A_4 be the alternating group of degree 4, and let $H = \langle (1\ 2\ 3) \rangle$. Find the right cosets of H in A_4.

5.1.4. As usual, $S_3 = \{e, (1\ 2\ 3), (1\ 3\ 2), (1\ 2), (1\ 3), (2\ 3)\}$ is the symmetric group on three letters, and $D_8 = \{e, a, a^2, a^3, b, ab, a^2b, a^3b\}$ is the dihedral group of order 8. Let $G = S_3 \times D_8$. Also let $x = ((1\ 2), a)$, $y = ((1\ 2\ 3), ab)$, and $z = (e, a)$ be elements of G.
   (a) What is $o(x)$?
   (b) What is $\langle y \rangle$?
   (c) What is $|G : \langle z \rangle|$?
   (d) Give three distinct right cosets of $\langle z \rangle$ in G.
   (e) What is the $\mathbf{Z}(G)$?

5.1.5. Let G be a group, and let $H \leq G$. Assume that $a \in G - H$. Can we have $aH = Ha^2$?

5.1.6. Let G be a group, and let $H \leq G$ with $|G : H| = 2$.
   (a) If K is a subgroup of G with at least one element not in H, show that $G = HK$.
   (b) Is it possible to find $y \in G$ such that $yH \neq Hy$?

5.1.7. Let $\mathrm{GL}(2, \mathbb{Z})$ be the group of 2×2 invertible integer matrices with determinant ± 1, and let $\mathrm{SL}(2, \mathbb{Z})$ be the subgroup of $\mathrm{GL}(2, \mathbb{Z})$ consisting of integer matrices with determinant 1. (See Problems 2.1.4 and 4.5.3.) Find $|\mathrm{GL}(2, \mathbb{Z}) : \mathrm{SL}(2, \mathbb{Z})|$, and describe the cosets of $\mathrm{SL}(2, \mathbb{Z})$ in $\mathrm{GL}(2, \mathbb{Z})$.

5.1.8. Suppose H is a subgroup of G. Show that $f(Ha) = a^{-1}H$ defines a bijection from the set of right cosets of H in G to the set of left cosets of H in G. (This shows—see the map in the proof of Lemma 5.12—that the set of left cosets has the same cardinality as the set of right cosets.) Instead of f, could we have used g, which is defined by $g(Ha) = aH$?

5.1.9. Let G be a group (not necessarily finite), let H be a subgroup of G, and let $x \in G$. Prove that H and xHx^{-1} have the same cardinality.

5.1.10. Let G be a group, and let $H \leq G$. Recall Definition 4.24 of $\mathbf{N}_G(H)$, the normalizer of H in G. Show that

$$\mathbf{N}_G(H) = \{x \in G \mid xHx^{-1} = H\} = \{x \in G \mid xH = Hx\}.$$

5.1.11. Let G be a group, and let H and K be subgroups of G. Assume that $\mathbf{N}_G(K) = G$. Use Problem **2.6.33** to show that HK is a subgroup of G.

5.1.12. Strengthen the result in Problem 5.1.11, by showing that you can replace the condition $\mathbf{N}_G(K) = G$ with the condition $H \leq \mathbf{N}_G(K)$.

5.1.13. Give an example of an infinite group G with a proper subgroup H such that $|G : H| < \infty$.

5.1.14. Let H and K be subgroups of a (not necessarily finite) group G. Assume that $|G : H|$ and $|G : K|$ are finite. Can $|G : H \cap K|$ be infinite? Prove your assertion.

5.1.15. Let G be a (possibly infinite) cyclic group, and let $\{e\} \neq H \leq G$. Show that $|G : H| < \infty$. In fact, if $G = \langle g \rangle$ and $H = \langle g^m \rangle$, then show that $H, Hg, \ldots, Hg^{m-1}$ are all of the right cosets of H in G. (See also Problem 11.7.19.)

5.1.16. Use Problem 5.1.15 to show that neither $\mathbb{Z} \times \mathbb{Z}$ nor $(\mathbb{Q}, +)$ are cyclic groups.

Does A_4 Have a Subgroup of Order 6? The alternating group of degree 4, A_4, is a group of order 12. Does it have a subgroup of order 6? You could try to answer this question by brute force and by trying all the possibilities. This would be very tedious and not that enlightening. Following Gallian [**Gal93**], in Problem **5.1.19** which depends on Problems 3.2.6 and **5.1.17**, you are asked to use cosets to answer the question. (Also see Brennan and MacHale [**BM00**] and Hogan [**Hog96**].)

In Problem 5.4.1, following Mackiw [**Mac96**], and again using Problem **5.1.17**, you are asked to show that $\mathrm{SL}(2,3)$—a group of order 24—does not have a subgroup of order 12.

5.1.17. Let G be a group, and let $H \leq G$. Assume that the number of elements in H is half of the number of elements in G.[1]
   (a) How many right cosets does H have in G?
   (b) If $x \in G$, show that $x^2 \in H$.
   (c) Let $g \in G$ with $o(g) = 3$. Show that $g \in H$.

---

[1] Adapted from Gallian [**Gal93**].

5.1.18. (a) Let G be a group, and let $H \leq G$ with $|G : H| = 2$. Generalizing Problem 5.1.17c and its proof, show that all elements of G of odd order are in H.

(b) Let G be a group of even order, and assume that more than half of the elements of G are of odd order. Show that G does not have a subgroup of index 2.[2]

5.1.19. Let $G = A_4$, the alternating group of degree 4.

(a) How many elements of order 3 does G have?

(b) Does G have a subgroup of order 6?

5.2. Lagrange's Theorem

The so-called Lagrange's theorem—which we will shortly state and prove—is now very easy to prove and yet has many unexpected consequences. This is because the theorem says that there is a limitation on the possible sizes of subgroups of a group. We will be able to use this limitation to prove many things about groups. The theorem is named for Lagrange who anticipated the study of groups with his work. However, he did not prove a theorem this general, and, in fact, the abstract concept of a group was developed after him.

Theorem 5.13 (Lagrange). *Let G be a finite group. Let $H \leq G$. Then*

$$|G| = |H| \cdot |G : H|.$$

In particular, $|H|$ divides $|G|$.

Proof. Let H act on G by translation. The orbits are the right cosets, each of these orbits have size $|H|$, and the orbits partition G. The number of orbits is $|G : H|$ and hence G is partitioned into $|G : H|$ sets, each having $|H|$ elements. □

Corollary 5.14. *Let $|G| < \infty$, a group, and let $a \in G$. Then $o(a)$ divides $|G|$.*

Proof. The cyclic group generated by a, that is $\langle a \rangle$, is a subgroup of G. Hence Lagrange's theorem implies that $|\langle a \rangle|$ divides $|G|$. But, $o(a) = |\langle a \rangle|$, and the proof is complete. □

The above corollary says that a group element is limited by virtue of the group that it is sitting in. For example, you may know one symmetry of a geometric object and wonder how many times you have to repeat it before you get the identity. The corollary says that the size of the whole symmetry group limits the possibilities for the answer. This is not an intuitively obvious result. Without the theorem, for example, it is not clear why a symmetry of order 5 cannot be a part of a symmetry group of order 12.

Corollary 5.15. *Let G be a group, and assume $|G| = p$, where p is a prime number. Let $e \neq x \in G$. Then $G = \langle x \rangle$. In particular, G is cyclic and isomorphic to $\mathbb{Z}/p\mathbb{Z}$.*

---

[2] Adapted from Brennan and MacHale [**BM00**] where, in addition to this result, various elementary proofs that A_4 does not have a subgroup of order 6 are gathered.

Proof. By Corollary 5.14, $o(x)$ divides $|G| = p$, a prime number. Hence $o(x)$ is either one or p, but $o(x)$ is greater than one—since $x \neq e$—and so $o(x)$ must be p. This means that $|\langle x \rangle| = o(x)$ is the same as $|G|$. Hence $G = \langle x \rangle$ is a cyclic group. Finally, by Theorem 2.50, we know that every cyclic group of order p is isomorphic to $\mathbb{Z}/p\mathbb{Z}$. $\qquad\square$

We now know that there is only one group of orders 1, 2, 3, 5, 7, 11, etc. In addition, such a group is cyclic, and so it is generated by just one of its elements. We further know that by starting with a non-identity element of the group and taking powers, we get every element of the group. All of these facts were not at all obvious before we constructed a theory. It should be clear now that our abstract constructions have power. They tell us things that we did not know before creating them. Moreover, we are just starting. There is much more to come.

Corollary 5.16. *Let G be a finite group with $|G| = n$. Let $a \in G$. Then $a^n = e$.*

Proof. By Corollary 5.14, $o(a)$ divides n. Hence, $n = ko(a)$. But then $a^n = a^{ko(a)} = (a^{o(a)})^k = e^k = e$. $\qquad\square$

If we are somewhat clever in using Lagrange's theorem, then we can, in fact, prove a variety of things.

Example 5.17. Assume that G is a group of order 169, and we are wondering whether G is abelian or not.

A priori, we need to check every element and see if it commutes with every other element. This involves lots of calculations. But that is before we use Lagrange's theorem.

Assume that we checked one non-identity element, and it did commute with every other element.

Now, instead of asking whether the group is abelian or not, we will ask what the center of the group is. The center of G, $\mathbf{Z}(G)$, is a subgroup, and hence, by Lagrange's theorem, its order must divide the order of the group. So $|\mathbf{Z}(G)| = 1, 13$, or 169. We know that this size is bigger than 1 (since both e and one other element are in the center). Thus there are either 13 or 169 elements in the center. This is already amazing. It says that we know for sure that there are 11 other elements that commute with every element. In fact, if we find a total of 14 in the center, then there must be 169 elements in the center. In other words the group is abelian.

Believe it or not, we can say more. Assume that there is an element x that was not in the center of G. What is the size of $|\mathbf{C}_G(x)|$? This is a subgroup again and hence its size is either 1, 13, or 169. But everything in the center must commute with x and hence it is in the centralizer of x. Thus there are at least 13 elements in the centralizer of x. But x also commutes with x. This makes for at least 14 elements that commute with x. But Lagrange's theorem now says that $|\mathbf{C}_G(x)| = 169$ and, in other words, $x \in \mathbf{Z}(G)$, a contradiction. Thus there cannot be any elements not in the center of G, and hence G is abelian!

Thus, in the case of a group of order 169, knowing that just one non-identity element commuted with everything, we were able to conclude that the group is abelian. This is just amazing. In fact, later we will prove that, in a group of order

169, there must be non-identity elements that commute with every element, and hence all groups of order 169 are abelian!

Remark 5.18. What made the above argument work was our insistence on using subgroups. There are two points here. First, in algebra, structures tell us more than elements, and hence, when possible, we want to think about algebraic structures and not individual elements. Second, in general, questions that have a yes or no binary answer are not the best kinds of questions. In many parts of mathematics, when forced with such a question, we try to reformulate the question and focus on understanding a structure that captures a range of possible answers. Analyzing such a structure will hopefully provide a deeper understanding of the original question. For example, is a group abelian? This question turns the world of groups into two categories, and even though sometimes it may be useful to do that, it is often better to change the question so that you get an spectrum of answers. We can ask what is the center of the group? Of course, if the center is the whole group, then we know that the group is abelian. However, we get more than knowing whether a group is abelian or not. We get a measure of "abelian-ness". The larger the center is, the closer the group is to being abelian. Or you may ask, is element x in the center? Again a binary yes-or-no question. A better question is what is the centralizer of x? Again this allows for a spectrum of answers, and, in addition, you are asking something about a subgroup, and hence our theorems about subgroups comes into play.

Corollary 5.19. *Let G be a finite group, and assume that $H \leq K \leq G$. Then*

$$|G : H| = |G : K| \, |K : H| \, .$$

Proof. Using Lagrange's theorem, we have

$$|G : K| \, |K : H| = \frac{|G|}{|K|} \frac{|K|}{|H|} = \frac{|G|}{|H|} = |G : H| \, . \qquad \square$$

Remark 5.20. Corollary 5.19 will be used often. In fact, instead of assuming G is a finite group, we only need $|G : H|$ to be finite for the conclusion to follow. Of course, our simple proof would not quite work in the more general case.

We want to use Corollary 5.19 to generalize the main argument in Example 5.17. It is convenient to talk about maximal subgroups first.

Definition 5.21 (Maximal subgroups). Let G be a group, and let H be a subgroup of G. The subgroup H is a *maximal subgroup* of G if $H < G$ (i.e., it is proper) and there exists no subgroup K with $H < K < G$.

Corollary 5.22. *Let G be a finite group, and let p be a prime number. Let H be a subgroup of G with $|G : H| = p$. Then H is a maximal subgroup of G.*

Proof. The subgroup H is a proper subgroup of G since $|G : H| > 1$. If H were not maximal, then there would exist a subgroup K with $H < K < G$. This would mean, by Corollary 5.19, that

$$p = |G : H| = |G : K| \, |K : H| \, .$$

Since $K > H$ and $G > K$, we have that $|G : K|$ and $|K : H|$ are both greater than 1, and so we have factored p into two non-trivial factors. This is impossible since p is a prime. The contradiction completes the proof. $\qquad\square$

The argument in Example 5.17 shows that $\mathbf{Z}(G)$ can never be a maximal subgroup.

Proposition 5.23. *Let G be a group, and let $\mathbf{Z}(G)$ denote—as usual—the center of G. Then $\mathbf{Z}(G)$ is not a maximal subgroup of G. In particular, $|G : \mathbf{Z}(G)|$ cannot be a prime number.*

Proof. If G is abelian, then $G = \mathbf{Z}(G)$ and $|G : \mathbf{Z}(G)| = 1$. Hence, assume G is not abelian, and let x be an element *not* in the center. Since x is not in the center, the centralizer in G of x is not the whole group. On the other hand, every element of the center of the group does commute with x, and hence $\mathbf{Z}(G) \leq \mathbf{C}_G(x)$. Now, x itself commutes with x, and so $x \in \mathbf{C}_G(x) - \mathbf{Z}(G)$. We conclude that the center of G is a *proper* subgroup of $\mathbf{C}_G(x)$. So, we have (this was actually Problem 2.6.5)

$$\mathbf{Z}(G) < \mathbf{C}_G(x) < G.$$

This means that $\mathbf{Z}(G)$ is not a maximal subgroup, and, by Corollary 5.22, $|G : \mathbf{Z}(G)|$ cannot be a prime number. $\qquad\square$

If H and K are subgroups of a finite group G, it may be that $H \leq K$, in which case, the various indices are related by Corollary 5.19. However, even if neither of the two subgroups are a subgroup of the other, there is much that we can say. In the Problems, you are asked to investigate the relationship between $|G : H \cap K|$, $|H : H \cap K|$, and $|K : H \cap K|$, and to prove:

Theorem 5.24. *Let G be a finite group, and let H and K be subgroups of G. Then*

(a) *(Problem **5.2.13**) The index $|G : K|$ is never smaller than $|H : H \cap K|$.*

(b) *(Problem **5.2.14**) We have $|G : K| = |H : H \cap K|$ if and only if $G = HK$.*

(c) *(Problem **5.2.15**) If $|G : H|$ and $|G : K|$ are relatively prime, then $G = HK$ and $|G : K| = |H : H \cap K|$.*

These results are somewhat unexpected and very powerful. We will come back to them in Chapter 9, but, for now, we limit ourselves to one example of their use.

Example 5.25. Assume that you have a group G of order 55 and that you have somehow found an element x of order 5 and an element y of order 11. The subgroup $\langle x \rangle = \{e, x, \ldots, x^4\}$ has five elements, and, likewise, the subgroup $\langle y \rangle = \{e, y, \ldots, y^{10}\}$ has 11 elements. If we find all elements of the form $x^i y^j$, for $0 \leq i \leq 4$ and $0 \leq j \leq 10$, we get 55 elements of the group G. Are all these elements distinct? In other words, do we have every element of G? After all, it may be possible that there is some repetition in the list.

By Lagrange's theorem, $|G : \langle x \rangle| = 11$ and $|G : \langle y \rangle| = 5$. These are relatively prime, and so, by Theorem 5.24(c), we have $\langle x \rangle \langle y \rangle = G$. We conclude that every element of G is of the form $x^i y^j$.

Problems

5.2.1. If H and K are subgroups of G of order 75 and 242 respectively, what can you say about $H \cap K$?

5.2.2. If G is a non-cyclic group of order 27, then for how many elements x of G do we have $x^9 = e$?

5.2.3. Suppose that a finite group G has an element g with order 7 and an element h with order 11. What is the minimum value of $|G|$?

5.2.4. (a) Let G be a non-cyclic group of order 121. How many subgroups does G have? Why?

   (b) Can you generalize your result of the previous part?

5.2.5. Let $D_{10} = \langle a, b \mid a^5 = b^2 = e, ba = a^4 b \rangle$ be the dihedral group of order 10. Assume x and y are two distinct elements of order 2 in D_{10}. Let $H = \langle x, y \rangle$. What can you say about $|H|$? Can x and y commute? Give your reasons.

5.2.6. Let G be a finite group, and let x and y be two distinct elements of order 2 that commute (i.e., $xy = yx$). Prove that the order of the group is divisible by 4.

5.2.7. Let G be a group of order 338. Assume x and y are two distinct elements of order 2 in G. Let $H = \langle x, y \rangle$. What are the possibilities for $|H|$? Can x and y commute? Give your reasons.

5.2.8. Let G be a group of order $143 = 11 \times 13$, and, as usual, let $\mathbf{Z}(G)$ denote the center of G. Assume that we have found an element $x \in \mathbf{Z}(G)$ with $x \neq e$. What are the possibilities for the $|\mathbf{Z}(G)|$? Prove any assertions you make.

5.2.9. **Subgroups of Q_8.** Find all the subgroups of Q_8, the quaternion group of order 8.

5.2.10. **Subgroups of $S_3 \times \mathbb{Z}/2\mathbb{Z}$.** Let $G = S_3 \times \mathbb{Z}/2\mathbb{Z}$. If L is a subgroup of G, what are the possible orders of L? Which groups could L be isomorphic to?

5.2.11. **Subgroups of A_4.** Let $G = A_4$, the alternating group of degree 4. If L is a subgroup of G, what are the possible orders of L? Which groups could L be isomorphic to?

Two Subgroups and Their Cosets. If a finite group G has two subgroups H and K, we can certainly find $|G : H|$ and $|G : K|$. However, $H \cap K$ is also a subgroup of G—as well as a subgroup of both H and K—and, hence, we can also find $|G : H \cap K|$, $|H : H \cap K|$, and $|K : H \cap K|$. Is there any meaningful relationship between these numbers?

In the easier case, when $K \leq H \leq G$, we have $H \cap K = K$, and Corollary 5.19 gives the relevant answer. In Problems 5.2.12 through **5.2.15** treat the general case, and prove the various parts of Theorem 5.24.

5.2.12. Give two examples of $U, V \leq G$ with $|G : U| < \infty$. Each example should satisfy one of the following:
  (a) $|V : V \cap U| < |G : U|$.
  (b) $|V : V \cap U| = |G : U|$.

5.2.13. **Proof of Theorem 5.24(a).** Let $U, V \leq G$ with $|G : U| < \infty$. Show that $|V : V \cap U| \leq |G : U|$.

5.2.14. **Proof of Theorem 5.24(b).** Let $U, V \leq G$ with $|G : U| < \infty$. Show that we have $|V : V \cap U| = |G : U|$ if and only if $VU = G$.

5.2.15. **Proof of Theorem 5.24(c).** Let $U, V \leq G$ with $|G : U|$ and $|G : V| < \infty$. Assume $\gcd(|G : U|, |G : V|) = 1$. Show $G = VU$.

5.2.16. Let G be a group of order 18. Assume that x, y, and z are elements of G, and we know that $|\langle x, y \rangle| = 9$ and that $o(z) = 2$. Prove that $G = \langle x, y, z \rangle$.

5.2.17. Let G be a finite group, and $H, K \leq G$. Assume that HK is a subgroup of G. Prove that
$$|HK| = \frac{|H|\ |K|}{|H \cap K|}.$$

5.3. Application to Number Theory*

Recall that we defined the elements of $\mathbb{Z}/n\mathbb{Z}$ to be $\{0, 1, 2, \ldots, n-1\}$, and we defined two operations $+$ and $\cdot$ for it. These were defined as the remainders of the usual sum and product when divided by n. We have seen that $(\mathbb{Z}/n\mathbb{Z}, +)$ is a cyclic group, and we have also proved that $((\mathbb{Z}/n\mathbb{Z})^\times, \cdot)$ is an abelian group of order $\phi(n)$. (Recall, from Definition 1.45, that $\phi(n)$ is the number of positive integers less than or equal to n that are relatively prime to n.) We will apply Corollary 5.16 to the group $((\mathbb{Z}/n\mathbb{Z})^\times, \cdot)$.

The basic idea is very straightforward. The group $G = ((\mathbb{Z}/n\mathbb{Z})^\times, \cdot)$ has $\phi(n)$ elements, and so if $a \in G$, then, by Corollary 5.16, $a^{\phi(n)} = e$. Now the identity of the group G is 1, the fact that $a \in G$ means that $1 \leq a < n$ and $\gcd(a, n) = 1$, and the group operation is multiplication mod n. Hence, if we take a an integer with $1 \leq a < n$ and with $\gcd(a, n) = 1$, then we have $a^{\phi(n)} \equiv 1 \mod n$. With a bit of care, we see that the condition $1 \leq a < n$ is quite unnecessary. The result is called Euler's theorem and a special case of it—when n is a prime p—is Fermat's Little Theorem.

In this section, we complete the above argument by first reviewing some elementary facts about congruences.

A useful way of organizing information about remainders is the mod notation which we will introduce now. Much can be done with congruences, but we will limit ourselves to the few facts that we need for Euler's theorem.

Definition 5.26 (mod n). Let n be a positive integer, and let a and b be integers. We write $a \equiv b \mod n$ if a and b have the same remainder when divided by n.

The following lemma is straightforward:

Lemma 5.27. *Let n be a positive integer, and let $a, b \in \mathbb{Z}$. Then the following are equivalent:*

(a) $a \equiv b \bmod n$,

(b) $n \mid a - b$,

(c) $b = nk + a$ for some $k \in \mathbb{Z}$.

Note that to say that $a \equiv b \bmod n$ is not the same as saying that b is a remainder of a when divided by n. If r is the remainder of a when divided by n, then, in addition to $a \equiv r \bmod n$, we have to have $0 \le r \le n - 1$.

Lemma 5.28. *Let n be a positive integer, and let $x, y, x', y' \in \mathbb{Z}$.*

If $x \equiv y \bmod n$ and $x' \equiv y' \bmod n$, then $xx' \equiv yy' \bmod n$.

Proof. We have $n \mid x - y$, and $n \mid x' - y'$. Thus $n \mid y(x' - y') + x'(x - y) = xx' - yy'$. $\qquad\qquad\qquad\qquad\qquad\qquad\qquad\qquad\qquad\qquad\qquad\qquad\qquad\qquad$ □

Corollary 5.29. *Let n and k be positive integers, and let $a, u \in \mathbb{Z}$.*

If $a \equiv u \bmod n$, then $a^k \equiv u^k \bmod n$.

Example 5.30. If you cube an integer and find its remainder when divided by 6, then what are the possible answers?

The above lemma and corollary imply that if we are interested in the remainder of n^3 when divided by 6, then the only determining factor is the remainder of n when divided by 6. By the division algorithm, the remainder n when divided by 6 is 0, 1, ..., or 5, and hence we can make a table as follows:

If n has remainder	0,	1,	2,	3,	4,	5	when divided by 6,
then n^2 has remainder	0,	1,	4,	3,	4,	1,	when divided by 6,
and then n^3 has remainder	0,	1,	2,	3,	4,	5,	when divided by 6.

Hence, we conclude that n and n^3 always have identical remainders when divided by 6.

Lemma 5.31. *Let n be a positive integer, and let a and u be integers. Assume that $\gcd(a, n) = 1$ and that $a \equiv u \bmod n$, then $\gcd(u, n) = 1$.*

Proof. Since $a \equiv u \bmod n$, we have $a = kn + u$ for some integer k. Now if u and n had some common prime divisor p, then $p \mid kn + u = a$, which would make p a common prime divisor of n and a. This is a contradiction since $\gcd(a, n) = 1$. The contradiction implies that u and n have no common prime divisors. $\qquad\qquad$ □

Now Euler's theorem of elementary number theory is an immediate consequence of Corollary 5.16:

Theorem 5.32 (Euler). *Let a be an integer, and let n be a positive integer. Assume $\gcd(a, n) = 1$. Then*

$$a^{\phi(n)} \equiv 1 \bmod n.$$

Proof. Let $0 \leq u \leq n-1$ with $a \equiv u \bmod n$. In other words, u is the remainder of a when divided by n. Now by Corollary 5.29, $a^{\phi(n)} \equiv u^{\phi(n)} \bmod n$, and it is enough to show $u^{\phi(n)} \equiv 1 \bmod n$.

But $0 \leq u \leq n-1$ and, by Lemma 5.31, $\gcd(u, n) = 1$. Thus u is an element of the group $(\mathbb{Z}/n\mathbb{Z})^{\times}$, and this group has $\phi(n)$ elements. Now, by Corollary 5.16, we have that $u^{\phi(n)}$ is the identity of $(\mathbb{Z}/n\mathbb{Z})^{\times}$, so $u^{\phi(n)} \equiv 1 \bmod n$. $\square$

The case when n is a prime is especially nice since clearly $\phi(p) = p - 1$ for primes p.

Corollary 5.33 (Fermat's Little Theorem). *Let a be an integer, and let p be a prime with $p \nmid a$. Then*

$$a^{p-1} \equiv 1 \bmod p.$$

Example 5.34. By Fermat's Little Theorem and with no calculation, we know that 10^{10} has remainder 1 when divided by 11.

To remove the condition that $p \nmid a$, we can reformulate Fermat's Little Theorem as follows:

Corollary 5.35. *Let p be a prime number, and let $a \in \mathbb{Z}$. Then*

$$a^p \equiv a \bmod p.$$

Proof. We do the proof in two cases. First, assume p divides a. This means that p also divides a^p and hence p divides $a^p - a$. We conclude that, in this case, $a^p \equiv a \bmod p$. For the second case, assume that p does not divide a. Then, by Corollary 5.33, $a^{p-1} \equiv 1 \bmod p$. On the other hand, we also have $a \equiv a \bmod p$. Now, putting these two together by Lemma 5.28, we get $a^p \equiv a \bmod p$. $\square$

Problems

5.3.1. (a) Suggest the following game to a friend: You agree on a method for picking a random integer n between 1 and 1000. Then you look at the remainder of n^2 when divided by 8. If the remainder is 1 or 4 you win, and if the reminder is 0, 2, 3, 5, 6, or 7 your friend will win. What is the probability that you will win this game?

   (b) You have an odd integer n. How is the remainder of n^3 when divided by 8 related to the remainder of n when divided by 8?

   (c) What are the possible remainders of n^4 when divided by 8?

5.3.2. Let a be a positive integer. Prove that a^{21} and a have the same remainder when divided by 15.

5.3.3. What is the remainder of $1835^{1910} + 1986^{2061}$ when divided by 7?[3]

5.3.4. What is the remainder of 47^{1994} when divided by 13?

---

[3]The three most recent appearances of Halley's comet were in the years 1835, 1910, and 1986; the next occurrence will be in 2061.

5.3.5. Let p be a prime. You come across the following three problems about p:
    (a) Solve the equation $x^2 = -1$ in $\mathbb{Z}/p\mathbb{Z}$?
    (b) Find all integers x such that $x^2 \equiv -1 \bmod p$.
    (c) Find all integers x such that p divides $x^2 + 1$.
Is there any difference between the three questions, or are all the same?
Answer all the questions for $p = 5$.

5.3.6. Assume that p is a prime number of the form $4n + 3$. Does $x^2 \equiv -1 \bmod p$ have any integer solutions?

5.3.7. Does $x^2 + y^2 = 3z^2$ have any integer solutions?

5.4. More Problems and Projects

Problems

5.4.1. **Does $\mathrm{SL}(2,3)$ have a subgroup of order 12?**[4] Let G be $\mathrm{SL}(2,3)$, the group of 2×2 matrices with determinant 1 and with entries in $\mathbb{Z}/3\mathbb{Z}$.
    (a) What is $|\mathrm{SL}(2,3)|$?
    (b) Find eight elements of order 3 in G.
    (c) Show that $\begin{bmatrix} -1 & 0 \\ 0 & -1 \end{bmatrix}$ is the only element of order 2 in G.
You could do this by checking the order of every element in G, or you could let A be a 2×2 matrix with determinant 1 such that $A^2 = I$, and use the following steps:
Step 1: Using the fact that $A^2 = I$, show that the only possibilities for eigenvalues for A are ± 1.
Step 2: Since A has determinant 1, show that A must have a repeated eigenvalue.
Step 3: Show that A is similar to a matrix of the form $\begin{bmatrix} -1 & b \\ 0 & -1 \end{bmatrix}$ or $\begin{bmatrix} 1 & b \\ 0 & 1 \end{bmatrix}$.
Step 4: Show that among the matrices in the previous step, only $\begin{bmatrix} -1 & 0 \\ 0 & -1 \end{bmatrix}$ has order 2.
Step 5: Show that $A = \begin{bmatrix} -1 & 0 \\ 0 & -1 \end{bmatrix}$.
    (d) Show that $\begin{bmatrix} -1 & 0 \\ 0 & -1 \end{bmatrix}$ is in the center of $\mathrm{SL}(2,3)$, and that the product of this element with any element of order 3 is an element of order 6.
    (e) Assume that H is a subgroup of order 12 in $\mathrm{SL}(2,3)$. Argue that H must contain an element of order 2, all eight elements of order 3, and eight elements of order 6, and hence H cannot exist.

---

[4] Adapted from Mackiw [**Mac96**].

The Classification of Groups of Order $2p$. Let p be a prime. We have seen (see Corollary 5.15) that the only group of order p is $\mathbb{Z}/p\mathbb{Z}$, the cyclic group of order p. We now want to classify all groups of order $2p$. If $p = 2$, then we know (see Problem *2.4.9*) that there are exactly two groups of order 4, namely $\mathbb{Z}/4\mathbb{Z}$ and $\mathbb{Z}/2\mathbb{Z} \times \mathbb{Z}/2\mathbb{Z}$. Hence, we concentrate on the case when p is an odd prime.

Proposition 5.36. *Let p be a prime greater than 2, and let G be a group of order $2p$. Then G is isomorphic to one of the following two groups:*

(a) *The cyclic group of order $2p$:*

$$\mathbb{Z}/2p\mathbb{Z} \cong \langle a \mid a^{2p} = e \rangle.$$

(b) *The dihedral group of order $2p$:*

$$D_{2p} = \langle a, b \mid a^p = b^2 = e, ba = a^{-1}b \rangle.$$

As an example, the proposition says that $\mathbb{Z}/6\mathbb{Z}$ and D_6 are the only groups of order 6. In particular, S_3 is isomorphic to D_6.

Outline of the Proof. We first show that G has an element a of order p. (This is a special case of a theorem known as Cauchy's theorem—Corollary 7.11—that asserts that if a prime p divides the order of a group, then there is an element of order p in the group. We will prove Cauchy's theorem in Chapter 7, but here, for the case of groups of order $2p$, we can give a direct proof.) We then assume that G is not the cyclic group of order $2p$, and we show that every element of G that is not in $\langle a \rangle$ has order 2. If b is such an element, we then show that $G = \langle a \rangle \langle b \rangle$, and $ba = a^{p-1}b$. We then argue that G and D_{2p} have the same multiplication table, and hence must be isomorphic. □

In the Problems, you are asked to fill in the details of the proof. This proof is adapted from Gallian [**Gal01**] with one notable difference. We use Theorem 5.24(a) to give a quick argument that every element outside of $\langle a \rangle$ has order 2.

5.4.2. Let p be a prime, and let G be a group of order $2p$. Show that G has an element of order p.

After dispensing with the case $p = 2$, you may find the following steps useful:

Step 1: What does Corollary 5.14 say about the orders of elements of G?

Step 2: Assume that every non-identity element of G had order 2. Apply Problem 2.2.2, and then construct a subgroup of order 4.

Step 3: Use Lagrange's theorem to finish your proof.

5.4.3. Let p be a prime greater than 2, and let G be a non-cyclic group of order $2p$. Assume that a is an element of order p in G. Let b be an element of G not in $\langle a \rangle$. Show that b has order 2.

You may find the following steps useful:

Step 1: What is $\langle a \rangle \cap \langle b \rangle$?

Step 2: Use Theorem 5.24(a)—with $U = \langle a \rangle$ and $V = \langle b \rangle$—to find $|\langle b \rangle|$.

5.4.4. Let p be a prime greater than 2, and let G be a non-cyclic group of order $2p$. Assume that a is an element of order p and b is an element of order 2. Show that $G = \langle a \rangle \langle b \rangle$ and that $ba = a^{-1}b$.

5.4.5. Using the proof outline and Problems 5.4.2–5.4.4 give a complete proof of Proposition 5.36.

5.4.6. **Another proof that A_4 does not have a subgroup of order 6.**[5] Using the classification of groups of order $2p$, Proposition 5.36, give another proof that A_4 does not have a subgroup of order 6.

You may find the following steps useful:

Step 1: Assume A_4 has a subgroup H of order 6, and use Proposition 5.36 to find a familiar group that is isomorphic to H.

Step 2: From the previous step conclude that H has three elements of order 2.

Step 3: How many elements of order 2 does A_4 have? Can you say exactly which three elements of order 2 must be in H?

Step 4: Verify that the three elements of order 2 and the identity element form a subgroup of order 4.

Step 5: Use Lagrange's theorem to get a contradiction.

Is a Group Abelian? A Random Test. A finite group G walks in the door, and we want to see if it is abelian. We randomly pick an element x from the group and check to see if x is in the center of G. If x is not in the center, then we know that G is not abelian. However, if x is in the center, then we cannot be sure that G is abelian. But what is the probability that x is in the center and yet G is not abelian? What if we repeat the experiment, and we repeatedly get elements in the center? When can we be reasonably sure that G is abelian?

Using Proposition 5.23, you are asked—Problems *5.4.7*, 5.4.8, and 5.4.9—to give some approximate answers to the above questions.

5.4.7. Let G be a finite non-abelian group. Show that at most $1/4$ of the elements of G can be in $\mathbf{Z}(G)$. Give an example of a group where exactly $1/4$ of its elements are in $\mathbf{Z}(G)$.

5.4.8. Let G be a finite group. Randomly and independently pick elements x_1, x_2, x_3, and x_4 from the group. (You are sampling with replacement. After picking x_1, return it to the group and choose another random element. Hence x_2 may be equal to x_1.) Assume that each of $x_1, \ldots, x_4$ are elements of the center of G. Based on this, you declare that G is abelian. Could you be mistaken? Prove that, regardless of the size of G, the probability that you could be wrong is less than 0.004 (four-tenths of one percent).

---

[5] Adapted from Brennan and MacHale [**BM00**].

5.4.9. Assume that G is a finite group of *odd* order. As in the previous problem, you randomly and independently choose four elements of the group and find that each is an element of the center of G. What can you say about the probability that G is non-abelian?

On the Probability that Two Random Elements of a Group Commute. A finite group G is given. You randomly pick an element x of G, put it back, and randomly choose another element y of the group. (This is sampling with replacement, and, hence, x could be equal to y.) What is the probability $p_c(G)$ that x and y commute? If we let $c(G) = |\{(x, y) \mid x, y \in G, xy = yx\}|$, then

$$p_c(G) = \frac{c(G)}{|G|^2}.$$

If the group G is abelian, then, of course, $p_c(G) = 1$. In what follows, you prove that if G is non-abelian—regardless of the size of G—then $p_c(G) \leq 5/8$. In other words, there is a *gap* in the values of $p_c(G)$. If $p_c(G)$ is less than 1, then it must actually be no more than 5/8. This result can be used to give another probabilistic test for deciding whether a group G is abelian.

5.4.10. Let G be a finite group, and let $c(G)$ be defined as above. Show that

$$c(G) = \sum_{x \in G} \mathbf{C}_G(x) = |\mathbf{Z}(G)|\,|G| + \sum_{x \in G - \mathbf{Z}(G)} \mathbf{C}_G(x).$$

5.4.11. Let G be a finite group, and let $p_c(G)$ be defined as above. Fill in the details, to show that the following calculation is correct:

$$
\begin{aligned}
p_c(G) &= \frac{|\mathbf{Z}(G)|}{|G|} + \frac{1}{|G|} \sum_{x \in G - \mathbf{Z}(G)} \frac{1}{|G : \mathbf{C}_G(x)|} \\
&\leq \frac{|\mathbf{Z}(G)|}{|G|} + \frac{1}{|G|} \sum_{x \in G - \mathbf{Z}(G)} \frac{1}{2} \\
&= \frac{1}{|G|} \left[|\mathbf{Z}(G)| + \frac{1}{2}|G - \mathbf{Z}(G)| \right] \\
&= \frac{1}{2} + \frac{1}{2\,|G : \mathbf{Z}(G)|} \\
&\leq \frac{5}{8}.
\end{aligned}
$$

5.4.12. We have a finite group G with over one billion elements, and we want to decide if G is abelian. We do the following experiment n times: Randomly choose one element x from G, put it back, and randomly choose another element y from G. Check to see if $xy = yx$. If at any time, we find two elements that do not commute, we declare that the group G is non-abelian. We would like to choose n large enough, such that if $xy = yx$ for each of n

experiments, then the probability that the group G is non-abelian is less than 0.001 (one-tenth of one percent). How large of an n do we need?

5.4.13. **Double cosets.** Let G be a group, and let H and K be two subgroups of G. Let $x \in G$. The set

$$HxK = \{hxk \mid h \in H, k \in K\}$$

is called a *double coset* with respect to H and K. Fix H and K and prove that the collection

$$\{HxK \mid x \in G\}$$

of double cosets with respect to H and K partitions G.

5.4.14. Let $G = D_8 = \langle a, b \mid a^4 = b^2 = e, ba = a^3 b \rangle$. Let $H = \langle b \rangle$, $K = \langle ab \rangle$, and $L = \langle a \rangle$. Find the double cosets of G once with respect to H and K, and then with respect to H and L.

5.4.15. Let G be a group, and let H and K be subgroups of G. Show that each double coset of G with respect to H and K is a union of left cosets of K (in G) as well as a union of right cosets of H.

Chapter 6

A Group Acts on Itself: Counting and the Conjugation Action

... where the fundamental counting principle, which is a relation between the sizes of orbits and stabilizers, is proved and then applied to a specific action—the conjugation action of a group on itself—to give the class equation, and to prove that, for every prime p, every group of order p^n has a non-trivial center, and every group of order p^2 is abelian.

6.1. The Fundamental Counting Principle

We now go back to the general situation of a group acting on a set. Whenever we have an action, we can look at orbits which partition the set and stabilizers which are subgroups of the group. In this section, we prove a fundamental relationship between the size of the orbit of a set element and the size of its stabilizer. In the following section, we will apply this general relationship to a specific action.

Theorem 6.1 (Fundamental Counting Principle, or FCP). *Let the group G act on the set Ω. Let $\alpha \in \Omega$ (and, as usual, $\mathcal{O}_\Omega(\alpha)$ and $\mathrm{Stab}_G(\alpha)$, respectively, denote the orbit of α in Ω and the stabilizer of α in G). Then*

$$|\mathcal{O}_\Omega(\alpha)| = |G : \mathrm{Stab}_G(\alpha)|.$$

Proof. Denote $\mathrm{Stab}_G(\alpha)$, the stabilizer of α in G, by H. We know that H is a subgroup of G, and we want to prove that the the orbit of α under the action of G has the same cardinality as the set of cosets (left or right cosets) of H in G. To do this, we construct an appropriate 1-1, onto map (see Figure 6.1).

We define

$$\psi : \{\, xH \mid x \in G \,\} \to \mathcal{O}_\Omega(\alpha),$$

by

$$\psi(xH) = x \cdot \alpha.$$

It appears that ψ is a map that for each left coset of H gives an element in the orbit of α. However, when defining maps on cosets, one has to be very careful. This is because a single coset may have many different "names", and we have to be careful that our map sends a coset to a well-defined destination regardless of the name of the coset. Thus the coset xH may also be yH, and, for our map to be well defined, we have to be sure that $x \cdot \alpha$ is the same as $y \cdot \alpha$. Otherwise, the map is sending the same coset to two different places. Thus, we have to show that the map is well defined, 1-1, and onto.

ψ *is well defined:* Assume $xH = yH$. We have to show that $x \cdot \alpha = y \cdot \alpha$.

From $xH = yH$ follows that $y \in yH = xH$, and this, in turn, means that $y = xh$ for some $h \in H$. Thus

$$y \cdot \alpha = xh \cdot \alpha = x \cdot (h \cdot \alpha) = x \cdot \alpha.$$

ψ *is 1-1:* Assume $\psi(xH) = \psi(yH)$. To show that $xH = yH$,

$$
\begin{aligned}
\psi(xH) = \psi(yH) &\Rightarrow& x \cdot \alpha = y \cdot \alpha \\
&\Rightarrow& x^{-1} \cdot (x \cdot \alpha) = x^{-1} \cdot (y \cdot \alpha) \\
&\Rightarrow& \alpha = x^{-1}y \cdot \alpha \\
&\Rightarrow& x^{-1}y \in H \\
&\Rightarrow& x^{-1}y = h \quad \text{for some} \quad h \in H \\
&\Rightarrow& y = xh \\
&\Rightarrow& y \in xH \\
&\Rightarrow& yH = xH.
\end{aligned}
$$

The last step follows since left cosets partition G and both $x, y \in xH$.

ψ *is onto:* Let $\beta \in \mathcal{O}_\Omega(\alpha)$. We have to find a left coset xH such that $\psi(xH) = \beta$.

If $\beta \in \mathcal{O}_\Omega(\alpha)$, then there exists $x \in G$ such that $x \cdot \alpha = \beta$. But then $\psi(xH) = x \cdot \alpha = \beta$. $\qquad\square$

Remark 6.2. The above proof actually proved more. We know that $\mathrm{Stab}_G(\alpha)$ consists of those elements of G that fix α. Now take $\beta \in \mathcal{O}_\Omega(\alpha)$. There will be an $x \in G$ such that $x \cdot \alpha = \beta$, and the proof shows that the set of all elements of G that send α to β is exactly the coset $x\mathrm{Stab}_G(\alpha)$. In other words, the elements in any of the left cosets of $\mathrm{Stab}_G(\alpha)$ are exactly those elements of G that send α to the same place (see Figure 6.1). This explains why the two sets have the same cardinality.

In the finite case, we can restate the FCP:

Corollary 6.3. *Let the finite group G act on the set Ω. Let $\alpha \in \Omega$. Then*

$$|G| = |\mathcal{O}_\Omega(\alpha)| \, |\mathrm{Stab}_G(\alpha)|.$$

In particular, $|\mathcal{O}_\Omega(\alpha)|$ divides $|G|$.

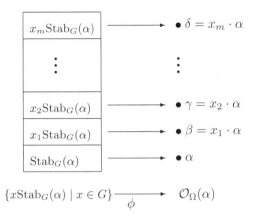

Figure 6.1. A bijection between the left cosets of the stabilizer and the elements in an orbit

Example 6.4. The dihedral group D_8 acts on the corners of the square $\Omega = \{1, 2, 3, 4\}$.

The orbit of $1 \in \Omega$ is $\mathcal{O}_\Omega(1) = \{1, 2, 3, 4\}$. Hence $|\mathcal{O}_\Omega(1)| = 4$ and so $|\text{Stab}_G(1)| = 2$. The element e certainly fixes 1, and hence we conclude—basically with no calculation—that, for each corner of the square, there is exactly one non-identity element of D_8 that fixes it. We know that each of the diagonal reflections fix two of the corners, and hence, these must be the only elements of D_8 that fix a corner.

Example 6.5. Recall that any group G acts on G by left multiplication (regular action). In other words for $g \in G$ and $x \in G$, we have $g \cdot x = gx$.

We have seen that the orbit of x is all of G (i.e., the action is transitive) and that the stabilizer of x is just the identity. Given the FCP Theorem 6.1, each of these two facts follows from the other.

Another Relative of the Regular Action. Let G be a finite group, and let H be a subgroup. Let Ω be the set of left cosets of H in G, i.e., $\Omega = \{xH \mid x \in G\}$. Note that $|\Omega| = |G : H|$. Now, G acts on Ω by $g \cdot xH = gxH$.

This action is transitive since if xH and yH are any two left cosets, then $(yx^{-1}) \cdot xH = yH$. Thus there is only one orbit: Ω.

The subgroup H itself is one of the elements of Ω. What does the FCP say about H? The orbit of H is all of Ω and, hence, $|\mathcal{O}_\Omega(H)| = |\Omega| = |G : H|$. On the other hand, what is the $\text{Stab}_G(H)$? We have

$$\begin{aligned}
\text{Stab}_G(H) &= \{g \in G \mid g \cdot H = H\} \\
&= \{g \in G \mid gH = H\} \\
&= H.
\end{aligned}$$

Thus, the FCP (in the case of this specific action) says $|G| = |H| \, |G : H|$. So for this action, the FCP is just Lagrange's theorem. For this reason we can think of the FCP as Lagrange's theorem generalized to arbitrary actions.

Question: Could we have done the above with right cosets?

Binomial Coefficients. As an example of the use of the FCP, we use it here to give a different proof for the familiar formula for binomial coefficients.

Definition 6.6. Let n be a positive integer. $[n] = \{1, 2, \ldots, n\}$ denotes a set with n elements, and $\left(\begin{smallmatrix} [n] \\ k \end{smallmatrix}\right)$ denotes the subsets of $[n]$ of size k. We denote the size of $\left(\begin{smallmatrix} [n] \\ k \end{smallmatrix}\right)$ by $\binom{n}{k}$. In other words, $\binom{n}{k}$ is the number of subsets of size k in a set with n elements. The expression $\binom{n}{k}$ is called a *binomial coefficient*.

Now let $G = S_n$ and $\Omega = \left(\begin{smallmatrix} [n] \\ k \end{smallmatrix}\right)$. The action of G on $[n]$ induces an action of G on $\left(\begin{smallmatrix} [n] \\ k \end{smallmatrix}\right)$ (every element of G acts on every element of a subset of size k producing another subset of size k—also see Example 4.3 and Problems 4.2.6 and 4.3.6). So for example, for $n = 5$, $k = 3$, we have

$$\begin{pmatrix} 2 & 3 & 5 \end{pmatrix} \cdot \{1, 3, 5\} = \{1, 5, 2\} = \{1, 2, 5\}.$$

Now the action is clearly transitive. Consider $\alpha = \{1, 2, \ldots, k\}$. What is the orbit and the stabilizer of α? The orbit of α is all of Ω, and hence the size of the orbit of α is $\binom{n}{k}$. The stabilizer of α consists of those permutations in S_n that are products of two smaller permutations. One that permutes $1, \ldots, k$ and another that permutes $k + 1, \ldots, n$. Thus $|\mathrm{Stab}_{S_n}(\alpha)| = |S_k \times S_{n-k}| = k!(n-k)!$. What does FCP tell us? It says that

$$n! = |S_n| = \binom{n}{k} k!(n-k)!.$$

We conclude that $\binom{n}{k} = \frac{n!}{k!(n-k)!}$, and we have a proof of the formula for binomial coefficients! We record this result for future use:

Lemma 6.7. *Let n and k be positive integers, with $n \geq k$. Then*

$$\binom{n}{k} = \frac{n!}{k!(n-k)!}.$$

Problems

6.1.1. Let G be a group of order 12, and let Ω be a set with five elements. Assume that G acts on Ω. Can the action be transitive? Either give an example or prove that it is impossible.

6.1.2. Let the finite group G act on the set Ω, and assume that the action is transitive. Further assume that $|\Omega| = |G|$. Is it possible to find a non-identity group element $g \in G$ and a set element $\alpha \in \Omega$ with $g \cdot \alpha = \alpha$? Why?

6.1.3. Let G be a group of order 121, and let Ω be a set with 16 elements. Assume that G acts on Ω. We are given that there is some orbit of size bigger than one. Let Ω_0 be the set of elements in Ω that are fixed by every element of G. What can you say about $|\Omega_0|$? Prove your assertions.

6.1.4. Recall Definition 4.14 of the conjugation action of a group on the set of its subgroups as well as Definition 4.24 of normalizers. Let $G = D_8$ act on the set of all of its subgroups by conjugation.

(a) Let $H = \langle b \rangle$. In Problem *4.3.3*, you found $\mathbf{N}_G(H)$. Use the FCP to find the size of the orbit of H. What are the elements in the orbit of H.

(b) Let $K = \langle a \rangle$. Find $\mathbf{N}_G(K)$, and the orbit of K.

6.2. The Conjugation Action

In this section we will explore the consequences of our previous results on a specific action. In general, as soon as we find a new action, we will ask, What are the stabilizers? What are the orbits? What does the FCP tell us? Later, we will learn some more general properties of actions, and thus we will be able to add new questions to the above list.

We already defined the family of conjugation actions on page 89. In this section, we further consider the conjugation action of a group on itself. We first recall the definition:

Definition 6.8 (The conjugation action). Let G be any group, and let $\Omega = G$. For $g \in G$ and $x \in \Omega$, define

$$g \cdot x = gxg^{-1}.$$

This defines an action of G on G called the *conjugation action*. In this action the orbits are called the *conjugacy classes*. The element gxg^{-1} is called a *conjugate* of x, and the conjugacy class of x (i.e., the orbit containing x) is denoted by $\mathrm{cl}_G(x)$.

In earlier Problems (e.g., 4.1.7, 4.4.6–4.4.16) you have been asked to become familiar with calculating conjugate elements and conjugacy classes. In Problem *2.3.13* you showed that two conjugate elements have the same order. We especially draw your attention to Problem **4.4.15**—and the discussion preceding it—where it is shown that conjugating in the symmetric group is particularly straightforward. This is one of the reasons why it is helpful to find a subgroup of S_n isomorphic to whatever group G you are interested in. Having done that, you can carry calculations related to conjugacy in the symmetric group.

Now that we have a new action, we know, for example, that the orbits (which are called conjugacy classes for this action) partition G and that the FCP applies. We will first look at an example of this action before exploiting its properties.

Example 6.9. Let G be the dihedral group of order 8: $\langle a, b \mid a^4 = b^2 = e, ba = a^3 b \rangle$.

To see the conjugation action of G on G, we draw the Cayley graph of this action with respect to the generating set $\{a, b\}$. (This was actually Problem 4.2.4.) We first have to see how a and b act. For $0 \le i \le 3$, we have

$$a \cdot a^i = a^i,$$

$$a \cdot a^i b = aa^i ba^3 = a^{i+2} b,$$

$$b \cdot a^i = ba^i b = a^{3i} = a^{-i},$$

$$b \cdot a^i b = ba^i bb = a^{3i} b = a^{-i} b.$$

Note that the action of all other elements follows from knowing the above. So for example,

$$(ab) \cdot a^3 b = a \cdot (b \cdot a^3 b)$$
$$= a \cdot (ab)$$
$$= a^3 b.$$

The Cayley graph of the conjugation action of D_8 on D_8 with respect to $S = \{a, b\}$ is given in Figure 6.2.

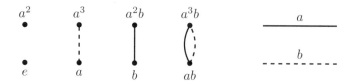

Figure 6.2. The Cayley graph of the conjugation action of D_8 on D_8 with $S = \{a, b\}$

The orbits of the action are precisely the connected components of the Cayley graph of the action:

$$\mathcal{O}_{D_8}(e) = \{e\},$$
$$\mathcal{O}_{D_8}(a) = \{a, a^3\} = \mathcal{O}_{D_8}(a^3),$$
$$\mathcal{O}_{D_8}(a^2) = \{a^2\},$$
$$\mathcal{O}_{D_8}(b) = \{b, a^2 b\} = \mathcal{O}_{D_8}(a^2 b),$$
$$\mathcal{O}_{D_8}(ab) = \{ab, a^3 b\} = \mathcal{O}_{D_8}(a^3 b).$$

As was mentioned before, the orbits of this action have the special name of conjugacy classes and are denoted by $\mathrm{cl}_G(x)$. In other words

$$\mathrm{cl}_G(x) = \{y \in G \mid \exists g \in G \text{ with } gxg^{-1} = y\}.$$

In our example we have found that the conjugacy classes of D_8 are

$$\{e\}, \{a^2\}, \{a, a^3\}, \{b, a^2 b\}, \{ab, a^3 b\}.$$

We also can find the stabilizers of this action by checking to see, for example, which group elements fix a given group element. But by Lemma 4.23—in fact, it is a straightforward argument—we know that, for this action, the stabilizer of a given group element is the centralizer of that element. In other words, for this action, $\mathrm{Stab}_G(x) = \mathbf{C}_G(x)$.

Now what does the FCP say about this specific action? The FCP gives a relation between the size of the orbits and the size of the stabilizers. If G acts on G by conjugation and $x \in G$, then the stabilizer of x is $\mathbf{C}_G(x)$ and the orbit of x is $\mathrm{cl}_G(x)$. Thus we have:

Corollary 6.10. *Let G be a group, and $x \in G$. Then*

$$|\mathrm{cl}_G(x)| = |G : \mathbf{C}_G(x)|.$$

In particular, if G is a finite group, we have $|G| = |\mathrm{cl}_G(x)||\mathbf{C}_G(x)|$, and the size of each conjugacy class of G divides the order of G.

Noting that an element x is in the center if and only if its centralizer is the whole group, we have the following corollary:

Corollary 6.11. *The following are equivalent:*

(a) $x \in \mathbf{Z}(G)$,

(b) $\mathbf{C}_G(x) = G$,

(c) $|\mathrm{cl}_G(x)| = 1$.

Example 6.12. Coming back to the example of D_8, we have already found all the conjugacy classes. From those and the FCP we know the size of the centralizer of each element. Since the center of D_8 is already in every centralizer and powers of every element commute with the element, we can quickly write down the centralizer of each element. These are, of course, the stabilizers of the conjugation action:

$$\mathbf{C}_{D_8}(e) = \mathbf{C}_{D_8}(a^2) = D_8, \quad \mathbf{C}_{D_8}(a) = \mathbf{C}_{D_8}(a^3) = \langle a \rangle,$$

$$\mathbf{C}_{D_8}(b) = \mathbf{C}_{D_8}(a^2 b) = \langle a^2, b \rangle, \quad \mathbf{C}_{D_8}(ab) = \mathbf{C}_{D_8}(a^3 b) = \langle a^2, ab \rangle.$$

Example 6.13. To show how we could use the relation between conjugacy classes and centralizers, we pose the following question:

Question 6.14. Assume that G is a group of order 8. What are the possibilities for $|\mathbf{Z}(G)|$?

If you ask someone off the street about this, they, not knowing any group theory, would have to say the answer is some integer between 1 and 8.

However, Lagrange's theorem says that, since $\mathbf{Z}(G)$ is a subgroup of G, the only possibilities are 1, 2, 4, and 8.

We can think of a number of groups of order 8: D_8, Q_8, $\mathbb{Z}/8\mathbb{Z}$, $\mathbb{Z}/4\mathbb{Z} \times \mathbb{Z}/2\mathbb{Z}$, and $\mathbb{Z}/2\mathbb{Z} \times \mathbb{Z}/2\mathbb{Z} \times \mathbb{Z}/2\mathbb{Z}$. The last three are abelian and hence their center has size 8. D_8 has a center with two elements. So we know that 2 and 8 are real possibilities for the answer. What about 1 and 4?

If $|\mathbf{Z}(G)| = 4$, then $|G : \mathbf{Z}(G)| = 2$, a prime number. This contradicts Proposition 5.23, and so $|\mathbf{Z}(G)| \neq 4$.

Now assume $|\mathbf{Z}(G)| = 1$. What are the sizes of conjugacy classes of G? These have to be numbers dividing 8 and hence they can be $1, 2, 4$ or 8. An element is in a conjugacy class of size 1 if and only if that element is in the center (Corollary 6.11). Thus there exists exactly one class of size 1. On the other hand, conjugacy classes partition G, and so the sum of the sizes of conjugacy classes must add up to 8. Can we have positive divisors of 8 greater than 1 add up to 7? The answer is no, since 7 is odd, and such a sum would necessarily be even. Thus $|\mathbf{Z}(G)| \neq 1$.

We thus conclude that a group of order 8 is either abelian or has a center of size 2!

Conjugate Elements in the Multiplication Table.[1] Given the multiplication table of a group, can we recognize conjugate elements easily and with no calculation? In fact, we can. In producing a multiplication table, we first have to choose an order for the elements. Assume that x is the ith element and y is the jth element. Then the (i, j) entry of the table is xy while the (j, i) entry of the table is yx. The point is that these two elements are conjugate, and, in fact, every pair of conjugate elements appears in this fashion. In other words, two elements of a group are conjugate if and only if they are situated, somewhere in the group multiplication table, symmetrically relative to the main diagonal.

Proposition 6.15. *Let G be a group, and let $a, b \in G$. Then a and b are conjugate elements if and only if there exists elements $x, y \in G$ with $xy = a$ and $yx = b$.*

Proof. First note that $yx = y(xy)y^{-1}$, and so, for all $x, y \in G$, xy and yx are conjugate elements.

We now have to show that if a and b are conjugate elements of G, then there exists $x, y \in G$ with $xy = a$ and $yx = b$.

If a and b are conjugate elements, then there exists $c \in G$ with $b = cac^{-1}$. Now let $x = c^{-1}$ and $y = ca$. Then $xy = a$ and $yx = cac^{-1} = b$.   □

Problems

6.2.1. As usual let $D_8 = \langle a, b \mid a^4 = b^2 = e,\ ba = a^3 b \rangle$ and $S_3 = \langle f, g \mid f^3 = g^2 = e,\ gf = f^2 g \rangle$. Let $G = D_8 \times S_3$.
   (a) Find $\mathbf{C}_G((b, e))$, $\mathbf{C}_G((a, f))$, and $\mathbf{Z}(G)$.
   (b) Find $\mathrm{cl}_G((b, e))$ and $\mathrm{cl}_G((a, f))$.
   (c) Let $H = \langle a \rangle \times \langle f \rangle$. H is a subgroup of G. Find $|G : H|$. Find the left and right cosets of H in G. What is $\mathbf{N}_G(H)$?

6.2.2. Let $G = D_8 \times \mathbb{Z}/4\mathbb{Z}$, and let a be an element of D_8 with order 4.
   (a) What is $|G|$?
   (b) What is $\langle (a^2, 3) \rangle$?
   (c) Is $\{(e, 0), (a^2, 2)\}$ a subgroup of G?
   (d) Let $A = \langle (a, 1) \rangle$, and $B = \langle (a^2, 3) \rangle$. Find two elements of AB that are in neither A nor B.
   (e) Find $\mathbf{C}_G((a, 1))$, the centralizer of $(a, 1)$ in G.
   (f) Find $|\mathrm{cl}_G((a, 1))|$, the number of elements in the conjugacy class of $(a, 1)$.

6.2.3. As usual, $S_3 = \{e, (1\,2\,3), (1\,3\,2), (1\,2), (1\,3), (2\,3)\}$ is the symmetric group on three letters, and $D_8 = \{e, a, a^2, a^3, b, ab, a^2 b, a^3 b\}$ is the dihedral group of order 8. Let $G = S_3 \times D_8$. Let $x = (\,(1\,2)\,, a)$ and $y = (\,(1\,2\,3)\,, ab)$ be elements of G. Also let $A = \langle x \rangle$ and $B = \langle y \rangle$.
   (a) What are the elements of B?
   (b) What is $|G : A|$?
   (c) Find two elements of AB that are in neither A nor B.

---

[1] From Golomb [**Gol96**].

(d) Find $\mathbf{C}_G(x)$, the centralizer of x in G.

(e) Find $|\mathrm{cl}_G(x)|$, the number of elements in the conjugacy class of x.

6.2.4. Let G be a group, and let $x \in G$. Define a map $\phi : G \to G$ by $\phi(g) = xgx^{-1}$. Is ϕ an isomorphism?

6.2.5. Assume G has a unique element x of order 2. Prove that $x \in \mathbf{Z}(G)$.

6.2.6. Write down the multiplication table for the quaternion group of order 8, Q_8. (See Definition 2.54.) Using Proposition 6.15, find the conjugacy classes of Q_8.

6.2.7. Let a, b, u, and v be elements of a group G. Assume that $uv = a$ and $vu = b$, show that $ubu^{-1} = a$.

6.2.8. Let G be a finite group, and let a and b be two conjugate elements of G. Assume that $a = gbg^{-1}$ for some $g \in G$, and let C denote $\mathbf{C}_G(a)$, the centralizer of a in G. Show that

$$\{(u, v) \in G \times G \mid uv = a, vu = b\} = \{(x, x^{-1}a) \mid x \in Cg\}.$$

Conclude that the number of pairs u, v of elements of G with $uv = a$ and $vu = b$ is the size of the centralizer of a in G.[2]

6.2.9. Let D_{2n} be the dihedral group of order $2n$:

$$D_{2n} = \langle a, b \mid a^n = b^2 = e, bab = a^{-1} \rangle.$$

Show that

(a) $a^i b a^{-i} = a^{2i} b$.

(b) If n is odd, then b is conjugate to ba.

(c) If n is even, then $z = a^{n/2} \in \mathbf{Z}(D_{2n})$, and

   (i) if $n/2$ is odd, then zb is conjugate to ba, and

   (ii) if $n/2$ is even, then zb is conjugate to b.

Conjugacy Classes in A_n. In Problem **4.4.15**—see also the discussion preceding the problem—you proved that two elements of S_n are conjugate if and only if they have the same cycle type. Consider the element $x = (1\ 2)(3\ 4) \in S_5$. The cycle type of this element is $1, 2, 2$—two 2-cycles and one, namely the unwritten (5), 1-cycle. So, we know that the conjugacy class of x consists precisely of those elements of S_5 that have cycle type $1, 2, 2$. We can count how many of these there are. There are five choices for the 1-cycle, and for each of these choices there are three choices for the rest. Hence, the size of the conjugacy class of x is 15. By Corollary 6.10, we now know that the size of the centralizer of x is $|S_5|/15 = 8$. So eight elements in S_5 commute with x.

But x is an even permutation, and so it is also an element of A_5. How big is the conjugacy class of x in A_5?

If y is conjugate to x in A_5, then y is also conjugate to x in S_5. But the converse may not be true. It may be that in S_5, whenever we had $y = gxg^{-1}$, the element g was an odd permutation. In such a case, x would not be conjugate to y in A_5. We

---

[2]Adapted from Golomb [**Gol96**].

conclude that some of the conjugacy classes of S_5 may split up to smaller conjugacy classes in A_5.

For the particular element x, we can decide the size of its conjugacy class in A_5 by first finding the size of its centralizer, and then using Corollary 6.10. We had that $|\mathbf{C}_{S_5}(x)| = 8$, but the size of A_5 is 60 and this is not divisible by 8. Hence, the number of elements in A_5 that commute with x must be ≤ 4 (we used Lagrange's theorem here). By Corollary 6.10 $|\mathrm{cl}_{A_5}(x)| \geq 15$. But, since the number of conjugates of x in S_5 is exactly 15, we now conclude that $|\mathrm{cl}_{A_5}(x)| = 15$ and, as a consequence, $|\mathbf{C}_{A_5}(x)| = 4$.

6.2.10. Let $n \geq 3$. Prove that $\mathbf{Z}(S_n) = \{e\}$.

6.2.11. Let $g = (1\ 2\ 3)$. How many elements of S_4 commute with g? What about S_5?

6.2.12. Let $n \geq 2$, and let t be some integer no larger than $n/2$. Assume $\sigma \in S_n$ is a product of t disjoint 2-cycles. What is the size of the conjugacy class of σ (in terms of t and n)? How many elements of S_n commute with σ?

6.2.13. Find the number of conjugacy classes of S_5 and the number of elements in each class.

6.2.14. Find the number of conjugacy classes of A_4 and the number of elements in each class.

6.2.15. Find the number of conjugacy classes of A_5 and the number of elements in each class.

6.2.16. Let $G = S_n$ be the symmetric group on n letters, and let $g \in G$. Assume that the cycle type of g is 1^{m_1}, 2^{m_2}, ..., n^{m_n}. In other words, for $1 \leq i \leq n$, g has—in its cycle decomposition—m_i cycles of size i. Find an expression for $|\mathbf{C}_G(g)|$ and $|\mathrm{cl}_G(g)|$ in terms of m_1, ..., m_n.

6.2.17. **Conjugacy in** $\mathrm{GL}(2,p)$**.** Let p be a prime, and let λ_1 and λ_2 be distinct non-zero elements in $\mathbb{Z}/p\mathbb{Z}$. Let

$$A = \begin{bmatrix} \lambda_1 & 0 \\ 0 & \lambda_2 \end{bmatrix} \in \mathrm{GL}(2,p).$$

(a) Find $\mathbf{C}_{\mathrm{GL}(2,p)}(A)$.
(b) Find $\left|\mathbf{C}_{\mathrm{GL}(2,p)}(A)\right|$.
(c) Use Corollary 6.10 (and Problem 1.4.4) to find the number of 2×2 matrices conjugate with A in $\mathrm{GL}(2,p)$.

6.3. The Class Equation and Groups of Order p^2

In this section we will explore some amazing applications of the results of the last section on the conjugation action. Let p be a prime number. We have proved that if the order of a group is p, then the group must be cyclic. We will prove in this section that if the order of the group is p^2, then the group must be abelian!

Furthermore, if the order of the group is any power of p, then the group must have elements other than identity that commute with every element (i.e., the center is not trivial).

You should also pay attention to the fact that, in the proof of these results, we will not resort to element-by-element calculations. We will depend on the limitations on the sizes of subgroups and sizes of conjugacy classes. The first two results follow directly from the fact that the size of the conjugacy class of an element is equal to the index of its centralizer, and the fact that conjugacy classes (being orbits of an action on the group) partition the group.

Corollary 6.16. *Let G be a finite group. Let $x_1, x_2, \ldots, x_n$ be representatives of the distinct conjugacy classes of G (i.e., pick one element from each conjugacy class). Then*

$$|G| = \sum_{i=1}^{n} |\mathrm{cl}_G(x_i)|$$

$$= \sum_{i=1}^{n} |G : \mathbf{C}_G(x_i)|.$$

Proof. The first formula says that the number of the elements of the group is the same as the sum of the number of elements in each conjugacy class. This is true since orbits of an action partition the set and conjugacy classes are the orbits of the conjugation action. The second formula follows from the first since, by the FCP, we have $|\mathrm{cl}_G(x_i)| = |G : \mathbf{C}_G(x_i)|$. □

Now we modify the above formula for $|G|$ a little bit. Among the conjugacy classes there are some that are of size 1. These are exactly the elements that are in the center. If we add the sizes of these classes, we get exactly the size of the center. This modification gives what is called the *class equation* for the group G.

Corollary 6.17 (Class equation). *Let $x_1, x_2, \ldots, x_n$ be representatives of distinct conjugacy classes of G that have more than one element (i.e., pick one element from each conjugacy class with more than one element). Then*

$$|G| = |\mathbf{Z}(G)| + \sum_{i=1}^{n} |\mathrm{cl}_G(x_i)|$$

$$= |\mathbf{Z}(G)| + \sum_{i=1}^{n} |G : \mathbf{C}_G(x_i)|. \quad □$$

We have built the theory a step at a time, and thus maybe it looks like we have not done much. But here is a somewhat surprising result. Every group whose order is a power of a prime has a non-identity element in its center!

Definition 6.18 (*p*-group). *Let p be a prime integer, and let n be a non-negative integer. Then a group of order p^n is called a *p-group*.*

As a warmup for the next proof, you may want to reread Question 6.14 and its solution.

Corollary 6.19. *Let G be a p-group. Then $|\mathbf{Z}(G)| > 1$.*

Proof. Assume $\mathbf{Z}(G) = \{e\}$. Let x_1, ..., x_n be representatives from conjugacy classes with more than one element. The class equation now becomes

$$(6.1) \qquad\qquad |G| = 1 + \sum_{i=1}^{n} |\mathrm{cl}_G(x_i)|.$$

Now for each i, $|\mathrm{cl}_G(x_i)|$ divides the order of G (since by the FCP, this size times the size of the centralizer is $|G|$), and hence it is a non-trivial power of p (it is non-trivial since we are assuming that all the conjugacy classes other than the identity have size bigger than one).

We have that p divides $|G|$ and p divides $|\mathrm{cl}_G(x_i)|$, for each i, and hence by the class equation, Corollary 6.17, p divides 1. This is a contradiction which proves that $|\mathbf{Z}(G)| > 1$. $\qquad\square$

Corollary 6.20. *Let G be a group of order p^2, where p is a prime. Then G is abelian.*

Proof. By the previous corollary, $|\mathbf{Z}(G)| > 1$. Thus by Lagrange's theorem $|\mathbf{Z}(G)| = p$ or p^2. If $|\mathbf{Z}(G)| = p$, then $|G : \mathbf{Z}(G)| = p$ which is a prime number. This contradicts Proposition 5.23, and so $|\mathbf{Z}(G)| \neq p$. The only other possibility is that the size of the center is p^2. But this means that every element is in the center, and so the group G must be abelian. $\qquad\square$

Groups of order p^2. It is now easy to classify all groups of order p^2. This is because the fundamental theorem of abelian groups (Theorem 13.12) says that every finite abelian group is a direct product of cyclic groups. Thus if G is a finite abelian group, then $G \cong \mathbb{Z}/n_1\mathbb{Z} \times \mathbb{Z}/n_2\mathbb{Z} \times \cdots \times \mathbb{Z}/n_k\mathbb{Z}$ where n_1, ..., n_k are positive integers. It follows then that any group of order p^2 is isomorphic to either $\mathbb{Z}/p^2\mathbb{Z}$ or $\mathbb{Z}/p\mathbb{Z} \times \mathbb{Z}/p\mathbb{Z}$. A direct proof of the fundamental theorem of abelian groups is given in (the optional) Chapter 13. The fundamental theorem of abelian groups is also a consequence of more general results on so-called modules.

Problems

6.3.1. Let X be a set with 47 elements. Choose 169 one-to-one and onto functions from X to X such that if you compose any two of the functions, you get another function in the set. Prove (one of the problems highlighted in the Preface) the following.
   (a) If f and g are two of your functions, then $f(g(x)) = g(f(x))$ for all $x \in X$.
   (b) If f is any of your functions and if you start composing f with itself, then after 169 iterations you will always get the identity function.

6.3.2. Let G be a non-abelian group of order 27. What are the possibilities for $|\mathbf{Z}(G)|$? Justify your answer.

6.3.3. Let G be a group of order p^3. Prove that either G is abelian or its center has exactly p elements.

6.3.4. A group P acts on a set Ω. We know that $|P| = 81$ and $|\Omega| = 98$. Let Ω_0 be the set of elements of Ω that are fixed by every element of P. In other words,

$$\Omega_0 = \{\alpha \in \Omega \mid \alpha \cdot g = \alpha \text{ for all } g \in P\}.$$

Show $|\Omega_0| = 3k + 2$ for some integer k with $0 \le k \le 32$.

6.3.5. Let p be a prime, and let P be a p-group. Assume that P acts on a finite set Ω, and let f be the number of elements of Ω that are fixed by every element of P. Show that $|\Omega| \equiv f \bmod p$.

6.4. More Problems and Projects

Problems

6.4.1. **Probability that two random elements of a group commute.** Recall that, on page 126, we defined, for a finite group G, the two quantities $c(G) = |\{(x, y) \mid x, y \in G, xy = yx\}|$ and $p_c(G) = c(G)/|G|^2$. The latter is the probability that two randomly chosen elements from the group G will commute. Prove that

$$p_c(G) = \frac{\text{the number of conjugacy classes of } G}{|G|}.$$

Use this to find $p_c(D_8)$ and $p_c(S_3)$.

Skew Centralizer and Reversing Symmetry Groups. Let G be a group, and let $a \in G$. In analogy with the centralizer of a in G, we define the *skew centralizer* of a in G by

$$B_G(a) = \{x \in G \mid xa = a^{-1}x\}.$$

We also define the *reversing symmetry group* of a in G by

$$E_G(a) = \mathbf{C}_G(a) \cup B_G(a).$$

In Problems 6.4.2–6.4.6—which are adapted from Goodson [**Goo99**]—you are asked to show that $E_G(a)$ is always a subgroup of G and that, if $B_G(a) \ne \emptyset$ and $a^2 \ne e$, then $|E_G(a) : \mathbf{C}_G(a)| = 2$. See Goodson [**Goo99**] for the relation of skew centralizers and reversing symmetry groups to dynamical systems.

6.4.2. Let G be a group, let $a \in G$, and let $B_G(a)$ be the skew centralizer of a in G.
  (a) Give an example where $B_G(a)$ is not a subgroup of G.
  (b) Show that $B_G(a)$ is not empty if and only if a is conjugate to a^{-1}.

6.4.3. Let G be a group, let $a \in G$, and let $B_G(a)$ be the skew centralizer of a in G. Assume that $B_G(a) \neq \emptyset$, then show that the following are equivalent:
  (a) $B_G(a)$ is a subgroup of G;
  (b) a is an involution, i.e., $a^2 = e$;
  (c) $B_G(a) = \mathbf{C}_G(a)$;
  (d) $B_G(a) \cap \mathbf{C}_G(a) \neq \emptyset$.

6.4.4. Let G be a group, let $a \in G$, let $B_G(a)$ be the skew centralizer of a in G, and let $E_G(a)$ be the reversing symmetry group of a. Show that $E_G(a)$ is always a subgroup of G.

6.4.5. Let G be a group, let $a \in G$, let $B_G(a)$ be the skew centralizer of a in G, and let $E_G(a)$ be the reversing symmetry group of a. Assume $B_G(a) \neq \emptyset$ and $a^2 \neq e$. Show that the cosets of $\mathbf{C}_G(a)$ in $E_G(a)$ are $\mathbf{C}_G(a)$ and $B_G(a)$, and conclude that $|E_G(a) : \mathbf{C}_G(a)| = 2$.

6.4.6. Let G be a finite group of odd order, and let $e \neq a \in G$. Prove that a cannot be conjugate to a^{-1}.

Chapter 7

Acting on Subsets, Cosets, and Subgroups: The Sylow Theorems

> ...where by further exploiting group actions, the Sylow theorems are proved, and it is shown that every finite group is guaranteed to have subgroups of certain sizes and elements of certain orders.

Let G be a group of order 24, and let H be a subgroup of G. By Lagrange's theorem, the possible orders of H are 1, 2, 3, 4, 6, 8, 12, and 24. We know that G will *always* have subgroups of order 1 and 24, but does it have to have subgroups of the other orders? Not necessarily, since, in general, the converse of Lagrange's theorem is not true. In other words, there is no guarantee that there are subgroups for every order dividing the order of G. In the problems we have seen examples of this. For example, A_4—a group of order 12—does not have a subgroup of order 6 (Problems **5.1.19** or 5.4.6), and $SL(2,3)$—a group of order 24—does not have a subgroup of order 12 (Problem 5.4.1).

It is surprising and one of the deeper results of elementary group theory that we can guarantee the existence of subgroups of all prime powers that divide the order of a group. In other words, if p is a prime, a a positive integer, and p^a divides the order of a finite group G, then G is guaranteed to have a subgroup of order p^a. The main part of this result—that a group is guaranteed to have a subgroup of size p^a if p^a is the *highest* power of p dividing the size of the group—is called *Sylow's existence theorem*, and is the subject of this chapter. As a quick consequence, we get that if a prime p divides the order of a group, then the group must have an element of order p. This is *Cauchy's theorem* and will be used often. As an example, after the results of this chapter, we know that any group of order 24 will be guaranteed to have subgroups of order 1, 2, 3, 4, and 8 (and, of course, 24) and elements of order 2 and 3.

The subgroups guaranteed by Sylow's theorem—those whose order is the highest power of a prime dividing the order of a group—are called Sylow subgroups of the group, and we can, in fact, say more about them. In this chapter, you are asked to complete an outline of the proof that all Sylow subgroups of a group for a given prime are conjugate and that every subgroup whose order is a power of a prime is inside a Sylow subgroup of the group. Other important results on Sylow subgroups will come after Chapter 10. In fact, a full appreciation of the power of Sylow theorems in analyzing the structure of finite groups will have to wait until Chapter 12.

All the main proofs in this chapter will utilize interesting actions. But before we begin, we need some preliminaries on binomial coefficients, which we will cover in the first section.

7.1. Binomial Coefficients mod p

Recall—see Definition 6.6—that the binomial coefficient $\binom{n}{k}$ denotes the number of subsets of size k of a set with n elements.

Lemma 7.1. *Let n and k be positive integers with $n \geq k$, and let x and y denote indeterminates. Then the binomial coefficients satisfy the following basic identities:*

$$(7.1) \qquad \binom{n}{k} = \frac{n!}{k!(n-k)!},$$

$$(7.2) \qquad \binom{n}{k} = \binom{n-1}{k} + \binom{n-1}{k-1},$$

$$(7.3) \qquad (x+y)^n = \sum_{k=0}^{n} \binom{n}{k} x^k y^{n-k}.$$

Proof. Let $[n] = \{1, 2, \ldots, n\}$ be our prototype of a set with n elements. Identity (7.1) was already proved in Lemma 6.7. For a more direct proof, count the number of k-permutations of $[n]$ (an ordering of k elements of $[n]$) in two different ways. First, pick k elements out of $[n]$ and then find all their permutations. This results in $\binom{n}{k} k!$. Second, build a k-permutation directly. There are n choices for the first element of the permutation, $n-1$ choices for the second one, and so on. This gives $n(n-1)(n-2)\cdots(n-k+1) = \frac{n!}{(n-k)!}$. Equating the two counts gives us (7.1).

The subsets of $[n]$ with k elements come in two varieties. Those that include n and those that do not. The number of subsets of size k that include n is $\binom{n-1}{k-1}$ since we have to pick $k-1$ elements from $[n-1]$. The number of subsets of size k that do not include n is $\binom{n-1}{k}$ since we have to pick k elements from $[n-1]$. Identity (7.2) now follows.

To prove (7.3), we expand $(x+y)^n = (x+y)(x+y)\cdots(x+y)$. A typical term of the product is gotten by picking either x or y from each parenthesis. If x is picked k times, then the resulting product will be $x^k y^{n-k}$. How many times does this term arise? From among the n parentheses we have to pick k—to choose the ones from which we are going to take x—and hence the answer is $\binom{n}{k}$. Thus after collecting like terms, a typical term will be $\binom{n}{k} x^k y^{n-k}$. $\qquad \square$

Identity (7.1) gives the formula for $\binom{n}{k}$, while (7.2) is known as Pascal's identity as it demonstrates that binomial coefficients are exactly the numbers in the so-called "Pascal's triangle".[1] Finally, (7.3) explains the word "binomial coefficient" by demonstrating that the binomial coefficients are exactly the numbers that appear when expanding a power of a "binomial". There are many identities involving binomial coefficients and most of them can be proved using one of the basic identities of Lemma 7.1.

Definition 7.2. Let $f(x) = a_0 + a_1 x + \cdots + a_n x^n$ and $g(x) = b_0 + b_1 x + \cdots + b_n x^n$ be polynomials of degree n with integer coefficients. We write

$$f(x) \equiv g(x) \bmod p$$

if, for $0 \leq i \leq n$, we have $a_i \equiv b_i \bmod p$.

Example 7.3.

$$7x^2 - 3x + 5 \equiv x^2 - 1 \bmod 3.$$

Lemma 7.4. *Let p be a prime number.*

(a) *The binomial coefficient $\binom{p}{k}$ is divisible by p, for $1 \leq k \leq p-1$.*

(b)

$$(x+1)^p \equiv (x^p + 1) \bmod p.$$

(c) *Let a be a positive integer. Then*

$$(x+1)^{p^a} \equiv (x^{p^a} + 1) \bmod p.$$

Proof. (a) We have

$$\binom{p}{k} = \frac{p!}{k!(p-k)!}.$$

For $1 \leq k \leq p-1$, the prime number p divides the numerator but it does not divide the denominator. Hence $p \mid \binom{p}{k}$.

(b) By Identity (7.3),

$$(x+1)^p = \sum_{k=0}^{p} \binom{p}{k} x^k.$$

By the previous part, except for the first and the last term, all other terms of this expansion have a coefficient that is divisible by p. Hence

$$(x+1)^p \equiv x^p + 1 \bmod p.$$

---

[1]In 1654, Blaise Pascal wrote *Traité du Triangle Arithmétique* which offered a systematic treatment of the triangle of binomial coefficients. In the early eighteenth century, French mathematicians began to refer to the triangle as "Pascal's triangle", and the designation has become standard. However, it should be noted that the binomial coefficients, their properties, and this specific triangle had been known to and used by mathematicians from China, India, Iran, the Islamic world, the Hebrew tradition, and northern Africa for hundreds of years before Pascal. One of the earliest known descriptions of the triangle and its properties is due to Abu Bekr Karaji (or al-Karaji if one prefers—as is common—to include the Arabic definite article as part of his name) who worked in Baghdad around the year 1000 CE. Berggren [**Ber86**, p. 58] suggests that the triangle "might with more justice be called al-Kariji's triangle".

(c) Using the previous part repeatedly, we have

$$(x+1)^{p^2} = [(x+1)^p]^p \equiv (x^p+1)^p \bmod p \equiv (x^{p^2}+1) \bmod p.$$

To get the desired identity, continue as above or, for a more formal proof, use induction on a. □

Lemma 7.5. *Let p be a prime number, and let m be a positive integer. Assume $n = p^a m$, where a is a non-negative integer. Then*

$$\binom{n}{p^a} \equiv m \bmod p.$$

In particular, if m is not divisible by p, then neither is $\binom{n}{p^a}$.

Proof. By identity (7.3), $\binom{n}{p^a}$ is the coefficient of x^{p^a} in the expansion of $(x+1)^n$. We are interested in the remainder of $\binom{n}{p^a}$ when divided by p. Hence, we should find the coefficient of x^{p^a} in $(x+1)^n \bmod p$. Using Lemma 7.4(c),

$$(x+1)^n = (x+1)^{p^a m} = ((x+1)^{p^a})^m \equiv (x^{p^a}+1)^m \bmod p,$$

and the coefficient of x^{p^a} in $(x^{p^a}+1)^m$ is $\binom{m}{1} = m$. Thus the coefficient of x^{p^a} in $(x+1)^n \bmod p$ is m, and the result follows. □

Example 7.6. In Lemma 7.5, let $p = 2$ and $n = 20$, and conclude that the binomial coefficient $\binom{20}{4}$ is odd.

Problems

7.1.1. Assume that p is a prime number and you know that it divides the sum

$$\sum_{k=0}^{2011} \binom{2011}{k} 10^k.$$

What can you say about p?

7.1.2. Is $\binom{238}{13}$ odd or even?

You could use the following steps:

Step 1: Write 238 as a sum of powers of 2. Use it to write $(1+x)^{238}$ as a product of terms of the form $(1+x)^a$ where a is a power of 2.

Step 2: Using the previous step, and Lemma 7.4(c), show that

$$(1+x)^{238} = (1+x^{128})(1+x^{64})(1+x^?)(1+x^?)(1+x^?)(1+x^2) \mod 2.$$

Step 3: Use the previous step to find the coefficient of x^{13} in $(1+x)^{238}$ mod 2.

Step 4: Argue that the coefficient of x^{13} in $(1+x)^{238}$ mod 2 gives us the desired answer, and finish the problem using the previous step.

7.1.3. Similarly to Problem 7.1.2, write 69 as a linear combination of powers of 5, and use this to find the coefficient of x^{37} in $(1+x)^{69}$ mod 5. What is $\binom{69}{37}$ mod 5?

7.1.4. **Lucas's theorem.** Using Problems 7.1.2 and 7.1.3 as templates, prove Lucas's theorem:

Theorem 7.7 (Lucas). *Let p be a prime, and let n and k be positive integers. Write $n = b_0 + b_1 p + \cdots + b_r p^r$ and $k = c_0 + c_1 p + \cdots + c_r p^r$, where, for $1 \leq i \leq r$, b_i and c_i are non-negative integers. Then*

$$\binom{n}{k} = \binom{b_r}{c_r}\binom{b_{r-1}}{c_{r-1}} \cdots \binom{b_1}{c_1}\binom{b_0}{c_0} \quad \text{mod } p.$$

7.2. The Sylow E(xistence) Theorem

Recall that a p-group is a group whose order is a power of a prime p. For each prime p, the largest p-subgroup of a group is called its Sylow p-subgroup.

Definition 7.8 (Sylow subgroups). Let G be a finite group with $|G| = p^a m$, where $p \nmid m$, and a is a non-negative integer. A subgroup $P \leq G$ with $|P| = p^a$ is called a *Sylow p-subgroup* of G.

The set of all Sylow p-subgroups of G is denoted by $\mathrm{Syl}_p(G)$.

Example 7.9. If G is a group of order 24, then a subgroup of order 3 would be one of its Sylow 3-subgroups and a subgroup of order 8 would be one of its Sylow 2-subgroups. The identity subgroup would be its Sylow 5-subgroup as well as its Sylow 11-subgroup. While we know that G has the identity subgroup, we need Sylow's theorem to know that G actually must have a subgroup of order 3 and a subgroup of order 8.

Theorem 7.10 (Sylow E(xistence) theorem). *If G is a finite group and p is a prime, then G is guaranteed to have a Sylow p-subgroup.*

Proof due to Wielandt. Write $|G| = p^a m$, where a is a non-negative integer and p does not divide m. We want an $H \leq G$ with $|H| = p^a$.

Let $\Omega = \{X \subseteq G \mid |X| = p^a\}$. So Ω is the set of all *subsets* of size p^a. Certainly, if a Sylow p-subgroup exists—and we are trying to prove that it does—it should be one of the elements of Ω. We will, however, produce this subgroup in a roundabout and somewhat surprising way.

By Lemma 7.5, $|\Omega| = \binom{n}{p^a} \equiv m \mod p$, and so p does not divide the size of Ω.

Let G act on Ω by $g \cdot X = gX$. It is clear that this does define an action. Hence the orbits of the action partition Ω. So the sum of the sizes of the orbits is $|\Omega|$, and $p \nmid |\Omega|$. Thus, there exists at least one orbit $\mathcal{O}$ with $p \nmid |\mathcal{O}|$.

Let $X \in \mathcal{O}$, so that $\mathcal{O}$ is the orbit, $\mathcal{O}_\Omega(X)$, of X in Ω.

CLAIM: $\mathrm{Stab}_G(X) \in \mathrm{Syl}_p(G)$.

PROOF OF THE CLAIM: The FCP implies that

$$|\mathcal{O}_\Omega(X)| = \frac{|G|}{|\mathrm{Stab}_G(X)|}.$$

Since $p \nmid |\mathcal{O}_\Omega(X)|$, we have $p^a \mid |\mathrm{Stab}_G(X)|$, so $p^a \leq |\mathrm{Stab}_G(X)|$.

Fix $x \in X$. If $g \in \mathrm{Stab}_G(X)$, then $g \cdot X = X$, and so $gx \in X$. We conclude that $\mathrm{Stab}_G(X)x \subseteq X$. So $|\mathrm{Stab}_G(X)| = |\mathrm{Stab}_G(X)x| \leq |X| = p^a$.

From the two opposite inequalities, we get that $|\mathrm{Stab}_G(X)| = p^a$. We conclude that $\mathrm{Stab}_G(X)$ is a subgroup of size p^a. Hence it is a Sylow p-subgroup of the group G. □

This theorem has many consequences. For now, we shall limit ourselves to one:

Corollary 7.11 (Cauchy). *Let G be a finite group, and let p be a prime number. Assume p divides $|G|$. Then there exists $g \in G$ with $o(g) = p$.*

Proof. Let $S \in \mathrm{Syl}_p(G)$. We know $|S| = p^a$ for some positive integer a. Let x be a non-identity element of S. Since $o(x)$ divides $|S|$ (Corollary 5.14 to Lagrange's theorem), we have $o(x) = p^b$ for some positive integer b. Now let $g = x^{p^{b-1}} \in G$. Then it follows that $o(g) = p$. □

Example 7.12. Let G be a group of order $500 = 2^2 \times 5^3$. Sylow's theorem tells us that G has subgroups of order 2^2 and 5^3. Cauchy's theorem tells us that there are elements (and hence subgroups) of order 2 and 5. Thus we know that a group of order 500 must have subgroups of orders 1, 2, 2^2, 5, and 5^3. In fact, later we shall prove that there also will be a subgroup of order 5^2.

Problems

7.2.1. Let p and q be two different primes, let G be a finite group, let $P \in \mathrm{Syl}_p(G)$, and let $Q \in \mathrm{Syl}_q(G)$. What can you say about $P \cap Q$?

7.2.2. Find all Sylow 2 and Sylow 3-subgroups of A_4. Find a familiar group that is isomorphic to a Sylow 2-subgroup of A_4.

7.2.3. (a) For each prime divisor p of $|S_4|$, find a Sylow p-subgroup of S_4. For each Sylow p-subgroup P find a well-known group that is isomorphic to P.
   (b) Do the same for S_5.

7.2.4. Let $G = \mathbb{Z}/10\mathbb{Z} \times \mathbb{Z}/10\mathbb{Z}$ be the direct product of $(\mathbb{Z}/10\mathbb{Z}, +)$ with itself. Find a Sylow 2 and a Sylow 5-subgroup of G. Find familiar groups that are isomorphic to the subgroups that you found.

7.2.5. Let G be an abelian group of order 338. Let t be the number of elements of order 2 in G. Cauchy's theorem says that $t \geq 1$. What are the possible values for t? Why?

7.2.6. Let G be a group of order $539 = 7^2 \times 11$. Assume that G acts on a set with ten elements and that there is some orbit of size bigger than 1.
   (a) What can you say about the orbit sizes? Why?
   (b) G is guaranteed to have subgroups of which sizes? Give reasons or proofs for your assertions.

7.2.7. Let p be a fixed prime. Let P be a Sylow p-subgroup of a finite group G.

(a) Let x be an arbitrary element of G. Under what conditions is xPx^{-1} also a Sylow p-subgroup of G?

(b) Assume that P is the unique Sylow p-subgroup of G. Show that, for every $x \in G$, we have $xP = Px$.

7.2.8. Let G be a group of order 21. Can G be abelian and yet not cyclic? Prove your assertion.

7.2.9. We have a group of order 21. What are the possibilities for $|\mathbf{Z}(G)|$? Give reasons for your assertions.

7.2.10. **Abelian groups of order** pq. Let p and q be distinct prime numbers, and assume that G is an abelian group of order pq. Prove that G is isomorphic to $\mathbb{Z}/pq\mathbb{Z}$.

7.2.11. Let $\sigma = (1\ 2\ 4\ 6)$, $\tau = (1\ 4)$, and $\delta = (3\ 5)$ be elements of S_6.

(a) Find a familiar group that is isomorphic to $\langle \sigma, \tau \rangle$.

(b) Find a familiar group that is isomorphic to $\langle \sigma, \tau, \delta \rangle$.

(c) Find a Sylow 2-subgroup of S_6 and give a familiar group that is isomorphic to it.

7.2.12. **Sylow p-subgroups of** $\mathrm{SL}(n,p)$ **and** $\mathrm{GL}(n,p)$.

(a) What is the order of a Sylow 11-subgroup of $\mathrm{SL}(2,11)$?

(b) Prove that the set of lower triangular 2×2 matrices with ones on the diagonal is a Sylow 11-subgroup of $\mathrm{SL}(2,11)$.

(c) Find a Sylow p-subgroup for $\mathrm{SL}(2,p)$.

(d) Generalize your example of the previous part to find a Sylow p-subgroup of $\mathrm{SL}(n,p)$.

(e) What is the order of a Sylow p-subgroup of $\mathrm{GL}(n,p)$? Can you find such a Sylow p-subgroup?

7.3. The Number and Conjugacy of Sylow Subgroups*

The existence of Sylow subgroups—which we have proved—is very useful in studying the structure of finite groups. A number of other related results—collectively called the *Sylow theorems*—give us additional information about the Sylow p-subgroups and about $\left|\mathrm{Syl}_p(G)\right|$, the number of Sylow p-subgroups of a group. These sets of theorems will complement each other and will prove very powerful. However, we will be able to appreciate the power of these theorems after we have studied normal subgroups in Chapter 10. In fact, we will come back to the Sylow theorems and their uses in Chapter 12. In this section, we will state most of the Sylow theorems, and give an outline of their proofs—each proof is another example of how one can utilize group actions. You are asked to complete the proofs in the Problems.

Completing this section, at this time, is optional. You can either complete the proofs of these theorems before you proceed—for the main theorems an outline is given in the text and you are asked to complete them in the Problems—or wait until Chapter 12 and come back to this section at that time. The advantage of waiting is that in Chapter 10 you will gain much experience with the normalizer subgroups which play a prominent role in the proofs in this section. Except for a few problems, the material in this section is not used before Chapter 12.

There are basically three Sylow theorems. We have already proved the first one—the Sylow existence theorem, often referred to as the Sylow E theorem—which asserts that every finite group G has a Sylow p-subgroup for each prime p. The second Sylow theorem is called the Sylow development theorem and is referred to as the Sylow D theorem. It asserts that given a p-subgroup Q—i.e., a subgroup Q with $|Q|$ a power of p—and a Sylow p-subgroup P, then $Q \leq gPg^{-1}$ for some $g \in G$. From this a number of results follow, including the fact that every two Sylow p-subgroups are conjugate (called the Sylow C theorem), and that the number of Sylow p-subgroups divides the order of G. The final result is the fact that the number of Sylow p-subgroups—in addition to dividing $|G|$—is 1 modulo p.

Theorem 7.13 (Sylow D(evelopment) theorem). *Let G be a finite group, let p be a prime number, and let P be a fixed Sylow p-subgroup of G. Assume $Q \leq G$ is a p-group (recall that a p-group is one whose order is a power of the prime p). Then, for some $x \in G$,*

$$Q \leq xPx^{-1}.$$

Outline of proof. We are given two subgroups of G: Q and P, where Q is a p-group and P is a Sylow p-subgroup of G. We want to show that there exists $x \in G$ such that $Q \leq xPx^{-1}$.

Let Ω be the set of left cosets of P in G. In other words, $\Omega = \{gP \mid g \in G\}$. Then Q acts on Ω by left multiplication (a relative of the regular action). We know orbits of this action partition the set Ω. Using the FCP and the fact that p does not divide $|\Omega| = |G : P|$, we can show that there is an orbit of size 1. If $\{xP\}$ is the orbit of size 1, then it can be shown that $Q \leq xPx^{-1}$. (You are asked to complete the details in Problem **7.3.1**.) □

Much follows from the Sylow D theorem. We organize some of the consequences in two corollaries:

Corollary 7.14 (Sylow C(onjugacy) theorem). *Let G be a finite group, let p be a prime number, and let P be a fixed Sylow p-subgroup of G. Then*

$$Q \in \mathrm{Syl}_p(G) \quad \text{if and only if} \quad Q = xPx^{-1} \text{ for some } x \in G.$$

As a consequence, if $P, Q \in \mathrm{Syl}_p(G)$, then P is isomorphic to Q.

Proof. ($\Leftarrow$) If $Q = xPx^{-1}$, then, by Problem **2.6.27**, $|Q| = \left|xPx^{-1}\right|$, and hence Q is a Sylow p-subgroup as well. (This actually was Problem **7.2.7**.)

($\Rightarrow$) If Q is a Sylow p-subgroup of G, then by the Sylow D theorem (Theorem 7.13), there exists $x \in G$ with $Q \leq xPx^{-1}$. But these two groups have the same size and hence must be equal. So Q is equal to a conjugate of P as claimed.

Finally, if P and Q are both Sylow p-subgroups of G, then $Q = xPx^{-1}$ for some $x \in G$ and hence the map $\phi : P \to Q$ defined by $\phi(g) = xgx^{-1}$ gives an isomorphism between P and Q. □

We now restate the Sylow C theorem (Corollary 7.14) using the language of the conjugation action of a group on its subgroups.

Corollary 7.15. *Let G be a finite group, let p be a prime number, and let $P \in$ $\mathrm{Syl}_p(G)$. The group G acts on $\Omega = \{H \mid H \leq G\}$, the set of subgroups of G, by conjugation—that is $g \cdot H = gHg^{-1}$.*

For this action, $\mathrm{Syl}_p(G)$ is the orbit of P and $\mathbf{N}_G(P)$ is the stabilizer of P. Hence

$$\left|\mathrm{Syl}_p(G)\right| = |G : \mathbf{N}_G(P)|,$$

and $\left|\mathrm{Syl}_p(G)\right|$, the number of Sylow p-subgroups of G, is a divisor of $|G|$. It further follows that

$$\mathbf{N}_G(P) = G \quad \text{if and only if } P \text{ is the unique Sylow } p\text{-subgroup of } G.$$

Proof. We already know that G acts on the set of its subgroups $\Omega = \{H \mid H \leq G\}$ by conjugation and that, for $H \leq G$, $\mathrm{Stab}_G(H)$ is called $\mathbf{N}_G(H)$ (Definition 4.14, Problem *4.1.5*, and Definition 4.24). Now if P is a Sylow p-subgroup of G, then, by the Sylow C theorem (Corollary 7.14), the orbit of P, $\mathcal{O}_\Omega(P)$, is the set of Sylow p-subgroups of G. The rest follows from the FCP (Theorem 6.1). □

The fact that the *number* of Sylow p-subgroups is a divisor of the group will end up being quite useful. However, we can say more about this integer. It also has to have remainder 1 when divided by p.

Theorem 7.16. *Let G be a finite group, and let p be a prime. Then*

$$\left|\mathrm{Syl}_p(G)\right| = 1 + kp, \quad \text{for some non-negative integer } k.$$

Outline of proof. We want to show that the number of Sylow p-subgroups—in addition to dividing $|G|$—is 1 modulo p. For this proof, we use yet another action—actually, just a modification of an already familiar one.

Let $\mathcal{S} = \mathrm{Syl}_p(G) = \{P_1, P_2, \ldots, P_s\}$ be the set of Sylow p-subgroups of G, and let $P = P_1$. We know that G acts on the set of subgroups of G by conjugation and that $\mathcal{S}$ is one of the orbits. This means that we can restrict the action of G to $\mathcal{S}$. Also, if a group acts on a set, so does each of its subgroups. Hence, we can restrict the conjugation action to an action of P on $\mathcal{S}$. In other words, P acts on $\mathcal{S}$ by $x \cdot Q = xQx^{-1}$, for all $x \in P$ and $Q \in \mathcal{S}$.

We analyze this action similarly to the way we obtained and used the Class Equation (Corollary 6.17). Since the orbits partition the set of Sylow p-subgroups, the number of Sylow p-subgroups is the sum of the sizes of orbits. Since the group acting is a p-group, by the FCP all orbits are of size 1 or a power of p. Hence, we ask how many of the Sylow subgroups are fixed by the action of P. It is straightforward to see that P itself is fixed by the action and hence is in an orbit of size 1. In fact, we can show—you are asked to prove this in Problems *7.3.5* and *7.3.7*—that P is the *only* Sylow p-subgroup fixed in this action. Hence the set $\mathcal{S}$ is partitioned into orbits, one of which has size 1 and all the others have sizes that are multiples of p. The result now follows. (You are asked to write up a complete proof in Problem *7.3.9*.) □

Example 7.17. Let G be a group of order $200 = 2^3 5^2$. By Sylow's existence theorem we know that G must have subgroups of order 8 and 25. By Cauchy's Theorem 7.11, G will also have subgroups of order 2 and 5. In fact, we will prove later (Theorem 12.1) that a group of order 8 must have a subgroup 4. Hence, G is guaranteed to have subgroups of order $1, 2, 4, 5, 8, 25$, and 200.

But we can do more. Let $s_5 = |\mathrm{Syl}_5(G)|$ be the number of Sylow 5-subgroups of G. By the Sylow theorems s_5 has remainder 1 when divided by five and yet, at the same time, has to divide $|G| = 200$. The only divisors of 200 that are not divisible by 5 are 1, 2, 4, and 8, and among these only 1 has remainder 1 when divided by 5. Thus $s_5 = 1$, and G has a unique Sylow 5-subgroup P. It follows (Corollary 7.15) that $\mathbf{N}_G(P) = G$. Let H be a subgroup of order 2. By Problem 5.1.11, HP will be a subgroup of G, and it is not hard to see that its order will be 50. In a similar fashion we can also show the existence of a subgroup of order 100.

Remark 7.18. Arguments such as those used in Example 7.17 will become clear and much more understandable after studying normal subgroups in Chapter 10. In fact, we postpone a discussion of the consequences and applications of Sylow theorems to Chapter 12 and especially Sections 12.3 and 12.4. We included all of the Sylow theorems in this early chapter because of their *proofs*. Each of these proofs illustrate the value of considering the different ways that a group can act and the orbits and the stabilizers of these actions. Hence, the problems in this section are focused on completing the proofs of the Sylow theorems. While, after this section, you can do many of the problems in Section 12.3, we prefer to wait to demonstrate the power of Sylow theorems in revealing the structure of a group until after introducing normal subgroups in Chapter 10 and after a deeper study of group homomorphisms in Chapter 11.

Problems

7.3.1. **Proof of Sylow D Theorem 7.13.** Using the outline in the text, prove that if G is a finite group, p a prime, $P \in \mathrm{Syl}_p(G)$, $Q \le G$, and $|Q|$ is a power of p, then $Q \le gPg^{-1}$ for some $g \in G$.

7.3.2. Let Q be a subgroup of G whose order is a power of p. Show that Q is a subgroup of some Sylow p-subgroup of G.

7.3.3. **Proof of Corollary 7.15.** Let G be a finite group and p a prime. By completing the outline in the text, write a complete proof of the fact that $\mathrm{Syl}_p(G)$ is exactly one orbit of the conjugation action of G on the set of subgroups of G, and that if $P \in \mathrm{Syl}_p(G)$, then

$$\left|\mathrm{Syl}_p(G)\right| = |G : \mathbf{N}_G(P)|.$$

Conclude that the number of Sylow p-subgroups is a divisor of the order of the group.

7.3.4. Find all Sylow 2-subgroups of S_3, confirm that they are all conjugate in G, and verify that the number of Sylow 2-subgroups is the index of the normalizer of any of the Sylow subgroups. Do the same for the Sylow 3-subgroups of S_3.

7.3.5. Let G be a finite group, $P \in \mathrm{Syl}_p(G)$, and $N = \mathbf{N}_G(P)$.
   (a) Show that $P \in \mathrm{Syl}_p(N)$.

(b) Show that the normalizer in N of P is N. In other words, $\mathbf{N}_N(P) = N$.

(c) Show that P is the unique Sylow p-subgroup of N. In other words, $\left|\mathrm{Syl}_p(N)\right| = 1$.

7.3.6. Let the group G act on a set S. Let T be a subset of S, and let O be one of the orbits of the action. Furthermore, let H be a subgroup of G.

(a) Using the same action as before, does G always act on T? Does G always act on O?

(b) Using the same action as before, does H always act on S? Does H always act on O?

7.3.7. Let G be a finite group, let $P \in \mathrm{Syl}_p(G)$, and let P act on $\mathrm{Syl}_p(G)$ by conjugation (see the Outline of Proof for Theorem 7.16). Assume that $Q \in \mathrm{Syl}_p(G)$ is fixed by this action, and let $N = \mathbf{N}_G(Q)$. Show that $P \leq N$. Using Problem 7.3.5, conclude that $Q = P$.

7.3.8. Let $G = A_4$, the alternating group of degree 4. What are the Sylow 3-subgroups of G? What is $|\mathrm{Syl}_3(G)|$? Let $P = \langle (1\ 2\ 3) \rangle$. Is P a Sylow 3-subgroup of G? Does P act on $\mathrm{Syl}_p(G)$ by conjugation? If so, draw the Cayley digraph of the action, and identify the orbits of the action. How many orbits of size 1 are there?

7.3.9. Proof of Theorem 7.16. Let G be a finite group and let p be a prime. Let $s = |\mathrm{Syl}_p(G)|$ be the number of $\mathrm{Syl}_p(G)$. By completing the outline in the text (and by using Problem 7.3.7), show that $s = 1 + kp$ for some non-negative integer k.

7.3.10. Let G be a group of order 35. What are the possibilities for $|\mathrm{Syl}_5(G)|$? What about $|\mathrm{Syl}_7(G)|$?

7.3.11. Let G be a finite group, and let p be a prime. Assume that P and Q are both Sylow p-subgroups of G. Using the outline provided in the proof of Corollary 7.14, write a complete proof of the fact that P and Q are isomorphic.

Chapter 8

Counting
the Number of Orbits*

... where the Cauchy–Frobenius counting lemma—that gives a formula for the number of orbits in the action of a finite group on a finite set—is proved, and applications to enumerative combinatorics are given.

8.1. The Cauchy–Frobenius Counting Lemma

If a group acts on a set, then we know that it can be fruitful to look at the orbits and the stabilizers. In this section, in the case when the set is finite, we focus on the *number* of orbits and prove a useful formula.

Definition 8.1. Let G be a group, and assume that G acts on a finite set Ω. For $g \in G$, we define fix(g)—or fix$_\Omega(g)$—to be the number of elements of Ω that are fixed by the action of g. In other words,

$$\text{fix}(g) = |\{\alpha \in \Omega \mid g \cdot \alpha = \alpha\}|.$$

Given an action of the group G on a finite set Ω, fix $: G \to \mathbb{Z}^{\geq 0}$ is a function on G. This function is also called the *permutation character* of the action and is sometimes denoted by χ.

Example 8.2. Let $D_8 = \langle a, b \mid a^4 = b^2 = e, ba = a^3 b \rangle$ act on the vertices of the square as usual. The identity fixes all four vertices, and each of the two reflections about diagonal axes fix two vertices. On the other hand, the vertical and horizontal reflections and the non-trivial rotations fix no vertex. Hence,

$$\text{fix}(e) = 4,$$
$$\text{fix}(ab) = \text{fix}(a^3 b) = 2,$$
$$\text{fix}(a) = \text{fix}(a^2) = \text{fix}(a^3) = \text{fix}(b) = \text{fix}(a^2 b) = 0.$$

Curiously, notice that the average of fix(g) over the whole group is an integer: 1.

Example 8.3. Let $\Omega = \{1, 2, 3, 4, 5, 6, 7, 8\}$, and let $G = \langle (2\ 3\ 4), (2\ 4) \rangle \leq S_8$. G acts on Ω in the usual way.

With a bit of calculation we see that $G = \{1, (2\ 3\ 4), (2\ 4\ 3), (2\ 4), (3\ 4), (2\ 3)\} \cong S_3$.

The orbits of the action are

$$\{1\}, \{2, 3, 4\}, \{5\}, \{6\}, \{7\}, \{8\}.$$

The stabilizers of the action are

$$\mathrm{Stab}_G(1) = \mathrm{Stab}_G(5) = \mathrm{Stab}_G(6) = \mathrm{Stab}_G(7) = \mathrm{Stab}_G(8) = G,$$
$$\mathrm{Stab}_G(2) = \langle (3\ 4) \rangle, \quad \mathrm{Stab}_G(3) = \langle (2\ 4) \rangle, \quad \mathrm{Stab}_G(4) = \langle (2\ 3) \rangle.$$

The values of the function fix are

$$\mathrm{fix}(e) = 8, \ \mathrm{fix}(2\ 3\ 4) = \mathrm{fix}(2\ 4\ 3) = 5, \ \mathrm{fix}(2\ 3) = \mathrm{fix}(3\ 4) = \mathrm{fix}(2\ 4) = 6.$$

In both examples, if we find the average of the values of the function fix, we get an integer. For the second example, we have

$$\frac{1}{6}(8 + 5 + 5 + 6 + 6 + 6) = 6.$$

In fact, the answer 6 is the number of orbits of the action. In other words, the number of orbits is the average of the values of the fix function. This identity is known as *Burnside's counting lemma* mainly because it appeared in Burnside's influential 1911 text in group theory [**Bur55**]. For many years, it was believed that Burnside had discovered this result, but in the late 1970s—see Neumann [**Neu79**] and Wright [**Wri81**]—it became clear that the result was known earlier and various versions of it are due to Cauchy (1845) and Frobenius (1887), and, in fact, Burnside himself attributed it to Frobenius. Many refer to the result as the *Cauchy–Frobenius counting lemma* while others continue to call it *Burnside's lemma*. Yet a third group calls the lemma "the lemma that is not Burnside's".

Theorem 8.4 (Cauchy–Frobenius; not due to Burnside). *Let a finite group G act on a finite set Ω, and let n be the number of orbits. Then*

$$n = \frac{1}{|G|} \sum_{g \in G} \mathrm{fix}(g).$$

Proof. Let $\mathcal{O}_1, \mathcal{O}_2, \ldots, \mathcal{O}_n$ be the orbits of the action. Make a table—see Table 8.1—whose rows are indexed by the elements of G and its columns are indexed by elements of Ω. List elements of Ω one orbit at a time. The entries of the table are either 1 or 0 and are defined as

$$(i, j) \text{ entry} = \begin{cases} 1 & \text{if } g_i \cdot \alpha_j = \alpha_j, \\ 0 & \text{otherwise.} \end{cases}$$

How many 1's are in the table? We can find the number of 1's by summing up all the entries, and we can do this in two different ways: one row at a time or one column at a time.

For the row associated with $g \in G$, the number of 1's is $\mathrm{fix}(g)$, and hence the total number of 1's in the table is

$$\sum_{g \in G} \mathrm{fix}(g).$$

Table 8.1. A group acts on a finite set. The set elements are listed one orbit at a time, and a 1 indicates that the set element corresponding to the column is fixed by the group element representing the row.

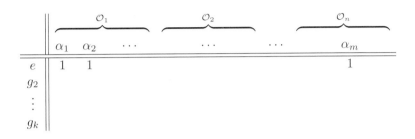

For the column associated with $\alpha \in \Omega$, the number of 1's is

$$|\mathrm{Stab}_G(\alpha)| = \frac{|G|}{|\mathcal{O}_\Omega(\alpha)|}.$$

If α and β are in the same orbit, then $|\mathcal{O}_\Omega(\alpha)| = |\mathcal{O}_\Omega(\beta)|$, and so their columns have exactly the same number of 1's. This means that the total number of 1's in the columns associated with the elements of one orbit $\mathcal{O}_i$ is

$$|\mathcal{O}_i| \frac{|G|}{|\mathcal{O}_i|} = |G|.$$

This number is independent of the orbit, and so, if there are n orbits, the total number of 1's in the table is

$$n\,|G|.$$

Equating the two expressions for the total number of 1's we get

$$n\,|G| = \sum_{g \in G} \mathrm{fix}(g),$$

and the proof is complete.   □

Problems

8.1.1. Let G be a finite group. Show that if we find the average of the sizes of centralizers of elements of G, we get the number of conjugacy classes of G. In other words, prove

$$\text{Number of conjugacy classes of } G = \frac{1}{|G|} \sum_{g \in G} |\mathbf{C}_G(g)|.$$

8.1.2. Let $s_n(k)$ be the number of elements of S_n that have exactly k fixed points. In other words, $s_n(k)$ is the number of permutations of n objects that leave exactly k objects fixed. Show that

$$\sum_{k=0}^{n} k s_n(k) = n!.$$

8.1.3. Let n be a fixed positive integer. We randomly pick an element $x \in S_n$, and we count the number of fixed points of x (e.g., if $x = (1\ 2\ 5) \in S_5$, then x fixes 3 and 4 and hence we record two fixed points). What is the expected value of the number of fixed points?

8.1.4. **Jordan's theorem.**[1] Assume a finite group G acts on a finite set Ω. Further assume that $|\Omega| > 1$ and that the action is transitive (i.e., the action has only one orbit). Show that there exists $g \in G$ with $\mathrm{fix}(g) = 0$.
An element $g \in G$ with $\mathrm{fix}(g) = 0$ is called a *derangement*.

8.1.5. Let G be a group, and let $H < G$ with $|G : H| < \infty$. Then G acts on the set of left cosets of H in G by left multiplication. Is this action transitive? By applying Jordan's theorem (see Problem 8.1.4) to this action, show that there exists a conjugacy class of G that is disjoint from H. Conclude that, in a finite group, if S is a set that contains at least one element from each conjugacy class, then S generates all of G.

8.2. Combinatorial Applications of the Counting Lemma

In enumerative combinatorics we often want to count the number of certain configurations. Consider a simple example: You may know that a cube has six faces and you want to count the number of edges. You could argue that the cube was constructed by gluing together six squares, and each of these squares has four sides. Hence, there are a total of $4 \times 6 = 24$ edges. This would not be quite right since it does not take into account that we glued the squares together along the edges, and so the original set of edges were paired up and every pair became one edge of the eventual cube. Hence the correct answer is $24/2 = 12$ edges. So, the original set of 24 edges was partitioned into 12 subsets—each subset consisting of the two later-identified edges—and we really wanted to count the number of subsets. We are thinking of the two elements of the subset as the same edge.

The Cauchy–Frobenius counting lemma allows us to count the number of subsets of an original set when the subsets are not necessarily of the same size, provided that we got the subsets as the orbits of a group action. Three examples will illustrate this common use of the counting lemma.

Coloring the Cube. Given m colors, how many "different" ways can we color a cube by assigning a color to each face? (This was one of the problems highlighted in the Preface.)

When are two cubes colored differently? If two cubes are colored in two ways such that one cube can be rotated to look like the other, then the two colorings are considered not different. For example, coloring the front face red while coloring all the other faces black is the same as coloring the top face red and all other faces black.

Consider small values of m. If $m = 1$ and we have one color, then there is exactly one way to color the cube. If $m = 2$ and the two colors are red and blue,

---

[1] For a wonderful discussion of Jordan's 1872 theorem and its implications in number theory and topology, see Serre [**Ser03**].

then let i be the number of faces colored red. For $i = 0$ and $i = 1$, we have one possible coloring of the cube. If $i = 2$, we can either color red two of the adjacent faces or two of the opposite faces. Hence, there are two such colorings. For $i = 3$, the three red faces either share one vertex or they do not and so there are two of these colorings. The case $i = 4$ is the same as $i = 2$—since four red faces means two blue faces—and $i = 5, 6$ are the same as $i = 1, 0$. We conclude that for $m = 2$, the total number of different ways that we can color the cube is $1 + 1 + 2 + 2 + 2 + 1 + 1 = 10$. Already the case $m = 3$ is too unwieldy for a brute force approach.

Now, let G be the group of rigid symmetries of the cube, and let Ω be the set of all m^6 possible colorings of the cube—m choices for each of the six faces of the cube gives m^6. However, many of these m^6 colorings are really not different from each other. For example, in this count, the coloring of the cube that consists of one red face and five blue faces is counted six times. On the other hand, the coloring with all red is counted exactly once.

Now G acts on Ω—since G is the group of symmetries of the cube—and each orbit represents one coloring of the cube. If two colorings are in two orbits, then we cannot get from one to the other by a rigid symmetry of the cube and, hence, they are different colorings. The number of orbits is the number of truly different colorings.

We will now list the rigid symmetries of the cube and calculate fix(g) for each of them and then use the counting lemma to find the number of orbits.

We have organized the information in Table 8.2. The group has one identity element and fix$(e) = |\Omega| = m^6$. Take an axis through the center of opposite faces. You can rotate the cube $90°$ forward or backward. Since there are three such axes—one for each pair of opposite faces—there are six of these rotations, and each has order 4—we do not actually need the order but it helps in making sure that we know which symmetry we are talking about. What is fix(g) when g is one of these rotations? As an example, take the axis of rotation to be through the faces on the left and right of the cube. For an element of Ω—the m^6 possible colorings of the cube—to be fixed by this rotation, it would have to have the same color on all the faces but the left and right ones. We have m choices for that color and another m choices of colors for each of the left and right squares. This means that fix$(g) = m^3$

Table 8.2. The cube has five types of rigid symmetries. For each type, the number of symmetries of that type, the order of those symmetries, and fix(g) is given.

type	Description	#	$o(g)$	fix(g)
1	identity	1	1	m^6
2	$90°$ rotation, axis through center of opposite faces	6	4	m^3
3	$180°$ rotation, axis through center of opposite faces	3	2	m^4
4	$120°$ rotation, axis through center of opposite vertices	8	3	m^2
5	$180°$ rotation, axis through center of opposite edges	6	2	m^3

if g is a 90° rotation with an axis through the middle of two opposite faces. Get an actual cube—a Rubik's cube would do—and go through each line of Table 8.2 and convince yourself that it makes sense.

The number of different colorings of the cube is exactly the number of orbits of Ω under the action of G. This is given by the Cauchy–Frobenius counting lemma:

$$
\begin{aligned}
n &= \frac{1}{24}(m^6 + 6m^3 + 3m^4 + 8m^2 + 6m^3) \\
&= \frac{m^2}{24}(m^4 + 3m^2 + 12m + 8).
\end{aligned}
$$

Not only was this formula not obvious at the start, it is not even clear, without further arguments, why you get an integer no matter what positive integer m you plug in.

Application to Chemistry. The NH_3 (ammonia) molecule is in the form of a pyramid with the three H's forming an equilateral base. The N forms an isosceles triangle with each pair of H's (see Figure 8.1). Assume that we can replace some of the H's with Cl or CH_3. How many visually distinct molecules can we get?

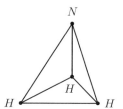

Figure 8.1. The ammonia molecule

Solution. We can "color" the vertices of the equilateral triangle with three colors, H, Cl, and CH_3, and we want to know the number of different colorings. We proceed as before, and first find the group of rigid symmetries. If we let r denote a 120 degree rotation, then the group of symmetries will be $G = \{e, r, r^2\}$. The set Ω of all possible colorings has 3^3 elements, and we have fix$(e) = 27$, fix$(r) =$ fix$(r^2) = 3$. Note that to be fixed by r, the three colors of the vertices must be the same. Thus the number of orbits is $n = 1/3(27 + 3 + 3) = 11$.

Among these 11 different colorings, there are two that have one of each H, Cl, and CH_3. These two molecules are called *optical isomers*. If we do not want to distinguish between them, then we should enlarge our symmetry group to include reflections of the base. Let b be one such reflection, then $G = \{e, r, r^2, b, rb, r^2b\}$, and we have fix$(b) =$ fix$(rb) =$ fix$(r^2b) = 9$. Thus the number of different non-isomer molecules is $n = \frac{1}{6}(27 + 3 + 3 + 9 + 9 + 9) = 10$. (We have not claimed that all of these molecules are actually possible in real life. That question and its answer belong to courses in chemistry. We have just counted the full range of possibilities.) $\square$

Switching Functions. We define a *switching function* to be a "black box" which has n binary inputs and one binary output. Each of the inputs can be a 0 or a 1, and hence there are 2^n possible messages that can enter a particular switching function. The output for each of these possible messages will also be a 0 or a 1. In other words, a switching function is a function $f : \{0,1\}^n \to \{0,1\}$. There are two choices—for the output—for each of the 2^n messages, and hence there are a total of $2^{(2^n)}$ possible switching functions.

Example 8.5. Let $n = 2$. Figure 8.2 gives a visualization of a switching function with two inputs and one output.

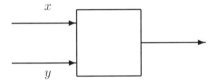

Figure 8.2. A switching function with two inputs

There are two lines—whose values are denoted by x and y—coming into the box, and thus there are 2^2 possible messages that can be entered. In Table 8.3, we have numbered these messages 1 through 4. Thus, for example, message 3 is achieved with inputs $x = 1$ and $y = 0$. Now each function gives an output of 0 or 1 for each of the messages. For example, f_6 gives the output 1 for the first two messages and 0 for the third and fourth messages. The number of possible functions is $2^{(2^2)} = 16$.

Table 8.3. With two inputs there are four possible messages and 16 different switching functions.

messages		x	y	f_1	f_2	f_3	f_4	f_5	f_6	$\cdots$	f_{16}
1	=	0	0	0	1	0	0	0	1	$\cdots$	1
2	=	0	1	0	0	1	0	0	1	$\cdots$	1
3	=	1	0	0	0	0	1	0	0	$\cdots$	1
4	=	1	1	0	0	0	0	1	0	$\cdots$	1

Consider functions f_3 and f_4 in Table 8.3. These are two different functions, but if we switch the two inputs—i.e., x and y—then f_3 becomes f_4. Hence, if we have already built the switching function f_3, then, by just rerouting the inputs, we can get the function f_4 as well. Hence, there is no point in manufacturing both f_3 and f_4. On the other hand, rerouting the inputs on the function f_{16} will give us the function f_{16} again. We want to know the minimum number of switching functions that we have to manufacture in order to get—by rerouting the inputs if necessary—the functionality of all 16 functions. We record the question:

Question 8.6. How many "different" switching functions are there if two functions are considered the same if rerouting the inputs of one yields the other?

We will solve this problem for the case $n = 2$ and ask you to do the case $n = 3$ as well as other variations in the problems. For $n = 2$, we could solve the problem by brute force and by listing all possibilities. But already for $n = 3$ this seems quite tedious. Here we use the Cauchy–Frobenius counting lemma.

Solution for Two Inputs. With two inputs, the only rerouting is to switch x and y. This results in switching the second and third messages in Table 8.3. So we can either do nothing or switch the second and third messages. These actions, in turn, will permute the set of 16 switching functions.

We let $\Omega = \{f_1, f_2, \ldots, f_{16}\}$, and we let $G = \{e, (2\ 3)\}$. We think of the group G as a subgroup of S_4 since a priori we had four messages, and we could have considered any permutation of these. Given that rerouting is our only possible action, out of all possible 24 permutations of four objects only the two in G are possible for us.

Now G acts on Ω. It does so by virtue of permuting the messages. For example, $(2\ 3) \cdot f_3 = f_4$, while $(2\ 3) \cdot f_{16} = f_{16}$. What we want is exactly the *number* of orbits of this action.

Clearly, and as always, $\mathrm{fix}(e) = |\Omega| = 16$. What is $\mathrm{fix}(2\ 3)$? The element $(2\ 3)$ will fix a function f_i if the output of f_i for message 2 is the same as the output of f_i for message 3. In other words, if $f_i(2) = f_i(3)$. How many of such functions do we have? We have two choices for each of $f_i(1)$, $f_i(2) = f_i(3)$, and $f_i(4)$, and so the total number of functions fixed by $(2\ 3)$ is $2 \times 2 \times 2 = 8$. Hence, the number of orbits is

$$n = 1/2(16 + 8) = 12.$$

We conclude that the number of truly different switching functions for two inputs is 12. $\qquad\qquad\qquad\qquad\qquad\qquad\qquad\qquad\qquad\qquad\qquad\qquad\qquad\qquad\quad\square$

Problems

8.2.1. Compute how many different ways there are to color the faces of a cube so that three faces are red, two are white, and one is blue.

8.2.2. How many different patchwork quilts, four patches long and three patches wide, can be made from five red and seven blue squares, assuming that the quilts cannot be turned over?

8.2.3. How many different ways can we color the vertices of an equilateral triangle with six colors. (Two triangles have identical colorings if they cannot be distinguished after being dropped on the floor.)

8.2.4. How many different ways are there to color the corners of a regular 5-gon (i.e., a pentagon) with the colors red, blue, and green? (Two 5-gons have identical colorings if they cannot be distinguished after being dropped on the floor.)

8.2.5. A necklace is made by placing a bead at each of the six corners of a regular hexagon. How many different necklaces can we make if we have m different colors of beads? (Note that two necklaces are considered the same if we cannot distinguish them after a number of rotations and flips.)

8.2.6. You want to make 2×3 rectangular patchwork quilts, and you can use one of ten colors for each of the patches. (Each of the six patches are squares and they are all of the same size. Different patches can have the same colors.) Two quilts are considered the same if you can get one from the other by turning or flipping it. How many different quilts are possible?

8.2.7. A certain kind of a ceramic tile has the form of a 4×4 board. Each of the 16 squares is colored white, blue, or yellow. How many different such ceramic tiles are there? Note that we cannot turn the tiles upside down.

8.2.8. Count the number of ways to color the edges of a cube with r colors. We, as usual, consider two colorings the same if one can be obtained from the other by an appropriate rotation—that is, a rigid symmetry—of the cube.

8.2.9. Find the number of distinct switching functions on three variables. As explained in the text, two switching functions are considered the same if we can get one from the other by rerouting the inputs.

8.2.10. Consider switching functions with two inputs. We have seen that there are 16 such functions but that, if we allow rerouting of the inputs, the number of "different" switching functions drops to 12.

   We can further reduce the number of different functions by applying inverters, or NOT gates, to the inputs. An inverter exchanges the two possible values for a bit, thus 1 becomes 0, and 0 becomes 1. Inverters can be applied to none, one, or both of the inputs. How many "different" switching functions do we have now? In solving this problem you will have a group acting on a set; identify the group.

8.2.11. Some transistor switching devices are sealed in a can with three input sockets at the vertices of an equilateral triangle. The three input wires are connected to a plug that will fit into the input sockets. The three inputs are either zero or one (off or on). How many different cans are needed to produce any function of three input variables? Note that the only way to reroute the inputs is to rotate the plug.

8.3. More Problems and Projects

The Theorem of Cameron and Cohen.[2] Let a finite group G act transitively on a finite set Ω. (Recall that a group action is transitive if all elements of Ω are in one orbit. In addition, if a group G acts—transitively or not—on a set, then it acts transitively on any of the orbits.) Assume $|\Omega| > 1$, and let G_0 be the set of elements of G that fix no element of Ω. In other words, $G_0 = \{g \in G \mid \text{fix}(g) = 0\}$.

---

[2]Adapted from Serre [**Ser03**].

As a measure of the relative size of G_0, let

$$c_0 = \frac{|G_0|}{|G|}.$$

Jordan's theorem (Problem 8.1.4) showed that $c_0 > 0$. In fact $c_0 \geq 1/|\Omega|$. We give an outline of this theorem of Cameron and Cohen here and ask you to complete the proof in Problem 8.3.2.

Theorem 8.7 (Cameron and Cohen 1992). *Let the finite group G act transitively on the finite set Ω. Assume $|\Omega| > 1$, let $G_0 = \{g \in G \mid \mathrm{fix}(g) = 0\}$, and let $c_0 = |G_0| / |G|$. Then*

$$c_0 \geq \frac{1}{|\Omega|}.$$

Outline of the Proof. To begin with, note that for $g \notin G_0$, we have $1 \leq \mathrm{fix}(g) \leq \Omega$, and so:

(8.1) $(\mathrm{fix}(g) - 1)(\mathrm{fix}(g) - |\Omega|) \leq 0$, for all $g \in G - G_0$.

Now, let

$$A = \frac{1}{|G|} \sum_{g \in G} (\mathrm{fix}(g) - 1)(\mathrm{fix}(g) - |\Omega|).$$

Expanding the product, and using the result of Problem 8.3.1 together with the Cauchy–Frobenius Theorem 8.4—remember that the action is transitive and so there is only one orbit—we can show that $A \geq 1$. On the other hand, using equation (8.1) and recalling that for $g \in G_0$, we have $\mathrm{fix}(g) = 0$, we can show that $A \leq c_0 |\Omega|$. Putting the two inequalities together gives the desired result. □

Problems

8.3.1. Let a finite group G act on a finite set Ω with $|\Omega| > 1$. Prove

$$\frac{1}{|G|} \sum_{g \in G} (\mathrm{fix}(g))^2 \geq 2.$$

You may find the following steps useful:

Step 1: Using the action of G on Ω define a natural action of G on $\Omega \times \Omega$, the Cartesian product of Ω with itself.

Step 2: Let $x, y \in \Omega$ with $x \neq y$. By considering the action of G on $(x, x), (x, y) \in \Omega \times \Omega$ show that the action of G on $\Omega \times \Omega$ has at least two orbits.

Step 3: For $g \in G$, if $\mathrm{fix}(g)$ denotes the number of fixed points of g on its action on Ω, show that $(\mathrm{fix}(g))^2$ gives the number of fixed points of the action of g on $\Omega \times \Omega$.

Step 4: Complete the proof using the Cauchy–Frobenius Theorem 8.4.

8.3.2. **Proof of the Cameron–Cohen Theorem 8.7.** Using the given outline, complete the proof of the Cameron–Cohen Theorem 8.7.

8.3.3. In the Cameron–Cohen Theorem 8.7, do we need the assumption that $|\Omega| > 1$?

8.3.4. Let a finite group G act on a finite set Ω with $|\Omega| > 2$. Prove

$$\frac{1}{|G|} \sum_{g \in G} (\text{fix}(g))^3 \geq 5.$$

Chapter 9

The Lattice of Subgroups*

...where we introduce the partially ordered set of subgroups of a group and its Hasse diagram, argue for using lattice diagrams—a visual representation of the subgroup structure of a group—and bring together useful facts for drawing partial lattice diagrams and for adding edge lengths to the diagrams.

As we have seen, in studying groups it is fruitful to understand the subgroups of a group, and as a result it will be helpful to somehow visualize these subgroups and their relations to each other. The subgroups of a group are a partially ordered set—see Definition 9.1—and showing the inclusion relations of the subgroups in a so-called Hasse diagram will be often productive. In this short chapter, we will define partially ordered sets (posets), Hasse diagrams of posets, and lattices. We will then focus on the lattice of subgroups of a group, and review and bring together a number of facts—already proved in the text or in the problems—that allow us to use subgroup lattices and their diagrams effectively.

9.1. Partially Ordered Sets, Hasse Diagrams, and Lattices

Posets. Let X be a set, and let $\sim$ be a relation on X (see Definition 4.25). Recall that (Definition 4.26) $\sim$ is said to be reflexive if $a \sim a$ for all $a \in X$, and transitive if, for all $a, b, c \in X$, $a \sim b$ and $b \sim c$ implies $a \sim c$. We make one further definition (compare with the definition of a symmetric relation in Definition 4.26):

Definition 9.1. Let X be a set, and let $\sim$ be a relation on X. If, for all $a, b \in X$, the two relations $a \sim b$ and $b \sim a$ imply that $a = b$, then we say that $\sim$ is *anti-symmetric*.

We can now define partially ordered sets or posets:

Definition 9.2 (Posets). A *partial order* on a non-empty set X is a relation that has reflexive, anti-symmetric, and transitive properties. In a partial order, if, for every $a, b \in X$, we have $a \sim b$ or $b \sim a$ (or both), then we call the relation a *total order*.

A set with a partial order is called a *partially ordered set* or a *poset*. In a poset we usually use the symbol $\leq$ instead of $\sim$. If X is a poset with $a, b \in X$, we write $a < b$ if $a \leq b$ *and* $a \neq b$.

Remark 9.3. Note that a relation is an equivalence relation if it is reflexive, symmetric, and transitive, while it is a poset if it is reflexive, *anti-symmetric*, and transitive. It may seem that equivalence relations and posets should have much in common, but that is not so.

Example 9.4. Let $\mathbb{Z}$ be the set of integers. Define $a \sim b$ if $a \leq b$. Then $\sim$ is a total order.

Example 9.5. Let $\mathbb{Z}$ be the set of integers. Define $a \sim b$ if $a < b$. Then $\mathbb{Z}$ is neither a partial order nor an equivalence relation.

Example 9.6. Let $\mathbb{Z}^{>0}$ be the set of positive integers. For $a, b \in \mathbb{Z}^{>0}$, define $a \sim b$ if $a \mid b$. This relation is a partial order.

Example 9.7. Let G be any group. Let X be the set of subgroups of G. Let H and K be elements of X. We say $H \sim K$ if $H \leq K$. This relation is a partial order, and the resulting poset is the object of this chapter.

Hasse Diagrams. Given a partial order on a set X, we can draw a representation of the set X by drawing its *Hasse diagram* which is a graph that allows us to "see" all the relations. In general, given a poset X, the vertices of this graph are the elements of the poset. We draw an edge between x and y if $x < y$ and there is no $z \in X$ with $x < z < y$. In other words we draw an edge only if x is immediately below y. Also if $x < y$, then y is drawn "above" x.

Example 9.8. Let $P = \{x, y, z, u, v\}$, and assume that we have the following relations among the elements of P:

$$x \leq y, \ y \leq z, \ x \leq z, \ x \leq u, \ y \leq u, \text{ and } w \leq w, \text{ for all } w \in P.$$

Then the Hasse diagram of P is given in Figure 9.1.

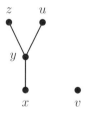

Figure 9.1. The Hasse diagram of the poset P of Example 9.8

Example 9.9. Let $\mathbb{Z}$ be the set of integers with the usual $\leq$ relation. The Hasse diagram of $(\mathbb{Z}, \leq)$ is given in Figure 9.2.

Figure 9.2. The Hasse diagram of $(\mathbb{Z}, \leq)$

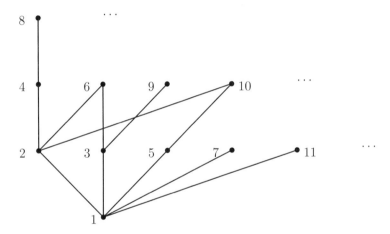

Figure 9.3. The (partial) Hasse diagram of $\mathbb{Z}^{>0}$ ordered by divisibility

Example 9.10. Let $\mathbb{Z}^{>0}$ be the set of positive integers. For $a, b \in \mathbb{Z}^{>0}$, define $a \sim b$ if $a \mid b$. Part of the Hasse diagram of $\mathbb{Z}^{>0}$ with this relation is given in Figure 9.3.

Example 9.11. Let $G = \mathbb{Z}/2\mathbb{Z} \times \mathbb{Z}/2\mathbb{Z}$ be the Klein 4-group, and let X be the set of all subgroups of G ordered by inclusion. The Hasse diagram for X is given in Figure 9.4.

Lattices. We will now define a special subclass of posets called lattices and point out that the poset of subgroups of a group is indeed a lattice.

Definition 9.12. Let P be a poset, and let $x, y \in P$. An element $z \in P$ is a *least upper bound* (or a *supremum*) for x and y if

(a) $z \geq x$, and $z \geq y$, and

(b) if $w \in P$ with $w \geq x$ and $w \geq y$, then $w \geq z$.

The *greatest lower bound* (or the *infimum*) is defined similarly.

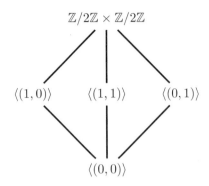

Figure 9.4. The Hasse diagram of subgroups of $(\mathbb{Z}/2\mathbb{Z} \times \mathbb{Z}/2\mathbb{Z}, +)$ ordered by inclusion

A partially ordered set is a *lattice* if every pair of elements have a least upper bound and a greatest lower bound.

Note that by definition, a least upper bound (or a greatest lower bound)—if it exists—is unique (if z and w were both least upper bounds, then we would have both $z \le w$ and $w \le z$ resulting in $w = z$).

Example 9.13. If the Hasse diagram of the poset P is given in Figure 9.5, then P is *not* a lattice. The elements 1, c, and d are all upper bounds of a and b, but none of them is a least upper bound for these. Likewise, c and d do not have a greatest lower bound.

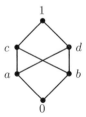

Figure 9.5. The Hasse diagram of a poset that is not a lattice

Going back to the poset of subgroups, recall that if H and K are subgroups of a group G, then $H \cap K$ is their intersection and $\langle H, K \rangle$ is the subgroup of G generated by all the elements of H and K. The following lemma is immediate:

Lemma 9.14. *Let G be a group, and let $\mathcal{P}$ be the poset of subgroups of G ordered by inclusion. Let H and K be subgroups of G. Then*

(a) $H \cap K$ *is the largest subgroup of G that is contained in both H and K. In other words, $H \cap K$ is the greatest lower bound of H and K in $\mathcal{P}$.*

(b) $\langle H, K \rangle$ *is the smallest subgroup of G that contains both H and K. In other words, $\langle H, K \rangle$ is the least upper bound of H and K in $\mathcal{P}$.*

In particular, $\mathcal{P}$ is a lattice.

Subgroup Lattice Diagrams

Definition 9.15. Let G be a group. The poset of subgroups of G ordered by inclusion will be called the *lattice of subgroups* or the *subgroup lattice* of G. The *lattice diagram* of G—or more precisely the *subgroup lattice diagram* of G—will refer to the Hasse diagram of this poset.

Remark 9.16. Subgroup lattice diagrams can be of great assistance in analyzing the properties of a group. We often go back and forth between information that we know about a group and what we can glean from parts of the lattice diagram. For a student beginning her study of group theory, it is worthwhile to "know" the lattice diagram of certain small groups well. These basic groups include the abelian groups, dihedral groups especially D_8 and D_{10}, the symmetric groups S_3 and S_4, the alternating groups A_4 and A_5, the quaternion group Q_8, and the special linear group $\mathrm{SL}(2,3)$.

Definition 9.17. Let P be a poset. A totally ordered subset of P is called a *chain*. In other words, a collection of elements $a_1, \ldots, a_k, \ldots$ form a chain in P if

$$a_1 \leq a_2 \leq \cdots \leq a_k \leq \cdots .$$

Example 9.18. The subgroup lattice diagram of $\mathbb{Z}/8\mathbb{Z}$ is given in Figure 9.6 and is a single chain.

$$\mathbb{Z}/8\mathbb{Z}$$

$$\langle 2 \rangle$$

$$\langle 4 \rangle$$

$$\{0\}$$

Figure 9.6. Subgroups of $(\mathbb{Z}/8\mathbb{Z}, +)$ ordered by inclusion

Example 9.19. Figure 9.7 gives the subgroup lattice diagram of $\mathbb{Z}/2\mathbb{Z} \times \mathbb{Z}/3\mathbb{Z} \times \mathbb{Z}/47\mathbb{Z}$.

Example 9.20. The subgroup lattice diagram of $D_8 = \langle a, b \mid a^4 = b^2 = e, ba = a^3 b \rangle$ is given in Figure 9.8.

Using the Subgroup Lattice Diagram. Often we know only parts of the lattice diagram, but often we have more information than just the lattice diagram. We may know the sizes of various subgroups or that certain elements commute with others. Here, we give several preliminary examples of how having the full lattice diagram can help in understanding a group.

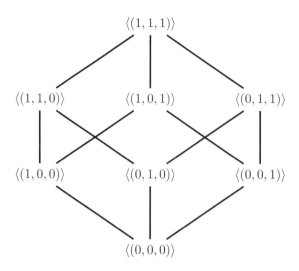

Figure 9.7. Subgroups of $\mathbb{Z}/2\mathbb{Z} \times \mathbb{Z}/3\mathbb{Z} \times \mathbb{Z}/47\mathbb{Z}$ ordered by inclusion

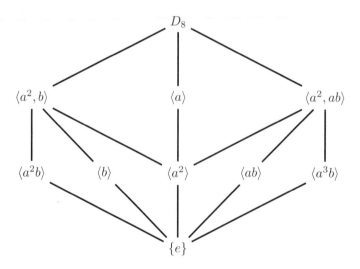

Figure 9.8. Subgroups of D_8 ordered by inclusion

Example 9.21. The subgroup lattice diagram of the dihedral group of order 8, D_8, is given in Figure 9.8. Here are several examples of what we can glean from it:

Using the fact that the least upper bound of two subgroups is the subgroup generated by them, we can conclude from the lattice diagram that the subgroup generated by $\langle a^2 \rangle$ and $\langle a^3b \rangle$ is $\langle a^2, ab \rangle$, while the subgroup generated by $\langle b \rangle$ and $\langle a^3b \rangle$ is all of D_8. Also, D_8 has three subgroups of order 4 and comparing the subgroups of these (in the lattice diagram) to those of $\mathbb{Z}/4\mathbb{Z}$ and $\mathbb{Z}/2\mathbb{Z} \times \mathbb{Z}/2\mathbb{Z}$, we conclude that D_8 has one subgroup isomorphic to $\mathbb{Z}/4\mathbb{Z}$ and two subgroups isomorphic to $\mathbb{Z}/2\mathbb{Z} \times \mathbb{Z}/2\mathbb{Z}$. In fact, we also see that these latter two subgroups intersect in $\langle a^2 \rangle$.

With the help of the lattice diagram, we argue that a^2 is in the center of D_8, while b is not. On one hand, we see that a^2 together with any other element of the group lives inside a subgroup of order 4 (e.g., a^2 and ab live in $\langle a^2, ab \rangle$), and we know—see Problem **2.4.9**—all groups of order 4 are abelian. Hence a^2 commutes with every element of D_8. On the other hand, b does not commute with ab since if it did, then these two elements of order 2, their product, and the identity would form a subgroup of order 4. However, according to the lattice diagram, there is no such subgroup. In fact, from the lattice diagram, we can tell that $D_8 = \langle b, ab \rangle$.

What is the centralizer of b? Every element commutes with itself, and a^2 is in the center of D_8. Hence, we know that $\langle a^2 \rangle$ and $\langle b \rangle$ are in the centralizer of b. Hence, the centralizer is a subgroup that contains both of these subgroups. By looking at the lattice diagram, we see that the centralizer is either $\langle a^2, b \rangle$ or D_8. It cannot be the latter since b is not in the center, and hence $\mathbf{C}_{D_8}(b) = \langle a^2, b \rangle$. We could have, of course, found this result directly. However, we did it here with hardly any calculation at all.

For a final example, assume that we want to find the normalizer in D_8 of $\langle ab \rangle$. Let $H = \langle ab \rangle$, and recall that $\mathbf{N}_{D_8}(H)$ is a subgroup of G consisting of elements $x \in D_8$ with $xHx^{-1} = H$ (see Definition 4.24). Clearly, elements of H itself and elements of the center of the group are in $\mathbf{N}_{D_8}(H)$. Hence, both $\langle ab \rangle$ and $\langle a^2 \rangle$ are in the normalizer. Looking at the subgroup lattice diagram, we see that this means that $\mathbf{N}_{D_8}(H)$ is either $\langle a^2, ab \rangle$ or D_8. Is $a \in \mathbf{N}_{D_8}(H)$? We have $a(ab)a^{-1} = a^2ba^3 = ba^5 = ba = a^3b \notin H$. Hence a is not in the normalizer and the normalizer cannot be all of D_8. We conclude that

$$\mathbf{N}_{D_8}(\langle ab \rangle) = \langle a^2, ab \rangle.$$

Example 9.22. The subgroup lattice diagram of $Q_8 = \langle i, j \mid i^4 = j^4 = e, i^2 = j^2, ij = -ji \rangle$ is given in Figure 9.9. (See Problem 5.2.9.) Which elements of Q_8 commute with i? The elements of $\langle i \rangle$ certainly commute with i, but are there others? From the subgroup lattice diagram, if the centralizer of $\langle i \rangle$ was any larger than $\langle i \rangle$ itself, then it would have to be all of Q_8 which would mean that i commutes with everything. But we know that $ij \neq ji$, and so we conclude that

$$\mathbf{C}_{Q_8}(i) = \langle i \rangle = \{1, i, -1, -i\}.$$

Problems

9.1.1. Give a complete argument that the lattice diagram of $\mathbb{Z}/2\mathbb{Z} \times \mathbb{Z}/3\mathbb{Z} \times \mathbb{Z}/47\mathbb{Z}$ is indeed the one given in Figure 9.7. Give examples of other groups that have exactly the same lattice diagram.

9.1.2. Give a complete argument that the lattice diagram of D_8 is given in Figure 9.8. For each subgroup of D_8, find its centralizer.

9.1.3. (a) Draw the lattice of subgroups of $\mathbb{Z}/6\mathbb{Z}$.
        (b) Repeat the above for the group S_3.

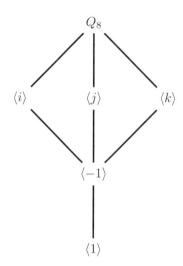

Figure 9.9. The subgroup lattice diagram of Q_8

9.1.4. Draw the full subgroup lattice diagram for $\mathbb{Z}/12\mathbb{Z}$ and for $\mathbb{Z}/6\mathbb{Z} \times \mathbb{Z}/2\mathbb{Z}$. Point out one difference between these two diagrams.

9.1.5. Verify that Figure 9.9 does give the lattice diagram of subgroups of Q_8. Find the normalizer of each of the subgroups of Q_8.

9.1.6. Give the full subgroup lattice diagram of $D_{10} = \langle a, b \mid a^5 = b^2 = e, ba = a^{-1}b\rangle$. For each subgroup of D_{10} give its centralizer.

9.1.7. The full subgroup lattice diagram of $\mathbb{Z}/8\mathbb{Z}$ is a single chain (see Figure 9.6). Can you suggest an infinite family of groups whose full subgroup lattice diagram is a single chain? Make a conjecture.

9.2. Edge Lengths and Partial Lattice Diagrams

We are able to add some extra information to a subgroup lattice by assigning a positive integer to each edge of the lattice diagram.

Edge Lengths. Let G be a finite group, and let H and K be subgroups of G. Assume $H \leq K$. Then the "length" of the path from H to K will be defined to be $|K : H| = |K| / |H|$.

If there is an edge from H to K, whenever possible, we draw this edge to have a *length* equal to $|K : H|$. Note that the total "height" of the lattice diagram of the group is $|G : \{e\}| = |G|$. We recall Corollary 5.19 that says that these numbers are combined multiplicatively:

Lemma 9.23. *Let G be a finite group, and assume that $H < K < L \leq G$. Then* $|L : H| = |L : K| \, |K : H|$.

Example 9.24. The subgroup lattice diagram of the poset of subgroups of $\mathbb{Z}/8\mathbb{Z}$—with edge lengths—is given in Figure 9.10.

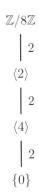

Figure 9.10. Subgroup lattice diagram with edge lengths for $\mathbb{Z}/8\mathbb{Z}$

Example 9.25. The lattice diagram of the alternating group of degree 4, A_4, is given in Figure 9.11. (See Problem 5.2.11.)

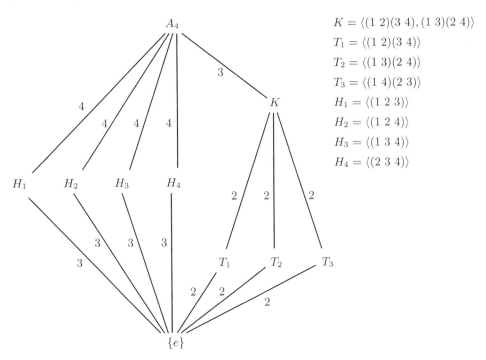

$$K = \langle (1\ 2)(3\ 4), (1\ 3)(2\ 4) \rangle$$
$$T_1 = \langle (1\ 2)(3\ 4) \rangle$$
$$T_2 = \langle (1\ 3)(2\ 4) \rangle$$
$$T_3 = \langle (1\ 4)(2\ 3) \rangle$$
$$H_1 = \langle (1\ 2\ 3) \rangle$$
$$H_2 = \langle (1\ 2\ 4) \rangle$$
$$H_3 = \langle (1\ 3\ 4) \rangle$$
$$H_4 = \langle (2\ 3\ 4) \rangle$$

Figure 9.11. The subgroup lattice diagram of A_4 with edge lengths.

Partial Lattice Diagrams. In general, the full Hasse diagram for the poset of subgroups of a group is too complicated and too large, and hence many times we will draw *partial lattice diagrams* (or *partial subgroup lattice diagrams*). This means that we will judiciously choose some of the subgroups of the group under consideration and draw them. Often, if we include two subgroups H and K, then we try to locate $H \cap K$ and $\langle H, K \rangle$ as well.

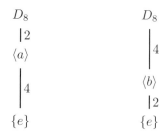

Figure 9.12. Partial lattice diagrams for D_8

Example 9.26. Two partial lattice diagrams for D_8 are given in Figure 9.12.

Two Subgroups in General Position. Let G be a group, and let H and K be two subgroups of G. If we know that $H \leq K$, then we draw the partial lattice diagram in Figure 9.13.

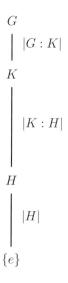

Figure 9.13. Two subgroups H and K with $H \leq K$

However, if H and K are subgroups of G and we do not know whether one is contained in the other, then we draw them in general position as in Figure 9.14.

When we draw subgroups in general position, we allow the possibility that two of them may be the same. For example, in Figure 9.14, if $K = H \cap K$, then we have the case when $K \leq H$. In this case, $|K : H \cap K| = 1$ and $H = \langle H, K \rangle$.

In Figure 9.14 we drew the edges as if $|\langle H, K \rangle : H|$ was bigger than $|K : H \cap K|$. This was not a coincidence because of the following—surprisingly useful—theorem:

Theorem 9.27. *Let G be a group, and let H and K be subgroups of G with $|G : H| < \infty$. Then*

$$|\langle H, K \rangle : H| \geq |K : H \cap K|.$$

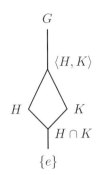

Figure 9.14. Two subgroups H and K in general position

Proof. This was stated in Theorem 5.24(a) (and you were asked to prove it in Problem **5.2.13**). Here we give the proof for completeness. We use Lemma 5.6— two cosets Hx and Hy are equal if and only if $y \in Hx$ if and only if $y = hx$ for some $h \in H$—repeatedly.

Let $L = \langle H, K \rangle$, and let $A = \{(H \cap K)x \mid x \in K\}$ be the set of right cosets of $H \cap K$ in K, and let $B = \{H\ell \mid \ell \in L\}$ be the right cosets of H in L. To show that $|A| \leq |B|$, it suffices to construct a 1-1 map $\theta \colon A \to B$.

Define $\theta \colon A \longrightarrow B$ by

$$\theta((H \cap K)x) = Hx \qquad \text{for } x \in K.$$

First, we show that θ is well defined by showing that, for $x, y \in K$, if $(H \cap K)x = (H \cap K)y$, then $Hx = Hy$. If $(H \cap K)x = (H \cap K)y$, then $y \in (H \cap K)x \subseteq Hx$. But $y \in Hx$ implies that $Hx = Hy$.

Next, we show that θ is 1-1 by showing that, for $x, y \in K$, whenever $\theta((H \cap K)x) = \theta((H \cap K)y)$, we have $(H \cap K)x = (H \cap K)y$. By the definition of θ, if $\theta((H \cap K)x) = \theta((H \cap K)y)$, then $Hx = Hy$. This, in turn, implies that $y = hx$ for some $h \in H$. But both x and y are in K and hence $h = yx^{-1}$ must be in K as well. So $h \in H \cap K$ and $y = hx \in (H \cap K)x$. The latter implies $(H \cap K)x = (H \cap K)y$, and the proof is complete. □

We know that $HK = \{hk \mid h \in H, k \in K\}$ is a subset of G that contains both H and K. If HK were a subgroup, then it would be a good candidate for $\langle H, K \rangle$. However, in general, HK is only a subset of G (see Problem 2.6.32).

We have the following useful proposition:

Proposition 9.28. *Let G be a group, and let H and K be subgroups of G. Assume $|G : H| \leq \infty$. Then the following are equivalent:*

(a) *HK is a subgroup of G,*

(b) *$HK = KH$,*

(c) *$HK = \langle H, K \rangle$,*

(d) *$|\langle H, K \rangle : H| = |K : H \cap K|$.*

Proof. (a) ⇔ (b). This is Problem *2.6.33*.

(a) ⇔ (c). This follows from the definition of $\langle H, K \rangle$ since HK is always a subset of $\langle H, K \rangle$.

The fact that part (d) is equivalent to the other parts was stated in Theorem 5.24(b) (and you were asked to give a proof in Problem **5.2.14**). We sketch the proof for completeness:

We proceed as in the proof of Theorem 9.27 and use the same notation. Recall that $L = \langle H, K \rangle$, $A = \{(H \cap K)x \mid x \in K\}$, and $B = \{H\ell \mid \ell \in L\}$. We defined $\theta : A \longrightarrow B$ by $\theta((H \cap K)x) = Hx$, for $x \in K$. We showed that θ is well defined and 1-1. It thus followed $|A| \leq |B|$.

We know that $|A| = |B|$ if and only if the map θ is onto.

(c) ⇒ (d). Assume $HK = L$. We need to show that $|A| = |B|$. We do this by showing that θ is onto.

From $L = HK$, we conclude that every element of L is of the form uv with $u \in H$, $v \in K$. To show θ is onto, pick an arbitrary element of B, say $H\ell$ with $\ell \in L$. But $\ell \in L$ means that $\ell = uv$ with $u \in H$, $v \in K$, and so $H\ell = Huv = Hv$, and this means that $\theta((H \cap K)v) = Hv = H\ell$. Thus we found an element of A that is mapped to $H\ell$, and so θ is onto.

(d) ⇒ (c). Now we assume θ is onto and show that $L = HK$. Since $L = \langle H, K \rangle$, we know that $HK \subseteq L$, and hence the proof will be complete when we show that $L \subseteq HK$.

Let $x \in L$ be arbitrary. We need to show that $x \in HK$. The map θ is onto which implies that θ maps some element of A, say $(H \cap K)y$, to Hx. Thus $Hx = \theta((H \cap K)y) = Hy$ for some $y \in K$. Now $Hx = Hy$ implies $x \in Hy$, which, in turn, means $x = uy$ for some $u \in H$ and $y \in K$. Hence $x \in HK$ and the proof is complete. □

Parallelograms. Assume that H and K are subgroups of a finite group G. Proposition 9.28 implies that if HK is a subgroup of G, then $|HK : H| = |K : H \cap K|$ and $|HK : K| = |H : H \cap K|$. To emphasize these relations among the edge lengths, we draw the partial lattice diagram including HK, H, K, and $H \cap K$ as a parallelogram. As a result, the relation among the edge lengths can be read from the fact that the opposite sides of a parallelogram have equal lengths. (See Figure 9.15.)

Example 9.29. Two partial lattice diagrams for D_8 are given in Figure 9.16. Note that in one of them the diagram is drawn as a parallelogram.

To be able to draw useful diagrams, it will be helpful to find additional conditions that guarantee that HK is a subgroup. Here, we give two such criteria (Propositions 9.30 and 9.32) which, in addition to Proposition 9.28, are quite useful. These propositions will be even more effective after learning about normal subgroups in Chapter 10.

Proposition 9.30. *Let G be a group and let H and K be subgroups of G. Assume that* $\gcd(|G : H|, |G : K|) = 1$. *Then $G = HK$.*

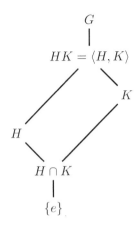

Figure 9.15. If HK is a subgroup, then we draw a parallelogram to emphasize that $|HK : H| = |K : H \cap K|$ and $|HK : K| = |H : H \cap K|$.

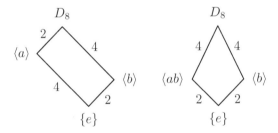

Figure 9.16. Two partial subgroup lattice diagrams for D_8

Proof. This was stated in Theorem 5.24(c) and you were asked to give a proof in Problem **5.2.15**.   □

Example 9.31. Let G be a group and let $H, K \leq G$. Assume that $|G : H| = 15$ and $|G : K| = 14$. Then, by Proposition 9.30, we have $G = HK$, and, thus by Proposition 9.28, we have that $|G : H| = |K : K \cap H|$. (See Figure 9.17.)

For the next proposition, recall that $\mathbf{N}_G(K)$ denotes the normalizer of K in G—Definition 4.24—and is the stabilizer of K in the action of G by conjugation on the set of the subgroups of G.

Proposition 9.32. *Let G be a group, and let H and K be subgroups of G. Assume that $H \leq \mathbf{N}_G(K)$, then HK is a subgroup of G.*

Proof. This was Problem **5.1.12**, but, for completeness, we give the proof here. To show that HK is a subgroup, by Problem **2.6.33** (restated in Proposition 9.28), it is enough to show that $HK = KH$.

Now if $x \in \mathbf{N}_G(K)$, then by definition, $xKx^{-1} = K$, and this means that $xK = Kx$ (Problem 5.1.10). We are given that every element of H is an element of the normalizer of K, and so we know that for all $h \in H$ we have $hK = Kh$. Now

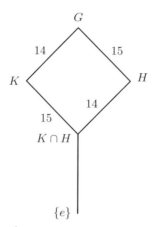

Figure 9.17. If $|G : K|$ and $|G : H|$ are relatively prime, then $G = HK$ and $|G : H| = |K : K \cap H|$.

let h and k be arbitrary elements of H and K, respectively. We have

$$hk \in hK = Kh \subseteq KH.$$

Hence $HK \subseteq KH$. Similarly $KH \subseteq HK$, and the proof is complete. □

More Than Two Subgroups. It is of course possible to draw partial lattice diagrams consisting of several subgroups. However, it is more difficult to do so while keeping track of parallelograms and edge lengths. We end this chapter with a couple of examples.

Example 9.33. A partial lattice diagram of D_8 consisting of several subgroups is given in Figure 9.18.

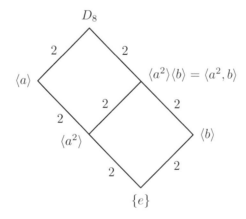

Figure 9.18. A partial lattice diagram for D_8

Example 9.34. A partial lattice diagram of S_5 is given in Figure 9.19.

Using this diagram, one can make a variety of arguments. For example, does S_4 have a subgroup of order 12 that contains S_3? If we just consider sizes, we will

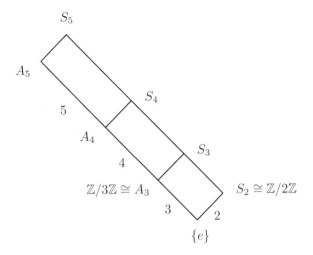

Figure 9.19. A partial lattice diagram for S_5

not arrive at a contradiction since S_3 has six elements and S_4 has 24 elements. However, if there were such a subgroup H, what would its intersection with A_4 be? Since $2 = |S_4 : A_4| \geq |H : H \cap A_4|$, we have to have $|H : H \cap A_4| = 2$. It is clear from the diagram that in this case, $H \cap A_4$ will be a subgroup of order 6 in A_4. But (by Problem **5.1.19**) A_4 does not have a subgroup of order 6, and thus S_4 does not have a subgroup of order 12 that contains S_3.

Remark 9.35. Problems 9.2.9 and 9.2.10 are often used when manipulating partial lattice diagrams of groups.

Problems

9.2.1. Verify that Figure 9.11 is indeed the lattice diagram of A_4, the alternating group of degree 4.

9.2.2. Assume that g_1 and g_2 are 3-cycles in S_4 (e.g., (1 2 3)). Let G be the subgroup of S_4 generated by g_1 and g_2. What are the possibilities for $|G|$? Which familiar groups can G be isomorphic to?

9.2.3. Let G be a group of order 200. Show that we can always find two nontrivial subgroups A and B such that $G = AB$ and $A \cap B = \{e\}$.

9.2.4. Let $G_1 = \mathbb{Z}/12\mathbb{Z}$ and $G_2 = \mathbb{Z}/6\mathbb{Z} \times \mathbb{Z}/2\mathbb{Z}$. Can you find two identical partial lattice diagrams, one for each of these groups? In other words, for $i = 1, 2$, find subgroups H_i and K_i of G_i such that the two partial lattice diagrams (one for $i = 1$ and one for $i = 2$) consisting of $\{e\}$, H_i, K_i, $H_i \cap K_i$, $\langle H_i, K_i \rangle$, and G_i are identical diagrams.

9.2.5. The group G has $720 = 2^4 \times 3^2 \times 5$ elements, N is a subgroup of G of order 60, and H is a Sylow 2-subgroup of G. Assume that $\mathbf{N}_G(N) = G$.
   (a) What can you say about $|NH|$? What about $|NH : H|$?

(b) Is $N \cap H$ a Sylow subgroup of N? Is it a Sylow subgroup of H?

9.2.6. The group G has 270 elements, and Q is a subgroup of G of order 9. Assume $\mathbf{N}_G(Q) = G$, and let P be a Sylow 3-subgroup of G. What can you say about $|PQ|$ and $|P \cap Q|$?

9.2.7. Let $G = D_{10} \times \mathbb{Z}/7\mathbb{Z}$, where, as usual, $D_{10} = \langle a, b \mid a^5 = b^2 = e, ba = a^4 b \rangle$. Let $H = \langle (b, 1) \rangle$ and $K = D_{10} \times \{0\}$.
  (a) Draw a partial lattice diagram of subgroups of G that includes the subgroups H and K. Include also $H \cap K$ and $\langle H, K \rangle$. Label the edges with appropriate numbers, and give reasons for what you have done.
  In what follows, you may refer back to your diagram.
  (b) What is $\mathbf{N}_G(H)$, the normalizer of H in G? Why?
  (c) Is $\mathbf{N}_G(K) = G$?
  (d) What is $|\mathrm{cl}_G(b, 1)|$? Note that $\mathrm{cl}_G(x)$ is the conjugacy class of x in G and that we are only interested in the order of this set.
  (e) Can you add the subgroup $L = \langle (a, 0) \rangle$ to your lattice diagram?

9.2.8. Let H and K be subgroups of a finite group G, and assume that $G = HK$. Can we find a set of elements $\{k_1, \ldots, k_\ell\}$ such that they are a set of coset reps for the right cosets of H in G *and* for the right cosets of $H \cap K$ in K?

9.2.9. Let H, K, and L be subgroups of a group G. Assume $HK = G$ and $H \leq L \leq G$ (see the diagram on the left of Figure 9.20). Show that $H(L \cap K) = L$.

  In other words, assume that a partial lattice diagram of subgroups of a group G is a parallelogram made of G, H, K, and $H \cap K$, and we have a subgroup L between H and G (see the diagram on the left of Figure 9.20). Then $L \cap K$ is a subgroup between $H \cap K$ and K, and we can draw the edge between L and $L \cap K$. You are asked to show that you can always draw this edge so as to make two parallelograms and that the information gleaned from the diagram on the right of Figure 9.20 is correct. Namely, $H(L \cap K) = L$.

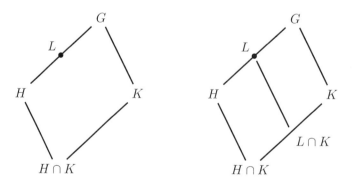

Figure 9.20. If $HK = G$, $H \leq L \leq G$, and we draw the edge between L and $L \cap K$, we get two parallelograms and $H(L \cap K) = L$.

9.2.10. Let H, K, and L be subgroups of a group G. Assume $HK = G$ and $H \cap K \leq L \leq K$ (see the diagram on the left of Figure 9.21). Further assume that LH is a subgroup of G and show that $LH \cap K = L$.

In other words, assume that a partial lattice diagram of subgroups of a group G is a parallelogram made of G, H, K, and $H \cap K$, and we have a subgroup L between $H \cap K$ and K (see the diagram on the left of Figure 9.21). Then LH is located between H and G and may or may not be a subgroup. If LH is a subgroup, we can draw the edge between L and LH, and the assertion is that we get two parallelograms and the information gleaned from the diagram on the right of Figure 9.21 is correct. Namely, $LH \cap K = L$.

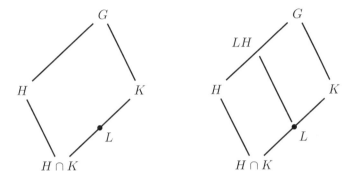

Figure 9.21. If $HK = G$, $H \cap K \leq L \leq K$, LH a subgroup, and we draw the edge between L and LH, we get two parallelograms and $LH \cap K = L$.

9.3. More Problems and Projects

Problems

9.3.1. **The group** $\mathrm{SL}(2,3)$. Recall—from Definition 1.63—that $\mathrm{SL}(2,3)$ is the group of 2×2 invertible matrices with determinant 1 over the field $\mathbb{Z}/3\mathbb{Z}$. In Theorem 1.64—proved in Problem *1.4.9*—we showed that this is a group of order 24. Problem 2.6.17 gave a cyclic subgroup of order 6 for $\mathrm{SL}(2,3)$, and Problem 5.4.1 asserted that $\mathrm{SL}(2,3)$ has a unique element of order 2 and no subgroup of order 12. The Hasse diagram of the lattice of subgroups of $\mathrm{SL}(2,3)$ is given in Figure 9.22.
   (a) If you haven't already done so, do Problems *1.4.9*, 2.6.17, and especially 5.4.1.
   (b) Add all the edge lengths in the Hasse diagram in Figure 9.22.
   (c) For each subgroup in Figure 9.22, find a set of generators.

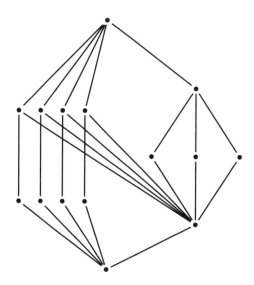

Figure 9.22. The lattice of subgroups of SL(2, 3)

(d) For each subgroup of SL(2, 3) find a familiar group to which it is isomorphic. Does the Hasse diagram help in identifying the subgroups? Which ones are the Sylow subgroups?

(e) What is $\mathbf{Z}(\mathrm{SL}(2,3))$, the center of SL(2, 3)?

(f) Let Ω be the set of subgroups of SL(2, 3). The group SL(2, 3) acts on Ω by conjugation. Identify all the orbits of the action.

(g) For each $H \leq \mathrm{SL}(2,3)$, find $\mathbf{N}_{\mathrm{SL}(2,3)}(H)$, the normalizer of H.

9.3.2. **Quasinormal subgroups.** Let H be a subgroup of a group G. We say that H is *quasinormal*[1] if $JH = HJ$ for all subgroups $J \leq G$. Prove that the following conditions are equivalent:

(a) H is quasinormal in G.

(b) $JH = HJ$ for every cyclic subgroup J of G.

(c) For every $g \in G$ and $h \in H$, there exists $r \in \mathbb{Z}$ and $h' \in H$ such that $hg = g^r h'$.

(d) For every subgroup $J \leq G$, we have $HJ = \langle H, J \rangle$.

If G is a finite group, then show that the above conditions are equivalent to the following

(e) For every subgroup J of G, $|\langle H, J \rangle : H| = |J : H \cap J|$.

9.3.3. **The necklace poset.** Let $\mathbf{2}^{[4]}$ denote the poset of subsets of $[4] = \{1, 2, 3, 4\}$ ordered by inclusion. Let $G = \langle (1\ 2\ 3\ 4) \rangle$ be a subgroup of S_4. Extend the action of S_4 on $[4]$ to an action of S_4 on $\mathbf{2}^{[4]}$. In other words, if $\sigma \in S_4$ and $A \subseteq [4]$, then to find $\sigma \cdot A$ just apply σ to every element of A (see Example 4.3 on page 86). Now since $G \leq S_4$, the group G also acts on $\mathbf{2}^{[4]}$.

(a) Find the orbits of the action of G on $\mathbf{2}^{[4]}$.

---

[1] Adapted from Hickerson, Stein, and Yamaoka [**HSY90**].

Let $\mathbf{2}^{[4]}/G$ denote the set of orbits of the action of G on $\mathbf{2}^{[4]}$. If $\mathcal{O}_1$, $\mathcal{O}_2 \in \mathbf{2}^{[4]}/G$, then say $\mathcal{O}_1 \leq \mathcal{O}_2$ if some element of $\mathcal{O}_1$ is a subset of some element of $\mathcal{O}_2$.

(b) Is the set of orbits, $\mathbf{2}^{[4]}/G$, together with the relation $\leq$ a partial order?

(c) Draw the Hasse diagram of $(\mathbf{2}^{[4]}/G, \leq)$.

Chapter 10

Acting on Its Subgroups:
Normal Subgroups
and Quotient Groups

... where normal subgroups are defined in a number of equivalent ways,
normalizers are explored, and quotient groups are introduced.

We now study a general and very important construction that is used in many
parts of algebra and mathematics. In some sense this a generalization of the idea
of division. When we divide 12 by 4, we get 3. This can be thought of as dividing a
pile of 12 objects into three sets of four objects each. In other words, a set X with
12 elements was partitioned into subsets each with four elements. Thinking of the
parts as individual objects, we have a new set with three elements.

To be able to carry out this construction in groups, we need the concept of
normal subgroups. This concept was first introduced by Galois and was used by
him to understand the solvability of equations of higher degree. In a sense, some
subgroups are better than others, and these are called normal subgroups. There
are a number of equivalent ways to define normal subgroups. We could first try
to define "division" for subgroups and extract the necessary definition from there.
Alternatively, we could first consider homomorphisms—we defined them in Definition 2.51, we will study them in some detail in the next chapter, and they play the
same role in group theory as linear transformations do in linear algebra—and then
define normal subgroups as the subgroups that occur as the kernels of these. In
this text, and in keeping with the theme of the book, we define normal subgroups
through group actions.

10.1. Normal Subgroups

We have seen the value of considering the center of a group. We defined the center
as the set of all elements of the group that commute with every element of the group

(Definition 2.72). Alternatively, we could have defined the center in terms of the conjugation action (Definition 4.13). Let the group G act on itself by conjugation (that is, $g \cdot x = gxg^{-1}$). Elements in the center are exactly those that are fixed by this action and have an orbit of size 1.

But we have seen another conjugation action as well. We are referring to the conjugation action of G on the set of subgroups of G. In this action, the group G acts on the set of all of its subgroups by $g \cdot H = gHg^{-1}$, and, in analogy with the central elements of the group, we may ask which subgroups are fixed by this action.

First, recall that if G is a group, H a subgroup of G, and $x \in G$, then xHx^{-1} is also a subgroup of G (Proposition 2.78(b)) that has the same cardinality as H (Problem **5.1.9**)—in fact, xHx^{-1} is isomorphic to H (Problem **2.6.27**)—and is called a *conjugate* of H. If we let Ω be the set of subgroups of G, then we have defined (Definition 4.14 and Problem *4.1.5*) the *conjugation action* of G on Ω by

$$x \cdot H = xHx^{-1} \quad \text{for } x \in G, \ H \in \Omega.$$

In this action, the stabilizer of a subgroup $H \in \Omega$ is called the *normalizer* of H in G and is denoted by $\mathbf{N}_G(H)$ (Definition 4.24).

Definition 10.1. Let G be a group, and let H be a subgroup of G. If, in the conjugation action of G on the set of all subgroups of G, the subgroup H is fixed (i.e., has an orbit of size 1), then H is called a *normal subgroup* of G. If H is a normal subgroup of G, we write $H \lhd G$, and we say H is normal in G.

Just as for the center, we could have defined normal subgroups without mentioning actions. We will now point out several equivalent definitions for normal subgroups. All of these are just a (superficial but useful) reformulation of the definition.

Directly from the definition, we have that $H \lhd G$ if and only if $xHx^{-1} = H$ for all $x \in G$.

Now $xHx^{-1} = H$ if and only if $xH = Hx$ (for one direction, we multiply on the right by x, and, for the converse we multiply on the right by x^{-1}). So, $H \lhd G$ if and only if $xH = Hx$ for all $x \in G$.

We can also restate the definition of a normal subgroup in terms of the stabilizer of the action. In the action of a group G on a set Ω, to say that an element $\alpha \in \Omega$ is fixed by the action is the same as saying that the orbit of α in Ω is of size 1. This is equivalent to saying that the stabilizer of α in G is all of G.

In the case of the conjugation action of a group on itself, the stabilizer of a group element is the centralizer of that element (Lemma 4.23). Hence—and this is clear directly anyway—a group element is in the center of G if and only if its centralizer is the whole group.

Let G be a group and let $H \leq G$. Recall that the normalizer of H in G is $\mathbf{N}_G(H) = \{x \in G \mid xHx^{-1} = H\}$ (see Definition 4.24). In the conjugation action of a group on its subgroups, the stabilizer of a subgroup is its normalizer. Thus—again this is clear also directly from the definition—a subgroup is normal in G if and only if its normalizer is the whole group.

We gather our conclusions in a lemma:

Lemma 10.2. *Let G be a group with $N \leq G$. Then the following are equivalent:*

(a) $N \lhd G$,

(b) $xNx^{-1} = N$ for all $x \in G$,

(c) $xN = Nx$ for all $x \in G$,

(d) $\mathbf{N}_G(N) = G$.

Remark 10.3. Each of the reformulations in Lemma 10.2 are useful. We defined normal subgroups in terms of an action. The condition $xNx^{-1} = N$ turns our focus to conjugate subgroups. (If N is a subgroup of G, then the set xNx^{-1} is also a subgroup of G, is called a *conjugate* of N, and has the same cardinality as N.) The condition $xN = Nx$ is about right and left cosets and allows us to bring in what we know about cosets. Finally, the last equivalence points out the importance of the subgroup $\mathbf{N}_G(N)$, and, in fact, prompts us to replace the question "Is N normal in G?" with the question "What is the normalizer of N in G?"

What is not clear so far is the significance of normal subgroups. Why are they important, what do they do for us, and how do they help us answer questions about groups? You will start seeing the answers in Section 10.3 and subsequent chapters.

Example 10.4. Let $G = D_8 = \langle a, b \mid a^4 = b^2 = e, ba = a^3b \rangle$, and let H be the subgroup $\langle a \rangle = \{e, a, a^2, a^3\}$. We claim that $H \lhd G$.

One way to show this is to show that, for every $x \in G$, we have $xH = Hx$. Let us try this for $x = b$.

We have $Hb = \{b, ab, a^2b, a^3b\}$ while $bH = \{b, ba, ba^2, ba^3\}$. Are these equal? Certainly, $ab \neq ba$, but that is not quite relevant. We need to see if the set Hb has exactly the same elements as the set bH. Rewriting the elements of bH, we get $bH = \{b, a^3b, a^2b, ab\}$, and these are exactly the elements of Hb. Hence, $Hb = bH$.

We can separately check every element of the group, or we can argue that for any $x \in G$, we have either $x \in H$ or $x \notin H$. In the first case, $xH = H = Hx$, and in the second case, both Hx and xH are cosets of H in G and consist of the elements of G not in H, and so they must be equal. We conclude that, for all $x \in G$, we have $xH = Hx$ and hence $H \lhd G$.

Now let $K = \langle b \rangle = \{e, b\} \leq G$. Note that $aK = \{a, ab\}$ while $Ka = \{a, ba = a^3b\}$. These sets are not the same and so K is not a normal subgroup of G.

Some subgroups are clearly normal. We gather some of these obvious facts here. These all follow directly from one of the conditions in Lemma 10.2, and the reader, by referring back to the lemma, should convince herself of their validity.

Lemma 10.5. *Let G be a group. Then*

(a) $\{e\} \lhd G$,

(b) $G \lhd G$,

(c) $\mathbf{Z}(G) \lhd G$, *and*

(d) *if G is abelian, then all subgroups of G are normal.*

In Example 10.4, we argued that $\langle a \rangle \lhd D_8$. In Problem **10.1.8** you are asked to generalize this to any subgroup of index 2 and prove:

Proposition 10.6. *Let* $H \leq G$ *be groups. If* $|G : H| = 2$, *then* $H \triangleleft G$. *In particular, for* $n > 1$, $A_n \triangleleft S_n$.

Example 10.7. Assume that among the subgroups of a group G there is only one subgroup of order 10. We claim that this subgroup must be normal. Call the subgroup H and use condition (b) of Lemma 10.2. Let $x \in G$. The set xHx^{-1} is also a subgroup and it also has ten elements. But H was the unique subgroup of order 10. Hence, $H = xHx^{-1}$ for all $x \in G$.

Remark 10.8. It should be clear by now that to tell whether a subgroup is a normal subgroup, it is not enough to know internal information about the subgroup. What matters is how the subgroup is situated in the big group. A group of order 8 could have a normal subgroup of order 2 as well as a non-normal subgroup of order 2. In fact in $D_8 = \langle a, b \mid a^4 = b^2 = e, ba = a^3b \rangle$, the subgroup generated by a^2 is normal while the subgroup generated by b is not. Both of these subgroups are isomorphic to $\mathbb{Z}/2\mathbb{Z}$, but one of them is normal while the other is not. "Being normal is not about who you are, but about your relation to the rest of the group."

Normal Sylow Subgroups. We should note that we have also already seen normal subgroups in the context of Sylow p-subgroups.[1] The same argument as the one in Example 10.7 shows that if a group G has a unique Sylow p-subgroup, then this subgroup must be normal. In fact, in Corollary 7.15, we proved the converse by proving that P is a unique Sylow p-subgroup if and only if $\mathbf{N}_G(P) = G$. The converse—that is if a Sylow p-subgroup is normal, then it must be unique—is not a straightforward fact and it followed from the Sylow D Theorem 7.13. We can now reword Corollary 7.15:

Corollary 10.9. *Let* G *be a finite group, let* p *be a prime number, and let* $P \in \mathrm{Syl}_p(G)$. *Then* $P \triangleleft G$ *if and only if* $\left|\mathrm{Syl}_p(G)\right| = 1$.

Lattice Diagrams and Normal Subgroups. We saw[2] in Chapter 9 that in drawing lattice diagrams and in analyzing the subgroups of a group, we are interested in knowing when the product of two subgroups is a subgroup. Normal subgroups are particularly nice in this regard.

Proposition 10.10. *Let* G *be a group, and let* H *and* K *be subgroups of* G. *Assume that* $H \triangleleft G$. *Then* HK *is a subgroup of* G.

Proof. From Proposition 2.78(c) (you gave a proof in Problem *2.6.33*), we know that HK is a subgroup if and only if $HK = KH$. For normal subgroups $Hx = xH$ for all $x \in G$, hence we have

$$HK = \bigcup_{x \in K} Hx = \bigcup_{x \in K} xH = KH. \qquad \square$$

Remark 10.11. Actually, previously we had used the same argument as in the proof of Proposition 10.10 to prove a more general fact! Proposition 9.32 proved that if H and K are subgroups of a group G and $K \leq \mathbf{N}_G(H)$, then HK is a

---

[1] If you have not done Chapter 7 yet, then skip this paragraph, and come back to it after doing the Sylow theorems.

[2] While Proposition 10.10 stands on its own, the rest of the discussion in this section assumes familiarity with the material in Chapter 9.

subgroup of G. (In fact, you had proved this even earlier in Problem **5.1.12**.) Now if $H \lhd G$, then $\mathbf{N}_G(H)$ is all of G, and hence the condition $K \leq \mathbf{N}_G(H)$ is automatically satisfied.

Remark 10.12. Because of Proposition 10.10, when we draw a partial lattice diagram involving two subgroups, and if we know one of them to be normal, then we draw a parallelogram. This is because when HK is a subgroup of G, we have that $|HK : H| = |K : K \cap H|$ (Theorem 9.28), and so the opposite sides of the quadrilateral with vertices H, K, $H \cap K$, and HK have equal lengths.

In addition, in drawing partial lattice diagrams, if possible and not too distracting, we put a double edge to signal that $H \lhd G$. See Figure 10.1.

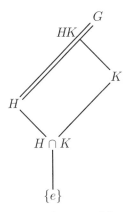

Figure 10.1. A partial lattice diagram with two subgroups when one subgroup is known to be normal

To illustrate how to use Proposition 10.10 and partial lattice diagrams as in Figure 10.1, we prove a fact about normal p-groups and Sylow p-subgroups:

Proposition 10.13. *Let G be a finite group, and let P be an arbitrary Sylow p-subgroup of G. Let Q be a normal subgroup of G and assume that the order of Q is a power of a prime p. Then Q is necessarily a subgroup of P.*

Proof. We begin by drawing a partial lattice diagram of G that includes P and Q. Since $Q \lhd G$, the diagram will be similar to Figure 10.1. In other words, QP is a subgroup of G and $|QP : P| = |Q : Q \cap P|$ as well as $|QP : Q| = |P : Q \cap P|$. See Figure 10.2.

We now use the diagram as a guide to argue as follows. The diagram reminds us that $|Q| = |Q \cap P| |Q : Q \cap P|$. Since $|Q|$ is a power of the prime p, then so is $|Q : Q \cap P|$. On the other hand, P is a Sylow p-subgroup, and hence $p \nmid |G : P|$. But—again we "see" this in the diagram—$|G : P| = |G : QP| |QP : P|$ and so p cannot divide $|QP : P|$. Now $|QP : P| = |Q : Q \cap P|$ ("the opposite sides of the parallelogram are of equal size"), and this integer is a power of p and yet is not divisible by p. Hence, $|QP : P| = |Q : Q \cap P| = 1$. This means that $Q \cap P = Q$, and so $Q \subseteq P$.

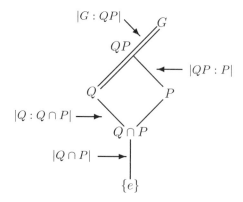

Figure 10.2. The subgroup Q is a normal p-group and P is a Sylow p-subgroup.

□

Problems

10.1.1. Find all normal subgroups of D_8 and of S_3.

10.1.2. Let $D_8 = \langle a, b \mid a^4 = b^2 = e, ba = a^3b \rangle$, and let S_3 be the symmetric group of degree 3. Let $G = D_8 \times S_3$. Let $H = \langle b \rangle \times \langle (1\ 2\ 3) \rangle$ and $K = \langle a \rangle \times \langle (1\ 2\ 3) \rangle$ be subgroups of G. Is H a normal subgroup of G? What about K?

10.1.3. Find all normal subgroups of D_{10}. Can you generalize your assertion to D_{2p} where p is a prime?

10.1.4. Find all normal subgroups of A_4.

10.1.5. Let $G = \mathrm{SL}(2,3)$ be the group of 2×2 matrices with determinant 1 over $\mathbb{Z}/3\mathbb{Z}$. Let H be the subgroup consisting of 2×2 lower triangular matrices with 1's on the diagonal. Is H a normal subgroup of G?

10.1.6. Let $G = D_8 \times A_5$ and $H = D_8 \times \{e\}$. Is H a normal subgroup of G? Can you generalize?

10.1.7. Is $\mathrm{SL}(2, \mathbb{Z}) \triangleleft \mathrm{GL}(2, \mathbb{Z})$? (See Problems 2.1.4 and 5.1.7.)

10.1.8. Proof of Proposition 10.6. Let G be a group. Assume $H \leq G$ and $|G : H| = 2$. Prove $H \triangleleft G$.

10.1.9. Find a group G, with subgroups H and K, such that $H \triangleleft K$, $K \triangleleft G$, but H not normal in G.

10.1.10. Assume that H and K are normal subgroups of the group G. Is $HK \triangleleft G$?

10.1.11. Let G be a finite abelian group with no non-trivial normal subgroups (i.e., $\{e\}$ and G are the only normal subgroups of G). What can you say about G?

10.1.12. Let G be a group of order p^5, where p is a prime. Show that G has a normal subgroup of order p.

10.1.13. A group G has come to talk to you. You find out that G has $100 = 2^2 \times 5^2$ elements and that G has exactly one subgroup of order 10 whose name is H.

    (a) Remind yourself, by rereading Example 10.7, why H must be a normal subgroup of G.

    (b) Let $P \in \mathrm{Syl}_5(G)$. What can you say about $|P|$, $|H \cap P|$, and $|\langle H, P \rangle|$? Prove your assertions.

    (c) The group G is guaranteed to have subgroups of which sizes?

10.1.14. Let G be a finite group, and let N and H be subgroups of G. Assume N is normal in G and $\gcd(|G : H|, |N|) = 1$. Show that $N \leq H$.

10.1.15. Give an example of a group that has several normal subgroups of the same size.

10.1.16. Let G be a finite group, and let $N \leq G$. Let $|N| = n$ and $|G : N| = m$. Assume that $\gcd(n, m) = 1$. Prove that $N \triangleleft G$ if and only if G has exactly one subgroup of order n, namely N. To illustrate the possibilities, give two examples, one where N is normal and another where it is not (in your examples, both n and m should be greater than 1).

10.1.17. Show that A_4 is the only subgroup of order 12 in S_4.

10.1.18. Let $G = S_n$. Assume that H is a subgroup of G, that H contains at least one odd permutation, and that $|H| > 2$. Show that H has some non-trivial normal subgroup.

10.2. The Normalizer

Let a group G act on the set of its subgroups by conjugation. Let H be a particular subgroup of G. We have seen that the stabilizer of H in G for this action is called the *normalizer* of H in G and is denoted by $\mathbf{N}_G(H)$. Thus

$$\mathbf{N}_G(H) = \{x \in G \mid x \cdot H = H\} = \{x \in G \mid xHx^{-1} = H\} = \{x \in G \mid xH = Hx\}.$$

By Lemma 10.2, $H \triangleleft G$ if and only if $\mathbf{N}_G(H) = G$. Thus instead of asking "Is H normal in G?" we could—and should—ask, "What is the normalizer of H in G?" In this section, we explore the normalizers of subgroups a bit further.

Example 10.14. Let $G = D_8$. Then $\mathbf{N}_G(\langle b \rangle) = \langle a^2, b \rangle$, while $\mathbf{N}_G(\langle a^2 \rangle) = D_8$. The latter follows since $\langle a^2 \rangle$ is the center of the group and so it is normal in D_8. The former follows since both b and a^2 are in the normalizer and $\langle b \rangle$ is not normal in D_8 and hence $\mathbf{N}_G(\langle b \rangle)$ cannot be the whole group.

Recall that if G is a group and $X \subseteq G$, then $\mathbf{C}_G(X) = \{g \in G \mid gx = xg, \forall x \in X\}$ is the set of elements of G that commute with every element of X. For reference, we gather some straightforward facts about the normalizer.

Lemma 10.15. *Let G be a group, and let $H \leq G$. Then*

(a) $\mathbf{N}_G(H) = \{x \in G \mid x^{-1}Hx = H\} = \{x \in G \mid xH = Hx\}$,

(b) $\{e\} \leq H \leq \mathbf{N}_G(H) \leq G$,

Figure 10.3. If $H \leq G$, then $H \lhd \mathbf{N}_G(H)$.

(c) $H \lhd \mathbf{N}_G(H)$, *and*

(d) $\mathbf{C}_G(H) \leq \mathbf{N}_G(H)$.

Proof. (a) This is immediate from the definition and had been noted before (see Problem 5.1.10).

(b) We know that $\mathbf{N}_G(H)$ is a subgroup of G since it is the stabilizer of an action. Hence the only thing that remains to show is that $H \leq \mathbf{N}_G(H)$. Let $h \in H$. We have $hH = H = Hh$, and so $h \in \mathbf{N}_G(H)$.

(c) If $x \in \mathbf{N}_G(H)$, then by definition $xH = Hx$. Hence $xH = Hx$ for all $x \in \mathbf{N}_G(H)$, and hence $H \lhd \mathbf{N}_G(H)$. (See Figure 10.3.)

(d) If $x \in \mathbf{C}_G(H)$, then $xh = hx$ for all $h \in H$, and hence $xH = Hx$. Thus $x \in \mathbf{N}_G(H)$. □

Lemma 10.16. *Let G be a group, let $H \leq G$, and let $x \in G$. Assume H is finite and $xHx^{-1} \leq H$. Then $xHx^{-1} = H$.*

Proof. We have $xHx^{-1} \leq H$ and $|xHx^{-1}| = |H|$. The result follows. □

For Lemma 10.16, we need H to be finite. For the case when $|H| = \infty$, we have:

Lemma 10.17. *Let G be a group, and let $H \leq G$. Assume $xHx^{-1} \leq H$ for all $x \in G$. Then $xHx^{-1} = H$ for all $x \in G$.*

Proof. Let $g \in G$. We want to show that $gHg^{-1} = H$. We know that $gHg^{-1} \subseteq H$. Letting $x = g^{-1}$, we also know $g^{-1}Hg \subseteq H$. We get

$$H = gg^{-1}Hgg^{-1} \subseteq gHg^{-1}.$$

The result now follows. □

We expand Lemma 10.2 to gather a number of equivalent formulations of normality—most we have seen already—in one place:

Proposition 10.18. *Let G be a group, and let $H \leq G$. Then the following are equivalent:*

(a) $H \triangleleft G$,

(b) $xH = Hx$ *for all* $x \in G$,

(c) $xHx^{-1} = H$ *for all* $x \in G$,

(d) $xHx^{-1} \subseteq H$ *for all* $x \in G$,

(e) $xhx^{-1} \in H$ *for all* $x \in G$, $h \in H$,

(f) $\mathbf{N}_G(H) = G$,

(g) H *is the union of some conjugacy classes of* G.

Proof. The equivalence of all the parts has been proved except for the equivalence of the last part. This is left to you in Problem *10.2.15*. Note that not every union of conjugacy classes is a subgroup. However, if such a union is a subgroup, then it is a normal subgroup. $\qquad\qquad\square$

Remark 10.19. We list a number of noteworthy results—in addition to Proposition 10.18(g)—that you are asked to prove in the problems.

Lemma 10.20 (Problem **10.2.4**). *If M and N are both normal subgroups of a group G, then so is $M \cap N$.*

Lemma 10.21 (Problem **10.2.5**). *If N is a normal subgroup of a group G and H is a subgroup of G, then $H \cap N$ is a normal subgroup of H.*

Proposition 10.22 (Problem **10.2.17**). *The alternating group of degree 5, A_5, has no non-trivial normal subgroups.*

Theorem 10.23 (Problem **10.2.21**). *Let M and N be normal subgroups of a group G. Assume that $M \cap N = \{e\}$. Then every element of M commutes with every element of N.*

Problems

10.2.1. (Re)draw the lattice of subgroups of D_8. For each subgroup indicate its normalizer.

10.2.2. Repeat the previous problem for A_4.

10.2.3. Let $G = A_4$. Find two subgroups H and K of G such that $HK = G$. What are the normalizers of H and K?

10.2.4. If M and N are normal subgroups of a group G, show that $M \cap N$ is also a normal subgroup of G.

10.2.5. Let G be a group, let $N \triangleleft G$, and let $H \leq G$. Draw a partial lattice diagram that includes G, N, H, NH, and $H \cap N$. Prove that $H \cap N \triangleleft H$.

10.2.6. **Kernels of homomorphisms.** Let G and H be groups, and let $\phi : G \rightarrow H$ be a group homomorphism. In Problem 2.6.19, you were asked to show that $\ker(\phi) = \{g \in G \mid \phi(g) = e_H\}$ is a subgroup of G. Show that $\ker(\phi)$ is a normal subgroup of G.

10.2.7. **Stabilizers of elements in the same orbit.** Let the group G act on the set Ω. Let $g \in G$, and let $\alpha, \beta \in \Omega$. Assume that $g \cdot \alpha = \beta$. Show that $\mathrm{Stab}_G(\beta) = g\mathrm{Stab}_G(\alpha)g^{-1}$. Conclude that two elements in the same orbit have isomorphic stabilizers.

10.2.8. **A normal subgroup of the affine group.** Let $G = \mathrm{Aff}(1, \mathbb{R})$ be the affine group of degree 1 over $\mathbb{R}$. Recall that G is the set of functions $T_{a,b} \colon \mathbb{R} \to \mathbb{R}$ of the form $T_{a,b}(x) = ax + b$ where $a, b \in \mathbb{R}$ and $a \neq 0$ (see Definition 2.80 and the ensuing problems). Let N be the set of "translations" in G. In other words, $N = \{T_{1,b} \in \mathrm{Aff}(1, \mathbb{R}) \mid b \in \mathbb{R}\}$ is the set of those functions for which $a = 1$. Show that $N \lhd G$.

10.2.9. **Does Lemma 10.16 apply to infinite groups?** As in the previous problem, let $G = \mathrm{Aff}(1, \mathbb{R})$. Let $H \subseteq G$ be the set of "integer translations". In other words, H consists of those $T_{a,b} \in G$ for which $a = 1$ and $b \in \mathbb{Z}$.
   (a) Show that H is a subgroup of G.
   (b) Find $\{\, g \in G \mid gHg^{-1} \leq H \,\}$.
   (c) Find $\mathbf{N}_G(H)$.
   (d) Is this a counterexample to Lemma 10.16?

10.2.10. Let G be a group of order 99.
   (a) Can we always find subgroups P and Q such that $G = PQ$?
   (b) Assume $G = PQ$ and $Q \leq \mathbf{N}_G(P)$. What can you say about $|\mathbf{N}_G(P)|$?

10.2.11. Let $G = D_{34} = \langle a, b \mid a^{17} = b^2 = e, ba = a^{16}b \rangle$. Let $H = \langle a \rangle$ and $K = \langle b \rangle$.
   (a) Draw a partial lattice diagram of subgroups of G that includes the subgroups H and K. Include also $H \cap K$ and $\langle H, K \rangle$. Label the edges with appropriate numbers, and give reasons for what you have done.
   In what follows, you may refer back to your diagram.
   (b) Is $H \lhd G$? What is $\mathbf{N}_G(H)$? What is $\mathbf{N}_G(K)$? Why? Is $K \lhd G$?
   (c) What is $|\mathrm{cl}_G(a)|$? What is $|\mathrm{cl}_G(b)|$? Why? Note that $\mathrm{cl}_G(x)$ is the conjugacy class of x in G.
   (d) What is $\mathbf{Z}(G)$? Why?

10.2.12. Let H be a subgroup of the group G.
   (a) What is another name for $\mathbf{C}_G(H) \cap H$?
   (b) Draw a partial lattice diagram of G that contains $\{e\}$, H, $\mathbf{C}_G(H)$, $\mathbf{N}_G(H)$, $H \cap \mathbf{C}_G(H)$, $\langle H, \mathbf{C}_G(H) \rangle$, and G.

10.2.13. Let H be a subgroup of the group G, $N = \mathbf{N}_G(H)$, and $h \in H$. Show that N acts on H by conjugation. In this action, what is the stabilizer in G of h?

10.2.14. Let G be a finite group, let p be a prime, and let $P \in \mathrm{Syl}_p(G)$. Assume that Q is a p-subgroup of G (that is, $|Q|$ is a power of p) such that $Q \leq \mathbf{N}_G(P)$. Show that $Q \leq P$.

10.2.15. **Proof of Proposition 10.18(g).** Let N be a subgroup of G. Prove that N is normal if and only if N is the union of some of the conjugacy classes of G.

10.2.16. **Another proof that A_4 has no subgroup of order** 6. Use Proposition 10.6 and Problems *10.2.15* and 6.2.14 to prove that A_4, the alternating group of degree 4, has no subgroup of order 6.

10.2.17. **The alternating group of degree** 5. Use Problems **6.2.15** and *10.2.15* and Lagrange's theorem to prove that A_5, the alternating group of degree 5, has no non-trivial normal subgroups.

10.2.18. Let H be a cyclic subgroup of a finite group G. Assume H is normal in G. Show that every subgroup of H is normal in G.

10.2.19. Let $H, K \leq G$. Assume $G = \langle H, K \rangle$, $N \triangleleft H$, and $N \triangleleft K$. Is $N \triangleleft G$?

10.2.20. Let H and K be groups, and let $G = H \times K$. Is $H \times \{e\} \triangleleft G$?

10.2.21. **Proof of Theorem 10.23.** G is a group. $M \triangleleft G$ and $N \triangleleft G$. Assume $M \cap N = \{e\}$. Show that for all $m \in M$ and $n \in N$, we have $mn = nm$. Does this mean that G is abelian?

10.2.22. The government has been investigating the group G for alleged illegal activities. In the investigation two facts have become clear: G has $242 = 2 \times 11^2$ members and G has a normal subgroup of order 2.
  (a) Let $P \in \mathrm{Syl}_{11}(G)$. What is $|P|$? Is $P \triangleleft G$?
  (b) For which integers k are we guaranteed to have a subgroup of G of order k? Give your reasons.
  (c) Draw a partial lattice diagram of subgroups for G that includes G, $\{e\}$, a Sylow 11-subgroup, and a Sylow 2-subgroup. Can you add any other subgroups to the diagram?
  (d) Among the subgroups of G, subgroups of which size are guaranteed to be abelian?

10.2.23. Let G be a finite group, and let $H \leq G$. Assume that $\bigcup_{g \in G} gHg^{-1} = G$.
  (a) Show that

$$|G| \leq 1 + |G : \mathbf{N}_G(H)| \, (|H| - 1).$$

  (b) Prove that $H = G$.
  (c) Can you give a different proof of the previous part using the result in Problem 8.1.5?

10.3. Quotient Groups

We are now ready to define a sort of "division" for groups. Given a group G and a normal subgroup N, we will be able to define a new group that can be thought of as the result of "dividing" G by N. This construction is a very important one, and analogs of it are present in much of modern mathematics. In the case of groups, we shall see that we can only "divide" by *normal* subgroups, and this explains the importance of normal subgroups. This whole topic is very much related to the notion of homomorphisms—which were defined in Chapter 2 but will be treated more systematically in the next chapter—and, we shall see there that normal subgroups, quotient groups, and homomorphisms are different ways of looking at the same phenomenon.

We begin with recalling the definition of G/N from Definition 5.7.

Definition 10.24 (G/N)**.** Let G be a group, and let $N \leq G$. The set $\{Nx \mid x \in G\}$, of right cosets of N in G, is called "G mod N" and is denoted by G/N.

Note that we have denoted $|G/N|$ by $|G : N|$ and, in particular, if G is finite, $|G/N| = \frac{|G|}{|N|}$.

The set G/N is a set of subsets of G and, at first, it is not clear why we are considering it. The point is that if N is normal, we will turn G/N into a group. This will be an unusual group since its elements are sets of elements of another group. To motivate the definition of this group, let us consider D_8 and its normal subgroup generated by a. Thus let $H = \langle a \rangle$, then D_8/H consists of two cosets : H and Hb. If you look at the multiplication table of D_8 and not pay much attention to details, you will see the pattern in Figure 10.4.

		H: $e\ a\ a^2\ a^3$	Hb: $b\ ab\ a^2b\ a^3b$
H	e, a, a^2, a^3	H	Hb
Hb	b, ab, a^2b, a^3b	Hb	H

Figure 10.4. A pattern in the multiplication table of D_8

The parts of the table that are labeled H consist only of elements of H, and likewise the parts of the table that are labeled Hb consist only of elements of the coset Hb. We can think of the pattern in Figure 10.4 as a kind of multiplication of cosets. We would write this multiplication table as:

	H	Hb
H	H	Hb
Hb	Hb	H

We recognize the above table as the multiplication table for the group $\mathbb{Z}/2\mathbb{Z}$. So we would like to be able to say that D_8/H is a group isomorphic to $\mathbb{Z}/2\mathbb{Z}$. We will now give the precise definitions.

Definition 10.25 (An operation on G/N)**.** Let G be a group, and let $N \lhd G$. Define $*$ on G/N by

$$(Nx) * (Ny) = Nxy.$$

We have defined an operation on cosets, and so we have to make sure that it is well defined. One more time, the worry is that, since each coset has a number of different names, it may be that the result of the operation depends on the particular name chosen. If this were possible, then we really have not defined an operation on cosets.

Lemma 10.26. *Let G be a group, and let $N \lhd G$. Then the operation $*$ is a well defined operation on G/N.*

Proof. We need to show that the result of the product of Nx and Ny does not depend on the representatives of the two cosets. So assume $Nx = Nx'$ and $Ny = Ny'$. We must show $Nxy = Nx'y'$.

Since $Nx = Nx'$, we have $x' = n_1 x$ for some $n_1 \in N$. Likewise, from $Ny = Ny'$, we get $y' = n_2 y$ for some $n_2 \in N$. Now we know $Nn_1 = N = Nn_2$ since $n_1, n_2 \in N$, and we also know $Nx = xN$ since $N \lhd G$. Hence,

$$Nx'y' = Nn_1xn_2y = Nxn_2y = xNn_2y = xNy = Nxy. \qquad \square$$

Example 10.27. In this example we show that the normality condition in the above lemma is essential. Let $G = D_8$, and let $H = \langle b \rangle$. If we define a multiplication on cosets of H as before, we would have $(Ha) * (Ha) = Ha^2$. However, $Ha = Ha^3b$, and so $(Ha) * (Ha) = (Ha^3b) * (Ha^3b)$ should be $Ha^3ba^3b = H$. Thus our definition of $*$ is not well defined since the product of the same two cosets wants to be both H and Ha^2. The problem arose since H is not a normal subgroup of G.

In the case of the normal subgroup $\langle a \rangle$, we saw in Figure 10.4 that arranging the multiplication table according to the cosets of $\langle a \rangle$ creates a distinctive pattern. This will not happen with non-normal subgroups. See Figure 10.5.

		H		Ha		Ha^2		Ha^3	
		e	b	a	a^3b	a^2	a^2b	a^3	ab
H	e	e	b	a	a^3b	a^2	a^2b	a^3	ab
	b	b	e	a^3b	a	a^2b	a^2	ab	a^3
Ha	a	a	ab	a^2	b	a^3	a^3b	e	a^2b
	a^3b	a^3b	a^3	a^2b	e	ab	a	b	a^2
Ha^2	a^2	a^2	a^2b	a^3	ab	e	b	a	a^3b
	a^2b	a^2b	a^2	ab	a^3	b	e	a^3b	a
Ha^3	a^3	a^3	a^3b	e	a^2b	a	ab	a^2	b
	ab	ab	a	b	a^2	a^3b	a^3	a^2b	e

Figure 10.5. The multiplication table of D_8 organized according to the cosets of $H = \langle b \rangle$, a subgroup that is not normal. The table does not exhibit a pattern similar to Figure 10.4.

Remark 10.28. Note that the operation $*$ is the usual multiplication of subsets of a group. In other words to find $(Nx) * (Ny)$, we can just multiply every element of Nx by every element of Ny. To see this, note that since N is normal we have $Nx = xN$, and hence

$$(Nx)(Ny) = N(xN)y = NNxy = Nxy = (Nx) * (Ny).$$

Theorem 10.29. *Let G be a group, let $N \lhd G$, and let the operation $*$ on G/N be defined as above. Then $(G/N, *)$ is a group.*

Proof. We have already proved that $*$ is a well-defined multiplication. Since the product of two right cosets is defined to be another right coset, the operation is closed. Associativity follows from the associativity of the group operation

$$(Nx * Ny) * Nz = (Nxy) * Nz = N(xy)z = Nx(yz) = Nx * (Nyz) = Nx * (Ny * Nz).$$

The identity element is the coset N, since, by definition, $N * Nx = Nx$ and $Nx * N = Nx$. Finally, the inverse of Nx is Nx^{-1}, and the proof is complete. $\qquad\square$

Definition 10.30 (Quotient group). Let G be a group, let $N \lhd G$, and let the operation $*$ on G/N be defined as above. $(G/N, *)$ is called a *quotient group* or a *factor group* of G.

As we have done with all groups, we will usually drop the $*$ and write $NxNy$ for $(Nx) * (Ny)$.

Example 10.31. $Z = \langle a^2 \rangle$ is a normal subgroup of D_8. The right cosets of Z in D_8—which are the elements of D_8/Z—are

$$Z = \{e, a^2\}, \ Za = \{a, a^3\}, \ Zb = \{b, a^2 b\}, \ Zab = \{ab, a^3 b\}.$$

We can read off the multiplication table for D_8/Z from Figure 10.6. Note that D_8 has eight elements, while D_8/Z has four elements. Each element of D_8/Z is a right coset but, even so, we have been able to define a multiplication for these cosets. We recognize the multiplication table in Figure 10.6 as the multiplication table for the Klein 4-group. (Recall that, up to isomorphism, there are only two groups of order 4: the cyclic group of order 4 and the Klein 4-group.)

		Z $e \quad a^2$	**Z**a $a \quad a^3$	**Z**b $b \quad a^2 b$	**Z**ab $ab \quad a^3 b$
Z	e a^2	**Z**	**Z**a	**Z**b	**Z**ab
Za	a a^3	**Z**a	**Z**	**Z**ab	**Z**b
Zb	b $a^2 b$	**Z**b	**Z**ab	**Z**	**Z**a
Zab	ab $a^3 b$	**Z**ab	**Z**b	**Z**a	**Z**

Figure 10.6. The multiplication table of D_8 organized according to the cosets of the center.

Thus so far we have seen that $D_8/\langle a \rangle \cong \mathbb{Z}/2\mathbb{Z}$ and $D_8/\langle a^2 \rangle \cong \mathbb{Z}/2\mathbb{Z} \times \mathbb{Z}/2\mathbb{Z}$.

Example 10.32. Let $G = \mathbb{Z}/12\mathbb{Z}$. G is abelian and hence all of its subgroups are normal. Let $H = \{0, 4, 8\}$. What is G/H?

The elements of G/H are the cosets of H in G. The operation in $\mathbb{Z}/12\mathbb{Z}$ is written additively and hence the cosets are $H = \{0, 4, 8\}$, $H + 1 = \{1, 5, 9\}$,

$H + 2 = \{2, 6, 10\}$, and $H + 3 = \{3, 7, 11\}$. Thus $G/H = \{H, H + 1, H + 2, H + 3\}$. We see that $G/H = \langle H + 1 \rangle$, and hence $G/H \cong \mathbb{Z}/4\mathbb{Z}$.

Example 10.33. This final example shows the reason for the notation that we have used for the cyclic groups $\mathbb{Z}/n\mathbb{Z}$. Let $G = \mathbb{Z}$ be the group of integers under addition. Let $H = n\mathbb{Z}$ be the subgroup consisting of multiples of n. Then $G/H = \{H, H + 1, H + 2, \ldots, H + (n - 1)\} \cong \mathbb{Z}/n\mathbb{Z}$. Thus the group $\mathbb{Z}/n\mathbb{Z}$ is really $\mathbb{Z}$ mod $n\mathbb{Z}$.

Quotient Groups and Lattice Diagrams. When we draw a partial lattice diagram of a group and some of its subgroups, there is a way to think about quotient groups that will sometimes be helpful. If G is a finite group and $N \lhd G$, then we have already labeled the edge from N to G with $|G : N|$, the index of N in G. This is the size of the quotient group G/N. In fact, it will be useful to think of the part of the diagram between N and G as a partial lattice diagram for G/N. It is actually true that if you draw N, G, and all subgroups between them (i.e., those subgroups of G that contain N), then you get a lattice identical to the lattice of subgroups of G/N! This is an interesting and very useful result which will be the main aim of the so-called "homomorphism theorems" of the next chapter. For now, we will just use the heuristic that the part of the lattice diagram above N is the lattice diagram for G/N. You can imagine that when you create G/N, you really just chop off anything below N, and of course you are making N into the new identity.

Example 10.34. We have seen that $D_8/\mathbf{Z}(D_8)$ is isomorphic to $\mathbb{Z}/2\mathbb{Z} \times \mathbb{Z}/2\mathbb{Z}$. We could have guessed this by comparing the top portion of the appropriate lattice diagrams. Figure 10.7 shows the partial lattice diagram of D_8 consisting of those subgroups that contain a^2 as well as the lattice diagram of $\mathbb{Z}/2\mathbb{Z} \times \mathbb{Z}/2\mathbb{Z}$.

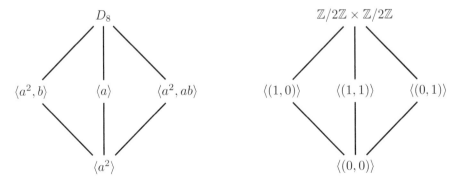

Figure 10.7. The subgroups of D_8 that contain $\langle a^2 \rangle$ form a lattice identical to the lattice of subgroups of $\mathbb{Z}/2\mathbb{Z} \times \mathbb{Z}/2\mathbb{Z}$.

The Import of Normal Subgroups and Quotient Groups. A long standing project in group theory has been to study groups by understanding their normal subgroups and quotient groups. The thinking goes as follows: Let a group G be given. If G has a non-trivial normal subgroup N, then N and G/N are smaller groups, and hence should be studied earlier. It should be then possible to translate information about N and G/N into information about G. There will be some straightforward ways in which the structure of N and/or G/N is reflected in the structure of G, and we will see some such examples. However, there remain many subtle ways in which one can glean information about G from these smaller structures. Given this outlook, a number of questions arise. How do we find normal subgroups? What if a group does not have any non-trivial normal subgroups? What are the ways that groups N and G/N can be "reassembled" to give a group G? For now we will make a definition and then some comments about the second question.

Definition 10.35 (Simple groups). Let G be a group. G is called a *simple group* if G has no non-trivial normal subgroups.

Given the discussion above, simple groups are the building blocks of group theory much in the same way that the prime numbers are the building blocks of integers. If a group is not simple, then we can at least try the project outlined above. However, to be successful, we have to have a good understanding of the simple groups. The first question is what are the finite simple groups?

Lemma 10.36. *Let G be a finite abelian group. Then G is simple if and only if $G \cong \mathbb{Z}/p\mathbb{Z}$, where p is a prime.*

Proof. If G is abelian, then all subgroups of G are normal. Hence G is simple if and only if it has no non-trivial subgroups. Thus the subgroup generated by any non-identity element will have to be the whole group, and hence G is cyclic. Furthermore, in a cyclic group for every divisor of $|G|$, we have a subgroup of that order. Hence, $|G|$ must be prime. (Also see Problem **2.6.24**.)   □

Proposition 10.37. *Let n be an integer greater than or equal to 5. The alternating group of degree n, A_n, is simple.*

Proof. You were asked to prove that A_5 is simple in Problem **10.2.17**. The idea of the proof was as follows: A normal subgroup is a union of conjugacy classes (Proposition 10.18(g)). On the other hand, by Lagrange's theorem, the size of a normal subgroup must divide the order of the group. For A_5 we found (Problem **6.2.15**) that the sizes of the conjugacy classes were 1, 15, 20, 12, and 12, and there is no way to add a subset of these numbers to get a divisor of 60 (other than 1 and 60). Hence, A_5 is simple.

The proof that the higher degree alternating groups are simple uses the fact that A_5 is simple and is relegated to Chapter 14 (see Theorem 14.20).   □

The Classification of Finite Non-abelian Simple Groups. Because of Lemma 10.36, the real problem is classifying finite non-abelian simple groups. Life would be much easier (and much more boring) if there were no such objects. However (see Proposition 10.37), for $n \geq 5$, the alternating groups A_n are an infinite family of finite non-abelian simple groups, and there are other such groups. From the beginning of group theory in the mid-nineteenth century until 1980, this classification problem was the most important problem in group theory. The classification was mainly finished in 1980 through the works of hundreds of mathematicians from around the globe. This work is known as "the classification" among group theorists and consists of two parts. The first is a list of finite non-abelian simple groups. It consists of a number of infinite families of groups and 26 "sporadic" simple groups. The second part is a proof that this list is complete. This proof originally took more than 15,000 pages in research journals and involved the work of more than 100 mathematicians from many different countries. As of this writing, no one person has claimed to have read all of this proof, and in fact there are major projects under way in an attempt to simplify this proof. This gigantic effort was a good example of how mathematicians rely on each other's work and how cooperative the mathematical enterprise is. One important step toward the proof of the classification was a theorem by John Thompson and Walter Feit. At the end of the nineteenth century, William Burnside had conjectured that all non-abelian finite simple groups have even order. In other words, no group of odd order is non-abelian simple. This theorem was proved by Feit and Thompson in 1962 [**FT63**], and is known as the Feit–Thompson odd order theorem. The 255-page proof of the odd order theorem opened the way for the attack on the classification problem. Even after more than 50 years, the proof of the odd order theorem remains impenetrable to all but the most sophisticated readers. One interesting aspect of the list of finite non-abelian simple groups is the existence of the 26 "sporadic groups" that are not a part of any infinite family of simple groups. The largest of these groups is a group called *the monster*[3]—or the *Friendly Giant*—and its order is

$$808,017,424,794,512,875,886,459,904,961,710,757,005,754,368,000,000,000.$$

Problems

10.3.1. Let $G = S_4$ and let $K = \{1, (1\ 2)(3\ 4), (1\ 3)(2\ 4), (1\ 4)(2\ 3)\}$. K is a normal subgroup of G. What is the order of G/K? Which group is G/K isomorphic to?

10.3.2. Let $G = (\mathbb{Z}/18\mathbb{Z}, +)$ be a cyclic group of order 18.
  (a) Find a subgroup H of G with $|H| = 3$.
  (b) What are the elements of G/H?
  (c) Find a familiar group that is isomorphic to G/H.

10.3.3. Let $G = \mathbb{Z}/4\mathbb{Z} \times \mathbb{Z}/6\mathbb{Z}$, and let H be the subgroup of G generated by $(2, 2)$.
  (a) What are the elements of H?
  (b) What are the elements of G/H?
  (c) Find a familiar group that is isomorphic to G/H.

10.3.4. Let $G = \mathbb{Z}/4\mathbb{Z} \times \mathbb{Z}/4\mathbb{Z}$, and let $H = \langle (0, 2), (2, 0) \rangle$. Find a familiar group that is isomorphic to G/H.

10.3.5. Let G be the group $S_3 \times \mathbb{Z}/2\mathbb{Z}$. Let $x = (1\ 2\ 3) \in S_3$, and let $H = \langle x \rangle \times \mathbb{Z}/2\mathbb{Z}$.
  (a) What is $o(x)$, $|H|$, $|G : H|$?

[3]For an expository account of the classification and the monster aimed at non-mathematicians, see Ronan [**Ron06**]. For a well written exposition aimed at mathematicians, see Solomon [**Sol01**].

     (b) Let $g = ((1\ 2), 1) \in G$. Find gH.

     (c) Is $H \lhd G$? If so, what group is G/H isomorphic to?

10.3.6. Let $G = D_{12} = \langle a, b \mid a^6 = b^2 = e, ba = a^5 b \rangle$.

     (a) Is $a^3 \in \mathbf{Z}(G)$?

     (b) Find $\mathbf{C}_G(a)$ and $\mathbf{C}_G(b)$. What are they isomorphic to?

     (c) Find $\mathbf{Z}(G)$.

     (d) Find $\mathbf{N}_G(\langle b \rangle)$ and $\mathbf{N}_G(\mathbf{C}_G(b))$.

     (e) Find a familiar group that is isomorphic to $G/\mathbf{Z}(G)$.

10.3.7. As usual let $D_8 = \langle a, b \mid a^4 = b^2 = e, ba = a^3 b \rangle$ and $D_6 = \langle f, g \mid f^3 = g^2 = e, gf = f^2 g \rangle$. Let $G = D_6 \times D_8$. Let $x = (f, a) \in G$, and let $H = \{e\} \times D_8 \le G$.

     (a) What is $o(x)$?

     (b) Find Hx. What is $|G : H|$?

     (c) Is $H \lhd G$? If the answer is yes, find a familiar group that is isomorphic to G/H?

     (d) Let $y = (e, a)$. What is $|\mathbf{C}_G(y)|$? What is $|\mathrm{cl}_G(y)|$?

     (e) Let $P \in \mathrm{Syl}_2(G)$. What is $|P|$? What is $|G : P|$?

     (f) Find one Sylow 3-subgroup of G.

10.3.8. Assume G is cyclic and $N \lhd G$. Is G/N cyclic?

10.3.9. Let G be a group, and let $N \lhd G$. Assume that $|G : N| = m$. Let $x \in G$. Prove that $x^m \in N$.

10.3.10. Let G be a group and, as usual, $\mathbf{Z}(G)$ denotes the center of G. Can $G/\mathbf{Z}(G)$ be a non-trivial cyclic group? Either prove that it cannot be or give an example where it is.

10.3.11. Assume that N is a normal subgroup of a group G. Assume E is a subgroup of G/N. Thus E is a collection of right cosets of N in G. Let K be the union of all the elements of E. In other words, K is a subset of G consisting of all the elements in the right cosets in E. Prove that K is a subgroup of G that contains N. What is $|K|$?

10.3.12. Assume that N is a normal subgroup of a group G. Assume E is a subgroup of G/N. As in Problem *10.3.11*, let K be the union of all the cosets in E. Show that K is normal in G if and only if E is normal in G/N.

10.3.13. Let G be a non-abelian group of order 27. Assume $x, y \in G$. Prove that there exists $z \in \mathbf{Z}(G)$ such that $yx = zxy$.

10.3.14. Let G be a group of order 27. Let $x \in G$. Prove that $x^3 \in \mathbf{Z}(G)$.

10.3.15. Let G be a group of order p^3, where p is a prime. Show that G has a normal subgroup of order p^2.

10.3.16. Can you generalize Problems 10.1.12 and 10.3.15?

10.3.17. Let n be an odd integer, and let D be a group of order $2n$. Assume D has a subgroup H of order n, and further assume that for all $h \in H$ and $x \in D - H$ we have

$$xh = h^{-1}x.$$

Prove that H is abelian, and that every element in $D - H$ is of order 2. In addition, give an example of such a group D.

You may find the following steps useful:

Step 1: Let b be an element of order 2 in D, and let h_1 and h_2 be elements of H. Explain and complete the following string of algebraic manipulations:

$$bh_1 h_2 = (h_1 h_2)^{-1} b = \; ?? \; = (h_2^{-1} b) \; ?? \; (h_1^{-1} b) = \; ?? \; = bh_2 h_1.$$

Step 2: Conclude that H is abelian.

Step 3: Let $x \in D - H$, and assume that $o(x)$ is not 2. Show that $x^2 \in H$, and that $x^4 \neq e$.

Step 4: If we let $h = x^2$, then show that $xh \neq h^{-1} x$.

Step 5: Conclude that every element of D not in H has order 2.

10.4. More Problems and Projects

Problems

10.4.1. Normalizers and the group multiplication table. Let G be a group of order n, and let H be a subgroup of G of order k. We will, in this problem and following Johnson [**Joh00**], exhibit a way of organizing the multiplication table of G so that the elements and the size of the normalizer of H in G can be gleaned easily.

Let $s = |G : H| = n/k$, and choose $a_1, \ldots, a_s \in G$, such that $Ha_1, \ldots, Ha_s$ are the distinct right cosets of H in G. Organize the column headings of the multiplication table of G by first listing all elements of Ha_1, then all elements of Ha_2, and so on. Organize the row headings by first listing all the elements of $a_1^{-1} H$, then all the elements of $a_2^{-1} H$, and so on, ending with all the elements of $a_s^{-1} H$. In this way the multiplication table will have been split into blocks. We are particularly interested in the blocks down the diagonal. The ith diagonal block consists of the products of elements from $a_i^{-1} H$ with those of Ha_i. (See Figure 10.8.)

(a) Show that $a_1^{-1} H, a_2^{-1} H, \ldots, a_s^{-1} H$ are the set of distinct left cosets of H in G. Conclude that by organizing the column and row headings as prescribed, we have indeed listed every element of G exactly once as a column heading and exactly once as a row heading. Hence, we do have a multiplication table for G.

(b) Let K be a subgroup of G. Show that K is a conjugate of H if and only if the elements of K are exactly the elements of G that occur in one of the blocks on the diagonal of the multiplication table of G.

(c) Show that the normalizer $\mathbf{N}_G(H)$ is the union of the right cosets Ha_i for which $a_i^{-1} Ha_i = H$.

	Ha_1	Ha_2	$\dots$	Ha_s
$a_1^{-1}H$	$a_1^{-1}Ha_1$	$a_1^{-1}Ha_2$	$\dots$	$a_1^{-1}Ha_s$
$a_2^{-1}H$	$a_2^{-1}Ha_1$	$a_2^{-1}Ha_2$	$\dots$	$a_2^{-1}Ha_s$
$\vdots$	$\vdots$	$\vdots$	$\ddots$	$\vdots$
$a_s^{-1}H$	$a_2^{-1}Ha_1$	$a_s^{-1}Ha_2$	$\dots$	$a_s^{-1}Ha_s$

Figure 10.8. We order the column and row headings of the multiplication table of G according to the cosets of H and get a partition of the multiplication table into blocks.

(d) Let m be the number of diagonal blocks that consist of only elements of H. Show that $|\mathbf{N}_G(H)| = mk$, $|\mathbf{N}_G(H) : H| = m$, and $|G : \mathbf{N}_G(H)| = s/m$.

Solvable groups

Definition 10.38 (Solvable groups)**.** A group G is called *solvable* if we can find subgroups of G,

$$\{e\} = N_0, N_1, N_2, \dots, N_k = G,$$

such that

(a) $N_i \leq N_{i+1}$, for $i = 0, \dots k - 1$.

(b) $N_i \lhd N_{i+1}$ for $i = 0, \dots, k - 1$.

(c) $|N_{i+1} : N_i|$ is a prime number for $i = 0, \dots k - 1$.

10.4.2. Which one(s) of the following are solvable: $\mathbb{Z}/n\mathbb{Z}$, D_8, and S_4.

10.4.3. Show that neither A_5 nor S_5 are solvable.

10.4.4. Prove that a subgroup of a solvable group is solvable.

On Subgroups of Prime Index. Proposition 10.6 stated that a subgroup of index 2 is automatically normal. The problems to follow are adapted from Lam [**Lam04**]: they give some necessary and some sufficient conditions for normality which are then used to prove that a subgroup of index p, where p is the smallest prime divisor of the group, is always normal. This theorem can also be proved using more sophisticated machinery (see Corollary 12.6 and its proof), but the proof here is elementary. In addition, the conditions can be used to prove normality in other cases as well.

10.4.5. Let G be a group, and let H be a subgroup of index k in G. Consider the following conditions on H and G:
 (a) H is normal in G.
 (b) For every $a \in G - H$, $a^k \in H$.
 (c) For every $a \in G - H$, $a^n \in H$ for some positive integer $n = n(a)$ with no prime divisor less than k.
 (d) For every $a \in G - H$, $a^2, \ldots, a^{k-1} \notin H$.
 Prove that (a) $\Rightarrow$ (b) and (c) $\Rightarrow$ (d) $\Rightarrow$ (a). Furthermore, prove that, if k is a prime, then (b) $\Rightarrow$ (c).

10.4.6. Let $G = A_4$, and let $H = \{e, (1\ 2)(3\ 4), (1\ 3)(2\ 4), (1\ 4)(2\ 3)\}$. Clearly, H is a subgroup of G.
 (a) Convince yourself that every element in $G - H$ is a 3-cycle.
 (b) Choose an appropriate $n(a)$ in condition 10.4.5(c) to show that H is a normal subgroup of G with no further calculation.

10.4.7. Let $G = S_4$, and let P be a Sylow 2-subgroup of G. Show that P cannot contain every transposition of G. Now, let $a \in G$ be a transposition not in P. Show that condition 10.4.5(b)—and condition 10.4.5(d)—fail for a. Conclude—with hardly any calculation—that P is not normal in G.

10.4.8. Use Problem 10.4.5 to prove that if the index of a subgroup is the smallest prime dividing the order of the group, then the subgroup is normal.

10.4.9. Assume that a group G of order 1081 has a normal subgroup of order 23. Prove that G is cyclic.

10.4.10. **Quasinormal versus normal.** Let H be a subgroup of a group G. Recall from Problem 9.3.2 that H is said to be *quasinormal* if $JH = HJ$ for all subgroups $J \leq G$. Prove[4]
 (a) if H is normal in G, then H is quasinormal in G, and
 (b) if $|G : H|$ is a prime and H is a quasinormal subgroup of G, then H is normal in G.

---

[4]Adapted from Hickerson, Stein, and Yamaoka [**HSY90**] where it is proved that if H is a quasinormal subgroup of G with $|G : H|$ a squarefree integer or twice a squarefree integer, then H is normal in G.

Chapter 11

Group Homomorphisms

> ... where homomorphisms, homomorphism theorems, and their relation
> to normal subgroups and quotient groups are studied, homomorphism
> diagrams are introduced, actions are used to produce homomorphisms,
> and, in an optional section, the relation of inner automorphisms to the
> center of the group and the so-called N/C theorem are investigated.

Recall that a mapping ϕ from a group G to another group G' is an isomorphism
if ϕ is 1-1 and onto and that $\phi(ab) = \phi(a)\phi(b)$ for all $a, b \in G$. The existence of an
isomorphism tells us that the two groups are really the same as groups—in fact, by
rearranging the elements of one of the two groups, we will have identical multiplication tables—and all group theoretic properties of one can be translated to the other.
We will now study maps that are like group isomorphisms except that they may not
be 1-1 and onto. These maps, which are called homomorphisms—Definition 2.51
already introduced them—preserve some information about the group product and
end up providing a very useful language for attacking group theoretic problems.
One reason for this importance is that, as we shall later see, homomorphisms are
just another way of talking about quotient groups and normal subgroups.

The idea of focusing on maps that preserve the operations of an algebraic
structure is not useful only for groups. In your linear algebra course, you may have
noticed the importance of linear transformations which are exactly the maps that
preserve the operations of a vector space (i.e., addition and scalar multiplication).
In fact, it can be argued—and this is formalized in more abstract approaches to
algebra—that the study of such maps is at the core of understanding an algebraic
structure. We shall see that it would have made sense to even start with homomorphisms and then define normal subgroups and quotient groups.

We will first recall the definition of homomorphisms, then study some of their
properties (some of the elementary properties of homomorphisms have already appeared in Problems but will be repeated here for completeness), and then—keeping

with the theme of the text—show that group actions provide a method for constructing homomorphisms.

11.1. Definitions, Examples, and Elementary Properties

Definition 11.1 (Homomorphisms). Let G and H be groups. The mapping $\phi \colon G \to H$ is a group *homomorphism* if, for all $a, b \in G$,

$$\phi(ab) = \phi(a)\phi(b).$$

A homomorphism $\phi : G \to H$ is called the *trivial homomorphism* if $\phi(g) = e_H$ for all $g \in G$.

Homomorphisms are so important that many special kinds of homomorphisms have their own names. In this text we will usually say that a map is a homomorphism and then enumerate its other properties, but sometimes we may use one of the following common expressions.

Definition 11.2. Let G and H be groups, and assume that $\phi : G \longrightarrow H$ is a homomorphism. Then

- if ϕ is onto, then ϕ is called an *epimorphism*,
- if ϕ is a 1-1, then ϕ is called a *monomorphism*,
- if ϕ is 1-1 and onto, then ϕ is called an *isomorphism*, and
- if ϕ is an isomorphism and $G = H$, then ϕ is called an *automorphism*.

Example 11.3. Let $G = (\mathbb{R}, +)$ and $G' = (\mathbb{R}^{>0}, \cdot)$. Define $\phi \colon G \to G'$ by

$$\phi(x) = e^x.$$

We have $\phi(x + y) = e^{x+y} = e^x e^y = \phi(x)\phi(y)$, and so ϕ is a homomorphism. In addition, ϕ is 1-1 and onto. Thus ϕ is an isomorphism and $(\mathbb{R}, +) \cong (\mathbb{R}^{>0}, \cdot)$.

Example 11.4. Let $G = (\mathbb{Z}, +)$ and $G' = (\{1, -1\}, \times)$.

Define $\phi \colon G \to G'$ by

$$\phi(m) = \begin{cases} 1 & \text{if } m \text{ is even,} \\ -1 & \text{if } m \text{ is odd.} \end{cases}$$

Then ϕ is easily seen to be an epimorphism.

Example 11.5. Let $G = D_8 = \langle a, b \mid a^4 = b^2 = e, ba = a^3 b \rangle$, and let $Z = \mathbf{Z}(G) = \langle a^2 \rangle$. Define a map $\phi : D_8 \longrightarrow \mathbb{Z}/2\mathbb{Z} \times \mathbb{Z}/2\mathbb{Z}$ as follows:

$$\phi(x) = \begin{cases} (0,0) & \text{if } x \in Z, \\ (1,0) & \text{if } x \in Za, \\ (0,1) & \text{if } x \in Zb, \\ (1,1) & \text{if } x \in Zab. \end{cases}$$

Since $G/Z = \{Z, Za, Zb, Zab\}$ is the set of cosets of Z in G, and cosets of a subgroup partition the group, we have defined the map on all of G. The map ϕ is clearly onto but not 1-1. In calculating with ϕ, do not get confused by the fact that the operation in G is written multiplicatively and the one for $\mathbb{Z}/2\mathbb{Z} \times \mathbb{Z}/2\mathbb{Z}$ is written additively. To check that ϕ is a homomorphism, notice that

$$\phi(a^2) = (0,0) = (1,0) + (1,0) = \phi(a) + \phi(a),$$

$$\phi(b^2) = (0,0) = (0,1) + (0,1) = \phi(b) + \phi(b),$$

and, in fact, it is straightforward to see that $\phi(a^i) = i\phi(a)$ and $\phi(b^j) = j\phi(b)$. Also

$$\phi(ab) = (1,1) = (1,0) + (0,1) = \phi(a) + \phi(b).$$

Now the reader can finish the proof that ϕ is a homomorphism. We conclude that ϕ is an epimorphism.

Example 11.6. Let G be any group and choose a fixed element $x \in G$.

Define $\phi_x \colon G \to G$ by

$$\phi_x(g) = xgx^{-1}.$$

To show that ϕ_x is a homomorphism, let g_1 and g_2 be arbitrary element of G, and calculate

$$
\begin{aligned}
\phi_x(g_1 g_2) &= x g_1 g_2 x^{-1} \\
&= x g_1 (x^{-1} x) g_2 x^{-1} \\
&= (x g_1 x^{-1})(x g_2 x^{-1}) \\
&= \phi_x(g_1)\phi_x(g_2).
\end{aligned}
$$

Is ϕ_x 1-1?

Assume that $\phi_x(g_1) = \phi_x(g_2)$. By the definition of ϕ_x, we have $x g_1 x^{-1} = x g_2 x^{-1}$. By multiplying on the left by x^{-1} and on the right by x, we get that $g_1 = g_2$. Thus ϕ_x is 1-1.

Is ϕ_x onto?

Let $h \in G$. Is there $g \in G$ with $\phi_x(g) = h$. This would mean $h = xgx^{-1}$, which translates to $g = x^{-1}hx$. Now that we have a candidate for g, it is easy to check that $\phi_x(g) = h$. Thus ϕ_x is onto.

So ϕ_x is an automorphism of the group G. Thus, starting with an element $x \in G$ and using this process, we get an automorphism of G. These automorphisms need not be distinct and, for example, if G is abelian, then every one of the ϕ_x is the identity automorphism. The automorphism ϕ_x is called an *inner automorphism* of G.

Definition 11.7. Let G be a group, and let $x \in G$ be fixed. The map $\phi_x : G \to G$ defined by

$$\phi_x(g) = xgx^{-1}$$

is an automorphism of G and is called an *inner automorphism* of G. The collection of all inner automorphisms of G is denoted by $\operatorname{Inn}(G)$.

We will now look at some elementary properties of homomorphisms. Recall the following notation:

Definition 11.8 ($f(C)$ and $f^{-1}(D)$). If $f : A \to B$ is a map and $C \subseteq A$, then $f(C) = \{f(c) \mid c \in C\}$ is a subset of B. Also if $D \subseteq B$, then $f^{-1}(D) = \{x \in A \mid f(x) \in D\}$ is a subset of A.

Remark 11.9. Note that as long as $f : A \to B$ is a map and $D \subseteq B$, the set $f^{-1}(D)$ is defined. This is even though the function f may not have an inverse.

Lemma 11.10. *Let G and G' be groups, and assume $\phi : G \longrightarrow G'$ is a homomorphism. Then*

(a) $\phi(e_G) = e_{G'}$.

(b) *For all $a \in G$, $(\phi(a))^{-1} = \phi(a^{-1})$.*

(c) *If $H \leq G$, then $\phi(H) \leq G'$.*

(d) *If $H' \leq G'$, then $\phi^{-1}(H') \leq G$.*

Proof. (a) This was Lemma 2.52, but it is easy enough to reproduce the proof:

$$\phi(e_G) = \phi(e_G e_G) = \phi(e_G)\phi(e_G).$$

Now G' is a group, and hence we have cancellation. Canceling $\phi(e_G)$ from both sides of the equation, we get $e_{G'} = \phi(e_G)$.

(b) The result follows from

$$e_{G'} = \phi(e_G) = \phi(aa^{-1}) = \phi(a)\phi(a^{-1}).$$

(c) We know that H is a subgroup of G, and we want to show that $\phi(H)$ is a subgroup of G'. (This was actually Problem 2.6.18.) First, $e_{G'} = \phi(e_G) \in \phi(H)$ and hence $\phi(H)$ is not empty. Next, we need to show that $\phi(H)$ is closed under the multiplication in G' and under taking inverses. Let $x, y \in \phi(H)$. Then $x = \phi(h_1)$ and $y = \phi(h_2)$, for some $h_1, h_2 \in H$. Thus $xy = \phi(h_1)\phi(h_2) = \phi(h_1 h_2) \in \phi(H)$. Also $x^{-1} = (\phi(h))^{-1} = \phi(h^{-1}) \in \phi(H)$, and the proof is complete.

(d) This is similar to the previous part (and was Problem 2.6.20). Recall that to check whether an element x is in $\phi^{-1}(H')$, we check $\phi(x)$. The element $x \in \phi^{-1}(H')$ if and only if $\phi(x) \in H'$. We know that $\phi(e_G) = e_{G'} \in H'$ and so $e_G \in \phi^{-1}(H')$ and so $\phi^{-1}(H')$ is not empty. Now let x and y be elements of $\phi^{-1}(H')$. We have to show that so are xy and x^{-1}. By the definition, we have that $\phi(x)$ and $\phi(y)$ are elements of H'. Thus $\phi(xy) = \phi(x)\phi(y) \in H'$ which implies that $xy \in \phi^{-1}(H')$. Also $\phi(x^{-1}) = \phi(x)^{-1} \in H'$ and so $x^{-1} \in \phi^{-1}(H')$.

$\square$

Remark 11.11. In addition to the problems in this section, the reader may want to go back to earlier sections where elementary properties of group homomorphisms were considered. In particular, Problems 2.4.12–2.4.17, 2.5.13–2.5.15, and 2.6.18–2.6.20 are all about homomorphisms.

Problems

11.1.1. Let $\mathbb{Z}$ denote the integers, and let

$$\mathbf{H}_3(\mathbb{Z}) = \{ \begin{bmatrix} 1 & a & b \\ 0 & 1 & c \\ 0 & 0 & 1 \end{bmatrix} \mid a, b, c \in \mathbb{Z} \}.$$

Under usual matrix multiplication, $\mathbf{H}_3(\mathbb{Z})$ is a group (see Problem 2.7.1) and is called the *discrete Heisenberg group*. Define $\phi : \mathbf{H}_3(\mathbb{Z}) \longrightarrow \mathrm{GL}(3, \mathbb{R})$ by

$$\phi \left(\begin{bmatrix} 1 & a & b \\ 0 & 1 & c \\ 0 & 0 & 1 \end{bmatrix} \right) = \begin{bmatrix} 1 & a & 0 \\ 0 & 1 & 0 \\ 0 & 0 & 1 \end{bmatrix}.$$

Show that ϕ is a homomorphism. Is it 1-1 or onto?

11.1.2. Define $\phi : (\mathbb{Z}/8\mathbb{Z}, +) \to (\mathbb{Z}/8\mathbb{Z}, +)$ by $\phi(x) = 2x$. Is ϕ a homomorphism? If so, what is $\phi^{-1}(\{0\})$? Answer the same questions for $\theta : (\mathbb{Z}/8\mathbb{Z}, +) \to (\mathbb{Z}/8\mathbb{Z}, +)$ defined by $\theta(x) = x^2$.

11.1.3. Let $G = (\mathbb{R}^{>0}, \cdot)$ be the group of positive real numbers under multiplication. Let r be a fixed real number, and define $\phi_r : G \to G$ by

$$\phi_r(x) = x^r.$$

For which values of r is ϕ an automorphism of G?

11.1.4. Let $G = \mathrm{SL}(2, 3)$, and define $\phi : G \to G$ by $\phi(A) = A^2$. Is ϕ a homomorphism?

11.1.5. Let $G = \mathrm{GL}(3, \mathbb{R})$, and define $\phi : G \to (\mathbb{R}^{\times}, \cdot)$ by $\phi(A) = \det(A)$. Is ϕ a homomorphism? If so, what is $\phi^{-1}(1)$?

11.1.6. Consider D_8, the dihedral group of order 8. For $x \in D_8$, let ϕ_x be the inner automorphism defined by x (see Definition 11.7). In other words, $\phi_x : D_8 \longrightarrow D_8$ and is defined by $\phi_x(g) = xgx^{-1}$ for all $g \in D_8$. Explicitly write down the inner automorphism defined by a, an element of order 4. How many *different* inner automorphisms does D_8 have?

11.1.7. How many different inner automorphisms does $(\mathbb{Z}/16\mathbb{Z}, +)$ have?

11.1.8. Find all inner automorphisms of Q_8, the quaternion group of order 8 (see Definition 2.54), and identify the group $\mathrm{Inn}(Q_8)$.

11.1.9. Let A_4 be the alternating group of degree 4. Use Example 11.5 as a model to produce a non-trivial homomorphism $\phi : A_4 \to \mathbb{Z}/3\mathbb{Z}$.

11.1.10. Let $\phi : G \longrightarrow H$ be an onto homomorphism.
   (a) Assume that G is abelian. Does this imply that H is abelian? What about the converse?
   (b) What if we replaced abelian by cyclic in the above question.

11.1.11. Let $\theta : G \longrightarrow H$ be a homomorphism.
   What can you say about the relation of $\theta(\mathbf{Z}(G))$ and $\mathbf{Z}(\theta(G))$?

11.1.12. Let G be a group, and assume that H and K are normal subgroups of G with trivial intersection. Further assume that $G = HK$. Prove that $G \cong H \times K$.

11.2. The Kernel and the Image

We will now consider homomorphisms and ask whether or not they are 1-1 and/or onto. Like other questions in algebra, we prefer to rephrase the question so that the answer is not just a yes or no. We rather have a mathematical structure which gives us a range of possible answers, and, just as for linear transformations in linear algebra, one will get much mileage by considering the kernel and the image of a group homomorphism.

Definition 11.12 (The kernel and the image). Let G and H be groups and assume that $\phi : G \longrightarrow H$ is a group homomorphism. Then

(a) the set $\phi^{-1}(e_H) = \{a \in G \mid \phi(a) = e_H\}$ is called the *kernel* of ϕ and is denoted by $\ker(\phi)$, and

(b) the set $\phi(G) = \{b \in H \mid \exists\, a \in G \text{ with } \phi(a) = b\}$ is called the *image* of ϕ and is denoted by $\mathrm{Im}(\phi)$.

The following is an immediate corollary of Lemma 11.10:

Corollary 11.13. *Let G and H be groups, and let $\phi : G \longrightarrow H$ be a group homomorphism. Then $\ker(\phi)$ is a subgroup of G and $\mathrm{Im}(\phi)$ is a subgroup of H.*

The kernel and the image are subgroups, and hence we can use group theoretical methods (e.g., Lagrange's theorem) for studying them. It is clear that a map is onto if and only if the image is all of the target group (i.e., $\phi(G) = H$). It is also clear that if ϕ is 1-1 then $\ker(\phi) = \{e_G\}$. What is remarkable is that the converse of this last statement is also true and hence kernels give us complete information about whether group homomorphisms are 1-1 just in the same way as the image of a map tells us whether a map is onto. We will now see how the kernel tells us more than what goes to the identity. The next proposition says that if you know one element that is mapped by a group homomorphism to some element in the target group, then using the kernel you can find all other elements that are sent to that element.

Proposition 11.14. *Let G_1 and G_2 be groups, and let $\phi : G_1 \longrightarrow G_2$ be a group homomorphism. Let $K = \ker(\phi)$, and let $x \in G_1$. Let $y = \phi(x) \in G_2$. Then*

$$\{\, g \in G_1 \mid \phi(g) = y \,\} = Kx.$$

In other words for $g, h \in G_1$, we have $\phi(g) = \phi(h)$ if and only if $Kg = Kh$.

Proof. We want to show that the set of all elements of G_1 that are mapped to y is the same as the coset Kx. To show that these sets are the same, we need to show that every element of each is also a member of the other.

For one direction, Let $g \in G_1$ such that $\phi(g) = y$. We want to show that $g \in Kx$. We have $\phi(gx^{-1}) = \phi(g)\phi(x)^{-1} = yy^{-1} = e_{G_2}$. So $gx^{-1} \in \ker(\phi) = K$. It follows that $g \in Kx$.

For the other direction, let $g \in Kx$. We have to show that $\phi(g) = y$. We know that $g \in Kx$, and so $g = kx$ for some $k \in K$. Now $\phi(g) = \phi(kx) = \phi(k)\phi(x) = e_{G_2}y = y$. Thus $g \in \{\, g \in G_1 \mid \phi(g) = y \,\}$. $\qquad\square$

The above proposition says that, for a group homomorphism, all the elements of one coset of the kernel are always mapped to the same element, and, conversely, the set of elements of domain that are mapped to one specific element of the target group is exactly one coset of the kernel. Of course, there may be elements of G_2 that are not hit by anything, but if an element is hit, then its inverse image is a whole coset. Thus if $\phi : G_1 \longrightarrow G_2$ is a group homomorphism, then we can think of the map in terms of Figure 11.1.

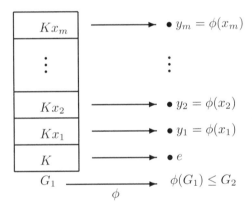

Figure 11.1. In a homomorphism, the cosets of the kernel correspond to the elements of the image.

Note that a homomorphism cannot say anything about the part of the target group that is not in the image. If $\phi : G_1 \longrightarrow G_2$ is a group homomorphism, then $\phi(G_1)$ is a subgroup of G_2, and we can think of ϕ as a map from G_1 onto $\phi(G_1)$, i.e., $\phi : G_1 \longrightarrow \phi(G_1) \le G_2$. This new map carries all the information of the old map and, in addition, is onto.

In the above proposition, what happens if we know that $\ker(\phi) = \{e_{G_1}\}$? In such a situation, every coset of the kernel would have exactly one element, and, hence, the map would be 1-1. Thus, by knowing that the identity is the only element mapped to identity, we would know that all other elements of the target group are hit by either one or zero elements. We have thus proved the only non-trivial part of the following corollary:

Corollary 11.15. *Let G_1 and G_2 be groups, and let $\phi\colon G_1 \to G_2$ be a homomorphism. Then*

(a) *the map ϕ is onto if and only if $\mathrm{Im}(\phi) = G_2$; and*

(b) *the map ϕ is 1-1 if and only if $\ker(\phi) = \{e_{G_1}\}$.*

The previous proposition also tells us that the elements in the image are in 1-1 correspondence with the cosets of the kernel, and hence we have:

Corollary 11.16. *Let G_1 and G_2 be groups, and let $\phi\colon G_1 \to G_2$ be a homomorphism. Then $|\phi(G_1)| = |G_1 : \ker(\phi)|$.*

In the next section, we will find a much stronger version of this result. We can remember the above corollary by drawing the diagram in Figure 11.2. The length of the edge from $\ker(\phi)$ to G_1 denotes $|G_1 : \ker(\phi)|$, and this is equal to the length of the edge from $\{e_{G_2}\}$ to $\phi(G_1)$ which denotes $|\phi(G_1)|$. This is exactly what Corollary 11.16 says.

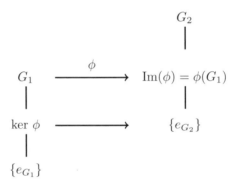

Figure 11.2. Given a homomorphism $\phi : G_1 \to G_2$, from this diagram we can "see" that $|\phi(G_1)| = |G_1 : \ker(\phi)|$.

Problems

11.2.1. Let $\phi : (\mathbb{Z}/12\mathbb{Z}, +) \to (\mathbb{Z}/12\mathbb{Z}, +)$ be defined by $\phi(x) = 4x$. Verify that ϕ is a homomorphism, and find its kernel and image. Which elements are mapped to 4? Can you draw a version of Figure 11.1 specifically for this map?

11.2.2. Repeat the previous problem for $\psi : (\mathbb{Z}/12\mathbb{Z}, +) \to (\mathbb{Z}/12\mathbb{Z}, +)$ defined by $\psi(x) = 5x$. Can you come up with a conjecture that accounts for the differences between ψ and the map ϕ of the last problem?

11.2.3. Let $\mathbb{Z}$ denote the integers, and let

$$\mathbf{H}_3(\mathbb{Z}) = \{ \begin{bmatrix} 1 & a & b \\ 0 & 1 & c \\ 0 & 0 & 1 \end{bmatrix} \mid a, b, c \in \mathbb{Z}\}$$

be the discrete Heisenberg group (Problem 2.7.1). In Problem 11.1.1 we defined a homomorphism $\phi : \mathbf{H}_3(\mathbb{Z}) \longrightarrow GL(3, \mathbb{R})$ by

$$\phi\left(\begin{bmatrix} 1 & a & b \\ 0 & 1 & c \\ 0 & 0 & 1 \end{bmatrix} \right) = \begin{bmatrix} 1 & a & 0 \\ 0 & 1 & 0 \\ 0 & 0 & 1 \end{bmatrix}.$$

What is the $\ker(\phi)$? What is the image of ϕ?

11.2.4. Let A_4 be the alternating group of degree 4. In Problem 11.1.9, we constructed a non-trivial homomorphism $\phi : A_4 \to \mathbb{Z}/3\mathbb{Z}$. Find the kernel and the image of your map, and verify Corollary 11.16.

11.2.5. Recall that $(\mathbb{Z}/n\mathbb{Z})^\times = \{\, x \in \mathbb{Z}/n\mathbb{Z} \mid \gcd(x, n) = 1 \,\}$ is a group under multiplication mod n. We have a non-trivial homomorphism $\theta \colon (\mathbb{Z}/9\mathbb{Z})^\times \to (\mathbb{Z}/11\mathbb{Z})^\times$. Find all of the elements in $\ker(\theta)$ and in $\mathrm{Im}(\theta)$.

11.3. Homomorphisms, Normal Subgroups, and Quotient Groups

In this section, we will explore the relation between kernels of homomorphisms, normal subgroups, and quotient groups. We have seen that the kernel is a subgroup. In fact, the kernel is a normal subgroup:

Theorem 11.17. *Let G and H be groups, and let $\phi : G \longrightarrow H$ be a group homomorphism. Then $\ker(\phi) \lhd G$.*

Proof. This was Problem 10.2.6, but here is the proof. Let $K = \ker(\phi)$. To show that K is a normal subgroup of G we need, by Proposition 10.18(e), to show that $xkx^{-1} \in K$ for all $x \in G$, and $k \in K$.

So fix $x \in G$ and $k \in K$, and let $g = xkx^{-1}$. We want to show that $g \in K$. We have

$$
\begin{aligned}
\phi(g) &= \phi(xkx^{-1}) \\
&= \phi(x)\phi(k)\phi(x^{-1}) \\
&= \phi(x)e_H\phi(x^{-1}) \\
&= \phi(xx^{-1}) = \phi(e_G) = e_H,
\end{aligned}
$$

and so g is in K, the kernel of ϕ. $\qquad\square$

Because of this theorem, one way to construct normal subgroups is to find a homomorphism from your group to some other group and then calculate the kernel. In fact—as we shall see later—every normal subgroup is the kernel of some homomorphism, and, hence, all normal subgroups can be found using homomorphisms.

We had seen that the number of cosets of the kernel in the group G is the same as the number of elements in the image. The next important theorem says that the cosets of the kernel have more than cardinality in common with the image. In fact, the group $G/\ker(\phi)$ is isomorphic as a group to the group $\phi(G)$.

Theorem 11.18. *Let G and H be groups, and let $\phi : G \longrightarrow H$ be a group homomorphism. Then*

$$ G/\ker(\phi) \cong \phi(G). $$

Proof. Let $K = \ker(\phi)$. To show the required isomorphism, we will need a group isomorphism $\psi \colon G/K \to \phi(G)$. In other words, every coset of K in G needs to be sent to an appropriate element in the image of ϕ. If you look back at Figure 11.1, then it is easy to guess the definition of ψ. We define ψ by

$$ \psi(Ka) = \phi(a). $$

Before we prove that this is an isomorphism, we have to show that the map is well defined. Then we prove that ψ is a group homomorphism and that it is 1-1 and onto.

ψ is well defined: The element $\phi(a)$ is clearly an element of $\phi(G)$, but ψ is a map on cosets, and hence we have to show that it is well defined. Let $Ka = Kb$. We need to show that $\phi(a) = \phi(b)$. Since $Ka = Kb$, then $b = ka$ for some $k \in K$. This means that $\phi(b) = \phi(ka) = \phi(k)\phi(a) = \phi(a)$.

ψ is a homomorphism: We need to show that $\psi(KaKb) = \psi(Ka)\psi(Kb)$. Now K is normal, and so $KaKb = Kab$. Thus, using the fact that ϕ is a homomorphism, we have

$$\psi(KaKb) = \psi(Kab) = \phi(ab) = \phi(a)\phi(b) = \psi(Ka)\psi(Kb).$$

ψ is 1-1: Since we have already shown ψ to be a group homomorphism, we only need to show that the kernel of ψ consists of the identity element of G/K. Assume $\psi(Ka) = e_H \in H$. We need to prove $Ka = K$. We have $\phi(a) = \psi(Ka) = e_H$ and so $a \in \ker(\phi) = K$. Thus $Ka = K$.

ψ is onto: Let $y \in \phi(G)$. So, by definition of $\phi(G)$, there exists $x \in G$ such that $\phi(x) = y$. But then $\psi(Kx) = \phi(x) = y$ as required.

We have shown that ψ is a group isomorphism between G/K and $\phi(G)$. Hence $G/K \cong \phi(G)$. □

Remark 11.19. Let $\phi : G \to H$ be a homomorphism, and let $K = \ker(\phi)$. We have proved that $G/K \cong \phi(G)$. This means that these two groups are identical as abstract groups and, hence, for example, have identical subgroup lattice diagrams. As a consequence of the homomorphism theorems, Theorem 11.38, we shall see that the lattice of subgroups of G/K is identical to the part of the lattice of subgroups of G between K and G. In other words, if we consider a partial lattice diagram of G consisting of only those subgroups that contain K, then we *exactly* get the full lattice of the subgroups of G/K.

This is remarkable, very useful, and will be proved in Theorem 11.38, but it also provides a heuristic for visualizing various facts about homomorphisms and quotient groups. Figure 11.3 is a repeat of Figure 11.2. The only difference is that we have now labelled the part of the partial lattice diagram above K and below G as G/K. This heuristic allows us to remember Theorem 11.18 as saying that the opposite sides of the rectangle are isomorphic. Whenever we have a homomorphism $\phi : G \to H$, we begin by drawing Figure 11.3. If the homomorphism $\phi : G \to H$ is onto, then $H = \phi(G)$ and the diagram will become the simpler one of Figure 11.4.

Example 11.20. Let $G = \mathbb{Z}/10\mathbb{Z}$ be the cyclic group of order 10, and define $\phi : G \longrightarrow G$ by $\phi(x) = 2x$. ϕ is a group homomorphism since

$$\phi(x + y) = 2(x + y) = 2x + 2y = \phi(x) + \phi(y).$$

For this map, $\text{Im}(\phi) = \{0, 2, 4, 6, 8\} = \langle 2 \rangle$, and $\ker(\phi) = \{0, 5\} = \langle 5 \rangle$. By Theorem 11.18, $G/\langle 5 \rangle \cong \langle 2 \rangle \cong \mathbb{Z}/5\mathbb{Z}$. The diagram in Figure 11.5 records much of the information.

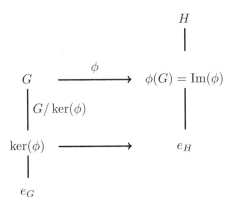

Figure 11.3. Given a homomorphism $\phi : G \to H$, the lattice diagram of $G/\ker(\phi)$ is identical to the partial lattice diagram of G consisting only of subgroups that contain $\ker(\phi)$.

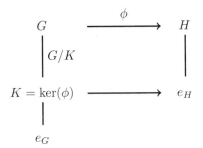

Figure 11.4. The diagram for an *onto* homomorphism $\phi : G \to H$

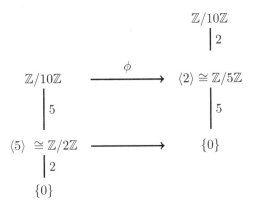

Figure 11.5. $\phi : \mathbb{Z}/10\mathbb{Z} \to \mathbb{Z}/10\mathbb{Z}$ is defined by $\phi(x) = 2x$.

We have seen that kernels are normal subgroups. We now want to show that every normal subgroup is the kernel of some homomorphism.

Definition 11.21 (The canonical homomorphism). Let G be a group, and let $N \lhd G$. Define

$$\pi : G \longrightarrow G/N$$

by

$$\pi(g) = Ng.$$

The map π is called the *canonical homomorphism* from G onto G/N.

Of course, to justify the name, we do have to show that π is a homomorphism.

Theorem 11.22. *Let G be a group, and let $N \lhd G$. Let $\pi : G \longrightarrow G/N$ be the canonical homomorphism. Then π is an onto homomorphism and* $\ker(\pi) = N$.

Proof. We have $\pi(g_1 g_2) = N g_1 g_2 = N g_1 N g_2 = \pi(g_1)\pi(g_2)$, and so π is a group homomorphism. It is clearly onto.

Now $x \in \ker(\pi)$ if and only if $\pi(x) = N$. The latter is equivalent to $Nx = N$ which, in turn, is equivalent to $x \in N$. $\qquad\square$

Because of Theorem 11.22, whenever we have $N \lhd G$, we can draw the diagram in Figure 11.6.

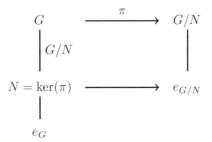

Figure 11.6. If $N \lhd G$, then the canonical homomorphism $\pi : G \to G/N$ is defined by $\pi(x) = Nx$.

So an arbitrary normal subgroup N is the kernel of the canonical homomorphism from G to G/N. Thus by focusing on homomorphisms and their kernels, we do not lose sight of any normal subgroups. We record this fact as a corollary:

Corollary 11.23. *Every normal subgroup of a group is the kernel of some homomorphism. Every quotient group of a group is the image of some homomorphism.*

Using the canonical homomorphism, we can see that the subgroups of G/N are what we expect:

Proposition 11.24. *Let G be a group, and let $N \lhd G$. Assume K is a subgroup of G/N. Then K is equal to H/N where H is some subgroup of G containing N.*

Proof. Let $\pi \colon G \to G/N$ be the canonical homomorphism. Let $H = \pi^{-1}(K)$. We know—by Lemma 11.10(d) since π is a homomorphism—that H is a subgroup of G containing N. Since π is onto, we also know that $\pi(H) = K$. Now recall Definition 11.21 of the canonical homomorphism π. For $x \in G$, we have $\pi(x) = Nx$. Hence

$$K = \pi(H) = \{ Nx \mid x \in H \} = H/N. \qquad\square$$

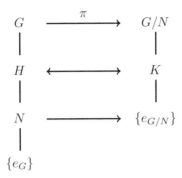

Figure 11.7. Every subgroup of G/N is of the form H/N, where H is a subgroup of G containing N.

Remark 11.25. In the proof above, what are the elements of H? The subgroup K of G/N consists of (some) cosets of N in G. The group H consists exactly of the elements of these cosets.

In the next section, we use group actions to construct homomorphisms and normal subgroups.

Problems

11.3.1. As usual, let $D_8 = \langle a, b \mid a^4 = b^2 = e, ba = a^{-1}b \rangle$. Is it possible to find a homomorphism ϕ from D_8 to some group G such that the kernel of ϕ is exactly $\langle b \rangle$? Either find such an example or prove that it is impossible.

11.3.2. As usual, let S_3 denote the symmetric group of degree 3. Assume that you have a mystery homomorphism $\theta \colon S_3 \to G$ where G is also a mystery group. What are the possibilities for $\ker(\theta)$?

11.3.3. (a) Verify that the mapping $f \colon (\mathbb{R}, +) \longrightarrow (\mathbb{C}^\times, \cdot)$ given by $f(x) = \cos(2\pi x) + i\sin(2\pi x)$ is a homomorphism.
(b) We know that $\mathbb{R}/\ker(f) \cong \mathrm{Im}(f)$. Explicitly find $\ker(f)$ and $\mathrm{Im}(f)$.

11.3.4. By finding an appropriate homomorphism $\phi : \mathrm{GL}(n, \mathbb{R}) \to ?$, show that $\mathrm{SL}(n, \mathbb{R}) \triangleleft \mathrm{GL}(n, \mathbb{R})$, and find a familiar group that is isomorphic to $\mathrm{GL}(n, \mathbb{R})/\mathrm{SL}(n, \mathbb{R})$.

11.3.5. Let D_8 and S_3, as usual, be the dihedral group of order 8 and the symmetric group of degree 3, respectively. Assume $\phi : D_8 \to S_3$ is a homomorphism. What are the possibilities for $|\ker(\phi)|$ and $|\mathrm{Im}(\phi)|$? For each possibility, give an explicit example.

11.3.6. Let S_3 be the symmetric group on three letters, and let $\mathbb{Z}/15\mathbb{Z}$ be the cyclic group of order 15. The map $\theta : S_3 \to \mathbb{Z}/15\mathbb{Z}$ is a group homomorphism. What can you say about $\ker(\theta)$ and $\mathrm{Im}(\theta)$?

11.3.7. Explicitly find all non-trivial group homomorphisms $\phi : (\mathbb{Z}/15\mathbb{Z}, +) \to (\mathbb{Z}/12\mathbb{Z}, +)$.

11.3.8. Consider the group $G = D_8 \times \mathbb{Z}/6\mathbb{Z}$, and let $N = D_8 \times \{0\}$. Show that $N \lhd G$ and that $G/N \cong \mathbb{Z}/6\mathbb{Z}$. Now $\mathbb{Z}/6\mathbb{Z}$—and, hence, G/N—has a unique subgroup of order 3. Let K be the subgroup of order 3 in G/N. Explicitly, find a subgroup H of G such that $K = H/N$.

11.3.9. Let G and H be groups, and let $\phi : G \to H$ be a group homomorphism. Let $N = \ker(\phi)$, and let the map $\pi : G \to G/N$ denote the canonical homomorphism. Give the definition of a homomorphism $\psi : G/N \to H$ such that the diagram in Figure 11.8 commutes.

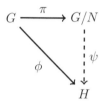

Figure 11.8. The homomorphism $\phi : G \to H$ can be "factored" as $\phi = \psi \circ \pi$. We say that ϕ is factored through G/N.

Commutators and the Commutator Subgroup

Definition 11.26 (Commutators). Let G be a group. Let $x \in G$. If x can be written as $aba^{-1}b^{-1}$ for some $a, b \in G$, then we call x a *commutator* in G.

Definition 11.27 (The commutator subgroup). Let G be a group. The subgroup of G generated by all the commutators of G is called the *commutator* subgroup of G and is denoted by G'.

11.3.10. Let G be a group. Prove that G is abelian if and only if $G' = \{e\}$.

11.3.11. Find the commutator subgroup of D_8 and of S_3.

11.3.12. Let $\phi : G \to H$ be an onto homomorphism. Prove that the following are equivalent:
   (a) The group H is abelian.
   (b) Every commutator of G is in $\ker(\phi)$.
   (c) The commutator subgroup G' of G is a subgroup of $\ker(\phi)$.

11.3.13. Let G be a group, and let N be a normal subgroup of G. Prove that G/N is abelian if and only if $G' \leq N$.

11.3.14. Let N and M be two normal subgroups of G. Assume that G/N and G/M are abelian. Prove that $G/(N \cap M)$ is abelian.

11.3.15. Let G be a group. Prove that G' is normal in G, and G/G' is abelian.

11.4. Actions and Homomorphisms

We like finding homomorphisms of groups. One reason is that the kernels of homomorphisms are a good source of normal subgroups. But how do we construct homomorphisms? One powerful method is through group actions! We already saw (see Lemma 4.5) that every group action gives a permutation for every group element. In other words, as soon as you have a group action, you have a map from your group to a symmetric group. We now show that this map is actually a homomorphism of groups. As a consequence, the image (i.e., the set of permutations given by group elements) is a subgroup of a symmetric group and the kernel will be a normal subgroup of the group. Recall that the elements of $\mathrm{Perm}(\Omega)$ are 1-1 and onto functions from Ω to Ω and $\mathrm{Perm}(\Omega)$ is a group with function composition as the group operation.

Theorem 11.28. *Let the group G act on the set Ω. For $g \in G$ define the map $f_g : \Omega \to \Omega$ by $f_g(\alpha) = g \cdot \alpha$. Then $f_g \in \mathrm{Perm}(\Omega)$.*

Define a map $\theta : G \longrightarrow \mathrm{Perm}(\Omega)$ by $\theta(g) = f_g$. Then θ is a group homomorphism.

In addition, $\ker(\theta)$ consists of those elements of G that fix every element of Ω.

Proof. We already proved in Lemma 4.5 that $f_g \in \mathrm{Perm}(\Omega)$. To show that θ is a homomorphism, we need to prove that $\theta(gh) = \theta(g)\theta(h)$. Since $\theta(gh) = f_{gh}$, $\theta(g) = f_g$, and $\theta(h) = f_h$, we want to show that $f_{gh} = f_g f_h$. Now f_{gh} and $f_g f_h$ are both maps from Ω to Ω. To show that these are the same maps, we will show that they both have the same effect on $\alpha \in \Omega$. So let $\alpha \in \Omega$. We have

$$f_{gh}(\alpha) = gh \cdot \alpha = g \cdot (h \cdot \alpha)$$
$$= g \cdot f_h(\alpha) = f_g(f_h(\alpha))$$
$$= f_g f_h(\alpha).$$

Now if $g \in \ker(\theta)$, then $f_g = \theta(g)$ must be the identity map in $\mathrm{Perm}(\Omega)$. This means that f_g fixes every element of Ω, i.e., $f_g(\alpha) = \alpha$ for all $\alpha \in \Omega$. But $f_g(\alpha) = g \cdot \alpha$, and so we conclude that if $g \in \ker(\theta)$, then g fixes every element of Ω. □

Definition 11.29. Let the group G act on the set Ω. Then the set of those elements of G that fix every element of Ω is called the *kernel* of the action.

Remark 11.30. Whenever a group G acts on a set Ω, by Theorem 11.28 we have a homomorphism $\theta : G \longrightarrow \mathrm{Perm}(\Omega)$ and the homomorphism diagram in Figure 11.9. If Ω is a finite set of size n, then this gives a map into the group S_n. Assuming that the action is not trivial, meaning that not all of the elements of G fix every element of Ω, we calculate the kernel of the action. We have one of two possibilities. Either the kernel is not trivial, in which case we have found a non-trivial normal subgroup of G, or else the kernel is trivial, in which case G is isomorphic to $\theta(G)$ which is a subgroup of $\mathrm{Perm}(\Omega)$. In the latter case, we have found a copy of our group inside the group $\mathrm{Perm}(\Omega)$. Thus, this is a win-win situation. We either have a non-trivial normal subgroup, or we know that our group lives in some specific symmetric group.

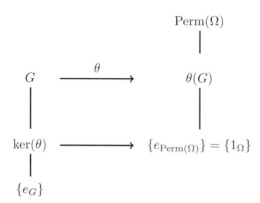

Figure 11.9. An action of the group G on the set Ω gives a homomorphism $\theta : G \to \mathrm{Perm}(\Omega)$.

Remark 11.31. In the proof of $f_{gh} = f_g f_h$, we did not use Lemma 4.5, and we only used the properties of an action. In fact, it is easy to check that f_e is the identity permutation and hence we get that $f_g f_{g^{-1}} = f_{gg^{-1}} = f_e$. This means that f_g and $f_{g^{-1}}$ are inverses of each other and hence both are 1-1 and onto maps from Ω to Ω. In other words $f_g \in \mathrm{Perm}(\Omega)$, and we have given a different (and shorter) proof of Lemma 4.5.

Example 11.32. The group D_8 acts on the corners of the square $\Omega = \{1, 2, 3, 4\}$. By Theorem 11.28 this gives a homomorphism $\theta : G \to S_4$. Now the only element of D_8 that fixes every corner of the square is the identity element, and hence $\ker(\theta) = e_{D_8}$. We have the homomorphism diagram of Figure 11.10. This means that $D_8 \cong \theta(D_8) \leq S_4$. So D_8 is isomorphic to a subgroup of S_4. In fact, it is easy to write down this subgroup. It is just the collection of permutations of Ω given by the action of elements of D_8 (see Example 4.6 on page 87):

$$\{1, \begin{pmatrix} 1 & 2 & 3 & 4 \end{pmatrix}, \begin{pmatrix} 1 & 3 \end{pmatrix}\begin{pmatrix} 2 & 4 \end{pmatrix}, \begin{pmatrix} 1 & 4 & 3 & 2 \end{pmatrix},$$
$$\begin{pmatrix} 1 & 2 \end{pmatrix}\begin{pmatrix} 3 & 4 \end{pmatrix}, \begin{pmatrix} 1 & 3 \end{pmatrix}, \begin{pmatrix} 1 & 4 \end{pmatrix}\begin{pmatrix} 2 & 3 \end{pmatrix}, \begin{pmatrix} 2 & 4 \end{pmatrix}\}.$$

We could have written down these elements before (and in fact we did on page 87). However, we now know, *without any calculation*, that these elements form a subgroup of S_4 that is isomorphic to D_8. Since $24 = 8 \times 3$, we also know that this is a Sylow 2-subgroup of S_4, and so S_4 has a Sylow 2-subgroup isomorphic to D_8.

Example 11.33. Assume that we had a group G of order 12 and we knew that G acted on a set with four elements in such a way that no element of G fixed every element of the set. What can we say about G? Since we have an action of G on a set with four elements, we immediately get a homomorphism ϕ from G into S_4. On the other hand, no element of G fixes every element of the set and so the kernel of this map is trivial. Hence,

$$G \cong G/\ker(\phi) \cong \phi(G) \leq S_4.$$

So G is a subgroup of S_4. But S_4 has only one subgroup of order 12, namely A_4 (see Problem *10.1.17*). We conclude that

$$G \cong A_4.$$

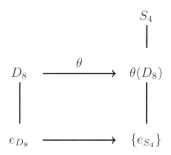

Figure 11.10. The action of D_8 on the corners of the square produces a homomorphism.

This example shows how even just one action—if on a small set and with no kernel—can reveal much about the structure of a group. *Representation theory* takes this point of view to the extreme. It uses *all* actions of a group (and generalizations thereof) to study the structure of a group.

Example 11.34. Let G be the group of the rigid symmetries (i.e., rotations) of the cube. To find the size of G, consider one face of the cube. A rigid symmetry takes this face to one of the other six faces. After deciding the position of this face, you have four choices for one of the faces adjacent to it. Thus so far, you have $6 \times 4 = 24$ choices. However, after fixing the place of two adjacent faces, you cannot rotate the cube anymore, and hence $|G| = 24$ (see Table 8.2 on page 160 for a list of the rigid symmetries of the cube). But which group of order 24 is G? We could consider the action of G on the six faces of the cube. This would tell us that G is isomorphic to a subgroup of S_6. But we can do better:

Let Ω be the four diagonals of the cube. G acts on Ω, and $|\Omega| = 4$. This gives a homomorphism $\theta : G \to S_4$.

What is the kernel of θ? Which group elements fix all four diagonals of the cube? Convince yourself that the only way a non-trivial symmetry could send each diagonal to itself is to flip every diagonal. But this would mean that every point of the cube should go to its "antipodal" point on the cube. Such a map is a symmetry but it is not a rigid symmetry (it is not a rotation and cannot be performed in the usual three-dimensional space by moving a cube around). Thus $\ker(\theta) = e_G$, and we have the homomorphism diagram of Figure 11.11. But $|S_4| = 24$, and $\theta(G) \cong G$ has order 24 also. Thus θ is onto, and $G \cong S_4$. So the group of rigid symmetries of the cube is S_4.

Using Theorem 11.28, we can actually show that *every* group is the subgroup of some symmetric group.

Theorem 11.35 (Cayley). *Let G be any group. Then there exists a set Ω such that G is isomorphic to a subgroup of* $\mathrm{Perm}(\Omega)$.

In particular, if $|G| < \infty$, then $G \cong H$ where $H \leq S_n$ for some n.

Proof. Let $\Omega = G$. Then G acts on Ω by the regular action. (In other words, the action of G on Ω is defined by $g \cdot x = gx$.) By Theorem 11.28, we have a homomorphism $\theta : G \to \mathrm{Perm}(\Omega) = \mathrm{Perm}(G)$.

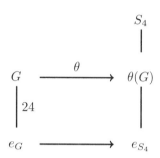

Figure 11.11. The homomorphism gotten from the action of the rigid symmetries of the cube on the diagonals.

We claim that θ is 1-1. To prove this, let $g \in \ker(\theta)$. This means that g fixes every element of Ω in the regular action of G on Ω. Thus, $g \cdot x = x$ for all $x \in G$. But this means that $gx = x$ which gives $g = e_G$.

Thus, $\theta : G \to \theta(G)$ is a 1-1, onto homomorphism. This means that $G \cong \theta(G)$ which is a subgroup of $\mathrm{Perm}(\Omega)$. $\square$

Because of Cayley's Theorem 11.35, for every group G we have the homomorphism diagram of Figure 11.12.

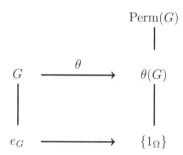

Figure 11.12. Every group G is isomorphic to a subgroup of $\mathrm{Perm}(G)$.

Problems

11.4.1. G is a subgroup of S_4 and is generated by (1 2 3 4) and (1 2 3). G acts on $\Omega = \{1, 2, 3, 4\}$ in the natural way. This action gives rise to a homomorphism $\theta : G \longrightarrow S_4$. Find the order of $\theta((1\ 2\ 3\ 4)^2)$.

11.4.2. According to Cayley's Theorem 11.35, every finite group is isomorphic to a subgroup of some S_n. Find the smallest n such that $(\mathbb{Z}/6\mathbb{Z}, +)$ is isomorphic to a subgroup of S_n.

11.4.3. Let $D_8 = \langle a, b \mid a^4 = b^2 = e, ba = a^{-1}b \rangle$, and let $H = \langle b \rangle$. Then H acts on D_8 by left multiplication (see Example 4.11). According to Theorem 11.28, this action gives rise to a homomorphism $\theta : H \to \mathrm{Perm}(D_8) \cong S_8$.

First, label each element of D_8 as 1, ..., 8, and then explicitly give $\theta(b)$ as an element of S_8.

11.4.4. Let $D_8 = \langle a, b \mid a^4 = b^2 = e, ba = a^{-1}b \rangle$ act on $\Omega = D_8$ by conjugation. According to Theorem 11.28. this action gives rise to a homomorphism $\theta : D_8 \to S_8$. What is the kernel of θ? Label each element of Ω as 1, ..., 8, and then explicitly give $\theta(a)$ as an element of S_8. What is $o(\theta(a))$?

11.4.5. Let S_3 act on $\Omega = S_3$ by conjugation, and let $\theta : S_3 \to S_6$ be the resulting homomorphism (see the previous problem). Label each element of Ω with 1, ..., 6, and explicitly give $\theta((1\ 2\ 3))$. Could you have done this problem more easily if you had used the Cayley digraph of this action given in Figure 4.4? Use the Cayley digraph to give $\theta((1\ 2))$.

11.4.6. Let D_8 act on the set Ω of subgroups of D_8 by conjugation. This action gives rise to a homomorphism θ from D_8 to some symmetric group S_n. For this action, what is n? After labeling elements of Ω with $1, \ldots, n$, explicitly give $\theta(a)$ and $\theta(b)$.

11.4.7. Let $G = S_4$ and let $\Omega = \{(1\ 2)(3\ 4), (1\ 3)(2\ 4), (1\ 4)(2\ 3)\} \subset S_4$. G acts on Ω by conjugation (i.e., $g \cdot x = gxg^{-1}$ for $g \in G$ and $x \in \Omega$). This action gives rise to a homomorphism

$$\theta : G \longrightarrow S_?.$$

(a) What is the "?" in $S_?$? What is Im θ? What is ker θ?
(b) Let $K = \Omega \cup \{e_{S_4}\}$. Is $K \leq G$? Why? Is $K \lhd G$? Why? If K is a normal subgroup of G, then what is a familiar group isomorphic to G/K? Why?

11.4.8. Identify the group of the rigid symmetries of a regular tetrahedron (i.e., find a well known group that is isomorphic to this group).

11.4.9. In how many different ways can we color the faces of a regular tetrahedron using m colors? (Note that we color an entire face with one color, we can color multiple faces with the same color, and two colorings are considered the same if they cannot be distinguished after a number of rotations of the two tetrahedra.)

11.5. The Homomorphism Theorems

We will now continue to prove a number of general theorems about homomorphisms. It is customary to single out three of these theorems as the *isomorphism theorems*. Usually Theorem 11.18 is called the *first isomorphism theorem*, and Theorem 11.43 and Corollary 11.41—which we will prove in this section—are called, respectively, the *second* and *third isomorphism theorems*. We will, however, focus on a more general statement—Theorem 11.38 below—and derive the others as corollaries. This theorem proves the heuristic that was the basis of our homomorphism diagrams of the previous sections. In other words, we see that when we mod out by a normal subgroup, the subgroup structure of the resulting quotient group can be gleaned from the subgroup structure of the original group. In fact, if G is a group and $N \lhd G$, then the lattice of subgroups of G/N is identical to the partial lattice of subgroups of G that contains N. It is helpful to first formulate the whole idea in terms of homomorphisms and kernels, and then apply it—using the canonical homomorphism—to quotient groups.

We first formalize a familiar notion.

Definition 11.36. If X and Y are two sets and $f : X \longrightarrow Y$ is a map, then we can extend this map to a map from *subsets* of X to *subsets* of Y. In other words, let 2^X be the collection of subsets of X and, likewise, let 2^Y denote the set of all subsets of Y. Using f, we define a new map—which we also denote by f—from 2^X to 2^Y. So we define

$$f : 2^X \to 2^Y$$

by

$$A \longmapsto f(A) = \{f(a) \mid a \in A\} \subseteq Y, \qquad \text{for } A \subseteq X.$$

We are used to using $f(A)$ to mean $\{f(a) \mid a \in A\}$, and hence using f both for the function $f : X \to Y$ and the extended function $f : 2^X \to 2^Y$ will not cause any confusion.

Of course, we can apply f to any collection of subsets of X and not necessarily to the collection of *all* subsets of X.

Example 11.37. Let $X = \mathbb{R}$, $Y = [-1, 1]$, and $f : X \to Y$ be defined by $f(x) = \sin x$. Then we have—as defined above—a function $f : 2^X \to 2^Y$. So for example, if $A = \{\frac{n}{2}\pi \mid n \in \mathbb{Z}\}$ and $B = \{\pm\pi/4\}$, then $f(A) = \{-1, 0, 1\}$ and $f(B) = \{\pm\frac{\sqrt{2}}{2}\}$. Both $f(A)$ and $f(B)$ are subsets of $[-1, 1]$ as expected.

Theorem 11.38. *Let G and H be groups, and let $\phi \colon G \to H$ be a homomorphism. Let $N = \ker(\phi)$. Define $\mathcal{S}$ to be the set of subgroups of G that contain N, and define $\mathcal{T}$ to be the set of subgroups of $\phi(G)$. In other words,*

$$\mathcal{S} = \{K \leq G \mid N \leq K\} \qquad and \qquad \mathcal{T} = \{H' \leq H \mid H' \leq \phi(G)\}.$$

Then the map ϕ extends to a map $\phi : \mathcal{S} \longrightarrow \mathcal{T}$ by

$$U \overset{\phi}{\longmapsto} \phi(U).$$

This map—which is also denoted by ϕ—is a bijection from $\mathcal{S}$ to $\mathcal{T}$. Furthermore, if $U \in \mathcal{S}$ and $V \in \mathcal{T}$ with $V = \phi(U)$, then

(a) *$U \lhd G$ if and only if $V \lhd \phi(G)$,*

(b) *$|G : U| = |\phi(G) : V|$, and*

(c) *if $U \lhd G$, then $G/U \cong \phi(G)/V$.*

Remark 11.39. Let $\phi : G \to H$ be a group homomorphism, and let $N = \ker(\phi)$. Theorem 11.38 is basically saying that not only the lattice of subgroups of the image $\phi(G)$ is identical to the lattice of subgroups of G that contain N, but edge lengths, normal subgroups, and quotient groups in the two lattices correspond as well. If you limit yourself to those subgroups of G that contain N, then what you see is exactly a view of the subgroups of $\phi(G)$. We have already proved—Theorem 11.18—that $G/N \cong \phi(G)$, and two isomorphic groups will have identical subgroup structures. What this theorem really does is to show that the subgroup structure of G/N is identical to the lattice of subgroups of G that contain N. Note also that, in the proof, we do not use Theorem 11.18, and, yet if we let $U = N$ in Theorem

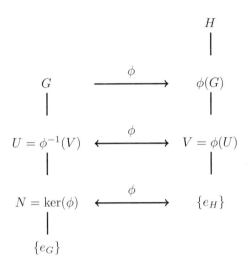

Figure 11.13. A 1-1 correspondence between subgroups of G that contain N and subgroups of $\phi(G)$

11.38, we do get that $G/N \cong \phi(G)$. Hence, we are reproving Theorem 11.18 here. The homomorphism diagram in Figure 11.13 captures some of the statements of Theorem 11.38.

Proof. As discussed in Definition 11.36, the map ϕ extends to a map from all subsets of G to all subsets of H. Now if you restrict the domain to subgroups of G, then the codomain will be subgroups of $\phi(G)$, and hence if you further restrict the domain to those subgroups of G that contain N, the kernel of (the original) ϕ, then the codomain can continue to be the subgroups of $\phi(G)$. Hence, ϕ does give a map from $\mathcal{S}$ to $\mathcal{T}$.

There are many things to show, and, so, the proof is long. However, no part of the proof is surprising or requires a deep insight. We urge the reader to attempt the proofs of the claims before reading the explanations. First, we show that (this extended) ϕ is a bijection from $\mathcal{S}$ to $\mathcal{T}$.

CLAIM: ϕ is 1-1.

PROOF OF CLAIM: Assume $\phi(U_1) = \phi(U_2)$, where U_1, U_2 are two subgroups of G that contain N. We first show that $U_1 \subseteq U_2$.

Let $x \in U_1$. We have $\phi(x) \in \phi(U_1) = \phi(U_2)$. Thus there exists $y \in U_2$ such that $\phi(y) = \phi(x)$. But this means—by Proposition 11.14—that x and y are in the same coset of N. Hence $x \in Ny \subseteq U_2$.

Similarly, $U_2 \subseteq U_1$, and so $U_1 = U_2$, and ϕ is 1-1.

CLAIM: ϕ is onto.

PROOF OF CLAIM: Let $V \in \mathcal{T}$. We need a subgroup U with $N \leq U \leq G$ such that $\phi(U) = V$. We let $U = \phi^{-1}(V) = \{g \in G \mid \phi(g) \in V\}$. Now, by the definition of $\mathcal{T}$, $V \subseteq \phi(G)$, and so every element of V has an inverse image in G. As a result, $\phi(U) = V$. Since, for all $n \in N$, $\phi(n) = e_H \in V$, we have $N \subseteq U$. Hence $U \in \mathcal{S}$ and $\phi(U) = V$.

Now assume that $U \in \mathcal{S}$, $V \in \mathcal{T}$, and $\phi(U) = V$.

CLAIM: If $U \lhd G$, then $V \lhd \phi(G)$.

PROOF OF CLAIM: We will show that $hVh^{-1} \subseteq V$ for all $h \in \phi(G)$. Let $v \in V$ and $h \in \phi(G)$ be arbitrary. Since $h \in \phi(G)$, we have $h = \phi(g)$ for some $g \in G$. Also $v \in V = \phi(U)$ and so $v = \phi(u)$ for some $u \in U$. We know $U \lhd G$ and hence $gug^{-1} \in U$, and so

$$hvh^{-1} = \phi(g)\phi(u)\phi(g)^{-1} = \phi(gug^{-1}) \in \phi(U) = V,$$

as claimed.

CLAIM: If $V \lhd \phi(G)$ then $U \lhd G$.

PROOF OF CLAIM: To show $U \lhd G$, we show that $gUg^{-1} \subseteq U$ for all $g \in G$. So fix $g \in G$ and $u \in U$. The proof will be complete when we show that $gug^{-1} \in U$. Let $h = \phi(g)$ and $v = \phi(u) \in V$. We know $hvh^{-1} \in V$ since $V \lhd \phi(G)$, and so

$$(11.1) \qquad \phi(gug^{-1}) = \phi(g)\phi(u)\phi(g)^{-1} = hvh^{-1} \in V.$$

Now $\phi(gug^{-1}) \in V$ means that $gug^{-1} \in \phi^{-1}(V) = U$, and the proof is complete.

Now let $G/U = \{Ug \mid g \in G\}$ denote the right cosets of U in G, and, likewise, let $\phi(G)/V = \{Vh \mid h \in \phi(G)\}$ denote the right cosets of V in $\phi(G)$. Note that we are *not* assuming that $U \lhd G$ or $V \lhd \phi(G)$.

Define a map $\psi : G/U \longrightarrow \phi(G)/V$ by

$$\psi(Ug) = V\phi(g).$$

To complete the proof, we show that ψ is a bijection and that it is an isomorphism if $U \lhd G$. However—as ψ is defined on cosets—we first have to show that ψ is well defined.

CLAIM: ψ is well defined.

PROOF OF CLAIM: Assume $Ug_1 = Ug_2$ for $g_1, g_2 \in G$. We then have to show that $V\phi(g_1) = V\phi(g_2)$.

$Ug_1 = Ug_2$ means that $g_2 \in Ug_1$, and hence $g_2 = ug_1$ for some $u \in U$. Note that $\phi(u) \in V$, and hence $V\phi(u) = V$. We now have

$$V\phi(g_2) = V\phi(ug_1) = V\phi(u)\phi(g_1) = V\phi(g_1).$$

CLAIM: ψ is 1-1.

PROOF OF CLAIM: Assume $\psi(Ug_1) = \psi(Ug_2)$, and we need to show that $Ug_1 = Ug_2$. From the assumption and by definition of the map, it follows that $V\phi(g_1) = V\phi(g_2)$. This means that $\phi(g_2) \in V\phi(g_1)$, and hence $\phi(g_2) = v\phi(g_1)$ for some $v \in V = \phi(U)$. Hence $v = \phi(u)$ for some $u \in U$, and $\phi(g_2) = v\phi(g_1) = \phi(ug_1)$. Thus we have that $g_2 \in Nug_1 \subseteq Ug_1$ since $Nu \subseteq U$. But $g_2 \in Ug_1$ implies that $Ug_1 = Ug_2$.

CLAIM: ψ is onto.

PROOF OF CLAIM: Let $Vh \in \phi(G)/V$, then $h \in \phi(G)$, and so $h = \phi(g)$ for some $g \in G$. We then have
$$\psi(Ug) = V\phi(g) = Vh,$$
and the map is onto.

CLAIM: If $U \lhd G$, then ψ is an isomorphism.

PROOF OF CLAIM: It remains to show that ψ is a homomorphism, since $U \lhd G$, $V \lhd \phi(G)$, and both G/U and $\phi(G)/V$ are groups with well-defined multiplications. We have
$$\psi(Ug_1 Ug_2) = \psi(Ug_1 g_2) = V\phi(g_1 g_2)$$
$$= V\phi(g_1)\phi(g_2) = V\phi(g_1)V\phi(g_2) = \psi(Ug_1)\psi(Ug_2). \qquad \square$$

Example 11.40. Let $G = D_8$ and $\mathbf{Z} = \mathbf{Z}(D_8) = \langle a^2 \rangle$. We know that
$$D_8/\mathbf{Z} = \{\mathbf{Z}, \mathbf{Z}a, \mathbf{Z}b, \mathbf{Z}ab\} \cong \mathbb{Z}/2\mathbb{Z} \times \mathbb{Z}/2\mathbb{Z} = \{(0,0), (1,0), (0,1), (1,1)\}.$$
One isomorphism between $D_8/\mathbf{Z}$ and $\mathbb{Z}/2\mathbb{Z} \times \mathbb{Z}/2\mathbb{Z}$ is given by
$$\psi : \begin{cases} \mathbf{Z} & \longmapsto (0,0), \\ \mathbf{Z}a & \longmapsto (1,0), \\ \mathbf{Z}b & \longmapsto (0,1), \\ \mathbf{Z}ab & \longmapsto (1,1). \end{cases}$$
We also have the canonical homomorphism $\pi : D_8 \longrightarrow D_8/\mathbf{Z}$, given by $\pi(x) = \mathbf{Z}x$ for every $x \in D_8$. Hence, we have the homomorphism diagram of Figure 11.14.

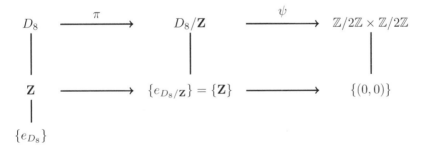

Figure 11.14. The canonical homomorphism $\pi : D_8 \to D_8/\mathbf{Z}$ and the isomorphism $\psi : D_8/\mathbf{Z} \to \mathbb{Z}/2\mathbb{Z} \times \mathbb{Z}/2\mathbb{Z}$

Now let
$$A = \{(0,0), (1,0)\} = \langle (1,0) \rangle \leq \mathbb{Z}/2\mathbb{Z} \times \mathbb{Z}/2\mathbb{Z}.$$
Now A corresponds to a subgroup of $D_8/\mathbf{Z}$ which in turn—by the homomorphism theorem, Theorem 11.38—corresponds to a subgroup of D_8 that contains $\mathbf{Z}$. We can draw the more refined homomorphism diagram of Figure 11.15. Note that the elements of $\langle a \rangle$ are exactly the union of the elements in the two cosets $\mathbf{Z}$ and $\mathbf{Z}a$, and that π maps elements of $\langle a \rangle$ onto $\{\mathbf{Z}, \mathbf{Z}a\}$ which is a subgroup of order 2 of $D_8/\mathbf{Z}$. We actually know exactly where π sends each element:
$$\pi(e) = \pi(a^2) = \mathbf{Z} \quad \text{and} \quad \pi(a) = \pi(a^3) = \mathbf{Z}a.$$

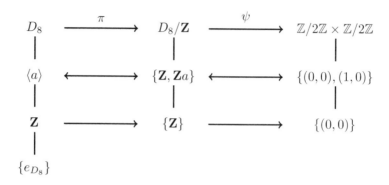

Figure 11.15. The subgroup $A = \{(0,0), (1,0)\}$ corresponds to $\{\mathbf{Z}, \mathbf{Z}a\}$ and to $\langle a \rangle$.

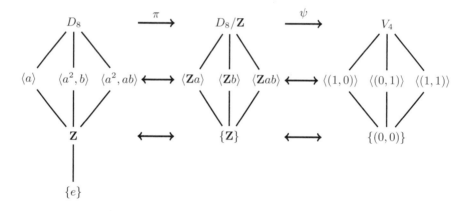

Figure 11.16. The lattice of subgroups of $V_4 = \mathbb{Z}/2\mathbb{Z} \times \mathbb{Z}/2\mathbb{Z}$ and $D_8/\mathbf{Z}(D_8)$ correspond to the subgroups of D_8 that contain $\mathbf{Z}(D_8)$.

In fact, the lattice of subgroups of $V_4 = \mathbb{Z}/2\mathbb{Z} \times \mathbb{Z}/2\mathbb{Z}$ is exactly the same as the lattice of subgroups of $D_8/\mathbf{Z}$, and this, in turn—by the homomorphism theorem, Theorem 11.38—is the same as the lattice of subgroups of D_8 that contain $\mathbf{Z}$. See Figure 11.16.

Corollary 11.41. *Let G be a group, and assume $N \lhd G$ and $N \le M \lhd G$. Then*

$$M/N \lhd G/N \qquad and \qquad (G/N)/(M/N) \cong G/M.$$

Proof. Let $\pi : G \to G/N$ be the canonical homomorphism. See the homomorphism diagram in Figure 11.17. We apply the homomorphism theorem, Theorem 11.38, directly. Since $M \lhd G$, we have $M/N = \pi(M) \lhd G/N$ (by Theorem 11.38(a)), and $G/M \cong (G/N)/(M/N)$ (by Theorem 11.38(c)). In the homomorphism diagram in Figure 11.17 we think of this as "the two sides of the upper rectangle are isomorphic". □

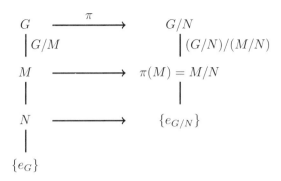

Figure 11.17. The canonical homomorphism $\pi : G \to G/N$, and $N \le M \lhd G$.

Example 11.42. We illustrate the use of the homomorphism theorems in understanding the structure of a group. Let $G = S_4$, the symmetric group of degree 4. Let $K = \{e, x = (1\ 2)(3\ 4), y = (1\ 3)(2\ 4), xy = (1\ 4)(2\ 3)\}$. It is straightforward to check that K is closed under multiplication and hence is a subgroup of S_4. Each non-identity element of K has order 2 and so K is isomorphic to $\mathbb{Z}/2\mathbb{Z} \times \mathbb{Z}/2\mathbb{Z}$, the Klein 4-group. The subgroup K is also the union of two conjugacy classes of S_4: $\{e\}$ and the conjugacy class consisting of all elements with cycle type $2, 2$. (Recall that two elements in S_n are conjugate if and only if they have the same cycle type.) By Theorem 10.18(g) we conclude that $K \lhd S_4$.

Now S_4/K is a group of order 6. Is it abelian? We have

$$K(1\ 2)K(1\ 3) = K(1\ 3\ 2) \quad \text{and} \quad K(1\ 3)K(1\ 2) = K(1\ 2\ 3).$$

Now $K(1\ 2\ 3) = K(1\ 3\ 2)$ if and only if $(1\ 2\ 3)(1\ 3\ 2)^{-1} \in K$. But $(1\ 2\ 3)(1\ 3\ 2)^{-1} = (1\ 3\ 2) \notin K$. Hence, we conclude that S_4/K is a non-abelian group of order 6. By Proposition 5.36, $S_4/K \cong S_3 = \langle f, g \mid f^3 = g^2 = e, gf = f^2g \rangle$. Knowing this and since elements of the same order have to be mapped to each other, it is straightforward to write down an actual isomorphism $\psi : S_4/K \to S_3$:

$$\psi : \begin{cases} K & \longmapsto e, \\ K(1\ 2\ 3) & \longmapsto f, \\ K(1\ 3\ 2) & \longmapsto f^2, \\ K(1\ 2) & \longmapsto g, \\ K(1\ 3) & \longmapsto fg, \\ K(2\ 3) & \longmapsto f^2g. \end{cases}$$

Hence, as a first step, we draw the homomorphism diagram of Figure 11.18. We can then add detail to the diagram. A partial lattice diagram for S_3 is given in Figure 11.19, and hence there will be a corresponding partial diagram for subgroups of S_4 that contain K. See Figure 11.20.

Now let U and V be the subgroups of S_4 containing K that correspond, in S_3, to $\langle f \rangle$ and $\langle g \rangle$, respectively. Which subgroups are these?

The subgroup V is a subgroup of order 8 containing K. In fact, $V = \phi^{-1}(\psi^{-1}\langle (1\ 2) \rangle)$ and so $V/K = \{K, K(1\ 2)\}$, and $V = K \cup K(1\ 2)$. We have

$$V = \{e, (1\ 2)(3\ 4), (1\ 3)(2\ 4), (1\ 4)(2\ 3), (1\ 2), (3\ 4), (1\ 3\ 2\ 4), (1\ 4\ 2\ 3)\}.$$

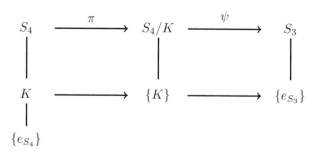

Figure 11.18. K consisting of the identity and all elements of cycle type $2, 2$ is a normal subgroup of S_4.

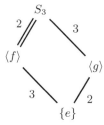

Figure 11.19. A partial lattice diagram for S_3

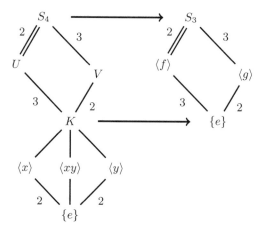

Figure 11.20. Subgroups of S_4 that contain K correspond to the subgroups of S_3

Evidently $V \cong D_8$. Hence S_4 has a subgroup isomorphic to D_8. In fact, since S_3 has three subgroups of order 2, S_4 also has three subgroups isomorphic to D_8 all of which contain K. Note that 8 is the highest power of 2 dividing 24 and hence V is a Sylow 2-subgroup of S_4.

In Figure 11.20, we have a partial lattice diagram of V. Hence this must correspond to a partial lattice diagram of D_8 as well. One such diagram is shown in Figure 11.21.

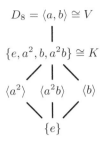

$$D_8 = \langle a, b \rangle \cong V$$

$$\{e, a^2, b, a^2 b\} \cong K$$

$$\langle a^2 \rangle \quad \langle a^2 b \rangle \quad \langle b \rangle$$

$$\{e\}$$

Figure 11.21. The partial lattice diagram of D_8 corresponding to the subgroup V of Figure 11.20

As for U, this is a subgroup of order 12 in S_4 that contains K. Again, since U corresponds to the group generated by $(1\ 2\ 3)$ in S_3, we conclude that $U = K \cup K(1\ 2\ 3) \cup K(1\ 3\ 2)$. Elements of K as well as 3-cycles are even permutations and hence U consists of 12 even permutations. We conclude that $U \cong A_4$.

As a byproduct of this example, we now have two (maximal) chains of subgroups of S_4 going from $\{e\}$ to S_4. These are given in Figure 11.22.

We had seen in Theorem 9.28 that, for two subgroups H and K of a group G, whenever KH is a subgroup then $|KH : K| = |H : K \cap H|$. We had also seen that if N is a normal subgroup of G, then NH is automatically a subgroup of G, and hence $|NH : N| = |H : N \cap H|$ for any subgroup H of G. In fact, in this situation and in Problem **10.2.5**, you were asked to show that $N \cap H \lhd H$. Here, we strengthen this result.

Theorem 11.43 (Direct diamond). *Let $N \lhd G$ and $H \le G$. Then*

$$(N \cap H) \lhd H \qquad and \qquad NH/N \cong H/(N \cap H).$$

The best way to remember this theorem is to recreate Figure 11.23.

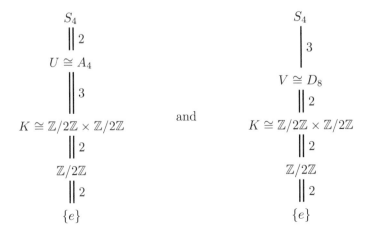

Figure 11.22. Two chains of subgroups for S_4

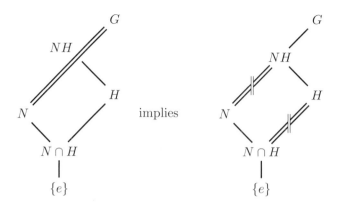

Figure 11.23. If $N \lhd G$ then $NH/N \cong H/(N \cap H)$

Proof. Define $\psi \colon H \to NH/N$ by $\psi(h) = Nh$.

CLAIM: ψ is an onto homomorphism.

PROOF OF CLAIM: We have $\psi(h_1 h_2) = Nh_1 h_2 = Nh_1 Nh_2 = \psi(h_1)\psi(h_2)$, and so ψ is a homomorphism. To show that ψ is onto, let $Nx \in NH/N$. We need $h \in H$ with $\psi(h) = Nx$. Now, $Nx \in NH/N$ means that $x \in NH$. Thus $x = nh$ for some $n \in N$ and $h \in H$. Hence,

$$\psi(h) = Nh = Nnh = Nx.$$

CLAIM: $\ker \psi = N \cap H$.

PROOF OF CLAIM: Let $h \in H$. The element h is in the kernel of ψ if and only if $\psi(h) = e$, which in turn is true if and only if $Nh = N$. In other words, $h \in H$ is in the kernel of ψ if and only if $h \in N$. Hence $\ker \psi = H \cap N$.

We conclude that $H \cap N \lhd H$ since $H \cap N$ is the kernel of a homomorphism with domain H. In addition, by Theorem 11.18, $H/\ker \psi \cong \operatorname{Im}\psi$. The theorem now follows. $\qquad\qquad\qquad\qquad\qquad\qquad\qquad\qquad\qquad\qquad\qquad\qquad\qquad\square$

Problems

11.5.1. Let G be a group, and let $f : G \longrightarrow \mathbb{Z}/12\mathbb{Z}$ be an onto homomorphism. We know that $|\ker(f)| = 5$.
   (a) What can you say about $|G|$?
   (b) Consider $H = \{0, 4, 8\}$ a subgroup of $\mathbb{Z}/12\mathbb{Z}$. What can you say about $|f^{-1}(H)|$?

11.5.2. Let G be a group, and let $\theta : G \to \mathbb{Z}/24\mathbb{Z}$ be an onto group homomorphism. We know that $|\ker(\theta)| = 5$.
   (a) What can you say about $|G|$?
   Let $H = \langle 3 \rangle = \{0, 3, 6, 9, 12, 15, 18, 21\} \leq \mathbb{Z}/24\mathbb{Z}$, and let $L = \theta^{-1}(H)$.
   (b) What is $|L|$?
   (c) Find a familiar group that is isomorphic to $L/\ker(\theta)$.

(d) Is $L \lhd G$? If the answer is yes, find a familiar group that is isomorphic to G/L.

11.5.3. Let G be a group, and let S_3 be the symmetric group of degree 3. The map $\theta : G \to S_3$ is an onto group homomorphism. We know that $|\ker(\theta)| = 4$.
   (a) What can you say about $|G|$?
   Let $H = \langle (1\ 2\ 3) \rangle \leq S_3$, and let $L = \theta^{-1}(H)$.
   (b) What can you say about $|L|$? Can you find a familiar group that is isomorphic to $L/\ker(\theta)$?
   (c) Is $L \lhd G$? If the answer is yes, then find a familiar group that is isomorphic to G/L.

11.5.4. Let G be a group. D_8 is the dihedral group of order 8, and a is an element of order 4 in D_8. The map $\theta : G \to D_8$ is an onto group homomorphism. We know that $|\ker(\theta)| = 4$.
   (a) What can you say about $|G|$?
   Let $H = \langle a^2 \rangle \leq D_8$, and let $L = \theta^{-1}(H)$.
   (b) What can you say about $|L|$? Can you find a familiar group that is isomorphic to $L/\ker(\theta)$?
   (c) Is $L \lhd G$? Why? If the answer is yes, then find a familiar group that is isomorphic to G/L.

11.5.5. Let $G = (\mathbb{Z}/24\mathbb{Z}, +)$.
   (a) Find a subgroup H of G that is isomorphic to $(\mathbb{Z}/4\mathbb{Z}, +)$. How many choices for such an H are there?
   (b) Find a familiar group that is isomorphic to G/H.
   (c) Find a subgroup K of G/H of order 3.
   (d) Find a familiar group that is isomorphic to $(G/H)/K$.

11.5.6. Let $K \leq N \leq G$ with K and N normal in G. Define the map $\phi : G/K \to G/N$ by $\phi(Kx) = Nx$. Is ϕ a well defined map? Is it a homomorphism? What is the image of ϕ? What is the kernel of ϕ?

11.5.7. The group G is of order 30, and it acts on a set Ω of size 3. We know that exactly five elements of G fix every element of Ω, i.e.,

$$|\{g \in G \mid g \cdot \alpha = \alpha \ \forall \ \alpha \in \Omega\}| = 5.$$

   (a) Complete the homomorphism diagram of Figure 11.9 that arises from this action as best as you can.
   (b) Using the homomorphism theorem, Theorem 11.38, and the subgroup lattice diagram of S_3, draw a partial lattice diagram of G.

11.5.8. **Normal maximal subgroups.** A proper subgroup H of a group G is said to be *maximal*, if there is no subgroup K with $H < K < G$. (See Definition 5.21.) Let G be a group, and let N be a proper normal subgroup of G. Prove that N is maximal in G if and only if $G/N \cong \mathbb{Z}/p\mathbb{Z}$ for some prime p.

11.5.9. **Maximal normal subgroups.** A proper normal subgroup H of a group G is said to be a *maximal normal* subgroup if there is no normal subgroup K with $H < K < G$. Let G be a group, and let N be a proper normal subgroup of G. Prove that N is maximal normal in G if and only if G/N is a simple group.

11.5.10. Assume that H and K are distinct maximal normal subgroups of a group G (see Problem 11.5.9). Prove that $H \cap K$ is a maximal normal subgroup of H.

11.5.11. Let G and H be finite groups, and let $\phi : G \longrightarrow H$ be a homomorphism with $K = \ker(\phi)$. Assume $L \leq G$. (The subgroup L does not necessarily contain nor is it contained in K).
  (a) Show $\phi(L) = \phi(KL)$.
  (b) How are the numbers $|L : K \cap L|$, $|KL : K|$, and $|\phi(L)|$ related?

11.5.12. Let $\theta : G \to H$ be a group homomorphism, and let $g \in G$. Show that
$$o(g) = o(\theta(g)) \cdot |\langle g \rangle \cap \ker(\theta)| \, .$$

11.5.13. **The group** $\mathrm{PSL}(2,3)$. Recall that the special linear group $\mathrm{SL}(n,p)$ is the group of $n \times n$ invertible matrices with determinant 1 over $\mathbb{F}_p = (\mathbb{Z}/p\mathbb{Z}, +, \cdot)$. The *Projective Special Linear Group*, $\mathrm{PSL}(n,p)$ is defined to be the quotient group $\mathrm{SL}(n,p)/\mathbf{Z}(\mathrm{SL}(n,p))$.[1] Find an already familiar group that is isomorphic to $\mathrm{PSL}(2,3)$.

  You may find the following steps helpful.
Step 1: If you have done Problem 9.3.1, then you may be able to guess the answer to this problem by looking at the relevant part of the lattice diagram for $\mathrm{SL}(2,3)$.
Step 2: Let $\mathbb{F}_3 = (\mathbb{Z}/3\mathbb{Z}, +, \cdot)$ be the field with three elements, and let $(\mathbb{F}_3)^2 = \{(a,b) \mid a, b \in \mathbb{F}_3\}$. Show that $(\mathbb{F}_3)^2$ is a two dimensional vector space over $\mathbb{F}_3$.
Step 3: Let Ω be the set of one dimensional subspaces of $(\mathbb{F}_3)^2$. What is $|\Omega|$?
Step 4: Define a (natural) action of $\mathrm{SL}(2,3)$ on Ω.
Step 5: Show that the kernel of this action is $\mathbf{Z}(\mathrm{SL}(2,3))$.
Step 6: Use Theorem 11.28 to identify $\mathrm{PSL}(2,3)$.

11.6. Automorphisms and Inner-automorphisms*

In this optional section we will use some of the vocabulary that we have developed to look at "symmetries" of a group. Recall that if Ω is a set, then $\mathrm{Perm}(\Omega)$ is the set of 1-1 and onto maps from Ω to Ω. These maps tell us the different ways that elements of Ω can be rearranged, and if Ω is only a set and does not have any additional structure, then these bijections can be thought of as the symmetries of the set Ω. However, if the set has any more structure (e.g., it is a group), then we would like its symmetries to preserve the structure as well. Thus, for example, for a group G we would like to consider maps $\phi : G \to G$ such that ϕ is a bijection and also satisfies $\phi(ab) = \phi(a)\phi(b)$ for all a and b in G. Such maps are, as we know, the automorphisms of G. Hence, automorphisms of a group play the role of the symmetries of the group. They tell us the ways of relabeling the elements of the group such that the multiplication table of the group remains valid after relabeling

---

[1] The groups $\mathrm{PSL}(2,p)$ were considered by Évariste Galois—See Part 3 and especially page 453—in the 1830s. In his last letter, written two days before his death at the age of 21, Galois observed that except for $p = 2$ and 3, the groups $\mathrm{PSL}(2,p)$ are simple.

all the row and column headings as well as all the entries. Since a group is only a set with a group multiplication, we do not expect anything more from its symmetries.

The set of all automorphisms of a group G will be denoted by $\mathrm{Aut}(G)$. But how do we find automorphisms of a group G? We saw in Example 11.6 (and Definition 11.7) one way of constructing automorphisms of G by using elements of G. If $x \in G$, then the map $\phi_x : G \to G$ defined by $\phi_x(g) = xgx^{-1}$ was shown (see Example 11.6) to be an automorphism of G, and it was called an inner automorphism of G. Recall that the set of inner automorphisms of a group G is denoted by $\mathrm{Inn}(G)$. We shall show a bit later that $\mathrm{Inn}(G)$ is a subgroup of $\mathrm{Aut}(G)$. We record the various groups of maps on G:

Definition 11.44 ($\mathrm{Perm}(G)$, $\mathrm{Aut}(G)$, $\mathrm{Inn}(G)$). Let G be a group, then

$$\mathrm{Perm}(G) = \{\phi : G \to G \mid \phi \text{ is 1-1 and onto}\},$$

$$\mathrm{Aut}(G) = \{\phi \in \mathrm{Perm}(G) \mid \phi \text{ is a homomorphism}\},$$

$$\mathrm{Inn}(G) = \{\phi \in \mathrm{Aut}(G) \mid \exists\, x \in G \text{ with } \phi(g) = xgx^{-1}\ \forall g \in G\}.$$

It is clear that $\mathrm{Inn}(G)$ is a subset of $\mathrm{Aut}(G)$ which, in turn, is a subset of $\mathrm{Perm}(G)$.

We know that (with function composition as the operation) $\mathrm{Perm}(G)$ is a group. Since the composition of two homomorphisms is a homomorphism and the inverse of an isomorphism is an isomorphism, we have

Lemma 11.45. $\mathrm{Aut}(G)$ *is a subgroup of* $\mathrm{Perm}(G)$.

We use actions to show, albeit indirectly, that $\mathrm{Inn}(G)$ is a subgroup.

Let G be any group. We know that G acts on G by conjugation: For $g \in G$ and $x \in G$, we have $g \cdot x = gxg^{-1}$. By Theorem 11.28 we know that this means that we have a homomorphism $\theta : G \longrightarrow \mathrm{Perm}(G)$ defined by $\theta(g) = f_g$ where $f_g \in \mathrm{Perm}(G)$ and $f_g(x) = g \cdot x = gxg^{-1}$. Note that θ sends every element of G to a map from G to G. In fact, $\theta(g)$ is exactly the inner automorphism of G induced by g. Hence the image of θ is $\mathrm{Inn}(G)$ which is a subset of $\mathrm{Aut}(G)$. What is the kernel of θ? The kernel of θ consists of those elements of G that act as the identity element. In other words, $g \in \ker(\theta)$ if and only if $gxg^{-1} = x$ for all $x \in G$. This is the same as saying that $gx = xg$ for all $x \in G$. We conclude that $\ker(\theta) = \mathbf{Z}(G)$ is the center of G. We have the homomorphism diagram of Figure 11.24, and using the fact that $G/\ker(\theta) \cong \theta(G)$, we get that $G/\mathbf{Z}(G) \cong \mathrm{Inn}(G)$. We record what we have proved:

Corollary 11.46. *Let G be a group, and let $\mathrm{Perm}(G)$, $\mathrm{Aut}(G)$, and $\mathrm{Inn}(G)$ be the groups of bijections, automorphisms, and inner automorphisms of G, respectively. Then*

$$\mathrm{Inn}(G) \leq \mathrm{Aut}(G) \leq \mathrm{Perm}(G)$$

and

$$G/\mathbf{Z}(G) \cong \mathrm{Inn}(G).$$

Since the center of a group is always a normal subgroup, we can always form $G/\mathbf{Z}(G)$. As we have just proved, this will always be isomorphic to the group of inner automorphisms of G. Of course, if the group is centerless (i.e., if $\mathbf{Z}(G) = \{e\}$),

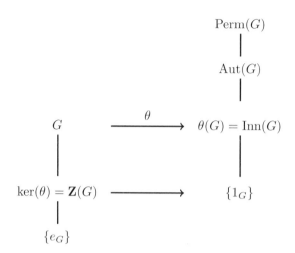

Figure 11.24. The action of G on G by conjugation gives a homomorphism with $\text{Inn}(G)$ as the image and $\mathbf{Z}(G)$ as the kernel.

which is what happens to non-abelian simple groups, then $\text{Inn}(G) \cong G/\mathbf{Z}(G) \cong G$. In general, a group G is called *capable* if there exists a group Γ such that $G \cong \Gamma/\mathbf{Z}(\Gamma) \cong \text{Inn}(\Gamma)$. In other words, capable groups are exactly the groups that appear as the inner automorphism groups of other groups. We have seen that non-abelian simple groups are capable. The Klein-4 group, $\mathbb{Z}/2\mathbb{Z} \times \mathbb{Z}/2\mathbb{Z}$, is also capable since $D_8/\mathbf{Z}(D_8) \cong \mathbb{Z}/2\mathbb{Z} \times \mathbb{Z}/2\mathbb{Z}$. However it is easy to see (see Problem 10.3.10) that non-trivial cyclic groups are *not* capable. The author, in his first research paper [**Sha87**], proved that if the quaternion group Q_8 is the normal subgroup of a group G, then G *cannot* be capable.

Theorem 11.47 (N/C theorem). *Let G be a group, and let $H \leq G$. Then*

$$\mathbf{C}_G(H) \vartriangleleft \mathbf{N}_G(H),$$

and $\mathbf{N}_G(H)/\mathbf{C}_G(H)$ is isomorphic to a subgroup of $\text{Aut}(H)$.

Proof. We already know—by Lemma 10.15—that $H \vartriangleleft \mathbf{N}_G(H)$ and that $\mathbf{C}_G(H)$ is a subgroup of $\mathbf{N}_G(H)$. We let $\mathbf{N}_G(H)$ act on H by conjugation. We have an action, since H is normal in the normalizer, and, hence, conjugating elements of H by elements of the normalizer results in other elements of H.

This action automatically gives us a homomorphism from the normalizer into $\text{Perm}(H)$, the bijections on H. However, the image of every element of $\mathbf{N}_G(H)$—in addition to being 1-1 and onto—is an automorphism of H. (Why?) Note that even though the action is conjugation, the image of every element of $\mathbf{N}_G(H)$ may not be an inner automorphism of H—inner automorphisms of H are the maps given by conjugating with an element of H. In any case, we have the homomorphism diagram in Figure 11.25.

What is the kernel of this homomorphism? Using the fact that $\mathbf{C}_G(H) \leq \mathbf{N}_G(H)$ we have

$$\ker(\psi) = \{n \in \mathbf{N}_G(H) \mid nhn^{-1} = h \quad \forall h \in H\} = \mathbf{C}_{\mathbf{N}_G(H)}(H) = \mathbf{C}_G(H).$$

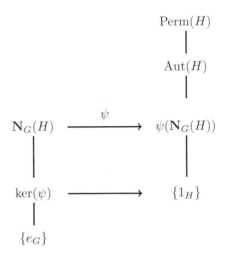

Figure 11.25. The action of $\mathbf{N}_G(H)$ on H gives a homomorphism ψ : $\mathbf{N}_G(H) \to \mathrm{Perm}(H)$.

Hence $\mathbf{C}_G(H) \lhd \mathbf{N}_G(H)$, and—by Theorem 11.18—we have

$$\mathbf{N}_G(H)/\mathbf{C}_G(H) \cong \psi(\mathbf{N}_G(H)) \leq \mathrm{Aut}(H). \qquad \square$$

We also conclude that, in general and for *every subgroup of any group*, we have the partial lattice diagram in Figure 11.26. To appreciate the power of the N/C theorem, do Problem *11.6.13*. To be able to use the full power of the N/C theorem, it is helpful to know $\mathrm{Aut}(G)$ for groups G that appear often. In Problem 11.6.6, which is based on Problem *11.6.5*, you are asked to show that the automorphism group of the cyclic group of order n is $(\mathbb{Z}/n\mathbb{Z})^{\times}$. In Problems 11.6.8 and 11.6.11, you are led to finding the automorphism groups of S_3 and D_8, respectively.

Problems

11.6.1. Let $G = D_8 \times D_8$, and let $H = \langle b \rangle \times D_8$ be a subgroup of G.
  (a) Draw the partial lattice diagram in Figure 11.26 for G and H, identify all the relevant subgroups, and include edge lengths.
  (b) Find a familiar group that is isomorphic to $\mathrm{Inn}(H)$.
  (c) Find a familiar group that is isomorphic to $\mathbf{N}_G(H)/\mathbf{C}_G(H)$.
  (d) Find a familiar group that is isomorphic to $\mathrm{Inn}(G)$.

11.6.2. Repeat Problem 11.6.1 for $G = D_8 \times D_8$ and $H = \langle b \rangle \times \langle a \rangle$. Can you explicitly write down an automorphism of H that is not an inner automorphism?

11.6.3. Is D_8 capable?

11.6.4. If $G = (\mathbb{Z}/12\mathbb{Z}, +)$, then identify all the groups in the homomorphism diagram of Figure 11.24.

11.6.5. Consider the cyclic group $(\mathbb{Z}/12\mathbb{Z}, +)$, and let $\sigma \in \mathrm{Aut}(\mathbb{Z}/12\mathbb{Z})$.

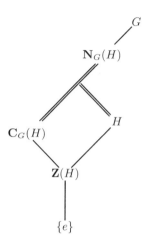

Figure 11.26. A partial lattice diagram for any $H \leq G$

   (a) What is $o(\sigma(1))$?

   (b) Must $\sigma(1) \in (\mathbb{Z}/12\mathbb{Z})^{\times}$?

   (c) What are the possibilities for $\sigma(1)$?

   (d) Show that, for $k \in \mathbb{Z}/12\mathbb{Z}$, $\sigma(k) = \underbrace{\sigma(1) + \cdots + \sigma(1)}_{k} = k\sigma(1)$.

   (e) Let $a \in (\mathbb{Z}/12\mathbb{Z})^{\times}$. Show that the map $\phi_a : \mathbb{Z}/12\mathbb{Z} \to \mathbb{Z}/12\mathbb{Z}$ defined by $\phi_a(k) = \underbrace{a + \cdots + a}_{k} = ka$ is an automorphism of $\mathbb{Z}/12\mathbb{Z}$.

11.6.6. Let C be a cyclic group of order n. Prove

$$\mathrm{Aut}(C) \cong (\mathbb{Z}/n\mathbb{Z})^{\times}.$$

11.6.7. Find a familiar group that is isomorphic to $\mathrm{Aut}(\mathbb{Z}/2\mathbb{Z} \times \mathbb{Z}/2\mathbb{Z})$.

11.6.8. Let S_3 be the symmetric group of degree 3.

   (a) Show that $\mathrm{Inn}(S_3) \cong S_3$.

   (b) Let $\sigma \in \mathrm{Aut}(S_3)$. Show that there are at most two possibilities for $\sigma((1\ 2\ 3))$ and at most three possibilities for $\sigma((1\ 2))$.

   (c) Show that $|\mathrm{Aut}(S_3)| \leq 6$.

   (d) Show that $\mathrm{Aut}(S_3) \cong S_3$.

11.6.9. Assume that a group G has a normal abelian subgroup H. Given this assumption, what does the partial lattice diagram in Figure 11.26 look like?

11.6.10. Assume that the order of a group G is $11m$, where m is an odd number not divisible by 5. Further assume, that G has a normal subgroup H of order 11. Prove that H must be in the center of G.

You may find the following steps useful:

Step 1: Draw the partial lattice diagram in Figure 11.26 for the subgroup H of G. You may want to do Problem 11.6.9 first.

Step 2: Use Theorem 11.47 and Problem 11.6.6 to limit the possibilities for $|G : \mathbf{C}_G(H)|$.

Step 3: Use the given information about $|G|$ to show $\mathbf{C}_G(H) = G$. Are you done?

11.6.11. Let $G = D_8 = \langle a, b \mid a^4 = b^2 = e, ba = a^{-1}b \rangle$.
    (a) Show that there exists a $\sigma \in \mathrm{Aut}(G)$ with $\sigma(a) = a^3$ and $\sigma(b) = ab$. What is the order of σ as an element of $\mathrm{Aut}(G)$?
    (b) Show that $|\mathrm{Aut}(G)|$ is 8.
    (c) Find every element of $\mathrm{Aut}(G)$, and, for every element, find its order.
    (d) Find a familiar group that is isomorphic to $\mathrm{Aut}(D_8)$.

11.6.12. Find a familiar group that is isomorphic to $\mathrm{Aut}(\mathbb{Z}/2\mathbb{Z} \times \mathbb{Z}/4\mathbb{Z})$.

11.6.13. Assume that a group G has a proper normal subgroup H isomorphic to D_8. Further assume that the only elements of G that commute with every element of H are the elements of the center of H. Prove that $|G| = 16$.

11.7. More Problems and Projects

Exact Sequences

Definition 11.48 (Exact sequence). Suppose $\{G_i \mid i = 1, \ldots, n\}$ is a collection of groups and, for $i = 1, \ldots, n-1$, we have homomorphisms $f_i : G_i \longrightarrow G_{i+1}$. Thus we have

$$\cdots \longrightarrow G_{i-1} \xrightarrow{f_{i-1}} G_i \xrightarrow{f_i} G_{i+1} \longrightarrow \cdots .$$

This sequence is called *exact* if $\mathrm{Im} f_{i-1} = \ker f_i$, for $i = 2, \ldots, n$.

Let $\phi : G \to F$ be a group homomorphism. Note that ϕ is onto if and only if $G \xrightarrow{\phi} F \longrightarrow \{e\}$ is exact (the second homomorphism is not identified since only the map that sends everything to $\{e\}$ is a possibility), and ϕ is 1-1 if and only if $\{e\} \longrightarrow G \xrightarrow{\phi} F$ is exact.

Exact sequences provide an alternative to the homomorphism diagrams. As Problems 11.7.3 and 11.7.4 show, we can encode a homomorphism, its kernel, and its image in a short exact sequence.

Problems

11.7.1. Assume that the sequence

$$V \xrightarrow{f} W \xrightarrow{g} U \xrightarrow{h} X$$

is exact. If f is onto, does h have to be 1-1? Either prove that it is or construct a counterexample.

11.7.2. Assume that

$$V \xrightarrow{f} W \xrightarrow{k} U \xrightarrow{h} X \xrightarrow{g} Y$$

is exact. Further assume that f is onto and g is 1-1. Show that $U = \{e\}$.

11.7.3. Let G and H be groups, and let $\phi : G \to H$ be a group homomorphism. Let $K = \ker(\phi)$ and $F = \phi(G)$. Show that there is an exact sequence of the form

$$\{e\} \longrightarrow K \longrightarrow G \xrightarrow{\phi} F \longrightarrow \{e\}.$$

Make sure that you identify all the maps.

11.7.4. Let G, K, and F be groups, and assume that we have the following short exact sequence:

$$\{e\} \xrightarrow{\gamma} K \xrightarrow{\alpha} G \xrightarrow{\beta} F \xrightarrow{\delta} \{e\}.$$

Show that G has a normal subgroup N such that
(a) $N \cong K$, and
(b) $G/N \cong F$.

11.7.5. Suppose that the diagram of groups and group homomorphisms in Figure 11.27 is commutative (see Definition 1.13) and has exact rows. Show that if α and γ are 1-1, then so is β.

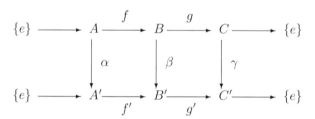

Figure 11.27. A commutative diagram of groups and homomorphisms with exact rows

11.7.6. Let $\mathbb{R}^\times$ denote the group of non-zero real numbers under multiplication. Give all the maps needed to make the following short exact sequence:

$$\{I_2\} \to \mathrm{SL}(2,\mathbb{R}) \to \mathrm{GL}(2,\mathbb{R}) \to \mathbb{R}^\times \to \{1\}.$$

11.7.7. Let $\mathrm{GL}(2,\mathbb{Z})$ be the group of invertible integer matrices with determinant ± 1, and let $\mathrm{SL}(2,\mathbb{Z})$ be the subgroup of $\mathrm{GL}(2,\mathbb{Z})$ consisting of integer matrices with determinant 1. (See Problems 2.1.4, 5.1.7, and 10.1.7.) Give all the maps and the mystery group to make a short exact sequence:

$$\{I_2\} \to \mathrm{SL}(2,\mathbb{Z}) \to \mathrm{GL}(2,\mathbb{Z}) \to ?? \to \{1\}.$$

Divisible Abelian Groups

Definition 11.49. Let G be an abelian group written additively. If $x \in G$ and k is a positive integer, then, as usual, $kx = \underbrace{x + \cdots + x}_{k}$, and $kG = \{kx \mid x \in G\}$. The group G is called *divisible* if $mG = G$ for all positive integers m.

11.7.8. Show that $(\mathbb{Q}, +)$ is a divisible group.

11.7.9. Assume that G_1 and G_2 are both divisible. Show that so is $G_1 \times G_2$.

11.7.10. Show that no finite (abelian) group is divisible.

11.7.11. Let G be a (abelian) divisible group. Prove that G does not have a maximal subgroup. (Maximal subgroups are defined in Definition 5.21.) You may find the following steps useful:
Step 1: Write G additively, and assume M is a maximal subgroup of G. Show (this is Problem 11.5.8) that $G/M \cong \mathbb{Z}/p\mathbb{Z}$ for some prime p.
Step 2: Show $p(x + M) = M$ for all $x \in G$.
Step 3: Show that $pG \subseteq M$, and conclude that $M = G$.[2]

The Affine and Generalized Stochastic Groups. Let F be a field (e.g., $\mathbb{Q}$, $\mathbb{R}$, $\mathbb{C}$, or $\mathbb{Z}/p\mathbb{Z}$ for p a prime), and let n be a positive integer.

Recall that $\mathrm{Aff}(n, F)$, the affine group of degree n over F, was defined (see Definition 2.80) as

$$\mathrm{Aff}(n, F) = \{T : F^n \to F^n \mid T(x) = Ax + b, \text{ where } A \in \mathrm{GL}(n, F), \text{ and } b \in F^n\}.$$

The group operation in $\mathrm{Aff}(n, F)$ is a function composition, and F^n stands for the set of $n \times 1$ column vectors with entries in F. (Problems 2.7.3–2.7.7, 10.2.8, and 10.2.9 were about the affine groups.)

Also recall, from Problem 4.5.2, that a generalized stochastic matrix is a matrix whose column sums is 1, and that the generalized stochastic group $S(n, F)$ consists of $n \times n$ invertible stochastic matrices over F (with matrix multiplication as group multiplication).

Here, following Poole [**Poo95**], we ask you to prove:

Theorem 11.50. *Let F be any field, and let $n \geq 2$ be an integer. Then*

$$S(n, F) \cong \mathrm{Aff}(n - 1, F).$$

Proof Sketch. Clearly, $S(n, f)$ is a subgroup of $\mathrm{GL}(n, F)$. Let $A(n, F)$ be the set of $n \times n$ invertible matrices over F whose last row is $(0\ 0\ \cdots\ 0\ 1)$. We first show that $\mathrm{Aff}(n - 1, F)$ is isomorphic to $A(n, F)$, which is also a subgroup of $\mathrm{GL}(n, F)$. We then complete the proof by showing that $S(n, F)$ and $A(n, F)$ are conjugate subgroups of $\mathrm{GL}(n, F)$. In other words, we exhibit a $Q \in \mathrm{GL}(n, F)$ such that $QS(n, F)Q^{-1} = A(n, F)$. $\qquad\square$

---

[2]Adapted from Malcolmson and Okoh [**MO00**].

11.7.12. Let $n \geq 2$, and let $\mathbf{0}_k$ denote the $k \times 1$ column vector of zeros (and $\mathbf{0}_k^t$ is the transpose of $\mathbf{0}_k$). Define $\phi : \mathrm{Aff}(n-1, F) \to \mathrm{GL}(n, F)$ by

$$\phi(T) = \begin{bmatrix} A & b \\ \mathbf{0}_{n-1}^t & 1 \end{bmatrix},$$

where $T : F^{n-1} \to F^{n-1}$ is given by $T(x) = Ax + b$ with $A \in \mathrm{GL}(n-1, F)$, $b \in F^{n-1}$. Show that ϕ is a 1-1 group homomorphism. Conclude that if $A(n, F)$ is the set of $n \times n$ invertible matrices over F whose last row is $(0\ 0\ \ldots\ 0\ 1)$, then

$$\mathrm{Aff}(n-1, F) \cong A(n, F) \leq \mathrm{GL}(n, F).$$

11.7.13. Let $n \geq 2$, let I_{n-1} denote the $(n-1) \times (n-1)$ identity matrix, let $\mathbf{j}$ denote the $(n-1) \times 1$ column vector of all ones, and let $\mathbf{0}_k$ be the $k \times 1$ column vector of all zeros. Finally, let F be any field. Define

$$Q = \begin{bmatrix} I_{n-1} & \mathbf{0}_{n-1} \\ \mathbf{j}^t & 1 \end{bmatrix}.$$

(a) Show that $Q^{-1} = \begin{bmatrix} I_{n-1} & \mathbf{0}_{n-1} \\ -\mathbf{j}^t & 1 \end{bmatrix}$.

(b) If $M \in S(n, F)$, then partition M as

$$M = \begin{bmatrix} N & u \\ v^t & \lambda \end{bmatrix},$$

where N is an $(n-1) \times (n-1)$ matrix, $u, v \in F^{n-1}$, and $\lambda \in F$. Show that

$$\mathbf{j}^t N + v^t = \mathbf{j}^t \text{ and } \mathbf{j}^t u + \lambda = 1,$$

and

$$QMQ^{-1} = \begin{bmatrix} N - u\mathbf{j}^t & u \\ \mathbf{0}_{n-1}^t & 1 \end{bmatrix}.$$

(c) Let $A(n, F)$ be the set of $n \times n$ invertible matrices over F whose last row is $(0\ 0\ \cdots\ 0\ 1)$, and $P = \begin{bmatrix} A & b \\ \mathbf{0}_{n-1}^t & 1 \end{bmatrix} \in A(n, F)$. Show $A \in \mathrm{GL}(n-1, F)$, and

$$Q^{-1}PQ = \begin{bmatrix} A + b\mathbf{j}^t & b \\ \mathbf{j}^t - \mathbf{j}^t A - \mathbf{j}^t b\mathbf{j}^t & 1 - \mathbf{j}^t b \end{bmatrix}.$$

11.7.14. Continuing with the notation and using the results of the previous problem, prove that

$$QS(n, F)Q^{-1} = A(n, F).$$

11.7.15. **Proof of Theorem 11.50.** Using the Proof Sketch and results of Problems 11.7.12–11.7.14, give a complete proof of Theorem 11.50. In other words, if F is any field and $n \geq 2$ is an integer, then prove that

$$S(n, F) \cong \mathrm{Aff}(n-1, F).$$

11.7.16. **An unlikely group.** Let n be an integer greater than 1, and let k be a positive integer relatively prime to n. As usual, $(\mathbb{Z}/n\mathbb{Z})^{\times}$ is the group of units of $\mathbb{Z}/n\mathbb{Z}$ under multiplication mod n. Let

$$G = k(\mathbb{Z}/n\mathbb{Z})^{\times} = \{kx \in \mathbb{Z} \mid x \in \mathbb{Z} \text{ with } 1 \le x < n, \text{ and } \gcd(x, n) = 1\}.$$

The binary operation on the set G is multiplication mod kn. (Note that the elements in G are not reduced mod n and the operation in G is *not* multiplication mod n.) In Problem 2.7.11, you were asked to prove that G (with multiplication mod kn) is a group, and if k' is the inverse of k mod n, then kk' is the identity of G. To understand the structure of G,[3] define the following three maps:

$\phi : (\mathbb{Z}/n\mathbb{Z})^{\times} \to G$ defined by $\phi(x) = kx$ for all $x \in (\mathbb{Z}/n\mathbb{Z})^{\times}$,

$\theta : (\mathbb{Z}/n\mathbb{Z})^{\times} \to G$ defined by $\theta(x) = kk'x \bmod kn$ for all $x \in (\mathbb{Z}/n\mathbb{Z})^{\times}$,

$\psi : G \to (\mathbb{Z}/n\mathbb{Z})^{\times}$ defined by $\psi(x) = x \bmod n$ for all $x \in G$.

    (a) Show that, in general, ϕ is *not* a group homomorphism.
    (b) Show that both θ and ψ are group isomorphisms, and that, in fact, they are inverses of each other.

In particular, conclude that $G \cong (\mathbb{Z}/n\mathbb{Z})^{\times}$.

11.7.17. **Central automorphisms.** Let G be a group, and, as usual, $\mathbf{Z}(G)$ denotes the center of G. Let α be an automorphism of G. The automorphism α is called a *central automorphism* of G if $x^{-1}\alpha(x) \in \mathbf{Z}(G)$ for all $x \in G$. In other words, a central automorphism will multiply every element by some central element. The set of all central automorphisms of G is denoted by $\mathrm{Aut}_c(G)$. Show that $\mathrm{Aut}_c(G)$ is a normal subgroup of $\mathrm{Aut}(G)$, and find $\mathrm{Aut}_c(D_8)$.

11.7.18. **A complex torus.** Consider $\mathbb{C}$, the additive group of complex numbers. Let $L = \{m_1 + m_2 i \mid m_1, m_2 \in \mathbb{Z}\}$.
    (a) Show that L is a subgroup of $\mathbb{C}$.
    (b) Let $X = \mathbb{C}/L$ and $\pi : \mathbb{C} \to X$ be the canonical homomorphism. Consider the set of points $R = \{a + bi \mid 0 \le a < 1, 0 \le b < 1\}$ in $\mathbb{C}$. Show that the map π restricted to R is 1-1 and onto. Convince yourself that this is another way of saying that R is a set of distinct representatives for the cosets of L in $\mathbb{C}$. In other words,

(11.2) $$X = \{r + L \mid r \in R\}.$$

    (c) Given the description of X in equation (11.2), there is a 1-1 correspondence between elements of X and points in R. We would rather replace R with $R' = \{a + bi \mid 0 \le a \le 1, 0 \le b \le 1\}$ (if you have had analysis, then you would know that R' is closed and bounded and hence compact). However, the map π restricted to R' is not 1-1 anymore. We can fix this, if we identify both pairs of opposite sides

---

[3]Adapted from Berger [**Ber05**]

of R'. In other words, we think of each point of the form $a + 0i$ as the same as the point $a + i$, and likewise each point of the form $0 + bi$ as the same as the point $1 + bi$. We denote by $\bar{R}$, the set R' with the opposite sides identified as described. Show that there is a 1-1 correspondence between X and $\bar{R}$.

(d) Explain why one may call $\bar{R}$—and hence X—a (complex) torus.

11.7.19. **Subgroups of infinite cyclic groups.** In Theorem 2.75 we proved that subgroups of cyclic groups—whether finite or infinite—are cyclic groups. In Problem 5.1.15, you showed that a subgroup of a (finite or infinite) cyclic group has a finite index. (See also Problems 5.1.13 and 5.1.16.) In fact, in the case of infinite groups, the converse is also true. Lanski [**Lan01**] has an elementary (but not short) proof of the following theorem of Fedorov [**Fed51**] from 1951.

> **Theorem.** *An infinite group is cyclic if and only if each of its non-identity subgroups has finite index.*

Find a copy of the paper, decipher the proof, and write an exposition of it.

11.7.20. **The complexes and the unit circle.** This is for students familiar with introductory set theory and basics of cardinal arithmetic. Let $G = (\mathbb{C}^\times, \cdot)$ be the multiplicative group of non-zero complex numbers, and let $H = \mathbb{S}^1$ be the proper subgroup of G consisting of points on the unit circle. A theorem of Clay [**Cla69**] from 1969 states that $G \cong H$. Azad and Laradji [**AL00**] give an elementary (and short) proof of this result. Their proof relies on using basic ideas of cardinality to show that there exists an isomorphism $f : (\mathbb{R}, +) \to (\mathbb{C}, +)$ with the property that $f(q) = q$ for all $q \in \mathbb{Q}$. Find a copy of the paper, decipher the proof, and write an exposition of it.

Chapter 12

Using Sylow Theorems to Analyze Finite Groups*

... where Sylow theorems are recalled and their power in analyzing the structure of finite groups is demonstrated.

It can be shown that there are only five non-abelian simple groups of order less than 1000. These groups have orders 60, 168, 360, 504, and 660. If we believe this result, then we know that, for example, a group of order 90 cannot be simple. But how we would prove this and similar facts?

If we know the size of a finite group, then, by Lagrange's theorem, the sizes of its subgroups are restricted to be the divisors of the size of the group. But for which of the divisors of the order of the group are we guaranteed to have a subgroup? We proved in Chapter 7 the existence of Sylow subgroups. But can we say more?

This chapter is the tip of an iceberg. We will recall the Sylow theorems and prove additional theorems on existence of subgroups and normal subgroups. These will be used to illustrate how we can analyze the structure of finite groups using scant information. The methods and techniques introduced here can be (and have been) refined much further. Our purpose is not to be exhaustive, but rather to convince the reader that much can be done with the theory that we have developed.

This is a good time to go back and (re)read Section 7.3. So far in the text we have not used the Sylow theorems extensively. After two brief sections—including one on p-groups—we will remind the reader of the statements of the Sylow theorems and proceed to illustrate their power and use.

12.1. p-groups

Recall that a group is a p-group if its order is a power of the prime p. For example, $\mathbb{Z}/2\mathbb{Z} \times \mathbb{Z}/2\mathbb{Z}$ and D_8 are 2-groups, while any group of order 27 is a 3-group.

We have shown that a group of order p must be cyclic (Corollary 5.15), a group of order p^2 must be abelian (Corollary 6.20), and any p-group has a non-trivial center (Corollary 6.19). These results—which if announced right after the definition of an abstract group would seem quite extraordinary—already show that p-groups are quite a special collection of groups. We now show that the converse of Lagrange's theorem holds for p-groups.

Theorem 12.1. *Let p be a prime, and let G be a p-group. Assume $|G| = p^a$ where a is a non-negative integer. Let b be an integer with $0 \le b \le a$. Then G has a subgroup of order p^b.*

Proof. We prove the result by induction on a. If $a = 1$, then G is a group of order p and its trivial subgroups—of order 1 and order p—already give all the needed subgroups.

Assume that the result has been proved for all groups G for which $a < m$. We want to prove the result for $a = m$. Thus, let G be a group of order p^m, and assume that b is an integer with $0 \le b \le m$. We need to show that G has a subgroup of order p^b. Since the identity subgroup has size 1, we can assume that $b \ge 1$.

By Corollary 6.19 and since G is a p-group, we know that $\mathbf{Z}(G)$ is non-trivial. By Lagrange's Theorem 5.13 the order of $\mathbf{Z}(G)$ is a power of p, and, by Cauchy's Theorem 7.11, $\mathbf{Z}(G)$ has an element z of order p.

Let $Z = \langle z \rangle$. We have that $|Z| = p$, and $Z \lhd G$, since Z is a subgroup of the center of G.

Now let $\pi \colon G \to G/Z$ be the canonical homomorphism. The group G/Z has order p^{m-1}, and so, by the inductive hypothesis, it has a subgroup of order p^{b-1} (note that $b \ge 1$ implies that $b - 1 \ge 0$, and so we have $0 \le b - 1 \le m - 1$). Let this subgroup be called $\overline{H}$.

Now we apply the Homomorphism Theorem 11.38. Consider $H = \pi^{-1}(\overline{H})$ (see Figure 12.1). By the homomorphism theorem $|H : Z| = |\overline{H}| = p^{b-1}$. But then $|H| = |H : Z|\,|Z| = p^b$, and we conclude that H is a subgroup of G of order p^b as desired. $\qquad\square$

Remark 12.2. The proof of Theorem 12.1 is typical of many proofs that take advantage of an existing normal subgroup. By modding out by the normal subgroup, one is in a smaller, more manageable, setting and may be able to use induction. The Homomorphism Theorem 11.38 provides the passage back and forth between the original group and the smaller factor group.

The group A_4 is a group of order 12 with no subgroup of order 6 (Problem **5.1.19**). This shows that the converse of Lagrange's theorem does not hold. However, combining Theorem 12.1 with Sylow's E theorem, Theorem 7.10, we now obtain a partial converse to Lagrange's theorem.

Corollary 12.3. *Let G be a finite group, let p be a prime, and let b be a non-negative integer. If p^b divides $|G|$, then G has a subgroup of order p^b.*

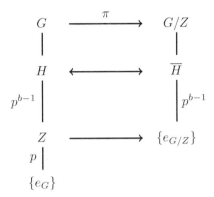

Figure 12.1. The subgroup $\overline{H}$ of G/Z corresponds to H a subgroup of G that contains Z.

Proof. By the Sylow E theorem, Theorem 7.10, G has a Sylow p-subgroup S, and $|S| = p^e$ with $e \geq b$. Now S is a p-group, and so it has a subgroup of order p^b. The latter will also be a subgroup of G.   □

Problems

12.1.1. Let G be a group of order 243. Show that there exists normal subgroups $Z_1, \ldots, Z_4$ of G such that
(a) $\{e\} = Z_0 < Z_1 < Z_2 < Z_3 < Z_4 < Z_5 = G$,
(b) for $1 \leq i \leq 4$, $|Z_i| = 3^i$, and
(c) for $1 \leq i \leq 4$, $Z_i/Z_{i-1} \leq \mathbf{Z}(G/Z_{i-1})$.

12.1.2. Generalize the previous problem to p-groups. As a corollary, conclude that p-groups are solvable. (See Definition 10.38.)

12.1.3. Let G be a group of order 11×13^5. Assume that G has a normal subgroup of order 11. For which integers m are we guaranteed a subgroup H of G with $|H| = m$?

12.1.4. Let G be a group of order 11×13^5. Assume that G has a normal subgroup of order 11. Prove that G is solvable. (See Definition 10.38.)

12.1.5. Let G be a group of order $11^2 \times 13^3 \times 17^4$. Assume that G has normal subgroups of order 11^2, 13^3, and 17^4.
(a) Prove that G has a normal subgroup of order $11^2 \times 13^3 \times 17$.
(b) Prove that G is solvable.

12.1.6. A group G is called *nilpotent* if every Sylow subgroup of G is normal in G.[1] Prove that every nilpotent group is solvable.

12.1.7. **Normal subgroups of p-groups.** Let P be a p-group. We showed in Corollary 6.19 that P has at least one non-identity element in its center.

---

[1] There are a number of equivalent definitions of nilpotent groups. See Section 14.2.

Let $\{e\} \neq N \triangleleft P$. Strengthen Corollary 6.19 by showing that $N \cap \mathbf{Z}(P) \neq \{e\}$.

12.2. Acting on Cosets and Existence of Normal Subgroups

To analyze a group, it is quite helpful to find a non-trivial normal subgroup. If we have such a normal subgroup N, we may be able to glean information about G using what we know about the smaller subgroups N and G/N. However, often—for example, using the Sylow E theorem, Theorem 7.10—we can only guarantee the existence of a (not necessarily normal) subgroup. In such cases, the next theorem may possibly yield a non-trivial normal subgroup.

Theorem 12.4. *Let G be a group, and let $H \leq G$. Assume that $|G : H| = n$. Then there exists a subgroup N, with $N \triangleleft G$, such that*

(a) *N is a subgroup of H, and*

(b) *$|G : N|$ divides $n!$.*

Note that the two conditions on N guaranteed by Theorem 12.4 may help in making sure that N is not trivial. The fact that $N \leq H$ means that N cannot be G (unless $H = G$), and the restriction on $|G : N|$ makes it less likely (and sometimes impossible) for N to be $\{e\}$.

Proof. We use a common strategy. We want to construct a normal subgroup N. This will be done by constructing a homomorphism on G and defining N to be the kernel of the homomorphism. But how do we create a homomorphism using the scant information that we have about G? The answer is that we make G act appropriately!

Let G act on $\Omega = \{xH \mid x \in G\}$, the set of left cosets of H in G, by $g \cdot xH = gxH$. We know that $|\Omega| = n$. As we have seen in Theorem 11.28, an action of a group G on a set with n elements, automatically gives a homomorphism $\theta : G \longrightarrow S_n$. See Figure 12.2.

Let $N = \ker(\theta)$. From Figure 12.2 (or Theorem 11.18) it is clear that $|G : N| = |\theta(G)|$. Now, $\theta(G)$ is a subgroup of S_n, and hence its order divides $|S_n| = n!$. Thus, $|G : N|$ divides $n!$ as claimed.

All that remains to be shown is that $N \subseteq H$.

Let $n \in N$. We need to show that $n \in H$. Recall, from Theorem 11.28, that $N = \ker(\theta)$ is the kernel of the action and consists exactly of those elements of the group that fix every element of Ω (i.e., act trivially). Hence $n \in N$ means that $n \cdot xH = xH$ for every left coset of H. In particular, $n \cdot H = H$ which means $nH = H$ and hence $n \in H$ and the proof is complete. $\square$

Example 12.5. Let G be a group of order 24. We show that G must have a normal subgroup of order 4 or of order 8. In particular, S cannot be simple.

Let $S \in \mathrm{Syl}_2(G)$. Then $|S| = 8$, and hence $|G : S| = 3$.

By Theorem 12.4, there exists $N \leq S$, such that $N \triangleleft G$, and $|G : N|$ divides 6.

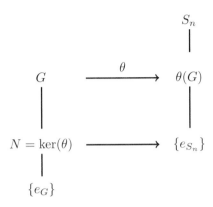

Figure 12.2. The action of G on the left cosets of H gives a homomorphism from G to S_n.

We know that $N \leq S$, and hence, by Lagrange's theorem, $|N|$ is 1, 2, 4, or 8. This means that the possibilities for $|G : N|$, respectively, are 24, 12, 6, or 3. However, $|G : N|$ must divide 6, and only the last two possibilities allow for that. Hence, $|N|$ is 4 or 8.

We know that if $H \leq G$ and $|G : H| = 2$, then $H \triangleleft G$. We now generalize this result. (A more elementary—and less conceptual—proof of the same result was given in a set of problems culminating in Problem 10.4.8.)

Corollary 12.6. *Let G be a finite group, and let $H \leq G$. Assume that $|G : H| = p$, and let p be the smallest prime divisor of $|G|$. Then $H \triangleleft G$.*

Proof. By Theorem 12.4 there exists $N \subset H$ with $N \triangleleft G$ such that $|G : N|$ divides $p!$.

But p is the smallest prime divisor of G, and hence $|G : N|$—which divides $p!$—must divide p. In particular, $|G : N| \leq p$.

On the other hand, $N \leq H$ and hence $|G : N| \geq |G : H| = p$. We conclude that $|G : N| = p$. This means $|N| = |H|$. But $N \leq H$, and so $N = H$. We conclude that $H = N \triangleleft G$ as claimed. $\qquad \square$

Corollary 12.7. *Let G be a p-group, and let $H \leq G$ with $|G : H| = p$. Then $H \triangleleft G$.*

Furthermore, if $|G| = p^k$, then G has subgroups H_0, H_1, ..., H_k such that $|H_i| = p^i$, and

$$\{e\} = H_0 \triangleleft H_1 \triangleleft \cdots \triangleleft H_{k-1} \triangleleft H_k = G.$$

In particular, every p-group is solvable. (See Definition 10.38.)

Proof. We already know, by Theorem 12.1, that G has a subgroup H of order p^{k-1}. By Corollary 12.6, H is a normal subgroup of G. Now repeat the same argument for H to get the chain of normal subgroups. $\qquad \square$

Corollary 12.8. *Let $|G| = pq$, where p and q are both prime. Assume that $p < q$. Then G has a normal subgroup H of order q.*

Proof. Let $Q \in \mathrm{Syl}_q(G)$. Then $|G : Q| = p$, and p is the smallest prime divisor of G. Hence, by Corollary 12.6, $Q \lhd G$. $\qquad\qquad\qquad\qquad\qquad\qquad\qquad\qquad\qquad\qquad\qquad\square$

Problems

12.2.1. Prove that a group of order 192 is not simple.

12.2.2. Prove that a group of order $216 = 2^3 \times 3^3$ is not simple.

12.2.3. A group of order 60 acts non-trivially (i.e., at least one of the group elements moves one of the set elements) on a set with four elements. Prove that the group cannot be simple.

12.2.4. A group G of order 72 has a subgroup H of order 24. Prove that either H is normal in G or H has a subgroup of order 12 which is normal in G.

12.2.5. G is a group of order $1155 = 3 \times 5 \times 7 \times 11$. N is a normal subgroup of G of order 55. K is a subgroup of G of order 35.
  (a) What can you say about $|\langle N, K\rangle|$ and $|N \cap K|$?
  (b) Does NK have to be a normal subgroup of G?

12.2.6. Let G be a group of order $3^6 \times 7$. Prove that G cannot be simple.

12.2.7. Let G be a group of order $2{,}025 = 3^4 \times 5^2$, and let P be a subgroup of order 3^4 (such a subgroup must exist by the Sylow E theorem). Assume that we know that $|\mathbf{N}_G(P)| > |P|$. Prove that G cannot be simple.

12.2.8. Let G be a group of order $168 = 2^3 \times 3 \times 7$. Assume that an element of order 7 is in the normalizer of a Sylow 2-subgroup of G. Prove that G is not simple.

12.2.9. Let G be an infinite group, and let H be a subgroup of G. Assume $|G : H| < \infty$. Prove that G must have a normal subgroup N with $N \le H$ and $|G : N| < \infty$.

12.2.10. Let G be a group, and let H be a subgroup of G. The group G acts on the left cosets of H by left multiplication (see proof of Theorem 12.4). Let N be the kernel of the action. Then show that
  (a) $N = \bigcap_{g \in G} gHg^{-1}$, and
  (b) if M is a normal subgroup of G that is contained in H, then $M \le N$.
  The subgroup N is the largest normal subgroup of G that is contained in H and it is called the *core* of H in G and is denoted by $\mathrm{core}_G(H)$.

12.3. Applying the Sylow Theorems

We proved Sylow's existence theorem—the so-called Sylow E theorem—in Section 7.2. Various other Sylow theorems were discussed in Section 7.3, and you were asked to complete their proofs in the problems. So far—even though, these important theorems are essential in analyzing finite groups—we have not seen many applications. We remedy this in this section, when we see a number of examples of how they can be used. We begin by recalling the statements of the Sylow theorems, and we urge the reader to (re)read Chapter 7 and especially Section 7.3 concurrently with this material.

Let p be a prime, let G be a finite group, and let $P \leq G$. Recall that P is a Sylow p-subgroup of G if P is a p-group and its order is the highest power of p that divides the order of G. The set of Sylow p-subgroups of G is denoted by $\mathrm{Syl}_p(G)$. (See Definition 7.8.)

Theorem 12.9 (The Sylow theorems). *Let G be a finite group, and let p be a prime number. Then*

(a) *Sylow E(xistence):*
$$\left|\mathrm{Syl}_p(G)\right| \geq 1.$$

Let P be a fixed Sylow p-subgroup of G. Then

(b) *Sylow D(evelopment): If $Q \leq G$ and Q a is p-group, then $Q \leq xPx^{-1}$ for some $x \in G$.*

And, as a consequence, we have

(c) *Sylow C(onjugacy):*
$$Q \in \mathrm{Syl}_p(G) \quad \text{if and only if} \quad Q = xPx^{-1} \text{ for some } x \in G,$$

and

(d) *The group G acts on $\Omega = \{H \mid H \leq G\}$, the set of subgroups of G, by conjugation—that is $g \cdot H = gHg^{-1}$. For this action, $\mathrm{Syl}_p(G)$ is the orbit of P, and $\mathbf{N}_G(P)$ is the stabilizer of P. Hence,*
$$\left|\mathrm{Syl}_p(G)\right| = \left|G : \mathbf{N}_G(P)\right|,$$

and $\left|\mathrm{Syl}_p(G)\right|$, the number of Sylow p-subgroups of G, is a divisor of $|G|$. Hence,

(e)
$$P \lhd G \quad \text{if and only if} \quad \left|\mathrm{Syl}_p(G)\right| = 1.$$

In addition,

(f)
$$\left|\mathrm{Syl}_p(G)\right| = 1 + kp, \quad \text{for some non-negative integer } k.$$

Proof. Part (a) was Theorem 7.10 and part (b) was Theorem 7.13. Parts (c), (d), and (e) were Corollaries 7.14, 7.15, and 10.9. Finally, part (f) was Theorem 7.16.   □

Remark 12.10. To analyze a group of order n, we usually start with its prime divisor and its Sylow subgroups. For each prime factor of $|G|$ we try to find the possibilities for $\left|\mathrm{Syl}_p(G)\right|$. Often the fact that this number divides the order of G and, at the same time, has remainder 1 when divided by p reduces the number of possibilities drastically. If we can conclude that there is only one Sylow p-subgroup, then we know that this subgroup is normal, and, in particular, the group is not simple.

Example 12.11. Let G be a group of order $405 = 3^4 \times 5$, and let $P \in \mathrm{Syl}_3(G)$. By the Sylow theorems $|\mathrm{Syl}_3(G)|$ is a divisor of $|G|$ and has remainder 1 when divided by 3. Hence, the number of Sylow 3-subgroups must be a divisor of 5, and so it is either 1 or 5. But 5 does not have remainder 1 when divided by 3, and so the only possibility is $|\mathrm{Syl}_3(G)| = 1$. This means, again by the Sylow theorems, that $P \lhd G$. We conclude that G is not simple and must have a normal subgroup of order 81.

With the same argument, the number of Sylow 5-subgroups is a divisor of 3^4 and has remainder 1 when divided by 5. The only possibilities are 1 or 81. Hence, G either has only one subgroup of order 5 or 81 such subgroups.

Before we do more applications of the Sylow theorems, we recall Theorem 10.23:

Theorem 12.12. *Let G be a group, and assume that M and N are normal subgroups of G with $M \cap N = \{e\}$. Then $mn = nm$ for every $m \in M$ and $n \in N$.*

Proof. You were asked to prove this in Problem **10.2.21**, but here is the proof: Let $m \in M$ and $n \in N$, and consider $x = mnm^{-1}n^{-1}$. Since M is normal in G, we have $nm^{-1}n^{-1} \in M$ and so $x = m(nm^{-1}n^{-1}) \in M$. But N is also normal in G, and so $mnm^{-1} \in N$. This means that $x = (mnm^{-1})n^{-1} \in N$. As a result, $x \in M \cap N = \{e\}$. From $mnm^{-1}n^{-1} = e$, we conclude that $mn = nm$. $\square$

Corollary 12.13. *Let M and N be normal subgroups of a group G. Assume that $M \cap N = \{e\}$, $MN = G$, and that both M and N are abelian. Then G is abelian.*

Proof. Since M is abelian, every element of M commutes with every element of M. By Theorem 12.12 every element of M also commutes with every element of N. Hence $M \subseteq \mathbf{Z}(G)$. Likewise, $N \subseteq \mathbf{Z}(G)$, and so $G = MN \subseteq \mathbf{Z}(G)$. Thus G is abelian. $\square$

Groups of order pq and pqr

Example 12.14. Let G be a group of order 15. Let $P \in \mathrm{Syl}_5(G)$ and $Q \in \mathrm{Syl}_3(G)$. By the Sylow theorems, $|\mathrm{Syl}_5(G)|$ divides 15 and is $1 \bmod 5$. The only possibility is $|\mathrm{Syl}_5(G)| = 1$. Similarly, $|\mathrm{Syl}_3(G)| = 1$. Hence, both P and Q are normal subgroups of G.

We can go further. The group P is a proper subgroup of PQ, and yet $|G : P| = 3$, and so there does not exist any subgroup H with $P < H < G$. Hence $PQ = G$. The sizes of P and Q are relatively prime, and so, by Lagrange's theorem, $P \cap Q = \{e\}$. Now, by Corollary 12.13, G is abelian. Let $P = \langle x \rangle$ and $Q = \langle y \rangle$. Then, since x and y commute, xy is an element of order $\mathrm{lcm}(o(x), o(y)) = 15$ (see Problem **2.3.16**). Hence, $G = \langle xy \rangle$ is a cyclic group of order 15.

We conclude that, up to isomorphism, there is only one group of order 15, namely $\mathbb{Z}/15\mathbb{Z}$.

We generalize Example 12.14:

Theorem 12.15. *Let G be a group of order pq, where p and q are primes, $p < q$ and p does not divide $q - 1$. Then $G \cong \mathbb{Z}/pq\mathbb{Z}$.*

Proof. Let $Q \in \mathrm{Syl}_q(G)$, then $|G : Q| = p$ and this is the smallest prime divisor of $|G|$. Hence, $Q \lhd G$ by Corollary 12.6. (This also follows from the Sylow theorems directly.)

The integer $\left|\mathrm{Syl}_p(G)\right|$ must divide pq and be 1 mod p, and so it is either 1 or q. If $q = 1 + kp$, then $kp = q - 1$ and $p \mid q - 1$, which is a contradiction. Hence $q \not\equiv 1 \mod p$ and so $\left|\mathrm{Syl}_p(G)\right| = 1$. A unique Sylow p-subgroup must be normal and so, if $P \in \mathrm{Syl}_p(G)$, then $P \lhd G$.

Since $P \lhd G$, PQ is a subgroup of G and its size is bigger than $|P| = p$. We know that $|G : P| = q$ a prime, and hence the only subgroup that contains P is all of G. We conclude that $G = PQ$. We also know, by Lagrange's theorem, that $P \cap Q = \{e\}$. By Corollary 12.13, G is abelian.

Let $P = \langle x \rangle$ and $Q = \langle y \rangle$. Then the order of xy is $\mathrm{lcm}(o(x), o(y)) = pq$ since $xy = yx$. Hence, $G = \langle xy \rangle$ is a cyclic group of order pq. We conclude that, if p does not divide $q - 1$, up to isomorphism, there is only one group of order pq, namely $\mathbb{Z}/pq\mathbb{Z}$. $\qquad\square$

Remark 12.16. If G is a group of order pq where p and q are primes, $p < q$, and $p \mid q - 1$, then it can be shown that G is isomorphic to one of two groups. The cyclic group of order pq or a non-abelian group of order pq that has a normal Sylow q-subgroup and q Sylow p-subgroups.

Example 12.17. Let G be a group of order $30 = 2 \times 3 \times 5$. Given the constraints of the Sylow theorems, we get that $|\mathrm{Syl}_5(G)|$ is 1 or 6, and $|\mathrm{Syl}_3(G)|$ is 1 or 10.

Assume that $|\mathrm{Syl}_5(G)| = 6$ and $|\mathrm{Syl}_3(G)| = 10$. Every Sylow 5-subgroup has four elements of order 5 and every two such subgroups intersect only in the identity element. Hence, there are $6 \times 4 = 24$ elements of order 5 in G. Likewise, there are $2 \times 10 = 20$ elements of order 3 in G. Thus we must have at least $24 + 20 + 1$ elements in G. But this is impossible in a group of order 30. We conclude that G has either a normal Sylow 5-subgroup or a normal Sylow 3-subgroup. Hence G cannot be simple.

We again generalize the argument in the previous example:

Theorem 12.18. *Let G be a group of order pqr, where p, q, and r are primes and $p < q < r$. Then G is not simple.*

Proof. Suppose G is simple. This means that G cannot have a unique Sylow subgroup for any of the primes p, q, or r. Given the constraints of the Sylow theorems, we get:

$|\mathrm{Syl}_r(G)| = pq$. This is because both p and q are bigger than 1 and less than $r + 1$, and hence cannot be equal to 1 mod r.

$\left|\mathrm{Syl}_q(G)\right| = r$, or pr. This number cannot be p since $p < 1 + q$.

Finally $\left|\mathrm{Syl}_p(G)\right| = q$, r or qr.

Now we count the number of elements of order p, q, and r. The number of elements of order p is $\left|\mathrm{Syl}_p(G)\right|(p - 1) \geq q(p - 1)$, the number of elements of order q is $\left|\mathrm{Syl}_q(G)\right|(q - 1) \geq r(q - 1)$, and the number of elements of order r is $|\mathrm{Syl}_r(G)|(r - 1) = pq(r - 1)$. Thus, the total number of elements of G is at least

$q(p-1)+r(q-1)+pq(r-1)+1 = pqr+(r-1)(q-1) > pqr$. This is a contradiction, and G cannot be simple. $\qquad\square$

Example 12.19. Assume that G is a group of order $72 = 2^3 \times 3^2$. We prove that G cannot be simple. If G has a unique Sylow 3-subgroup, then this subgroup will be normal in G, and G will not be simple. Assume that G has more than one Sylow 3-subgroup, then, given the restrictions of the Sylow theorems, we have $|\mathrm{Syl}_3(G)| = 4$. Let $P \in \mathrm{Syl}_3(G)$, then $4 = |\mathrm{Syl}_3(G)| = |G : \mathbf{N}_G(P)|$. Hence $\mathbf{N}_G(P)$ is a subgroup of G of order 18. Now apply Theorem 12.4 to get a normal subgroup N such that $N \leq \mathbf{N}_G(P)$ and $|G : N|$ divides 24. The former means that $N \neq G$, and the latter means that $N \neq \{e\}$. Hence N is a non-trivial normal subgroup of G and G is not simple.

Example 12.20. Let G be a group of order $90 = 2 \times 3^2 \times 5$. We prove that G cannot be simple.

Assume G is simple, and apply the Sylow theorems to get that $|\mathrm{Syl}_5(G)| = 6$ and $|\mathrm{Syl}_3(G)| = 10$. (Convince yourself that counting the number of elements will not produce a contradiction.)

Let $P \in \mathrm{Syl}_5(G)$. We have $|G : \mathbf{N}_G(P)| = 6$, and hence $\mathbf{N}_G(P)$ is a subgroup of order 15. (Convince yourself that Theorem 12.4 does not produce a non-trivial normal subgroup.) We have shown that the only group of order 15 is the cyclic group of order 15 (see Example 12.14). Hence $\mathbf{N}_G(P)$ is a cyclic group of order 15.

Now let Ω be the set of left cosets of $\mathbf{N}_G(P)$ in G. G acts on Ω by multiplication (i.e., $g \cdot x\mathbf{N}_G(P) = gx\mathbf{N}_G(P)$). Since $|\Omega| = 6$, the action gives a homomorphism $\theta : G \to S_6$. Since G is simple, the kernel of this action is $\{e\}$. See Figure 12.3.

Now $G \cong G/\ker\theta \cong \theta(G) \leq S_6$. Hence, S_6 has a subgroup isomorphic to G. Now, $\mathbf{N}_G(P)$ is a cyclic group of order 15 in G, and hence S_6 must have a cyclic subgroup of order 15. But the largest order of an element in S_6 is 6. The contradiction shows that G is not simple.

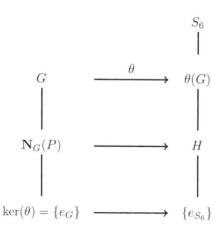

Figure 12.3. The action of G on the left cosets of $\mathbf{N}_G(H)$ gives a homomorphism from G to S_6.

Remark 12.21. Two big theorems that eliminate many group sizes as candidates for the sizes of simple groups are the odd order theorem and the $p^a q^b$ theorem. The $p^a q^b$ theorem, originally proved by Burnside, states that no group of order $p^a q^b$ where p and q are primes can be simple. This theorem can be proved using representation theory, and the proof is beyond the scope of this volume. The odd order theorem proved by Walter Feit and John Thompson [**FT63**] in 1962 states that no group of odd order can be simple (in fact, it says that every group of odd order is solvable).

If you recall how simple the definition of a group was, then the consequences of that definition are quite remarkable. One of our purposes has been to show the reader that the theory can take us quite far and that we can extract quite detailed information about groups, even when we start with as little information as only the size of the group.

Problems

12.3.1. Let G be a group of order $585 = 3^2 \times 5 \times 13$.
  (a) Prove that G is not simple.
  The group G continues to have 585 elements. Let $\mathcal{S} = \mathrm{Syl}_3(G)$ be the set of Sylow 3-subgroups of G. Let $P \in \mathcal{S}$. Assume that P is not normal in G.
  (b) What can you say about $|\mathcal{S}|$?
  (c) What can you say about $|\mathbf{N}_G(P)|$?
  (d) The group G acts on the set $\mathcal{S}$ by conjugation (i.e., for $x \in G$ and $Q \in \mathcal{S}$ we define $x \cdot Q = xQx^{-1}$). What can you say about the size of the orbit of P?

12.3.2. Let p be an odd prime, and let G be a group order $4p^2$. Assume that a Sylow p-subgroup of G is not normal in G. Prove that $|G| = 36$.

12.3.3. Let p be a prime, and let G be a group of order $4p^2$. Prove that G cannot be simple.

12.3.4. Prove that a group of order 45 must be abelian. Does it have to be cyclic?

12.3.5. Prove that a group of order 56 cannot be simple.

12.3.6. Let G be a group of order 105.
  (a) Prove that G cannot be simple.
  (b) Prove that G must have a subgroup of order 35.

12.3.7. Our friend G is a group of order 63.
  (a) The group G is guaranteed to have subgroups of which sizes? Why?
  (b) Assume that H is a subgroup of G of order 21. Prove that $H \lhd G$.

12.3.8. The group G is of order $66 = 2 \times 3 \times 11$.
  (a) Prove that G has a cyclic normal subgroup of order 33.
  (b) What is the minimum number of elements of order 33 in G?
  (c) What are the possibilities for $|\mathrm{Syl}_3(G)|$?
  (d) G is guaranteed to have subgroups of which orders?

12.3.9. Let G be a group of order 135.
   (a) Show that every Sylow subgroup of G is normal in G.
   (b) For which integers m is G guaranteed to have a subgroup of order m? Prove your assertions.

12.3.10. Let G be a group of order 99. Show that $G = PQ$ where P is a Sylow 3-subgroup and Q is a Sylow 11-subgroup.

12.3.11. Does there exist a non-abelian group of order 99?

12.3.12. Let G be a group of order 30. Does G have to have a subgroup of order 15? Does G have to have an element of order 15?

12.3.13. Let G be a group of order 30. Show that G has a normal Sylow 3-subgroup *and* a normal Sylow 5-subgroup.

12.3.14. The groups $\mathbb{Z}/30\mathbb{Z}$, D_{30}, $D_{10} \times \mathbb{Z}/3\mathbb{Z}$, and $D_6 \times \mathbb{Z}/5\mathbb{Z}$ are all groups of order 30. For each of them find the number of Sylow 2-subgroups.[2]

12.3.15. You are introduced to a mysterious group G. All you know is that G has $8{,}225 = 5^2 \times 7 \times 47$ elements.
   (a) What are the possibilities for $|\mathrm{Syl}_5(G)|$, $|\mathrm{Syl}_7(G)|$, and $|\mathrm{Syl}_{47}(G)|$?
   (b) Let $P \in \mathrm{Syl}_{47}(G)$ and $Q \in \mathrm{Syl}_7(G)$. Is PQ necessarily a subgroup, a normal subgroup, an abelian subgroup, or a cyclic subgroup?
   (c) If k divides $|G|$, then is G guaranteed to have a subgroup of order k? Why?
   (d) If $R \in \mathrm{Syl}_5(G)$, then is $PQR = G$? Why?
   (e) What is the smallest value possible for $|\mathbf{Z}(G)|$? Why?
   (f) What is the size of the largest cyclic subgroup that G is guaranteed to have? Why?

12.3.16. Let G be a group of order $455 = 5 \times 7 \times 13$.
   (a) What are the possibilities for $|\mathrm{Syl}_5(G)|$, $|\mathrm{Syl}_7(G)|$, $|\mathrm{Syl}_{13}(G)|$?
   (b) Can G be simple?
   (c) How many elements of order 13 does G have?
   (d) Let $P \in \mathrm{Syl}_{13}(G)$ and $Q \in \mathrm{Syl}_5(G)$. Is PQ necessarily a group? Why? Assume that PQ is a group, then what are the possibilities for $|\mathbf{Z}(PQ)|$?
   (e) Let $x \in G$ with $o(x) = 13$. What are the possibilities for $|\mathbf{C}_G(x)|$?
   (f) What are the possibilities for $|\mathbf{Z}(G)|$?
   (g) Are there any other comments that you wish to make regarding the group G?

12.3.17. Let G be a group of order $5{,}145 = 3 \times 5 \times 7^3$, and assume that H is a subgroup of order $1{,}029$. Show that $H \lhd G$.

12.3.18. Prove that a group of order 132 cannot be simple.

12.3.19. Let p and q be prime numbers, and assume that the group G has $p^2 q$ elements. Show that G has a normal Sylow p-subgroup or a normal Sylow q-subgroup.

12.3.20. Let G be a finite group, let p be a prime, let $N \leq G$, let $P \in \mathrm{Syl}_p(N)$, and let $H = \mathbf{N}_G(P)$.

---

[2]In fact, these are all of the groups of order 30 up to isomorphism.

(a) Let $g \in G$. Is gPg^{-1} necessarily a Sylow p-subgroup of N? Either prove that it is or give a counterexample.

(b) Let $g \in G$ and $x \in N$, and assume $gPg^{-1} = xPx^{-1}$. Prove that $g \in NH$.

(c) Assume that N is a normal subgroup of G and that $g \in G$. Is $gPg^{-1} \in \mathrm{Syl}_p(N)$?

12.3.21. Assume that G is a finite group, let p be a prime, let $N \triangleleft G$, let $P \in \mathrm{Syl}_p(N)$, and let $H = \mathbf{N}_G(P)$.

(a) Prove that $G = NH$.

(b) Draw a partial lattice diagram that includes $\{e\}$, N, H, P, $N \cap H$, and G. Where is $\mathbf{N}_N(P)$?

(c) Show that $|H|$ is divisible by both $|P|$ and $|G:N|$.

12.3.22. Let G be a group of odd order. Assume N is a normal subgroup of G of order 17. Prove that $N \leq \mathbf{Z}(G)$.

You may find the following steps helpful.

STEP 1: Let Q be a Sylow 17-subgroup of G. First show $N \leq Q$, then use Problem 12.1.7 and show $N \leq \mathbf{Z}(Q)$.

STEP 2: Let P be a Sylow p-subgroup of G where $p \neq 17$ is a prime divisor of $|G|$. Consider the subgroup NP, show that $P \triangleleft NP$, and use Theorem 10.23 to show that $P \leq \mathbf{C}_G(N)$.

STEP 3: Show that $|G|$ divides $|\mathbf{C}_G(N)|$, and complete the proof.

12.3.23. Let G be a group of order 180. Prove that G is not simple.

If you assume that G *is* simple, then you may find the following steps helpful.

STEP 1: Show that the number of Sylow 3-subgroups of G cannot be 4. Conclude that the number of Sylow 3-subgroups of G must be 10.

STEP 2: Show that if the number of Sylow 5-subgroups of G is 36, then at least two of the Sylow 3-subgroups must intersect non-trivially.

STEP 3: Let P and Q be two Sylow 3-subgroups of G, and assume that they intersect non-trivially. Define $R = P \cap Q$, and $L = \langle P, Q \rangle$. Show that $R \triangleleft L$, and so $L < G$. Further (by drawing a partial lattice diagram and using Theorem 9.27), show that $|L| \geq 36$.

STEP 4: Use the subgroup L in the previous part to show that the number of Sylow 5-subgroups of G cannot be 36. Conclude that the number of Sylow 5-subgroups of G must be 6.

STEP 5: Do Problem 12.3.12, and apply your results to the normalizer of a Sylow 5-subgroup.

STEP 6: Show that G must be isomorphic to a subgroup of S_6.

STEP 7: Arrive at a contradiction.

12.3.24. While waiting for your next class at the student lounge in the mathematics building, you overhear a frustrated student:

The group $G = \mathrm{GL}(2,3)$ has $48 = 2^4 \times 3$ elements. Right? So, a Sylow 2-subgroup has order 16. Look at the element $\alpha = \begin{bmatrix} 0 & 1 \\ 1 & 1 \end{bmatrix}$. I believe that α is an element of order 8, and so $\langle \alpha \rangle$ is a subgroup of order 8, and hence by the Sylow D Theorem,

$\langle\alpha\rangle$ is a subgroup of some Sylow 2-subgroup P of G. Moreover, $|P : \langle\alpha\rangle| = 2$, which means that $\langle\alpha\rangle \triangleleft P$. Are you with me? But—and here comes the twist—the element $x = \begin{bmatrix} 2 & 0 \\ 0 & 1 \end{bmatrix}$ is an element of order 2, and, as far as I can tell, $x\alpha x^{-1}$ is *not* an element of $\langle\alpha\rangle$, and so x does *not* normalize $\langle\alpha\rangle$. I am confused.

Check every statement of the student, and clear up the confusion. When you are done, determine the number of Sylow 2-subgroups of G.

12.3.25. Let G be a finite group, and let p be a fixed prime divisor of $|G|$. For each element $x \in G$, we define

$$N_p(x) = |\{P \in \mathrm{Syl}_p(G) \mid xPx^{-1} = P\}|.$$

In words, $N_p(x)$ is the number of Sylow p-subgroups P for which $xPx^{-1} = P$.

(a) Let $G = S_3$ be the symmetric group of degree 3, and let $p = 2$. First list all the Sylow 2-subgroups of G and then fill out the table:

x	e	$(1\ 2)$	$(1\ 3)$	$(2\ 3)$	$(1\ 2\ 3)$	$(1\ 3\ 2)$
$N_p(x)$						

(b) For a general finite group G and a fixed prime p, what can you say about the average of the numbers $N_p(x)$ as x ranges over G? In other words, make and prove a conjecture about

$$\frac{1}{|G|} \sum_{x \in G} N_p(x).$$

12.4. A_5 Is the Only Simple Group of Order 60

As another example of how to use our results to analyze the structure of groups, we prove that A_5, the alternating group of degree 5, is the only simple group of order 60. We will also find the isomorphism type of the subgroups of A_5.

Recall that we have already proved (Proposition 10.37) that A_5 is simple. We begin with two lemmas.

Lemma 12.22. *The alternating group of degree 5, A_5, is the only subgroup of order 60 in S_5.*

Proof. The set of even permutations in S_5 form the subgroup A_5. Now, assume that H was another subgroup of order 60 in S_5. Draw a partial lattice diagram of S_5 consisting of A_5, H, and their intersection. See Figure 12.4. Since A_5 has index 2 in S_5 and is normal in S_5, $S_5 = A_5 H$, and we have a parallelogram. Hence $2 = |G : H| = |A_5 : A_5 \cap H|$. This would mean that A_5 has a subgroup of index 2 which by necessity would be normal in A_5. But A_5 is simple, and hence we have a contradiction which proves that the subgroup H does not exist. $\square$

Lemma 12.23. *Let G be a simple group of order 60. Assume that G has a subgroup of order 12. Then $G \cong A_5$.*

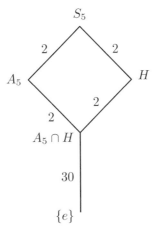

Figure 12.4. If S_5 had another subgroup of order 60, then A_5 would have a subgroup of index 2.

Proof. Let $H \leq G$ with $|G : H| = 5$. This means that H has five left cosets in G. Let $\Omega = \{xH \mid x \in G\}$ be the set of left cosets of H in G. We know that $|\Omega| = 5$. Now, G acts on Ω by $g \cdot xH = gxH$. As soon as we have an action of a group on a set with five elements, we know, by Theorem 11.28, that we have a homomorphism θ from G into S_5. Because G is simple, we know that the kernel of the action has to be the identity subgroup.

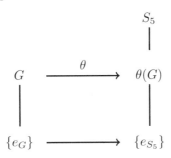

Thus

$$G \cong G / \ker \theta \cong \theta(G) \leq S_5.$$

Hence G is isomorphic to a subgroup of S_5 of order 60. By Lemma 12.22, A_5 is the only subgroup of S_5 of order 60, and hence $G \cong A_5$. $\qquad \square$

Theorem 12.24. *The group A_5 is the unique simple group of order 60. In other words, let G be a group of order 60. Then G is simple if and only if $G \cong A_5$.*

Furthermore, each non-trivial subgroup of A_5 is isomorphic to one of the following:

$$\mathbb{Z}/2\mathbb{Z}, \ \mathbb{Z}/3\mathbb{Z}, \ \mathbb{Z}/2\mathbb{Z} \times \mathbb{Z}/2\mathbb{Z}, \ \mathbb{Z}/5\mathbb{Z}, \ S_3, \ D_{10}, \ A_4.$$

In particular, A_5 does not have subgroups of order 15, 20, and 30. Moreover, the number of Sylow 2, 3, and 5-subgroups of A_5 is 5, 10, and 6, respectively, and, if, in A_5, x is any element of order 5 and y is any element of order 3, then $A_5 = \langle x, y \rangle$.

Proof. We have already proved (Proposition 10.37) that A_5 is a simple group of order 60. This also means that every fact that we prove for a simple group of order 60 will hold true for A_5. Of course, by the end of the proof, we will know that A_5 is the *only* simple group of order 60.

CLAIM 1: A simple group of order 60 does not have subgroups of order 15, 20, or 30.

PROOF OF CLAIM 1: Assume that G is a group of order 60 that is known to have a subgroup H of order 15. Then, by Theorem 12.4, there exists $N \leq H$ with $N \lhd G$, and $|G : N|$ a factor of $4! = 24$.

Now, $N \leq H$ means that $|N|$ can be one of 1, 3, 5, or 15, which, in turn, implies that $|G : N|$ will be 60, 20, 12, or 4, respectively. However, only the latter two divide 24. Hence the only possibilities for N are 5 and 15.

In particular, we conclude that G must have a normal subgroup of size 5 or 15 and cannot be simple. Hence, a simple group of order 60 does not have a subgroup of order 15.

An identical argument shows that a simple group of order 60 also does not have subgroups of order 20 or 30.

CLAIM 2: A simple group of order 60 has six Sylow 5-subgroups, 24 elements of order 5, ten Sylow 3-subgroups, and 20 elements of order 3.

PROOF OF CLAIM 2: Let G be a simple group of order 60, and let $n_p(G)$ denote $|\mathrm{Syl}_p(G)|$, the number of Sylow p-subgroups of G. If $P \in \mathrm{Syl}_p(G)$, we know that $n_p(G) = |G : \mathbf{N}_G(P)|$ divides $|G|$ and also has remainder one when divided by p. In addition, if p is a divisor of the order of the group, then $n_p(G) = 1$ means that the Sylow p-subgroup is unique and hence normal in G. This cannot happen since G is simple.

With these restrictions, the only possibilities are $n_5(G) = 6$ and $n_3(G) = 4$ or 10.

If $n_3(G) = 4$ and $P \in \mathrm{Syl}_3(G)$, then $|G : \mathbf{N}_G(P)| = 4$, and $\mathbf{N}_G(P)$ would be a subgroup of order 15, contradicting Claim 1. Hence, G has six Sylow 5-subgroups and ten Sylow 3-subgroups.

The six Sylow 5-subgroups are each a cyclic group of order 5, their pairwise intersection is just the identity element, and each has four elements of order 5 in addition to the identity element. In addition, every element of order 5 in G will generate a Sylow 5-subgroup. Hence, G has exactly $6 \times 4 = 24$ elements of order 5. Likewise, and with the same argument, all the elements of order 3 are contained in the ten Sylow 3-subgroups, and, hence, G has exactly $10 \times 2 = 20$ elements of order 3.

CLAIM 3: The number of Sylow 2-subgroups of a simple group of order 60 is either 5 or 15. In the latter case, at least two Sylow 2-subgroups intersect non-trivially.

PROOF OF CLAIM 3: Let G be a simple group of order 60. Because of the restrictions imposed by the Sylow theorems, $n_2(G)$, the number of Sylow 2-subgroups of G, is either 1, 3, 5, or 15. This number cannot be 1, since G is simple. On the other hand, if $n_2(G) = 3$ and $Q \in \mathrm{Syl}_2(G)$, then $|G : \mathbf{N}_G(Q)| = 3$, and $\mathbf{N}_G(Q)$ would be a subgroup of order 20, contradicting Claim 1. Hence, $n_2(G) = 5$ or 15.

If G has 15 Sylow 2-subgroups each of order 4, then could all their pairwise intersections be the trivial group? If that were so, then we would have $3 \times 15 = 45$ non-identity elements in these Sylow 2-subgroups. This is impossible, since, by Claim 2, we have already accounted for another $1 + 24 + 20 = 45$ elements of this group of order 60. Hence, at least two of the Sylow 2-subgroups intersect non-trivially.

CLAIM 4: A simple group of order 60 must have a subgroup of order 12, and, hence, it is isomorphic to A_5.

PROOF OF CLAIM 4: Let G be a simple group of order 60. By Lemma 12.23, if we prove that G has a subgroup of order 12, then $G \cong A_5$, and the proof is complete. Now, let $P \in \mathrm{Syl}_2(G)$. Because of Claim 3, we do two cases:

CASE 1: Assume that the number of Sylow 2-subgroups of G is 5. This means that $|G : \mathbf{N}_G(P)| = 5$, and, so, $\mathbf{N}_G(P)$ is a subgroup of G of order 12 as desired.

CASE 2: Assume that the number of Sylow 2-subgroups of G is 15. Then, by Claim 3, we can find two distinct Sylow 2-subgroups R and S of order 4, such that $|R \cap S| = 2$. Let x be the non-identity element of order 2 in $R \cap S$. What is $|\mathbf{C}_G(x)|$? Every group of order 4 is abelian, and, hence, both R and S are subgroups of $\mathbf{C}_G(x)$. This means that $|\mathbf{C}_G(x)|$ is divisible by 4 and is a divisor of 60. Thus, $|\mathbf{C}_G(x)|$ is 12, 20, or 60. We have already proved that G does not have a subgroup of order 20. Likewise, $|\mathbf{C}_G(x)| = 60$ means that x is a non-trivial element of the center of G. But G is simple and non-abelian, and so its center is trivial. We conclude that $|\mathbf{C}_G(x)| = 12$, and we have a subgroup of order 12 as desired.

CLAIM 5: The group A_5 has subgroups of order 2, 3, 4, and 5, and these are isomorphic to $\mathbb{Z}/2\mathbb{Z}$, $\mathbb{Z}/3\mathbb{Z}$, $\mathbb{Z}/2\mathbb{Z} \times \mathbb{Z}/2\mathbb{Z}$, and $\mathbb{Z}/5\mathbb{Z}$, respectively.

PROOF OF CLAIM 5: By Cauchy's theorem, A_5 must have elements of order 2, 3, and 5 (in fact, we can write examples of these easily: (1 2)(3 4), (1 2 3), and (1 2 3 4 5)), and these generate cyclic subgroups of orders 2, 3, and 5. A Sylow 2-subgroup of A_5 is of order 4, and, since A_5 has no element of order 4 (elements of order 4 in S_4 are odd permutations), these subgroups must be isomorphic to the Klein 4-group $\mathbb{Z}/2\mathbb{Z} \times \mathbb{Z}/2\mathbb{Z}$.

CLAIM 6: The group A_5 has subgroups of order 10, and any such subgroup is isomorphic to D_{10}.

PROOF OF CLAIM 6: By Claim 2, the number of Sylow 5-subgroups of A_5 is 6. This means that if $P \in \mathrm{Syl}_5(G)$, then $|G : \mathbf{N}_G(P)| = 6$. Hence the normalizer of a Sylow 5-subgroup is a subgroup of order 10. In Proposition 5.36, we saw that a group of order 10 is either cyclic or isomorphic to D_{10}. The group A_5 does not have elements of order 10, and so all subgroups of order 10 in A_5 must be isomorphic to D_{10}. This completes the proof of Claim 6, but we can also be more concrete: If you consider the natural definition of D_{10} as the symmetries of a pentagon, we see that it is generated by a rotation and a reflection. The former is a 5-cycle, while the latter is the product of two 2-cycles. Both of these generators are even permutations and hence elements of A_5.

CLAIM 7: The group A_5 has subgroups of order 6, and any such subgroup is isomorphic to S_3.

PROOF OF CLAIM 7: There is no element of order 6 in A_5, and, hence, if there is a subgroup of order 6, it would be isomorphic to $S_3 \cong D_6$. By Claim 2, if $Q \in \mathrm{Syl}_3(G)$, then $|\mathbf{N}_G(Q) : Q| = |\mathrm{Syl}_3(G)| = 10$. Hence $\mathbf{N}_G(Q)$ is a subgroup of order 6 isomorphic to S_3.

To see concretely what these subgroups of order 6 look like, consider the natural definition of S_3 as the permutations of $\{1, 2, 3\}$. This group is certainly not a subgroup of A_5, since its elements of order 2 (e.g., (1 2)) are odd permutations. However, a small modification gives an isomorphic group inside A_5. Consider the subgroup $H = \langle (1\ 2\ 3), (1\ 2)(4\ 5) \rangle$. Since the generators are even permutations, $H \leq A_5$. If we map $(1\ 2\ 3) \mapsto (1\ 2\ 3)$ and $(1\ 2)(4\ 5) \mapsto (1\ 2)$, we get an isomorphism from H onto S_3.

CLAIM 8: The subgroups of order 12 in A_5 are isomorphic to A_4.

PROOF OF CLAIM 8: By Claim 4, A_5 has a subgroup of order 12, and we also know that $A_4 \leq A_5$. Hence, A_5 does have subgroups of order 12 isomorphic to A_4. In fact, A_4 is the even permutations on four objects and hence there are at least five copies of A_4 in A_5 (one is even permutations of $\{1, 2, 3, 4\}$, another is the even permutations of $\{1, 2, 3, 5\}$, and so on).

To prove that *every* subgroup of order 12 in A_5 is isomorphic to A_4, we use Problem 12.4.1 where you are asked to prove that any group of order 12 is either isomorphic to A_4 or has a normal Sylow 3-subgroup. Now in A_5 there are ten Sylow 3-subgroups, and hence the normalizer of any of them is of size 6. So a subgroup of order 3 of A_5 cannot be a normal subgroup of a subgroup of order 12. From this we conclude that every subgroup of order 12 in A_5 is isomorphic to A_4.

CLAIM 9: The number of Sylow 2-subgroups in A_5 is 5.

PROOF OF CLAIM 9: By Claim 3, $|\mathrm{Syl}_2(G)|$ is either five or 15. Let

$$K = \langle (1\ 2)(3\ 4), (1\ 3)(2\ 4) \rangle \leq A_4 \leq A_5.$$

The group K is a Sylow 2-subgroup of A_5, since it has four elements. It is also a normal subgroup of A_4, since it is a union of conjugacy classes in A_4 (it contains all elements of A_4 of cycle type $2, 2$ together with the identity). Hence, the normalizer of K in A_5 contains A_4, and so the size of this normalizer is at least 12. Thus, $|\mathrm{Syl}_2(G)| = |A_5 : \mathbf{N}_{A_5}(K)| \leq 60/12 = 5$. Since the choices for this number were 5 or 15, we conclude that $|\mathrm{Syl}_2(G)| = 5$.

CLAIM 10: In A_5, any element of order 5 together with any element of order 3 generates the whole group.

PROOF OF CLAIM 10: Let x be any element of order 5, and let y be any element of order 3 in A_5. By Lagrange's theorem, the order of the subgroup $\langle x, y \rangle$ is at least 15. But by Claim 1, A_5 has no proper subgroups with 15 or more elements. Hence, the subgroup generated by x and y is the whole group. We have now completed our analysis of A_5 and simple groups of order 60. $\square$

Remark 12.25. A good problem to test your facility with the arguments in finite group theory is Problem 12.4.7. That problem asks you to prove that if G is a non-abelian simple group with $|G| < 168$, then $G \cong A_5$. After a graduate course covering finite groups, a similarly instructive, albeit difficult, problem is to show that if a G is a non-abelian simple group with $|G| \leq 1000$, then $|G|$ is 60, 168, 360, 504, or 660. Eliminating $|G| = 720$ is particularly tricky.

Problems

12.4.1. Let G be a group of order 12. Prove that either $G \cong A_4$ or G has a normal Sylow 3-subgroup.

12.4.2. Let n be an integer greater than 2, and let G be a finite simple group with more than two elements. Assume that G is a subgroup of S_n. Prove that G is a subgroup of A_n.

12.4.3. Prove that a group of order 112 cannot be simple.

12.4.4. Let G be a simple group of order 168.
   (a) How many Sylow 7-subgroups does G have?
   (b) Show that G must have a subgroup of order 21.
   (c) How many elements of order 7 does G have?

12.4.5. Let G be a simple group of order 168. Show that G is isomorphic to a subgroup of A_8.

12.4.6. Let G be a group of order 60. Assume that G does not have a normal Sylow 5-subgroup. Prove that G is simple. As a corollary give a new proof that A_5 is simple.

12.4.7. Assume that G is a non-abelian simple group and that $|G| < 168$. Prove that $G \cong A_5$.

Chapter 13

Direct
and Semidirect Products*

... where construction of groups using direct and semidirect products is investigated, the fundamental theorem of abelian groups is proved, and a list and classification of all groups of order ≤ 15 is presented.

How do we construct groups? Cyclic groups, dihedral groups, symmetric groups, and groups of invertible matrices over fields are some of the standard examples that we have seen. We can then construct new groups from the old ones. For example, new groups can be found by looking at subgroups of known groups—the alternating groups and the special linear groups are examples—or by finding direct products of familiar groups.

In this (optional) chapter, we first look at direct products and see how to recognize a given group as isomorphic to a direct product of other smaller groups. As a byproduct, in Section 13.2, we sketch a proof that all finite *abelian* groups are direct products of cyclic groups. This theorem—and some refinements of it—is called the *fundamental theorem of abelian groups*, and it allows us to give a complete list of the abelian groups of a given order up to isomorphism.

In Section 13.3, we turn our attention to a more general construction called the semidirect product. As an illustration and using this construction, we then proceed to classify, in Section 13.4, all groups of order no more than 15.

In group theory—as in much of discrete mathematics—when analyzing a given situation, often we can prove general statements as long as we exclude a few possible exceptions. Sometimes the exceptions are there because our proof techniques were not powerful enough to prove the most general statement possible. But, often, the exceptions are real. In such cases, we usually need to construct examples of groups to show that the exceptions actually do occur. In such circumstances, semidirect products—which include direct products as a special case—are indispensible.

13.1. Direct Products of Groups

We defined the direct product of two groups in Section 2.5. For completeness, we repeat the definition here. It should be clear to you that there is no reason to limit ourselves to just two groups. Hence, we will state the more general definition of the direct product of k groups.

Definition 13.1 (The direct product of a finite number of groups). Let H_1, H_2, ..., H_k be a collection of groups. Then the set $H_1 \times H_2 \times \cdots \times H_k$ is defined by

$$H_1 \times H_2 \times \cdots \times H_k = \{(h_1, h_2, \ldots, h_k) \mid h_1 \in H_1, h_2 \in H_2, \ldots, h_k \in H_k\}.$$

This set is made into a group by defining the following operation:

$$(h_1, h_2, \ldots, h_k)(h_1', h_2', \ldots, h_k') = (h_1 h_1', h_2 h_2', \ldots, h_k h_k').$$

This group is called the *(external) direct product* of H_1, H_2, ..., H_k.

Remark 13.2. If the groups H_1, ..., H_k are abelian, and we are using the additive notation, then an alternative expression for the direct product of H_1, ..., H_k is the *direct sum* of H_1, ..., H_k which is denoted by $H_1 \oplus H_2 \oplus \cdots \oplus H_k$. Thus $\mathbb{Z}/2\mathbb{Z} \oplus \mathbb{Z}/2\mathbb{Z}$ is the same object as what we have been calling $(\mathbb{Z}/2\mathbb{Z} \times \mathbb{Z}/2\mathbb{Z}, +)$.

Assume a group G is already given. Our main task in this section is to see if we can recognize G as isomorphic to a direct product of a number of (smaller) groups. We begin with the simpler case of groups isomorphic to the direct products of two groups. First, we identify a set of features of direct products, and then show that any group with these features must be isomorphic to the direct product of two groups. We then repeat the same procedure for the direct product of k groups.

Theorem 13.3. *Let H and K be groups, and define $G = H \times K$. Then G has two subgroups H_1 and K_1 such that*

(a) $H_1 \cong H$, *and*

(b) $K_1 \cong K$, *and*

(c) H_1 *and* K_1 *are normal subgroups of G, and*

(d) $H_1 \cap K_1 = \{e\}$, *and*

(e) $H_1 K_1 = G$.

Proof Sketch. We let $H_1 = H \times \{e_K\} = \{(h, e_K) \mid h \in H\}$ and $K_1 = \{e_H\} \times K = \{(e_H, k) \mid k \in K\}$. Then it is straightforward to see that all the conditions are satisfied. Problem 13.1.1 asks the reader to fill in the details. $\qquad \square$

Theorem 13.3 says that inside any direct product of two groups there are two normal subgroups whose product is the whole group and whose intersection is trivial. We now see that the converse of this is true as well and provides a method for recognizing that a group G is isomorphic to the direct product of two groups.

Theorem 13.4. *Let G be a group, and let H and K be subgroups of G. Assume that*

(a) H *and* K *are normal subgroups of G, and*

(b) $H \cap K = \{e\}$, *and*

(c) $G = HK$.

Then $G \cong H \times K$.

Proof. This was Problem 11.1.12 but we give the proof here. We construct the group $H \times K$, and define a map $f : H \times K \longrightarrow G$ by $f(h,k) = hk$.

Note that since both H and K are normal in G and $H \cap K = \{e\}$, by Theorem 12.12, elements of H commute with elements of K. The map f is a group homomorphism since

$$f((h,k)(h',k')) = f(hh',kk') = hh'kk' = hkh'k' = f(h,k)f(h',k').$$

Now, assume $f(h,k) = e$, then $hk = e$, which implies that $h = k^{-1}$. This would mean that $h \in H \cap K$, since $h = k^{-1} \in K$ and of course $h \in H$. But $H \cap K = \{e\}$ and so $h = k = e$. Thus the kernel of f is trivial, and so f is 1-1. The map f is onto, since $HK = G$. We conclude that f is an isomorphism, and we have $H \times K \cong G$. $\square$

Remark 13.5. In Problem 13.1.3, you are asked to show that, given the rest of the assumptions of Theorem 13.4, the condition that H and K be normal in G is equivalent to the condition that elements of H commute with elements of K.

Example 13.6. Let $D_{12} = \langle a,b \mid a^6 = b^2 = e, ba = a^{-1}b \rangle$ be the dihedral group of order 12. Let $H = \langle a^2,b \rangle$ and $K = \langle a^3 \rangle$ be subgroups of D_{12}.

Now $ba^3 = a^{-3}b = a^3b$. Hence, a^3 commutes with both a and b and is an element of order 2 in the center of D_{12}. Hence, $K \cong \mathbb{Z}/2\mathbb{Z}$ and $K \lhd D_{12}$.

We have $ba^2 = a^{-2}b$, and so $ba^2b = a^{-2}$. This means that b normalizes $\langle a^2 \rangle$. Hence, $H = \langle a^2,b \rangle = \langle a^2 \rangle \langle b \rangle$ is a non-abelian group of order 6. So $H \cong S_3$. Since $|D_{12} : H| = 2$, we have $H \lhd D_{12}$.

Clearly, $H \cap K = \{e\}$. Hence, HK is a subgroup of order 12, and we must have $HK = D_{12}$.

We conclude, by Theorem 13.4, that $D_{12} \cong H \times K$. In other words,

$$D_{12} \cong S_3 \times \mathbb{Z}/2\mathbb{Z}.$$

We now consider the more general case of the direct product of more than two groups. Again, we first identify a feature of the direct product, and then we show that we can use it to possibly identify a group as isomorphic to the direct product of some of its subgroups.

Theorem 13.7. *Let H_1, H_2, ..., H_k be groups, and let $G = H_1 \times H_2 \times \cdots \times H_k$.*

For $i = 1, \ldots, k$, define

$$\tilde{H}_i = \{e_{H_1}\} \times \{e_{H_2}\} \times \cdots \times \{e_{H_{i-1}}\} \times H_i \times \{e_{H_{i+1}}\} \times \cdots \times \{e_{H_k}\}.$$

Then, for $1 \leq i \leq k$,

(a) $\tilde{H}_i \cong H_i$, *and*

(b) $\tilde{H}_i \lhd G$, *and*

(c) $\tilde{H}_i \cap \tilde{H}_1 \cdots \tilde{H}_{i-1} \tilde{H}_{i+1} \cdots \tilde{H}_k = \{e_G\}$, *and*

(d) $\tilde{H}_1 \tilde{H}_2 \cdots \tilde{H}_k = G$.

Proof. This a generalization of Theorem 13.3, and the proof is straightforward. You are asked to write down the details in Problem 13.1.7. $\qquad\square$

Remark 13.8. Using the notation of Theorem 13.7, note that if $G = H_1 \times H_2 \times \cdots \times H_k$, then it is certainly true, and follows from part (c) of the theorem, that $\tilde{H}_i \cap \tilde{H}_j = \{e_G\}$ for $i \neq j$. However, as it turns out, this simpler condition is *not* enough for identifying a group as isomorphic to a direct product of its subgroups. (See Problem 13.1.8 for a simple example.)

Definition 13.9 (Internal direct product). Let G be a group, and let N_1, N_2, ..., N_k be subgroups of G. Assume that for $i = 1, \ldots, k$ we have

(a) $N_i \triangleleft G$, and

(b) $N_i \cap N_1 \cdots N_{i-1} N_{i+1} \cdots N_k = \{e_G\}$, and

(c) $G = N_1 N_2 \cdots N_k$.

Then we say that G is the *internal direct product* of N_1, ..., N_k.

The use of the term "direct product" is justified because of the following:

Theorem 13.10. *Assume that the group G is the internal direct product of subgroups $N_1, \ldots, N_k$, then*

$$G \cong N_1 \times N_2 \times \cdots \times N_k.$$

Proof Sketch. The proof is a straightforward generalization of the proof of Theorem 13.4. You are asked to write a complete proof in Problem 13.1.10. To show that the map $f : N_1 \times \cdots \times N_k \longrightarrow G$ defined by $f(n_1, \ldots, n_k) = n_1 n_2 \cdots n_k$ is a homomorphism, you again use the fact that, for $i \neq j$, by Theorem 12.12 applied to the group $N_i N_j$, the elements of N_i commute with elements of N_j. You use the same fact and condition (b) to show that the map is 1-1. The map is clearly onto, and hence an isomorphism. $\qquad\square$

Remark 13.11. Note the difference between external direct products and internal direct products. In Definition 13.1, we started with k groups and constructed a new group that was called their external direct product. In contrast, in Definition 13.9 and Theorem 13.10, we started with a group G and recognized that it is isomorphic to the direct product of k of its subgroups.

Problems

13.1.1. Complete the proof of Theorem 13.3.

13.1.2. Let G be a group. Assume H and K are subgroups of G with $G = HK$. Prove that the following are equivalent:
   (a) Every element $g \in G$ can be written *uniquely* as a product hk with $h \in H$ and $k \in K$.
   (b) $H \cap K = \{e\}$.

13.1.3. Let G be a group. Assume that H and K are subgroups of G with $HK = G$ and $H \cap K = \{e\}$. Prove that the following are equivalent:

(a) Every element of H commutes with every element of K.

(b) Both H and K are normal in G.

Conclude that if either of the above conditions hold, then $G \cong H \times K$.

13.1.4. Let $D_{20} = \langle a, b \mid a^{10} = b^2 = e, ba = a^{-1}b \rangle$ be the dihedral group of order 20. Likewise, $D_{10} = \langle a, b \mid a^5 = b^2 = e, ba = a^{-1}b \rangle$ denotes the dihedral group of order 10. Prove

$$D_{20} \cong D_{10} \times \mathbb{Z}/2\mathbb{Z}.$$

13.1.5. Let D_n denote the dihedral group of order n. Is $D_{24} \cong D_{12} \times \mathbb{Z}/2\mathbb{Z}$?

13.1.6. Let m and n be positive integers with $m < n$. Let A be a subset of size m of $[n] = \{1, \ldots, n\}$. Let G consists of elements $\sigma \in S_n$ with the property that $\sigma(A) = A$. Prove

$$G \cong S_m \times S_{n-m}.$$

13.1.7. Write down the details of the proof of Theorem 13.7.

13.1.8. Let $G = \mathbb{Z}/2\mathbb{Z} \times \mathbb{Z}/2\mathbb{Z}$ be the Klein 4-group. Let $H_1 = \langle (1,0) \rangle$, $H_2 = \langle (0,1) \rangle$, and $H_3 = \langle (1,1) \rangle$. Verify that

(a) $H_i \lhd G$ for $i = 1, 2, 3$, and

(b) $H_i \cap H_j = \{e_G\}$ for $i \neq j$, and

(c) $G = H_1 H_2 H_3$, and

(d) G is *not* isomorphic to $H_1 \times H_2 \times H_3$.

Reconcile this example with Theorem 13.10.

13.1.9. Let

$$G = \mathbb{Z}/90\mathbb{Z} \times \mathbb{Z}/10\mathbb{Z} \times \mathbb{Z}/5\mathbb{Z}.$$

What are all the Sylow subgroups of G? Is G the (internal) direct product of its Sylow subgroups?

13.1.10. Complete the proof sketched for Theorem 13.10.

13.1.11. Generalize Problems 13.1.2 and 13.1.3 to the case with more than two subgroups.

13.1.12. Assume that every Sylow subgroup of a finite group G is a normal subgroup of G. Prove that G is isomorphic to the direct product of its Sylow subgroups.

13.1.13. Let H, K, G, M, and N be groups. Assume that the group G is isomorphic to $H \times K$. Further assume that the group K itself is isomorphic to $M \times N$. Prove that G is isomorphic to $H \times M \times N$.

13.1.14. Let m and n be relatively prime integers, and let G be a abelian group of order mn. Let $H = \{x \in G \mid x^m = e\}$ and $K = \{x \in G \mid x^n = e\}$. Prove that G is isomorphic to $H \times K$.

13.2. Fundamental Theorem of Finite Abelian Groups

In this section, we prove the following theorem:

Theorem 13.12. *Let G be a finite abelian group. Then G is isomorphic to a direct product of cyclic groups.*

In other words, if G is a finite abelian group, then

$$G \cong \mathbb{Z}/n_1\mathbb{Z} \times \mathbb{Z}/n_2\mathbb{Z} \times \cdots \times \mathbb{Z}/n_k\mathbb{Z}$$

for some positive integers $n_1, n_2, \ldots, n_k$.

One can actually refine this theorem and be more specific about the list of positive integers $n_1, n_2, \ldots, n_k$. The theorem can also be generalized to include (finitely generated) infinite abelian groups (in such a case we need to allow for $\mathbb{Z}$ as a possible factor also). We will do neither refinement here. The theorem as stated (together with things we know about direct products of cyclic groups) will allow us to classify the finite abelian groups of any given order, and this will be powerful enough.

Before we begin, we remind the reader of the following:

Lemma 13.13. *Let G be a finite group. Let $x \in G$, and let $i > 0$. Then*

$$o(x^i) = \frac{o(x)}{\gcd(i, o(x))}.$$

Proof. This was Problem **2.3.18**, but we include its proof here for completeness. Let $y = x^i$, $r = o(x)$, $k = o(y)$, and $s = \gcd(i, r)$. Then $i = sa$ and $r = sb$, with a and b relatively prime. We want to prove that $k = b$.

We have $y^b = (x^i)^b = (x^{sa})^b = (x^{sb})^a = (x^r)^a = e$, and so $k = o(y) \mid b$.

On the other hand, $e = y^k = x^{ik}$. This means $r = o(x) \mid ik$ and so $sb \mid sak$. Hence, $b \mid ak$, and, since a and b are relatively prime, $b \mid k$. (This argument was really Problem 1.3.11.) We conclude that $k = b$.   □

We have chosen to turn one step of the proof of the main theorem into the following proposition:

Proposition 13.14. *Let G be a finite abelian group. Choose $a \in G$ such that $o(a)$ is as large as possible. Let $m = o(a)$, and let $A = \langle a \rangle$.*

Let $\pi : G \to G/A$ be the canonical homomorphism. Let $y \in G/A$ and assume that $o(y) = r$. Then there exists $x \in G$ such that $o(x) = r$ and $\pi(x) = y$.

Proof. Since $y \in G/A$, we can write $y = Ab$ for some $b \in G$. We know that the order of y is r, and so $A = e_{G/A} = y^r = Ab^r$. Hence, $b^r \in A = \langle a \rangle$. Thus $b^r = a^n$, for some $0 \le n < m$.

We write $n = qr + s$ for $0 \le s < r$, and define $x = a^{-q}b$. We want to show that x is the sought after element. Certainly, $x \in G$, and $\pi(x) = \pi(a^{-q})\pi(b) = \pi(b) = Ab = y$. We only need to prove that $o(x) = r$.

Now, $y^{o(x)} = \pi(x)^{o(x)} = \pi(x^{o(x)}) = e$, and hence $r \mid o(x)$. (This argument is really the solution to Problem *2.4.17*.)

We also have $x^r = a^{-qr}b^r = a^{n-qr} = a^s$. The proof would be complete, if we show $s = 0$. This is because if $s = 0$, then $x^r = a^s = e$, which means that $o(x) \mid r$.

Using proof by contradiction, assume $s > 0$, and let $d = \gcd(s, m)$. By Lemma 13.13

$$o(x^r) = o(a^s) = \frac{m}{d}.$$

But applying Lemma 13.13 directly to x^r and using the fact that $r \mid o(x)$, we have $o(x^r) = \frac{o(x)}{r}$. Hence, $o(x) = \frac{r}{d}m$. But $d \leq s < r$, and so $r/d > 1$ and $(r/d)m > m$. This contradicts the assumption that $o(a) = m$ was as large as possible. Hence, $s = 0$, and the proof is complete. $\square$

We are now ready to prove Theorem 13.12.

Proof of Theorem 13.12. We use strong induction on $|G|$. If $|G| = 1$, then the validity of the theorem is clear, and hence we assume that $|G| > 1$, and that the theorem is true for any finite abelian group whose order is less than $|G|$.

Now, let a_1 be an element of G of maximum order, and let $A_1 = \langle a_1 \rangle$. Let $\pi : G \to G/A_1$ be the canonical homomorphism.

Applying the inductive hypothesis to G/A_1, we get that G/A_1 is isomorphic to the direct product of finite cyclic groups. By Theorem 13.7, we have that G/A_1 has normal cyclic subgroups $L_2, L_3, \ldots, L_k$ such that $G/A_1 = L_2 L_3 \cdots L_k$, and such that the intersection of any of them with the product of the rest is trivial. Assume that $L_i = \langle y_i \rangle$ for $i = 2, \ldots, k$. By Proposition 13.14 for each $i = 2, \ldots, k$, we have elements $a_i \in G$ such that $\pi(a_i) = y_i$ and $o(a_i) = o(y_i)$. Let $A_i = \langle a_i \rangle$.

CLAIM 1: $G = A_1 A_2 \cdots A_k$.

PROOF OF CLAIM 1: Let $g \in G$. Now $\pi(g) \in G/A_1 = L_2 L_3 \cdots L_k$, and so $\pi(g) = y_2^{n_2} y_3^{n_3} \cdots y_k^{n_k}$ for some positive integers $n_1, \ldots, n_k$. Let $x = g a_2^{-n_2} a_3^{-n_3} \cdots a_k^{-n_k}$ be an element of G. We have $\pi(x) = e$, and hence $x \in \ker(\pi) = A_1$. Thus $g = x a_2^{n_2} \cdots a_k^{n_k} \in A_1 A_2 \cdots A_k$.

CLAIM 2: $A_1 \cap A_2 \cdots A_k = \{e\}$.

PROOF OF CLAIM 2: Let $x \in A_1 \cap A_2 \cdots A_k$. Then $x = a_2^{n_2} \cdots a_k^{n_k}$ for some positive integers $n_2, \ldots, n_k$. Apply π—remember that A_1 is the kernel of π—and get that $e = y_2^{n_2} \cdots y_k^{n_k}$. Now, fix $2 \leq i \leq k$, and get $y_i^{-n_i} = \prod_{\substack{j=2 \\ j \neq i}}^{k} y_j^{n_j}$. Hence, $y_i^{-n_i} \in L_i \cap \prod_{j \neq i} L_j = \{e\}$. Hence $o(a_i) = o(y_i) \mid n_i$, and $a_i^{n_i} = e$. Since this is true for every $2 \leq i \leq k$, we have $x = e$ as claimed.

CLAIM 3: For $2 \leq i \leq k$, $A_i \cap A_1 \cdots A_{i-1} A_{i+1} \cdots A_k = \{e\}$.

PROOF OF CLAIM 3: Assume $x \in A_i \cap A_1 \cdots A_{i-1} A_{i+1} \cdots A_k$. Then $x = a_i^{n_i} = \prod_{\substack{j=1 \\ j \neq i}}^{k} a_j^{n_j}$ for some positive integers $n_1, \ldots, n_k$. Apply the homomorphism π to get $\pi(x) = y_i^{n_i} = \prod_{\substack{j=2 \\ j \neq i}}^{k} y_j^{n_j}$. This implies that $y_i^{n_i} \in L_i \cap \prod_{j \neq i} L_j = \{e\}$. Hence, $o(a_i) = o(y_i) \mid n_i$. and so $x = a_i^{n_i} = e$ as claimed.

Now each A_i is a cyclic normal subgroup of G, and we conclude by Theorem 13.10 that G is isomorphic to $A_1 \times A_2 \times \cdots \times A_k$. $\square$

Corollary 13.15. *Let G be a finite abelian group. Then G is isomorphic to a direct product of cyclic p-groups.*

Sketch of the Proof. This follows from Theorem 13.12 and the fact that every cyclic group is a direct product of its Sylow subgroups. $\square$

Example 13.16. What are the abelian groups of order $72 = 2^3 \times 3^2$? Using Corollary 13.15, any abelian group of order 72 is isomorphic to one of the following:

$$\mathbb{Z}/9\mathbb{Z} \times \mathbb{Z}/8\mathbb{Z}, \qquad\qquad \mathbb{Z}/3\mathbb{Z} \times \mathbb{Z}/3\mathbb{Z} \times \mathbb{Z}/8\mathbb{Z},$$

$$\mathbb{Z}/9\mathbb{Z} \times \mathbb{Z}/4\mathbb{Z} \times \mathbb{Z}/2\mathbb{Z}, \qquad\qquad \mathbb{Z}/3\mathbb{Z} \times \mathbb{Z}/3\mathbb{Z} \times \mathbb{Z}/4\mathbb{Z} \times \mathbb{Z}/2\mathbb{Z},$$

$$\mathbb{Z}/9\mathbb{Z} \times \mathbb{Z}/2\mathbb{Z} \times \mathbb{Z}/2\mathbb{Z} \times \mathbb{Z}/2\mathbb{Z}, \qquad \mathbb{Z}/3\mathbb{Z} \times \mathbb{Z}/3\mathbb{Z} \times \mathbb{Z}/2\mathbb{Z} \times \mathbb{Z}/2\mathbb{Z} \times \mathbb{Z}/2\mathbb{Z}.$$

Remark 13.17. A different and more conceptual proof of the fundamental theorem of abelian groups—including the extension to finitely generated infinite abelian groups—can be obtained from theorems on modules. We will cover this alternative approach in the (hopefully) forthcoming second volume of this text.

Problems

13.2.1. Which abelian groups are *not* the direct product of two *proper* subgroups?

13.2.2. Let $G = (\mathbb{Z}/7\mathbb{Z} \times \mathbb{Z}/3\mathbb{Z} \times \mathbb{Z}/3\mathbb{Z} \times \mathbb{Z}/3\mathbb{Z}, +)$ and $a = (1, 1, 1, 0) \in G$. Let $A = \langle a \rangle$, and let $\pi : G \to G/A$ be the canonical homomorphism. Let $b = (0, 1, 2, 1) \in G$ and $y = Ab \in G/A$ (you may prefer to use the additive notation and write $A + b$ instead of Ab). Find an $x \in G$ with properties promised by Proposition 13.14.

13.2.3. List (up to isomorphism) all abelian groups of order 48.

13.2.4. List (up to isomorphism) all abelian groups of order 32.

13.2.5. Classify (up to isomorphism) all groups of order 45.

13.2.6. Using Theorem 13.12 and after doing Problems 13.1.12 and 13.1.13, write a complete proof of Corollary 13.15.

13.2.7. Use your solution to Problem 13.1.11 to streamline the Proof of Theorem 13.12, Claims 2 and 3.

13.2.8. **Invariant factors of a finite abelian group.** Let

$$G = \mathbb{Z}/25\mathbb{Z} \times \mathbb{Z}/9\mathbb{Z} \times \mathbb{Z}/9\mathbb{Z} \times \mathbb{Z}/7\mathbb{Z} \times \mathbb{Z}/7\mathbb{Z} \times \mathbb{Z}/5\mathbb{Z} \times \mathbb{Z}/3\mathbb{Z} \times \mathbb{Z}/3\mathbb{Z}.$$

Find positive integers $n_1, \ldots, n_k$ with $n_k \mid n_{k-1} \mid \cdots \mid n_1$, such that G is isomorphic to

$$\mathbb{Z}/n_1\mathbb{Z} \times \mathbb{Z}/n_2\mathbb{Z} \times \cdots \times \mathbb{Z}/n_k\mathbb{Z}.$$

The integers $n_1, \ldots, n_k$ are called the *invariant factors* of G. Is the list of invariant factors of G unique?

13.2.9. Let G be an abelian p-group, and let $x \in G$ have the largest possible order in G. Prove that G is isomorphic to $\langle x \rangle \times K$ for some subgroup K of G.

13.2.10. **Elementary abelian p-groups.** Let p be a prime. A group E is an *elementary abelian p-group* if it is abelian and $x^p = e$ for all $x \in E$. Prove that every finite elementary abelian p-group is isomorphic to the direct product of a number of copies of $\mathbb{Z}/p\mathbb{Z}$.

13.3. Semidirect Products

Given two groups H and K, we have seen that we can construct their direct product $G = H \times K$. This group has two *normal* subgroups $\tilde{H}$ and $\tilde{K}$ that are, respectively, isomorphic to H and K. In addition, $\tilde{H}\tilde{K} = G$ and $\tilde{H} \cap \tilde{K} = \{e\}$. In this section, we see how to construct a group G from groups N and H in such a way that again G has subgroups $\tilde{N}$ and $\tilde{H}$ isomorphic to N and H, respectively, and again $\tilde{N}\tilde{H} = G$ and $\tilde{N} \cap \tilde{H} = \{e\}$. However, in this new construction, only one of the subgroups, $\tilde{N}$, will necessarily be normal in G. The subgroup $\tilde{H}$ does not have to be normal in G. Of course, if $\tilde{H}$ is normal in G, then G will be isomorphic to $N \times H$, but the construction will allow us to construct groups that are *not* direct products of two of their subgroups. The group G constructed from N and H will be called the *semidirect product* or *split extension* of N by H and will be denoted by $N \rtimes H$.

Just as in the case of direct products, we distinguish between *constructing* a new group G as a semidirect product and *recognizing* an existing group as a semidirect product. The former is called an *external* semidirect product while the latter is called an *internal* semidirect product. We start with internal semidirect products.

Definition 13.18. Let G be a group. Assume G has two subgroups N and H that satisfy the following:

(a) $N \lhd G$, and

(b) $NH = G$, and

(c) $N \cap H = \{e\}$.

Then we say that G is the *internal semidirect product* of N by H, and will write $G = N \rtimes H$.

In other words, G is the semidirect product of N by H if and only if it has the partial lattice diagram of Figure 13.1.

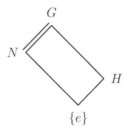

Figure 13.1. A partial lattice diagram for the semidirect product $N \rtimes H$

Example 13.19. Let $G = D_8 = \langle a, b \mid a^4 = b^2 = e, ba = a^{-1}b \rangle$. Let $N = \langle a \rangle$ and $H = \langle b \rangle$. Then $G = N \rtimes H$ is a semidirect product.

Example 13.20. Let $G = Q_8$ be the quaternion group of order 8. Then G is *not* a semidirect product of two of its proper subgroups. This is because $\mathbf{Z}(G)$ is a subgroup of order 2 that is contained in *all* non-trivial subgroups. Hence, there do not exist two non-trivial subgroups with trivial intersection. (See the lattice diagram in Figure 9.9.)

Example 13.21. Let $G = A_4$ be the alternating group of order 12. Using the lattice diagram in Figure 9.11 (on page 175), we let $K = \langle(1\ 2)(3\ 4), (1\ 3)(2\ 4)\rangle$ and $H = \langle(1\ 2\ 3)\rangle$. Then K is a normal subgroup of order 4 (K is normal since it consists of the identity element and *all* elements with cycle structure $2, 2$, and so it is a union of conjugacy classes), and H is a subgroup of order 3. In addition, $H \cap K = \{e\}$ and $KH = G$. Hence, $A_4 = K \rtimes H$ is the semidirect product of K by H.

Given two groups N and H, how do we construct a semidirect product $N \rtimes H$? This question is not well defined, since, unlike the construction of the direct product $N \times H$, we need more information than just the two groups N and H. For example, if $N = \mathbb{Z}/2\mathbb{Z} \times \mathbb{Z}/2\mathbb{Z}$ and $H = \mathbb{Z}/3\mathbb{Z}$, then there is more than one way to construct $N \rtimes H$. In fact, both A_4 and $\mathbb{Z}/2\mathbb{Z} \times \mathbb{Z}/2\mathbb{Z} \times \mathbb{Z}/3\mathbb{Z}$ are examples of such a semidirect product. (By our definition, the direct product $N \times H$ is an example of a semidirect product.)

If G is the internal direct product of N and H, then, in G, elements of N commute with elements of H. Hence, if we just know the groups N and H, we can reconstruct the group G and its multiplication table.

However, if H is not normal in G, then elements of N and H need not commute. To construct $N \rtimes H$, we need to know how to write nh (with $n \in N$ and $h \in H$) as $h'n'$ where $h' \in H$ and $n' \in N$. Now, since N is assumed to be normal in G, we have $hnh^{-1} = n' \in N$, from which we get $hn = n'h$. In other words, if we know the value of hnh^{-1} for all $h \in H$ and $n \in N$, then we have enough information to build $N \rtimes H$.

The most straightforward way to organize/generalize this information is through the concept of "action via automorphisms":

Definition 13.22. Let N and H be groups, and assume that H acts on N. Further assume that for every $h \in H$ and $n_1, n_2 \in N$ we have

$$h \cdot (n_1 n_2) = (h \cdot n_1)(h \cdot n_2).$$

Then we say that H *acts on N via automorphisms*.

We know that when H acts on N, every element of H gives a permutation of N. If H acts on N via automorphisms, then each element of H gives an *automorphism* of N. In fact, to have that H acts on N via automorphisms is equivalent to having a group homomorphism $\phi \colon H \to \operatorname{Aut}(N)$ (see Problem **13.3.1**).

It is straightforward to check that we do have an action via automorphisms in an internal semidirect product:

Lemma 13.23. *Assume $G = N \rtimes H$ is an internal semidirect product. In other words, N and H are subgroups of the group G, N is normal in G, $NH = G$, and $N \cap H = \{e\}$. Define an action of H on N by $h \cdot n = hnh^{-1}$ for $h \in H, n \in N$. Then H acts on N via automorphisms.*

Proof. Since N is normal in G, $hnh^{-1} \in N$, and conjugation does give an action of H on N. Furthermore, for $h \in H$ and $n_1, n_2 \in N$, we have

$$h \cdot (n_1 n_2) = hn_1 n_2 h^{-1} = hn_1 h^{-1} hn_2 h^{-1} = (h \cdot n_1)(h \cdot n_2). \qquad \square$$

We next show that, in an internal semidirect product, knowing the subgroups N, H, and the action via automorphisms of H on N is enough data to uniquely determine the group up to isomorphism.

Proposition 13.24. *Let the groups $G_1 = N_1 \rtimes H_1$ and $G_2 = N_2 \rtimes H_2$ be semidirect products. As in Lemma 13.23, for $i = 1, 2$, H_i acts on N_i by conjugation. Assume that $\psi \colon N_1 \to N_2$ and $\phi \colon H_1 \to H_2$ are isomorphisms such that*

$$(13.1) \qquad \psi(h \cdot n) = \phi(h) \cdot \psi(n) \quad \text{for all } n \in N_1, h \in H_1.$$

Then $G_1 \cong G_2$.

Proof. Since the action (by automorphisms) of H_i on N_i is conjugation, equation (13.1) translates to

$$(13.2) \qquad \psi(hnh^{-1}) = \phi(h)\psi(n)\phi(h)^{-1}.$$

Now define $\theta \colon G_1 \to G_2$ by $\theta(nh) = \psi(n)\phi(h)$ for all $n \in N_1$, $h \in H_1$. Since $G_1 = N_1 H_1$, every element of G_1 is of the form nh for some $n \in N_1$, $h \in H_1$, and since $N_1 \cap H_1 = \{e\}$ this representation of elements of G_1—as a product of elements of N_1 and elements of H_1—is unique (see Problem 13.1.2). Hence, θ is well defined on all of G_1.

Showing that θ is an isomorphism is straightforward. Let $g = nh$ and $g' = n'h'$ be two elements of G_1 with $n, n' \in N_1$ and $h, h' \in H_1$. Then, since $hn'h^{-1} \in N_1$ and using equation (13.2), we have

$$\begin{aligned}
\theta(gg') = \theta(nhn'h') &= \theta(nhn'h^{-1}hh') \\
&= \psi(nhn'h^{-1})\phi(hh') = \psi(n)\psi(hn'h^{-1})\phi(h)\phi(h') \\
&= \psi(n)\phi(h)\psi(n')\phi(h)^{-1}\phi(h)\phi(h') \\
&= \theta(nh)\theta(n'h') = \theta(g)\theta(g'),
\end{aligned}$$

and so θ is a homomorphism. Clearly θ is onto, and $\theta(nh) = e_{G_2}$ implies that $\psi(n)\phi(h) = e$. But this means that $\psi(n) = \phi(h)^{-1} \in N_2 \cap H_2 = \{e\}$. Now, since ψ and ϕ are 1-1, we get that $n = h = e_{G_1}$, and so θ is 1-1. The proof is now complete. $\square$

We are now ready to construct external semidirect products:

Theorem 13.25. *Let N and H be groups, and assume that H acts on N via automorphisms. Then there exists a group G (unique up to isomorphism) with subgroups $\tilde{N}$ and $\tilde{H}$ such that*

(a) *The group G is the internal semidirect product of $\tilde{N}$ by $\tilde{H}$. In other words, $\tilde{N} \triangleleft G$, $\tilde{N}\tilde{H} = G$, and $\tilde{N} \cap \tilde{H} = \{e\}$.*

(b) *There exists isomorphisms $\phi \colon H \to \tilde{H}$ and $\psi \colon N \to \tilde{N}$ such that, for all $h \in H$, $n \in N$, we have*

$$\psi(h \cdot n) = \phi(h)\psi(n)\phi(h)^{-1},$$

where the three elements on the right hand side are multiplied in the group G.

What the theorem says is that given N and H and an action via automorphisms of H on N, we can uniquely construct a group G that is the internal semidirect product of $\tilde{N}$ by $\tilde{H}$ in such a way that the action of H on N becomes the conjugation action of $\tilde{H}$ on $\tilde{N}$ (inside G).

Proof. Let $G = \{(n,h) \mid n \in N, h \in H\}$. We define the following multiplication on G:

$$(n_1, h_1)(n_2, h_2) = (n_1(h_1 \cdot n_2), h_1 h_2).$$

In other words, as a *set*, G is the same as $N \times H$, but we have defined a different product on the elements.

We leave to the reader, the (somewhat tedious but straightforward) details of showing that G with this multiplication is a group. Note that the element $(e_N, e_H) \in G$ is the identity element and $(h^{-1} \cdot n^{-1}, h^{-1})$ is the inverse of (n, h) in G.

Let $\tilde{N} = \{(n, e_H) \mid n \in N\}$, and let $\psi \colon N \to \tilde{N}$ be defined by $\psi(n) = (n, e_H)$. Likewise, $\tilde{H} = \{(e_N, h) \mid h \in H\}$, and $\phi \colon H \to \tilde{H}$ is defined by $\phi(h) = (e_N, h)$. Then ϕ and ψ are clearly isomorphisms, $\tilde{N}\tilde{H} = G$, and $\tilde{N} \cap \tilde{H} = (e_N, e_H) = e_G$.

Also, for $h \in H$ and $n \in N$, we have

$$
\begin{aligned}
\phi(h)\psi(n)\phi(h)^{-1} &= (e_N, h)(n, e_H)(e_N, h)^{-1} \\
&= (h \cdot n, h)(h^{-1} \cdot e_N, h^{-1}) \\
&= ((h \cdot n)(h \cdot e_N), hh^{-1}) \\
&= (h \cdot n, e_H) \\
&= \psi(h \cdot n).
\end{aligned}
$$

Finally, the same calculation showed that every element of $\tilde{H}$ normalizes $\tilde{N}$. Since $G = \tilde{N}\tilde{H}$, and $\tilde{N}$ certainly normalizes itself also, we conclude that $\tilde{N} \triangleleft G$. Uniqueness was already proved in Proposition 13.24, and so the proof is complete. $\square$

Remark 13.26. If H acts trivially on N, that is $h \cdot n = n$ for all $h \in H$ and $n \in N$, then $N \rtimes H = N \times H$. This can easily be seen from the construction, in the proof of Theorem 13.25, since the product in G is then identical to the one in the direct product.

Remark 13.27. The statement of Theorem 13.25 only declares that a group G with certain properties exist. The actual construction of the group is in the proof. The group G as a set is the Cartesian product of N and H, and then a certain multiplication is defined on it. However, as Proposition 13.24 shows, the information about the group G given in the statement of Theorem 13.25 is enough to determine G up to isomorphism. Hence, almost all the time, all we need to know is what is in the statement of Theorem 13.25.

Example 13.28. Let $N = \langle a \mid a^3 = e \rangle$ be the cyclic group of order 3 (written multiplicatively), and let $H = \langle b \mid b^4 = e \rangle$ be the cyclic group of order 4. To define an action of H on N, it suffices to specify the action of b on N. To define an action of H on N *via automorphisms*, it suffices to specify the action of b on a. We define

$$b \cdot a = a^{-1}.$$

This means that $b \cdot a^2 = (b \cdot a)(b \cdot a) = a^{-2} = a$ and $b \cdot a^3 = a^{-3} = e$. Now, $b^2 \cdot a = b \cdot (b \cdot a) = a$, and so b^2 fixes everything. It follows that b^3 acts as b. We do have an action of H on N via automorphisms, and hence we can construct $T = N \rtimes H \cong \mathbb{Z}/3\mathbb{Z} \rtimes \mathbb{Z}/4\mathbb{Z}$.

What do we know about the group T? The group T has 12 elements, a normal subgroup isomorphic to $\mathbb{Z}/3\mathbb{Z}$, and another subgroup isomorphic to $\mathbb{Z}/4\mathbb{Z}$. If x and y are the generators of these two groups, then every element of T is of the form $x^i y^j$ with $0 \le i \le 2$, $0 \le j \le 3$, and we have $yxy^{-1} = x^{-1}$. Hence,

$$T = \langle x, y \mid x^3 = y^4 = e, yx = x^{-1}y \rangle.$$

This is a group of order 12 that is not isomorphic to A_4 (since A_4 has a normal Sylow 2-subgroup, but T has a normal Sylow 3-subgroup) or D_{12} (since the Sylow 2-subgroup of D_{12} is isomorphic to $\mathbb{Z}/2\mathbb{Z} \times \mathbb{Z}/2\mathbb{Z}$, while T has a subgroup isomorphic to $\mathbb{Z}/4\mathbb{Z}$).

Note that we could not define $b \cdot a$ arbitrarily. The action of b has to coincide with the action of an automorphism of the group N. Now N has an automorphism that inverts every element, and hence the action of b on a as defined here does give an action of H on N via automorphisms.

Problems

13.3.1. Let H and N be groups. Show that to say that H acts on N via automorphisms is equivalent to saying that we have a homomorphism $\phi \colon H \to \mathrm{Aut}(N)$. In other words, given an action of H on N via automorphisms, construct a homomorphism $\phi \colon H \to \mathrm{Aut}(N)$, and vice versa.

13.3.2. Let $G = N \rtimes H$ be the internal semidirect product of N by H. Assume that every element of H commutes with every element of N. Show that $G = N \times H$.

13.3.3. Write down the complete details—including a proof of the associativity— that shows that the multiplication defined, in the proof of Theorem 13.25 on the set $G = N \times H$, does indeed make G a group.

13.3.4. Let p and q be distinct primes, and let G be a group of order pq. Show that G is a semidirect product of two of its proper subgroups.

13.3.5. Let $N = \langle a \mid a^5 = e \rangle$ be a cyclic group of order 5. Let $\phi \colon N \to N$ be defined by $\phi(x) = x^2$ for all $x \in N$. Is ϕ an automorphism of N? Let $H = \{\phi, \phi^2, \phi^3, \phi^4\}$, where $\phi^i = \phi \circ \phi \circ \cdots \circ \phi$ is the composition of ϕ with itself i times. Show that H is a group under function composition. Define an action of H on N by

$$\phi^i \cdot a^j = \phi^i(a^j).$$

Is this an action by automorphisms? If so, then describe the group $N \rtimes H$ by giving a set of generators and relations for it.

13.3.6. Let $N = \mathbb{Z}/4\mathbb{Z}$ and $H = \mathbb{Z}/3\mathbb{Z}$.
(a) Show that $\mathrm{Aut}(N) \cong \mathbb{Z}/2\mathbb{Z}$.

(b) Show that if $\phi\colon H \to \mathrm{Aut}(N)$ is a homomorphism, then $\ker(\phi) = H$.

(c) Show that any group of the form $\mathbb{Z}/4\mathbb{Z} \rtimes \mathbb{Z}/3\mathbb{Z}$ is isomorphic to $\mathbb{Z}/12\mathbb{Z}$.

13.3.7. Construct as many non-isomorphic non-abelian groups of order 16 as you can.

13.4. Groups of Very Small Order

As a reminder of some of what we have done, in this section we classify all groups of order up to 15. We can even go higher, but the process becomes quite tedious (especially for groups of order p^n where p is a prime and n is a large integer). Table 13.1 gives a list of all groups of order ≤ 15 (up to isomorphism).

Table 13.1. All groups of order no more than 15 up to isomorphism. The group T is the group in Example 13.28.

Order	Cyclic	Abelian but not Cyclic	Non-abelian
2	$\mathbb{Z}/2\mathbb{Z}$	$--$	$--$
3	$\mathbb{Z}/3\mathbb{Z}$	$--$	$--$
4	$\mathbb{Z}/4\mathbb{Z}$	$\mathbb{Z}/2\mathbb{Z} \times \mathbb{Z}/2\mathbb{Z}$	$--$
5	$\mathbb{Z}/5\mathbb{Z}$	$--$	$--$
6	$\mathbb{Z}/6\mathbb{Z}$	$--$	S_3
7	$\mathbb{Z}/7\mathbb{Z}$	$--$	$--$
8	$\mathbb{Z}/8\mathbb{Z}$	$\mathbb{Z}/4\mathbb{Z} \times \mathbb{Z}/2\mathbb{Z}, \mathbb{Z}/2\mathbb{Z} \times \mathbb{Z}/2\mathbb{Z} \times \mathbb{Z}/2\mathbb{Z}$	D_8, Q_8
9	$\mathbb{Z}/9\mathbb{Z}$	$\mathbb{Z}/3\mathbb{Z} \times \mathbb{Z}/3\mathbb{Z}$	$--$
10	$\mathbb{Z}/10\mathbb{Z}$	$--$	D_{10}
11	$\mathbb{Z}/11\mathbb{Z}$	$--$	$--$
12	$\mathbb{Z}/12\mathbb{Z}$	$\mathbb{Z}/3\mathbb{Z} \times \mathbb{Z}/2\mathbb{Z} \times \mathbb{Z}/2\mathbb{Z}$	D_{12}, A_4, T
13	$\mathbb{Z}/13\mathbb{Z}$	$--$	$--$
14	$\mathbb{Z}/14\mathbb{Z}$	$--$	D_{14}
15	$\mathbb{Z}/15\mathbb{Z}$	$--$	$--$

We have already met all the groups in Table 13.1, and except for the case of groups of order 8 and 12, we already knew that the groups in the table are the only groups for these orders. Here, we remind the reader of the results used to create the table:

Proposition 13.29. *Let p and q denote prime numbers with $p < q$, then:*

(a) (Corollary 5.15) *Every group of order p is isomorphic to $\mathbb{Z}/p\mathbb{Z}$.*

(b) (Corollary 6.20 and Theorem 13.12) *Every group of order p^2 is abelian, and hence isomorphic to either $\mathbb{Z}/p^2\mathbb{Z}$ or $\mathbb{Z}/p\mathbb{Z} \times \mathbb{Z}/p\mathbb{Z}$.*

(c) (Proposition 5.36) *Every group of order $2p$ is isomorphic to either $\mathbb{Z}/2p\mathbb{Z}$ or D_{2p}.*

(d) (Theorem 12.15) *Every group of order pq where p does not divide q − 1 is isomorphic to* $\mathbb{Z}/pq\mathbb{Z}$.

We close this section (and chapter) by classifying groups of order 8 and 12.

Groups of Order 8. Let G be a group of order 8. We know by the fundamental theorem of abelian groups, Theorem 13.12, that if G is abelian, then G is isomorphic to one of $\mathbb{Z}/8\mathbb{Z}$, $\mathbb{Z}/4\mathbb{Z} \times \mathbb{Z}/2\mathbb{Z}$, or $\mathbb{Z}/2\mathbb{Z} \times \mathbb{Z}/2\mathbb{Z} \times \mathbb{Z}/2\mathbb{Z}$. Hence, assume that G is non-abelian.

Because G is non-abelian, G cannot have an element of order 8, and not all of the elements of G can have order 2 (see Problem 2.2.3). So, let $a \in G$ with $o(a) = 4$. We have $|G : \langle a \rangle| = 2$, and so $\langle a \rangle \triangleleft G$, and a together with any element of G outside of $\langle a \rangle$ will generate G.

If G has an element b of order 2 not in $\langle a \rangle$, then $\langle a \rangle \langle b \rangle = G$ and $\langle a \rangle \cap \langle b \rangle = \{e\}$. Hence, $G = \langle a \rangle \rtimes \langle b \rangle$. What is bab^{-1}? Since $\langle a \rangle$ is normal in G, $bab^{-1} \in \langle a \rangle$. In addition, $o(bab^{-1}) = o(a) = 4$. The only choices are $bab^{-1} = a$ or $bab^{-1} = a^{-1}$. In the former case, G is, contrary to assumption, abelian. Hence, $bab^{-1} = a^{-1}$, and so

$$G = \langle a, b \mid a^4 = b^2 = e, ba = a^{-1}b \rangle \cong D_8.$$

We are left with the case when every element of G not in $\langle a \rangle$ is of order 4. Hence, the only element of order 2 in G is a^2. Let $b \in G - \langle a \rangle$. Then b^2 is an element of order 2, and so $b^2 = a^2$. Now, since $G = \langle a \rangle \langle b \rangle$, we have

$$G = \langle a, b \mid a^4 = b^4 = e, a^2 = b^2, ba = a^{-1}b \rangle \cong Q_8.$$

We conclude that there are five groups of order 8:

$$\mathbb{Z}/8\mathbb{Z}, \ \mathbb{Z}/4\mathbb{Z} \times \mathbb{Z}/2\mathbb{Z}, \ \mathbb{Z}/2\mathbb{Z} \times \mathbb{Z}/2\mathbb{Z} \times \mathbb{Z}/2\mathbb{Z}, \ D_8, \ Q_8.$$

Groups of Order 12. Let G be a group of order 12. Let $P \in \mathrm{Syl}_2(G)$ and $Q \in \mathrm{Syl}_3(G)$. The group P has order 4 and is isomorphic to either $\mathbb{Z}/4\mathbb{Z}$ or $\mathbb{Z}/2\mathbb{Z} \times \mathbb{Z}/2\mathbb{Z}$. The group Q has order 3 and is isomorphic to $\mathbb{Z}/3\mathbb{Z}$. The orders and indices of P and Q are relatively prime, and hence by Lagrange's theorem $P \cap Q = \{e\}$ and by Proposition 9.30, $PQ = G$.

By the Sylow theorems, $|\mathrm{Syl}_2(G)|$ is 1 or 3, and $|\mathrm{Syl}_3(G)|$ is 1 or 4.

If $|\mathrm{Syl}_2(G)| = |\mathrm{Syl}_3(G)| = 1$, then both P and Q are normal in G, and $G \cong P \times Q$ is abelian. In such a case, by Corollary 13.15, G is isomorphic to $\mathbb{Z}/4\mathbb{Z} \times \mathbb{Z}/3\mathbb{Z}$ or $\mathbb{Z}/2\mathbb{Z} \times \mathbb{Z}/2\mathbb{Z} \times \mathbb{Z}/3\mathbb{Z}$. Note that $\mathbb{Z}/4\mathbb{Z} \times \mathbb{Z}/3\mathbb{Z} \cong \mathbb{Z}/12\mathbb{Z}$.

If $|\mathrm{Syl}_3(G)| = 4$, then G has $4 \times 2 = 8$ elements of order 3, and the remaining four elements must constitute the unique subgroup of order 4. Hence, in this case, $|\mathrm{Syl}_2(G)| = 1$. Thus two cases remain: P is the unique Sylow 3-subgroup, or Q is the unique Sylow 2-subgroup. In the former case, $P \triangleleft G$, and in the latter case $Q \triangleleft G$.

Case 1: G is non-abelian and $Q \triangleleft G$.

In this case, G does not have a normal Sylow 3-subgroup. By Problem 12.4.1, G is isomorphic to A_4.

Case 2: G is non-abelian and $P \triangleleft G$.

In this case, $G = P \rtimes Q$. It remains to identify the action of Q on P via automorphisms. The group $P = \langle a \mid a^3 = e \rangle$ is a cyclic group of order 3, and every automorphism of P must send an element of order 3 to another element of order 3. Hence, the only non-trivial automorphism of P sends a to a^{-1}. In fact, $\mathrm{Aut}(P) \cong \mathbb{Z}/2\mathbb{Z}$.

Now Q is isomorphic to either $\mathbb{Z}/4\mathbb{Z}$ or $\mathbb{Z}/2\mathbb{Z} \times \mathbb{Z}/2\mathbb{Z}$. Writing multiplicatively, $Q \cong \langle b \mid b^4 = e \rangle$ or $Q \cong \langle c, d \mid c^2 = d^2 = e, cd = dc \rangle$.

If $Q \cong \langle b \mid b^4 = e \rangle$, to get a non-trivial action, we have to have $b \cdot a = a^{-1}$ and this determines the action completely. This gives the group T of Example 13.28.

If $Q \cong \langle c, d \mid c^2 = d^2 = e, cd = dc \rangle$, in a non-trivial action by automorphisms of Q on P at least one of c or d must send a to a^{-1}. This results in two from among c, d, and cd inverting a, and the remaining third element fixing a. Without loss of generality, assume c fixes a. Then, in the group $G = P \rtimes Q$, the elements c and a commute, and ca is an element of order 6. On the other hand, d is an element of order 2 and, in G, $d(ca)d^{-1} = c(dad^{-1}) = ca^{-1} = (ca)^{-1}$. Hence, letting $x = ca$ and $y = d$, we have

$$ G = \langle x, y \mid x^6 = y^2 = e, yx = x^{-1}y \rangle. $$

Evidently, in this case, G is isomorphic to D_{12}.

We conclude that any group of order 12 is isomorphic to one of $\mathbb{Z}/12\mathbb{Z}$, $\mathbb{Z}/3\mathbb{Z} \times \mathbb{Z}/2\mathbb{Z} \times \mathbb{Z}/2\mathbb{Z}$, A_4, D_{12}, or group T of Example 13.28.

Problems

13.4.1. Which groups in Table 13.1 are *not* the semidirect product of two proper subgroups?

13.4.2. In the classification of groups of order 8, we asserted that

$$ \langle a, b \mid a^4 = b^4 = e, a^2 = b^2, ba = a^{-1}b \rangle \cong Q_8. $$

Using Definition 2.54 of Q_8, prove this assertion.

13.4.3. Give explicit arguments to show that no two groups in Table 13.1 can be isomorphic.

13.4.4. Classify—up to isomorphism—all groups of order 20.

13.4.5. Let G be a finite group. Assume that $x, y \in G$ are distinct elements of order 2 and that $G = \langle x, y \rangle$. Let $n = o(xy)$. Prove that $G \cong D_{2n}$.

Chapter 14

Solvable
and Nilpotent Groups*

...where various characterizations of solvable and nilpotent groups are given, it is proved that A_n is simple for $n \geq 5$, the derived subgroup, characteristic subgroups, and chains of subgroups are studied, and the Jordan–Hölder theorem is proved.

14.1. Solvable Groups

Definition and Examples. From one vantage point—to become clear later—solvable groups are the groups that are put together from cyclic groups, and, hence, should be the easiest to study. In fact, abelian groups are a special case of solvable groups, and so the study of solvable groups maybe a good next step after understanding abelian groups. Solvable groups also play an important role in—and, in fact originated in the study of—Galois theory. For this reason, the reader may want to postpone studying this chapter until it is needed in Galois theory. We have already defined solvable groups in Definition 10.38. Here, we begin by giving an alternative definition. Later, we will show the equivalence of the two definitions. Recall that a chain of subgroups (Definition 9.17) is just a set of subgroups totally ordered by inclusion.

Definition 14.1 (Solvable groups)**.** Let G be a group. Then G is *solvable* if G has a finite chain of subgroups

$$\{e\} = G_0 \leq G_1 \leq \cdots \leq G_{n-1} \leq G_n = G,$$

such that, for $i = 0, \ldots, n-1$,

(a) $G_i \lhd G_{i+1}$, and

(b) G_{i+1}/G_i is an abelian group.

Example 14.2. If G is abelian, then the chain $\{e\} \leq G$ has the required properties, and hence G is solvable.

Example 14.3. Let $G = S_3$, and let $P = \langle (1\ 2\ 3) \rangle$. Then P is a subgroup of index 2 and is normal in G. We have

$$\{e\} \leq P \leq G, \text{ and } P \cong \mathbb{Z}/3\mathbb{Z}, \ G/P \cong \mathbb{Z}/2\mathbb{Z},$$

and so S_3 is solvable.

Example 14.4. Let $G = D_8 = \langle a, b \mid a^4 = b^2 = e, ba = a^3 b \rangle$ and $H = \langle a \rangle$. Then $\{e\} \leq H \leq G$, H is normal in G since it is of index 2, $H \cong \mathbb{Z}/4\mathbb{Z}$, and $G/H \cong \mathbb{Z}/2\mathbb{Z}$. Hence, D_8 is solvable.

Example 14.5. Consider the group S_4 and its subgroup A_4, the alternating group of degree 4 consisting of even permutations in S_4. Let

$$K = \{e, (1\ 2)(3\ 4), (1\ 3)(2\ 4), (1\ 4)(2\ 3)\} \leq A_4.$$

Then we have

$$\{e\} \lhd K \lhd A_4 \lhd S_4,$$

and $K \cong \mathbb{Z}/2\mathbb{Z} \times \mathbb{Z}/2\mathbb{Z}$, $A_4/K \cong \mathbb{Z}/3\mathbb{Z}$, and $S_4/A_4 \cong \mathbb{Z}/2\mathbb{Z}$. Hence, S_4 is solvable.

Example 14.6. In Proposition 10.37 we proved that A_5, the alternating group of degree 5, is simple. Since A_5 has no normal subgroups and is not abelian itself, it cannot satisfy the condition of Definition 14.1. Hence A_5 is *not* solvable. In fact, A_5 is the smallest non-solvable group.

The Commutator Subgroup. To study solvable groups, a particular subgroup of the group—the so-called commutator subgroup—and a particular chain of subgroups will be of much use. We already defined commutators and the commutator subgroup in Definitions 11.26 and 11.27, but we will repeat the definitions here.

Definition 14.7 (Commutators). Let G be a group. Let $x \in G$. If x can be written as $aba^{-1}b^{-1}$ for some $a, b \in G$, then we call x a *commutator* in G.

Definition 14.8 (The commutator subgroup). Let G be a group. The subgroup of G generated by all the commutators of G is called the *commutator subgroup* (or the *derived subgroup*) of G and is denoted by G'. We will denote the commutator subgroup of G' with G'' or $G^{(2)}$, and more generally, for $n > 1$, $G^{(n)}$ will denote $(G^{(n-1)})'$.

Finding commutator subgroups directly and using the definition can be cumbersome. We will see examples later, after developing some theoretical facts about commutator subgroups.

Lemma 14.9. *Let G and H be groups, and let $\phi : G \to H$ be an onto homomorphism. Then $\phi(G') = H'$.*

Proof. Since $\phi(xyx^{-1}y^{-1}) = \phi(x)\phi(y)\phi(x)^{-1}\phi(y)^{-1}$, we see that commutators are mapped into commutators. Hence, the commutator subgroup of G—that is generated by the commutators of G—is mapped into the commutator subgroup of H. On the other hand, since ϕ is onto, each commutator of H is the image of some commutator. So H' is in the image of G', and thus $\phi(G') = H'$. $\qquad\square$

As the next lemma shows, the commutator subgroup provides an alternative to the center for deciding whether a group is abelian.

Lemma 14.10. *Let G be a group. Then G is abelian if and only if $G' = \{e\}$.*

Proof. (This was Problem **11.3.10**.) If G is abelian, then clearly all the commutators—and hence the subgroup generated by them—are trivial. On the other hand, if $G' = \{e\}$, then for all $x, y \in G$, $xyx^{-1}y^{-1} = e$, and this means that $xy = yx$. $\square$

We can use the lemma to generalize it considerably. The bigger the center of a group is, the closer that group is to being abelian. In contrast, as the next corollary shows, a small commutator subgroup signifies a large abelian quotient group.

Corollary 14.11. *Let G be a group, and let $N \triangleleft G$. Then G/N is abelian if and only if $G' \leq N$.*

Proof. (This was Problem 11.3.13.) Let $\phi : G \to G/N$ be the canonical homomorphism. Then the kernel of ϕ is N. Now if G/N is abelian, then by Lemmas 14.9 and 14.10 $\phi(G') = (G/N)' = \{e\}$ which means that $G' \leq \ker \phi = N$.

On the other hand, if $G' \leq N$, then $(G/N)' = \phi(G') = \{e\}$ proving, by Lemma 14.10, that G/N is abelian. $\square$

Characteristic Subgroups. Is $G' \triangleleft G$? The answer is yes. In fact, G' has a property that is stronger than being normal.

A subgroup N of a group G is normal if $xNx^{-1} = N$. We could rephrase the definition by saying that $N \triangleleft G$ if $\phi(N) = N$ for all $\phi \in \mathrm{Inn}(G)$. In other words, normal subgroups are exactly the subgroups that are fixed—not necessarily element-wise but as a set—by every inner automorphism of G. We get a stronger condition if we consider subgroups that are fixed by every automorphism of G. Such a subgroup is called a characteristic subgroup and, by necessity, is a normal subgroup.

Definition 14.12. *Let G be a group, and let $H \leq G$. We say that H is a characteristic subgroup of G if $\phi(H) = H$ for all $\phi \in \mathrm{Aut}(G)$. If H is a characteristic subgroup of G, then we write H char G.*

We know that if $H \triangleleft K \triangleleft G$, then H may or may not be normal in G (Problem 10.1.9). However, we can conclude normality if H is not merely normal in K but it is a characteristic subgroup.

Lemma 14.13. *Let G be a group, and let H, K be subgroups of G. If H char $K \triangleleft G$, then $H \triangleleft G$. In particular, H char G implies $H \triangleleft G$.*

Proof. Let g be an arbitrary element of G, and define $\phi_g : G \to G$ by $\phi_g(x) = gxg^{-1}$. Then ϕ_g is an automorphism of G (in fact, ϕ_g is called an inner automorphism—see Definition 11.44 and Corollary 11.46). The restriction $\phi_g|_K$ of ϕ_g to K is an automorphism of K, and hence fixes H. We have $\phi_g(H) = H$ which means that $gHg^{-1} = H$. Since $g \in G$ was arbitrary, we conclude that H is normal in G. $\square$

When do you suspect that a subgroup is a characteristic subgroup? Often, if you can identify the subgroup as "the" (as opposed to "a") subgroup of the group with some group theoretic property, then that subgroup is characteristic. For example, "the" center of a group is a characteristic subgroup (Problem 14.1.13). We now show that "the" commutator subgroup of a group is a characteristic subgroup.

Lemma 14.14. *Let G be a group. Then G' char G. In particular, for all positive integers n, $G^{(n)} \triangleleft G$, and G/G' is the largest abelian quotient group of G.*

Proof. Let $\phi \in \text{Aut}(G)$. We have to show that $\phi(G') = G'$. But this follows from Lemma 14.9, since $\phi : G \to G$ is an onto homomorphism. A characteristic subgroup is normal, and, hence, G' is normal in G. Now G'' char $G' \triangleleft G$, which means that $G'' \triangleleft G$. Continue with the same argument to get $G^{(n)} \triangleleft G$.

By Corollary 14.11, G/G' is abelian, and if G/N is abelian, then $G' \leq N$. This shows that G/G' is the largest abelian quotient group of G.   □

Example 14.15. Let $G = S_3$. This group has only three normal subgroups: $\{e\}$, $P = \langle (1\ 2\ 3) \rangle$, and G. Both G/G and G/P are abelian while $G/\{e\} \cong G$ is not. Hence, $G' = P$. Now P itself is abelian, and so $G'' = P' = \{e\}$.

Equivalent Definitions of Solvability. We are now ready to prove various equivalent definitions for solvable groups.

Theorem 14.16. *Let G be a group. Then the following are equivalent:*

(a) *G is solvable.*

(b) *There exists a finite chain of subgroups*
$$\{e\} = H_0 \leq H_1 \leq \cdots \leq H_{n-1} \leq H_n = G,$$
such that, for $0 \leq i < n$, $H_i \triangleleft H_{i+1}$, and H_{i+1}/H_i is a cyclic group of prime order.

(c) *$G^{(n)} = \{e\}$ for some positive integer n.*

(d) *We have a finite chain of subgroups*
$$\{e\} = K_0 \leq K_1 \leq \cdots \leq K_n = G,$$
such that, for $0 \leq i < n$, $K_i \triangleleft G$, and K_{i+1}/K_i is an abelian group.

Remark 14.17. The theorem says that in a solvable group G we can find several kinds of finite chains of subgroups. We could insist that every subgroup is normal in the whole group, in which case we get that the factor groups are abelian. On the other hand, we can ask that each subgroup be just normal in the next subgroup in the series. In that case, we can insist that the factor groups be cyclic of prime order. Moreover, if there is a finite chain of subgroups where each subgroup is normal in the next one and the factor groups are merely abelian, then the group will also have both of the other kinds of chains of subgroups.

Proof. (a) $\Rightarrow$ (b) By definition of solvability, we already have a chain of subgroups
$$\{e\} = G_0 \triangleleft G_1 \triangleleft \cdots \triangleleft G_{n-1} \triangleleft G_n = G,$$
such that, for $i = 0, \ldots, n - 1$, G_{i+1}/G_i is an abelian group. We refine this chain to get the desired one. In particular, let p be a prime divisor of $|G_{i+1}/G_i|$, and let H be a subgroup of order p in G_{i+1}/G_i. The inverse image of H under the canonical homomorphism $\pi : G_{i+1} \to G_{i+1}/G_i$ is a group L with $G_i \leq L \leq G_{i+1}$

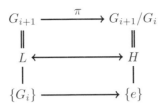

Figure 14.1. Refining the abelian factor groups

and with L/G_i a cyclic group of order p. (See Figure 14.1.) Since G_{i+1}/G_i was abelian, $H \lhd G_{i+1}/G_i$, and, by homomorphism theorems, $L \lhd G_{i+1}$. Hence, in the original chain of subgroups, we can add L between G_i and G_{i+1}. Continuing in this manner, we can make all factor groups be cyclic of prime order.

(b) $\Rightarrow$ (c) Given the chain of subgroups with the given properties, we know that G/H_{n-1} is abelian—in fact, cyclic—and so, by Lemma 14.11, $G' \leq H_{n-1}$. Similarly, since H_{n-1}/H_{n-2} is abelian, we have $H'_{n-1} \leq H_{n-2}$. Now, we argue

$$G' \leq H_{n-1} \quad \Rightarrow \quad G'' \leq H'_{n-1} \leq H_{n-2}.$$

Continuing with the same argument, we get $G^{(3)} \leq H_{n-3}$, $\ldots$, $G^{(n)} \leq H_0 = \{e\}$, and the proof is complete.

(c) $\Rightarrow$ (d) Consider the chain of subgroups (called the *derived series* of G)

$$G^{(n)} \leq G^{(n-1)} \leq \cdots \leq G'' \leq G' \leq G.$$

By hypothesis $G^{(n)} = \{e\}$. By Lemma 14.14 each of the subgroups in the chain is normal in G, and, by Corollary 14.11, each factor group is abelian.

(d) $\Rightarrow$ (a) This is clear, since the given chain of subgroups has stronger properties—namely that each subgroup is normal in the whole group—than what is required in the definition of solvability. $\qquad\square$

We will use the characterization of solvable groups given by Theorem 14.16 part (c) to prove the following useful facts:

Theorem 14.18. *Let G be a group.*

(a) *If G is solvable and $H \leq G$, then H is solvable.*

(b) *If G is solvable and $N \lhd G$, then G/N is solvable.*

(c) *If $N \lhd G$ and both N and G/N are solvable, then so is G.*

Proof. (a) If $H \leq G$, then $H' \leq G'$ and $H^{(n)} \leq G^{(n)}$. Now if G is solvable, then for some positive integer n we have $G^{(n)} = \{e\}$. Hence, $H^{(n)} \leq G^{(n)} = \{e\}$ proving that H is solvable.

(b) Let $\pi : G \to G/N$ be the canonical homomorphism. Then, by repeated application of Lemma 14.9, we have $(G/N)^{(n)} = \pi(G^{(n)})$. Now if G is solvable, then for some positive integer n, $G^{(n)} = \{e\}$. This results in $(G/N)^{(n)}$ being the identity, which in turn implies that G/N is solvable.

(c) Assume N and G/N are solvable and, again, let $\pi : G \to G/N$ be the canonical homomorphism. We know that for some positive integer k and ℓ, we have $N^{(k)}$ and $(G/N)^{(\ell)}$ are both trivial subgroups. We then get that $\pi(G^{(\ell)}) = (G/N)^{(\ell)} = \{e\}$ and hence $G^{(\ell)} \le \ker \pi = N$. But then $G^{(\ell+k)} \le N^{(k)} = \{e\}$ proving that G is solvable. $\qquad\square$

Remark 14.19. Earlier, we mentioned some famous (and difficult) theorems that assert non-simplicity of many groups. As a corollary of these, we get solvability results as well. For example, a theorem of Burnside asserts that every group of order $p^a q^b$ where p and q are primes is solvable. Hence, the order of a non-solvable group has to be divisible by at least three distinct primes. On the other hand, every group of order pqr where p, q, and r are distinct primes is also solvable. The odd order theorem of Feit and Thompson [**FT63**] asserts that every group of odd order is solvable. Based on these, the smallest order of a non-solvable group is $2^2 \times 3 \times 5 = 60$. Indeed, there is a non-solvable group of order 60, namely A_5.

Simplicity of Alternating Groups. Recall that a group is simple if it has no non-trivial normal subgroups. A non-abelian simple group cannot be solvable since we cannot find a chain of normal subgroups with the desired properties.

Proposition 10.37 asserted that the alternating groups A_n, for $n \ge 5$, are an infinite family of non-abelian simple groups. At that point, we gave the proof that A_5 was simple, but relegated the proof for $n > 5$ for later. Here we complete the gap. We warn the reader that to follow the proof, you need to be comfortable with the Direct Diamond Theorem 11.43, and the language and basic properties of group actions (see Chapter 4).

Theorem 14.20. *Let n be an integer greater than 4, then A_n, the alternating group of degree n, is a non-abelian simple group.*

Proof. We induct on n. The case $n = 5$ was proved in Proposition 10.37, and so let $n \ge 6$. Let $G = A_n$, and assume that $N \lhd G$.

For $\alpha \in [n] = \{1, \dots, n\}$, we let G_α stand for the elements of A_n that fix α. In other words, in the natural action of A_n on $[n]$, $G_\alpha = \mathrm{Stab}_G(\alpha)$. Now, for $\alpha = n$, the elements of G_α are exactly those of A_{n-1}, and clearly $G_\alpha \cong G_\beta$, for all $\alpha, \beta \in [n]$. Hence, $G_\alpha \cong A_{n-1}$, and by the inductive hypothesis, G_α is a non-abelian simple group. In addition, note that the action of A_n on $[n]$ is transitive (i.e., for every pair $\alpha, \beta \in [n]$, there exists $g \in A_n$ with $g \cdot \alpha = \beta$).

Since $N \lhd G$, we have, by the Direct Diamond Theorem 11.43, that $N \cap G_\alpha \lhd G_\alpha$. (See Figure 14.2.) But G_α is simple, and hence, for each $\alpha \in [n]$, $N \cap G_\alpha$ is either $\{e\}$ or G_α. Hence, either, for some $\alpha \in [n]$, we have $N \cap G_\alpha = G_\alpha$, or for every $\alpha \in [n]$, we have $N \cap G_\alpha = \{e\}$.

CASE 1: For some $\alpha \in [n]$, we have $N \cap G_\alpha = G_\alpha$.

PROOF OF THE THEOREM IN CASE 1: We will show that in this case $N = G$ is trivial. Let β be an arbitrary element of $[n]$, and choose $g \in G$ with $g \cdot \alpha = \beta$—such a g exists since the action of G on $[n]$ is transitive. Note that acting on both sides by g^{-1} gives $g^{-1} \cdot \beta = \alpha$.

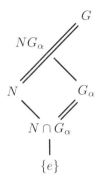

Figure 14.2. Since $N \lhd G$, then $N \cap G_\alpha \lhd G_\alpha$.

CLAIM: $gG_\alpha g^{-1} = G_\beta$.

PROOF OF CLAIM: Let $y \in gG_\alpha g^{-1}$. We want to show that $y \in G_\beta$. Now $y = gxg^{-1}$ where $x \in G_\alpha$, and

$$y \cdot \beta = (gxg^{-1}) \cdot \beta = (gx) \cdot (g^{-1} \cdot \beta) = gx \cdot \alpha = g \cdot \alpha = \beta.$$

Hence $y \in G_\beta$ as desired. This means that $gG_\alpha g^{-1} \subseteq G_\beta$. But $\left|gG_\alpha g^{-1}\right| = |G_\alpha| = |G_\beta|$, and so $gG_\alpha g^{-1} = G_\beta$.

CLAIM: $G_\beta \leq N$.

PROOF OF CLAIM: The condition $N \cap G_\alpha = G_\alpha$ means that $G_\alpha \subseteq N$, and it follows that $gG_\alpha g^{-1} \leq gNg^{-1}$. But $N \lhd G$, and so

$$G_\beta = gG_\alpha g^{-1} \leq gNg^{-1} = N.$$

CLAIM: $N = G$.

PROOF OF CLAIM: Let x be an arbitrary element of $G = A_n$. Elements of G are the even permutations, and hence x is a product of an even number of permutations. This means that $x = g_1 \cdots g_k$, where each g_i is a product of two transpositions. Then each $g_i \in G$, and since $n \geq 5$ (in fact $n \geq 6$), each g_i fixes some element of $[n]$ (a product of two transpositions only moves four elements of $[n]$). Hence each g_i is in some G_β for $\beta \in [n]$. Since, for every $\beta \in [n]$, $G_\beta \leq N$, we have $g_i \in N$ for all i. Hence, $x \in N$, and we have proved that $G \subseteq N$.

CASE 2: For every $\alpha \in [n]$, we have $N \cap G_\alpha = \{e\}$.

PROOF OF CASE 2: In this case—where a non-identity element of N does not fix any of the elements of $[n]$—we will prove that $N = \{e\}$ is trivial.

Let $a \in N$ with $a \neq e$. Then by assumption a moves every element of $[n]$. Without loss of generality, say that $a \cdot 1 = 2$, and $a \cdot 2 = \alpha \in [n]$ where α could possibly be 1. (Can write this as $a : 1 \mapsto 2 \mapsto \alpha$.) Choose $\beta \notin \{1, 2, \alpha\}$. Regardless of the value of $a \cdot \beta$, the set $\{1, 2, \alpha, \beta, a \cdot \beta\}$ has no more than five elements, and $n \geq 6$. Hence, pick $\gamma \notin \{1, 2, \alpha, \beta, a \cdot \beta\}$.

The 3-cycle $y = (1\ 2\ \beta)$ is an even permutation and, hence, is an element of G. So is $y^{-1} = (1\ \beta\ 2)$. Let $c = yay^{-1}a^{-1} \in G$, and consider the action of c on γ. We have

$$c \cdot \gamma = (ya) \cdot (y^{-1} \cdot (a^{-1} \cdot \gamma)).$$

We know that y^{-1} is a 3-cycle and only moves 1, 2, and β. Now $a^{-1} \cdot \gamma$ cannot be any of these, since $a^{-1} \cdot \gamma = \delta$ means $\gamma = a \cdot \delta$, and our choice of γ guaranteed that it is not any of $a \cdot 1 = 2$, $a \cdot 2 = \alpha$, or $a \cdot \beta$. Hence y^{-1} fixes $a^{-1} \cdot \gamma$. Also note that y fixes γ, since γ is not any of 1, 2, or β. Thus, we have

$$\begin{aligned}
c \cdot \gamma &= (ya) \cdot (y^{-1} \cdot (a^{-1} \cdot \gamma)) \\
&= (ya) \cdot (a^{-1} \cdot \gamma) \\
&= (yaa^{-1}) \cdot \gamma \\
&= y \cdot \gamma \\
&= \gamma.
\end{aligned}$$

We conclude that c fixes γ. We further note that $a \in N$, and since N is normal in G, we have $yay^{-1} \in N$ which means that $c = (yay^{-1})a^{-1} \in N$.

By assumption (for this case) the only element of N that fixes any element of $[n]$ is the identity. Hence $c = e$, which means that $yay^{-1} = a$. However, consider the action of yay^{-1} and a on 2. We have $a \cdot 2 = \alpha$, $yay^{-1} \cdot 2 = ya \cdot 1 = y \cdot 2 = \beta$, and $\beta \neq \alpha$. The contradiction proves that there are no $a \in N$ with $a \neq e$, and hence $N = \{e\}$.

In either case, we have proved that N is trivial and, hence, G is simple. $\square$

Corollary 14.21. *For $n \geq 5$, S_n is not solvable.*

Proof. If S_n were solvable, then, by Theorem 14.18, $A_n \leq S_n$ would be solvable. But, for $n \geq 5$, A_n is non-abelian simple and, hence, not solvable. $\square$

Problems

14.1.1. Give an example of a finite group that is neither simple nor solvable.

14.1.2. Let p be a prime, and let n be a positive integer. Prove that a group of order p^n is solvable.

14.1.3. Let H and K be two solvable groups. Prove that $H \times K$ is solvable.

14.1.4. Assume that a finite group G is the internal direct product (Definition 13.9) of P_1, ..., P_k, where each of the P_i is a Sylow p-subgroup of G. Prove that G is solvable.

14.1.5. Let $p < q < r$ be prime numbers. Prove that every group of order pqr is solvable.

14.1.6. Let $G = S_4$. Find $G^{(n)}$ for all positive integers n.

14.1.7. Let $G = S_5$. Find $G^{(n)}$ for all positive integers n.

14.1.8. If G is a non-abelian simple group, then what can you say about $G^{(n)}$?

14.1.9. Let p be a prime, and let G be a non-abelian group of order p^3. Show that $G' = \mathbf{Z}(G)$.

14.1.10. Let G be a group, and let H be a subgroup of G with $G' \leq H$. Prove that $H \triangleleft G$.

14.1.11. Let G be a group, and let N be a normal subgroup of G. Assume that $N \cap G' = \{e\}$. Prove that $N \leq \mathbf{Z}(G)$.

14.1.12. Let G be a finite group, and let H be the only subgroup of G that has order 25. Is H necessarily characteristic in G?

14.1.13. Let G be a group, and let $\mathbf{Z}(G)$ be the center of G. Show that $\mathbf{Z}(G)$ char G.

14.1.14. If N is a normal subgroup of a group G, then is $\mathbf{Z}(N)$ necessarily a normal subgroup of G?

14.1.15. Assume that a Sylow subgroup of a group is normal. Prove that this subgroup is actually characteristic. Conclude that if G is a finite group, N a normal subgroup of G, and $P \in \mathrm{Syl}_p(N)$ with $P \triangleleft N$, then $P \triangleleft G$.

14.1.16. Let G be a group, and let H and K be subgroups of G. Prove that if H char K char G, then H char G.

14.1.17. Give an example of a group with a normal subgroup that is not characteristic.

14.1.18. Let G be a simple group of odd order. The Feit–Thompson theorem states that G is a cyclic group of prime order. Use the Feit–Thompson theorem to prove that every group of odd order is solvable.

14.1.19. **Minimal normal subgroups of solvable groups.** Recall—see Problem 13.2.10—that a group E is an elementary abelian p-group, where p is a prime, if E is abelian and $x^p = e$ for all $x \in E$. Let G be a finite solvable group, and let N be a minimal normal subgroup of G. Prove that N is an elementary abelian p-group for some prime p.

  You may find the following steps helpful:
  STEP 1: Show that N' is a proper subgroup of N and normal in G.
  STEP 2: Show that N is abelian.
  STEP 3: Let p be a prime divisor of $|N|$, and show that $U = \{x \in N \mid x^p = e\}$ is a characteristic subgroup of N.
  STEP 4: Show that $U = N$, and complete the proof.

14.1.20. Let G be a group (not necessarily solvable), and let N be a minimal normal subgroup of G. Assume that N is a finite solvable group. Strengthen the result of the previous problem by showing that N is an elementary abelian p-group for some prime p.

14.1.21. **On derived length.** Let G be a solvable group. Then the *derived length* of G, denoted by $\mathrm{dl}(G)$, is the minimum non-negative integer n such that $G^{(n)} = \{e\}$.

  Let G be a solvable group and $N \triangleleft G$. Prove that $\mathrm{dl}(G) \leq \mathrm{dl}(G/N) + \mathrm{dl}(N)$.

14.2. Nilpotent Groups

Every abelian group is solvable but not vice versa. Nilpotent groups are a class of groups between abelian and solvable, and so every abelian group is nilpotent and every nilpotent group is solvable. Just as for solvable groups, there are a variety ways of characterizing nilpotent groups. We will choose one of these as the definition, proceed to show the relation of nilpotent groups to abelian groups, solvable groups, and p-groups, and then prove various equivalent characterizations.

Definition 14.22. Let G be a group. A chain of subgroups

$$\{e\} = N_0 \leq N_1 \leq N_2 \leq \cdots \leq N_k = G$$

is called a *central series* of G, if, for $0 \leq i \leq k - 1$,

(a) the subgroup N_i is normal in G, and

(b) the quotient group N_{i+1}/N_i is in the center of G/N_i.

If a group G has a central series, then it is called a *nilpotent* group.

Proposition 14.23. *Every abelian group is nilpotent, every nilpotent group is solvable, and every finite p-group is nilpotent.*

Proof. Let G be any abelian group, then let $N_0 = \{e\}$ and $N_1 = G$. Then $\{e\} = N_0 \leq N_1 = G$ is a central series, and so G is nilpotent.

In Definition 14.22 of a central series, each quotient N_{i+1}/N_i is in the center of G/N_i and hence is abelian. It follows from Theorem 14.16(d) that a nilpotent group is solvable.

Finally, let p be a prime, and let P be any finite p-group. Then we construct a central series for P as follows.

Let $N_0 = \{e\}$, and let $N_1 = \mathbf{Z}(P)$ be the center of P. If $|P| > 1$, then we know by Corollary 6.19 that N_1 is bigger than N_0. If N_1 happens to be all of P, then we are done and $N_0 \leq N_1$ is a central series.

Otherwise, consider the group P/N_1, and let $\pi: P \to P/N_1$ be the canonical homomorphism. Find $\mathbf{Z}(P/N_1)$ and let $N_2 = \pi^{-1}\left(\mathbf{Z}(P/N_1)\right)$. Again, by Corollary 6.19 we know that $|\mathbf{Z}(P/N_1)| > 1$, and so $N_2 > N_1$. In fact, by (the proof of) Proposition 11.24, $N_2/N_1 = \mathbf{Z}(P/N_1)$. If $N_2 = P$, then $N_0 < N_1 < N_2$ is the sought-after central series.

$$
\begin{array}{ccc}
P & \xrightarrow{\;\pi\;} & P/N_1 \\
\big| & & \big| \\
N_2 & \longleftrightarrow & \mathbf{Z}(P/N_1) \\
\big| & & \big| \\
N_1 = \mathbf{Z}(P) & \longrightarrow & \{e\} \\
\big| & & \\
N_0 = \{e\} & &
\end{array}
$$

If $N_2 \neq P$, we repeat the same process and find a subgroup N_3 with $N_2 < N_3 \lhd P$ and $N_3/N_2 = \mathbf{Z}(P/N_2)$. Continuing in this manner, since P is a finite group and each N_i is bigger than the previous N_{i-1}, the process will eventually end with some $N_k = P$.

This gives a central series for P and proves that every finite p-group is nilpotent. $\square$

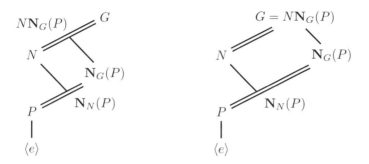

Figure 14.3. If $P \in \mathrm{Syl}_p(N)$ and $N \lhd G$, a priori we have the diagram on the left. The Frattini argument implies that, in fact, we have the diagram on the right.

A useful result that comes in handy when working with nilpotent groups is the *Frattini argument* which we shall prove next. (You were actually asked to give a proof in Problem 12.3.21.)

Theorem 14.24 (Frattini argument). *Let G be a finite group, let N be a normal subgroup of G, let p be a prime, and let $P \in \mathrm{Syl}_p(N)$. Then*

$$G = N\mathbf{N}_G(P).$$

Proof. The given information, and with no further assumptions, can be depicted in the partial lattice diagram on the left of Figure 14.3. We want to prove that, in fact, we have the partial lattice diagram on the right of Figure 14.3.

To prove that $G = N\mathbf{N}_G(P)$, it is sufficient to prove that $G \subseteq N\mathbf{N}_G(P)$. Let $g \in G$, and denote the subgroup gPg^{-1} by Q. Since $P \leq N$, we have $Q = gPg^{-1} \leq gNg^{-1} = N$. But $|Q| = |P|$ and hence we conclude that Q is a Sylow p-subgroup of N. Since P and Q are both Sylow p-subgroups of N, by the Sylow C Theorem 12.9(c), they are conjugate in N. Hence, there exists an element $n \in N$ such that $P = nQn^{-1} = ngP(ng)^{-1}$. But this means that $ng \in \mathbf{N}_G(P)$, and so $g \in n^{-1}\mathbf{N}_G(P) \subseteq N\mathbf{N}_G(P)$, and the proof is complete. $\square$

The following lemma will also be useful.

Lemma 14.25. *Let G be a finite group, let H be a subgroup of G, and let p be a prime. Assume that a Sylow p-subgroup of G is normal in G. Then a Sylow p-subgroup of H is normal in H, and if $H \lhd G$, then a Sylow p-subgroup of G/H is normal in G/H.*

Proof. Let $P \in \mathrm{Syl}_p(G)$ with $P \lhd G$. (In fact, P is the unique Sylow p-subgroup of G.)

By the Direct Diamond Theorem 11.43 since $P \lhd G$, we
have $H \cap P \lhd H$, and $|H : H \cap P| = |HP : P|$ divides
$|G : P|$ and is relatively prime to p. Now $H \cap P$ is a sub-
group of P and so $|H \cap P|$ is a power of p. We conclude
that $H \cap P$ is a normal Sylow p-subgroup of H (and hence
the only Sylow p-subgroup of H).

Likewise, when $H \lhd G$, again by the Direct Diamond The-
orem 11.43, $|HP : H| = |P : H \cap P|$ divides $|P|$ and is a
prime power, while $|G : HP|$ divides $|G : P|$ and is rela-
tively prime to p. Hence, HP/H is a Sylow p-subgroup of
G/H. Moreover, since both H and P are normal in G, so is
HP, and therefore, HP/H is a normal Sylow p-subgroup
of G/H.

□

We are now ready to give a number of characterizations of finite nilpotent
groups. These characterizations make it clear that the finite nilpotent groups—
while being a class of groups large enough to include all finite abelian groups and all
finite p-groups—are very special and have a number of useful properties. Recall—
see Definition 5.21—that a subgroup H of a group G is called a *maximal subgroup*
if H is a proper subgroup and there exists no subgroup K with $H < K < G$.

Theorem 14.26. *Let G be a finite group. Then the following are equivalent:*

(a) *G is nilpotent.*

(b) *Normalizers of proper subgroups grow. In other words, if H is a proper subgroup of G, then $\mathbf{N}_G(H) > H$.*

(c) *All maximal subgroups of G are normal in G.*

(d) *Every Sylow subgroup of G is normal in G.*

(e) *G is the (internal) direct product of $P_1, \ldots, P_k$, where each P_i, for $1 \leq i \leq k$,
is a Sylow subgroup of G.*

(f) *If N is a proper normal subgroup of G, then G/N has a non-trivial center.*

Proof. (a) $\Rightarrow$ (b). Let $H < G$, and let $\{e\} = N_0 < N_1 < \cdots < N_k = G$ be a central
series for G. We know that $N_0 \subseteq H$ and $N_k \subsetneq H$. So choose i so that $N_i \subseteq H$
and yet $N_{i+1} \subsetneq H$. We have that N_{i+1}/N_i is in the center of G/N_i, and hence
its elements commute with elements of H/N_i. Thus if x^{-1} and h^{-1} are arbitrary
elements of N_{i+1} and H, respectively, we have $x^{-1}N_i h^{-1}N_i = h^{-1}N_i x^{-1}N_i$. It
follows that $xhx^{-1}h^{-1}N_i = N_i$ and so $xhx^{-1}h^{-1} \in N_i \leq H$. Thus, for each
$x \in N_{i+1}$, $xhx^{-1} \in H$ for every $h \in H$. We have proved that $xHx^{-1} \leq H$ for every
$x \in N_{i+1}$. This means that $xHx^{-1} = H$ for every $x \in N_{i+1}$. (Why? See the proof
of Lemma 10.17.) Thus $N_{i+1} \leq \mathbf{N}_G(H)$. Since N_{i+1} has elements that are not in
H, we conclude that $\mathbf{N}_G(H)$ is a subgroup strictly bigger than H.

(b) $\Rightarrow$ (c). This is straightforward. If M is a maximal subgroup, then, by assumption, $M < \mathbf{N}_G(M)$ which forces $\mathbf{N}_G(M)$ to be G. This means that $M \lhd G$.

(c) $\Rightarrow$ (d). Let P be a Sylow p-subgroup of G. We assume that P is not normal in G, and arrive at a contradiction. Since P is not normal, $\mathbf{N}_G(P)$ is a proper subgroup of G that contains P. Let M be a maximal subgroup of G that contains $\mathbf{N}_G(P)$ (it is possible that $M = \mathbf{N}_G(P)$). By assumption M is normal in G and P is a Sylow p-subgroup of M (as well as of G), and we can apply the Frattini argument (Theorem 14.24). We conclude that $G = M\mathbf{N}_G(P)$. But $\mathbf{N}_G(P) \leq M$ and so $G = M\mathbf{N}_G(P) = M$, and this is a contradiction since maximal subgroups are proper by definition.

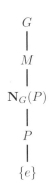

(d) $\Rightarrow$ (e). Assume that every Sylow subgroup of G is normal in G, and, so, for every prime p dividing $|G|$, there is a unique Sylow p-subgroup. Let $P_1, \ldots, P_k$ be the (non-trivial) Sylow subgroups of G. To show that G is the internal direct product of $P_1, \ldots, P_k$ (see Definition 13.9), we have to show three things: each P_i is normal in G, $G = P_1 P_2 \cdots P_k$, and, for $1 \leq i \leq k$, $P_i \cap P_1 \cdots P_{i-1} P_{i+1} \cdots P_k = \{e_G\}$. The first of these is true by assumption and the other two follow immediately. In fact, $P_1 \cap P_2 = \{e\}$, and since $P_1 \lhd G$, $P_1 P_2$ is a subgroup of G of order $|P_1|\,|P_2|$. Now, $P_3 \lhd G$ and so $P_1 P_2 P_3$ is a subgroup of G of order $|P_1|\,|P_2|\,|P_3|$. Repeating the same argument, we get that $|G| = |P_1|\,|P_2|\,\cdots\,|P_k| = |P_1 P_2 \cdots P_k|$, and so $G = P_1 P_2 \cdots P_k$. Likewise, $H_i = P_1 \cdots P_{i-1} P_{i+1} \cdots P_k$ and P_i are groups with relatively prime orders, and hence their intersection is $\{e\}$.

(e) $\Rightarrow$ (d). This is straightforward. Assume that G is the internal direct product of $P_1, \ldots, P_k$, with each P_i a Sylow subgroup of G. Let p be a prime and $P \in \mathrm{Syl}_p(G)$. If p does not divide the order of G, then $P = \{e\} \lhd G$. Otherwise, p divides $|G| = |P_1|\,|P_2|\cdots|P_k|$, and so p has to divide the order of P_i for some $1 \leq i \leq k$. This means that P_i is a Sylow p-subgroup of G. But $P_i \lhd G$ and so there is a unique Sylow p-subgroup of G. Hence, $P = P_i \lhd G$.

(d) $\Rightarrow$ (f). Assume that the Sylow subgroups of G are normal. We will show that G has a non-trivial center. Even though this is a special case of condition (f), it is all we have to show. If N is a normal subgroup of G, then Lemma 14.25 gives that all Sylow subgroups of G/N are also normal and so—applying what we are about to show to G/N—we conclude that $\mathbf{Z}(G/N) > \{e\}$.

Let p be a prime divisor of $|G|$, let $P \in \mathrm{Syl}_p(G)$, and let $Z = \mathbf{Z}(P)$ be the center of P. Since P is a p-group, by Corollary 6.19 $|Z| > 1$. We claim that $Z \leq \mathbf{Z}(G)$, which would complete the proof. To prove the claim, consider $C = \mathbf{C}_G(Z)$, the centralizer of Z in G. Certainly, $P \leq C$ and, hence, $|P|$ divides $|C|$. Now let Q be Sylow q-subgroup of G, where q is a prime distinct from p. Then, by assumption, both P and Q are normal subgroups of G, and $P \cap Q = \{e\}$ since $\gcd(|P|, |Q|) = 1$. We conclude (by Theorem 10.23) that every element of Q commutes with every element of P. This means that $Q \leq C$. Thus, $|C|$ is divisible by the size of *every* Sylow subgroup of G. Hence, $|C| \geq |G|$. We conclude that $C = G$, and so $\{e\} < Z \leq \mathbf{Z}(G)$.

(f) $\Rightarrow$ (a). If $G = \{e\}$, then G is certainly nilpotent. For $G \neq \{e\}$, define a series of subgroups as follows: $N_0 = \{e\}$, and for $i \geq 1$, $N_i = \pi^{-1}(\mathbf{Z}(G/N_{i-1}))$ where $\pi : G \to G/N_{i-1}$ is the canonical homomorphism. In other words, $N_1 = \mathbf{Z}(G)$, and, for $i > 1$, N_i is defined to be the subgroup of G such that $N_i/N_{i-1} = \mathbf{Z}(G/N_{i-1})$. Now, since $\mathbf{Z}(G/N_{i-1}) \lhd G/N_{i-1}$, then by the homomorphism theorems its inverse image is normal in G. Thus $N_i \lhd G$. As long as $N_{i-1} < G$, by hypothesis we have that $\mathbf{Z}(G/N_{i-1}) > \{e\}$ and so $N_i > N_{i-1}$. Since $|G| < \infty$, we eventually have $N_k = G$, for some k. Thus $\{e\} = N_0 \leq N_1 \leq \cdots \leq N_k = G$ is a central series for G and G is nilpotent. $\square$

Remark 14.27. We had showed that p-groups are nilpotent (Proposition 14.23), and we had proved that all p-groups have a non-trivial center (Corollary 6.19). In Theorem 14.26(f), we generalized the latter by showing that all nilpotent groups have a non-trivial center. In fact, the proof of Theorem 14.26(f) showed that in nilpotent groups, the center of each Sylow subgroup is contained in the center of the whole group. In a similar vain, in Problem 14.2.9, you are asked to show that every non-trivial normal subgroup of a nilpotent group intersects the center non-trivially. This strengthens Problem *12.1.7*, where you were asked to prove the same for p-groups.

Corollary 14.28. *Subgroups and quotient groups of nilpotent groups are nilpotent.*

Proof. This is immediate from characterization (d) of nilpotent groups in Theorem 14.26 and from Lemma 14.25. $\square$

Problems

14.2.1. Give an example of a non-abelian nilpotent group that is not a p-group. Give an example of a solvable group that is not nilpotent.

14.2.2. Give an example of a group G with a normal subgroup N such that both N and G/N are nilpotent but G is not.

14.2.3. Show that the implication (a) $\Rightarrow$ (b) of Theorem 14.26 works even if the group G is infinite.

14.2.4. Let p and q be primes, and let a and b be non-negative integers.
   (a) Let G be a group of order $p^a q^b$, let $P \in \mathrm{Syl}_p(G)$, and let $Q \in \mathrm{Syl}_q(G)$. Show that $G = PQ$.
   (b) Give an example of a group of order $p^a q^b$ such that the group is not nilpotent.

14.2.5. Show that the direct product of a finite number of finite nilpotent groups is nilpotent.

14.2.6. Let $G = A_5$, the alternating group of degree 5.
   (a) Can you find examples of $P \in \mathrm{Syl}_5(G)$, $Q \in \mathrm{Syl}_3(G)$, and $R \in \mathrm{Syl}_2(G)$ such that $G = PQR$?
   (b) Can you find examples of $P \in \mathrm{Syl}_5(G)$, $Q \in \mathrm{Syl}_3(G)$, and $R \in \mathrm{Syl}_2(G)$ such that $G \neq PQR$?

14.2.7. **Subnormal subgroups.** Let G be a group, and let H be a subgroup of G. Then H is called a *subnormal subgroup* of G if there exists subgroups

$H_1, H_2, \ldots, H_k$ with

$$H = H_0 \lhd H_1 \lhd H_2 \lhd \cdots \lhd H_k = G.$$

Prove that every subgroup of a finite group is subnormal if and only if the group is nilpotent.

14.2.8. Let G be a finite nilpotent group, and let $M < G$ be a maximal subgroup of G. Show that $G/M \cong \mathbb{Z}/p\mathbb{Z}$ for some prime p.

14.2.9. Let G be a finite nilpotent group, and let $\{e\} < N \lhd G$. Prove that $N \cap \mathbf{Z}(G) > \{e\}$.

The Frattini Subgroup

Definition 14.29 (The Frattini subgroup). Let G be a group. The intersection of all maximal subgroups of G is called the *Frattini subgroup* of G, and is denoted by $\Phi(G)$.

Definition 14.30 (Non-generators). Let G be a group. An element $g \in G$ is a *non-generator* if whenever $\langle X \cup \{g\} \rangle = G$ for some subset $X \subseteq G$, then $\langle X \rangle = G$. (In other words, if a set fails to generate G, then adjoining g to it will not help.)

In Theorem 14.18(c), we proved that if $N \lhd G$, and N and G/N are both solvable, then G is solvable. Hence, we can reduce the task of checking for solvability of a given group to smaller groups. If we replace solvable with nilpotent, then this statement becomes false (Problem 14.2.2). However, the Frattini subgroup $\Phi(G)$ provides an alternate way of checking for nilpotency. In the Problems, you are asked to prove—possibly by looking at the hints—a number of properties of the Frattini subgroup.

Proposition 14.31. *Let G be a finite group.*

(a) (Problems 14.2.12 and 14.2.15.) *The Frattini subgroup $\Phi(G)$ is a normal nilpotent subgroup of G.*

(b) (Problem 14.2.16.) *The group G is nilpotent if and only if $G/\Phi(G)$ is nilpotent. In fact, G is nilpotent if and only if $G/\Phi(G)$ is abelian.*

(c) (Problem 14.2.19.) *A normal subgroup N of G is nilpotent if and only if N', the commutator subgroup of N, is contained in $\Phi(G)$.*

(d) (Problem 14.2.14.) *The Frattini subgroup is exactly the set of non-generators of the group.*

14.2.10. Find the Frattini subgroup of each of D_8, S_3, $\mathbb{Z}/8\mathbb{Z}$, and $\mathbb{Z}/4\mathbb{Z} \times \mathbb{Z}/2\mathbb{Z}$.

14.2.11. Find the Frattini subgroup of each of A_4, S_4, and A_5.

14.2.12. Prove that the Frattini subgroup of a group G is characteristic in G.

14.2.13. Let G be a finite group. Assume H is a subgroup of G with $H\Phi(G) = G$. Show that $H = G$.

14.2.14. $\Phi(G)$ **and non-generators.** Let G be a finite group. Show that the Frattini subgroup of G equals the set of non-generators of G.

14.2.15. $\Phi(G)$ **is nilpotent.** Let G be a finite group. Show that the Frattini subgroup of G is nilpotent.

14.2.16. Let G be a finite group. Prove that the following are equivalent:
  (a) G is nilpotent.
  (b) $G' \leq \Phi(G)$ (recall that G' is the commutator subgroup of G).
  (c) $G/\Phi(G)$ is abelian.
  (d) $G/\Phi(G)$ is nilpotent.

14.2.17. Let G be a finite group, and let K be a normal subgroup of G that contains $\Phi(G)$. Strengthen the implication Problem 14.2.16(d) $\Rightarrow$ 14.2.16(a)—as well as Problem 14.2.15—by showing that if $K/\Phi(G)$ is nilpotent, then K is nilpotent.

14.2.18. Let G be a finite group, and $N \lhd G$. Then show that $\Phi(N) \leq \Phi(G)$.

14.2.19. Let G be a finite group, and let $N \lhd G$. Strengthen one part of Problem 14.2.16 by proving that N is nilpotent if and only if $N' \leq \Phi(G)$.

14.2.20. Let G be a finite group, and let K and N be normal subgroups of G with $N \leq K \cap \Phi(G)$. Strengthen Problem 14.2.17 by showing that if K/N nilpotent, then K is nilpotent.

14.2.21. **Frattini subgroup of a finite p-group.** Let G be a finite p-group, and, as usual, let G' denote the derived subgroup of G. Furthermore, define

$$G^p = \langle x^p \mid x \in G \rangle.$$

Prove that $\Phi(G) = G^p G'$.

   You may find the following steps useful:

   STEP 1: Let M be a maximal subgroup of G. Show $M \lhd G$ and $|G : M| = p$.

   STEP 2: Let $x \in G$. By considering the element $xM \in G/M$, show that $x^p \in M$. Conclude that $G^p \leq M$ for all maximal subgroups M.

   STEP 3: First show $G' \leq M$ for all maximal subgroups M. Then conclude that $G^p G' \leq \Phi(G)$.

   STEP 4: Show that $G^p G' \lhd G$ and that $G/G^p G'$ is an elementary abelian p-group (see Problem 13.2.10). Problem 14.1.10 and Corollary 14.11 may be relevant.

   STEP 5: Use Problem 13.2.10 to write $G/G^p G'$ as $\mathbb{Z}/p\mathbb{Z} \times \mathbb{Z}/p\mathbb{Z} \times \cdots \times \mathbb{Z}/p\mathbb{Z}$. Construct a collection of maximal subgroups of G whose intersection is $G^p G'$. Conclude that $\Phi(G) \leq G^p G'$, and finish the proof.

14.3. The Jordan–Hölder Theorem

In the study of any partially ordered set, the chains of elements—see Definition 9.17—play an important role. In the previous two sections, we defined nilpotent and solvable groups using certain chains of subgroups in the poset of subgroups of the group ordered by inclusion. In this section, we continue the study of chains of subgroups of a group. To make the section self-contained, we repeat some definitions, starting with a few general definitions for chains in posets.

Definition 14.32 (Chains and graded posets). Let P be a poset. A totally ordered subset of P is called a *chain*. In other words, if $\{a_1, \ldots, a_k, \ldots\}$ is a subset of P and

$$a_1 < a_2 < \cdots < a_k < \cdots,$$

then $\{a_1, \ldots, a_k, \ldots\}$ is a chain in P. A chain with k elements is called a chain of *size* k and of *length* $k - 1$. A chain is *maximal* if it cannot be enlarged while staying a chain. A poset where all the maximal chains have the same length is called *graded*.

Example 14.33. Let $G = A_4$, and let P be the poset of subgroups of G ordered by inclusion. If $H = \langle (1\ 2\ 3) \rangle$, $T = \langle (1\ 2)(3\ 4) \rangle$ and $K = \langle (1\ 2)(3\ 4), (1\ 3)(2\ 4) \rangle$. Then

$$\{e\} < H < A_4,$$
$$\{e\} < T < K < A_4$$

are both maximal chains (see Figure 9.11) and P is not a graded poset.

We recall a couple of definitions about subgroups:

Definition 14.34. Let G be a group.

A proper subgroup M of G is a *maximal normal subgroup* of G if $M \triangleleft G$ and there is no subgroup L with $M < L < G$ and $L \triangleleft G$.

Likewise, a non-trivial subgroup N of G is a *minimal normal subgroup* of G if $N \triangleleft G$ and there is no subgroup K with $\{e\} < K < N$ and $K \triangleleft G$.

A subgroup H of G is a *subnormal* subgroup of G, sometimes denoted by $H \triangleleft \triangleleft G$, if there exist subgroups $G_0, G_1, \ldots, G_k$ such that

$$H = G_0 \triangleleft G_1 \triangleleft \cdots \triangleleft G_{k-1} \triangleleft G_k = G.$$

Definition 14.35 (Subnormal series, composition series, normal series, and chief series). Let G be a group. A chain of subgroups

$$e = G_0 \triangleleft G_1 \triangleleft \cdots \triangleleft G_k = G$$

is called a *subnormal series of subgroups* in the group G. For $0 \leq i \leq k$, the subgroups G_i are called the *terms* of the subnormal series, while, for $1 \leq i \leq k$, the quotient groups G_i/G_{i-1} are called the *factors* of the subnormal series. If, for $0 \leq i \leq k-1$, G_i is a maximal normal subgroup of G_{i+1}, then the subnormal series is called a *composition series* of G. The factors of a composition series are called the *composition factors* of the series.

A chain of subgroups

$$e = H_0 \leq H_1 \leq \cdots \leq H_m = G$$

is called a *normal series* if, for $0 \leq i \leq m$, $H_i \triangleleft G$. If, for $0 \leq i \leq m-1$, H_i/H_{i-1} is a minimal normal subgroup of G/H_{i-1}, then the normal series is called a *chief series* of G.[1]

---

[1] A word of caution: Many authors use "normal series" for what we have called a "subnormal series". These authors usually do not have a particular name for what we have called a normal series. On the other hand, the terms composition series and chief series are standard.

Remark 14.36. Note that for a subnormal series, we only require that each term be normal in the next, while in a normal series, we require that each term be normal in the whole group. Hence, every term of a subnormal series is a subnormal subgroup of G, and every normal series is automatically a subnormal series. In a composition series, the condition that each term is a maximal normal subgroup of the next is equivalent—via the homomorphism theorems—to the factors being simple groups (Problem 11.5.9). In a chief series, we require that no normal subgroup of the group be properly sandwiched between two terms of the series. Note that in both composition and chief series, by definition, the terms are distinct and there are no repeats. While, by definition, composition and chief series have finite length, in studying infinite groups, one may consider subnormal and normal series of infinite length.

Remark 14.37. Let G be a group. Let P be the poset of subgroups of G ordered by inclusion. Let Q be the subposet of P consisting of all subnormal subgroups of G, and let R be the subposet of Q consisting of all normal subgroups of G.

The terms of any subnormal series are elements of Q, and a subnormal series is a composition series if and only if it is a maximal chain in Q. On the other hand, normal series correspond to chains in R, and a normal series is a chief series if and only if it is a maximal chain in R. (See Problems 14.3.5 and 14.3.12.)

We can now restate the definition of solvability in terms of series. The following follows directly from Definition 14.1 and Theorem 14.16.

Lemma 14.38. *Let G be a finite group. Then the following are equivalent:*

(a) *The group G is solvable.*

(b) *The group G has a subnormal series with abelian factors.*

(c) *The group G has a composition series with factors that are cyclic groups of order p.*

(d) *The group G has a normal series with abelian factors.*

(e) *The group G has a chief series with abelian factors.*

Example 14.39. The alternating group of degree 4, A_4, is a solvable group of order 12. If $T = \langle (1\ 2)(3\ 4) \rangle$ and $K = \langle (1\ 2)(3\ 4), (1\ 3)(2\ 4) \rangle$, then

$$\{e\} < T < K < A_4$$

is the only composition series for A_4, and

$$\{e\} < K < A_4$$

is the only chief series for A_4.

Motivated by the above lemma, we define a subclass of solvable groups.

Definition 14.40 (Supersolvable). A group is called *supersolvable* if it has a chief series with factors that are cyclic groups of a prime order.

You are asked, in Problem 14.3.3 to show that every nilpotent group is supersolvable. Hence, the class of supersolvable groups is one sandwiched between nilpotent and solvable. (See Figure 14.4.)

Figure 14.4. p-groups, abelian, nilpotent, supersolvable, and solvable groups

Remark 14.41. One could ask, How much of the group theoretic properties of a group are reflected in the poset of its subgroups? We mention two results in this direction: A 1941 theorem of Iwasawa [**Iwa41**] states that the poset of subgroups of a finite group ordered by inclusion is graded if and only if the group is supersolvable. Another such result is that a finite group is solvable if and only if it has a maximal chain of subgroups with the same length as one of its chief series. (Kohler [**Koh68**] proved the only-if direction in 1968, and Shareshian and Woodroofe [**SW12**] proved the if direction in 2012.)

A group may or may not have a composition series, and if it does have a composition series, such a series may be unique or not. However, what is remarkable is that, if a group has a composition series, then all composition series will have the same length (recall that the length of a series is the same as the number of factors for the series) and isomorphic composition factors, albeit possibly in a different order. The same result is also true for chief series. Hence, two chief series of a group have the same length and isomorphic factors, but the factors may occur in a different order. These results are called the Jordan–Hölder theorem. We prove the result about composition series and leave the one about chief series to the reader.[2]

Definition 14.42. Let G be a group, and let

$$\{e\} = H_0 \lhd H_1 \lhd \cdots \lhd H_{n-1} \lhd H_n = G, \text{ and}$$
$$\{e\} = K_0 \lhd K_1 \lhd \cdots \lhd K_{m-1} \lhd K_m = G$$

be two subnormal series of finite length for G. We say that the two series are *equivalent* if $n = m$, and, after a possible reordering, the list (with possible repeats) of factors of the two series are the same.

More precisely, for $1 \le i \le n$, let $\bar{H}_i = H_i/H_{i-1}$, and, for $1 \le j \le m$, let $\bar{K}_j = K_j/K_{j-1}$. Then the two subnormal series are equivalent if $n = m$, and there exists a permutation $\sigma \in S_n$ such that, for $1 \le i \le n$, the group $\bar{H}_i$ is isomorphic to the group $\bar{K}_{\sigma(i)}$.

Example 14.43. Let $D_8 = \langle a, b \mid a^4 = b^2 = e, ba = a^{-1}b \rangle$, and let S_3 be the symmetric group of degree 3. Define $G = D_8 \times S_3$, a group of order 48.

$$\{(e,e)\} \lhd \langle (a^2, e) \rangle \lhd \langle (a, e) \rangle \lhd \langle (a, e), (b, e) \rangle \lhd \langle (a, e), (b, e), (e, (1\ 2\ 3)) \rangle \lhd G$$

is a composition series for G, and the composition factors are $\mathbb{Z}/2\mathbb{Z}$, $\mathbb{Z}/2\mathbb{Z}$, $\mathbb{Z}/2\mathbb{Z}$, $\mathbb{Z}/3\mathbb{Z}$, and $\mathbb{Z}/2\mathbb{Z}$.

$$\{(e,e)\} \lhd \langle (e, (1\ 2\ 3)) \rangle \lhd \langle (a^2, (1\ 2\ 3)) \rangle \lhd \langle (a, (1\ 2\ 3)) \rangle \lhd \langle (a, (1\ 2\ 3)), (b, e) \rangle \lhd G$$

---

[2]It is not that hard to create a framework for proving both results at the same time. See Isaacs [**Isa94**].

is another composition series for G. This time the composition factors, in order, are $\mathbb{Z}/3\mathbb{Z}$, $\mathbb{Z}/2\mathbb{Z}$, $\mathbb{Z}/2\mathbb{Z}$, $\mathbb{Z}/2\mathbb{Z}$, and $\mathbb{Z}/2\mathbb{Z}$. The two series are equivalent, and in fact, both of these series are chief series as well.

Theorem 14.44 (Jordan–Hölder). *Let G be a finite group. Then any two composition series for G are equivalent.*

Proof. We use induction on $|G|$. The theorem is certainly true for a group of size 1. Now, for the inductive step, assume that the theorem is true for all groups with fewer elements than G.

Now assume that

$$\{e\} = H_0 \lhd H_1 \lhd \cdots \lhd H_{n-1} \lhd H_n = G, \text{ and}$$
$$\{e\} = K_0 \lhd K_1 \lhd \cdots \lhd K_{m-1} \lhd K_m = G$$

are two composition series for G. We have to show that the two series are equivalent.

First consider the case when $H_{n-1} = K_{m-1}$. Call this group H and note that by induction every two composition series for H are equivalent. Now $\{e\} = H_0 \lhd \cdots \lhd H_{n-1} = H$ and $\{e\} = K_0 \lhd \cdots \lhd K_{m-1} = H$ are two composition series for H, and so they are equivalent. In particular, $n - 1 = m - 1$ and hence $n = m$. Also the factors in these two composition series for H are isomorphic. Now, the two series for G have only one extra factor, and we have $G/H_{n-1} = G/H = G/K_{m-1}$. Hence, the two series of G are equivalent as well.

Now assume $H_{n-1} \neq K_{m-1}$, and let $L = H_{n-1} \cap K_{m-1}$. Both H_{n-1} and K_{m-1} are normal subgroups of G and, thus, so is L. Moreover, we claim that L is a maximal normal subgroup of both H_{n-1} and K_{m-1}. This was Problem 11.5.10 but the reasoning is straightforward: H_{n-1} is maximal normal in G, and so $H_{n-1}K_{m-1} = G$ and G/H_{n-1} is a simple group. Now by the Direct Diamond Theorem 11.43, $K_{m-1}/L \cong G/H_{n-1}$ is simple and so L is maximal normal in K_{m-1}. Similarly, L is maximal normal in H_{n-1}.

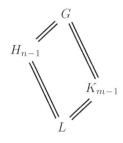

Now let

$$\{e\} = L_0 \lhd L_1 \lhd \cdots \lhd L_{r-1} \lhd L_r = L,$$

be a composition series of L. Hence, in addition to the original two composition series, we have two more composition series for G:

$$\{e\} = L_0 \lhd L_1 \lhd \cdots \lhd L_{r-1} \lhd L_r = L \lhd H_{n-1} \lhd H_n = G, \text{ and}$$
$$\{e\} = L_0 \lhd L_1 \lhd \cdots \lhd L_{r-1} \lhd L_r = L \lhd K_{m-1} \lhd K_m = G.$$

If we temporarily ignore the last factor for these four composition series, we get two composition series for H_{n-1} and two composition seires for K_{m-1}. Both H_{n-1} and K_{m-1} are groups smaller in size than G, and hence by the inductive hypothesis, the two composition series for each of these groups are equivalent. Hence,

$$\{e\} = L_0 \lhd L_1 \lhd \cdots \lhd L_r = L \lhd H_{n-1}, \text{ and}$$
$$\{e\} = H_0 \lhd H_1 \lhd \cdots \lhd H_{n-1}$$

are equivalent series. So $n-1 = r+1$, and the factors in the series of H's are, in some order, the factors in the composition series for L together with $H_{n-1}/L \cong G/K_{m-1}$.

Likewise,

$$\{e\} = L_0 \lhd L_1 \lhd \cdots \lhd L_r = L \lhd K_{m-1}, \text{ and}$$
$$\{e\} = K_0 \lhd K_1 \lhd \cdots \lhd K_{m-1}$$

are equivalent series, and so $m - 1 = r + 1$, and again the factors in the series of K's are, in some order, the factors in the composition series for L together with $K_{m-1}/L \cong G/H_{n-1}$.

We can now conclude that $n = r + 2 = m$, and the factors in the two original series for G are, in some order, the factors in the composition series for L together with $H_{n-1}/L \cong G/K_{m-1}$ and $G/H_{n-1} \cong K_{m-1}/L$. Thus, the two series are equivalent, and the proof is complete. $\square$

Proposition 14.45. (a) *A finite group always has a composition series.*

(b) *An abelian group has a composition series if and only if it is finite.*

Proof. (a) Begin with the subnormal series $\{e\} \lhd G$ and "refine" it. In other words, either G is simple, in which case we already have a composition series, or G has a normal subgroup N. We now have $\{e\} \lhd N \lhd G$. Again, either both N and G/N are simple and we have found our composition series or one or the other has a normal subgroup. We continue inserting new subgroups in this way. The process has to stop since the group is finite and will only stop if we have reached a composition series.

(b) If an abelian group has a composition series, the factors must be abelian simple groups. The only abelian simple groups are $\mathbb{Z}/p\mathbb{Z}$ where p is a prime. The number of factors is finite and the order of the group is the product of the sizes of the factors. Hence, the group is finite. $\square$

Problems

14.3.1. Write down a complete proof of Lemma 14.38.

14.3.2. Give an example of a solvable group that is not supersolvable. Give an example of a supersolvable group that is not nilpotent.

14.3.3. Prove that every nilpotent group is supersolvable.

14.3.4. Find all composition series of $\mathbb{Z}/12\mathbb{Z}$ and of D_{16}.

14.3.5. Let G be a group, and let Q be the poset of subnormal subgroups of G ordered by inclusion. Is every chain in Q a subnormal series in G? Is every subnormal series in G a chain in Q? Is every maximal chain in Q a composition series of G? Is every composition series of G, a maximal chain in Q? Is Q a graded poset?

14.3.6. Let $G = (\mathbb{Z}, +)$ be the additive group of integers. Find a subnormal series of G with five terms. What about one with ten terms? Does G have a composition series?

14.3.7. Let p be a prime, let n be a positive integer, and let G be a group of order p^n. What are the composition factors of G? What about its chief factors?

14.3.8. Let G be a finite group, and let $N \lhd G$. Show that G has a composition series that has N as one of its terms.

14.3.9. Give an example of two non-isomorphic groups with identical composition factors.

14.3.10. Find all chief series of $\mathbb{Z}/12\mathbb{Z}$ and of D_{16}.

14.3.11. Prove another version of the Jordan–Hölder theorem with "composition series" replaced with "chief series".

14.3.12. Let G be a group, and let R be the poset of normal subgroups of G ordered by inclusion. Is every chain in R a normal series in G? Is every maximal chain in R a chief series of G? Is R a graded poset?

14.3.13. Let G be a finite solvable group. Show that the chief factors of G are elementary abelian p-groups.

Part 2

(Mostly Commutative) Ring Theory

Chapter 15

Rings

... where the solution to Diophantine equations serves as a motivation for
the study of rings, and where rings, integral domains, division rings,
and fields are defined, and important examples of rings are given.

In this chapter, we begin our study of rings. Rings are an abstract set of
elements together with two operations—called "addition" and "multiplication"—
and where these operations follow a set of rules. In some ways, it may seem natural
to move from the study of groups (sets with *one* operation) to rings (sets with
two operations). You may, in fact, expect that "ring theory" will be a slightly
more complicated version of "group theory". This point of view can be somewhat
misleading.

Certainly, modern treatments of algebraic objects do bring out common themes
and approaches to both groups and rings. In studying both, we will concentrate
on *substructures* (rather than elements), and *maps* that preserve the operations—
group and ring homomorphisms—play a crucial role in both theories. In fact,
rings have within them an abelian group, and so what we know about groups can
sometimes be directly used.

However, we study different kinds of questions when we focus on rings, and in
many ways ring theory has a different flavor from group theory. Historically, also,
the need for studying abstract rings arose in a somewhat different context than
groups.

Up through the nineteenth century—and even now in the high school curri-
culum—algebra denoted the study of equations. Groups are the mathematician's
way of studying symmetries of geometric and mathematical objects, and first arose
as groups of permutations. In fact, the real power of group theory was first demonstrated in Galois theory (see Chapter 21), where groups were used to exploit the
relations among the permutations of roots of polynomials.

The study of rings, on the other hand, became necessary in studying a different
set of equations: Diophantine equations. Classically, a *Diophantine equation* refers

to an equation in a number of variables where we are interested in *integer* solutions. One of the more famous Diophantine equations is $x^n + y^n = z^n$. Fermat conjectured that, for n an integer greater than 2, the only integer solutions to this equation are the ones when at least one of x, y, or z is zero.[1] This conjecture was proved in 1995 by Andrew Wiles, who put the finishing work on 350 years of intense study. In studying Diophantine equations, it quickly becomes clear that one would like to see if different sets of numbers behave like integers. The study of integers is the domain of number theory, and integers have certain familiar properties. Prime numbers play an important role: we factor integers into primes, we talk about divisibility, and we find greatest common divisors. A *commutative ring* is a generalization of the ordinary integers. In rings, we can add, subtract, and multiply (but not necessarily divide). After defining them, we are interested in understanding which rings behave most like the integers.

In the first section of this chapter, we start with a few simple Diophantine equations to illustrate the need to understand factorization properties of various classes of numbers. We follow this with the definition of rings and various subclasses of rings. In the following chapters, we will develop the theory of rings. The questions we ask will be motivated by our discussion of Diophantine equations. We will return to Diophantine equations as examples of how rings can be used.

The basic theory of rings that we will develop in this part of the book will be used throughout (in Galois theory, module theory, representation theory, and commutative algebra). Having arisen from the study of Diophantine equations, rings are now ubiquitous in algebra.

15.1. Diophantine Equations and Rings

Notation. We have been using and will continue to use the following notation throughout:

$\mathbb{Z}$ is the set of *integers*.

$\mathbb{Z}^{\geq 0}$ (or $\mathbb{N}$) is the set of *natural numbers*: $0, 1, 2, \ldots$.

$\mathbb{Z}^{>0}$ is the set of *positive integers*: $1, 2, 3, \ldots$.

$\mathbb{Q}$ is the set of *rational numbers*.

$\mathbb{R}$ is the set of *real numbers*.

$\mathbb{C}$ is the set of *complex numbers*.

A *Diophantine equation* is a polynomial equation in several variables where we are interested in finding *integer* solutions. These problems have been around since ancient times and are quite natural. Looking at patterns of numbers, you may wonder, for example, if the difference of the cube of an integer and the square of another integer is ever equal to 2. In other words, does $z^3 - y^2 = 2$ have any integer solutions? Certainly $z = 3$ and $y = \pm 5$ are solutions, but are there others?

These problems can be notoriously hard to solve, as the example of Fermat's equation $x^n + y^n = z^n$ has shown. In fact, similar to Fermat's conjecture, Euler, in 1778, conjectured that, for $n > 2$, no n-th power can be a non-trivial sum of less

---
[1]This conjecture is known as Fermat's Last Theorem.

than n other nth powers. In other words, none of the following have non-trivial integer solutions:

$$x^4 + y^4 + z^4 = w^4$$
$$x^5 + y^5 + z^5 + w^5 = u^5$$
$$\cdots$$

These sets of problems, for a long time, seemed as intractable as Fermat's equation. However, in 1966, Lander and Parkin [**LP66, LP67**] using a CDC 6600 (one of the earliest supercomputers) found that

$$27^5 + 84^5 + 110^5 + 133^5 = 144^5$$

and, hence, disproved Euler's conjecture.[2] A number of years later, in 1988, Noam Elkies [**Elk88**] proved that there are infinitely many solutions to $x^4 + y^4 + z^4 = w^4$. Following this, Roger Frye's computer search showed that the smallest integer solution to $x^4 + y^4 + z^4 = w^4$ is

$$95{,}800^4 + 217{,}519^4 + 414{,}560^4 = 422{,}481^4,$$

and there are no other solutions with all the integers less than one million.

The fate of Euler's conjectures already shows the difficulty of the subject. A seemingly random combination of integers may be the only solution to a Diophantine equation. What kind of theory would unearth such examples, or would, much more modestly, allow us to conclude that for certain types of equations no integer solutions exist?

Here, we will look at three simple examples of Diophantine equations and their (possible) solutions. Our approach to these problems serves as a guide to the kinds of questions that we would want to ask in ring theory.

We approach all three problems similarly. First, we isolate one of the variables and factor the other side. We then argue that the factors are relatively prime. From this—and using a version of the fundamental theorem of arithmetic—we argue that each of the factors have to be a power (a cube in the first two problems and a square in the third problem). Writing one of the factors as the appropriate power, then we either show that there are no solutions or get enough constraints to find the solutions. The only part of the "proofs" that can possibly cause any trouble is the belief in unique factorization into primes (the fundamental theorem of arithmetic) for sets of numbers other than integers. In the first problem, we actually just use unique factorization for ordinary integers, and hence the proof is complete. In the second and third problems, we use unique factorization for a set of numbers larger than integers. It happens that unique factorization works in one case and *not* in the other case. The breakdown of our method provides a cautionary tale that leads us to the questions that will guide our study of commutative rings.

At this point, it is *not* necessary for the reader to master all the details of the arguments. We have given complete details for all three solutions in order to convince you that the belief in unique factorization is the only weak link in the arguments. Working through these examples is beneficial and will put our study of commutative rings in a context.

---

[2]Lander and Parkin's paper [**LP66**] consists of two sentences in five lines!

y² − z³ = 4, y odd. Are there any integer solutions to $z^3 = y^2 - 4$ when y is odd? In other words, can four plus the cube of an integer be the square of an odd integer?

The answer is no. To see why, first write the equation as $z^3 = (y-2)(y+2)$. Now, notice that if a prime number divided both $y - 2$ and $y + 2$, then it would have to divide their difference $(y+2) - (y-2) = 4$, and, so, it would have to be 2. Since $y - 2$ is odd, we conclude that $\gcd(y-2, y+2) = 1$. If the highest power of the prime p that divides z is p^a, then p^{3a} divides z^3, and—since $y - 2$ and $y + 2$ do not share any primes in their factorization—p^{3a} has to divide either $y - 2$ or $y + 2$. We conclude that both $y - 2$ and $y + 2$ are cubes.

If $y - 2 = a^3$ and $y + 2 = b^3$, then $b^3 - a^3 = 4$. But are there two cubes whose difference is 4? Look at Table 15.1, and it is clear that the answer is no. We can check the differences at the beginning of the table, and the differences only get much larger as the integers get larger.

Table 15.1. Cubes of small integers. The smallest difference of two positive cubes is 7.

n	0	1	2	3	4	5	$\cdots$
n^3	0	1	8	27	64	125	$\cdots$

We conclude that the Diophantine equation $z^3 = y^2 - 4$ has no integer solutions when y is odd.

z³ − y² = 2. Are there any integer solutions to $z^3 = y^2 + 2$? This looks very much like the previous example, but does a similar argument help us answer the question?

First note that if y or z are even, then both are even. This is impossible since, for y and z even, the remainder of z^3 when divided by 4 is 0 while the reminder of $y^2 + 2$ when divided by 4 is 2. Hence, for any solution to $z^3 = y^2 + 2$, both y and z are odd integers.

If we follow the same argument as the previous example, we would first factor the equation as $z^3 = (y + \sqrt{2}i)(y - \sqrt{2}i)$. Now the problem is that we were looking for integer solutions, and we have already introduced both $\sqrt{2}$ and i. In the previous solution, both $y + 2$ and $y - 2$ were ordinary integers, and we could consider their greatest common divisor and prime factors.

To be able to proceed as before, we need to consider the set of numbers of the form $a + b\sqrt{2}i$ where a and b are integers. We denote the set $\{a + b\sqrt{2}i \mid a, b \in \mathbb{Z}\}$ by $\mathbb{Z}[\sqrt{2}i]$ (this notation will be explained in the next section).

For the sake of argument, assume that we could treat $\mathbb{Z}[\sqrt{2}i]$ as an expanded set of integers. After all, this set is closed under addition, subtraction, and multiplication. If so, then, presumably, some of these numbers are "primes", and maybe we can speak about factorization into primes and greatest common divisors.

Before we continue, we need to introduce one computational tool. Let d be an integer (positive or negative) that is not divisible by the square of a prime,

and let $\mathbb{Z}[\sqrt{d}]$ denote the set $\{a + b\sqrt{d} \mid a, b \in \mathbb{Z}, (\sqrt{d})^2 = d\}$. (For example, when $d = -2$, we get $\mathbb{Z}[\sqrt{d}] = \mathbb{Z}[\sqrt{2}i]$.) Now, define the map $N \colon \mathbb{Z}[\sqrt{d}] \to \mathbb{Z}$ by $N(a + b\sqrt{d}) = a^2 - db^2$. Then, if $\alpha, \beta \in \mathbb{Z}[\sqrt{d}]$, a straightforward computation shows that $N(\alpha\beta) = N(\alpha)N(\beta)$.

We first claim that for any odd integer y, the greatest common divisor, in $\mathbb{Z}[\sqrt{2}i]$, of $y + \sqrt{2}i$ and $y - \sqrt{2}i$ is 1. (Note that $2 \pm \sqrt{2}i = \sqrt{2}i\left[\pm 1 - \sqrt{2}i\right]$, and so the claim would not have been true if y was allowed to be even.)

To prove the claim, assume that $\alpha = a_0 + b_0\sqrt{2}i$ is an element of $\mathbb{Z}[\sqrt{2}i]$ that divides both $y - \sqrt{2}i$ and $y + \sqrt{2}i$. We want to show that $\alpha = \pm 1$. We know α divides $(y + \sqrt{2}i) - (y - \sqrt{2}i) = 2\sqrt{2}i$.

Hence, we have $2\sqrt{2}i = \alpha\beta$, for some $\beta \in \mathbb{Z}[\sqrt{2}i]$. Applying the function N to both sides, we have

$$8 = N(2\sqrt{2}i) = N(\alpha)N(\beta).$$

Since $N(\alpha)$ is an ordinary integer, the only choices for $N(\alpha)$ are ± 1, ± 2, ± 4, and ± 8. But $N(\alpha) = a_0^2 + 2b_0^2$, and, hence the only possibilities for α are ± 1, ± 2, $\pm\sqrt{2}i$, and $\pm 2\sqrt{2}i$. Hence, if α is not ± 1, and since $(\sqrt{2}i)(\sqrt{2}i) = -2$, we conclude that z^3—which is the product of $y - \sqrt{2}i$ and $y + \sqrt{2}i$—is divisible by at least 2. But z^3 is odd, and hence α must be ± 1.

Now, as in the previous example, the product of $y - \sqrt{2}i$ and $y + \sqrt{2}i$ is a cube, and these two factors have no common factors. Can we conclude that each of these must be a cube as well? We can, if the elements of $\mathbb{Z}[\sqrt{2}i]$ like ordinary integers factor uniquely into "primes". Assuming this is so, we would argue that if p^a is the highest power of a prime p that divides z, then p^{3a} divides z^3, and p^{3a} would have to divide one or the other of $y + \sqrt{2}i$ and $y - \sqrt{2}i$ (but not both). This would mean that both $y + \sqrt{2}i$ and $y - \sqrt{2}i$ are cubes.

Now write $y + \sqrt{2}i = (a + b\sqrt{2}i)^3$, and expand the right-hand side to get

$$y + \sqrt{2}i = a^3 - 6ab^2 + (3a^2b - 2b^3)\sqrt{2}i.$$

Equating the real and imaginary part of the two sides, we have

$$y = a^3 - 6ab^2,$$
$$1 = b(3a^2 - 2b^2).$$

Now, a and b are ordinary integers, and, in the second equation, the product of two integers is 1. Hence, either $b = 1 = 3a^2 - 2b^2$ or $b = -1 = 3a^2 - 2b^2$. The latter has no solution, and the former gives $a = \pm 1$ and $b = 1$.

Hence, $y = \pm 5$ and $z = 3$ are the only integer solutions to $z^3 - y^2 = 2$.

The above solution is complete except for the fact that we do not know if $\mathbb{Z}[\sqrt{2}i]$ has primes and, even if it does, whether every element can uniquely (up to rearrangement) be written as a product of primes. One of the goals of upcoming chapters on ring theory is to clarify these issues thereby legitimizing solutions like the one above.

Now, if it were true that all such arguments always work, then our task would be merely to verify that nothing can go wrong. But things can go wrong!

$z^2 = y^2 + 5$. What are all the integral solutions to $z^2 = y^2 + 5$? The reader may be tired of going through the same argument for the third time, but this time something different happens.

There are actually several ways of solving this problem, but we want to see what happens if we follow the method of the previous example. In fact, our steps will be quite similar, and the reader can skim the argument.

Note that if y or z is divisible by 5, then both are, and this is impossible, since, in such a case, the remainder of z^2 when divided by 25 is zero, while the remainder of $y^2 + 5$ when divided by 25 is 5. Hence, neither y nor z are divisible by 5.

Write $z^2 = (y + \sqrt{5}i)(y - \sqrt{5}i)$, and work with numbers of the form $a + b\sqrt{5}i$ where a and b are integers. Denote $\{a + b\sqrt{5}i \mid a, b \in \mathbb{Z}\}$ by $\mathbb{Z}[\sqrt{5}i]$, and assume that $\alpha \in \mathbb{Z}[\sqrt{5}i]$ divides both $y + \sqrt{5}i$ and $y - \sqrt{5}i$. Again, α must divide the difference, and hence $2\sqrt{5}i = \alpha\beta$ for some $\beta \in \mathbb{Z}[\sqrt{5}i]$. Applying the function N, we get

$$20 = N(\alpha)N(\beta).$$

But if $\alpha = a_0 + b_0\sqrt{5}i$, then $N(\alpha) = a_0^2 + 5b_0^2$. The only integers of this form that divide 20 are 1, 4, 5, and 20. The only possibilities for α are ± 1, ± 2, $\pm\sqrt{5}i$ and $\pm 2\sqrt{5}i$.

If the common factor α of $y + \sqrt{5}i$ and $y - \sqrt{5}i$ is $\pm\sqrt{5}i$ or $\pm 2\sqrt{5}i$, then the product $z^2 = (y + \sqrt{5}i)(y - \sqrt{5}i)$ would be divisible by 5, and we had argued that z is not divisible by 5.

If $\alpha = 2$, then $y + \sqrt{5}i = 2\beta$ for some $\beta \in \mathbb{Z}[\sqrt{5}i]$. Applying the function N to both sides, we get $y^2 + 5 = 4N(\beta)$, which means that y is an odd integer. As a result z is an even integer. Now the square of an odd integer is of the form $(2k + 1)^2 = 4(k^2 + k) + 1$ and always has remainder 1 when divided by 4. The square of an even integer is divisible by 4 and has remainder 0 when divided by 4. So, $z^2 = 0 \mod 4$ while $y^2 + 5 = 2 \mod 4$. Hence, $z^2 \neq y^2 + 5$.

We conclude that $\alpha = \pm 1$, and hence $y + \sqrt{5}i$ and $y - \sqrt{5}i$ have no common divisors in $\mathbb{Z}[\sqrt{5}i]$. Arguing as before, we would like to conclude that since the product of $y + \sqrt{5}i$ and $y - \sqrt{5}i$ is a square and since these two factors are relatively prime, then each of these factors is a square. It would then follow that

$$y + \sqrt{5}i = (a + b\sqrt{5}i)^2 = a^2 - 5b^2 + 2ab\sqrt{5}i.$$

Equating the real and imaginary parts, we have

$$y = a^2 - 5b^2,$$
$$1 = 2ab.$$

But a and b are ordinary integers, and $2ab$ can never be equal to 1. Hence, we conclude that $z^2 = y^2 + 5$ has no integer solutions.

But this is wrong! In fact, $z = \pm 3$ and $y = \pm 2$ are solutions!

We could have easily found these solutions if we had followed a different path. From $z^2 = y^2 + 5$, we get $(z - y)(z + y) = 5$. Hence, either $z - y = \pm 1$ and $z + y = \pm 5$ or $z - y = \pm 5$ and $z + y = \pm 1$. Solving these is easy and gives the solutions $z = \pm 3$ and $y = \pm 2$.

What went wrong with our method? The key fact is that in $\mathbb{Z}[\sqrt{5}i]$ we have

$$3 \times 3 = 9 = (2 + \sqrt{5}i)(2 - \sqrt{5}i),$$

and neither 3 nor $2 \pm \sqrt{5}i$ can be further factored. In other words, in $\mathbb{Z}[\sqrt{5}i]$, the number 9 has two *different* factorizations. The fundamental theorem of arithmetic—which asserts that every ordinary integer is uniquely factorable into primes—does not hold in $\mathbb{Z}[\sqrt{5}i]$.

What about our proof that $y = \pm 5$ and $z = 3$ are the only integer solutions to $z^3 - y^2 = 2$? Is that wrong also? Actually, our method does work in that case since it can be proved that we have unique factorization into primes in $\mathbb{Z}[\sqrt{2}i]$.

So, we curiously realize that $\mathbb{Z}[\sqrt{2}i]$ is fundamentally different from $\mathbb{Z}[\sqrt{5}i]$. Sorting this out and clarifying the underlying concepts can be seen as a motivating question for the ring theory that we will develop.

Guiding Questions. Rings—which will be formally defined in the next section—are sets together with an addition and a multiplication that follow certain rules. We will mostly be concerned with *commutative* ring theory where both operations are commutative (as it turns out, addition always has to be commutative while multiplication does not), and our prototype will be the integers $\mathbb{Z}$. Just as the integers, every ring will be closed under addition, subtraction, and multiplication but not necessarily division.

Based on the examples we have seen of Diophantine equations, we have the following guiding questions for the study of rings:

Question 15.1. Can we define divisibility and primes for arbitrary rings? For ordinary integers, the fundamental theorem of arithmetic (Theorem 1.48) states that every integer is uniquely expressible as a product of primes. For which rings does this remain true? An ordinary prime can be defined as an integer that can only be factored trivially. It can also be defined as an integer p with the property that whenever p divides a product ab, then p must necessarily divide a or b. For which rings are these two concepts equivalent?

Early on, in Theorem 1.47 of Chapter 1, the division algorithm for integers was proved. It turns out that the special class of rings—called Euclidean domains—for which we have a division algorithm will be very important.

Question 15.2. For which commutative rings do we have an analog of the division algorithm?

The integers are a subset of rational numbers $\mathbb{Q}$, and, in fact, the rational numbers are constructed from the integers. Rational numbers are an example of a field. A field is a set in which we can add, subtract, multiply, and divide by non-zero elements (and the operations follow a number of reasonable rules).

Question 15.3. For which commutative rings can we construct a field that contains them?

These questions—as well as an examination of rings of polynomials that will lead us to construction of finite fields—will guide us throughout the chapters on ring theory.

Problems

15.1.1. Does $x^2 - 3y^2 = 0$ have any integer solutions? Prove your assertion. Can you generalize your claim? Make a conjecture.

15.1.2. In analyzing the possible solutions to $y^2 - z^3 = 4$, we used Table 15.1 to argue that the difference of two cubes cannot be 4. Prove this fact algebraically without referring to a table of cubes.

HW #1

15.1.3. Let d be an integer (positive or negative) not divisible by a square of a prime, and let $\mathbb{Z}[\sqrt{d}] = \{a + b\sqrt{d} \mid a, b \in \mathbb{Z}\}$. Let $N \colon \mathbb{Z}[\sqrt{d}] \to \mathbb{Z}$ be defined by $N(a + b\sqrt{d}) = a^2 - db^2$. Prove that, for $x, y \in \mathbb{Z}[\sqrt{d}]$, we have

$$N(xy) = N(x)N(y).$$

15.1.4. Without using ± 1 as one of the factors, can you factor 3 in $\mathbb{Z}[\sqrt{2}i]$? What about 5?

15.1.5. Show that, without using ± 1 as one of the factors, neither 3 nor $2 + \sqrt{5}i$ can be factored in $\mathbb{Z}[\sqrt{5}i]$.

15.1.6. Let $\mathbb{Z}[i]$ denote $\{a + bi \mid a, b \in \mathbb{Z}\}$, and accept without proof that the fundamental theorem of arithmetic generalizes to $\mathbb{Z}[i]$. Find all integer solutions to $y^3 = x^2 + 1$.

15.2. Rings, Integral Domains, Division Rings, and Fields

We now formally define a ring, introduce some vocabulary, prove some elementary properties, and see many examples.

Definition 15.4 (Ring). A *ring* $(R, +, \cdot)$ is a non-empty set R together with two binary operations $+$ and $\cdot$ such that:

every element commutative over +

(a) The set R together with the operation $+$ forms an abelian group.

(b) The operation $\cdot$ is associative on R. In other words, for all $a, b, c \in R$, we have $a \cdot (b \cdot c) = (a \cdot b) \cdot c$.

(c) The distributive laws hold. In other words, for all $a, b, c \in R$,

For +

$$a \cdot (b + c) = a \cdot b + a \cdot c,$$
$$(b + c) \cdot a = b \cdot a + c \cdot a.$$

We write 0 for the identity element of the group $(R, +)$. As in groups, most of the time, we drop $\cdot$, and write ab for $a \cdot b$.

For mult.

Definition 15.5 (Commutative ring). The ring $(R, +, \cdot)$ is a *commutative ring* if $\cdot$ is a commutative operation on R.

Definition 15.6 (Ring with identity). The ring $(R, +, \cdot)$ is a *ring with identity* if R contains an identity for the operation $\cdot$. We denote such a multiplicative identity by 1. Other names used for a ring with identity are a *ring with unity*, a *unit ring*, a *unital ring*, or a *unitary ring*.

For ·

Remark 15.7. Many authors define rings to be what we have called rings with identity. This causes no problem since most interesting rings used in algebra³ have an identity. However, you have to be careful that if all rings must have an identity, then most ideals—defined later—will not be subrings.

You may wonder why we required that one of the operations be commutative. The following proposition shows that, for rings with identity, we had no choice.

Proposition 15.8. *Let* $(R, +, \cdot)$ *satisfy all axioms of a ring except possibly* $a + b = b + a$ *for all* $a, b \in R$. *Further, assume that* R *has a multiplicative identity* 1. *Then* $(R, +, \cdot)$ *is a ring.*   ↖ *shows ring with identity is inherently commutative*

Proof. Calculate $(1 + a)(1 + b)$ in two ways and get $1 + a + b + ab = 1 + b + a + ab$. Thus $a + b = b + a$.   □

Proposition 15.9. *Let* $(R, +, .)$ *be a ring. Then*

(a) *The zero element is unique, and, for every* $a \in R$, $-a$ *is unique.*

(b) *For all* $a \in R$, *we have* $a \cdot 0 = 0 = 0 \cdot a$.

(c) *If* R *is a ring with identity, then* 1 *is unique.*

(d) *If* R *is a ring with identity and* $|R| \geq 2$, *then* $1 \neq 0$.

(e) *For* $a, b \in R$, *we have* $-(a + b) = (-a) + (-b)$ *and* $-(-a) = a$.

(f) *For* $a, b \in R$, *we have* $-(ab) = (-a)b = a(-b)$ *and* $ab = (-a)(-b)$.

Proof. (a) These follow since $(R, +)$ is an abelian group.

(b) $a \cdot 0 = a(0 + 0) = a \cdot 0 + a \cdot 0$, and thus $a \cdot 0 = 0$.

(c) If 1 and v are both 1's, then $1 = 1v = v$.

(d) If $1 = 0$, then we have $0 = r \cdot 0 = r \cdot 1 = r$ for all elements $r \in R$. This would mean that $R = \{0\}$.

(e) This is true since $(R, +)$ is a group.

(f) $a(-b) + ab = a(-b + b) = a \cdot 0 = 0$, and so $a(-b) = -ab$. Similarly $(-a)b = -ab$, and now we have $(-a)(-b) = -(-a)b = (-(-a))b = ab$.   □

A ring can have just one element $\{0\}$. This a commutative ring with identity— called the *trivial ring*—and in this ring $0 = 1$. But for all other rings with identity, $0 \neq 1$.

Note that since a ring has two operations, there are many possibilities for a ring. For example, if you have a ring R with five elements, then we know from group theory that $(R, +)$ is a group of order 5, and, hence, it must be isomorphic, as a group, to $(\mathbb{Z}/5\mathbb{Z}, +)$. This does *not* mean that $(R, +, \cdot) \cong (\mathbb{Z}/5\mathbb{Z}, +, \cdot)$. In other words, while the addition in R will coincide with the addition in $\mathbb{Z}/5\mathbb{Z}$, it is possible that the multiplication in R will be quite different than the one in $\mathbb{Z}/5\mathbb{Z}$.

Now note that if R is a ring with more than one element, then 0 does not have an inverse, and, hence, $(R, \cdot)$ cannot be a group.

---

³There are a number of important rings without identity in analysis.

Definition 15.10 (Unit). Let $(R, +, \cdot)$ be a ring with identity, and let $u \in R$. u is called a *unit* if there is an element $u^{-1} \in R$ such that $uu^{-1} = u^{-1}u = 1$. The element u^{-1} is called the *(multiplicative) inverse* of u. The set of units in R is denoted by $R^{\times}$ and is called the *group of units* of R.

Recall that we have already used the notation $(\mathbb{Z}/n\mathbb{Z})^{\times}$ to denote the elements of $\mathbb{Z}/n\mathbb{Z}$ that have a multiplicative inverse (Definition 1.51). We have also used the notation $\mathbb{C}^{\times}$ to denote the non-zero complex numbers (Example 2.13). Since $(\mathbb{Z}/n\mathbb{Z}, +, \cdot)$ and $(\mathbb{C}, +, \cdot)$ are rings, these notations are consistent with our general definition of $R^{\times}$ as the set of units of a ring R.

Proposition 15.11. *Let $(R, +, \cdot)$ be a ring with identity. Then $(R^{\times}, \cdot)$ is a group.*

Proof. The binary operation $\cdot$ is already known to be associative, and since R is a ring with identity, $(R^{\times}, \cdot)$ has an identity, namely the element 1. The set $R^{\times}$ is closed under $\cdot$ since the product of two invertible elements is invertible. To see this, assume a and b are invertible, then $(a \cdot b) \cdot (b^{-1} \cdot a^{-1}) = 1$ showing that $(a \cdot b)^{-1} = b^{-1} \cdot a^{-1}$. Finally, by definition, every element u of $R^{\times}$ has an inverse u^{-1}, and, since u^{-1} is also invertible, we have $u^{-1} \in R^{\times}$.   □

Definition 15.12. Let n be a positive integer. Let R be a ring, and let $a \in R$. We define $0_{\mathbb{Z}}a = 0_R$, $na = \underbrace{a + a + \cdots + a}_{n \text{ terms}}$, and $(-n)a = -(na)$.

We further define $a^1 = a$, $a^n = aa^{n-1}$. If R has an identity, then we define $a^0 = 1$. If the element a has a multiplicative inverse a^{-1}, then we define $a^{-n} = (a^{-1})^n$.

Lemma 15.13. *Let R be a ring, and let $a \in R$. Let $m, n \in \mathbb{Z}$. Then*

(a) $ma + na = (m + n)a$,

(b) $m(na) = (mn)a$,

(c) $a^m a^n = a^{m+n}$,

(d) $(a^n)^{-1} = a^{-n}$,

(e) $(a^m)^n = a^{mn}$.

Proof. We prove item (d) and leave the rest to the reader. We have

$$a^n \cdot (a^{-1})^n = \underbrace{a \cdots (a}_{n \text{ times}} \underbrace{a^{-1}) \cdots a^{-1}}_{n \text{ times}} = 1.$$

So $(a^n)^{-1} = (a^{-1})^n$. But the latter is defined to be a^{-n}. Hence, $(a^n)^{-1} = a^{-n}$.   □

The nicest kinds of a ring are those for which $(R - \{0\}, \cdot)$ is an abelian group. These are fields.

Definition 15.14 (Field). Let $(R, +, \cdot)$ be a ring. $(R, +, \cdot)$ is a *field* if $(R - \{0\}, \cdot)$ is an abelian group.

Definition 15.15 (Division ring). Let $(R, +, \cdot)$ be a ring. $(R, +, \cdot)$ is a *division ring* if $(R - \{0\}, \cdot)$ is a group.

Note that both division rings and fields have an identity and must have more than one element. The only difference between a field and a division ring is that in a field multiplication is commutative. In both division rings and fields, every non-zero element has a multiplicative inverse. Hence, if R is a division ring and $ab = 0$, then either $a = 0$ or $b = 0$. This is a very useful property, and, in fact, $\mathbb{Z}$, the ring of integers, also has this property even though $\mathbb{Z}$ is not a field.

Definition 15.16 (Zero-divisor). Let $(R, +, \cdot)$ be a ring, and let $0 \neq a \in R$. The element a is called a *left zero-divisor* if there exists $b \in R - \{0\}$ with $a \cdot b = 0$. The element a is called a *right zero-divisor* if there exists $b \in R - \{0\}$ with $b \cdot a = 0$. The element a is called a *zero-divisor* if it is a left or a right zero-divisor.

Definition 15.17 (Integral domain). Let $(R, +, \cdot)$ be a ring. The ring $(R, +, \cdot)$ is an *integral domain* if $(R, +, \cdot)$ is a commutative ring with identity, $|R| \geq 2$, and R has no zero-divisors. In other words, in an integral domain $a \cdot b = 0$ implies that either a or b is zero. ↖ No zero divisors

Proposition 15.18. *A field is an integral domain.*

Proof. If we have $ab = 0$ and $a \neq 0$, then $b = a^{-1}(ab) = a^{-1}0 = 0$. □

Figure 15.1 summarizes the relationship between fields, divisions rings, integral domains, and rings.

$$\begin{array}{ccccc}
\text{Fields} & \Rightarrow & \text{Integral Domains} & \Rightarrow & \text{Commutative Rings} \\
\Downarrow & & \Downarrow & & \Downarrow \\
\text{Division Rings} & \Rightarrow & \text{Rings with identity} & \Rightarrow & \text{Rings}
\end{array}$$

Figure 15.1. The hierarchy among fields, division rings, integral domains, and rings

Remark 15.19. Some mathematicians include the 0 element among zero-divisors—in such texts, you will see the expression "non-zero zero-divisor" often. And some mathematicians consider the ring with one element $\{0\}$ to be an integral domain—in such texts, you will see the expression "non-trivial integral domain".

Examples of Rings. Rings are everywhere, and, before proceeding any further, we will look at many examples of rings.

The Ring of Integers $(\mathbb{Z}, +, \cdot)$. The ring of integers with ordinary addition and multiplication is our prototype of a commutative ring with identity. In fact, $(\mathbb{Z}, +, \cdot)$ is an integral domain.

The only invertible elements in $\mathbb{Z}$ are ± 1, and the group of units of this ring is $\mathbb{Z}^{\times} = \{\pm 1\} \cong \mathbb{Z}/2\mathbb{Z}$.

In addition to being an integral domain, $\mathbb{Z}$ has a number of additional properties. Chief among these are the division algorithm (Theorem 1.47) and the fundamental theorem of arithmetic (Theorem 1.48). In Chapter 18, we will consider integral domains that have one of these properties. As a part of this investigation we will use the division algorithm to give a proof of the fundamental theorem.

Some subsets of $\mathbb{Z}$ are also rings (*subrings* of $\mathbb{Z}$). Let $n > 1$ be a fixed positive integer, then $(n\mathbb{Z}, +, \cdot) = \{na \mid a \in \mathbb{Z}\}$ is a commutative ring *without* identity.

Quadratic Integer Rings. If a is a positive integer, then, as usual, $\sqrt{a}$ denotes the unique *positive* number whose square is a, and $\sqrt{-a}$ denotes $i\sqrt{a}$. Let d be an integer (positive or negative) that is not divisible by the square of a prime. As in previous section, define

$$\mathbb{Z}[\sqrt{d}] = \{a + b\sqrt{d} \mid a, b \in \mathbb{Z}\}.$$

Then $(\mathbb{Z}[\sqrt{d}], +, \cdot)$ is an integral domain. In Section 15.1, while solving Diophantine equations, we saw the value of understanding divisibility, "primes", and unique factorization in $\mathbb{Z}[\sqrt{d}]$.

As a specific example, if $d = -1$, then $\mathbb{Z}[i] = \{a + bi \mid a, b \in \mathbb{Z}\}$ is called the ring of *Gaussian integers*.

If d is a square-free integer as before and, in addition, $d = 1 \bmod 4$, then we have the slightly larger integral domain:

$$\mathbb{Z}[\frac{1 + \sqrt{d}}{2}] = \{a + b\frac{1 + \sqrt{d}}{2} \mid a, b \in \mathbb{Z}\}.$$

You have to do a short calculation (Problem 15.2.17) to see that, when $d = 1 \bmod 4$, $\mathbb{Z}[\frac{1+\sqrt{d}}{2}]$ is closed under multiplication.[4]

The Ring of Integers mod n. Let n be a positive integer. Let $\mathbb{Z}/n\mathbb{Z} = \{0, 1, 2, \ldots, n-1\}$. With addition and multiplication mod n, $\mathbb{Z}/n\mathbb{Z}$ is a commutative ring with identity. The ring $(\mathbb{Z}/n\mathbb{Z}, +, \cdot)$ is called the *ring of integers mod n*.

We repeat some of what we know (Theorem 1.55 and Corollary 1.57) about the ring of integers mod n:

Theorem 15.20. *The ring of integers* mod n, $(\mathbb{Z}/n\mathbb{Z}, +, \cdot)$ *is a commutative ring with identity. If p is a prime number, then $\mathbb{Z}/p\mathbb{Z}$ is a field, and if n is composite, then $\mathbb{Z}/n\mathbb{Z}$ is* not *an integral domain. More generally,*

$$(\mathbb{Z}/n\mathbb{Z})^{\times} = \{a \in \mathbb{Z}/n\mathbb{Z} \mid \gcd(a, n) = 1\}.$$

Notation 15.21. Recall that if $p > 0$ is a prime integer, then the field $(\mathbb{Z}/p\mathbb{Z}, +, \cdot)$ will often be denoted by $\mathbb{F}_p$.

Polynomial Rings. Let R be any commutative ring with identity. Let $R[x]$ be the set of polynomials in the variable x and with coefficients in R. Thus an element of $R[x]$ is of the form $a_0 + a_1 x + a_2 x^2 + \cdots + a_n x^n$ for some non-negative integer n. Let $+$ and $\cdot$ denote the ordinary addition and multiplication of polynomials. Then $(R[x], +, \cdot)$ is a commutative ring with identity and is called the *ring of polynomials over R*.

Now, if R is a commutative ring with identity, then $S = R[x]$ is also a commutative ring with identity, and so we can construct $S[y]$. The ring $S[y] = R[x][y]$ is also a commutative ring with identity, and it consists of polynomials in y whose coefficients are polynomials in x. If we multiply these out, we just get polynomials in two variables x and y and with coefficients in R. Hence, $S[y]$ is the same as the ring of polynomials in two variables x and y and with coefficients in R. We usually denote $S[y] = R[x][y]$ by $R[x, y]$. We can construct a ring $R[x_1, \ldots, x_k]$ of polynomials in k variables in the same fashion.

---

[4]See Problem 15.2.22 for a reason for considering the admittedly strange looking $\mathbb{Z}[\frac{1+\sqrt{d}}{2}]$.

Polynomial rings often reflect the properties of the coefficient ring R. For example, $R[x_1, \ldots, x_k]$ is an integral domain if and only if R is.

Rings of Matrices. Let R be a commutative ring with identity. Let $M_{n \times n}(R)$ be the set of $n \times n$ matrices with entries in R. Let $+$ and $\cdot$ denote the usual matrix addition and multiplication. Then $(M_{n \times n}(R), +, \cdot)$ is a ring with identity. The identity matrix is the identity of the ring, and the ring is not necessarily commutative. In fact, rings of matrices provide a rich set of examples of non-commutative rings.

Note that in $M_{2 \times 2}(\mathbb{Z})$ we have

$$\begin{bmatrix} 1 & 2 \\ 2 & 4 \end{bmatrix} \begin{bmatrix} 4 & -6 \\ -2 & 3 \end{bmatrix} = \begin{bmatrix} 0 & 0 \\ 0 & 0 \end{bmatrix}.$$

Hence, the product of two non-zero elements can be zero, and $M_{2 \times 2}(\mathbb{Z})$ is not a division ring.

If F is a field, then $M_{n \times n}(F)^{\times}$, the group of units of $M_{n \times n}(F)$, is the familiar $\mathrm{GL}(n, F)$. For R, a general commutative ring with identity, $M_{n \times n}(R)^{\times}$ is also denoted by $\mathrm{GL}(n, R)$ and is the set of $n \times n$ invertible matrices with entries in R whose inverse is also in $M_{n \times n}(R)$.

Thus, for example, $\mathrm{GL}(n, \mathbb{Z}) = M_{n \times n}(\mathbb{Z})^{\times}$ is the group consisting of all invertible $n \times n$ integer matrices whose inverse is also an integer matrix. These are precisely the $n \times n$ integer matrices whose determinant is ± 1. (See Problem 2.1.4.)

Rings of Functions. There are many ways to construct rings whose elements are functions. We give two examples here.

Let X be any set, and let $S = \{ f \mid f \colon X \to \mathbb{R} \text{ is a function} \}$ be the set of real valued maps on X. Define addition and multiplication pointwise:

$$(f + g)(x) = f(x) + g(x),$$
$$(fg)(x) = f(x)g(x).$$

Then $(S, +, \cdot)$ is a commutative ring with identity. If $|X| \geq 2$, then S is *not* an integral domain.

For our second example, let A be an abelian group (written additively), and let $\mathrm{End}(A)$ be the set of endomorphisms of A (an endomorphism of A is a group homomorphism from A to A). For $f, g \in \mathrm{End}(A)$ define, for all $a \in A$,

$$(f + g)(a) = f(a) + g(a),$$
$$(fg)(a) = f(g(a)).$$

Then $(\mathrm{End}(A), +, \cdot)$ is a (not necessarily commutative) ring with identity.

Examples of Fields. As we discussed in Section 1.4, the rational numbers $(\mathbb{Q}, +, \cdot)$, the real numbers $(\mathbb{R}, +, \cdot)$, and the complex numbers $(\mathbb{C}, +, \cdot)$ are all (infinite) fields. We also have seen (Theorem 15.20) that, for p a prime, $\mathbb{F}_p = (\mathbb{Z}/p\mathbb{Z}, +, \cdot)$, where addition and multiplication are mod p, is a finite field.

In fact, if p is a prime and n is a positive integer, then there is exactly one (up to isomorphism) finite field of order p^n, and these are the only finite fields. The proof of this fact and a study of finite fields will be relegated to Chapter 27.

Quadratic Number Fields. Just as the ring of integers $\mathbb{Z}$ is a subset of the field of rational numbers $\mathbb{Q}$, the quadratic integer rings $\mathbb{Z}[\sqrt{d}]$ and, for $d = 1 \bmod 4$, $\mathbb{Z}[\frac{1+\sqrt{d}}{2}]$ live inside fields that play an important role in (algebraic) number theory.

Let d be an integer (positive or negative) that is not divisible by the square of a prime. Define

$$\mathbb{Q}[\sqrt{d}] = \{a + b\sqrt{d} \mid a, b \in \mathbb{Q}\}.$$

It is clear that $\mathbb{Q}[\sqrt{d}]$ is closed under addition, subtraction, and multiplication. To see that we can divide by non-zero elements, note that

$$\frac{1}{a + b\sqrt{d}} = \frac{a - b\sqrt{d}}{(a + b\sqrt{d})(a - b\sqrt{d})} = \frac{a}{a^2 - db^2} - \frac{b}{a^2 - db^2}\sqrt{d} \in \mathbb{Q}[\sqrt{d}].$$

It follows that $\mathbb{Q}[\sqrt{d}]$ is a field. It is called a *quadratic number field*.

Problems

15.2.1. We know that $\mathbb{F}_3 = (\mathbb{Z}/3\mathbb{Z}, +, \cdot)$ is a field. Can you find a ring with three elements that is not a field?

15.2.2. Explicitly give the addition and multiplication table of all fields with two and three elements.

HW #1

15.2.3. **Direct product of rings.** Let R and S be rings, and let $T = R \times S = \{(r, s) \mid r \in R, s \in S\}$ be the direct product of R and S. As usual, we use the operations in R and S to define addition and multiplication in T. Thus $(a, b) + (a', b') = (a + a', b + b')$ and $(a, b) \cdot (a', b') = (a \cdot a', b \cdot b')$, where the operations in the first coordinate are the operations of R and the operations in the second coordinate are the operations of S.
  (a) Show that T is a ring.
  (b) If R and S are commutative rings, then show that T is a commutative ring.
  (c) If R and S are rings with identity, then show that T is a ring with identity.

HW #)

15.2.4. Let R and S be integral domains, and let $T = R \times S$ be the direct product of R and S. Is T necessarily an integral domain? Can T ever be an integral domain?

15.2.5. Find all the units of $\mathbb{Z}/3\mathbb{Z} \times \mathbb{Z}/5\mathbb{Z}$.

15.2.6. Is $\mathbb{Z}/2\mathbb{Z} \times \mathbb{Z}/2\mathbb{Z}$ a field? Is $\mathbb{Z}/4\mathbb{Z}$ a field? Can you find a field with four elements? If so, give its addition and multiplication tables explicitly.

15.2.7. Let $\mathbb{F}_2 = (\mathbb{Z}/2\mathbb{Z}, +, \cdot)$, and define $E = \{\begin{bmatrix} a & b \\ b & a+b \end{bmatrix} \mid a, b \in \mathbb{F}_2\}$. How many elements does E have? With the usual matrix addition and multiplication, is E a field?

15.2.8. With the usual addition and multiplication of real numbers, are either of the following sets of real numbers a ring?

(a)
$$\{a + b\sqrt{47} \mid a, b \in \mathbb{R}, \ a^2 + b^2 \leq 1\}.$$

(b)
$$\left\{\frac{n}{47^m} \mid n \text{ is an integer and } m \text{ is a non-negative integer}\right\}.$$

15.2.9. Define $+$ and $\odot$ on $\mathbb{R}^2$ as follows:
$$(a, b) + (c, d) = (a + c, b + d),$$
$$(a, b) \odot (c, d) = (ac, ad + bc).$$

Is $(\mathbb{R}^2, +, \odot)$ a ring? Is it commutative? Does it have an identity? Is it an integral domain?

15.2.10. Define $+$ and $\odot$ on $\mathbb{R}^2$ as follows:
$$(a, b) + (c, d) = (a + c, b + d),$$
$$(a, b) \odot (c, d) = (ac - bd, ad + bc).$$

Is $(\mathbb{R}^2, +, \odot)$ a ring? Is it commutative? Does it have an identity? Is it an integral domain? Is it a field? Have you seen this set and these operations elsewhere?

15.2.11. Let X be a non-empty set, and recall (Definition 2.20) that 2^X is the set of all subsets of X, and for A and B subsets of X, their *symmetric difference* is denoted by $\triangle$ and is defined by
$$A \triangle B = (A - B) \cup (B - A).$$

Show that $(2^X, \triangle, \cap)$ is a commutative ring with identity. Is it an integral domain?

HW #1

15.2.12. Let F be a field, and define
$$R = \left\{ \begin{bmatrix} a & -b \\ b & a \end{bmatrix} \mid a, b \in F \right\}.$$

Show that, with the usual matrix addition and multiplication, R is a commutative ring with identity. Decide if R is a field, for each of the following choices for F: $\mathbb{Q}$, $\mathbb{R}$, $\mathbb{C}$, $\mathbb{F}_5 = \mathbb{Z}/5\mathbb{Z}$, and $\mathbb{F}_7 = \mathbb{Z}/7\mathbb{Z}$.

15.2.13. Find the group of units of $\mathbb{Z}/5\mathbb{Z}$, $\mathbb{Z}/6\mathbb{Z}$, $\mathbb{Z}/12\mathbb{Z}$, and $\mathbb{Z}/24\mathbb{Z}$.

15.2.14. Let R be a ring with identity. Let n be a positive integer and assume that $n1 = \underbrace{1 + 1 + \cdots + 1}_{n} = 0$. Show that $nx = 0$ for every $x \in R$.

15.2.15. Let D be an integral domain. Assume that there exists a positive integer n such that $n1 = 0$. Prove that the smallest positive integer n with $n1 = 0$ is a prime number.

15.2.16. Is $\{a + b\frac{1+\sqrt{5}}{2} \mid a, b \in \mathbb{Z}\}$ an integral domain? What about $\{a + b\frac{1+\sqrt{7}}{2} \mid a, b \in \mathbb{Z}\}$?

HW #1

15.2.17. Let d be a square-free integer with remainder 1 when divided by 4. Show that $\mathbb{Z}[\frac{1+\sqrt{d}}{2}]$ is an integral domain. Is the condition $d = 1 \bmod 4$ necessary?

15.2.18. Let $R = \{\frac{a+b\sqrt{-3}}{2} \mid a, b \in \mathbb{Z}, a \text{ and } b \text{ both even or both odd}\}$. Is R a commutative ring with identity? Is it an integral domain? Identify $R^\times$, the group of units of R.

15.2.19. Find a familiar group that is isomorphic to $GL(2, \mathbb{Z}/6\mathbb{Z})$, the group of units of $M_{2\times 2}(\mathbb{Z}/6\mathbb{Z})$.

15.2.20. Consider the ring $M_{2\times 2}(\mathbb{Z}/8\mathbb{Z})$. Can you find a unit in this ring with determinant 3?

15.2.21. Let d be a square-free integer. Show that the field $\mathbb{Q}[\sqrt{d}]$—in addition to being a field—is a vector space over $\mathbb{Q}$ (to be a vector space *over* $\mathbb{Q}$ means that the scalars are from $\mathbb{Q}$). What is its dimension?

15.2.22. Let d be a square-free integer. Which elements of $\mathbb{Q}[\sqrt{d}]$ are roots of monic polynomials with integer coefficients? (A polynomial is *monic* if the coefficient of its highest degree is 1.) By looking at examples, make a conjecture. Is the case when $d = 1 \bmod 4$ different than the others?[5]

15.2.23. Let $a < b$ be two real numbers, and let $R = \mathbf{C}([a, b], \mathbb{R})$ be the ring of continuous functions from $[a, b]$ to $\mathbb{R}$. Functions in this ring are added and multiplied pointwise. This means that if f and g are in R, then $f + g$ and fg are both functions from $[a, b]$ to $\mathbb{R}$ and are defined by

$$(f + g)(x) = f(x) + g(x), \quad (fg)(x) = f(x)g(x) \quad \text{for all } x \in [a, b].$$

Is R a commutative ring with identity? Is it an integral domain?

15.2.24. Assume $F_1, F_2, \ldots, F_n, \ldots$ is an infinite sequence of fields with

$$F_1 \subseteq F_2 \subseteq \cdots \subseteq F_n \subseteq \cdots.$$

Is $\bigcup_{i=1}^{\infty} F_i$ a field?

The Quaternion Ring. Let $Q_8 = \{\pm 1, \pm i, \pm j, \pm k\}$ be the quaternion group of order 8. Recall that 1 is the identity, -1 gets multiplied as usual, and the rest are multiplied similar to the cross product of vectors, i.e.,

$$ij = k, jk = i, ki = j, ji = -k, kj = -i, ik = -j, i^2 = j^2 = k^2 = -1.$$

Now let F be any field for which $-1 \neq 1$, and define

$$Q = \{a + bi + cj + dk \mid a, b, c, d \in F\}.$$

We can add two elements of Q by adding the like terms:

$$(a + bi + cj + dk) + (a' + b'i + c'j + d'k) = (a + a') + (b + b')i + (c + c')j + (d + d')k.$$

We can also multiply two elements of Q by extending the product in Q_8 to all of Q. So for example,

$$(1 + 2i + 3j)(i + k) = i + k + 2i^2 + 2ik + 3ji + 3jk = -2 + 4i - 2j - 2k.$$

---

[5]A complex number that is a root of a monic polynomial with integer coefficients is called an *algebraic integer*.

Q with the above addition and multiplication is called the *ring of quaternions over F*. In fact, Q is a division ring (see Problem *15.2.25*).

15.2.25. Let F be a field with $-1 \neq 1$. Let Q be the ring of quaternions over F. That is, $Q = \{a + bi + cj + dk \mid a, b, c, d \in F\}$. Find the multiplicative inverse of a typical non-zero element. Conclude that Q is a division ring.

15.2.26. As usual, let $\mathbb{R}$ denote the real numbers, and let Q be the ring of quaternions over $\mathbb{R}$. Show that there are an infinite number of solutions for $x^2 = -1$ in Q.

15.3. Finite Integral Domains

Focusing on integral domains, we may want to start with finite integral domains. In this short section, we show that the only finite integral domains are fields. It is also true—although harder to prove—that the only finite division rings are also fields. Hence, in the study of ring theory—unlike our treatment of groups—the emphasis will be on infinite rings.

Definition 15.22 (Cancellation laws). Let $(R, +, \cdot)$ be a ring. We say that R has *left cancellation* if whenever $ab = ac$ and $a \neq 0$ we can conclude that $b = c$. Right cancellation is defined similarly.

Proposition 15.23. *Let $(R, +, \cdot)$ be a ring.*

(a) *If R has left (right) cancellation, then R has no left (right) zero-divisors.*

(b) *Assume R is a commutative ring with identity. Then R is an integral domain if and only if R has left (and therefore right) cancellation.*

Proof. (a) Let a be a non-zero element of R, and assume $ab = 0$. We have $ab = a0$, and, if R has left cancellation, we can cancel a to get $b = 0$. This proves that a is not a left zero-divisor.

(b) One direction follows from the previous part. Now assume R is an integral domain and $ab = ac$ with $a \neq 0$. We have $ab - ac = 0$, and hence $a(b - c) = 0$. Since R has no zero-divisors and $a \neq 0$, we must have $b - c = 0$. Thus $b = c$ as desired. $\qquad \square$

Theorem 15.24. *Let D be a finite integral domain. Then D is a field.*

Proof. Let r be an arbitrary non-zero element of D. To show that D is a field, we need to find a multiplicative inverse for r. Consider $r, r^2, r^3, \ldots, r^k, \ldots$. Since the ring is finite, there must be repeats in this list. Suppose for $i < j$, we have $r^i = r^j$. Thus

$$\underbrace{r \cdot r \cdot \ldots \cdot r}_{i} \cdot 1 = \underbrace{r \cdot r \cdot \ldots \cdot r}_{j}.$$

Use left cancellation repeatedly and get $r^{j-i} = 1$. Now let $s = r^{j-i-1}$, and we have $rs = sr = 1$ and thus $s = r^{-1}$. $\qquad \square$

Theorem 15.25 (Wedderburn's theorem). *Every finite division ring is a field.*

Proof. The proof of this theorem is not as straightforward as the previous one. We will only be able to prove it much later using more sophisticated techniques. See Problem 27.4.10.   □

Problems

15.3.1. Let D be an integral domain, and let $x \in D$. Assume that $x^2 = 1$. Show that $x = \pm 1$.

Idempotent and Nilpotent Elements

Definition 15.26 (Idempotent element). An element b in a ring R is called an *idempotent* if $b^2 = b$.

Definition 15.27 (Nilpotent element). An element a of a ring R is called *nilpotent* if $a^n = 0$ for some natural number n.

15.3.2. Determine all idempotents in the rings $\mathbb{Z}/6\mathbb{Z}$ and $\mathbb{Z}/8\mathbb{Z}$.

15.3.3. What are the possibilities for the number of idempotents in an integral domain?

15.3.4. Let X be a non-empty set, and consider the ring $(2^X, \triangle, \cap)$ of Problem 15.2.11.
   (a) Determine all the idempotents in the ring $(2^X, \triangle, \cap)$.
   (b) When is the ring $(2^X, \triangle, \cap)$ an integral domain?

15.3.5. Let x be a nilpotent element of a ring with identity. Prove that $1 + x$ is a unit.

15.3.6. Let R be a ring with identity containing elements a and b with $ab = b$ and $b^2 = a$. Prove that there is a unit $u \in R$ such that $ub = bu = a$.[6]

---

[6]Adopted from Peck [**Pec02**].

Chapter 16

Homomorphisms, Ideals, and Quotient Rings

... where subrings, ring homomorphisms, ideals, quotient rings, the characteristic of a ring, and prime subfields are introduced, and homomorphism theorems are proved.

In this chapter, we follow a common pattern for studying algebraic objects. We define *subrings* just as we defined subgroups for groups. In group theory, we saw that some subgroups are more important than others. Normal subgroups were exactly those subgroups that are kernels of group homomorphisms, and, given a normal subgroup, we could construct quotient groups. Exactly the same happens for rings. Subrings that are kernels of ring homomorphisms are especially important and are called *ideals*, and, given ideals, we can construct *quotient rings*.

In ring theory the concept of the *characteristic of a ring* helps us distinguish and contextualize rings, integral domains, and fields. As a consequence, we see that each integral domain has a subring isomorphic to either $(\mathbb{Z}, +, \cdot)$ or to $(\mathbb{Z}/p\mathbb{Z}, +, \cdot)$ for some prime p. Each field will have a subfield isomorphic to $(\mathbb{Q}, +, \cdot)$ or $\mathbb{F}_p = (\mathbb{Z}/p\mathbb{Z}, +, \cdot)$ for some prime p.

16.1. Subrings, Homomorphisms, and Ideals

The material in this section (and the next) will follow the development of group theory closely, and most of the proofs will not be surprising. Subrings, homomorphisms, ideals (which play the role of normal subgroups) will be defined in this section, and quotient rings and homomorphism theorems will be discussed in the next section. We ask the reader (who has already seen the corresponding theorems and proofs in group theory) to go through this material quickly.

Subrings. Subrings are defined as you would expect. However, it turns out that the concept of an ideal—to be defined later—will be more central.

Definition 16.1. Let $(R, +, \cdot)$ be a ring, and let S be a non-empty subset of R. Then S is a *subring* of R if $(S, +, \cdot)$ is itself a ring.

The following characterization of subrings is immediate:

Proposition 16.2. *Let $(R, +, \cdot)$ be a ring, and let S be a non-empty subset of R. Then S is a subring of R if and only if*

(a) *For all $s_1, s_2 \in S$, $s_1 + s_2$ and $s_1 s_2$ are in S, and*

(b) *for all $s \in S$, the element $-s$ is in S.*

In other words, a non-empty subset is a subring if it is closed under addition, multiplication, and subtraction.

Example 16.3. We have $\mathbb{Z} \subseteq \mathbb{Q} \subseteq \mathbb{R} \subseteq \mathbb{C}$, and each is a subring of the next one.

If $(S, +, \cdot)$ is a subring of $(R, +, \cdot)$, then $(S, +)$ is a subgroup of the abelian group $(R, +)$ (written additively). Hence, 0_R is also 0_S. The situation with the multiplicative identity is much less predictable. $(R, +, \cdot)$ may not even have an identity, but even it does, there is no reason for $(S, +, \cdot)$ to have an identity. In fact, it is possible for both R and S to be rings with identity and $1_R \neq 1_S$.

Example 16.4. The set $3\mathbb{Z}$ consisting of integer multiples of 3 is a subring of $\mathbb{Z}$. Note that $\mathbb{Z}$ has an identity while $3\mathbb{Z}$ does not.

Example 16.5. The set $\{0, 3\}$, with addition and multiplication mod 6, is a subring of $\mathbb{Z}/6\mathbb{Z}$. Both of these rings are commutative rings with identity. However, the identity for $\{0, 3\}$ is 3, while the identity for $\mathbb{Z}/6\mathbb{Z}$ is 1.

Integral domains behave better in regards to their identities as the next lemma shows.

Lemma 16.6. *Assume that R is an integral domain and S is a non-trivial subring of R with an identity. Then $1_R = 1_S$.*

Proof. We know that $1_S 1_S = 1_S$. But since $1_S \in R$ and 1_R is the identity of R, we have $1_S 1_R = 1_S$. So $1_S 1_S = 1_S 1_R$. In integral domains—and, in particular, in R—we can cancel non-zero elements on the left. We have assumed that S is non-trivial—meaning that it has more than one element—and so $1_S \neq 0_S$. Hence, starting from $1_S 1_S = 1_S 1_R$, we can cancel 1_S on the left, and get $1_S = 1_R$. $\qquad\square$

We define subfields as well. Since fields are integral domains, we know that a subfield must contain the 0 and the 1 of the bigger field.

Definition 16.7. Let $(F, +, \cdot)$ be a field. Let K be a non-empty subset of F. Then K is a *subfield* of F if $(K, +, \cdot)$ is a field itself. In other words, a subset K of F is a subfield if 0_F and 1_F are in K, and, for all $a, b \in K$, we have that $a + b$, $a - b$, and ab are in K, and if $a \neq 0$, then $1/a \in K$.

Ring Homomorphisms. In the study of groups, we saw the importance of homomorphisms. We also saw that normal subgroups—that allowed us to define quotient groups—were precisely those subgroups which were kernels of homomorphisms. For rings we begin with homomorphisms, consider the subrings that occur as kernels of ring homomorphisms, and then construct quotient rings.

Definition 16.8 (Ring homomorphism). Let R and S be rings. A map $\phi : R \to S$ is a *ring homomorphism* if, for all $a, b \in R$,

(a) $\phi(a + b) = \phi(a) + \phi(b)$, and

(b) $\phi(ab) = \phi(a)\phi(b)$.[1]

Definition 16.9 (Isomorphisms, endomorphisms, automorphisms). Let R and S be rings, and let $\phi : R \to S$ be a ring homomorphism.

- If $R = S$, then ϕ is called a ring *endomorphism*.

- If ϕ is 1-1 and onto, then ϕ is called a ring *isomorphism*. In such a case we say R and S are *isomorphic* and write $R \cong S$.

- If $R = S$ and ϕ is 1-1 and onto, then ϕ is called a ring *automorphism*.

Example 16.10. Let $\phi : \mathbb{Z} \to \mathbb{Z}/n\mathbb{Z}$, where $\phi(m)$ is the remainder of m when divided by n. It follows from the basics of modulo arithmetic that ϕ is an onto homomorphism. It is not 1-1.

Example 16.11. If R and S are rings, then the map $\phi \colon R \to S$, defined by $\phi(r) = 0_S$ for all $r \in R$, is a ring homomorphism. This homomorphism is usually called the *trivial* homomorphism or the *zero* homomorphism.

Theorem 16.12. *Let $(R, +, \cdot)$ and $(S, \oplus, \odot)$ be rings, and let $\phi : R \to S$ be a ring homomorphism. Let $a \in R$. Then*

(a) *The map $\phi : (R, +) \to (S, \oplus)$ is a group homomorphism. In particular, $\phi(0) = 0$ and $\phi(-a) = -\phi(a)$.*

(b) *If 1 is the identity of R, then $\phi(1)$ is the identity of $\phi(R)$.*

(c) *If a^{-1} exists, then $\phi(a^{-1})$ is the inverse of $\phi(a)$ in $\phi(R)$.*

(d) *If R' is a subring of R, then the image of R', $\phi(R')$, is a subring of S.*

(e) *If S' is a subring of S, then the inverse image of S', $\phi^{-1}(S')$, is a subring of R.*

Proof. Part (a) follows directly from the definition. For part (b), note that $\phi(1) \odot \phi(r) = \phi(1 \cdot r) = \phi(r)$ and, similarly, $\phi(r) \odot \phi(1) = \phi(r)$. Hence, $\phi(1)$ is the identity of $\phi(R)$. For part (c) we see that $\phi(a^{-1}) \odot \phi(a) = \phi(a^{-1} \cdot a) = \phi(1)$ and, similarly, $\phi(a) \odot \phi(a^{-1}) = \phi(1)$. So $\phi(a^{-1})$ is indeed the inverse of $\phi(a)$ in $\phi(R)$. The proofs of parts (d) and (e) are straightforward and are left to the reader (Problem 16.1.4). $\qquad\square$

Example 16.13. Let $\phi : \mathbb{Z}/2\mathbb{Z} \to \mathbb{Z}/6\mathbb{Z}$ be defined by $\phi(0) = 0$ and $\phi(1) = 3$. Then ϕ is a ring homomorphism as can directly be checked. Note that 3 is the identity of $\phi(\mathbb{Z}/2\mathbb{Z}) = \{0, 3\}$ and *not* the identity of $\mathbb{Z}/6\mathbb{Z}$. Also 3 is the (multiplicative) inverse of 3 in $\phi(\mathbb{Z}/2\mathbb{Z}) = \{0, 3\}$ and *not* the inverse of 3 in $\mathbb{Z}/6\mathbb{Z}$.

Kernels of homomorphisms always play a special role.

---

[1] A word of caution: Authors who only consider rings with identity, often, in the definition of a ring homomorphism, also assume $\phi(1_R) = 1_S$.

Definition 16.14 (Kernels). Let R and S be rings, and let $\phi : R \to S$ be a ring homomorphism. The *kernel* of ϕ is denoted by $\ker(\phi)$ is $\phi^{-1}(\{0_S\}) = \{r \in R \mid \phi(r) = 0_S\}$.

If $(R, +, \cdot)$ and $(S, +, \cdot)$ are rings and $\phi : R \to S$ is a ring homomorphism, then $\phi : (R, +) \to (S, +)$ is a group homomorphism, and the kernel of the ring homomorphism is the same as the kernel of the group homomorphism. Thus we can use properties of group kernels and homomorphisms. In particular,

Proposition 16.15. *Let R and S be rings, and let $\phi : R \to S$ be a ring homomorphism. Then ϕ is 1-1 if and only if $\ker(\phi) = \{0_R\}$.*

Example 16.16. Let $\phi : \mathbb{Z} \to \mathbb{Z}/n\mathbb{Z}$ where $\phi(m)$ is the remainder of m when divided by n. Then $\ker(\phi) = n\mathbb{Z}$. We have the homomorphism diagram of Figure 16.1.

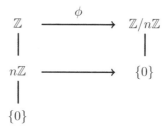

Figure 16.1. For $m \in \mathbb{Z}$, $\phi(m) = m \bmod n$.

Ideals. We first prove an essential property of kernels of homomorphisms and then use it to identify which subrings can be kernels. Precisely these special subrings will be called *ideals*.

Theorem 16.17. *Let R and S be rings, and let $\phi : R \to S$ be a ring homomorphism. Let $r \in \ker(\phi)$, then, for all $x \in R$, xr and rx are elements of $\ker(\phi)$.*

Proof. To prove that xr (or rx) are in the kernel, we show that their image is zero. This is straightforward: $\phi(xr) = \phi(x)\phi(r) = \phi(x) \cdot 0 = 0$, and $\phi(rx) = \phi(r)\phi(x) = 0$. $\qquad\square$

If ϕ is a ring homomorphism with the ring R as its domain, then $\ker(\phi)$ is a subring of R. Hence, if we multiply (or add or subtract) two elements of $\ker(\phi)$, we get an element of $\ker(\phi)$. We, however, proved that when you multiply an element of $\ker(\phi)$ by *any* element of the ring, then we get something in $\ker(\phi)$.

Definition 16.18 (Ideals). Let $(R, +, \cdot)$ be a ring, and let $I \subseteq R$. I is a *left ideal* of R if

(a) I is a subring of R, and

(b) $r \cdot s$ is in I for all $r \in R$ and $s \in I$.

Right ideals are defined analogously. I is an *ideal* (or a *two-sided ideal*) of R if it is both a left and a right ideal.

In other words, I is a left ideal of R if $rI \subseteq I$ for all $r \in R$, and I is a right ideal of R if $Ir \subseteq I$ for all $r \in R$.

The set $\{0\}$, consisting of only the zero element, and the whole ring R are called the *trivial* ideals of R. If I is an ideal of a ring R, we sometimes denote this by writing $I \leq R$. (Some authors use the notation $I \lhd R$.)

Corollary 16.19. *If $\phi : R \to S$ is a ring homomorphism, then $\ker(\phi)$ is an ideal of R.*

Proof. An immediate consequence of Theorem 16.17. $\qquad\square$

Example 16.20. Let n be a positive integer, then the set of multiples of n, $n\mathbb{Z}$, is an ideal of $\mathbb{Z}$. This is because the multiples of n form a subring, and if you multiply a multiple of n by *any* integer, then you get a multiple of n. In Example 16.16, we saw that $n\mathbb{Z}$ is indeed the kernel of a homomorphism with $\mathbb{Z}$ as its domain.

Definition 16.21 (The ideal generated by a subset)**.** Let R be a ring. If X is a subset of R, then $\langle X \rangle$, the *ideal generated by* X, is the smallest ideal of R that contains X.

For the ideal generated by X, we use the notation $\langle X \rangle$. An alternate notation, used by many authors, is (X).

Remark 16.22. Recall (see Definition 2.66) that to say $\langle X \rangle$ is the *smallest* (or the *unique minimal*) ideal of R that contains X means that $\langle X \rangle$ is a subset of R with all of the following properties:

(a) $\langle X \rangle$ is an ideal of R,

(b) $X \subseteq \langle X \rangle$, and

(c) if I is another ideal of R that contains X, then $\langle X \rangle \subseteq I$.

Thus, we actually need to prove that $\langle X \rangle$ exists. A priori, it is possible that there would be two ideals of R containing X such that neither contains the other. We need to show that in such a case there is a third ideal that is contained in both of the original ideals and still contains X. This is not hard since the intersection of two ideals is still an ideal. In fact, the smallest ideal containing X will be the intersection of all ideals that contain X. See the Problems.

In the case of commutative rings with identity, an ideal generated by a finite set of elements is the collection of *linear combinations* of these elements (where the scalars are the ring elements).

Lemma 16.23. *Let R be a commutative ring with identity. Let $a_1, \dots, a_n \in R$, then $\langle a_1, \dots, a_n \rangle = \{r_1 a_1 + \cdots + r_n a_n \mid r_1, \dots, r_n \in R\}$.*

Proof. Let $I = \{r_1 a_1 + \cdots + r_n a_n \mid r_1, \dots, r_n \in R\}$. Then I is a subring since it is closed under addition, subtraction, and multiplication. It is an ideal since if you multiply any element of I by any element of R, then you get an element of I. Now, $a_1, \dots, a_n$ are all elements of I, and, hence, I is an ideal containing $a_1, \dots, a_n$.

Further, any ideal that contains $a_1, \ldots, a_n$ will, by definition of an ideal, contain $r_1a_1, \ldots, r_na_n$ for any $r_1, \ldots, r_n \in R$. Hence, it will also contain $r_1a_1 + \cdots + r_na_n$. We conclude that any ideal containing $a_1, \ldots, a_n$ will contain all of I.

We conclude that I is the smallest ideal of R containing $a_1, \ldots, a_n$, and, hence, it is the ideal generated by these. $\square$

Example 16.24. Let $R = \mathbb{Z}$. In this ring, $\langle 3 \rangle$ consists of all integer multiples of 3. Now what is $\langle 4, 6 \rangle$, the ideal generated by 4 and 6? It consists, by Lemma 16.23, of all integer linear combinations of 4 and 6. A moment's reflection will convince you that $\langle 4, 6 \rangle = \langle 2 \rangle$. Note the similarity between this and the fact that the greatest common divisor of 4 and 6 is 2. As you may realize, this is not a coincidence. (In fact, many authors denote both $\langle 4, 6 \rangle$ and $\gcd(4, 6)$ by $(4, 6)$.)

Example 16.25. Let $R = \mathbb{Z}[x]$ be the ring of polynomials with integer coefficients. Let $I = \langle 2, x \rangle$ be the ideal generated by 2 and x. Then the elements of I are all polynomials of the form $2p + xq$ where p and q are arbitrary polynomials with integer coefficients. These are exactly the set of polynomials with integer coefficients where the constant term is even.

Example 16.26. Let $R = 3\mathbb{Z}$. What is $\langle 9 \rangle$? Since R does not have a unity, we have to be careful. The ideal generated by 9 is not $9R$, the multiples of 9 in R, since, in R, the element 9 itself is not a multiple of 9! In fact, the ideal generated by 9 is $9R + 9\mathbb{Z} = 9\mathbb{Z}$. (See Problem *16.1.19*.)

The role of cyclic groups in group theory is taken up by principal ideals:

Definition 16.27 (Principal ideals). Let R be a ring, and let I be an ideal of R. I is called a *principal ideal* if $I = \langle a \rangle$ for some $a \in R$.

Definition 16.28 (Principal ideal rings, principal ideal domains). Let R be a ring. R is called a *principal ideal ring* if every ideal of R is principal.

If R is an integral domain *and* a principal ideal ring, then R is called a *principal ideal domain*. Often, we use the abbreviation PID for a principal ideal domain.

Example 16.29. What are the ideals of $\mathbb{Z}/6\mathbb{Z} = \{0, 1, 2, 3, 4, 5\}$?
$$\langle 0 \rangle = \{0\}$$
$$\langle 1 \rangle = \langle 5 \rangle = \mathbb{Z}/6\mathbb{Z}$$
$$\langle 2 \rangle = \langle 4 \rangle = \{0, 2, 4\}$$
$$\langle 3 \rangle = \{0, 3\}$$
And these are the only ideals, as can easily be checked. Hence, $\mathbb{Z}/6\mathbb{Z}$ is a principal ideal ring but not a PID since it is not an integral domain.

Example 16.30. Let $a < b$ be two real numbers, and let $R = \mathbf{C}([a, b], \mathbb{R})$ be the ring of continuous functions from $[a, b]$ to $\mathbb{R}$. Addition and multiplication of functions is pointwise (see Problem 15.2.23). The ring R is a commutative ring with identity. The function $\mathbf{1}: [a, b] \to \mathbb{R}$ defined by $\mathbf{1}(x) = 1$, for all $x \in [a, b]$, is the identity element.

Now, let c be a real number with $a < c < b$, and define $I = \{f \in R \mid f(c) = 0\}$. The set I is easily seen to be an ideal of R. We claim that I is not a principal ideal.

Assume $I = \langle h \rangle$ for $h \in R$. Let $f \in I$ be a straight line through $(c, 0)$ with a non-zero slope. Then $f = gh$ for some $g \in R$. Since $f(x) \neq 0$ for $x \neq c$, we have $h(x) = 0$ if and only if $x = c$. Since h is continuous, we also must have $\lim_{x \to c} h(x) = h(c) = 0$.

Now note that $h^{1/3} \in I$, and, hence, $h^{1/3} = gh$ for some $g \in R$. This means that $g = h^{-2/3}$ for $x \neq c$ (since for these values of x, we have $h(x) \neq 0$). But then $\lim_{x \to c} g$ does not exist. This means that $g \notin R$, which is a contradiction.

We conclude that R is not a principal ideal ring.

In ring theory—just as in the study of any algebraic object—we tend to translate questions of interest to questions about substructures. As an example, we characterize fields based on their ideal structures. More examples of this approach will be seen in future chapters.

Definition 16.31. Let R be a ring. Then $\{0\}$ and R are called the *trivial* ideals of R. A ring is *simple* if it has exactly two ideals namely the trivial ones.

Proposition 16.32. *Let R be a commutative ring with identity. Then R is a field if and only if R is simple.*

Proof. ($\Rightarrow$) Assume that R is a field. Let J be an ideal of R, and assume that J contains at least one non-zero element r. Since R is a field, r is a unit, and so $r^{-1} \in R$. Now when we multiply any element of the ring with an element in the ideal, we get something in the ideal. Hence $1 = r^{-1}r \in J$. Now if s is any arbitrary element of R, then $s1 \in J$—again something in R times something in J is in J—which means that every element of R is in J. Hence $J = R$ is a trivial ideal.

($\Leftarrow$) Assume that R is simple. To show that R is a field, we have to show that $(R - \{0\}, \cdot)$ is an abelian group. We already know that the product is associative and commutative and we have a 1. Hence, we only need closure and inverses. Since $(R, \cdot)$ is closed, the only way closure could fail in $(R - \{0\}, \cdot)$ is if the product of two non-zero elements was zero (in other words, if we had zero-divisors). But if $xy = 0$ and x is invertible, then we have $y = x^{-1}xy = 0$. Hence, all we have to show is that every non-zero element in R has a (multiplicative) inverse in R. Now let r be a non-zero element of R. Consider the ideal $\langle r \rangle = \{rs \mid s \in R\}$. Since R is simple and $\langle r \rangle$ is not the zero ideal, we have $\langle r \rangle = R$. Since $1 \in R$, we have $1 \in \langle r \rangle$, and, for some $s \in R$, we have $rs = 1$. This means that r has an inverse, and hence R is a field. $\square$

Problem 16.1.24 generalizes one direction of Proposition 16.32 to non-commutative rings.

We will now define maximal ideals. These will be important in the future chapters.

Definition 16.33 (Maximal ideals). Let R be a ring. A proper ideal M of R is said to be a *maximal* ideal of R if, other than M itself, there exists no proper ideal of R that contains M.

Remark 16.34 (Axiom of choice and maximal ideals). Is an ideal of a ring always contained in a maximal ideal? To properly answer this question, we would have to veer off into the foundations of set theory and discuss the so-called *axiom of choice* or its equivalent the *Kuratowski–Zorn lemma* (almost universally known as Zorn's Lemma). The axiom of choice is an axiom about infinite sets that basically says that if you have a family of non-empty sets—even an uncountable family of non-empty sets—then you can create a new set by choosing one element from each of the sets in the family. On the face of it, this seems like a reasonable axiom, and maybe even provable from other axioms. This is not so. The axiom of choice is independent of, and consistent with, the usual axioms of set theory (the so-called Zermelo–Fraenkel axioms), and hence it—or its negation—can be adopted as a new axiom. There are a number of other axioms that end up being equivalent to the axiom of choice. The Well-Ordering Theorem—not discussed here and not to be confused with the well-ordering principle, Axiom 1.39—and the Kuratowski–Zorn lemma are two examples. In this text, we assume the axiom of choice when needed, but, in algebra, often the equivalent Kuratowski–Zorn lemma is more readily useful. We state it as an axiom. Recall that a *partially ordered set* is a set together with a relation that is transitive, reflexive, and anti-symmetric (Definition 9.2) and a *chain* in a poset is a totally ordered subset of the poset (Definition 9.17).

Axiom 16.35 (Kuratowski–Zorn lemma a.k.a. Zorn's Lemma). Let P be a non-empty partially ordered set. If every chain in P has an upper bound, then P has at least one maximal element.

One consequence of the Kuratowski–Zorn lemma is the existence of maximal ideals.

Corollary 16.36. *Every proper ideal of a non-trivial ring with identity is contained in a maximal ideal.*

Proof. Let R be a non-trivial ring with identity, and let I be a proper ideal of R. Let P be the poset of (two-sided) proper ideals of R that contain I, ordered by inclusion. If $I_1 \subseteq I_2 \subseteq \cdots$ is a chain of ideals in P, then let $J = \bigcup_{i=1}^{\infty} I_i$. Then J is an ideal containing I and an upper bound for the chain of ideals. (If $x, y \in J$, then $x \in I_n$ and $y \in I_m$ for some positive integers n and m, and that x, y, and $x + y$ are elements of $I_\ell \subseteq J$ where $\ell = \max(n, m)$. Similarly, for $r \in R$, both rx and xr are elements of $I_n \subseteq J$.) Hence, by the Kuratowski–Zorn lemma, P will have a maximal element. That maximal element is a maximal ideal containing I. $\qquad\square$

Another consequence of the Kuratowski–Zorn lemma is that all vector spaces have bases (see Problem 16.1.30). While many mathematicians accept the axiom of choice when need be, its acceptance does result in counterintuitive results. For example, using the axiom of choice, one can prove that a solid three-dimensional ball can be partitioned into a finite number of pieces which can then be reassembled by just translating and rotating the pieces in such a way as to result in *two* new solid balls identical to the original![2]

---

[2]This is the amazing Banach–Tarski paradox. See Wagon [**Wag93**]. For more on the axiom of choice and its uses in elementary ring theory, see Becker and Weispfenning [**BW93**, pp. 141-149].

Problems

16.1.1. If D is an integral domain and R is a subring of D with at least two elements, then is R necessarily an integral domain? Either prove that it is, or give an example where it is not.

16.1.2. Let $(G, +)$ be an abelian group. Can you define a multiplication for elements of G in such a way that $R = (G, +, \cdot)$ is a ring and so that the ideals of the ring R are exactly the additive subgroups of the original abelian group G?

16.1.3. The *center* of the ring R is $\mathbf{Z}(R) = \{x \in R \mid xr = rx \text{ for all } r \in R\}$. Show that $\mathbf{Z}(R)$ is a commutative subring of R, and R is commutative if and only if $\mathbf{Z}(R) = R$.

16.1.4. **Proof of Theorem 16.12(d) and 16.12(e).** Let R and S be rings, and let $\phi \colon R \to S$ be a ring homomorphism. Let R' and S' be subrings, respectively, of R and S. Prove that $\phi(R')$ and $\phi^{-1}(S')$ are subrings, respectively, of S and R.

16.1.5. Let R be a ring, and let

$$\mathrm{Aut}(R) = \{\phi : R \to R \mid \phi \text{ is a ring automorphism}\}.$$

Let $\circ$ denote function composition, and prove that $(\mathrm{Aut}(R), \circ)$ is a group.

16.1.6. Let R be a ring with identity, and let D be an integral domain. Let $\phi \colon R \to D$ be a non-trivial ring homomorphism. Show that $\phi(1_R)$ is the identity of D.

16.1.7. Let R be a ring with identity. How many ring homomorphisms $\phi \colon \mathbb{Z} \to R$ are there with $\phi(1) = 1_R$?

16.1.8. Let F and E be fields, and let $\phi \colon F \to E$. Show that ϕ is a non-trivial ring homomorphism if and only if, for all $\alpha, \beta \in F$,
  (a) $\phi(0_F) = 0_E$,
  (b) $\phi(1_F) = 1_E$,
  (c) $\phi(\alpha + \beta) = \phi(\alpha) + \phi(\beta)$,
  (d) $\phi(\alpha - \beta) = \phi(\alpha) - \phi(\beta)$,
  (e) $\phi(\alpha\beta) = \phi(\alpha)\phi(\beta)$, and
  (f) $\phi(\alpha/\beta) = \phi(\alpha)/\phi(\beta)$ as long as $\beta \neq 0$.

16.1.9. The elements of the set $\mathbb{Z}_{(5)}$ are rational numbers. A rational number q is in the set $\mathbb{Z}_{(5)}$ if and only if q can be written as a/b where a and b are integers, $\gcd(a, b) = 1$, and b is not divisible by 5. Is $\mathbb{Z}_{(5)}$ a ring? Is it a field? Can you find a non-trivial ideal of $\mathbb{Z}_{(5)}$? If the answer is yes, explicitly construct such an ideal, and if the answer is no, give a reason.

16.1.10. Let $R = \mathbb{Q}[\sqrt{2}]$ and $S = \mathbb{Q}[\sqrt{3}]$. Show that the only ring homomorphism from R to S is the trivial one. In particular, conclude that R and S are not isomorphic rings. In other words, assume $f : R \to S$ is a ring homomorphism. Show that $f(r) = 0$ for all $r \in R$.

16.1.11. Let R be a commutative ring.
  (a) If I_1 and I_2 are ideals of R, show that $I_1 \cap I_2$ is also an ideal of R.

(b) Let $X \subseteq R$. Show that $\langle X \rangle$ exists. In other words, show that there is a *smallest* ideal of R that contains X.

16.1.12. Let I be the ideal generated by 3 in $\mathbb{Z}/36\mathbb{Z}$. Show that as a group I is isomorphic to $(\mathbb{Z}/12\mathbb{Z}, +)$. Show that the ring I is not isomorphic to the ring $(\mathbb{Z}/12\mathbb{Z}, +, \cdot)$. Similarly show that in $\mathbb{Z}/36\mathbb{Z}$, $\langle 6 \rangle$ is not isomorphic as a ring to $\mathbb{Z}/6\mathbb{Z}$.

16.1.13. (a) In $(\mathbb{Z}/12\mathbb{Z}, +, \cdot)$ is the ideal generated by 6 isomorphic to $(\mathbb{Z}/2\mathbb{Z}, +, \cdot)$?
(b) In $(\mathbb{Z}/12\mathbb{Z}, +, \cdot)$ is the ideal generated by 4 isomorphic to $(\mathbb{Z}/3\mathbb{Z}, +, \cdot)$?

16.1.14. Let $R = M_{2\times2}(\mathbb{R})$ be the ring of two-by-two matrices with real entries. Let $S = \{ \begin{bmatrix} a & b \\ 0 & a \end{bmatrix} \mid a, b \in \mathbb{R} \}$ and $T = \{ \begin{bmatrix} 0 & b \\ 0 & 0 \end{bmatrix} \mid b \in \mathbb{R} \}$.
(a) Are T and S subrings of R?
(b) Is T an ideal of S? Is T an ideal of R? Is S an ideal of R?

16.1.15. Can you find a proper non-trivial ideal of $\mathbb{Z}/3\mathbb{Z} \times \mathbb{Z}/5\mathbb{Z}$?

16.1.16. Let $\mathbb{Q}$ be the field of rational numbers, and let $R = \mathbb{Q}[x]$. Is the ideal generated by 2 and x in R a principal ideal?

16.1.17. Let $\mathbb{Z}$ be the ring of ordinary integers, and let $R = \mathbb{Z}[x]$. Is the ideal generated by 2 and x in R a principal ideal?

16.1.18. Let R be a (not necessarily commutative) ring with identity, and let $a \in R$. Show that
$$\langle a \rangle = \{ras \mid r, s \in R\}.$$

16.1.19. Let R be a commutative ring, and let $d \in R$. Assume R does *not* have an identity. Show that $\langle d \rangle = \{rd + nd \mid r \in R, n \in \mathbb{Z}\}$.

16.1.20. Let R be a ring with identity, and let J be an ideal of R. Assume that J contains a unit of R. Prove that $J = R$.

16.1.21. Let R be a commutative ring with identity. Show that R is a field if and only if $\{0\}$ is a maximal ideal.

16.1.22. Let R be a ring, and let F be a field. Let $\phi: F \to R$ be a ring homomorphism. Assume ϕ is not trivial. In other words, there exists $x \in F$ with $\phi(x) \neq 0$. Show that ϕ must be 1-1.

16.1.23. Let R be a commutative ring with identity. Let I be a non-trivial ideal. Assume that I contains no zero-divisors of R. Show that R is an integral domain.

16.1.24. Let R be a simple ring with identity (not necessarily commutative). Show that the center of R is a field. (This generalizes one direction of Proposition 16.32.)

16.1.25. **Chinese Remainder Theorem.** Let m and n be relatively prime positive integers, and let $N = nm$. Define $f \colon \mathbb{Z}/N\mathbb{Z} \to \mathbb{Z}/n\mathbb{Z} \times \mathbb{Z}/m\mathbb{Z}$ by $f(x) = (x \mod n, x \mod m)$.
(a) Show that f is a ring homomorphism.
(b) Show that $\ker(f) = \{0\}$.
(c) Show that f is a ring isomorphism.
(d) I have a mystery number and I am willing to tell you its remainders when divided by 5 and by 47. Based on that information, will you be

able to determine the remainder of the mystery number when divided by 235?

16.1.26. Let A and B be commutative rings with identity, and let $R = A \times B$.

    (a) Let I and J be ideals of A and B respectively. Show that $I \times J$ is an ideal of R.

    (b) Let K be an ideal of R. Show that there exists ideals I and J, of A and B respectively, such that $K = I \times J$.

16.1.27. Let D be a division ring, and let $a \in D$. Define the *centralizer* of a in D to be $\mathbf{C}_D(a) = \{b \in D \mid ab = ba\}$.

    (a) Show that $\mathbf{C}_D(a)$ is a division ring itself (a *subdivision ring* of D).

    (b) Show that $\mathbf{Z}(D)$, the center of D, is a subfield of D.

16.1.28. Let R be a commutative ring with identity, and let I be a proper ideal of R. Then show that I is a maximal ideal if and only if for all $a \in R - I$, there exists $r \in R$ and $b \in I$ with $b + ar = 1$.

16.1.29. In Problem 11.7.11, you showed that a divisible abelian group does not have any maximal subgroups. Starting from this fact, construct a ring with no maximal ideals.[3] Does this contradict Corollary 16.36?

16.1.30. Let F be a field, and let $V \neq \{0\}$ be a vector space over F. Use the Kuratowski–Zorn lemma (Axiom 16.35) and show that V has a basis.

16.2. Quotient Rings and Homomorphism Theorems

If $(R, +, \cdot)$ is a ring and I is an ideal of R, then $(R, +)$ is an abelian group, and $(I, +)$ is a normal subgroup of $(R, +)$. Hence, we can construct the cosets of I in R, and the set of cosets, R/I, is an abelian group with addition of cosets as its operation. In this section, we will show that we can also define a multiplication on R/I. Abusing notation, we call these operations (on cosets) $+$ and $\cdot$ also, and we will have constructed the *quotient ring* $(R/I, +, \cdot)$. Of course, $(R/I, +)$ is an abelian group, even if I is just a subring (as opposed to an ideal). However, only when I is an ideal, will $(R/I, +, \cdot)$ be a ring.

Recall the definition of a coset.

Definition 16.37 (Cosets). Let $(R, +, \cdot)$ be a ring, let I be an ideal of R, and let $r \in R$. Then the set $r + I = \{r + a \mid a \in I\}$ is called a *coset* or the *residue class of r modulo I*.

We emphasize that this is *not* a new definition. The set $r + I$ is a coset of the subgroup I in the abelian group $(R, +)$—written additively since the operation is $+$—and, hence, these cosets have the same properties as cosets of any subgroup of an abelian group. We recall some of the properties of cosets in the following lemma:

Lemma 16.38. *Let R be a ring, and let I be an ideal of R. Then*

(a) *The set of cosets of I in R partition R.*

---

[3]See Malcolmson and Okoh [**MO00**] (and Problem 19.7.11) where other examples of rings with no maximal ideals are given.

(b) *Let $r, s \in R$, then the following are equivalent:*
    (i) $r + I = s + I$,
    (ii) $s - r \in I$,
    (iii) $s = r + a$ *for some $a \in I$.*

Proof. These results repeat Corollary 5.2 and Lemma 5.6 for the case of a subgroup of an abelian group written additively.   □

Definition 16.39 (Quotient rings). Let $(R, +, \cdot)$ be a ring, and let I be a (two-sided) ideal of R. Recall that $R/I = \{r + I \mid r \in R\}$, and $(R/I, +)$—where $(r + I) + (s + I)$ is defined to be $(r + s + I)$—is an abelian group. Make $(R/I, +, \cdot)$ into a ring by defining

$$(r + I) \cdot (s + I) = rs + I.$$

$(R/I, +, \cdot)$ is called the *quotient ring* of R by I or the *factor ring* of R by I or the *residue class ring* of R modulo I.

Lemma 16.40. *The multiplication of cosets defined in Definition* 16.39 *is well defined.*

Proof. Since each coset has many aliases, we have to show that our definition of coset multiplication does not depend on the particular coset representative chosen. To this end, assume that $r + I = r' + I$ and $s + I = s' + I$. We have to show that $rs + I = r's' + I$. From $r + I = r' + I$ and $s + I = s' + I$, we get that $r' = r + x$ and $s' = s + y$ with $x, y \in I$. Now

$$r's' + I = (r + x)(s + y) + I = rs + ry + xs + xy + I.$$

Now, since I is a two-sided ideal and $x, y \in I$, we have that ry, xs, and xy are all in I. Hence, $rs + ry + xs + xy + I = rs + I$ and $r's' + I = rs + I$.   □

Theorem 16.41. *Let R be a ring, and let I be an ideal of R. The set R/I with addition and multiplication defined in Definition* 16.39 *is a ring.*

Proof. Since, from group theory, we already know that $(R/I, +)$ is an abelian group, we only need to show that multiplication of cosets is associative and that the distributive laws hold. These all follow directly from the definition of the product of two cosets and are left to the reader. (See Problem 16.2.1.)   □

    We have seen that kernels of homomorphisms are ideals. Just as in groups, every ideal is the kernel of some ring homomorphism.

Lemma 16.42. *Let R be a ring, and let I be a (two-sided) ideal of R. Let $\phi : R \to R/I$ be defined by $\phi(r) = r + I$. Then ϕ is an onto ring homomorphism and $\ker(\phi) = I$. This homomorphism is called the* canonical *(ring) homomorphism from R to R/I.*

Proof. Thinking of $(I, +)$ as a normal subgroup of the abelian group $(R, +)$, the map ϕ is the canonical group homomorphism from R to R/I (Definition 11.21). This map is an onto group homomorphism with kernel I (Theorem 11.22). All of this remains true when we think of I as an ideal of the ring R. Hence, the only thing to show is that $\phi(rs) = \phi(r)\phi(s)$, for all $r, s \in R$. This is straightforward since, for $r, s \in R$,

$$\phi(rs) = rs + I = (r + I)(s + I) = \phi(r)\phi(s). \qquad \square$$

Theorem 16.43. *Let R and S be rings, and let $\phi : R \to S$ be a ring homomorphism. Then*

$$R/\ker(\phi) \cong \phi(R).$$

Proof. Let $K = \ker(\phi)$, and define $\psi : R/K \to \phi(R)$ by $\psi(x + K) = \phi(x)$, for $x \in R$. We need to show that ψ is a *ring* isomorphism. Again, to shorten our proof, we appeal to our work in group theory. Thinking as $(R, +)$ and $(K, +)$ as abelian groups, we have shown (proof of Theorem 11.18) that ψ is a group isomorphism. Hence, we have already showed that ψ is well defined, 1-1, onto, and that it preserves addition. It only remains to show that ψ preserves multiplication. This is a straightforward calculation:

$$\psi((x + K)(y + K)) = \psi(xy + K) = \phi(xy) = \phi(x)\phi(y) = \psi(x + K)\psi(y + K). \quad \square$$

Remark 16.44. As in group theory, if $\phi \colon R \to S$ is a ring homomorphism, then we draw the homomorphism diagram of Figure 16.2. We think of $R/\ker(\phi)$ as the portion of the vertical line between $\ker(\phi)$ and R.

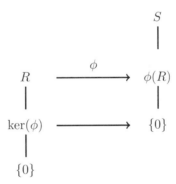

Figure 16.2. A ring homomorphism diagram for a homomorphism $\phi \colon R \to S$

Just as in groups, if $\phi \colon R \to S$ is a ring homomorphism, then not only $R/\ker(\phi) \cong \phi(R)$ but the whole lattice of ideals of R that contain $\ker(\phi)$ is the same as the lattice of ideals of S. We now state this homomorphism correspondence theorem. As we have seen in the last two proofs (of Lemma 16.42 and Theorem 16.43), in proving statements about homomorphisms of rings, we can use the corresponding statements about groups as a starting point. We leave it to the reader to fill in what is needed to complete the proof.

Theorem 16.45 (Homomorphism theorems). *Let $\phi : R \to S$ be a ring homomorphism. Let $K = \ker(\phi)$. Let*

$$\mathcal{I} = \{\text{ideals of } R \text{ containing } K\} \text{ and } \mathcal{J} = \{\text{ideals of } \phi(R)\}.$$

Then the map ϕ extends to a map $\phi : \mathcal{I} \longrightarrow \mathcal{J}$ by

$$I \overset{\phi}{\longmapsto} \phi(I).$$

This map—which is also denoted by ϕ—is a bijection from $\mathcal{I}$ to $\mathcal{J}$. In particular, if I is an ideal of R containing K, then $\phi(I)$ is an ideal of $\phi(R)$, and if J is an ideal of $\phi(R)$, then $\phi^{-1}(J)$ is an ideal of R containing K.

Furthermore, if $I \in \mathcal{I}$ and $J \in \mathcal{J}$ with $\phi(I) = J$, then $I/K \cong J$ and $R/I \cong \phi(R)/J$.

In particular, $R/K \cong \phi(R)$.

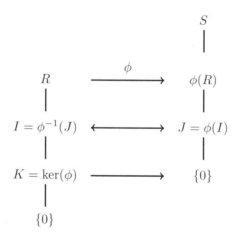

Figure 16.3. The ideals of R that contain the kernel are in 1-1 correspondence with the ideals of $\phi(R)$.

Proof. The homomorphism diagram of Figure 16.3 helps us visualize the statement of the theorem. The reader is asked in Problem 16.2.16 to use and mimic the proof of Theorem 11.38 to write a complete proof.   □

Example 16.46. Let $\phi \colon \mathbb{Z} \to \mathbb{Z}/12\mathbb{Z}$ be defined by $\phi(m) = m$ mod 12. Then ϕ is a ring homomorphism and $\ker(\phi) = 12\mathbb{Z}$. Now, in the ring $\mathbb{Z}/12\mathbb{Z}$, the ideal generated by 4, $\langle 4 \rangle$, is $\{0, 4, 8\}$, and $\phi^{-1}\langle 4 \rangle = 4\mathbb{Z}$. We draw the homomorphism diagram in Figure 16.4. By the homomorphism theorem 16.45, $4\mathbb{Z}/12\mathbb{Z} \cong \langle 4 \rangle \cong \mathbb{Z}/3\mathbb{Z}$ and $(\mathbb{Z}/12\mathbb{Z})/\langle 4 \rangle \cong \mathbb{Z}/4\mathbb{Z}$.

Problems

16.2.1. **Proof of Theorem 16.41.** Write a complete proof of Theorem 16.41. In other words, show that if R is a ring and I is an ideal, then with coset addition and multiplication the set of cosets R/I is a ring.

16.2.2. Let $R = \mathbb{Z}/36\mathbb{Z}$. Find $R/\langle 5 \rangle$.

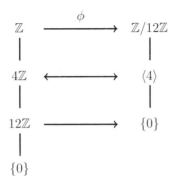

Figure 16.4. The ring homomorphism $\phi\colon \mathbb{Z} \to \mathbb{Z}/12\mathbb{Z}$ is defined by $\phi(m) = m \bmod 12$.

16.2.3. Is $15\mathbb{Z}$ an ideal of $5\mathbb{Z}$? If the answer is yes, find a familiar ring that is isomorphic to $5\mathbb{Z}/15\mathbb{Z}$.

16.2.4. Let $R = \mathbb{Z}[x]$, and let I be the ideal generated by $x^3 + 2x$. Find integers a, b, and c such that in R/I we have $I + x^5 + x^6 = I + a + bx + cx^2$.

16.2.5. Let $R = \mathbb{Z}[x]$, and let I be the ideal generated by $x^2 - 2x$.
   (a) Can you find a zero-divisor in R/I?
   (b) Can you find an ideal J with $I \subsetneq J \subsetneq R$?
   (c) Can you find integers a and b such that in R/I we have $I + x^5 = I + a + bx$?

16.2.6. Let $R = (\mathbb{Z}/2\mathbb{Z})[x, y]$ be the ring of polynomials in x and y and with coefficients in $\mathbb{Z}/2\mathbb{Z}$. Let $I = \langle x^2, xy, y^2 \rangle$ be an ideal of R, and let $S = R/I$. Show S has eight elements. Write down its multiplication table.

16.2.7. Recall that the Gaussian integers are denoted by $\mathbb{Z}[i]$ and are defined by
$$\mathbb{Z}[i] = \{a + bi \mid a, b \in \mathbb{Z}, i^2 = -1\}.$$
Let $I = \langle 1 + 3i \rangle$ be the ideal generated by $1 + 3i$ in $\mathbb{Z}[i]$, and define $R = \mathbb{Z}[i]/I$.
   (a) $I + i$ and $I + 3$ are two elements of R. Are they equal? What about $I + 9$ and $I + 1$?
   (b) How many elements does R have?
   (c) Can you find a familiar ring that is isomorphic to R?

16.2.8. Let I be the ideal generated by $2 + 5i$ in the ring of Gaussian integers $\mathbb{Z}[i]$. Find a familiar ring that is isomorphic to $R = \mathbb{Z}[i]/I$. Is R a field?

16.2.9. Let R be a commutative ring with identity. Using the homomorphism theorem (Theorem 16.45) and Proposition 16.32 show that an ideal M of R is maximal if and only if R/M is a field.

16.2.10. Let $R = \mathbb{Z}/10\mathbb{Z}$. Find all maximal ideals of R. For each maximal ideal J, find a familiar field that is isomorphic to R/J.

16.2.11. Let
$$\mathbb{Z}_{(2)} = \{r \in \mathbb{Q} \mid r = \frac{a}{b}, \text{ with } a, b \in \mathbb{Z}, \gcd(a, b) = 1, \text{ and } b \text{ odd}\}.$$

In other words, any rational number whose denominator, when written in reduced form, is odd is in $\mathbb{Z}_{(2)}$. The operations for $\mathbb{Z}_{(2)}$ are the usual addition and multiplication of rational numbers.

(a) Is $\mathbb{Z}_{(2)}$ a ring? An integral domain? A field?

(b) What are the units of $\mathbb{Z}_{(2)}$?

(c) What is $\langle 3 \rangle$? What is $\langle \frac{1}{3} \rangle$? What is $\langle 2 \rangle$?

(d) Can you find a maximal ideal in $\mathbb{Z}_{(2)}$? Give a proof that the ideal that you are suggesting is actually maximal.

(e) Can you identify $\mathbb{Z}_{(2)}/\langle 2 \rangle$?

16.2.12. Let I be an ideal in a commutative ring R. Prove that $I[x]$ is an ideal in $R[x]$. Prove that $R[x]/I[x] \cong (R/I)[x]$.

16.2.13. Let A and B be commutative rings with identity and let $R = A \times B$. Let I and J be ideals of A and B respectively, and define $\phi \colon A \times B \to A/I \times B/J$ by $\phi(a, b) = (a+I, b+J)$. Is ϕ a ring homomorphism? What is the image? What is the kernel?

16.2.14. Let A and B be commutative rings with identity, $R = A \times B$, and I and J ideals of A and B respectively. Prove

$$(A \times B)/(I \times J) \cong (A/I) \times (B/J).$$

16.2.15. Let R and S be rings. Assume that R and S have no elements in common. Further assume that S has a subring isomorphic to R. Show that there exists a ring T isomorphic to S that contains R as a subring.

16.2.16. **Proof of Theorem 16.45.** Look back at the proof of Theorem 11.38, and then write down a complete proof of Theorem 16.45.

16.3. Characteristic of Rings with Identity, Integral Domains, and Fields

The concept of the *characteristic* of a ring with identity does *not* have a direct (useful) analog in groups and is a useful parameter for rings with identity. Recall that if n is a positive integer and a is an element of a ring, then na is defined to be $\underbrace{a + a + \cdots + a}_{n \text{ terms}}$. Note that na does not denote a product in the ring, rather it is a shorthand for adding a to itself n times. We begin with a lemma in order to build a subring starting with the identity element.

Lemma 16.47. *Let R be a ring with identity, and let s and t be two positive integers. Then*

$$(s + t)1 = (s1) + (t1),$$
$$(st)1 = (s1)(t1).$$

Proof. The first one follows since

$$(s1) + (t1) = (\underbrace{1 + 1 + \cdots + 1}_{s}) + (\underbrace{1 + 1 + \cdots + 1}_{t}) = \underbrace{1 + 1 + \cdots + 1}_{s+t} = (s + t)1.$$

The second follows from the distributive laws since

$$(s1)(t1) = \underbrace{(1 + 1 + \cdots + 1)}_{s}\underbrace{(1 + 1 + \cdots + 1)}_{t} = \underbrace{1 + 1 + \cdots + 1}_{st} = (st)1. \qquad \square$$

Proposition 16.48. *Let R be a ring with identity. Let*

$$S = \{m1_R \mid m \in \mathbb{Z}\}.$$

Then S is a subring of R.

Proof. The subset S is non-empty since it contains 1 and is closed under addition and multiplication by Lemma 16.47. In addition, $-(m1_R) = (-m)1_R$, and so S is closed under taking additive inverses. Hence, S is a subring of R. $\qquad \square$

We will now define the characteristic of a ring. Later we will come back—in the case of commutative rings with identity—to the subring of the multiples of 1 and see the connection.

Definition 16.49 (The characteristic of a ring)**.** Let R be a ring. The *characteristic* of R is the smallest $n \in \mathbb{Z}^{>0}$ such that $nx = 0$ for all x in R. If no such integer exists, we say R has characteristic 0.

Note that the characteristic is an integer and not an element of the ring under consideration.

Example 16.50. The characteristic of $\mathbb{Z}$ is 0, while the characteristic of $\mathbb{Z}/n\mathbb{Z}$ is n.

Proposition 16.51. *Let R be a ring with identity, and let n be a positive integer. Then*

(a) *The characteristic of R is n if and only if 1 has order n under addition. That is, if and only if n is the smallest positive integer with $n1 = 0$.*

(b) *The characteristic of R is 0 if and only if 1 has infinite order under addition. That is, there is no positive integer n with $n1 = 0$.*

Proof. You were asked to provide the main argument in Problem **15.2.14** but, for completeness, we will repeat it here.

($\Rightarrow$) For both parts, this direction is clear from the definition.

($\Leftarrow$) If 1 has infinite order, then there is no positive integer n such that $n1 = 0$, so R has characteristic zero.

Now suppose that 1 has additive order n. Then $n1 = 0$ and n is the least positive integer with this property. So for any x in R we have

$$nx = \underbrace{x + \cdots + x}_{n \text{ times}} = \underbrace{(1 + \cdots + 1)}_{n \text{ times}} x = (n1)x = 0x = 0.$$

Thus R has characteristic n. $\qquad \square$

Theorem 16.52. *The characteristic of an integral domain is 0 or a prime number. In particular, the characteristic of a finite field is a prime number.*

Proof. This was Problem 15.2.15. We will also give the argument here.

Assume that the characteristic of an integral domain is $n \neq 0$. By Proposition 16.51, this means that n is the additive order of 1, and we need to show that n is a prime.

Assume to the contrary that $n = st$ where $1 < s$ and $t < n$. Then

$$0 = n1 = (st)1 = (s1)(t1).$$

Since we are in an integral domain, either $s1$ or $t1$ must be zero. But this contradicts the fact that 1 has order n. Thus n could not be factored, and so it is a prime.

In the case of a finite field—recall that every finite integral domain is a field—1 cannot have infinite order and, hence, the characteristic is a prime number.   □

We are now ready to identify (up to isomorphism) the subring $S = \{m1_R \mid m \in \mathbb{Z}\}$ based on the characteristic of the ring.

Theorem 16.53. *Let R be a ring with identity, and let $S = \{m1_R \mid m \in \mathbb{Z}\}$. If* $\mathrm{char}(R) = 0$, *then S is isomorphic to $\mathbb{Z}$, and if* $\mathrm{char}(R) = n$, *then S is isomorphic to $\mathbb{Z}/n\mathbb{Z}$.*

In particular, every ring with identity has a subring isomorphic to either $\mathbb{Z}$ or $\mathbb{Z}/n\mathbb{Z}$ for some positive integer n. Every integral domain has a subring isomorphic to $\mathbb{Z}$ or $\mathbb{Z}/p\mathbb{Z}$ for some prime number p.

Proof. We define a map $\phi : \mathbb{Z} \to R$ by $\phi(k) = k1_R$. In other words,

$$\phi(k) = \begin{cases} 0, & \text{if } k = 0; \\ \underbrace{1 + 1 + \cdots + 1}_{k}, & \text{if } k > 0; \\ -(\underbrace{1 + 1 + \cdots + 1}_{-k}), & \text{if } k < 0. \end{cases}$$

It is straightforward to check that ϕ is a ring homomorphism. Clearly, S is $\phi(\mathbb{Z}) = \mathrm{Im}(\phi)$. There are two cases:

CASE 1: $\mathrm{char}(R) = 0$.

In this case we have $\ker(\phi) = \{0\}$, and so $\mathbb{Z} \cong \mathbb{Z}/\ker(\phi) \cong \mathrm{Im}(\phi) = S \subseteq R$. So S is isomorphic to $\mathbb{Z}$, and R has a subring S isomorphic to $\mathbb{Z}$.

CASE 2: $\mathrm{char}(R) = n$.

In this case, by the characterization of the characteristic in Proposition 16.51, n is the smallest positive integer such that $\phi(n) = 0$. Hence $n \in \ker(\phi)$. We claim that $\ker(\phi) = n\mathbb{Z}$.

It is clear that if m is a multiple of n, then $\phi(m) = \phi(nh) = \phi(n)\phi(h) = 0$, and hence $m \in \ker(\phi)$. On the other hand, if $m \in \ker(\phi)$, then, using the division algorithm for integers, write $m = nq + r$ with $0 \leq r < n$. Now, $0 = \phi(m) = \phi(n)\phi(q) + \phi(r) = 0 + \phi(r) = \phi(r)$. Now n was the smallest positive integer in the $\ker(\phi)$ and so r could not be a positive number. Hence $r = 0$ and $m = nq \in n\mathbb{Z}$. Using the isomorphism theorem, we get $\mathbb{Z}/n\mathbb{Z} = \mathbb{Z}/\ker(\phi) \cong \mathrm{Im}(\phi) = S \subseteq R$. Thus R has a subring S isomorphic to $\mathbb{Z}/n\mathbb{Z}$.   □

We can say a bit more about fields.

Definition 16.54. Let K be a field, and let P be the intersection of all subfields of K. Then P is called the *prime subfield* of K.

Corollary 16.55. *Let K be a field, and let P be the prime subfield of K. Then $P \cong \mathbb{Q}$ or $P \cong \mathbb{F}_p$ $(= (\mathbb{Z}/p\mathbb{Z}, +, \cdot))$ for some prime number p.*

Proof. Any subfield of K will have to contain $S = \{m1_K \mid m \in \mathbb{Z}\}$. By Theorem 16.53, S is isomorphic to either $\mathbb{Z}$ or $\mathbb{Z}/p\mathbb{Z}$ for some prime number p. In the latter case, we are done. In the former case, since P is a field, it will have to include the inverses of elements of S as well. Hence P will contain $\{\frac{m1_K}{n1_K} \mid m, n \in \mathbb{Z}, n \neq 0\}$, which is easily seen to be isomorphic to $\mathbb{Q}$. $\square$

Problems

16.3.1. Can you find a commutative ring with identity with characteristic 5 that is not an integral domain? Either give an example or prove that it is not possible.

16.3.2. Is it possible to find an infinite ring with characteristic 3? Either give an example or prove that it is not possible.

16.3.3. Find the characteristic of $(\mathbb{Z}/2\mathbb{Z})[x]$.

16.3.4. Let $R = \mathbb{Z}/3\mathbb{Z} \times \mathbb{Z}/5\mathbb{Z}$. (Note that addition and multiplication are mod 3 in the first coordinate and mod 5 in the second coordinate.)
   (a) Is R a commutative ring with identity? What is the characteristic of R?
   (b) Find a subring S of R with characteristic 5.
   (c) Is the ring S an ideal of R? Does S have an identity?

16.3.5. Let D be an integral domain, and let R be a subring of D with an identity. Must D and R have the same characteristic? Either prove that they do, or give an example that shows they do not have to.

16.3.6. Let R be a ring with identity, and let n be the characteristic of R. Assume that $mx = 0$ for a particular positive integer m and a non-zero element $x \in R$. Does m have to be a multiple of n? Either prove that it is or give an example where it is not. What if R was an integral domain?

16.3.7. Let $R = \mathbb{Z}[i]$ be the ring of Gaussian integers, and let $I = \langle 2 + 3i \rangle$. What is the characteristic of the ring R/I?

16.3.8. Let R be a ring without identity. Prove that there exists a ring T with identity such that
   (a) T has the same characteristic as R, and
   (b) T has an ideal isomorphic (as a ring) to R.

16.4. Manipulating Ideals*

There are many ways that we can find new ideals from given ones. In this (optional) section, we will introduce some of the most important constructions.

Definition 16.56 (New ideals from old). Let R be a commutative ring with identity, and let I, J be ideals of R. Then

(a) **Sum.**
$$I + J = \{x + y \mid x \in I, y \in J\}.$$

(b) **Intersection.**
$$I \cap J = \{x \in R \mid x \in I, x \in J\}.$$

(c) **Product.**
$$IJ = \{\sum_{i=1}^{n} x_i y_i \mid x_i \in I, y_i \in J \text{ for } 1 \leq i \leq n\}.$$

(d) **Quotient.**
$$I : J = \{r \in R \mid rx \in I, \ \forall x \in J\}.$$

(e) **Radical.**
$$\sqrt{I} = \{r \in R \mid r^n \in I, \text{ for some positive integer } n\}.$$

Beware that IJ, the product of ideals I and J, does *not* consist only of products of elements of I with elements of J. Rather, to make sure that we get an ideal, we have to include all *sums* of such products.

Definition 16.57 (Radical ideals). An ideal I is *radical* if $I = \sqrt{I}$.

Lemma 16.58. *Let R be a commutative ring with identity, and let I, J be ideals of R. Then $I + J$, $I \cap J$, IJ, $I : J$, and $\sqrt{I}$ are all ideals of R. Furthermore, $\sqrt{I}$ is a radical ideal that contains I.*

Proof. The reader is asked to prove these in the Problems. □

Example 16.59. Let R be the ring of integers, and let $I = \langle 360 \rangle$ and $J = \langle 45 \rangle$. Then the reader can verify that:
$$I + J = \langle 45 \rangle,$$
$$I \cap J = \langle 360 \rangle,$$
$$IJ = \langle 16200 \rangle,$$
$$I : J = \langle 8 \rangle,$$
$$\sqrt{I} = \langle 30 \rangle.$$

Definition 16.60 (Nilradical of a ring). Recall that an element r of a commutative ring is nilpotent if we have $r^n = 0$ for some positive integer n. Let R be a commutative ring with identity. The ideal consisting of all nilpotent elements of R is the same as $\sqrt{\{0_R\}}$ and is called the *nilradical* of R.

The operations that we have defined satisfy many properties. When studying commutative algebra and algebraic geometry, these relations become useful. We list (without proof) a number of these here. You are asked to prove some of them in the Problems (none are that difficult, you could indeed prove all of them).

Proposition 16.61. *Let R be a commutative ring with identity, and let I, J, and K be ideals of R. Then*

(a) $(I \cap K) + (J \cap K) \subseteq (I + J) \cap K$,

(b) $(I : K) + (J : K) \subseteq (I + J) : K$,

(c) $I : (J + K) = (I : J) \cap (I : K)$,

(d) $I \subseteq J \Rightarrow \sqrt{I} \subseteq \sqrt{J}$,

(e) $\sqrt{I^k} = \sqrt{I}$,

(f) $\sqrt{I} \cap \sqrt{J} = \sqrt{\sqrt{I} \cap \sqrt{J}}$,

(g) $\sqrt{I + J} = \sqrt{\sqrt{I} + \sqrt{J}}$,

(h) $\sqrt{IJ} \supseteq \sqrt{I}\sqrt{J}$,

(i) $\sqrt{I : J} = \sqrt{I} : \sqrt{J}$.

Definition 16.62 (Annihilator of an ideal). Let R be a ring, and let U be an ideal of R. Let $r(U) = \{x \in R \mid xu = 0 \text{ for all } u \in U\}$. Then $r(U)$ is called the *(left) annihilator* of U.

Proposition 16.63. *Let R be a ring, and let U be an ideal of R. Then $r(U)$, the annihilator of U, is an ideal.*

Proof. This is Problem 16.4.5. □

Partial Lattice Diagrams of Ideals. Let R be a commutative ring with identity. In Problem 16.4.4, you are asked to show that the poset of ideals of R ordered by inclusion is a lattice. More precisely, you show that if I and J are two ideals of R, then $I + J$ is the smallest ideal that contains both of them, and $I \cap J$ is the largest ideal contained in both I and J. Hence, just as in groups (see Chapter 9), we often draw *partial* lattice diagrams of the ideals of R. Usually in our diagram, if we include two ideals I and J, we also include the ideals $I + J$ and $I \cap J$. Ideals are analogs of normal subgroups—both ideals and normal subgroups are kernels of homomorphisms—and, just as for normal subgroups, we draw the part of a lattice diagram that includes I, J, $I + J$, and $I \cap J$ as a parallelogram. Problem 16.4.17—which is the analog of the direct diamond theorem, Theorem 11.43, for groups—says that the opposite sides of the parallelogram are isomorphic.

Example 16.64. Let $\mathbb{Z}$ be the ring of integers, let $I = \langle 245 \rangle$ and $J = \langle 189 \rangle$. Note that $245 = 5 \times 7^2$ and $189 = 3^3 \times 7$. Elements of I are multiples of 245, while elements of J are multiples of 189. Hence, $I \cap J$ consists of integers that are multiples of $3^3 \times 5 \times 7^2$—that is, the least common multiple of 245 and 189. So $I \cap J = \langle 6615 \rangle$. Elements of $I + J$ are integers of the form $245a + 189b$, where a and b are integers. A moment's thought—this will become obvious in later sections and does follow from Theorem 1.49—gives that these are precisely the multiples of $\gcd(245, 189) = 7$. Hence, $I + J = \langle 7 \rangle$. Figure 16.5 is a partial lattice diagram of ideals $\mathbb{Z}$ that includes I, J, $I + J$, and $I \cap J$.

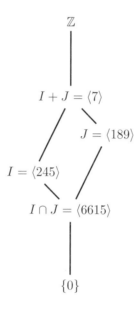

Figure 16.5. A partial lattice diagram of ideals I, J, $I + J$, and $I \cap J$ in $\mathbb{Z}$

Problem 16.4.17 implies that the opposite sides of the parallelogram are isomorphic as rings. Here, as an example, we have $\langle 7 \rangle / \langle 245 \rangle \cong \langle 189 \rangle / \langle 6615 \rangle$. In fact, each of these factor rings are isomorphic to $\mathbb{Z}/35\mathbb{Z}$.

Continuing the example further, elements of IJ are sums of products of elements of I and J, and hence $IJ = \langle 46305 \rangle$. Elements of $I : J$ are those integers that if multiplied by an element of J give an element of I. These are exactly the multiples of 35, and so $I : J = \langle 35 \rangle$. Finally, elements of $\sqrt{I}$ are those integer that raised to some power are in I. These are the multiples of 35. Likewise, $\sqrt{J} = \langle 21 \rangle$. In Figure 16.6, we have included these ideals as well.

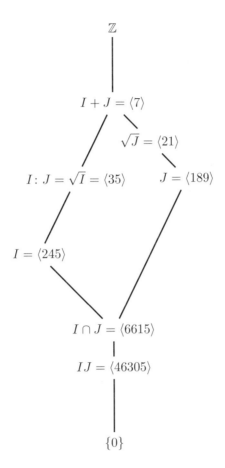

Figure 16.6. A partial lattice diagram in $\mathbb{Z}$ of some ideals related to $I = \langle 245 \rangle$ and $J = \langle 189 \rangle$

Problems

16.4.1. Let $R = \mathbb{Z}$, and let $I = \langle 18000 \rangle$. Find $\sqrt{I}$, and $\sqrt{\sqrt{I}}$.

16.4.2. Let $\mathbb{Z}$ be the ring of integers, let $I = \langle 1800 \rangle$ and $J = \langle 315 \rangle$. Find $I + J$, $I \cap J$, IJ, $I \colon J$, $J \colon I$, $\sqrt{I}$, $\sqrt{J}$, $\sqrt{I + J}$, and $\sqrt{I} + \sqrt{J}$.

16.4.3. Let $R = \mathbb{Z}[\sqrt{5}] = \{a + b\sqrt{5} \mid a, b \in \mathbb{Z}\}$. Let $I = \langle 2 \rangle$ and $J = \langle 3 \rangle$ be, respectively, the ideals generated by 2 and 3. Explicitly describe $I + J$ and $I \cap J$.

16.4.4. Let R be a commutative ring with identity, and let I and J be ideals of R. Show that $I \cap J$ is the largest ideal that is contained in both I and J. Likewise, show that $I + J$ is the smallest ideal that contains both I and J. Conclude that the poset of ideals of R ordered by inclusion is a lattice.

16.4.5. (a) Let $R = \mathbb{Z}/36\mathbb{Z}$ and $U = \langle 6 \rangle$. Find $r(U)$, the annihilator of U.
      (b) Let R be an integral domain, and let U be an ideal of R. What can you say about $r(U)$?
      (c) Let R be a ring, and let U be an ideal of R. Prove that the annihilator of U, $r(U)$, is an ideal of R.

16.4.6. Let R be a commutative ring with identity, and let I and J be ideals of R. Show that IJ and $I : J$ are ideals of R.

16.4.7. Let R be a commutative ring with identity, and let I be an ideal of R. Show that $\sqrt{I}$ is a radical ideal of R that contains I.

16.4.8. Let R be a commutative ring with identity, and let I be an ideal of R. Show that I is a maximal ideal of R if and only if $I + \langle x \rangle = R$ for all $x \in R - I$. How does this result compare with that of Problem 16.1.28?

16.4.9. Let R be a commutative ring with identity.
      (a) Let $I = \langle a \rangle$ and $J = \langle b \rangle$ be principal ideals of R. Show that $IJ = \langle ab \rangle$.
      (b) Let $R = \mathbb{Z}[x]$, $I = \langle x, 2 \rangle$, and $J = \langle 3 \rangle$. Describe the elements of IJ, and give a set of generators for it.

16.4.10. Consider the ring of integers $\mathbb{Z}$. Let a and b be integers, and let $I = \langle a \rangle$ and $J = \langle b \rangle$. Give a condition on a and b that assures $I : J = \langle a/b \rangle$.

16.4.11. Consider the ring of integers $\mathbb{Z}$. Let a be an integer, and let $I = \langle a \rangle$. Could we ever have $\sqrt{I} = \langle \sqrt{a} \rangle$? Under what conditions?

16.4.12. Let $I = \langle 245 \rangle$ and $J = \langle 189 \rangle$ be ideals of the ring of integers. In the partial lattice diagram of Figure 16.6, add the ideals $\sqrt{I} \cap J$, $I \cap \sqrt{J}$, and $\sqrt{I} \cap \sqrt{J}$. What is $I + (\sqrt{I} \cap \sqrt{J})$? Could you have guessed the answer from the diagram?

16.4.13. Let R be a commutative ring with identity, and let I and J be ideals of R.
      (a) Show that $IJ \subseteq I \cap J$.
      (b) Give an example where $IJ \neq I \cap J$.
      (c) Prove that if $I + J = R$, then $IJ = I \cap J$.

16.4.14. Let R be a commutative ring with identity, and let I, J, and K be ideals of R. Show
      (a) $(I + J) \cap K \supseteq (I \cap K) + (J \cap K)$,
      (b) $(I + J) : K \supseteq (I : K) + (J : K)$,
      (c) $I : (J + K) = (I : J) \cap (I : K)$,
      (d) $\sqrt{I^k} = \sqrt{I}$.

16.4.15. Let R be a commutative ring with identity.
      (a) Let I and J be ideals of R. Prove that

$$\sqrt{I} \cap \sqrt{J} = \sqrt{\sqrt{I} \cap \sqrt{J}}.$$

      (b) Prove that the intersection of two radical ideals is radical.

16.4.16. Let R be a commutative ring with identity. Let I and J be ideals of R. Assume, for some $k \geq 1$,

$$I^k \subseteq J \subseteq I.$$

How are $\sqrt{I}$ and $\sqrt{J}$ related?

16.4.17. **Direct diamond for ideals.** Let R be a commutative ring with identity. Let $\{0\} \neq S$ be a subring of R, and let I be a proper ideal of R. First, convince yourself that $S \cap I$ is a proper ideal of S, $S + I$ is a subring of R that contains S and I, and that I is a proper ideal of the ring $S + I$. Then, prove

$$S/(S \cap I) \cong (S + I)/I.$$

Chapter 17

Field of Fractions and Localization

...where generalizing the construction of rational numbers from the integers, we start with an integral domain or a commutative ring with identity R, and construct a bigger ring in which some judiciously chosen elements of R will become invertible.

The field of rational numbers $\mathbb{Q}$ is constructed from the ring of integers $\mathbb{Z}$. After all, the rational numbers are just ratios of integers. Can we mimic this process and build other fields starting from other rings?

The construction of $\mathbb{Q}$ has to be a bit more subtle than just "take the set of ratios of integers and then you get a field called the rational numbers." If the set $\mathbb{Q}$ was just the set of ratios of integers (with the provision that zero cannot be in the denominator), then $\frac{1}{2}$ and $\frac{3}{6}$ would be two different elements of $\mathbb{Q}$. Evidently, $\mathbb{Q}$ is *not* just the set of "ratios" of integers.

This problem is solved by defining the rational numbers, not as ratios of integers, but as equivalence classes of ratios of integers. The whole set $\{\frac{1}{2}, \frac{2}{4}, \frac{3}{6}, \dots\}$ will be one element of $\mathbb{Q}$, and we define addition and multiplication for these equivalence classes. Since every class has an infinite number of representatives—to make sure that the addition and multiplication of elements of $\mathbb{Q}$ are well defined—it is necessary to prove that the operations are independent of the choice of representatives. For example, we have to make sure that

$$\frac{1}{2} + \frac{1}{3} = \frac{3}{6} + \frac{5}{15}.$$

In the first section of this chapter, we want to rigorously construct rational numbers from the integers, and to generalize the construction to all integral domains. In other words, starting with R an integral domain, we want to build a field F, called its *field of fractions*, and we want this field to have certain properties. In particular, F is a field that contains a copy of R, it is the smallest such field, and after constructing it, every element of it can be represented as the quotient of two elements in R.

In F, the field of fractions of an integral domain R, every element of R—except 0—will have an inverse. We will generalize this further and identify other integral domains S with $R \subseteq S \subseteq F$, such that, in S, a judiciously chosen set of elements of R are invertible. In other words, given M—a so-called "multiplicative system" in R— in the integral domain S, every element of M will be a unit. In addition, every element of S will be of the form rm^{-1} with $r \in R$ and $m \in M$. In other words, we will have a bigger integral domain, in which we have some fractions, namely those with denominators in M. The ring S will be called the localization of R at M.[1]

In the second section of this chapter, we will be more ambitious, and replace integral domains with general commutative rings with identity. The second section *is* more general in scope and does repeat many of the arguments of the first section, but, in the first pass through abstract algebra, the reader can safely skip the more general construction.

17.1. Field of Fractions and Localization of an Integral Domain

Field of Fractions. Let D be an integral domain. In this section we build a field F that contains a copy of D and such that every element of F is equal to $rs^{-1} = \frac{r}{s}$, where r and s are elements of D (and s is not zero). In other words, the elements of F are "fractions" with elements of D in the numerator and the denominator. The construction will mimic the construction of the rational numbers from the integers. More precisely, we have:

Theorem 17.1. *Let D be an integral domain. Then there exists a field F, with the following properties:*

(a) *F has a subring D_0 isomorphic to D;*

(b) *every element of F is equal to rs^{-1} for some $r, s \in D_0$ with $s \neq 0$; and*

(c) *if K is another field with a subring D_1 that is isomorphic to D, then there exists a subfield E of K such that $D_1 \subseteq E$ and $E \cong F$.*

The last condition is a uniqueness (up to isomorphism) condition (see Problem 17.1.8) saying that F is the smallest field that contains a copy of D.

Definition 17.2 (Field of fractions). Let D be an integral domain. The field F guaranteed by Theorem 17.1 is called the *field of fractions*, or the *field of quotients*, or the *quotient field* of D.

The Construction of the Field of Fractions. Let D be an integral domain, and let $M = D - \{0\}$ be the set of non-zero elements of D. Consider $D \times M = \{(r, m) \mid r \in D, m \in M\}$. Akin to rational numbers, $D \times M$ consists of pairs of elements of D with the second element never being zero. So, heuristically, you should think of the element (r, m) as $\frac{r}{m}$. However, for rational numbers, $\frac{1}{2} = \frac{8}{16}$. Hence, we have to identify some elements of $D \times M$ with each other. We do this by defining a relation

---

[1]Technically, localization also refers to even more general situations when the constructed new ring S does not (and cannot) contain a copy of the original ring R. We will clarify these issues in Section 17.2.

on $D \times M$. Remembering the rational numbers again, for (a, b), $(c, d) \in D \times M$, we define $(a, b) \sim (c, d)$ if and only if $ad = bc$.

CLAIM: $\sim$ is an equivalence relation.

Proof. It is clear that $\sim$ is reflexive and symmetric. To prove the transitive property, assume $(a, m) \sim (b, n)$ and $(b, n) \sim (c, l)$. Thus $an = bm$ and $bl = cn$. So, we have $anl = bml = cnm$. Thus, $n(al - cm) = 0$. The element n is non-zero and is not a zero divisor, and so $al = cm$. Hence, $(a, m) \sim (c, l)$. $\qquad\square$

The equivalence classes of the equivalence relation $\sim$ partition $D \times M$. Denote the equivalence class that contains (r, m) by $[r, m]$, and define

$$(17.1) \qquad\qquad F = \{[r, m] \mid r \in D, m \in M\}.$$

We will turn the set F, consisting of the equivalence classes, into a ring by defining an addition and a multiplication. We then show that F with these operations is indeed a field.

Mimicking the rational numbers, we define addition and multiplication on the set F—defined in (17.1)—by:

$$(17.2) \qquad \begin{aligned} [r_1, m_1] + [r_2, m_2] &= [r_1 m_2 + r_2 m_1, m_1 m_2], \\ [r_1, m_1][r_2, m_2] &= [r_1 r_2, m_1 m_2]. \end{aligned}$$

The product of two non-zero elements is non-zero, and so for m_1, $m_2 \in M$, we have $m_1 m_2 \in M$. Hence, in our definitions, it is legitimate to have $m_1 m_2$ in the denominator. Since these operations are defined on equivalence classes, we have to prove that they are well defined. In other words, we have to show that the results of addition and multiplication are independent of the choice of representatives.

Before we begin, note that follows from the definition of the equivalence relation—and is predictable given what we know of rational numbers—that for $r \in D$ and $m, m' \in M$, we have $[r, m] = [rm', mm']$.

CLAIM. The addition and multiplication defined in (17.2) are well defined.

Proof. We will give the proof for the addition operation and leave the (easier) proof for multiplication to the reader. Assume that $[r_1, m_1] = [r_1', m_1']$ and $[r_2, m_2] = [r_2', m_2']$. We have to show that

$$[r_1 m_2 + r_2 m_1, m_1 m_2] = [r_1' m_2' + r_2' m_1', m_1' m_2'].$$

From $[r_1, m_1] = [r_1', m_1']$, we have $r_1 m_1' = m_1 r_1'$. Similarly, $r_2 m_2' = m_2 r_2'$. Using these—and the fact that $[r, m] = [rm', mm']$ for $m' \in M$—we have

$$\begin{aligned} [r_1' m_2' + r_2' m_1', m_1' m_2'] &= [(r_1' m_2' + r_2' m_1') m_1 m_2, m_1' m_2' m_1 m_2] \\ &= [r_1' m_1 m_2' m_2 + r_2' m_2 m_1' m_1, m_1' m_2' m_1 m_2] \\ &= [r_1 m_1' m_2' m_2 + r_2 m_2' m_1' m_1, m_1' m_2' m_1 m_2] \\ &= [m_1' m_2' (r_1 m_2 + r_2 m_1), m_1' m_2' m_1 m_2] \\ &= [r_1 m_2 + r_2 m_1, m_1 m_2]. \qquad\square \end{aligned}$$

We are now ready to prove Theorem 17.1 by showing that the set F defined in (17.1) together with the addition and multiplication defined in (17.2) satisfies the conclusions of the Theorem.

Proposition 17.3. *Let D be an integral domain. Let $M = D - \{0\}$ and construct the set F of (17.1). Define addition and multiplication on F via equations (17.2). Then*

(a) *F is a field;*

(b) *the element $[1, 1]$ is the identity of F;*

(c) *if $r \in D$ is not zero, then the inverse of $[r, m]$ is $[m, r]$;*

(d) *the set of elements $D_0 = \{[r, 1] \mid r \in D\}$ is a subring of F isomorphic to D;*

(e) *every element $[r, m]$ of F is equal to $[r, 1]([m, 1])^{-1}$, and $[r, 1], [m, 1] \in D_0$; and*

(f) *F is the smallest field containing D, meaning that if K is another field that contains a copy of D, then K must contain a copy of F.*

Proof. Most of the assertions are routine calculations and are left to the reader. Make sure to prove that the two operations given in (17.2) are associative. Note that, for every $m \in M$, $[0, 1] = [0, m]$ and $[1, 1] = [m, m]$. It is easy to check that the element $[0, 1]$ is the zero (i.e., the additive identity), $-[a, m] = [-a, m]$, $[1, 1]$ is the (multiplicative) identity, and $[m, r]$ is the (multiplicative) inverse of $[r, m]$ (for $r \neq 0$).

Define a map $\phi : D \to F$ by $\phi(r) = [r, 1]$. It is straightforward to show that ϕ is a well defined, 1-1, ring homomorphism, and $\phi(D) = D_0$. Hence, D_0 is a subring of F isomorphic to D.

It is also clear that every element of F can now be written as a "fraction" using elements of D_0. This is because, for $r \in D$, $m \in M$, we have

$$[r, m] = [r, 1][1, m] = [r, 1]([m, 1])^{-1}.$$

It remains to show that F is the *smallest* field containing a copy of D. Let K be a field, and assume that $D_1 \subseteq K$ is an integral domain with $D_1 \cong D$. Let $\psi : D \to D_1$ be a ring isomorphism. We want to show that there is a copy of F contained in K. Define a map $\theta : F \to K$ by

$$\theta([r, m]) = \psi(r)\psi(m)^{-1}, \text{ for } r \in D, m \in M.$$

Note that, if $r \in D$ and $m \in M = D - \{0\}$, then $\psi(r) \in D_1$ and $\psi(m) \in D_1 - \{0\}$. Hence, $\psi(m)^{-1} \in K$, and we can find the product $\psi(r)\psi(m)^{-1}$ in K.

We leave it to the reader to show that θ is a well defined, 1-1, ring homomorphism. This means that $F \cong \theta(F) \subseteq K$. Hence, $\theta(F)$ is a subfield of K that contains D_1 and is isomorphic to F. □

To summarize, let D be an integral domain, and let F be its field of fractions. Then F is a field that contains a copy of D, it is the smallest such field, and after constructing it, every element of it can be represented as the quotient of two elements in D.

Example 17.4. If we start with the integers as our integral domain, then, in this section, we have seen a rigorous construction of the rational numbers $\mathbb{Q}$. In other words, the field of rational numbers is the field of fractions of the integers.

Example 17.5. The ring $\mathbb{Z}[\sqrt{2}] = \{a + b\sqrt{2} \mid a, b \in \mathbb{Z}\}$ is an integral domain. The field of rational numbers $\mathbb{Q}$ does not contain all of $\mathbb{Z}[\sqrt{2}]$ and hence it is *not* its field of fractions. On the other extreme, the real numbers $\mathbb{R}$ certainly contain all of $\mathbb{Z}[\sqrt{2}]$, and, hence, we can find the field of fractions inside of $\mathbb{R}$. The elements of the field of fractions are real numbers that can be written as fractions with the numerator and denominator coming from $\mathbb{Z}[\sqrt{2}]$. We have

$$\frac{a + b\sqrt{2}}{c + d\sqrt{2}} = \frac{(a + b\sqrt{2})(c - d\sqrt{2})}{c^2 - 2d^2} = \frac{ac - 2bd}{c^2 - 2d^2} + \frac{bc - ad}{c^2 - 2d^2}\sqrt{2}.$$

This means that all ratios of two elements in $\mathbb{Z}[\sqrt{2}]$ belong to the quadratic number field $\mathbb{Q}[\sqrt{2}] = \{\alpha + \beta\sqrt{2} \mid \alpha, \beta \in \mathbb{Q}\}$. On the other hand, clearly every element of $\mathbb{Q}[\sqrt{2}]$ can be written as a ratio of two elements of $\mathbb{Z}[\sqrt{2}]$. Hence, the field of fractions of $\mathbb{Z}[\sqrt{2}]$ is $\mathbb{Q}[\sqrt{2}]$.

Example 17.6. The ring of polynomials with rational coefficients, $\mathbb{Q}[x]$, is an integral domain. Its field of fractions is the set of all rational functions with rational coefficients (that is, quotients f/g with $f, g \in \mathbb{Q}[x]$ and with $g \neq 0$). This field is denoted by $\mathbb{Q}(x)$.

Remark 17.7. If D is an integral domain, we often write $D \subseteq F$ where F is its field of fractions. The elements of D are not literally elements of F—for example, the integer 3 is not an equivalence class of pairs of integers and hence is not an element of $\mathbb{Q}$—but there is an isomorphic copy of D contained in F. If $D' \subseteq F$ and $\phi\colon D \to D'$ is an isomorphism, then, to avoid clutter, we identify $d \in D$ with $\phi(d) \in D'$. This means that we think of d and $\phi(d)$ as the same element. For example, if $r \in D$, we may ask "is r a unit in F?" More precisely, we should have asked "is $\phi(r)$ a unit in F?" Since ϕ is a ring isomorphism, we can always use ϕ to translate ring theoretic statements back and forth between D and D', and, hence, our identifying elements of D and D' will not cause any problems. This is not really a new idea. When we say $\mathbb{Z} \subseteq \mathbb{Q}$, given our construction of the rationals, we really mean that $\mathbb{Q}$ has a subring isomorphic to $\mathbb{Z}$ and we identify $\mathbb{Z}$ with its copy in $\mathbb{Q}$. For example, we identify the integer 3 with the rational number 3/1 (which after all is an equivalence class of pairs of integers and is certainly *not* an integer).

Localization. Let D be an integral domain, and let F be its field of fractions. In F every non-zero element of D is a unit and has an inverse. What if we just need some elements of D to be invertible? We will now find integral domains S that are contained in F and contain (a copy of) D, and such that in S, some elements of D are invertible. If a and b are elements of D and are chosen to be units of a bigger integral domain S, then in S, $b^{-1}a^{-1}$ would be the inverse of ab. Hence, ab—whether we like it or not—will be a unit of S also. This can be the motivation for the following definition.

Definition 17.8 (Multiplicative system). Let R be a commutative ring with identity. Let $M \subseteq R$. The set M is called a *multiplicative system* in R if

(a) $0_R \notin M$, $1_R \in M$, and

(b) if $m, n \in M$ then $mn \in M$.

We are now ready to define localizations for integral domains.

Definition 17.9. Let D be an integral domain, let F be the field of fractions of D, and let M be a multiplicative system in D. Define the set $D[M^{-1}]$,[2] the *localization of D at M*, by

$$D[M^{-1}] = \{rs^{-1} \in F \mid r \in D, s \in M\}.$$

Proposition 17.10. *Let D be an integral domain, and let M be a multiplicative system in D. Then $D[M^{-1}]$, the localization of D at M, is an integral domain containing (a copy of) D. Moreover, (after identifying D with its isomorphic copy in $D[M^{-1}]$), every element of M is a unit in $D[M^{-1}]$, and every element of $D[M^{-1}]$ has the form rm^{-1} for some $r \in R$ and $m \in M$.*

Proof. The proof is left to the reader. See Problem 17.1.11. $\qquad\square$

Example 17.11. Let D be an integral domain, and let $M = D - \{0\}$. Then M is a multiplicative system in D, and $D[M^{-1}]$ is exactly the field of fractions of D. From this point of view, localization is a generalization of the construction of field of fractions.

Example 17.12. Let $p \in \mathbb{Z}$ be a prime number, and let $M = \mathbb{Z} - \langle p \rangle$ be the set of integers that are not multiples of p. The set M is a multiplicative system in the ring of integers $\mathbb{Z}$, since M is closed under multiplication, $1 \in M$, and $0 \notin M$. What is $\mathbb{Z}[M^{-1}]$? It consists of fractions $\frac{a}{b}$ where $a \in \mathbb{Z}$ and $b \in M$. In other words, $\mathbb{Z}[M^{-1}]$ is the set of fractions that, when written in reduced form, have a denominator not divisible by p. This integral domain is often denoted by $\mathbb{Z}_{(p)}$.

$$\mathbb{Z}_{(p)} = \mathbb{Z}[M^{-1}] = \{\frac{a}{b} \in \mathbb{Q} \mid \text{when } \gcd(a,b) = 1, \text{ then } b \text{ is not divisible by p}\}.$$

This is the smallest ring containing $\mathbb{Z}$ in which every integer that is not divisible by p has an inverse. (Also see Problems 16.1.9 and 16.2.11 as well as the Problems in Section 18.6.)

Example 17.13. Consider the ring of polynomials $\mathbb{Z}[x]$, and let $M = \{1, x, x^2, \ldots\}$. Then M is a multiplicative system and $\mathbb{Z}[x][M^{-1}]$ consists of rational functions of the form $\frac{p(x)}{x^k}$ where $p \in \mathbb{Z}[x]$ and k is a non-negative integer. Splitting the terms up, we see that a typical element of this ring is of the form

$$\frac{a_{-m}}{x^m} + \frac{a_{-m+1}}{x^{m-1}} + \cdots + \frac{a_{-1}}{x} + a_0 + a_1 x + \cdots + a_n x^n,$$

where n and m are non-negative integers, and the coefficients $a_{-m}, \ldots, a_n$ are integers. The ring $\mathbb{Z}[x][M^{-1}]$ is called the *ring of Laurent polynomials* over $\mathbb{Z}$, and is denoted by $\mathbb{Z}[x, \frac{1}{x}]$.

---

[2]Some authors use the notation $M^{-1}D$.

Problems

17.1.1. What is the field of fractions of the Gaussian integers $\mathbb{Z}[i] = \{a + bi \mid a, b \in \mathbb{Z}\}$?

17.1.2. What are the elements of the field of fractions of $\mathbb{Z}[\sqrt{3}] = \{a + b\sqrt{3} \mid a, b \in \mathbb{Z}\}$?

17.1.3. Let $\mathbb{Z}_{(5)} = \{\frac{a}{b} \in \mathbb{Q} \mid \gcd(a, b) = 1,\ b$ is not divisible by $5\}$. What is the field of fractions of $\mathbb{Z}_{(5)}$?

17.1.4. What is the field of fractions of $\mathbb{Z}[x]$, the ring of polynomials with integer coefficients?

17.1.5. The field of fractions of $\mathbb{C}[x]$ is denoted by $\mathbb{C}(x)$. Is $\frac{x^2+1}{x-1} \in \mathbb{C}(x)$? What does a typical element of $\mathbb{C}(x)$ look like?

17.1.6. Let $R_1 = \mathbb{Z}[\sqrt{5}] = \{a + b\sqrt{5} \mid a, b \in \mathbb{Z}\}$, and let $R_2 = \mathbb{Z}[\frac{1+\sqrt{5}}{2}] = \{a + b\frac{1+\sqrt{5}}{2} \mid a, b \in \mathbb{Z}\}$. (See Problem 15.2.17.) Describe the elements in the field of fractions of each of R_1 and R_2. Are these two fields the same? Is one contained in the other?

17.1.7. Let $R_1 = \mathbb{Z}[\sqrt{-3}] = \{a + b\sqrt{-3} \mid a, b \in \mathbb{Z}\}$, and let $R_2 = \{\frac{a+b\sqrt{-3}}{2} \mid a, b \in \mathbb{Z}, a$ and b both even or both odd$\}$. (See Problem 15.2.18.) Describe the elements in the field of fractions of each of R_1 and R_2. Are these two fields the same? Is one contained in the other?

17.1.8. Let D be an integral domain. Assume that F_1 and F_2 are two fields satisfying the conclusions of Theorem 17.1 and that D_1 and D_2 are their respective subrings isomorphic to D. Then show that there exists a field isomorphism $\phi\colon F_1 \to F_2$ such that the restriction of ϕ to D_1 gives a ring isomorphism from D_1 to D_2. Conclude that field of fractions of D is unique up to isomorphism.

17.1.9. Write a complete proof of Proposition 17.3.

17.1.10. Let D be an integral domain, and let K be the field of fractions of D. As usual identify D with the subring of K isomorphic to D so that we can assume $D \subseteq K$. Let F be a field and assume that $\phi\colon D \to F$ is a 1-1 ring homomorphism.
   (a) Show that there exists a unique ring homomorphism $\Phi\colon K \to F$ such that $\Phi\mid_D$, the restriction of Φ to D, is the map ϕ.
   (b) Was the assumption that ϕ is 1-1 necessary?

17.1.11. **Proof of Proposition 17.10.** Let R be an integral domain, and let M be a multiplicative system in R.
   (a) Show that $R[M^{-1}]$ is an integral domain that has a subring isomorphic to R.
   (b) Which elements of $R[M^{-1}]$ are units?
   (c) Is every element of $R[M^{-1}]$ of the form rm^{-1} where $r \in R$ and $m \in M$? If so, explain what this means exactly and why it is true.

17.1.12. Let R be an integral domain, and let a be a non-zero element of R. Let $M = \{1, a, a^2, a^3, \ldots\}$. Is M a multiplicative system in R? If so, what are the elements of the ring $R[M^{-1}]$?

17.1.13. Let $R = (\mathbb{Z}/5\mathbb{Z})[x]$, and let $M = \{1, x+1, (x+1)^2, \ldots\}$. Is M a multiplicative system in R? If so, what are the elements of the ring $R[M^{-1}]$? What is the characteristic of this ring? What are the units of this ring? Give one element of $R[M^{-1}]$ that is not a unit.

17.1.14. Let $R = \mathbb{Z}$, and let $M = \langle 5 \rangle \cup \{1\} - \{0\}$. Is M a multiplicative system? If the answer is yes, what are the elements of $S = R[M^{-1}]$? Can you find a maximal ideal in S? Can you find the field of fractions of S?

17.1.15. Let $R = \mathbb{Z}$, the ring of integers, and let $M = \{1, 5, 5^2, 5^3, \ldots\}$. Let $S = R[M^{-1}]$ be the localization of R at M. Let I and J be the ideals generated by 7 in R and S, respectively. What are the elements of J and how is $J \cap R$ related to I? Answer the same questions if we replaced 7 by 5.

17.2. Localization of Commutative Rings with Identity*

For R, a general commutative ring with identity, we will not be able to construct a field of fractions, but we will be able to construct a localization of R at M when M is a particularly nice multiplicative system.

The advantage of integral domains was that since we constructed, once and for all, the field of fractions, we could then work within this field to "find" the localizations. For R, a general commutative ring with unity, we have to build a bigger ring S that will contain a copy of R (and we have to repeat the construction for every multiplicative system). For convenience, we will give a name to this situation.

Definition 17.14 (Unitary overrings)**.** Let R and S be rings with identity. Assume that S has a subring R' such that $R \cong R'$ and that $1_{R'} = 1_S$. Then S is called a *unitary overring* of R. Often, we identify R and R'. Hence, when we speak of elements of R in S, we are referring to the corresponding elements of R' in S.

Example 17.15. If D is an integral domain and if M is a multiplicative system in D, then $D[M^{-1}]$, the localization of D at M, is a unitary overring of D. In particular, the field of fractions of D is a unitary overring of D.

Let R be a commutative ring with identity. We want a unitary overring of R in which some or all of the non-zero elements of R are units. If R is an integral domain, then it is possible to have a unitary overring in which *all* non-zero elements of R are units. However, if R is not an integral domain, then it has zero-divisors, and these cannot be invertible in the unitary overring of R:

Lemma 17.16. *Let R be a commutative ring with identity, and assume S is a unitary overring of R. Let z be a zero-divisor of R, then z cannot be a unit in S.*

Proof. If z is a zero-divisor in R, then there exists $0 \neq r \in R$ with $zr = 0$. We still have $zr = 0$ in S. (Technically, if R' is the subring of S isomorphic to R and

$\phi\colon R \to R'$ is a ring isomorphism, we have $\phi(z)\phi(r) = 0$. See Remark 17.7.) If z was invertible in S, then we would have $z^{-1}(zr) = 0$ which would imply $r = 0$. The contradiction proves that z cannot be a unit in S. $\square$

Let R be a commutative ring with identity, and let $M \subseteq R$. Recall (from Definition 17.17) that M is a *multiplicative system* if $0 \notin M$, $1 \in M$, and whenever $a, b \in M$, then $ab \in M$. We want the elements of a multiplicative system M to be units in a unitary overring of R. Because of Lemma 17.16, M cannot contain any zero divisors. (We will see in the problems that if M contains zero-divisors, then we still can construct a suitable S, but then the constructed ring S will not contain a copy of R.) We state the following definition.

Definition 17.17 ($\mathcal{Z}(M)$)**.** Let R be a commutative ring with identity, and let M be a multiplicative system in R. Define

$$\mathcal{Z}(M) = \{r \in R \mid rm = 0 \text{ for some } \overset{.}{m} \in M\}.$$

In the case of integral domains we have the following.

Lemma 17.18. *Let R be a commutative ring with idenity, and let $M = R - \{0\}$. Then the following are equivalent:*

(a) *The ring R is an integral domain.*

(b) *The subset M is a multiplicative system, and $\mathcal{Z}(M) = \{0\}$.*

(c) *The subset M is a multiplicative system.*

Proof. The proof follows directly from the definitions and is left to the reader. (See Problem 17.2.1.) $\square$

Example 17.19. Let $R = (\mathbb{Z}/10\mathbb{Z}, +, \cdot)$.

(a) Let $M = \{1, 3, 7, 9\}$. The set M is a multiplicative system and $\mathcal{Z}(M) = \{0\}$.

(b) Let $M' = \{1, 2, 4, 6, 8\}$. The set M' is also a multiplicative system and $\mathcal{Z}(M') = \{0, 5\}$.

The subset $\mathcal{Z}(M)$ is an ideal and signals the existence of zero-divisors in M.

Lemma 17.20. *Let R be a commutative ring with identity, let $M \subseteq R$ be a multiplicative system. Then $\mathcal{Z}(M)$ is a proper ideal of R, its non-zero elements are (some of the) zero-divisors of R, and if $\mathcal{Z}(M) \neq \{0\}$, then M contains a zero-divisor of R, and we cannot construct a unitary overring of R in which every element of M is a unit.*

Proof. By definition, the non-zero elements of $\mathcal{Z}(M)$ are zero-divisors of R. The existence of even one non-zero element in $\mathcal{Z}(M)$ means that some element of M is also a zero-divisor, and, by Lemma 17.16, this means that there is no unitary overring of R in which every element of M has an inverse.

It remains to show that $\mathcal{Z}(M)$ is a proper ideal of R. The identity of the ring R is not a zero-divisor and cannot be a member of $\mathcal{Z}(M)$. On the other hand $0 \in \mathcal{Z}(M)$. Hence $\mathcal{Z}(M)$ is a non-empty proper subset of R. To show that $\mathcal{Z}(M)$ is an additive subgroup of R, let $x, y \in \mathcal{Z}(M)$, and let $m, n \in R$ with $xm = 0 = yn$.

Now $(x - y)(mn) = 0$, and, since $mn \in M$, we conclude that $x - y \in \mathcal{Z}(M)$. To show that $\mathcal{Z}(M)$ is an ideal, let $r \in R$, $x \in \mathcal{Z}(M)$, and $m \in M$ with $xm = 0$. Then $(rx)m = 0$ and, hence $rx \in \mathcal{Z}(M)$. This completes the proof. $\qquad\square$

A non-trivial $\mathcal{Z}(M)$ is the only obstruction to constructing the desired unitary overring.

Theorem 17.21. *Let R be a commutative ring with identity, and let M be a multiplicative system in R. Assume that $\mathcal{Z}(M) = \{0\}$. Then there exists a unique (up to isomorphism) unitary overring S of R such that (after identifying R with its isomorphic copy in S) every element of M is a unit in S, and every element of S has the form rm^{-1} for some $r \in R$ and $m \in M$.*

Definition 17.22 (Localization). Let R be a commutative ring with identity, and let M be a multiplicative system in R. Assume $\mathcal{Z}(M) = \{0\}$. Then the ring S, guaranteed in Theorem 17.21, is called the *localization* of R at M and is denoted by $R[M^{-1}]$ or $M^{-1}R$.

We prove Theorem 17.21 by explicitly constructing the overring $R[M^{-1}]$. If you are given a commutative ring with identity R and a multiplicative system M of R and you want to construct $R[M^{-1}]$, you can certainly consult the construction given below. However, most often Theorem 17.21 suffices. It tells you that the ring $R[M^{-1}]$ consists of all the elements of the form $\frac{r}{m} = rm^{-1}$ with $r \in R$ and $m \in M$. Hence,

$$R[M^{-1}] = \{\frac{r}{m} \mid r \in R, m \in M\}.$$

We mimic the construction of a field of fractions.

Construction of $R[M^{-1}]$. Let R be a commutative ring with identity, and let M be a multiplicative system in R. Further assume that $\mathcal{Z}(M) = \{0\}$. Consider $R \times M = \{(r, m) \mid r \in R, m \in M\}$. Define a relation on $R \times M$ by $(a, b) \sim (c, d)$ if and only if $ad = bc$.

The relation $\sim$ is an equivalence relation and the proof is very similar to the corresponding proof for the field of fractions. The equivalence classes of the equivalence relation $\sim$ partition $R \times M$. Denote the equivalence class that contains (r, m) by $[r, m]$, and define

(17.3) $S = \{[r, m] \mid r \in R, m \in M\}.$

Mimicking the addition and multiplication of fractions (and our construction of fraction fields), we define addition and multiplication on the set S by

(17.4)
$$[r_1, m_1] + [r_2, m_2] = [r_1 m_2 + r_2 m_1, m_1 m_2],$$
$$[r_1, m_1][r_2, m_2] = [r_1 r_2, m_1 m_2].$$

The proof that these operations are well defined is very similar to the case when R is an integral domain and is left to the reader. Next we state the properties of $R[M^{-1}]$, promised by Theorem 17.21. The proof is again very similar to the corresponding proofs for field of fractions.

Proposition 17.23. *Let R be a commutative ring with identity, and let M be a multiplicative system in R. Assume that $\mathcal{Z}(M) = \{0\}$. Then*

(a) $R[M^{-1}]$ is a commutative ring with identity and a unitary overring of R.

(b) The subset $R_0 = \{[r, 1] \mid r \in R\}$ is a subring of $R[M^{-1}]$ isomorphic to R.

(c) For $m \in M$, the element $[1, m]$ is the inverse of $[m, 1]$ in $R[M^{-1}]$.

(d) For $r \in R$, $m \in M$, we have $[r, m] = [r, 1]([m, 1])^{-1}$.

(e) Let T be a commutative ring with identity. Assume that T is a unitary overring of R in which every element of M has an inverse. Then T has a subring S that is isomorphic to $R[M^{-1}]$. In other words, $R[M^{-1}]$ is the smallest unitary overring of R in which elements of M are invertible.

Theorem 17.21 now follows.

Example 17.24. Let $R = (\mathbb{Z}/10\mathbb{Z}, +, \cdot)$, and let $M = \{1, 3, 7, 9\}$. Then M is a multiplicative system and $\mathcal{Z}(M) = \{0\}$. Hence, we can construct $R[M^{-1}]$.

$$R[M^{-1}] = \{\frac{a}{b} \mid a \in \mathbb{Z}/10\mathbb{Z}, b \in M\}.$$

We have already proved that this is a commutative ring with identity. It is also clear that this ring contains a copy of $\mathbb{Z}/10\mathbb{Z}$ and that elements of M are invertible in this overring of R.

But note that elements of M were already invertible in R. Namely, in $\mathbb{Z}/10\mathbb{Z}$, we have $\frac{1}{1} = 1$, $\frac{1}{3} = 7$, $\frac{1}{7} = 3$, and $\frac{1}{9} = 9$. Hence, fractions with denominator in M are equivalent to elements of $\mathbb{Z}/10\mathbb{Z}$. For example, $\frac{2}{7} = 2 \times \frac{1}{7} = 2 \times 3 = 6$. Hence, $R[M^{-1}] = R$.

An alternative argument would be to say that R is already a unitary overring of R in which all elements of M are invertible. This means $R[M^{-1}] = R$, since $R[M^{-1}]$ is the *smallest* unitary overring of R in which all elements of M are invertible.

Remark 17.25. We constructed $R[M^{-1}]$ when R was a commutative ring with identity, M was a multiplicative system, and $\mathcal{Z}(M) = \{0\}$. It is possible, with some care, to generalize this construction in a number of directions.

(a) If $\mathcal{Z}(M)$ has non-zero elements, then, by Lemma 17.20, we *cannot* construct a ring that has a copy of R as a subring, and in which every element of M has an inverse. What we can do is to construct a ring S that has a copy of $R/\mathcal{Z}(M)$ as a subring, and in which elements of the form $m + \mathcal{Z}(M)$ where $m \in M$ have inverses. You are asked to do this in Problem 17.2.10.

(b) It is possible to relax the condition that a multiplicative system necessarily contains 1. You are asked to do this in Problem 17.2.8.

(c) With much more care, it is possible to generalize the construction of $R[M^{-1}]$ to non-commutative rings R.

Problems

17.2.1. **Proof of Lemma 17.18.** Let R be a commutative ring with identity, and let $M = R - \{0\}$. Show that the following are equivalent:

(a) R is an integral domain,

    (b) M is a multiplicative system and $\mathcal{Z}(M) = \{0\}$,

    (c) M is a multiplicative system.

17.2.2. Let $R = (\mathbb{Z}/4\mathbb{Z})[x]$ be the ring of polynomials with coefficients in $\mathbb{Z}/4\mathbb{Z}$. Let $M = \{1, x, x^2, \ldots\}$. Is R an integral domain? Is M a multiplicative system in R. What is $\mathcal{Z}(M)$? Is it possible to construct the ring $R[M^{-1}]$? If so, what are its elements?

17.2.3. Let $R = \mathbb{Z} \times \mathbb{Z}$ and let $M = \{(1, 2^n) \mid n \in \mathbb{Z}^{\geq 0}\}$. Is R an integral domain? Is M a multiplicative system in R. What is $\mathcal{Z}(M)$? Is it possible to construct the ring $R[M^{-1}]$? If so, what are its elements? Pick two arbitrary elements of $R[M^{-1}]$ and explicitly find their sum and their product.

17.2.4. Let R be a commutative ring with identity, and let M be a multiplicative system in R. Assume R is not an integral domain. Is it possible for $\mathcal{Z}(M)$ to contain every zero-divisor of R? Either give an example or prove that it is impossible.

17.2.5. **The Construction of $R[M^{-1}]$.** Let R be a commutative ring with identity, and let M be a multiplicative system in R. Assume $\mathcal{Z}(M) = \{0\}$. Complete the details of the construction of $R[M^{-1}]$ on page 362. In particular, show that the relation $\sim$ is an equivalence relation and that the addition and multiplication defined on the equivalence classes are well defined.

17.2.6. **Proof of Proposition 17.23.** By mimicking the case of an integral domain, write a complete proof of Proposition 17.23.

17.2.7. **Universal Property.** Let A be a commutative ring with identity, and let M be a multiplicative system in A with $\mathcal{Z}(M) = \{0\}$. Let R be $A[M^{-1}]$, the localization of A at M. The ring R has a subring isomorphic to A. Let $\theta\colon A \to R$ be a 1-1 ring homomorphism. (In other words, $\theta(A)$ is the subring of R isomorphic to A and θ gives the isomorphism.) Now let B be a commutative ring with identity, and assume $\phi\colon A \to B$ is a ring homomorphism with $\phi(1_A) = 1_B$. Assume that, for all $m \in M$, $\phi(m)$ is a unit of B. Prove that there exists a unique ring homomorphism $\psi\colon R \to B$ such that the following diagram of rings and homomorphisms commutes.

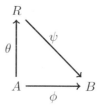

17.2.8. Generalize Theorem 17.21 to the case when M is not assumed to contain 1. In other words, let R be a commutative ring with identity. Let $M \subseteq R$. Assume

    (a) M is non-empty, and

    (b) if $m, n \in M$, then $mn \in M$, and

    (c) $0 \notin M$, and

    (d) if $m \in M$, $r \in R$, and $mr = 0$, then $r = 0$.

Show that there exists a ring S such that

    (a) S has a subring R_0 isomorphic to R, and if we identify R_0 and R then we have $R \subseteq S$, and

    (b) $1_S = 1_R$, and

    (c) for all $m \in M$, m^{-1} exists in S, and

    (d) for all $s \in S$, we can write $s = m^{-1}r$ with $m \in M$ and $r \in R$.

17.2.9. Let R be a commutative ring with identity, and let $M \subseteq R$ be a multiplicative system. Let $I = \mathcal{Z}(M)$, and let $\overline{R} = R/I$. Let $\pi : R \to \overline{R}$ be the canonical homomorphism (i.e., π is defined by $\pi(r) = I + r$). Show that $\pi(M)$ is a multiplicative system in $\overline{R}$ and $\mathcal{Z}(\pi(M)) = \{0_{\overline{R}}\}$.

17.2.10. Let R be a commutative ring with identity, and let $M \subseteq R$ be a multiplicative system. Show that there exists a ring S and a homomorphism $\theta : R \to S$ such that

    (a) $\ker(\theta) = \mathcal{Z}(M)$,

    (b) $\theta(1_R) = 1_S$,

    (c) elements of $\theta(M)$ have inverses in S, and

    (d) every element $s \in S$ is of the form $s = \theta(a)\theta(m)^{-1}$ for some $a \in R$ and $m \in M$.

Chapter 18

Factorization, EDs, PIDs, and UFDs

... where irreducibles, primes, greatest common divisors, maximal ideals, prime ideals, Euclidean domains (ED), unique factorization domains (UFD), and noetherian domains are defined, and where it is proved that ED ⇒ PID ⇒ UFD.

18.1. Factorization in Commutative Rings

In Section 15.1, we argued that we should study commutative rings in order to use various rings of numbers in solving Diophantine equations. We are now ready to develop the basic notions necessary for doing number theory in rings. We will define divisibility, primes, factorization, and so on. In doing so, we will *not* depend on what we know of these in the ordinary ring of integers. Hence, in the process, we will also be rigorously proving the properties of ordinary integers.

Definition 18.1 (Units, associates, irreducibles, and primes). Let R be a commutative ring with identity. Let $a, b \in R$.

If a is a non-zero element, then a *divides* b if there exists an element c of R with $b = ca$. This is denoted by $a \mid b$. Hence, $v \in R$ is a *unit* (that is, v has a multiplicative inverse in R) if and only if $v \mid 1$. We say that a and b are *associates* if $a = bv$ for some unit $v \in R$.

Let $p \in R$. The element p is called a *prime* if $p \neq 0$, p is not a unit, and whenever $p \mid ab$, then $p \mid a$ or $p \mid b$. The element p is called *irreducible* if $p \neq 0$, p is not a unit, and whenever $p = ab$, then either a is a unit or b is a unit.

Example 18.2. In the ring of integers $\mathbb{Z}$, the only units are ± 1. So, for example, 15 and -15 are associates. In this ring being a prime is the same as being an irreducible, and, given our definition, if p is a prime, then so is $-p$. Hence, a list of primes (and irreducibles) can begin with ± 2, ± 3, ± 5, ± 7, ... (When we use the integers as an example for our definitions, we rely on your prior familiarity with them. However, most facts about integers—for example, that, for the integers, primes are the same as irreducibles—will be proved as part of the development of ring theory.)

Remark 18.3. Before studying rings, you were probably accustomed to a prime being defined as an integer greater than 1 that is divisible by only itself and 1. We are now calling an ordinary integer with this property an *irreducible*. While being an irreducible and a prime is the same for integers, it is not so in general, and, in fact, the distinction is an important one.

In the case of integers, we are also allowing for primes to be negative, and hence both 47 and -47 are primes in $\mathbb{Z}$. The two integers 47 and -47 are associates, and neither has any other associates (in $\mathbb{Z}$). However, other commutative rings with identity may have many more units, and, hence, a prime may have a multitude of associates. While, for the ordinary integers, we may have a preference for positive integers, there is no natural way, in a commutative ring with identity, to privilege any one of the associates over the others.

The existence of associates will require us to be careful in wording theorems. For example, if we just say "an integer can be uniquely factored into a product of primes", then we are incorrect. After all, 6 is 2×3, but it is also $(-2) \times (-3)$ (not to mention, 3×2).

Example 18.4. Consider $\mathbb{Z}[\sqrt{2}] = \{a + b\sqrt{2} \mid a, b \in \mathbb{Z}\}$. The elements $3 + 2\sqrt{2}$ and $3 - 2\sqrt{2}$ are both units of this ring since $(3 + 2\sqrt{2})(3 - 2\sqrt{2}) = 1$. Hence, in this ring, 5, -5, $15 + 10\sqrt{2}$, and $15 - 10\sqrt{2}$ are all associates. This means that, for example,

$$2 \times 5 = 10 = (6 - 4\sqrt{2})(15 + 10\sqrt{2}),$$

but this does *not* show that 10 has two distinct factorizations in $\mathbb{Z}[\sqrt{2}]$. We have just replaced the factors 2 and 5 by their associates.

We will be mostly studying integral domains and, in fact, more restricted subclasses of integral domains where we have more control over primes and irreducibles. The next example shows why.

Example 18.5. Let $R = (\mathbb{Z}/6\mathbb{Z}, +, \cdot)$. The ring R is a commutative ring with identity, but it is *not* an integral domain. The multiplication table of R is given in Table 18.1. The units of R are $\pm 1 = 1, 5$. The elements 2 and $4 = -2$ are associates,

Table 18.1. The multiplication table of $(\mathbb{Z}/6\mathbb{Z}, +, \cdot)$

$\times$	0	1	2	3	4	5
0	0	0	0	0	0	0
1	0	1	2	3	4	5
2	0	2	4	0	2	4
3	0	3	0	3	0	3
4	0	4	2	0	4	2
5	0	5	4	3	2	1

as are the units 1 and 5. The elements 0 and 3 do not have any associates other than themselves. Since $2 = 2 \times 4$, $3 = 3 \times 3$, and $4 = 2 \times 2$, this ring has *no* irreducibles. On the other hand 2, 3, and 4 are all primes. For example, to see that 2 is a prime, note that 2 only divides 0, 2, and 4. In the multiplication table, whenever we see 0, 2, or 4, then either the row or the column heading is one of 0, 2, or 4. This means that if $2 \mid ab$, then $2 \mid a$ or $2 \mid b$.

The fact that there are no irreducibles makes factoring unusual. For example, we have

$$4 = 2 \times 2,$$
$$4 = 2 \times 2 \times 2 \times 2,$$
$$4 = 2 \times 2 \times 2 \times 2 \times 2 \times 2.$$

Lemma 18.6. *Let R be a commutative ring with identity, and let a, b, c, and v be non-zero elements of R. Then*

(a) *Every non-zero element of R divides 0.*

(b) *If v is a unit, then v^{-1} is a unit.*

(c) *The element v is a unit if and only if v and 1 are associates.*

(d) *If $a \mid b$ and $b \mid c$, then $a \mid c$.*

(e) *If a and b are associates, then $a \mid b$ and $b \mid a$.*

(f) *Assume R is an integral domain. If $a \mid b$ and $b \mid a$, then a and b are associates.*

Proof. Most assertions follow trivially from the definitions. We only give a proof that, in an integral domain, if $a \mid b$ and $b \mid a$, then a and b are associates. From $a \mid b$ and $b \mid a$, we get that $b = au$ and $a = bv$ for some ring elements u and v. Thus $b = au = bvu$. Since R is an integral domain, we can cancel b and get $1 = vu$. Hence, v and u are units, and a and b are associates. $\qquad\square$

Example 18.7. We give an example to show that if R is *not* an integral domain, then $a \mid b$ and $b \mid a$ does *not* necessarily imply that a and b are associates.

Let $R = \mathcal{C}([-1, 1], \mathbb{R})$ be the ring of all real valued continuous functions on the interval $-1 \le x \le 1$ (with pointwise addition and multiplication). Let

$$f(x) = \begin{cases} 2x + 1 & \text{for } -1 \le x \le -1/2 \\ 0 & \text{for } -1/2 \le x \le 1/2 \\ 2x - 1 & \text{for } 1/2 \le x \le 1, \end{cases}$$

and let $g(x) = |f(x)|$. Then f and g are both in R, and we can show (see Problem 18.1.17) that, in R, $f \mid g$ and $g \mid f$. However, f and g are not associates.

Algebra is about structures and substructures. Just as in group theory, where we preferred to translate our questions to ones about subgroups, in ring theory, we translate most notions to ones about ideals. Recall that $\langle a \rangle$ is the ideal generated by a.

Proposition 18.8. *Let R be a commutative ring with identity, and let $a, b \in R - \{0\}$. Then*

(a) *The element a divides the element b if and only if $\langle b \rangle \subseteq \langle a \rangle$.*

(b) *If a and b are associates, then $\langle a \rangle = \langle b \rangle$.*

(c) *If R is an integral domain and $\langle a \rangle = \langle b \rangle$, then a and b are associates.*

(d) *The element a is a unit if and only if $\langle a \rangle = R$.*

Proof. By definition a divides b if and only if $b = av$ for some $v \in R$. But $b = av$ is equivalent to $b \in \langle a \rangle$, which in turn is equivalent to $\langle b \rangle \subseteq \langle a \rangle$.

Parts (b) and (c) follow from (a) and the corresponding statements of Lemma 18.6.

The element a is a unit if and only if a is an associate of 1. Hence, by (b), a is a unit if and only if $\langle a \rangle = \langle 1 \rangle = R$. $\square$

Remark 18.9. Because of Proposition 18.8(a), if I and J are ideals and $I \supseteq J$, then we may think of this as I "divides" J. Hence, remember that "to contain is to divide".

We want to translate questions about irreducibles and primes to ideals also. To do this, we recall Definition 16.33 of a maximal ideal and then introduce prime ideals.

Definition 18.10 (Maximal ideals). *Let R be a ring, and let M be an ideal of R. Then M is a maximal ideal of R if*

(a) *$M \neq R$, and*

(b) *if there exists an ideal I with $M \subseteq I \subseteq R$, then $I = M$ or $I = R$.*

Definition 18.11 (Prime ideals). *Let R be a commutative ring with identity. Let P be an ideal of R. Then P is a prime ideal if*

(a) *$P \neq R$, and*

(b) *for all $a, b \in R$, if $ab \in P$, then $a \in P$ or $b \in P$.*

It is often useful to go back and forth between properties of elements, of ideals, and of quotient rings. The following theorem brings together a number of useful connections. In the statement of the theorem, ID stands for *integral domain*, PID stands for a *principal ideal domain*, and UFD stands for a *unique factorization domain*. The latter will be defined later in Section 18.3.

Theorem 18.12. *Let R be a commutative ring with identity, and let a be a nonzero element of R. Then*

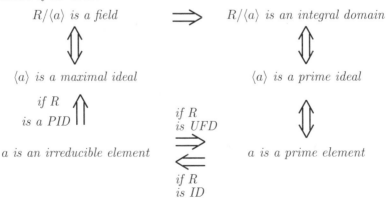

Note that other implications follow from the ones in the diagram. For example, if $\langle a \rangle$ is a maximal ideal and R is an integral domain, then a is an irreducible element. Also, $\langle a \rangle$ a maximal ideal implies that $\langle a \rangle$ is a prime ideal.

We prove this theorem by proving various implications. In some cases, we will, in fact, prove more general statements. Only one of the implications—that, in a unique factorization domain, an irreducible element is prime—requires additional ideas and will be proved in a later section (Section 18.3).

If the ring R is a principal ideal domain, then—even without proving that in a UFD every irreducible is prime—all of the statements in the theorem become equivalent. Hence, by the end of the section, we will have a complete proof of

Corollary 18.13. *Let R be a principal ideal domain (PID), and let a be a non-zero element of R. Then the following are equivalent:*

(a) *The ideal $\langle a \rangle$ is a maximal ideal.*

(b) *The ideal $\langle a \rangle$ is a prime ideal.*

(c) *The element a is a prime element.*

(d) *The element a is an irreducible element.*

(e) *The quotient ring $R/\langle a \rangle$ is a field.*

(f) *The quotient ring $R/\langle a \rangle$ is an integral domain.*

We now begin establishing results that will eventually give the proof of Theorem 18.12. Recall the straightforward argument that if a left (or a right) ideal of a ring with identity contains a unit of the ring, then it will also contain 1, and as a result will be the whole ring (Problem **16.1.20** or Proof of Proposition 16.32). It follows (Proposition 16.32) that a commutative ring with identity is a field if and only if its only ideals are $\{0\}$ and R.

Theorem 18.14. *Let R be a ring, and let M be an (two-sided) ideal of R. Then the following are equivalent:*

(a) *The ideal M is a maximal ideal.*

(b) *The quotient ring R/M contains no proper non-trivial ideals.*

Furthermore, if R is a commutative ring with identity, then each of the above is equivalent to

(c) *The quotient ring R/M is a field.*

Proof. This is Problem **16.2.9**, but we repeat the argument here: Let $\phi : R \to R/M$ be the canonical homomorphism—that is, $\phi(r) = M + r$. We know that ϕ is an onto ring homomorphism and its kernel is M. (See the homomorphism diagram of Figure 18.1.) Now, the homomorphism theorem (Theorem 16.45) says that there is a 1-1 correspondence between ideals of R containing $\ker(\phi)$ and ideals of R/M. It follows that $M = \ker(\phi)$ is a maximal ideal of R if and only if R/M has no non-trivial ideals.

In case of a commutative ring with 1, by Proposition 16.32, R/M has no non-trivial ideals if and only if R/M is a field, and the result follows. $\qquad \square$

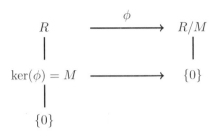

Figure 18.1. The canonical homomorphism $\phi\colon R \to R/M$

Example 18.15. Since $\mathbb{Z}/3\mathbb{Z}$ is a field and $\mathbb{Z}/6\mathbb{Z}$ is not, we conclude that $3\mathbb{Z}$ is a maximal ideal of $\mathbb{Z}$ while $6\mathbb{Z}$ is not.

We now turn to prime ideals. Recall that in our definition of an integral domain, we—unlike some other authors—had insisted that an integral domain have more than one element, and hence we do not consider the trivial ring $\{0\}$ an integral domain.

Theorem 18.16. *Let R be commutative ring with identity, and let I be an ideal of R. Then I is a prime ideal if and only if R/I is an integral domain.*

Proof. ($\Rightarrow$) Assume that I is a prime ideal. We want to show that R/I is an integral domain. Assume that $x, y \in R/I$ and $xy = 0_{R/I}$. This means that $x = r + I$ and $y = s + I$, where $r, s \in R$, and $rs + I = (r + I)(s + I) = xy = 0_{R/I}$. The zero element in R/I is I, and, hence, we have $rs + I = I$. Thus $rs \in I$. But I is a prime ideal, and, so, $r \in I$ or $s \in I$. Without loss of generality say $r \in I$, then $x = r + I = I$. This means that x is the zero of the ring. We conclude that R/I has no zero divisors. Since $I \neq R$, we have that R/I has more than one element and is non-trivial. Hence, R/I is an integral domain.

($\Leftarrow$) Assume that R/I is an integral domain. We want to show that I is a prime ideal. First, by definition, an integral domain has more than one element, and so $I \neq R$. Now, assume that $ab \in I$, then $(a + I)(b + I) = ab + I = I = 0_{R/I}$ and so either $a + I = I$ or $b + I = I$. Thus, $a \in I$ or $b \in I$. So, I is a prime ideal. $\square$

Corollary 18.17. *Let R be a commutative ring with identity, and let I be a maximal ideal of R. Then I is a prime ideal.*

Proof. The ideal I is a maximal ideal. This means, by Theorem 18.14 that R/I is a field. Every field is an integral domain (Proposition 15.18), and so R/I is an integral domain. We conclude, by Theorem 18.16, that I is a prime ideal. $\square$

Example 18.18. Not every prime ideal is maximal. For example, consider the ideal generated by x in $\mathbb{Z}[x]$, the ring of polynomials with integer coefficients. Define the map

$$\phi : \mathbb{Z}[x] \to \mathbb{Z}$$

by $\phi(p(x)) = p(0)$. This map is an onto ring homomorphism and its kernel is $\langle x \rangle$. Hence

$$\mathbb{Z}[x]/\langle x \rangle \cong \mathbb{Z}.$$

Now $\mathbb{Z}$ is an integral domain and yet it is not a field. Hence, $\langle x \rangle$ is a prime ideal but not a maximal ideal.

We could have argued that $\langle x \rangle$ is prime and not maximal directly. If $p(x)q(x) \in \langle x \rangle$, then both $p(x)$ and $q(x)$ cannot have a non-zero constant term since $p(x)q(x)$ has an x as a factor. Hence, at least one of $p(x)$ or $q(x)$ is in $\langle x \rangle$. This proves $\langle x \rangle$ is a prime ideal. On the other hand, the set of polynomials with an even constant term is a proper ideal of $\mathbb{Z}[x]$ that contains $\langle x \rangle$. Hence, $\langle x \rangle$ is not a maximal ideal.

According to Theorem 18.12 or Corollary 18.13, this example means that $\mathbb{Z}[x]$ is not a principal ideal domain. In fact, the set of polynomials with an even constant term is an ideal of $\mathbb{Z}[x]$ generated by 2 and x. It can be shown (Problem *16.1.17*) that this ideal, namely $\langle 2, x \rangle$, is not a principal ideal.

Proposition 18.19. *Let R be a commutative ring with identity, and let $p \in R$. Then p is a prime if and only if $\langle p \rangle$ is a non-zero prime ideal.*

Proof. This proof is straightforward and follows from the definitions of prime elements and prime ideals. The details are left to the reader (Problem *18.1.11*). □

Proposition 18.20. *Let D be an integral domain. Let $p \in D$ be a prime element. Then p is irreducible.*

Proof. Assume $p = ab$. We want to show that either a or b is a unit. If $p = ab$, then $p \mid ab$, which—since p is a prime element—means that $p \mid a$ or $p \mid b$. Without any loss of generality, assume $p \mid a$. In other words, $a = px$, for some $x \in D$. We now have $p = ab = pxb$. Since D is an integral domain and $p \neq 0$, we can cancel p from both sides and get $1 = xb$. This means b is a unit, and the proof is complete. □

Theorem 18.21. *Let R be a principal ideal domain (PID), and let $a \in R$. Assume a is an irreducible element. Then $\langle a \rangle$ is a maximal ideal.*

Proof. By definition, a is not a zero or a unit. Since a is not a unit, we have $1 \notin \langle a \rangle$ and hence $\langle a \rangle \neq R$. We now assume $\langle a \rangle \subseteq I \subseteq R$, and we have to show that either $I = \langle a \rangle$ or $I = R$.

The ring R is a PID, and hence $I = \langle b \rangle$, for some $b \in R$. We have $\langle a \rangle \subseteq \langle b \rangle$, and so $a \in \langle b \rangle$ which means $a = bx$, for some $x \in R$. But the element a is irreducible, and hence either b or x is a unit. If b is a unit, then $I = \langle b \rangle = R$. If x is a unit, then a and b are associates, and, hence, $\langle a \rangle = \langle b \rangle$. We conclude that $\langle a \rangle$ is a maximal ideal. □

Proof of Theorem 18.12. Except for the assertion that, in a unique factorization domain, primes and irreducibles are the same, every other assertion in Theorem 18.12 has now been proved. In particular, Corollary 18.13 has been completely proved. Theorem 18.40 of Section 18.3 will provide the final missing link. □

Example 18.22. Continuing with Example 18.4, consider $\mathbb{Z}/6\mathbb{Z}$, the ring of integers modulo 6 whose multiplication table is given in Table 18.1. In this ring, 1 and 5 are units and so $\langle 1 \rangle = \mathbb{Z}/6\mathbb{Z} = \langle 5 \rangle$. We have $\langle 2 \rangle = \langle 4 \rangle = \{0, 2, 4\}$, and

$(\mathbb{Z}/6\mathbb{Z})/\langle 2 \rangle$ is a ring with two elements isomorphic to $\mathbb{Z}/2\mathbb{Z}$. The latter is a field, and hence $\langle 2 \rangle$ is both a maximal and a prime ideal, and both 2 and 4 are prime elements. However, as noted in Example 18.4, neither 2 nor 4 are irreducible. Likewise, $\langle 3 \rangle = \{0, 3\}$, $(\mathbb{Z}/6\mathbb{Z})/\langle 3 \rangle$ is isomorphic to $\mathbb{Z}/3\mathbb{Z}$. This is also a field and hence $\langle 3 \rangle$ is also a maximal ideal, and 3 is a prime. Again, 3 is not irreducible, as noted in Example 18.4.

Constructing Fields. Because of its importance, we take a short detour to see how the above results—particularly Theorems 18.14 and 18.21—are used to construct fields. In what follows we accept two facts (these will be fully argued in Chapter 19): The first is that if F is a field, then $F[x]$, the ring of polynomials over F, is a PID (Corollary 19.22). The second—which is quite straightforward—is that in $F[x]$, where F is a field, a polynomial is irreducible if it cannot be factored into two non-constant polynomials (Proposition 19.30). We should have really waited until Chapter 19, but we wanted to whet your appetite.

Consider the ring $\mathbb{R}[x]$, the ring of polynomials over the reals $\mathbb{R}$. If we find a maximal ideal M of this ring, then the quotient ring $\mathbb{R}[x]/M$ will be a field (Theorem 18.14). The ring $\mathbb{R}[x]$ is a principal ideal domain (PID), and hence we have to look only for ideals generated by one element.

First, let us consider the ideal $\langle x + a \rangle$. Is this ideal maximal? Define $\phi : \mathbb{R}[x] \to \mathbb{R}$ by $\phi(p(x)) = p(-a)$. This is an onto ring homomorphism with $\ker(\phi) = \langle x + a \rangle$. Thus $\mathbb{R}[x]/\langle x + a \rangle \cong \mathbb{R}$. The latter is a field, and hence, by Theorem 18.14, $\langle x + a \rangle$ is a maximal ideal. This argument was actually the opposite of what we promised. Here we knew that $\mathbb{R}$ is a field and used it to show that $\langle x + a \rangle$ is a maximal ideal.

Now consider the ideal $\langle x^2 - 1 \rangle$. Is this ideal maximal? The answer is no, since $x^2 - 1 = (x-1)(x+1)$, and, hence, $\langle x^2 - 1 \rangle \subsetneq \langle x - 1 \rangle \subsetneq \mathbb{R}[x]$.

In fact, since $\mathbb{R}[x]$ is a PID, by Corollary 18.13, $\langle p(x) \rangle$ is maximal if and only if $p(x)$ is irreducible, and, in $\mathbb{R}[x]$, a polynomial is irreducible if it cannot be factored into two non-constant polynomials. Now, $x^2 + 1$ cannot be factored in $\mathbb{R}[x]$, and so $x^2 + 1$ is irreducible. Hence, $\langle x^2 + 1 \rangle$ is maximal, and $\mathbb{R}[x]/\langle x^2 + 1 \rangle$ must be a field.

What is $F = \mathbb{R}[x]/\langle x^2 + 1 \rangle$? Let $I = \langle x^2 + 1 \rangle$. Then a typical element of F is of the form $p(x) + I$ where $p(x)$ is a polynomial in $\mathbb{R}[x]$. But, by dividing $p(x)$ by $x^2 + 1$, we get $p(x) = q(x)(x^2 + 1) + r(x)$ where $r(x)$ is a polynomial of degree 1 or 0. This is useful since $q(x)(x^2 + 1) \in I$, and, hence $p(x) + I = r(x) + I$. We conclude that

$$F = \{a + bx + I \mid a, b \in \mathbb{R}\}.$$

Now, note that $1 + I$ is the multiplicative identity in F since $(1 + I)(a + bx + I) = a + bx + I$. The other element of note is $x + I$. In fact,

$$a + bx + I = a(1 + I) + b(x + I),$$

and, hence, every element of F is a linear combination of $1 + I$ and $x + I$. Further note that

$$(x + I)^2 = x^2 + I = -1 + x^2 + 1 + I = -1 + I = -(1 + I).$$

Since $1 + I$ is the identity of F, we may decide to just use 1 for it. Given this, the element $x + I$ has the property that when squared you get -1. Hence, we may choose to denote $x + I$ with i. Then F consists of linear combinations of 1 and i:

$$F = \{a + bi \mid a, b \in \mathbb{R}, i^2 = -1\}.$$

Evidently, F is the field of complex numbers, and we just gave a construction of the field $(\mathbb{C}, +, \cdot)$.

To summarize, starting with the field of real numbers $\mathbb{R}$, we first constructed the ring of polynomials $\mathbb{R}[x]$. We then found $x^2 + 1$, an irreducible polynomial of degree 2, and formed the field $F = \mathbb{R}[x]/\langle x^2 + 1 \rangle$. If we so desire, we can take this to be the definition of the field of complex numbers.

Incidentally, in the newly constructed field F, the polynomial $x^2 + 1$ has a root, namely $x + I$. Moreover, the set $R_0 = \{a + I \mid a \in \mathbb{R}\}$ is a subfield of F isomorphic to $\mathbb{R}$. Hence, starting with the real numbers, we have constructed a field that has a subfield isomorphic to $\mathbb{R}$, and, in addition, in this new field the polynomial $x^2 + 1$ has a root. This approach will be taken up in Chapter 22 when we begin a systematic study of roots of polynomials in fields.

Units and Irreducibles in Quadratic Integer Rings. As we saw in Section 15.1, the quadratic integer rings $\mathbb{Z}[\sqrt{d}]$ play an especially important role in analyzing Diophantine equations. Here, we recall an important tool—namely the *norm* map— that can be used to find units and irreducibles in quadratic integer rings.

Definition 18.23. Let d be an integer (possibly negative) not divisible by a perfect square. Recall that

$$\mathbb{Z}[\sqrt{d}] = \{a + b\sqrt{d} \mid a, b \in \mathbb{Z}\}.$$

Define $N : \mathbb{Z}[\sqrt{d}] \to \mathbb{Z}$ by $N(a + b\sqrt{d}) = a^2 - db^2$. The integer $N(x)$ is called the *norm* of x.

For the ring $\mathbb{Z}[\sqrt{d}]$, the norm map is very useful in finding units and irreducibles because of the following theorem:

Theorem 18.24. *Let d be an integer not divisible by a perfect square, and let*

$$N : \mathbb{Z}[\sqrt{d}] \to \mathbb{Z}$$

be the norm map defined by $N(a + b\sqrt{d}) = a^2 - db^2$. Then

(a) $N(xy) = N(x)N(y)$ *for all* $x, y \in \mathbb{Z}[\sqrt{d}]$.

(b) *The element* $x \in \mathbb{Z}[\sqrt{d}]$ *is a unit if and only if* $N(x) = \pm 1$.

(c) *If $N(x)$ is irreducible in $\mathbb{Z}$, then x is irreducible in $\mathbb{Z}[\sqrt{d}]$.*

Proof. (a) This is a straightforward calculation. Calculate both sides, and do a bit of algebra to see that they are the same. (You were asked to do this in Problem **15.1.3**.)

(b) Assume $x = a + b\sqrt{d}$ is a unit. Let y be the inverse of x, and we have $xy = 1$. This implies that

$$N(x)N(y) = N(xy) = N(1) = 1.$$

The only way the product of two ordinary integers is 1 is if both of those integers are 1 or -1. Hence, $N(x) = \pm 1$.

Conversely, assume $N(x) = \pm 1$. Regardless of the value of $N(x)$ we have

$$\frac{1}{a + b\sqrt{d}} = \frac{a - b\sqrt{d}}{a^2 - b^2 d} = \frac{a - b\sqrt{d}}{N(x)}.$$

Thus, if $N(x) = \pm 1$, we have that $\pm(a - b\sqrt{d})$ is the inverse of x in $\mathbb{Z}[\sqrt{d}]$ and x is a unit.

(c) Assume that $x = yz$ where y and z where $y, z \in \mathbb{Z}[\sqrt{d}]$. Then $N(x) = N(y)N(z)$. Now we know that $N(x)$ is irreducible in $\mathbb{Z}$, and, hence, one of $N(y)$ or $N(z)$ must be a unit of $\mathbb{Z}$. From the previous part, we conclude that y or z must be a unit of $\mathbb{Z}[\sqrt{d}]$, and hence x is irreducible. $\qquad\square$

Example 18.25. Consider the quadratic integer ring $R = \mathbb{Z}[\sqrt{-3}]$. What are the units of R? If $x = a + b\sqrt{-3} \in R$ is a unit, then $N(x) = a^2 + 3b^2$ must be ± 1. Since $a^2 + 3b^2$ is always positive, $N(x)$ can never be -1. Likewise if $b \neq 0$, then $N(x)$ will be no less than 3, and hence not equal to 1. We conclude that $a = \pm 1$ and $b = 0$ are the only solutions to $a^2 + 3b^2 = \pm 1$. Hence ± 1 are the only units in R.

The elements $2+3\sqrt{-3}$, $2-3\sqrt{-3}$, $-2+3\sqrt{-3}$, and $-2-3\sqrt{-3}$ are all irreducible in R, since the norm of each of them is 31, and 31 is an irreducible element of $\mathbb{Z}$.

The integer 5 is irreducible in $\mathbb{Z}$. Does it remain irreducible in R? We cannot directly use Theorem 18.24(c) since $N(5) = 25$ is not irreducible in $\mathbb{Z}$. First note that $a^2 + 3b^2 = 5$ has no integer solutions, and so if $x \in R$, then $N(x) \neq 5$. Now assume $5 = xy$ with $x, y \in R$. This means that $25 = N(5) = N(xy) = N(x)N(y)$. Now, $N(x)$ and $N(y)$ are ordinary integers, and neither can be 5. Hence, one of them is 1 and the other 25. We conclude that one of x or y is a unit, and hence 5 continues to be irreducible in $\mathbb{Z}[\sqrt{-3}]$.

The situation for 7 is different. The equation $7 = a^2 + 3b^2$ does have solutions, namely $a = \pm 2$ and $b = \pm 1$. Hence, the norm of $\pm 2 \pm \sqrt{-3}$ is equal to 7. Similar to the calculation for 5, if $7 = xy$ in R and neither x nor y are units, then $N(x) = N(y) = 7$. The candidates for x and y are few, and we can check them. We get that

$$7 = (2 + \sqrt{-3})(2 - \sqrt{-3})$$

and conclude that 7 "splits" in $\mathbb{Z}[\sqrt{-3}]$ and is not irreducible. By Theorem 18.24(c), both $2 + \sqrt{-3}$ and $2 - \sqrt{-3}$ *are* irreducible in R, since the norm of each is 7, and 7 is irreducible in $\mathbb{Z}$.

Problems

18.1.1. Let R be a commutative ring with identity. Assume that u and v are both units in the ring R. Are u and v necessarily associates? What are the associates of 0?

18.1.2. Let R be a commutative ring with identity, and let $a, b \in R$. Assume ab is a unit in R. Do a and b have to be units?

18.1.3. In Definition 18.1, we defined divisibility only for commutative rings with identity. We could use the same definition for commutative rings without identity. Let $R = 2\mathbb{Z}$ be the ring consisting of the integer multiples of 2. In this ring, what elements divide 2? Does 2 divide 2?

18.1.4. Find all the units, irreducibles, and primes in the ring $(\mathbb{Z}/10\mathbb{Z}, +, \cdot)$.

18.1.5. Let R be a commutative ring with identity. Show that R is an integral domain if and only if $\{0\}$ is a prime ideal.

18.1.6. Is $\langle x^3 \rangle$ a prime ideal in $\mathbb{Z}[x]$?

18.1.7. Find all prime and maximal ideals of $(\mathbb{Z}/12\mathbb{Z}, +, \cdot)$.

18.1.8. Let R be a commutative ring with identity. Is x an irreducible element of $R[x]$? Either prove that it is or give a counterexample. If the answer is no, then give a condition on R that would assure that x is irreducible in $R[x]$.

18.1.9. Is $6\mathbb{Z}$ a maximal ideal of $\mathbb{Z}$? What about $5\mathbb{Z}$? Can you generalize?

18.1.10. Does $\mathbb{Z}/36\mathbb{Z}$ have a prime ideal that is not maximal? Can you generalize?

18.1.11. Proof of Proposition 18.19. Let R be a commutative ring with identity, and let $p \in R$. Show that p is a prime if and only if $\langle p \rangle$ is a non-zero prime ideal.

18.1.12. What are the prime ideals of $\mathbb{Z}$, the ring of integers?

18.1.13. Let R be a commutative ring with identity.
(a) Show that $R[x]/\langle x \rangle \cong R$.
(b) Assume $R[x]$ is a PID. Show that R is a field.

18.1.14. The ring R is a commutative ring with identity. Let U be the set of all units of R, and let I be a proper ideal of R. Assume $I \cup U = R$. Prove that I is a maximal ideal of R.

18.1.15. Give an example of an irreducible element in an integral domain such that the ideal generated by this element is not a maximal ideal.

18.1.16. Let $R = \mathbb{C}[x, y]$ be the ring of polynomials in two variables over $\mathbb{C}$. Let $P = \langle x \rangle$ be the ideal of R generated by x. Define a map $\theta \colon R \to \mathbb{C}[y]$ by $\theta(p(x, y)) = p(0, y)$. In other words, given a polynomial in two variables x and y, we plug in 0 for x to get a polynomial in y.
(a) Is θ a ring homomorphism? What is the kernel and the image of θ?
(b) Is P a prime ideal? Is P a maximal ideal?
(c) Is $\mathbb{C}[x, y]$ a PID?

18.1.17. **Completing Example 18.7.** Let $R = \mathcal{C}([-1, 1], \mathbb{R})$ be the ring of all continuous functions on the interval $-1 \leq x \leq 1$ (with pointwise addition and multiplication).[1] Let

$$f(x) = \begin{cases} 2x + 1 & \text{for } -1 \leq x \leq -1/2 \\ 0 & \text{for } -1/2 \leq x \leq 1/2 \\ 2x - 1 & \text{for } 1/2 \leq x \leq 1, \end{cases}$$

---

[1] This problem is from Isaacs [**Isa94**, Problem 16.9].

and let $g(x) = |f(x)|$. Graph f and g. Are f and g elements of R? Show that, in R, $f \mid g$ and $g \mid f$ but that there is no unit $u \in R$ with $g = uf$.

18.1.18. Show that 5 is not irreducible in $\mathbb{Z}[i]$, the ring of Gaussian integers.

18.1.19. (a) What are the units of $\mathbb{Z}[i]$?
(b) Find all the associates of $2 + 3i$ in $\mathbb{Z}[i]$ and in $\mathbb{C}$.

18.1.20. Let $R = \mathbb{Z}[\sqrt{3}]$. Find 14 different units and 10 different irreducible elements of R. How is finding integer solutions to $x^2 - 3y^2 = 1$ related to finding units of R?

18.1.21. Let $R = \mathbb{Z}[\sqrt{-5}]$.
(a) Find all units of R.
(b) Is $1 + \sqrt{-5}$ irreducible in R?
(c) Is 2 a prime in R?

18.1.22. Let $\mathbb{Z}$ be the ring of integers. Define $\phi : \mathbb{Z} \times \mathbb{Z} \to \mathbb{Z}$ by $\phi(r, s) = r$.
(a) Is ϕ a ring homomorphism?
(b) Let $K = \ker(\phi)$. What are the elements of K? Is K a prime ideal? Is K a maximal ideal?
(c) Let $I = \langle 5 \rangle \subseteq \mathbb{Z}$. Let $L = \phi^{-1}(I)$. Describe L. Is L prime? Is L maximal?
(d) Find a familiar ring that is isomorphic to $(\mathbb{Z} \times \mathbb{Z})/L$.

18.1.23. Let $B = \{(4x, y) \mid x, y \in \mathbb{Z}\}$. Is B an ideal of $\mathbb{Z} \times \mathbb{Z}$? Is B prime and/or maximal? If B is not maximal, find a maximal ideal of $\mathbb{Z} \times \mathbb{Z}$ that contains B.

18.1.24. Let $I = \{(2m, 3n) \mid m, n \in \mathbb{Z}\}$. Verify that I is an ideal in $\mathbb{Z} \times \mathbb{Z}$. Is this ideal principal, prime, and/or maximal?

18.1.25. Let $R = \mathbb{Z}[x]$, and let $I = \langle x^3 + x \rangle$ be the ideal generated by $x^3 + x$. Is I a prime ideal or a maximal ideal? Is R/I an integral domain? Is it a field? In R/I do we have $I + x^5 = I + x$?

18.1.26. Define an onto homomorphism $f : (\mathbb{Z}/36\mathbb{Z})[x] \to \mathbb{Z}/36\mathbb{Z}$ such that $\ker(f) = \langle x \rangle$.
(a) Is $\langle x \rangle$ prime and/or maximal in $(\mathbb{Z}/36\mathbb{Z})[x]$?
(b) Let $A = \langle 3 \rangle$ be an ideal of $\mathbb{Z}/36\mathbb{Z}$. Find $f^{-1}(A)$. Do the same for $B = \langle 6 \rangle$.
(c) Find a familiar ring that is isomorphic to $(\mathbb{Z}/36\mathbb{Z})[x]/f^{-1}(A)$. Do the same for $f^{-1}(A)/\langle x \rangle$, and $f^{-1}(B)/\langle x \rangle$.
(d) Find a ring of the form $(\mathbb{Z}/36\mathbb{Z})/??$ that is isomorphic to $(\mathbb{Z}/36\mathbb{Z})[x]/f^{-1}(B)$.
(e) Find two maximal ideals in $(\mathbb{Z}/36\mathbb{Z})[x]$.

18.1.27. Let $R = (\mathbb{Z}/30\mathbb{Z})[x]$.
(a) Is R an integral domain? Does R have an identity?
Define $\phi : R \to \mathbb{Z}/30\mathbb{Z}$ by $\phi(p(x)) = 15p(0)$.
(b) Is ϕ a ring homomorphism?
(c) What is $\ker(\phi)$? What is $\operatorname{Im}(\phi)$?
Let $K = \ker(\phi)$.
(d) Find a familiar ring that is isomorphic to R/K.
(e) Is K a maximal ideal of R? Is K a prime ideal of R?

18.1.28. Let $R = (\mathbb{Z}/100\mathbb{Z})[x]$, and let I be the ideal of R generated by x. Is I a maximal ideal of R? Either prove that it is, or find a maximal ideal M of R such that $I < M < R$.

Is I a prime ideal? Why?

18.1.29. Let $f = x^2 + x + 1$.
   (a) Is the ring $\mathbb{F}_7[x]/\langle f \rangle$ an integral domain?
   (b) Show that $\mathbb{Z}[x]/\langle 7 \rangle \cong \mathbb{F}_7[x]$.
   (c) Is $\langle f, 7 \rangle$ a maximal ideal of $\mathbb{Z}[x]$? Is it a prime ideal?

18.1.30. Let $R = \mathbb{Z}[x]$, $f = x^2 - x + 1$, $I = \langle f, 13 \rangle$, and $J = \langle f, 17 \rangle$. Is I a maximal ideal of R? Is J a maximal ideal of R?

18.1.31. Let A and B be commutative rings with identity, and let $R = A \times B$.
   (a) Show that R is an integral domain if and only if one of A or B is the trivial ring $\{0\}$ and the other is an integral domain.
   (b) Let P be a prime ideal of A. Show that $P \times B$ is a prime ideal of R.
   (c) Let S be a prime ideal of R. Show that either $S = P \times B$ or $S = A \times Q$ where P and Q are prime ideals of A and B respectively.

18.1.32. Let R be a commutative ring with identity, and let $a, b \in R$. Let $I = \langle a \rangle$ and $J = \langle b \rangle$. Show that $a + J$ is a prime element of R/J if and only if $b + I$ is a prime element of R/I.

18.2. Ascending Chain Condition and Noetherian Rings

Let R be a ring, and let $\mathcal{I}$ be the set of ideals of R ordered by inclusion. $\mathcal{I}$ is a partially ordered set (poset) of great interest to us. (See page 347 for partial lattice diagrams of this poset.) After a couple of definitions for general posets, we introduce the class of noetherian rings. This is a rich collection of rings that includes all principal ideal domains.

Definition 18.26 (Ascending and descending chain). Let P be a partially ordered set. A set $a_1, a_2, \ldots$ of (not necessarily distinct) elements of P is called an *ascending chain* in P if

$$a_1 \leq a_2 \leq \cdots \leq a_n \leq \cdots.$$

ACC/DCC

A *descending chain* is defined similarly.

Definition 18.27 (The ascending and descending chain conditions). Let P be a poset. We say that P satisfies the *ascending chain condition* (or ACC) if every ascending chain eventually becomes constant. In other words, if $a_1 \leq a_2 \leq \cdots$ is an ascending chain, then there exists a positive integer n such that $a_n = a_{n+1} = \cdots$.

The *descending chain condition* (or DCC) is defined similarly.

Example 18.28. If P is a finite poset, then P satisfies ACC and DCC.

Example 18.29. The set of positive integers, $\mathbb{Z}^{>0}$, ordered by divisibility satisfies DCC but not ACC. Look at the Hasse diagram of the poset in Figure 9.3 to convince yourself that every descending chain will have to eventually become constant. On the other hand $1, 2, 2^2, 2^3, \ldots$ is an ascending chain with no repeats.

Example 18.30. The poset of ideals of $\mathbb{Z}$ ordered by inclusion satisfies ACC but not DCC. We shall see soon that $\mathbb{Z}$ is a principal ideal domain, and, hence, all of its ideals are generated by one element. Also recall that $a \mid b$ if and only if $\langle a \rangle \supseteq \langle b \rangle$. (We had suggested that you remember "to contain is to divide".) Hence, the Hasse diagram of the poset of ideals of $\mathbb{Z}$ is exactly the upside down version of the Hasse diagram of the poset of the Example 18.29. Every ascending chain has to become constant, and $\langle 1 \rangle \supset \langle 2 \rangle \supset \langle 2^2 \rangle \supset \langle 2^3 \rangle \cdots$ is a descending chain with no repeats.

Noetherian

Definition 18.31 (Noetherian and artinian ring). Let R be a commutative ring with identity, and let $\mathcal{I}$ be the poset of ideals of R ordered by inclusion. Then R is called *noetherian* if $\mathcal{I}$ satisfies the ascending chain condition. The ring R is called *artinian* if $\mathcal{I}$ satisfies the descending chain condition.

While they have similar sounding definitions, noetherian and artinian rings are quite different. By Example 18.30, our prototype ring, the ring of integers, is noetherian but not artinian. Hence, in this text, we only treat noetherian rings. Shortly, we give another characterization of noetherian rings.

Definition 18.32 (Finitely generated). Let R be a ring, and let I be an ideal of R. The ideal I is *finitely generated*, if there exists a finite set X with $I = \langle X \rangle$.

Lemma 18.33. *Let R be a commutative ring with identity. Let*

$$I_1 \subseteq I_2 \subseteq I_3 \subseteq \cdots$$

be an ascending chain of ideals of R, and let

$$I = \bigcup_{i=1}^{\infty} I_i.$$

Then

(a) *The set I is an ideal of R.*

(b) *If I is finitely generated, then $I = I_n$ for some positive integer n.*

Proof. (a) Let $x, y \in I$ and $r \in R$. Then, for some positive integers i and j, $x \in I_i$ and $y \in I_j$. Without loss of generality assume $j \geq i$. Since $I_i \subseteq I_j$, we have that x and y are both in I_j. Hence, since I_j is an ideal, $x + y$, rx, and $-x$ are elements of I_j. Since $I_j \subseteq I$, we conclude that $x + y$, rx, and $-x$ are in I, and I is an ideal.

(b) Assume that $I = \langle Y \rangle$ and $Y = \{y_1, \ldots, y_m\}$. For $1 \leq i \leq m$, we have $y_i \in I_{\ell(i)}$ for some positive integer $\ell(i)$. Let $n = \max\{\ell(1), \ell(2), \ldots, \ell(m)\}$. Then every one of the elements of Y is in I_n. But this means that $I = \langle Y \rangle \subseteq I_n$. But, by the definition of I, we also have $I_n \subseteq I$. Hence, we conclude that $I = I_n$. $\square$

Theorem 18.34. *Let R be a commutative ring with identity. Then R is noetherian if and only if all ideals of R are finitely generated.*

Proof. ($\Rightarrow$) Assume R is noetherian, and let I be an ideal of R. We need to prove that I is finitely generated. If $I = \langle 0 \rangle$, then it is certainly finitely generated. Otherwise, pick a non-zero element a_1 of I. If $I = \langle a_1 \rangle$, then again I is finitely generated, and we are done. Otherwise, pick $a_2 \in I - \langle a_1 \rangle$. Again if $I = \langle a_1, a_2 \rangle$, then I is finitely generated. Otherwise continue this process. At some point, after a finite number of steps, you will have a set of finite generators for I, proving that I is finitely generated. If this were not the case, the process of picking new elements for the generating set would continue indefinitely, and we would have

$$\langle a_1 \rangle \subsetneq \langle a_1, a_2 \rangle \subsetneq \langle a_1, a_2, a_3 \rangle \cdots,$$

contradicting the ascending chain condition.

($\Leftarrow$) Assume that all ideals of R are finitely generated. In order to show that R is noetherian, let

$$I_1 \subseteq I_2 \subseteq I_3 \subseteq \cdots$$

be an ascending chain of ideals. Let $I = \bigcup I_i$. By Lemma 18.33(a), I is an ideal, and since I is finitely generated—in fact, all ideals of R are assumed to be finitely generated—by Lemma 18.33(b), we have $I = I_n$ for some positive integer n. Now, for all $m > n$, we have $I_m \subseteq I = I_n \subseteq I_m$, concluding that $I_m = I_n$. Hence, the ring R satisfies the ascending chain condition, and is noetherian. $\qquad\square$

The following is now immediate:

Corollary 18.35. *If R is a principal ideal domain, then R is noetherian.*

↑ PID =r noetherian

Problems

18.2.1. Let R be a commutative ring with identity, and assume that I is an ideal of R. Prove that if R is noetherian, then so is R/I.

18.2.2. Let R and S be integral domains with $R \subseteq S$. Assume S is artinian. Does R have to be artinian?

18.2.3. Let P be a poset, and let $S \subset P$. An element $x \in S$ is called a *maximal element* of S if there is no $y \in S$ with $y > x$. The poset P satisfies the *maximal condition* if every non-empty subset of P has a maximal element.

Show that a poset P satisfies the ascending chain condition if and only if it satisfies the maximal condition.

Note: You may not notice it, but one direction of your proof will use the axiom of choice.

18.2.4. By mimicking Problem 18.2.3, define posets with minimal conditions and prove that a poset P satisfies the descending chain condition if and only if it satisfies the minimal condition.

18.2.5. Let K be a field, and let $R = K[x_1, x_2, \ldots]$ be the ring of polynomials in an infinite sequence of variables over the field K. (The elements of R are polynomials with a finite number of terms, but the number of variables is infinite.)

(a) Convince yourself that R is an integral domain.

(b) Is R noetherian?

18.2.6. Let $\mathcal{C}[0,1] = \{f : [0,1] \to \mathbb{R} \mid f \text{ is continuous}\}$ denote the ring of real valued continuous functions with pointwise addition and multiplication $((f+g)(x) = f(x) + g(x)$ and $(fg)(x) = f(x)g(x))$.

    (a) Let $0 \leq a \leq b \leq 1$ be real numbers, and define $J = \{f \in \mathcal{C}[0,1] \mid f(x) = 0$ for $a \leq x \leq b\}$. Is J an ideal of $\mathcal{C}[0,1]$?

    (b) Is $\mathcal{C}[0,1]$ noetherian?

18.2.7. Assume that R is an artinian ring and I is an ideal of R. Show that R/I is artinian as well.

18.2.8. Prove that R is an artinian integral domain if and only if R is a field.

18.2.9. Let R be a noetherian ring and let I be an ideal of R. Show that there exists a finite number of prime ideals, P_1, P_2, ..., P_k, of R such that $P_1 P_2 \cdots P_k \subseteq I$. (See Definition 16.56 for the definition of a product of ideals.)

18.3. A PID is a UFD

The fundamental theorem of arithmetic says that every ordinary integer can be written uniquely as a product of primes. We are now ready to consider rings for which (a properly worded version of) this theorem is true. We call such a ring, a unique factorization domain (or a UFD). In this section, we define UFDs, and prove that every PID is a UFD. We had shown that in a principal ideal domain an irreducible element is also prime (Corollary 18.13). Here, we strengthen this to show that, in fact, in every UFD, every irreducible element is prime. Incidentally, this also completes the proof of Theorem 18.12.

UFD

Definition 18.36 (Unique factorization domains). Let D be an integral domain. Then D is a *unique factorization domain* (or a UFD) if it satisfies the following two conditions:

UF1 If a is a non-zero element of D that is not a unit, then $a = p_1 p_2 \cdots p_n$ where n is a positive integer and each p_i is irreducible in D.

UF2 For every element a of D, if $a = p_1 p_2 \cdots p_n = q_1 q_2 \cdots q_m$, where every p_i and q_j are irreducible elements of D and n and m are positive integers, then we have $n = m$, and, after rearranging the terms, p_i is associate to q_i, for $1 \leq i \leq n$.

We begin with showing that every non-zero non-unit of a noetherian ring can be written as a product of irreducibles. In other words, we show that every noetherian ring satisfies UF1.

Theorem 18.37. *Let R be a noetherian integral domain, then R satisfies UF1. In other words, if $a \in R$, then $a = 0$ or a is a unit or a is a product of a finite number of irreducible elements.*

Proof. Assume $a \neq 0$, a is not a unit, and a is not a product of a finite number of irreducibles.

As a result, a is not irreducible, and so $a = a_1 b_1$ with neither a_1 nor b_1 a unit (or zero). This means that $\langle a \rangle \subseteq \langle a_1 \rangle$. Furthermore, these ideals cannot be equal. If $\langle a_1 \rangle = \langle a \rangle$, then a and a_1 are associates. This would mean that $a = a_1 u$, where u is a unit. But then $a_1 u = a = a_1 b_1$, which would mean $b_1 = u$ is a unit.

Since $a = a_1 b_1$, both a_1 and b_1 cannot be irreducible elements, since otherwise a will be a product of irreducibles. Without loss of generality, assume a_1 is not irreducible (and we also know that a_1 is not a unit nor zero). So $a_1 = a_2 b_2$. We continue repeating the same argument to get

$$\langle a \rangle \subsetneq \langle a_1 \rangle \subsetneq \langle a_2 \rangle \cdots,$$

contradicting the ascending chain condition. $\qquad\square$

Since PIDs are noetherian, we now know that PIDs satisfy UF1. Our next goal is to prove that an integral domain satisfying UF1 also satisfies UF2 if and only if every irreducible is prime. Since we know that in PIDs every irreducible is prime (see Corollary 18.13), the proof that every PID is a UFD will then follow. We need two lemmas:

Lemma 18.38. *Let D be an integral domain. Let p be irreducible, and let u be a unit. Then up is irreducible.*

Proof. Assume that $up = xy$ then $p = (u^{-1}x)y$. Since p is irreducible, we have either y or $u^{-1}x$ is a unit. In the latter case, we get that $x = u(u^{-1}x)$ is a unit. Hence, either y or x are units, and we have proved that up is irreducible. $\qquad\square$

Lemma 18.39. *Let D be an integral domain. Then the following are equivalent:*

UF1   *If a is a non-zero element of D that is not a unit, then $a = p_1 p_2 \cdots p_n$, where n is a positive integer and each p_i is irreducible in D.*

UF1′   *If a is a non-zero element of D, then $a = u p_1 p_2 \cdots p_n$, where u is a unit, each p_i is irreducible, and n is a non-negative integer.*

Proof. This follows directly from the previous lemma. $\qquad\square$

Theorem 18.40. *Assume D is a unique factorization domain (UFD), and let $p \in D$. Then p is irreducible if and only if p is prime.*

Proof. In Proposition 18.20, we showed that, in any integral domain, primes are irreducible. For the other direction, assume that p is irreducible. To show that p is prime, assume $p \mid ab$. We need to show that $p \mid a$ or $p \mid b$.

We have $p \mid ab$ which means $ab = pc$ for some $c \in D$. We can assume a and b are non-zero (since $p \mid 0$) and so $c \neq 0$. Since D is a UFD, we use condition UF1′ of Lemma 18.39 to write:

$$a = u_1 p_1 \cdots p_n,$$
$$b = u_2 q_1 \cdots q_m,$$
$$c = u_3 r_1 \cdots r_s,$$

where u_1, u_2, and u_3 are units, and the p's, q's, and r's are irreducible in D. From $ab = pc$ we conclude that

$$p(u_3 r_1) r_2 \cdots r_s = (u_1 p_1) p_2 \cdots p_n (u_2 q_1) q_2 \cdots q_m.$$

By Lemma 18.38, $u_3 r_1$, $u_1 p_1$, and $u_2 q_1$ are irreducible. In addition, we know every other factor (i.e., p, the r's, the p's, and the q's) are also irreducible. Invoking property UF2 of a unique factorization domain, we now conclude that the irreducible element p is an associate of a p_i or a q_j (if p is an associate of $u_1 p_1$ or $u_2 q_1$, then it is also an assoicate of p_1 or q_1). If p is an associate of p_i, then $p \mid p_i \mid a$. Likewise, if p is an associate of q_j, then $p \mid q_j \mid b$. So p either divides a or b, and the proof is complete.   □

We now show that the difference between primes and irreducibles is at the heart of determining which rings are unique factorization domains.

Theorem 18.41. *Let D be an integral domain, and assume D satisfies UF1. Then UF2 is satisfied if and only if every irreducible in D is prime in D.*

Proof. ($\Rightarrow$) Assume that D—in addition to being an integral domain satisfying UF1—satisfies UF2. Then D is a unique factorization domain, and so, by Theorem 18.40, every irreducible in D is prime in D.

($\Leftarrow$) Now, assume that every irreducible in D is a prime. We need to show that UF2 is satisfied. Assume

$$p_1 p_2 \cdots p_n = q_1 q_2 \cdots q_m,$$

where n and m are positive integers and, for $1 \leq i \leq n$ and $1 \leq j \leq m$, p_i and q_j are irreducible elements. We want to show that $m = n$ and, after possibly rearranging the q's, that each p_i is an associate of q_i. We induct on n.

For the base case, assume $n = 1$. Then $p_1 = (q_1 q_2 \cdots q_{m-1}) q_m$. Since p_1 is irreducible, if $m \geq 2$, then, by definition of irreducibility, either q_m or $q_1 \cdots q_{m-1}$ is a unit. If u is the inverse of $q_1 \cdots q_{m-1}$, then $q_1 \cdots q_{m-1} u = 1$, which means that q_1 is a unit. Hence either q_m or q_1 is a unit. This is a contradiction since they are both irreducibles and, hence, $m = 1$ and $p_1 = q_1$. The proof is complete in this case.

For $n \geq 2$, assume that the claim is true for $n = k$, and we would want to prove it for $n = k + 1$. We have

$$p_1 p_2 \cdots p_k p_{k+1} = q_1 q_2 \cdots q_m,$$

where all the p's and q's are irreducible. Note that the same argument as the one for the base case assures that $m \geq 2$. We are assuming that every irreducible is prime, and so all the p's and q's are prime elements. Now $p_{k+1} \mid q_1 \cdots q_m$ and p_{k+1} is a prime. So $p_{k+1} \mid q_j$ for some j. This means that $q_j = p_{k+1} r$. But q_j is irreducible and hence r must be a unit. We conclude that p_{k+1} and q_j are associates and hence $p_{k+1} u = q_j$ for some unit $u \in D$. We now have

$$p_1 p_2 \cdots p_k p_{k+1} = q_1 q_2 \cdots q_{j-1} q_{j+1} \cdots q_m p_{k+1} u.$$

Cancelling the p_{k+1}, we get

$$p_1 p_2 \cdots p_k = (u q_1) q_2 \cdots q_{j-1} q_{j+1} \cdots q_m.$$

Now, by Lemma 18.38, $u q_1$ is an irreducible element, and on the left hand side we have a product of k irreducible elements. Hence we can apply the inductive hypothesis, and conclude that $k = m - 1$—hence $m = k + 1$—and each p_i is the associate of one of the q_j.   □

Theorem 18.42. *Every principal ideal domain is a unique factorization domain. In other words,*

$$PID \Rightarrow UFD.$$

Proof. Let D be a PID. Then D is noetherian by Corollary 18.35 and hence satisfies UF1 by Theorem 18.37. In addition, every irreducible in D is a prime by Corollary 18.13, and, hence, D satisfies UF2 by Theorem 18.41. Thus D is a UFD. $\square$

Remark 18.43. In the proof of Theorem 18.37—that a noetherian ring satisfies UF1—when producing an infinite chain of ideals, we used the axiom of choice. In this text, we have assumed an intuitive understanding of basic set theory, and we do not detail our use of the various axioms, and, on several occasions, we do use the axiom of choice. For example, our result here that every PID is a UFD is dependent on axiom of choice. (Also see Remark 16.34.)

Example 18.44. In the ring $\mathbb{Z}[\sqrt{-5}]$, we have

$$21 = 3 \times 7 = (1 + 2\sqrt{-5})(1 - 2\sqrt{-5}).$$

Does this immediately show that $\mathbb{Z}[\sqrt{-5}]$ is not a UFD? In order to show that 21 does not factor uniquely into irreducibles, we have to show that each of the factors are irreducible and that the factors in one factorization are not the associates of the factors in the other factorization. In the case of quadratic integer rings, the norm map will be very helpful in answering these questions (see Theorem 18.24 and Example 18.25). For $\mathbb{Z}[\sqrt{-5}]$, if $x = a + b\sqrt{-5}$, then $N(x) = a^2 + 5b^2$ is a non-negative integer.

For example, to show that 3 is irreducible in $\mathbb{Z}[\sqrt{-5}]$, assume $3 = xy$, with neither x nor y a unit. Then $N(x)N(y) = N(xy) = N(3) = 9$. The only factorizations of 9 into positive integers is 1×9, 9×1, and 3×3. Neither x nor y is a unit and hence neither $N(x)$ nor $N(y)$ can be 1. This means that $N(x) = N(y) = 3$. However, $N(x) = a^2 + 5b^2$, and this cannot be equal to 3. The contradiction shows that 3 is irreducible in $\mathbb{Z}[\sqrt{-5}]$.

We can also show that 3 is not an associate of either $1 + 2\sqrt{5}$ or $1 - 2\sqrt{5}$. Indeed if $3 = (1 \pm 2\sqrt{5})u$ where u was a unit, then we would have $9 = N(3) = N(1 \pm 2\sqrt{5})N(u) = N(1 \pm 2\sqrt{5}) = 21$, a contradiction. Alternatively, we could first find the units of $\mathbb{Z}[\sqrt{-5}]$. An element $u \in \mathbb{Z}[\sqrt{-5}]$ is a unit if and only if $N(u) = \pm 1$. Hence, if $u = a + b\sqrt{-5}$ and u is a unit, we need $a^2 + 5b^2 = \pm 1$. This is only possible if $b = 0$ and $a = \pm 1$. We conclude that ± 1 are the only units of $\mathbb{Z}[\sqrt{-5}]$. Hence, the only associates of 3 are ± 3 (and not $1 \pm 2\sqrt{5}$).

In Problem 18.3.5, you are asked to complete the rest of the details. As a result, we declare that 21 does not factor irreducibly in $\mathbb{Z}[\sqrt{-5}]$, and so $\mathbb{Z}[\sqrt{-5}]$ is not a UFD or a PID.

Problems

18.3.1. Let $R = (\mathbb{Z}/6\mathbb{Z}, +, \cdot)$. Is R noetherian? Let a be a non-zero, non-unit element of R. Is a necessarily a product of irreducibles? Do your answers contradict Theorem 18.37?

18.3.2. Let R be a commutative ring with identity. Can a unit in R be a product of irreducibles?

18.3.3. In Definition 18.36 of a unique factorization domain, in condition UF2, we stipulated that n and m be positive integers. Instead, could we have n and m be non-negative integers with the understanding that a product of 0 terms (i.e., the empty product) is 1? If the answer is yes, then redo the proof of Theorem 18.41 by streamlining the proof of the base case of induction.

18.3.4. Let D be a noetherian integral domain. Show that D is a unique factorization domain if and only if every irreducible in D is a prime.

18.3.5. Complete Example 18.44, and show that in $\mathbb{Z}[\sqrt{-5}]$, 21 does not factor uniquely as a product of irreducibles. Conclude that $\mathbb{Z}[\sqrt{-5}]$ is not a UFD or a PID.

18.3.6. In $\mathbb{Z}[\sqrt{-5}]$, does 5 have a unique factorization into irreducibles? Prove your assertion.

18.3.7. Can you find an element of $\mathbb{Z}[\sqrt{-5}]$ that is irreducible but not prime? If the answer is yes, find one explicitly.

18.3.8. In Problem 18.3.5 you showed that $\mathbb{Z}[\sqrt{-5}]$ is *not* a principal ideal domain. Can you explicitly find an ideal that is not principal?

18.3.9. Can 6 be factored uniquely in $\mathbb{Z}[\sqrt{-6}]$? Is $\mathbb{Z}[\sqrt{-6}]$ a UFD?

18.3.10. In $\mathbb{Z}[\sqrt{-10}]$, does 14 have a unique factorization into irreducible elements? Is $\mathbb{Z}[\sqrt{-10}]$ a UFD?

18.3.11. Show that $\mathbb{Z}[\sqrt{5}]$ is not a UFD. Find an irreducible element of $\mathbb{Z}[\sqrt{5}]$ that is not prime.

18.3.12. Let R be a UFD. Assume that every non-zero prime ideal of R is maximal. Prove that every prime ideal of R is principal.

18.3.13. Let
$$R = \{p \in \mathbb{Q}[x] \mid p(0) \in \mathbb{Z}\}.$$
In other words, R consists of those polynomials with rational coefficients whose constant term is an integer.
(a) Is R an integral domain?
(b) What are the units of R?
(c) Is x irreducible in R? Is x a product of irreducibles? Why?
(d) Is x prime in R? Is R a UFD? Why?
(e) What are the elements of $\langle x \rangle$? Describe them explicitly.
(f) What are the elements of $R/\langle x \rangle$? Can you find an explicit zero divisor in this ring?
(g) Is R noetherian? Why?

18.3.14. Let R be a UFD, let M be a multiplicative system in R, and let $R[M^{-1}]$ be the localization of R at M. For $a \in R$ and $m \in M$, we write a/m for the element $am^{-1} \in R[M^{-1}]$. Assume that a is an irreducible element of R, and show that $a/1$ is an irreducible element of $R[M^{-1}]$.

18.3.15. **Localization of a UFD.** Let R be a UFD, and let M be a multiplicative system in R. Show that $R[M^{-1}]$, the localization of R at M, is a UFD.

18.4. Euclidean Domains

In this section, we turn to the division algorithm, a fundamental property of the integers. For the integers, we proved the division algorithm early on in Theorem 1.47—which says that given two integers a and b, as long as $b \neq 0$, you can divide a by b, get a quotient q and a remainder r such that q and r are integers and $0 \leq r < |b|$. Here, we consider integral domains where an appropriate version of the division algorithm holds. Such an integral domain will be called a *Euclidean domain* (ED). We will prove that Euclidean domains are principal ideal domains. This will mean that Euclidean domains are also unique factorization domains. In fact, the most straightforward way to show that an integral domain is a unique factorization domain is to show that the integral domain is a Euclidean domain. Since we already know that the ring of integers is a Euclidean domain, this will also prove that $\mathbb{Z}$ is a principal ideal domain and a unique factorization domain. These facts certainly can (and should) be proved directly. However, our results will provide a proof, and, in fact, if you go back through the various proofs and specialize them to the integers, you will find a fairly common proof that the integers are a UFD and a PID.

Definition 18.45. Let D be an integral domain. Then D is a *Euclidean domain* (ED) if there exists a function

$$d : D - \{0\} \to \mathbb{Z}^{\geq 0},$$

such that

- $d(a) \leq d(ab)$ for all $a, b \in D - \{0\}$, and
- if $a, b \in D$ and $b \neq 0$, then there exists $q, r \in D$ such that
$$a = bq + r \quad \text{where } r = 0 \text{ or } d(r) < d(b).$$

A function d with these properties is called a *degree function* (or a *Euclidean valuation*).

Example 18.46. The ring of integers $\mathbb{Z}$ is a Euclidean domain with $d(a) = |a|$ for all $a \in \mathbb{Z}$. In other words, the ordinary absolute value function provides a degree function for $\mathbb{Z}$.

The proof that the absolute value function satisfies the conditions of a degree function is straightforward. If a and b are non-zero integers, we clearly have that $|a| \leq |a||b| = |ab|$. Now if a is any integer and b a non-zero integer, then by the division algorithm (Theorem 1.47) there exist unique integers q and r such that

$$a = qb + r \quad \text{and} \quad 0 \leq r < |b|.$$

Example 18.47. The field of rational numbers $\mathbb{Q}$ is a Euclidean domain with $d(a) = 1$ for all $a \in \mathbb{Q}$. This constant valued function is a degree function since, for all a and all non-zero b, we have $a = bq + 0$. In fact, the same argument shows that every field is a Euclidean domain.

Example 18.48. In the next chapter we will prove that if F is a field, then the ring of polynomials over F, $F[x]$, is a Euclidean domain with $d(p(x))$ equal to the degree of the polynomial $p(x)$. This explains why the function d is called a *degree* function (and also why, in the definition of the degree function, we excluded 0 from the domain).

Example 18.49. The ring of Gaussian integers $\mathbb{Z}[i]$ is a Euclidean domain with $d(a + bi)$ equal to the norm of $a + bi$, namely $a^2 + b^2$. You are asked to show that this is indeed the case in Problem 18.4.7. We will come back to Gaussian integers and explore this property further in Chapter 20. We should note that the norm function in quadratic integer rings sometimes gives a degree function and sometimes does not. In Example 18.44 (and Problem 18.3.5) we showed that $\mathbb{Z}[\sqrt{-5}]$ is not a UFD and, hence, it is not a PID. In the next theorem, we show that every ED is a PID. As a result, $\mathbb{Z}[\sqrt{-5}]$ is not a Euclidean domain and cannot have any degree function.

ED⇒PID **Theorem 18.50.** *Every Euclidean domain is a principal ideal domain.*

Proof. Let D be an ED with degree function d, and let I be an ideal of D. If $I = \{0\}$, then I is a principal ideal. So, we can assume $I \neq \{0\}$. Consider

$$V = \{d(a) \mid a \in I - \{0\}\}.$$

The set V consists of non-negative integers and hence it must have a smallest element. Choose $x \in I$ with the property that $d(x)$ is the smallest integer in V.

We claim that $I = \langle x \rangle$. Since $x \in I$, we have $\langle x \rangle \subseteq I$, and, hence, we need to only to show that every element of I is a multiple of x.

Let y be an arbitrary element of I. Since D is an ED and $x \neq 0$, we have $y = xq + r$ with $r = 0$ or $d(r) < d(x)$. We will be done when we show that $r = 0$, and so assume that $r \neq 0$ and, hence, $d(r) < d(x)$. Now $r = y - xq$ and both y and xq are elements of I. Hence, r is a non-zero element of I, and, by our choice of x, we have $d(x) \leq d(r)$. This contradicts $d(r) < d(x)$, and so $r = 0$ and y is a multiple of x and an element of I. The proof is complete. $\square$

Corollary 18.51. *Every Euclidean domain is a principal ideal domain, and every principal domain is a unique factorization domain. In other words,*

$$ED \Rightarrow PID \Rightarrow UFD.$$

Proof. This is just the combination of Theorems 18.42 and 18.50. $\square$

We have proved the division algorithm for the ring of integers (Theorem 1.47). From this it follows that $\mathbb{Z}$ is a Euclidean domain. Hence, we now have a proof of the fundamental theorem of arithmetic for the integers. We also know that the integers are a principal ideal domain. For completeness, we record this particular corollary of Corollary 18.51:

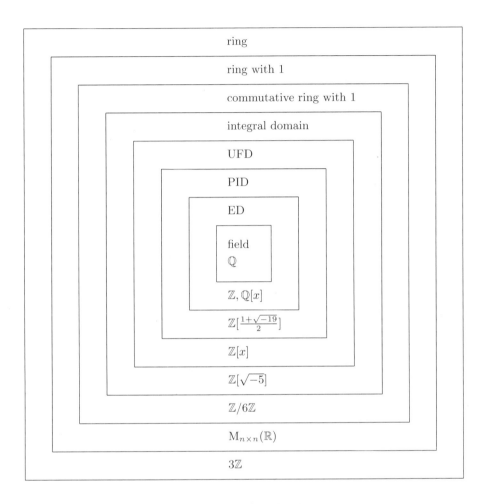

Figure 18.2. A hierarchy of rings

Corollary 18.52. *Let $(\mathbb{Z}, +, \cdot)$ be the ring of ordinary integers. The ring $\mathbb{Z}$ is a Euclidean domain, a principal ideal domain, and a unique factorization domain.*

Remark 18.53. There are certainly integral domains that have unique factorization without being a principal ideal domain. Likewise there are principal ideal domains that are not Euclidean domains. (See Figure 18.2.) Even so, often to prove that a ring is a UFD, we attempt to prove that it is an ED, and to prove that a ring is not an ED, we attempt to prove that it is not a UFD.

Remark 18.54. In the proof that every ED is a PID, we did not use the full force of a degree function. It was not necessary to assume that $d(a) \leq d(ab)$ for all non-zero elements a and b. In fact, the condition $d(a) \leq d(ab)$, while handy in determining the units of an ED, is superfluous in the definition of an ED. Let D be an integral domain, and assume that there exists a function

$$d \colon D - \{0\} \to \mathbb{Z}^{\geq 0},$$

such that if $a, b \in D$ and $b \neq 0$, then there exists $q, r \in D$ with the property that

$$a = bq + r \quad \text{where } r = 0 \text{ or } d(r) < d(b).$$

Then—even with the weaker condition on the function d—the ring D continues to be a Euclidean domain (as defined in Definition 18.45). The reason is that we can define a new function $d^* : D - \{0\} \to \mathbb{Z}^{\geq 0}$ as follows: For $a \in D - \{0\}$, $d^*(a)$ will be the minimum value of d among the non-zero elements of the ideal generated by a. Then it can be shown that d^* satisfies both conditions of a degree function in Definition 18.45, and hence D will be a Euclidean domain.[2]

Most authors keep the condition $d(a) \leq d(ab)$ in the definition of an ED, since showing that it is superfluous detracts from the main issues, and the condition itself simplifies the task of finding the units in a Euclidean domain.[3] We record the relevant result in the next lemma (which should be compared to Theorem 18.24(b)) and ask the reader to give the proof in the problems.

Lemma 18.55 (Problem *18.4.1*). *Let D be a Euclidean domain with a degree function d. Let $u \in D$. Then u is a unit of D if and only if $d(u) = d(1)$.*

Problems

18.4.1. **Proof of Lemma 18.55.** Let D be a Euclidean domain with degree function d. Assume $u \in D$. Show that u is a unit if and only if $d(u) = d(1)$.

18.4.2. Show that, in Definition 18.45, the first condition for a degree function is equivalent to the following:

for all $a, b \in D - \{0\}$, if $a \mid b$, then $d(a) \leq d(b)$.

18.4.3. Let D be a Euclidean domain with degree function d. Let $a \in D - \{0\}$. Show the following.
   (a) If $a = bc$ with b and c non-units, then $d(b) < d(a)$.
   (b) If $d(a) = 0$, then a is a unit.
   (c) If $d(a) = 1$, then a is either a unit or irreducible.

18.4.4. In Example 18.46, we showed that using the absolute function as the degree function, the ring of integers $\mathbb{Z}$ satisfies Definition 18.45 and is a Euclidean domain.
   (a) If $a = 22$ and $b = 4$, does $22 = 5(4) + 2$ satisfy the requirements of Definition 18.45? What about $22 = 6(4) - 2$?
   (b) In Definition 18.45, could/should we have insisted that q and r, the quotient and the remainder, be unique? Comment.
   (c) In the early version of Division algorithm for the integers, Theorem 1.47, we did have a unique quotient and remainder. What is the difference?

---

[2] See Rogers [**Rog71**].
[3] A condition weaker than the degree function of a Euclidean domain that still results in PIDs is discussed on page 401.

18.4.5. Let D be an integral domain. In the definition of a degree function (Definition 18.45), we insisted that for all non-zero elements $a \in D$, $d(a)$ is a *non-negative* integer. Assume that D has a "degree function" that satisfies all the conditions of a degree function, except that its codomain is the set of integers. In other words, $d : D - \{0\} \to \mathbb{Z}$. Prove that D continues to be a PID.

18.4.6. Let $R = \mathbb{Z}[i]$, and for $x \in R$, define $d(x) = N(x)$ where $N(x)$ is the norm of x (Definition 18.23 page 375).
   (a) For $x, y \in R - \{0\}$, show that $d(x) \leq d(xy)$.
   (b) For the following values of x and y find q and r such that $x = yq + r$, and either $r = 0$ or $d(r) < d(y)$:
       (i) $x = 2 + 10i$, $y = 5 + 3i$.
       (ii) $x = 5 + 11i$, $y = 2 + 8i$.

18.4.7. Prove that $\mathbb{Z}[i]$ is a Euclidean domain.

18.4.8. Let $R = \mathbb{Z}[\sqrt{-10}]$, and let $N : R \to \mathbb{Z}^{\geq 0}$ defined by $N(a + b\sqrt{-10}) = a^2 + 10b^2$ be the usual norm function. Prove that the norm function is *not* a degree function for R.

18.4.9. Show that $\mathbb{Z}[\sqrt{5}]$ is not an ED while $\mathbb{Z}[\frac{1+\sqrt{5}}{2}]$ is.

18.5. The Greatest Common Divisor⋆

In Definition 1.42, we defined the greatest common divisor (gcd) of two ordinary integers. We now want to develop the same concept in commutative rings.

Definition 18.56 (The greatest common divisor). Let R be a commutative ring with identity, and let $a, b \in R - \{0\}$. The element $d \in R$ is a *greatest common divisor* of a and b, written $\gcd(a, b)$, if

(a) the element d divides both a and b, and

(b) whenever, for $c \in R$, we have $c \mid a$ and $c \mid b$, then $c \mid d$.

In an integral domain, if a greatest common divisor of two non-zero elements is 1, then we say that the two elements are *relatively prime*.

   We shall see that the more we know about the ring, the more we can say about greatest common divisors. Let R be a commutative ring, and let a and b be non-zero elements of R. In general commutative rings, greatest common divisors may or may not exist. If R is an integral domain, and if d is a greatest common divisor for a and b, then the associates of d are exactly all of the greatest common divisors of a and b. If R is a UFD, then greatest common divisors exist. If R is a PID, then $\gcd(a, b)$ will be a linear combination of a and b. Finally, if R is an ED, then we will have an algorithm—the Euclidean algorithm—for finding the greatest common divisor.

Example 18.57. Let $R = \mathbb{Z}$, then both 2 and -2 are the greatest common divisors of 2 and 4.

Example 18.58. We could have used the same definition for a greatest common divisor in rings without identity. As an example, let $R = 2\mathbb{Z}$. What is $\gcd(2, 4)$? The greatest common divisor of 2 and 4 must divide 2. However, in $2\mathbb{Z}$ *nothing* divides 2! In fact, since $2\mathbb{Z}$ does not have an identity, 2 itself does not divide 2. Hence, in $2\mathbb{Z}$, the elements 2 and 4 do not have a greatest common divisor.

Example 18.59. Let $R = \mathbb{Q}$, then every non-zero rational number satisfies the conditions for $\gcd(6, 8)$. In fact, in any field F, every non-zero element of the field is a greatest common divisor for any pair of non-zero elements of the field.

Example 18.60. Let $R = \mathbb{Z}[\sqrt{-5}]$. Then R is an integral domain, but not a unique factorization domain. In R we have

$$9 = 3 \times 3 = (2 + \sqrt{-5})(2 - \sqrt{-5}).$$

Now, what is $\gcd(9, 6 + 3\sqrt{-5})$? Both 3 and $2 + \sqrt{-5}$ divide both elements, and yet neither divides the other. Hence, 9 and $6 + 3\sqrt{-5}$ do not have a greatest common divisor.

We first see that, in an integral domain, greatest common divisors—if they exist—are unique up to associates. Hence, in an integral domain, we often talk about *the* greatest common divisor (thinking that all associates are essentially the same element).

Lemma 18.61. *Let D be an integral domain, and let $a, b \in D - \{0\}$. Assume that x is a greatest common divisor for a and b. Then $y \in D$ is a greatest common divisor of a and b if and only if x and y are associates.*

Proof. ($\Rightarrow$) Assume y is a greatest common divisor of a and b. Then $y \mid a$ and $y \mid b$, and, since x is a gcd of a and b, we have that $y \mid x$. With the same argument, we also get that $x \mid y$. Since D is an integral domain, $x \mid y$ and $y \mid x$ imply that x and y are associates.

($\Leftarrow$) Now assume that y is an associate of x. Hence, $y = ux$ with u a unit. We know $x \mid a$, and, hence, $a = dx$ for some $d \in D$. Substituting $u^{-1}y$ for x, we have $a = u^{-1}dy$, which means $y \mid a$. Similarly, $y \mid b$. So y is a common divisor of a and b. It remains to show that it is a *greatest* common divisor. So, assume $c \mid a$ and $c \mid b$. Since x is a greatest common divisor, we must have that $c \mid x$. Hence, $x = ce$ for some $e \in D$. But then $y = uce$, and $c \mid y$. We have proved that y is a gcd of a and b. $\qquad\square$

Theorem 18.62. *Let D be a unique factorization domain, and let $a, b \in D - \{0\}$. Then $\gcd(a, b)$ exists.*

Proof. Since D is a unique factorization domain, we can write

$$a = u p_1^{r_1} p_2^{r_2} \cdots p_w^{r_w},$$
$$b = v p_1^{s_1} p_2^{s_2} \cdots p_w^{s_w},$$

where u and v are units, $w \geq 0$, r_i and s_j are non-negative integers for $1 \leq i, j \leq w$, and the p's are irreducible elements of D. For $1 \leq i \leq w$, let $n_i = \min(r_i, s_i)$, and define

$$d = p_1^{n_1} \cdots p_w^{n_w}.$$

(If $w = 0$, then d is 1.) Now clearly d divides both a and b. If c also divides both a and b, then c cannot have any irreducible factors other than $p_1, \ldots, p_w$. In addition, for $1 \leq i \leq w$, $p_i^{n_i}$ is the largest power of p_i that can divide c. Hence, c divides d, and it follows that d is $\gcd(a, b)$. $\qquad \square$

Theorem 18.63. *Let D be a principal ideal domain, and let $a, b \in D - \{0\}$. Then $\gcd(a, b) = sa + tb$ for some $s, t \in D$. In particular,*

$$\langle \gcd(a, b) \rangle = \langle a, b \rangle.$$

Proof. Let I be the ideal of D generated by a and b. In other words, $I = \langle a, b \rangle = \{ya + zb \mid y, z \in D\}$. The ring D is a principal ideal domain and so I is generated by one element. Let $I = \langle d \rangle$ for $d \in D$. Since d is an element of I, we have $d = sa + tb$ for some $s, t \in D$. We claim that d is actually a greatest common divisor of a and b. Both a and b are in I, and I is generated by d. Hence, $d \mid a$ and $d \mid b$. On the other hand, if $c \mid a$ and $c \mid b$, then $c \mid sa + tb = d$. Hence $d = \gcd(a, b)$ and a linear combination of a and b.

We actually proved that a generator for the ideal $\langle a, b \rangle$ is a greatest common divisor of a and b. Since all greatest common divisors are associates and two associate elements generate the same ideal, we can write $\langle \gcd(a, b) \rangle = \langle a, b \rangle$. $\qquad \square$

Example 18.64. In the ring of ordinary integers, $\gcd(6, 8) = 2$. As predicted by the theorem, 2 is an integer linear combination of 6 and 8. In fact, $2 = -6 + 8$.

Corollary 18.65. *Let D be a principal ideal domain, and let $a, b \in D - \{0\}$. Then a and b are relatively prime if and only if there exist u and v in D with $au + bv = 1$.*

Remark 18.66. We have shown that, in unique factorization domains, a greatest common divisors exists. The converse is not true. There are integral domains where every pair of non-zero elements has a greatest common divisor and yet unique factorization fails. This larger class of integral domains—those in which every pair of non-zero elements has a greatest common divisor—are called *gcd-domains*. Likewise, integral domains in which the greatest common divisor of every pair of non-zero elements is a linear combination of the two elements are called *Bézout domains* (Theorem 18.63 is sometimes called Bézout's theorem as well). We have shown that a UFD is a gcd-domain, and a PID is a Bézout domain. In fact, it is true that a ring is a PID if and only if it is a UFD and a Bézout domain. (You are asked to prove this in Problem 18.6.26.)

The proof of Theorem 18.62 provides a method for finding greatest common divisors in unique factorization domains. However, given a and b, to find $\gcd(a, b)$, we first need to factor a and b into a product of irreducible elements. Factoring is not easy even for ordinary integers. In fact, the RSA algorithm for public key cryptography, used in internet security, depends on the difficulty of factoring large integers into primes. Turning to Euclidean domains, we will be able to give an algorithm—that only depends on the division algorithm—for finding greatest common divisors. We begin with a lemma that generalizes a familiar fact about ordinary integers to unique factorization domains.

Lemma 18.67. *Let D be a unique factorization domain, and let $a, b, c \in D - \{0\}$. Assume $a \mid bc$ and $\gcd(a, b) = 1$. Then $a \mid c$.*

Proof. The reader is asked to provide the proof in Problem 18.5.2. □

Lemma 18.68. *Let D be a Euclidean domain, and let $a, b \in D - \{0\}$. If $a = bq + r$ with $r \neq 0$, then $\gcd(a, b) = \gcd(b, r)$.*

Proof. Let $d = \gcd(a, b)$. Then $d \mid a$ and $d \mid b$ which means that $d \mid a - bq = r$. Thus d is a common divisor of b and r. On the other hand, let c be another common divisor of b and r. We have $c \mid bq + r = a$, and so c is a common divisor of b and a. Since $d = \gcd(a, b)$, we conclude that c divides d. Hence, d is the greatest common divisor of b and r. □

We are now ready to present and prove the Euclidean algorithm for finding greatest common divisors.

Theorem 18.69. *Let D be a Euclidean domain with degree function d. Let a and b be non-zero elements of D. Define a sequence of elements of D recursively as follows: Let $r_0 = a$, $r_1 = b$, and, for $k > 1$, as long as $r_{k-1} \neq 0$, write*

$$r_{k-2} = q_k r_{k-1} + r_k,$$

where r_k is either 0 or $d(r_k) < d(r_{k-1})$. If r_{k-1} is zero, then let r_k be zero as well. Then the sequence $r_0, r_1, \ldots, r_k, \ldots$ will terminate in a string of zeros, and its last non-zero element will be the greatest common divisor of a and b.

In other words, if $b \mid a$, then $\gcd(a, b) = b$. Otherwise, write

$$a = q_2 b + r_2 \quad r_2 \neq 0$$
$$b = q_3 r_2 + r_3 \quad r_3 \neq 0$$
$$r_2 = q_4 r_3 + r_4 \quad r_4 \neq 0$$
$$\vdots$$
$$r_{k-2} = q_k r_{k-1} + r_k \quad r_k \neq 0$$
$$r_{k-1} = q_k r_k.$$

Then $r_k = \gcd(a, b)$ in D.

Proof. Since d is a degree function, as long as the remainders are non-zero, we have $d(b) > d(r_2) > d(r_3) > \cdots$. This sequence of *non-negative* integers cannot decrease forever, and so, eventually, one of the remainders will be zero. Hence, the sequence terminates in a string of zeros.

By repeated use of Lemma 18.68,

$$\gcd(a, b) = \gcd(b, r_2) = \gcd(r_2, r_3) = \cdots = \gcd(r_{k-1}, r_k) = r_k.$$

The proof is complete. □

Example 18.70. Let D be the ring of polynomials over the reals, $\mathbb{R}[x]$. In Chapter 19, we will show that D is a Euclidean domain. Hence, we can use the Euclidean algorithm to find $\gcd(x^4 - 1, x^6 - 1)$:

$$x^4 - 1 = 0(x^6 - 1) + x^4 - 1$$
$$x^6 - 1 = x^2(x^4 - 1) + x^2 - 1$$
$$x^4 - 1 = (x^2 + 1)(x^2 - 1) + 0.$$

We conclude that $\gcd(x^4 - 1, x^6 - 1) = x^2 - 1$.

We close this section with the definition of a *least common multplier*. The proofs of the accompanying results are left to the reader.

Definition 18.71. Let R be a commutative ring with identity, and let $a, b \in R - \{0\}$. Then $c \in R$ is a *least common multiple* (lcm) of a and b, written $\text{lcm}(a, b)$, if

- the elements a and b both divide c, and
- for any $d \in R$, if $a \mid d$ and $b \mid d$, then $c \mid d$.

In Problems 18.5.3 and 18.5.5, you are asked to recast the notion of least common multipliers in terms of ideals and to show that in every gcd-domain—an integral domain in which every pair of non-zero elements has a greatest common divisor—every pair of non-zero elements has a least common multiplier. We record the results here for completeness.

Lemma 18.72 (Problem 18.5.3). *Let D be an integral domain with $a, b \in D - \{0\}$, and let $c \in D$. Then $c = \text{lcm}(a, b)$ if and only if $\langle c \rangle = \langle a \rangle \cap \langle b \rangle$.*

Lemma 18.73 (Problem 18.5.5). *Let D be an integral domain in which every pair of non-zero elements have a greatest common divisor. Then every pair of non-zero elements of D also has a least common multiplier. In particular, in a UFD every pair of non-zero elements has a least common multiplier.*

Problems

18.5.1. What is the greatest common divisor of $1 + 3i$ and $1 - 3i$ in $\mathbb{Z}[i]$?

18.5.2. **Proof of Lemma 18.67.** Let D be a UFD, and let a, b, and c be non-zero elements of D. Assume that $a \mid bc$ and $\gcd(a, b) = 1$. Show that $a \mid c$.

18.5.3. **Proof of Lemma 18.72.** Let D be an integral domain with $a, b \in D - \{0\}$ and let $c \in D$. Then show that $c = \text{lcm}(a, b)$ if and only if $\langle c \rangle = \langle a \rangle \cap \langle b \rangle$.

18.5.4. Let D be an integral domain in which any two elements have a gcd, and let $0 \neq d \in D$. Then prove that $d = \gcd(da, db)$ if and only if $\gcd(a, b) = 1$.

18.5.5. **Proof of Lemma 18.73.** Let D be an integral domain in which any two elements have a gcd. Then show that any two elements have an lcm. If $a, b \in D$, then what is $\gcd(a, b)\text{lcm}(a, b)$?

18.5.6. Let R be a commutative ring with identity, and let $a, b \in R$.
  (a) Assume that $\langle a \rangle + \langle b \rangle = \langle d \rangle$ for some $d \in R$. Then show that d is a greatest common divisor of a and b.
  (b) By considering the ideal $\langle 2, x \rangle$ in $\mathbb{Z}[x]$, show that, in general, the converse is not true.

18.5.7. Let D be a PID, let $a, b \in D$, and let $d = \gcd(a, b)$. Prove that
$$\langle a \rangle + \langle b \rangle = \langle d \rangle.$$

18.5.8. Let $p, q \in \mathbb{Z}[x]$. Show that p and q are relatively prime in $\mathbb{Q}[x]$ if and only if the ideal they generate in $\mathbb{Z}[x]$ contains an integer.

18.5.9. Recall (Definition 16.56) that if I and J are two ideals in R, a commutative ring with identity, then

$$I : J = \{ r \in R \mid rx \in I \ \forall x \in J \}.$$

Assume that D is a UFD, and let $a, b \in D$. Prove

$$\langle a \rangle : \langle b \rangle = \langle a / \gcd(a, b) \rangle.$$

18.6. More Problems and Projects

Local and Discrete Valuation Rings. As we have seen, maximal ideals of a ring carry much information. Rings that have just one maximal ideal are called *local* and play an important role in algebraic number theory. We limit our discussion to commutative rings with identity.

Definition 18.74 (Local ring). Let R be a commutative ring with identity. Then R is called a *local ring* if it has a unique maximal ideal. If R is a local ring and M is its unique maximal ideal, then the field R/M is called the *residue field* of R.

A field is a local ring since $\{0\}$ is its only proper ideal. Problem 18.6.1 gives several equivalent conditions for being a local ring. The term *local* refers to an attempt to focus on "local behavior". In analysis, one often is interested in the local behavior of a function near a point. This can be formalized in the notion of the "germ" of a function near a point, and these germs give rise to a local ring. (See Problem 18.6.14.) There is also a connection between local rings and "localization" of Section 17.2. Problem 18.6.5 explores the connection.

Discrete valuation rings (abbreviated as DVR) are a well behaved class of local rings sandwiched between Euclidean domains and fields. Among their many properties is that they have a unique prime element. DVRs allow one to focus on the unique prime—number theorists think of this as "local" behavior—and, as a result, they are useful in algebraic number theory. As we shall see in Propositions 18.76 and 18.77, there are many ways to define a DVR.

Definition 18.75. A *discrete valuation ring* or *DVR* is a local ring that is a PID but not a field.[4]

Proposition 18.76. *Let R be a PID. Then the following are equivalent:*

(a) *R is a discrete valuation ring.*

(b) *R has a unique non-zero maximal ideal.*

(c) *R has a unique non-zero prime ideal.*

(d) *R has a unique (up to associates) irreducible element.*

(e) *R has a unique (up to associates) prime element.*

---

[4]In addition to the discussion and the ensuing problems here, see Problem 19.7.8 for more on DVRs.

Proof. You are asked to prove this in Problem 18.6.6.   □

Proposition 18.77. *Let D be an integral domain. Then the following are equivalent:*

(a) *D is discrete valuation ring.*

(b) *D has a unique irreducible element (up to associates) and satisfies UF1.*

In particular, a UFD (or a noetherian integral domain) with a unique irreducible element is a DVR.

Proof. You are asked to prove this in Problem *18.6.7*.   □

We defined DVRs as principal ideal domains that are not a field but have a unique maximal ideal. Proposition 18.77 shows that instead we could have defined a DVR as a UFD with a unique irreducible element. In fact, a DVR is always a Euclidean domain.

Proposition 18.78. *A discrete valuation ring is a Euclidean domain.*

Proof. You are asked to prove this in Problem 18.6.12.   □

Example 18.79. Let p be a prime, and define

$$\mathbb{Z}_{(p)} = \{r \in \mathbb{Q} \mid r = \frac{a}{b} \text{ with } a, b \in \mathbb{Z}, \ \gcd(a,b) = 1, \text{ and } p \nmid b\}.$$

In other words, $\mathbb{Z}_{(p)}$ consists of those rational numbers that, in reduced form, have a denominator not divisible by p. We have encountered $\mathbb{Z}_{(p)}$ before (see Problems 16.1.9 and 16.2.11, and Example 17.12 where $\mathbb{Z}_{(p)}$ was defined as a localization of $\mathbb{Z}$ at $M = \mathbb{Z} - \langle p \rangle$), and in Problem 18.6.8 you are asked to show that $\mathbb{Z}_{(p)}$ is a discrete valuation ring.

Problems

18.6.1. Let R be a non-trivial commutative ring with identity. Show that the following are equivalent:
  (a) R is local.
  (b) The set of all non-units of R forms an ideal of R.
  (c) The sum of any two non-units in R is a non-unit.
  (d) If $x \in R$, then x or $1 - x$ is a unit.

18.6.2. Let R be a local commutative ring with identity, and assume that the maximal ideal in R is a principal ideal. Let a be an irreducible element of R. Show that $\langle a \rangle$ is maximal.

18.6.3. Let S be a local integral domain. Let $0 \neq p \in S$, and assume that $I = \langle p \rangle$ is the unique maximal ideal of S. Is p a prime in S? If q is an associate of p in S, then is q a prime of S? Could S have any primes other than p and its associates?

18.6.4. Consider $\mathbb{Z}$, the ring of integers, and let $M = \mathbb{Z} - \langle 47 \rangle$. Define $\mathbb{Z}_{(47)}$ to be $\mathbb{Z}[M^{-1}]$, the localization of $\mathbb{Z}$ at M (see Definition 17.9).

(a) Convince yourself that this definition of $\mathbb{Z}_{(47)}$ is consistent with the definition in Example 18.79.

(b) What are the units of $\mathbb{Z}_{(47)}$? What is $\text{char}(\mathbb{Z}_{(47)})$? Is $47 \in \mathbb{Z}_{(47)}$? What is the field of fractions of $\mathbb{Z}_{(47)}$?

(c) Let I be the ideal generated by 47 in $\mathbb{Z}$, and let J be the ideal generated by 47 in $\mathbb{Z}_{(47)}$. Is $I = J$? Describe the elements of I and J.

(d) Let $x \in \mathbb{Z}_{(47)} - J$. Is x invertible in $\mathbb{Z}_{(47)}$? Use your answer to prove that J is a maximal ideal in $\mathbb{Z}_{(47)}$. Does $\mathbb{Z}_{(47)}$ have any other maximal ideals?

(e) Is 47 a prime in $\mathbb{Z}_{(47)}$? Apart from 47 and its associates, are there any other primes in $\mathbb{Z}_{(47)}$?

(f) Is $\mathbb{Z}_{(47)}$ a DVR?

18.6.5. Let R be an integral domain, and let P be a prime ideal of R. Let $M = R - P$, and let $S = R[M^{-1}]$. (See Definition 17.9.) Let J be the ideal of S generated by P.

(a) Show that $J = \{ab^{-1} \mid a \in P, b \in M\}$.

(b) Show that J is the unique maximal ideal of S and hence S is a local ring.

18.6.6. **Proof of Proposition 18.76.** Prove the equivalence of the various conditions on a PID given in Proposition 18.76.

18.6.7. **Proof of Proposition 18.77.** Prove that, for an integral domain, being a DVR is equivalent to satisfying UF1 and having a unique irreducible element (up to associates).

18.6.8. As in Example 18.79 and Problem 18.6.4, let p be a prime, and define

$$\mathbb{Z}_{(p)} = \{r \in \mathbb{Q} \mid r = \frac{a}{b} \text{ with } a, b \in \mathbb{Z}, \ \gcd(a, b) = 1, \text{ and } p \nmid b\}.$$

Prove that $\mathbb{Z}_{(p)}$ is a discrete valuation ring.

18.6.9. The ring $\mathbb{Z}_{(p)}$ is defined as in Example 18.79. Describe *all* the ideals of $\mathbb{Z}_{(p)}$. Is $\mathbb{Z}_{(p)}$ an artinian ring? (See Definition 18.31.) What is the field of fractions of $\mathbb{Z}_{(p)}$?

18.6.10. The ring $\mathbb{Z}_{(p)}$ is defined as in Example 18.79. What is the residue field of $\mathbb{Z}_{(p)}$ (see Definition 18.74)? Prove your assertion.

18.6.11. Let D be a DVR, and let x be an irreducible element of D. Show that every non-zero element of D can be written as ux^i where u is a unit and i a non-negative integer.

18.6.12. **Proof of Proposition 18.78.** Prove that a discrete valuation ring is a Euclidean domain.

18.6.13. Let $M = \{1, 70, 70^2, \ldots\}$. Let $R = \mathbb{Z}[M^{-1}]$, the localization of $\mathbb{Z}$ at M, and $S = \mathbb{Z}/\langle 70 \rangle$. Find all the prime ideals of R and all the prime ideals of S. Make a comment on the relation between the prime ideals of $\mathbb{Z}$ and the prime ideals of R and S. Is either R or S a local ring?

18.6.14. Let $\mathcal{F}$ denote the set of continuous real valued functions defined on some open interval around 0.[5] In other words, $f \in \mathcal{F}$ if $f \colon X \to \mathbb{R}$ is continuous, where $X \subseteq \mathbb{R}$ is an open interval containing 0. (Note that different functions in $\mathcal{F}$ can have different domains.) If $f \colon X \to \mathbb{R}$ and $g \colon Y \to \mathbb{R}$ are elements of $\mathcal{F}$, then we define addition and multiplication pointwise (on a possibly smaller domain):

$$f + g \colon X \cap Y \to \mathbb{R} \quad \text{by} \quad (f + g)(\alpha) = f(\alpha) + g(\alpha),$$
$$fg \colon X \cap Y \to \mathbb{R} \quad \text{by} \quad (fg)(\alpha) = f(\alpha)\, g(\alpha).$$

(a) Convince yourself that $\mathcal{F}$ with pointwise addition and multiplication is a commutative ring with identity.

We are only interested in the "local" behavior of the functions in $\mathcal{F}$ near zero, and so we define a relation $\sim$ among the elements of $\mathcal{F}$: For $f, g \in \mathcal{F}$, say $f \sim g$ if there exists a possibly very small interval $I \subseteq \mathbb{R}$ such that $0 \in I$ and $f(x) = g(x)$ for all $x \in I$. In other words, two functions in $\mathcal{F}$ are related if they agree on an open interval near zero.

(b) Show that $\sim$ is an equivalence relation on the set $\mathcal{F}$.
(c) Are the functions $y = 0$ and $y = x^2$ (defined on the real line) equivalent? Give three functions that are equivalent to $y = x^2$.

The equivalence class of $f \in \mathcal{F}$ is denoted by $\widetilde{f}$ and is called the *germ of f at 0*. The set of equivalence classes (i.e., the germs of real valued continuous functions near 0) will be denoted by $\widetilde{\mathcal{F}}$. For germs $\widetilde{f}, \widetilde{g} \in \widetilde{\mathcal{F}}$, we define

$$\widetilde{f} + \widetilde{g} = \widetilde{f + g},$$
$$\widetilde{f}\widetilde{g} = \widetilde{fg}.$$

(d) Show that the addition and multiplication on $\widetilde{\mathcal{F}}$ is well-defined. In other words, show that the result of the addition and multiplication does not depend on the choice of representatives for the germs.
(e) Show that $\widetilde{\mathcal{F}}$ with this addition and multiplication is a commutative ring with identity.
(f) Show that $\widetilde{f} \in \widetilde{\mathcal{F}}$ is a unit if and only if $f(0) \neq 0$.
(g) Show that $\widetilde{\mathcal{F}}$ is a local ring. What is the unique maximal ideal?
(h) Can differentiability at zero and Taylor polynomials and series at zero be defined for germs? In other words, will two functions with the same germ have the same derivative at zero (if differentiable) and have the same Taylor series (if analytic)?

Radical, Prime, and Semiprime Ideals. We have already defined prime ideals in commutative rings with identity (Definition 18.11). That definition, while straightforward, does not work well for non-commutative rings, mainly because it is a statement about the multiplication of *elements*. Rewritten in terms of

---

[5]A familiarity with continuous functions (and introductory real analysis) is assumed for this problem.

ideals, it can serve as a definition for prime ideals for both commutative and non-commutative rings.

Definition 18.80. Let R be a (not necessarily commutative) ring with identity, and let P be proper ideal of R. Then P is called a *prime ideal* of R if whenever $AB \subseteq P$ for ideals A and B, then at least one of A or B is a subset of P.

Remark 18.81. Recall (Remark 18.9) that we suggested the mantra "to contain is to divide". The above (new) definition of a prime ideal says that a proper ideal P of R is prime if, for ideals A and B, $P \supseteq AB$ implies $P \supseteq A$ of $P \supseteq B$. Do you see the connection with the definition of a prime element (Definition 18.1)?

In Problem 18.6.18, you are asked to show that, in the case of commutative rings with identity, this new definition is equivalent to the older definition (Definition 18.11). These two definitions are *not* equivalent in general.

Recall that an ideal I of R, a commutative ring with identity, is *radical* if $I = \sqrt{I}$ where $\sqrt{I} = \{x \in R \mid x^n \in I, \text{ for some } n \in \mathbb{Z}^{>0}\}$ (Definition 16.57). In Problem 18.6.16, you are asked to show that all prime ideals are radical.

We will also define for ideals a condition—that of being *semiprime*—weaker than being prime. Problem 18.6.15 asks you to translate this condition (just as we had done for maximal and prime ideals) to one about factor rings.

Definition 18.82 (Semiprime ideal)**.** Let R be a commutative ring with identity. A proper ideal I of R is called *semiprime* if for all $x \in R$ such that $x^2 \in I$, we have $x \in I$.

18.6.15. Let R be a commutative ring with identity. Recall that $x \in R$ is called nilpotent if $x^n = 0$ for some $n \in \mathbb{Z}^{>0}$.
   (a) Are all prime ideals semiprime? What about the converse?
   (b) Show that I is a semiprime ideal of R if and only if R/I has no non-zero nilpotent elements.

18.6.16. Let R be a commutative ring with identity. If I is a prime ideal, then show that I is a radical ideal. (See Definition 16.57.)

18.6.17. Let R be a commutative ring with identity. Assume I is the intersection of a finite number of prime ideals. Show that I is a radical ideal.

18.6.18. Let R be a commutative ring with identity. Let P be a proper ideal of R. Show that the following are equivalent:
   (a) $R - P$ is a multiplicative system;
   (b) P is a prime ideal of R (using Definition 18.11);
   (c) if $AB \subseteq P$ for ideals A and B, then $A \subseteq P$ or $B \subseteq P$;
   (d) there do not exist ideals $A \supsetneq P$ and $B \supsetneq P$ such that $AB \subseteq P$.

18.6.19. Let R be a commutative ring with identity, and let P be a proper ideal of R. In Problem 18.6.18, you showed that P is a prime ideal if and only if $R - P$ is multiplicative system. Let $R = \mathbb{Z}$ and $M = \{1, 7, 7^2, \ldots\}$. The set M is evidently a multiplicative system in $\mathbb{Z}$. Is the set $\mathbb{Z} - M$ a prime ideal of $\mathbb{Z}$? Comment.

18.6.20. Let R be a commutative right with identity. Let I_1, I_2, ..., I_k be ideals of R, and let P be a prime ideal of R. Assume $\bigcap_{i=1}^{k} I_i \subseteq P$. Prove that, at least for one $1 \leq i \leq k$, P contains the ideal I_i.

18.6.21. Let R be a PID, and let P, $Q_1, \ldots, Q_k$ be non-zero prime ideals of R. Assume $P \supseteq Q_1 Q_2 \cdots Q_k$. Show that $P = Q_i$ for some $1 \leq i \leq k$. Comment on the mantra "to contain is to divide" in this context.

18.6.22. Let $R = \mathbb{Z}[\sqrt{-5}]$. Let $P = \langle 2 \rangle$, $Q_1 = \langle 2, 1 + \sqrt{-5} \rangle$, and $Q_2 = \langle 2, 1 - \sqrt{-5} \rangle$ be ideals of R.
  (a) Are Q_1 and Q_2 *proper* ideals of R?
  (b) Is $P \subseteq Q_1 Q_2$ and/or $Q_1 Q_2 \subseteq P$?
  (c) Is P a prime ideal in R?
  (d) Is R a PID?

Dedekind–Hasse Norms. In Remark 18.54, we stated that the full force of the definition of the degree function was not used to prove that Euclidean domains are principal ideal domains.(Also see Problem 18.4.5.) In fact, there are principal ideal domains that are not Euclidean domains. It would be helpful to have a weaker condition on integral domains that would result in a PID.

Definition 18.83. Let R be a commutative ring with identity, and let $N \colon R - \{0\} \to \mathbb{Z}^{\geq 0}$. Assume that, for all $a, b \in R$ with $b \neq 0$,

$$a \in \langle b \rangle \quad \text{or} \quad \exists \, 0 \neq r \in \langle a \rangle + \langle b \rangle \text{ with } N(r) < N(b).$$

Then N is called a *Dedekind–Hasse norm* on R.

In the Problems, we will also refer to Bézout domains.

Definition 18.84. An integral domain in which the greatest common divisor of every pair of non-zero elements is a linear combination of the two elements is called a *Bézout domain*. (See Remark 18.66.)

Theorem 18.63 proved that every PID is a Bézout domain.

18.6.23. Let R be a commutative ring with identity, and let $N \colon R - \{0\} \to \mathbb{Z}^{\geq 0}$. Let $a, b \in R$ with $b \neq 0$. Show that the following two conditions are equivalent:
  (a) $a \in \langle b \rangle$ or there exists $0 \neq r \in \langle a \rangle + \langle b \rangle$ with $N(r) < N(b)$.
  (b) $a = tb$ for $t \in R$ or there exists $r, s, t \in R$ with $sa = tb + r$, $r \neq 0$, and $N(r) < N(b)$.
  In particular, conclude that the degree function in a Euclidean domain is a Dedekind–Hasse norm.

18.6.24. Consider $\mathbb{Z}$, the ring of integers. The absolute value function is a degree function (of a Euclidean domain) for $\mathbb{Z}$. (See Example 18.46.) Now, define $N \colon \mathbb{Z} - \{0\} \to \mathbb{Z}^{\geq 0}$ by $N(n)$ equal to the number of prime factors (including multiplicity) in a prime factorization of n. In other words, if

$n = p_1^{\ell_1} p_2^{\ell_2} \dots p_k^{\ell_k}$ where $p_1, \dots, p_k$ are primes, then $N(n) = \ell_1 + \ell_2 + \dots + \ell_k$.

    (a) Is N a degree function of a Euclidean domain?

    (b) Is N a Dedekind–Hasse norm?

18.6.25. Let R be a commutative ring with identity. Assume that R has a Dedekind–Hasse norm. Show that R is a principal ideal ring.

18.6.26. Let R be an integral domain. Show that the following are equivalent:

    (a) R is a PID.

    (b) R is a UFD and a Bézout domain.

    (c) R is noetherian and a Bézout domain.

    (d) R has a Dedekind–Hasse norm.

18.6.27. Let R be an integral domain. Prove that R is a PID if and only if every prime ideal of R is principal.

18.6.28. Let R be a UFD. Prove that R is a PID if and only if every non-zero prime ideal of R is maximal.

Chapter 19

Polynomial Rings

... where rings of polynomials are studied in some detail, Gauss's lemma and the Schönemann-Eisenstein irreducibility criterion are proved, and where it is proved that K a field $\Rightarrow K[x]$ an ED, R UFD $\Rightarrow R[x]$ UFD, and R noetherian $\Rightarrow R[x]$ noetherian.

In this chapter we explore polynomial rings. These rings provide a rich set of examples, play a fundamental role in algebraic geometry and Galois theory, and often appear in applications of the subject. To study polynomial rings, we will use extensively the general theory of commutative rings that we developed in Chapter 18. However, a few additional tools—specific to polynomial rings—will be introduced as well.

19.1. Polynomials

Definition 19.1 (Polynomial). Let R be a ring, and let x be some symbol. Then a *polynomial over R in the indeterminate x* is an expression of the form

$$a_n x^n + a_{n-1} x^{n-1} + \cdots + a_1 x + a_0,$$

where $n \in \mathbb{Z}^{\geq 0}$, and, for $0 \leq i \leq n$, $a_i \in R$.

Remark 19.2. A polynomial is not the same as a function. Formally, a polynomial is an ordered sequence of elements of R such that only finitely many are non-zero together with a symbol x.

Definition 19.3 (Ring of polynomials). Let R be a ring. The set of all polynomials over R in x is denoted by $R[x]$. With the usual polynomial addition and multiplication, $R[x]$ becomes a ring and is then called the *ring of polynomials in the indeterminate x with coefficients in R*.

Example 19.4. $(\mathbb{Z}/4\mathbb{Z})[x]$ is the collection of polynomials with coefficients in $\mathbb{Z}/4\mathbb{Z}$. We add and multiply these elements using polynomial addition and multiplication except that all the operations on the coefficients are performed in the ring $\mathbb{Z}/4\mathbb{Z}$. So, for example, $(2x + 1)(2x) = 2x$.

Definition 19.5 (Degree of a polynomial). Let $f \in R[x]$, then $f = a_0 + a_1 x + \cdots + a_n x^n$ for some $a_0, a_1, \ldots, a_n \in R$ and some non-negative integer n.

The *degree* of f, denoted by $\deg(f)$ is the largest exponent of x with a non-zero coefficient. If $\deg(f) = n$, then the (non-zero) coefficient a_n is called the *leading coefficient* of f and is denoted by $\mathrm{lc}(f)$. A polynomial is called *monic* if its leading coefficient is 1. A polynomial of degree 0 is called a *constant polynomial*. The polynomial with all of its coefficients equal to zero is called the *zero polynomial* and has no degree.

The following is straightforward.

Lemma 19.6. *Let R be a ring. Then*

(a) $R[x]$ *is a ring,*

(b) *the constant polynomials form a subring of $R[x]$ isomorphic to R,*

(c) $\deg(f + g) \leq \max\{\deg(f), \deg(g)\}$, *and*

(d) $\deg(fg) \leq \deg f + \deg g$.

Recall from Definition 17.14 that S is called a *unitary overring* of R if R and S are rings with identity, S has a subring R' isomorphic to R, and that $1_{R'} = 1_S$. To avoid clutter, we identify R and R', and hence we speak of R as a subset of S. So, for example, if $r \in R$, $s \in S$, and S is a unitary overring of R, then we can find the product rs. By this we really mean the product $\phi(r)s$, where $\phi \colon R \to R'$ is the ring isomorphism (that we used to identify R and R') and R' is an actual subset of S.

Definition 19.7 ($f(\alpha)$). Let S be a unitary overring of R, and let $f = a_0 + a_1 x + \cdots + a_n x^n \in R[x]$. Assume $\alpha \in S$. We define $f(\alpha) = a_0 + a_1 \alpha + \cdots + a_n \alpha^n$.

The map

$$\epsilon_\alpha : R[x] \to S$$

defined by

$$\epsilon_\alpha(f(x)) = f(\alpha)$$

is called the *evaluation map* at α.

Remark 19.8. Note that $f(\alpha)$ is not a polynomial. It is an element of S. Also, for this definition, there was no real need to assume that $1_R = 1_S$. In other words, we could have assumed that S is just an overring—as opposed to a *unitary* overring—of R. However, in such a case, we have to be careful. For example, if $f(x) = x$, then $f(\alpha) = 1_R \cdot \alpha$. If we want this to be equal to α, then we have to have $1_S = 1_R$.

The reader can provide the proof of the following.

Lemma 19.9. *Let S be a unitary overring of R, and fix $\alpha \in S$. Then the evaluation map $\epsilon_\alpha : R[x] \to S$ is a ring homomorphism.*

Since ϵ_α is a ring homomorphism, its image is a subring of S. The image consists of all the elements of S that can be written as a polynomial—with coefficients in R—in α. We formalize the definition of this important ring:

Definition 19.10. Let S be a unitary overring of R, and let $\alpha \in S$. Then $R[\alpha]$ is the image of the evaluation map ϵ_α. In other words,

$$R[\alpha] = \{g(\alpha) \mid g \in R[x]\}.$$

$R[\alpha]$ is the smallest subring of S that contains both R and α. It is called the subring of S generated by R and α. We read $R[\alpha]$ as "R joined α".

If $R \subseteq S$ are both rings with $1_R = 1_S$ and $\alpha \in S$, then $R[\alpha]$ is the collection of elements of S that can be written using elements of R, α, and the three operations of addition, subtraction, and multiplication.

Example 19.11. Consider $\mathbb{Q} \subseteq \mathbb{C}$. Now both i and $\sqrt[3]{2}$ are elements of $\mathbb{C}$, and

$$\mathbb{Q}[i] = \{a + bi \mid a, b \in \mathbb{Q}\},$$
$$\mathbb{Q}[\sqrt[3]{2}] = \{a + b\sqrt[3]{2} + c\sqrt[3]{4} \mid a, b, c \in \mathbb{Q}\}.$$

Example 19.12. Let d be an integer (positive or negative) that is not divisible by the square of a prime. We had already defined (see page 320) the quadratic integer rings:

$$\mathbb{Z}[\sqrt{d}] = \{a + b\sqrt{d} \mid a, b \in \mathbb{Z}\}.$$

The definition of $\mathbb{Z}[\sqrt{d}]$ is a special case of the general construction of rings $R[\alpha]$.

Remark 19.13. If $F \subseteq E$, both E and F are fields, and $\alpha \in E$, then we can certainly construct $F[\alpha]$. The ring $F[\alpha]$ will be an integral domain, but it is possible that $F[\alpha]$ is not a field. For example, if the multiplicative inverse of α cannot be written as a polynomial in α, then α will not have an inverse in $F[\alpha]$, and $F[\alpha]$ will not be a field. However, since $F[\alpha]$ is an integral domain, it will have a field of fractions, and since E is already a field containing $F[\alpha]$, the field E will have a subfield containing $F[\alpha]$ that is isomorphic to this field of fractions. This field will be denoted by $F(\alpha)$, is the smallest subfield of E that contains both F and α, and its elements are ratios of elements of $F[\alpha]$ (with the denominator not allowed to be zero). The field $F(\alpha)$ will play an important role in field theory and will be (re)introduced and studied in Chapter 22.

Our construction of polynomials in one variable with arbitrary coefficient rings allows us to construct—without much extra effort—rings of polynomials in several variables. Here, we confine ourselves to commutative rings with identity.

Definition 19.14. Let R be a commutative ring with identity. As usual $R[x]$ is the ring of polynomials with coefficients from R. Now use this new ring as the ring of coefficients and form $(R[x])[y]$. The ring $(R[x])[y]$ is denoted by $R[x, y]$, and its elements are *polynomials in two indeterminates with coefficients in R*. The ring $R[x_1, x_2, \ldots, x_n]$ is defined similarly.

Example 19.15. The polynomial $3xy^2 + 5x^2y^3$ is a member of $\mathbb{Z}[x, y]$, while the polynomial $\frac{1}{\pi}w - \sqrt{2}u + z^2$ is a member of $\mathbb{R}[w, u, z]$. Of course, $3xy^2 + 5x^2y^3$ is also a member of $(\mathbb{Z}/7\mathbb{Z})[x, y]$, as well as $\mathbb{R}[x, y]$ and $\mathbb{Q}[x, y, z, w]$.

The following lemma and theorem are straightforward.

Lemma 19.16. *Let R be a commutative ring with identity, and let $f, g \in R[x] - \{0\}$. Assume $\deg(f) = m$, $\deg(g) = n$, the leading coefficient of f is a, and the leading coefficient of g is b.*

Then if $ab \neq 0$, then $\deg(fg) = m + n$, and the leading coefficient of fg is ab. In particular,

(a) *if R is an integral domain, then so is $R[x]$, and*

(b) *if f and g are monic, then so is fg.*

Theorem 19.17. *Let R be a ring. Then*

(a) *R commutative $\Rightarrow R[x_1, \ldots, x_n]$ commutative.*

(b) *R has an identity $\Rightarrow R[x_1, \ldots, x_n]$ has an identity.*

(c) *R an integral domain $\Rightarrow R[x_1, \ldots, x_n]$ is an integral domain.*

Theorem 19.17, while not particularly profound, is typical of many theorems about polynomial rings. These theorems show how properties of the coefficient ring R are reflected onto the ring of polynomials $R[x]$ or $R[x_1, \ldots, x_n]$. Along the same lines, the following basic properties of polynomial rings will be proved in the course of this chapter:

Theorem 19.18. (a) *If F is a field, then $F[x]$ is a Euclidean domain.*

(b) *If R is a unique factorization domain, then $R[x]$ is also a unique factorization domain.*

(c) *(The Hilbert basis theorem) If R is noetherian, then $R[x]$ is also noetherian.*

Remark 19.19. One can repeatedly apply the last two parts of Theorem 19.18 to get that if R is a UFD, then so is $R[x_1, \ldots, x_n]$, and if R is noetherian, then so is $R[x_1, \ldots, x_n]$.

Problems

19.1.1. Let $R = \mathbf{M}_{2 \times 2}(\mathbb{R})$ be the ring of 2×2 matrices with real entries. Let $A = \begin{bmatrix} 2 & 1 \\ 3 & 0 \end{bmatrix}$ and $B = \begin{bmatrix} 0 & 1 \\ 1 & 0 \end{bmatrix}$, and let I_2 denote the 2×2 identity matrix. Let $p(x) = Ax^2 + Bx + I_2$ and $q(x) = I_2 x + B$ be two elements of $R[x]$. Find $p(x) + q(x)$, $p(x)q(x)$, and $q(x)p(x)$. Is $R[x]$ commutative? Does it have an identity? Does it have any zero divisors?

19.1.2. Let S be a unitary overring of R, and let $\alpha \in S$. In Definition 19.10, it is asserted that $R[\alpha]$ is "the smallest subring of S that contains both R and α". Give an argument for this assertion.

19.1.3. Recall that if α is any complex number, then $\mathbb{Z}[\alpha] = \{p(\alpha) \mid p(x) \in \mathbb{Z}[x]\}$. Let $R = \mathbb{Z}[\frac{1}{5}]$.
   (a) Show that $12 + \frac{3}{5} + \frac{4}{25} + \frac{1}{125} \in R$.
   (b) Is $\langle 2, \frac{1}{5} \rangle$, the ideal generated by 2 and $1/5$, a principal ideal? What about $\langle 2, \frac{3}{5} \rangle$?

19.1.4. Let $\omega = e^{\frac{2\pi i}{3}} = \frac{-1+i\sqrt{3}}{2} \in \mathbb{C}$. Can every element of $\mathbb{Z}[\omega]$ be written as an integer linear combination of 1 and ω?

19.1.5. Let $R = (\mathbb{Z}[\sqrt{3}])[x]$, the ring of polynomials with coefficients in $\mathbb{Z}[\sqrt{3}]$. Let K be the field of fractions of R. What are the elements of K?

19.1.6. Let $M = \{1, 5, 5^2, \ldots\}$, and let $R = \mathbb{Z}[M^{-1}]$ be the localization of $\mathbb{Z}$ at M (see Definition 17.9). Is R the same as $\mathbb{Z}[\frac{1}{5}]$? Is one contained in the other? Can you generalize?

19.1.7. Let $M = \{1, x, x^2, \ldots\}$, and let $R = \mathbb{Z}[x][M^{-1}]$ be the localization of $\mathbb{Z}[x]$ at M (see Example 17.13). Define $S = \mathbb{Z}[x, \frac{1}{x}] = \{q(x, \frac{1}{x}) \mid q(x, y) \in \mathbb{Z}[x, y]\}$ and $T = \mathbb{Z}[\frac{1}{x}] = \{p(\frac{1}{x}) \mid p(x) \in \mathbb{Z}[x]\}$.
   (a) Are S and/or T commutative rings with identity? Are they integral domains? Are either a field?
   (b) From among R, S, and T, are any two the same? Are any of them contained in another?
   (c) If possible, find the field of fractions for each of R, S, and T.

19.1.8. Let $M = \mathbb{Z} - \langle 5 \rangle$, and let $\mathbb{Z}_{(5)} = \mathbb{Z}[M^{-1}]$ be the localization of $\mathbb{Z}$ at M. Describe the elements of $\mathbb{Z}_{(5)}[\sqrt{2}]$ and $\mathbb{Z}_{(5)}[\frac{1}{2}]$.

19.1.9. Describe the elements of $(\mathbb{Z}[\frac{1}{2}])[\sqrt{2}]$ and of $(\mathbb{Z}[\frac{1}{2}])[\frac{1}{5}]$. Is the latter any different from $\mathbb{Z}[\frac{1}{2}, \frac{1}{5}] = \{p(\frac{1}{2}, \frac{1}{5}) \mid p \in \mathbb{Z}[x, y]\}$?

19.1.10. Let $R = \mathbb{Z} \times \mathbb{Z}$, and let $M = \{(2^n, 3^m) \mid n, m \in \mathbb{Z}^{\geq 0}\}$. Is $R[M^{-1}]$, the localization of R at M (see Problem 17.2.3), isomorphic to $\mathbb{Z}[\frac{1}{2}] \times \mathbb{Z}[\frac{1}{3}]$?

19.2. K a field $\Rightarrow K[x]$ an ED

The purpose of this section is to prove that if K is a field, then, $K[x]$, the ring of polynomials with coefficients in K, is a Euclidean domain. This will mean also that $K[x]$ is a principal ideal domain and a unique factorization domain. All of these facts will be extremely useful. To show that $K[x]$ is a Euclidean domain, we need a degree function, and the obvious candidate is the function that gives the degrees of the polynomials. We begin by showing that this does *not* work in the case of $\mathbb{Z}[x]$. This can be seen as a reason why we need a stronger condition on the coefficient ring. In fact, we will also show that for $R[x]$ to even be a principal ideal domain, then R has to be a field.

Example 19.20. Let $R = \mathbb{Z}[x]$, $f(x) = x^2 + 5$, and $g(x) = 3x + 2$. Can we find q and r in $\mathbb{Z}[x]$ with

$$f = qg + r \text{ such that } r = 0 \text{ or } \deg r < \deg g?$$

The answer is *no* since when dividing f by g, we cannot find an element of $\mathbb{Z}[x]$ that when multiplied by $3x$ gives x^2. Hence, the polynomial r will by necessity have degree 2 or larger.

However, all is not lost. Even if R is not a field, a simple condition can guarantee the division algorithm.

Theorem 19.21 (Division algorithm for polynomials). *Let R be an integral domain, and let $f, g \in R[x]$. Assume that the leading coefficient of g is a unit in R. Then there exists unique polynomials $q, r \in R[x]$ such that*

$$f = qg + r \text{ such that } r = 0 \text{ or } \deg r < \deg g.$$

Proof. Induct on $n = \deg f$.

If $\deg f = \deg g = 0$, then g is equal to its leading coefficient b_0 and is invertible. Hence $f = g(b_0^{-1} f) + 0$, and the theorem holds. On the other hand, if $\deg g > \deg f$—regardless of whether $\deg f = 0$—then we can write $f = 0g + f$ and the theorem holds.

Now assume that the theorem has been proved for all polynomials f of degree less than or equal to $n - 1$. We want to prove the statement for $\deg f = n$ and $\deg g \leq n$.

So let $f(x) = a_0 + a_1 x + \cdots + a_n x^n$ and $g(x) = b_0 + b_1 x + \cdots + b_m x^m$ with $0 < m \leq n$. Let $h = a_n b_m^{-1} x^{n-m} g$, and let $f_1 = f - h$. Then $\deg f_1 \leq n - 1$.

By the inductive hypothesis applied to f_1 and g, we get $f_1 = gq_1 + r_1$ with $r_1 = 0$ or $\deg r_1 < \deg g$. Now going back to the original f, we have $f - h = gq_1 + r_1$. Hence,

$$f = h + gq_1 + r_1 = (a_n b_m^{-1} x^{n-m} + q_1)g + r_1,$$

and the proof of the existence of the appropriate q and r is complete. We now turn to uniqueness.

Suppose $f = qg + r = q_0 g + r_0$, where $r = 0$ or $\deg r < \deg g$ and $r_0 = 0$ or $\deg r_0 < \deg g$. In particular, $\deg(r_0 - r) < \deg g$.

Rewriting $qg + r = q_0 g + r_0$, we get $(q - q_0)g = r_0 - r$. If $q - q_0$ is not the zero polynomial, then $\deg(r_0 - r) = \deg(q - q_0)g \geq \deg g$. The contradiction shows that

$$q - q_0 = 0, \text{ and therefore } r_0 - r = 0. \qquad \square$$

Corollary 19.22. *Let F be a field. Then $F[x]$ is a Euclidean domain. As a consequence, $F[x]$ is a principal ideal domain and a unique factorization domain.*

Proof. This follows directly from Theorem 19.21. You are asked to write up the proof in Problem 19.2.5. $\qquad \square$

In Example 19.20, we saw that the usual degree function does not turn $\mathbb{Z}[x]$ into a Euclidean domain. But maybe a different degree function would. We now prove a converse to Corollary 19.22 which shows that $\mathbb{Z}[x]$ cannot be a Euclidean domain no matter what degree function we use. In fact, $\mathbb{Z}[x]$ cannot even be a principal ideal domain.

Theorem 19.23. *If the polynomial ring $R[x]$ is a principal ideal domain, then R is a field.*

Proof. This was Problem 18.1.13 but here is the argument. Assume $R[x]$ is a PID. Then $R[x]$ is an integral domain which implies that R is an integral domain. Now let $\phi \colon R[x] \to R$ be defined by $\phi(p(x)) = p(0)$. The map ϕ is easily seen to be an onto ring homomorphism with kernel $\langle x \rangle$, the ideal generated by x. By the homomorphism theorems, $R[x]/\langle x \rangle$ is isomorphic to R which is an integral domain.

Thus by Theorem 18.16, $\langle x \rangle$ is a prime ideal. Since $R[x]$ is a PID, by Theorem 18.12 $\langle x \rangle$ is a maximal ideal. But this means, by Theorem 18.14, that $R[x]/\langle x \rangle$ is a field. But then $R \cong R[x]/\langle x \rangle$ is a field, and the proof is complete. $\qquad \square$

Problems

19.2.1. In each of the following a field F and two polynomials f and g in $F[x]$ are given. Divide f by g and find the remainder and quotient. In other words, find polynomials q and r in $F[x]$ such that

$$f = qg + r, \text{ and } r = 0 \text{ or } \deg(r) < \deg(g).$$

(a) The field F is $\mathbb{Q}$ the field of rational numbers, $f = x^7 - 2x^3 + 6$, and $g = x^3 + 13$.
(b) The field F is $\mathbb{Q}$, $f = x^2 + 1$, and $g = x^2 - 1$.
(c) The field F is $\mathbb{R}$, the field of real numbers, $f = 5x^3 - 47x^2 + x - 13$, and $g = 3x + 5$.
(d) The field F is $\mathbb{Z}/3\mathbb{Z}$, $f = x^3 + 2x^2 - x + 1$, and $g = x + 2$.
(e) The field F is $\mathbb{Z}/7\mathbb{Z}$, $f = x^7 - 5x^6 + 2x^3 - 3x + 6$, and $g = 3x^3 - 3$.

19.2.2. Use the Euclidean algorithm (Theorem 18.69) to find the greatest common divisor of $x^4 + 2x^3 - x - 2$ and $3x^6 - 6x^5 - 3x - 6$ in $\mathbb{Q}[x]$.

19.2.3. Let $\mathbb{F}_5 = \mathbb{Z}/5\mathbb{Z}$, and find the greatest common divisor of $3x^4 + x^3 + 2x^2 + 1$ and $x^2 + 4x + 2$ in $\mathbb{F}_5[x]$.

19.2.4. Let K be a field, and let $f, g, h \in K[x]$. Assume $\gcd(f, g) = 1$. If both f and g divide h, show that fg divides h as well.

19.2.5. **Proof of Corollary 19.22.** Using Theorem 19.21, write a complete proof of Corollary 19.22. In other words, let K be a field, and prove that $K[x]$ is a Euclidean domain.

19.2.6. Is $\mathbb{Q}[x]$ a PID? Is $\mathbb{Z}[x]$ an ED? Prove your assertions.

19.2.7. Let R be a commutative ring with identity. Is it possible for $R[x]$ to be a PID without being a Euclidean domain?

19.2.8. Let $\mathbb{F}_7 = \mathbb{Z}/7\mathbb{Z}$ be the field with seven elements, and let $f, g \in \mathbb{F}_7[x]$ be defined by $f(x) = x^5 + x^3 + x + 1$ and $g(x) = x^2 + x$. Is f in the ideal generated by g? If the answer is no, then find $h \in \mathbb{F}_7[x]$ with degree less than two such that $f - h$ is in the ideal generated by g.

19.2.9. Let $\mathbb{F}_5$ denote the field $\mathbb{Z}/5\mathbb{Z}$, and let $p(x) = 3x^2 + 4x + 3 \in \mathbb{F}_5[x]$. Does $3x + 2$ divide $p(x)$? What about $4x + 1$? In two distinct ways, write $p(x)$ as a product of two linear factors. Does this mean that $\mathbb{F}_5[x]$ is not a unique factorization domain? Explain.

19.2.10. Let R and S be integral domains with $R \subseteq S$, and let $f, g \in R[x]$. Assume that the leading coefficient of f is a unit of R and that there exists $h \in S[x]$ with $g = fh$. Does h have to be in $R[x]$?

19.2.11. Let $F \subseteq E$ be fields, and let $f, g \in F[x]$. Prove that the greatest common divisor of f and g in $F[x]$ is the same as the greatest common divisor of f and g in $E[x]$.

19.2.12. Let $f, g \in (\mathbb{Z}[\frac{1}{2}])[x]$ be defined by $f = 3x^4 - 2x^3 + x^2 + 7$, and let $g = 4x^2 - x + 1$. Can you find $q, r \in (\mathbb{Z}[\frac{1}{2}])[x]$ with $f = qg + r$ and with $r = 0$ or $\deg r < \deg g$? If so, find them. What if $g = 3x^2 - x + 1$?

19.2.13. Let $M = \mathbb{Z} - \langle 5 \rangle$, and let $\mathbb{Z}_{(5)} = \mathbb{Z}[M^{-1}]$ be the localization of $\mathbb{Z}$ at M. Let $f, g \in \mathbb{Z}_{(5)}[x]$ be defined by $f = 3x^4 - 2x^3 + x^2 + 7$, and let $g = 4x^2 - x + 1$. Can you find $q, r \in \mathbb{Z}_{(5)}[x]$ with $f = qg + r$ and with $r = 0$ or $\deg r < \deg g$? If so, find them.

19.2.14. Let $p \in \mathbb{R}[x]$ be a fixed polynomial in x with real coefficients. Let $I = \langle y - p \rangle$ be the ideal generated by $y - p$ in $A = \mathbb{R}[x, y]$.
  (a) Is $A/I = \{q + I \mid q \in \mathbb{R}[x]\}$? Is $A/I = \{q + I \mid q \in \mathbb{R}[y]\}$? In both cases, either give a reason or a counterexample.
  (b) Using your answer to the previous part, can you decide if A/I is an integral domain? Is I a prime ideal? Is I a maximal ideal?
  (c) Define $\phi : \mathbb{R}[x, y] \longrightarrow \mathbb{R}[x]$ by $\phi(f(x, y)) = f(x, p(x))$. Is ϕ a ring homomorphism? What is the image? What is the kernel? (Make sure that you justify your answers.) Do your answers provide an alternative way of answering the questions: Is A/I an integral domain? Is I a prime ideal? Is I a maximal ideal?

19.3. Roots of Polynomials and Construction of Finite Fields

Historically, one of the original threads of algebra has been the attempt to find roots of polynomials. We will treat this topic fully when we develop Galois theory starting in Chapter 21. In this section, using the division algorithm for polynomials, we formalize the relation between roots of polynomials and linear factors—something that should be familiar to you from high school algebra—and, as a result, we identify irreducible polynomials of degrees 1, 2, and 3. To whet your appetite then, we give an example of how to construct a finite field with four elements. We will fully develop the theory of finite fields in Chapter 27, but we can see the basic outline of the story here.

Definition 19.24 (Roots of f). Let $f \in R[x]$, let S be a unitary overring of R, and let $\alpha \in S$. Then the element α is called a *root* of f in S if $f(\alpha) = 0$.

From the division algorithm for polynomials, Theorem 19.21, we can immediately decide whether a polynomial has a linear factor.

Corollary 19.25 (Factor theorem). *Let R be an integral domain. Let $f \in R[x]$ and $a \in R$. Then the following hold.*

(a) *The remainder of f when divided by $x - a$ is $f(a)$. In other words, there exists a polynomial $q(x) \in R[x]$ such that*

$$f(x) = q(x)(x - a) + f(a).$$

(b) *The element a is a root of f if and only if $x - a$ is a factor of $f(x)$. In other words, $x - a \mid f(x)$ if and only if $f(a) = 0$.*

Proof. The second statement is immediate from the first. To prove the first statement, use the division algorithm for polynomials, Theorem 19.21, to write

$$f(x) = q(x)(x - a) + r(x),$$

where $r(x) = 0$ or $\deg(r(x)) < \deg(x - a) = 1$. (Note that we could apply the theorem since the leading coefficient of $x - a$ is 1 which is a unit of R.) We conclude that—whether $r(x)$ is zero or not—$r(x)$ is a constant. Now $r(a) = f(a)$, and so $r(x)$, the remainder of f when divided by $x - a$, is $f(a)$. □

Definition 19.26 (Multiplicity of a root). Let R be an integral domain, and let $f \in R[x]$. The element $\alpha \in R$ is called a root of f of *multiplicity m* if $(x - \alpha)^m \mid f$ and $(x - \alpha)^{m+1} \nmid f$.

Corollary 19.27. *Let R be an integral domain, and let $f \in R[x]$ with $\deg(f) = n$. Then f has at most n roots (counting multiplicities) in R.*

Proof. If the distinct roots of f are α_1, ..., α_k, and the multiplicities of these roots, respectively, are n_1, ..., n_k, then $(x - \alpha_1)^{n_1}(x - \alpha_2)^{n_2} \cdots (x - \alpha_k)^{n_k}$ divides f. This means that the n, the degree of f, is at least $n_1 + n_2 + \cdots + n_k$. □

A Plan for Constructing Fields. We will review and then carry out the plan for constructing fields which was already discussed within Section 18.1 (see page 374). We know that if R is a commutative ring with identity, and M a maximal ideal of R, then R/M is a field (Theorem 18.14). We also know that, in a principal ideal domain, $\langle a \rangle$ is maximal if a is irreducible (Theorem 18.21). Now—by Theorem 19.22—we know that, if F is a field, then $F[x]$ is a Euclidean domain and hence a principal ideal domain. So, if we can find an irreducible polynomial $p(x)$, in $F[x]$, then $F[x]/\langle p(x) \rangle$ will be a field. We record this observation.

Proposition 19.28. *Let F be a field, and let $p(x)$ be an irreducible polynomial in $F[x]$. Then $F[x]/\langle p(x) \rangle$ is a field.*

To carry out this project, we first need to determine the units and irreducibles of $F[x]$.

Lemma 19.29. *Let R be an integral domain. Then the units of $R[x]$ are exactly the units of R. In other words, $R[x]^\times = R^\times$.*

Proof. Clearly, the units of R remain invertible in $R[x]$. If $p(x) \in R[x]$ and $\deg(p(x)) > 0$, then multiplying by other polynomials will not bring down the degree—since R is an integral domain—and hence $p(x)$ cannot be a unit. □

Hence, if F is a field, then the non-zero elements of F are precisely all the units of $F[x]$. Now if $p(x) \in F[x]$ is reducible, then $p(x) = f(x)g(x)$ with neither f nor g a unit. Hence f or g cannot be of degree 0. The following is now clear:

Proposition 19.30. *Let F be a field, and let $p(x) \in F[x]$ have degree greater than zero. Then the polynomial $p(x)$ is irreducible in $F[x]$ if and only if it cannot be factored into two polynomials of lower degree.*

+	$\bar{0}$	$\bar{1}$	$\overline{x}$	$\overline{x+1}$
$\bar{0}$	$\bar{0}$	$\bar{1}$	$\overline{x}$	$\overline{x+1}$
$\bar{1}$	$\bar{1}$	$\bar{0}$	$\overline{x+1}$	$\overline{x}$
$\overline{x}$	$\overline{x}$	$\overline{x+1}$	$\bar{0}$	$\bar{1}$
$\overline{x+1}$	$\overline{x+1}$	$\overline{x}$	$\bar{1}$	$\bar{0}$

$\times$	$\bar{0}$	$\bar{1}$	$\overline{x}$	$\overline{x+1}$
$\bar{0}$	$\bar{0}$	$\bar{0}$	$\bar{0}$	$\bar{0}$
$\bar{1}$	$\bar{0}$	$\bar{1}$	$\overline{x}$	$\overline{x+1}$
$\overline{x}$	$\bar{0}$	$\overline{x}$	$\overline{x+1}$	$\bar{1}$
$\overline{x+1}$	$\bar{0}$	$\overline{x+1}$	$\bar{1}$	$\overline{x}$

Figure 19.1. The addition and multiplication tables for $E = (\mathbb{Z}/2\mathbb{Z}[x])/(x^2 + x + 1)$

Corollary 19.31. *Let F be a field, and let $p(x) \in F[x]$. Then*

(a) *if $\deg(p(x)) = 1$, then $p(x)$ is irreducible in $F[x]$,*

(b) *if $\deg(p(x)) = 2$ or 3, then $p(x)$ is irreducible in $F[x]$ if and only if $p(x)$ has no roots in F.*

Proof. If, in $F[x]$, you can factor a polynomial of degree 2 or 3, then one of the factors must be of degree 1. By the factor theorem, Theorem 19.25, a linear factor of degree 1 corresponds to a root of the polynomial. □

A Field with Four Elements. Let $\mathbb{F}_2 = (\mathbb{Z}/2\mathbb{Z}, +, \cdot)$. Then, in $\mathbb{F}_2[x]$, there are exactly four polynomials of degree 2:

$$x^2, \ x^2 + 1, \ x^2 + x, \ x^2 + x + 1.$$

Since $\mathbb{F}_2$ has two elements, it is trivial to check if these polynomials have roots, and, hence, whether they factor. We have that

$$x^2 = xx, \ x^2 + 1 = (x + 1)^2, \ x^2 + x = x(x + 1), \text{ and } x^2 + x + 1 \text{ is irreducible.}$$

So, the polynomial $x^2 + x + 1$ is irreducible in $\mathbb{F}_2[x]$, and $I = \langle x^2 + x + 1 \rangle$ is a maximal ideal of $\mathbb{F}_2[x]$. Hence, $\mathbb{F}_2[x]/I$ is a field.

The elements of $\mathbb{F}_2[x]/I$ are of the form $p(x) + I$ with $p(x) \in \mathbb{F}_2[x]$. However, $x^2 + x + 1 + I = I$, and—remembering that $-1 = 1$ in $\mathbb{Z}/2\mathbb{Z}$—we have $x^2 + I = -x - 1 + I = x + 1 + I$. So, every occurrence of $x^2 + I$ can be replaced with $x + 1 + I$. This means that $x^n + I$ where $n \geq 2$ can also be replaced with an element of the form $ax + b + I$, where a and b are elements of $\mathbb{F}_2$. For example,

$$x^3 + I = (x + I)(x^2 + I) = (x + I)(x + 1 + I) = x^2 + x + I = 1 + I.$$

We conclude that

$$\mathbb{F}_2[x]/I = \{a + bx + I \mid a, b \in \mathbb{F}_2\}.$$

Denote $\mathbb{F}_2[x]/I$ by E. Then E has four elements:

$$\bar{0} = I, \ \bar{1} = 1 + I, \ \overline{x} = x + I, \ \overline{x+1} = x + 1 + I.$$

We can write down the addition and multiplication tables for E, a field with four elements. (See Figure 19.1.)

We have constructed a field E with four elements. Note that this is *not* $(\mathbb{Z}/4\mathbb{Z}, +, \cdot)$ or $\mathbb{Z}/2\mathbb{Z} \times \mathbb{Z}/2\mathbb{Z}$, the direct product of $(\mathbb{Z}/2\mathbb{Z}, +, \cdot)$ with itself. Neither of these are even an integral domain since $2 \cdot 2 = 0$ in $\mathbb{Z}/4\mathbb{Z}$ and $(1, 0)(0, 1) = (0, 0)$ in $\mathbb{Z}/2\mathbb{Z} \times \mathbb{Z}/2\mathbb{Z}$. In fact, you can see that the addition and multiplication are quite distinct. (See Figures 19.2 and 19.3.)

+	$(0,0)$	$(1,0)$	$(0,1)$	$(1,1)$
$(0,0)$	$(0,0)$	$(1,0)$	$(0,1)$	$(1,1)$
$(1,0)$	$(1,0)$	$(0,0)$	$(1,1)$	$(0,1)$
$(0,1)$	$(0,1)$	$(1,1)$	$(0,0)$	$(1,0)$
$(1,1)$	$(1,1)$	$(0,1)$	$(1,0)$	$(0,0)$

$\times$	$(0,0)$	$(1,0)$	$(0,1)$	$(1,1)$
$(0,0)$	$(0,0)$	$(0,0)$	$(0,0)$	$(0,0)$
$(1,0)$	$(0,0)$	$(1,0)$	$(0,0)$	$(1,0)$
$(0,1)$	$(0,0)$	$(0,0)$	$(0,1)$	$(0,1)$
$(1,1)$	$(0,0)$	$(1,0)$	$(0,1)$	$(1,1)$

Figure 19.2. The addition and multiplication tables for $\mathbb{Z}/2\mathbb{Z} \times \mathbb{Z}/2\mathbb{Z}$

+	0	1	2	3
0	0	1	2	3
1	1	2	3	0
2	2	3	0	1
3	3	0	1	2

$\times$	0	1	2	3
0	0	0	0	0
1	0	1	2	3
2	0	2	0	2
3	0	3	2	1

Figure 19.3. The addition and multiplication tables for $\mathbb{Z}/4\mathbb{Z}$

We note that $(E, +) \cong \mathbb{Z}/2\mathbb{Z} \times \mathbb{Z}/2\mathbb{Z}$ while $(E - \{0\}, \times) \cong \mathbb{Z}/3\mathbb{Z}$. In fact, $\overline{x+1} = \overline{x}^2$, and hence, if we denote $\overline{x}$ with α, then we can write

$$E = \{0, 1, \alpha, \alpha^2\} \quad \text{where } \alpha + 1 = \alpha^2, \text{ and } \alpha^3 = 1.$$

The addition and multiplication tables, with this notation, are given in Figure 19.4. Note that when we think of elements of E as powers of α, then writing the multiplication table becomes easy. However, when writing the addition table, we have to use the relations.

+	0	1	α	α^2
0	0	1	α	α^2
1	1	0	α^2	α
α	α	α^2	0	1
α^2	α^2	α	1	0

$\times$	0	1	α	α^2
0	0	0	0	0
1	0	1	α	α^2
α	0	α	α^2	1
α^2	0	α^2	1	α

Figure 19.4. The addition and multiplication tables for $E = \{0, 1, \alpha, \alpha^2\}$

Note that E has a subfield consisting of $\{0, 1\}$. This is isomorphic to the original field $\mathbb{F}_2$. In other words, E is a bigger field that contains a copy of $\mathbb{F}_2$. We express this by saying that E is a field extension of $\mathbb{F}_2$. The original irreducible polynomial $x^2 + x + 1$ was a polynomial in $\mathbb{F}_2[x]$ but can also be thought of as a polynomial in $E[x]$. However, it is not irreducible any more. In fact, it now has a root! This is because $\alpha^2 + \alpha + 1 = 0$. In other words, we started with a field of two elements and a polynomial of degree 2 that had no roots in this field. We then constructed a field extension in which the polynomial has roots. In fact, in $E[x]$, we have $x^2 + x + 1 = (x - \alpha)(x - \alpha^2)$. Finally, we can prove that, up to isomorphism, E is the *only* field of order 4. As such, it will be denoted by $\mathbb{F}_4$. These ideas will be considered in detail in the chapters on fields and on Galois theory. In particular, in Chapter 27, we will construct and classify all finite fields.

Remark 19.32. Recall that, if p is a prime integer, we use the notation $\mathbb{F}_p$ to denote the field $(\mathbb{Z}/p\mathbb{Z}, +, \cdot)$.

Problems

19.3.1. As usual, let $\mathbb{C}$ denote the field of complex numbers. Let $f(x) \in \mathbb{C}[x]$, and let $a \in \mathbb{C}$. Corollary 19.25(a) stated that there exists a polynomial $q(x) \in \mathbb{C}[x]$ such that

$$f(x) - f(a) = q(x)(x - a).$$

Prove that $q(a)$ is equal to $f'(a)$, the derivative of $f(x)$ at a.

19.3.2. Factor $x^5 + x^3 + 6x^2 + 6$ into irreducible factors in $\mathbb{F}_7[x]$.

19.3.3. Construct a field E with eight elements. Start with $\mathbb{F}_2$, find an irreducible polynomial of degree 3, and mod out by the ideal generated by the polynomial. Give the addition and multiplication table of the field. Find familiar groups that are isomorphic to $(E, +)$ and $(E - \{0\}, \times)$.

19.3.4. Construct a field E with nine elements. Start with $\mathbb{F}_3$, find an irreducible polynomial of degree 2, and mod out by the ideal generated by the polynomial. Give the addition and multiplication tables of the field. Find two familiar groups that are isomorphic to $(E, +)$ and $(E - \{0\}, \times)$.

19.3.5. Let $p(x) = x^3 - x + 1 \in \mathbb{F}_2[x]$. Let $I = \langle p(x) \rangle$, and let $R = \mathbb{F}_2[x]/I$. Let $\alpha = I + x \in R$.
   (a) Is $p(x)$ irreducible in $\mathbb{F}_2[x]$?
   (b) How many elements does R have?
   (c) Is α a unit in R?
   (d) What is the additive order of α? What is the multiplicative order of α?
   (e) What is the characteristic of R?

19.3.6. In $\mathbb{F}_3[x]$ find a reducible polynomial with no roots.

19.3.7. Find all the irreducible polynomials of degrees 3 or 4 in $\mathbb{F}_2[x]$.

19.3.8. Let $f = x^3 + 3x + 2$. Is f irreducible in $\mathbb{F}_7[x]$? What about $\mathbb{F}_{19}[x]$? What about $\mathbb{Z}[x]$?

19.3.9. (a) Factor $x^5 + x + 1$ into irreducible factors in $\mathbb{F}_2[x]$.
   (b) Do the same for $x^5 + x^2 + 1$.

19.3.10. Factor $x^5 + 1$ into irreducible factors in $\mathbb{F}_2[x]$.

19.3.11. Let $R = \mathbb{F}_3[x]$, the ring of polynomials with coefficients in $\mathbb{F}_3$.
   (a) How many polynomials of degree 1 does R have? Does each of these have a root in $\mathbb{F}_3$?
   (b) Does $x^3 - x - 1$ have a root in $\mathbb{F}_3$? Is $x^3 - x - 1$ irreducible?
   (c) Show that $L = R/\langle x^3 - x - 1 \rangle$ is a field. How many elements does L have?

19.3.12. Factor $x^4 + 1 \in \mathbb{F}_3[x]$ into irreducible factors.

19.3.13. Let $f = x^7 - x \in \mathbb{F}_7[x]$.
  (a) What are all the roots of f?
  (b) Write f as a product of irreducible elements of $\mathbb{F}_7[x]$.

19.3.14. Let $p(x) = x(x - 1)(x + 1) \in (\mathbb{Z}/6\mathbb{Z})[x]$. Find all the roots of $p(x)$ in $\mathbb{Z}/6\mathbb{Z}$. Can you give a different factorization of $p(x)$?

Euler's Formula and Complex Roots. In finding roots of some polynomials in the complex numbers, it is helpful to use Euler's formula,

$$e^{ix} = \cos(x) + i\sin(x).$$

The formula allows for switching to a polar coordinate representation of complex numbers and a use of the usual rules for the manipulation of exponents. Every non-zero complex number $a + bi \in \mathbb{C}$ can be written as $re^{i\theta}$ with $r \geq 0$ and $0 \leq \theta < 2\pi$. The formula also says that, if k is an integer, then $e^{2k\pi i} = 1$. We can use this to find all solutions to $x^3 = 7$, for example. The polynomial $x^3 - 7$ has one real root that we denote by $\sqrt[3]{7}$ but it also has two complex roots. To find these, we first write $x^3 = 7 = 7 \times 1 = 7e^{2k\pi i}$ and then take cube roots of both sides. We get $x = \sqrt[3]{7}e^{\frac{2k\pi}{3}i}$. But, if the rules of exponentiation do indeed work, then this quantity should be a solution to $x^3 = 7$ regardless of the value of $k \in \mathbb{Z}$. Plugging in $k = 0, 1$, and 2 gives

$$x = \sqrt[3]{7},$$

$$x = \sqrt[3]{7}e^{\frac{2\pi i}{3}} = \sqrt[3]{7}\left(\cos(\frac{2\pi}{3}) + i\sin(\frac{2\pi}{3})\right) = \left(-\frac{1}{2} + \frac{\sqrt{3}}{2}i\right)\sqrt[3]{7},$$

$$x = \sqrt[3]{7}e^{\frac{4\pi i}{3}} = \sqrt[3]{7}\left(\cos(\frac{4\pi}{3}) + i\sin(\frac{4\pi}{3})\right) = \left(-\frac{1}{2} - \frac{\sqrt{3}}{2}i\right)\sqrt[3]{7}.$$

Other values of k repeat these three roots over and over.

But where does Euler's formula come from? To be rigorous, you first have to decide on your *definition* of complex exponentiation. What does an expression such as 2^i mean? In fact, one feasible approach—not the most common—is to take Euler's formula as the definition of complex exponentiation and derive other facts from it. Another is to start with the Taylor series for e^x at $x = 0$ and prove that it converges for all $x \in \mathbb{C}$. Then one uses the Taylor series to define e^x when $x \in \mathbb{C}$ and proceed to prove Euler's formula using Taylor series expansions of sine and cosine. We leave the details and the subtleties to a course in complex analysis but will come back to the roots of polynomials of the form $x^n - a$ in Section 27.3 and to the general solvability of polynomials in $\mathbb{C}[x]$ in Chapter 28.

19.3.15. Find all the roots of $x^4 + 1$ in $\mathbb{C}$. Is $x^4 + 1$ irreducible in $\mathbb{R}[x]$? Why?

19.3.16. Find all the roots of $x^5 - 3$ in $\mathbb{C}$.

19.3.17. Find all the roots of $x^4 - x^2 + 1$ in $\mathbb{C}$.

19.3.18. Find all the roots of $x^8 + x^4 + 1$ in $\mathbb{C}$.

19.3.19. Find the number of roots of $x^3 - x$ in $\mathbb{R}$, in $\mathbb{F}_5$, and in $\mathbb{Z}/6\mathbb{Z}$.

19.3.20. Let $f = x^4 + 8 \in \mathbb{Q}[x]$. Is f irreducible in $\mathbb{Q}[x]$? Is f irreducible in $\mathbb{R}[x]$?

19.3.21. Let $a, b \in \mathbb{R}$ and $\alpha = a + bi \in \mathbb{C}$.
   (a) Show that α has a square root in $\mathbb{C}$. In other words, show that $x^2 - \alpha = 0$ has a root in $\mathbb{C}$.
   (b) Find a square root of α and write it as $c + di$ where c and d are given in terms of a and b.

19.3.22. Let $p \in \mathbb{Z}$ be a positive prime with $p = 3 \pmod 4$. Let $f = x^2 + 1 \in \mathbb{F}_p[x]$. Show that f is irreducible.

19.3.23. Let $p \in \mathbb{Z}$ be a positive prime, and let $f = x^2 + x + 1 \in \mathbb{F}_p[x]$.
   (a) Factor f into a product of irreducible polynomials if $p = 3$.
   (b) Factor f into product of irreducible polynomials if $p = 7$.
   (c) If $p = 2 \pmod 3$, show that f is irreducible.

19.3.24. Let F be a finite field. Does $F[x]$ contain irreducible polynomials of arbitrarily large degree?

19.4. R **UFD** $\Rightarrow$ $R[x]$ **UFD and Gauss's Lemma**

As we have seen, roots of polynomials are intimately connected with irreducible polynomials. In $\mathbb{C}[x]$, for example, by the celebrated fundamental theorem of algebra (Theorem 26.11), every polynomial of positive degree has a root. This means that the only irreducible polynomials in $\mathbb{C}[x]$ are linear polynomials. It also means that the strategy of the previous section for creating a bigger field that contains $\mathbb{C}$ is not going to work. For other fields, the situation was different. Intuitively, in a finite field, the number of possible roots is finite and hence there is a better chance for finding polynomials with no roots. In fact, in $\mathbb{F}_p[x]$ there are irreducible polynomials of arbitrarily large degree (see Problem **19.3.24**). As motivation for our treatment of Gauss's lemma, we begin with the rational roots theorem which, for $R = \mathbb{Z}$ and $F = \mathbb{Q}$, could be familiar from high school algebra:

Theorem 19.33 (Rational roots theorem). *Let R be a UFD, and let F be its field of fractions. Let $f(x) = a_0 + a_1 x + \cdots + a_n x^n \in R[x]$ be a polynomial of degree n. Let $\beta \in F$ be a root of f. Write $\beta = rs^{-1}$ with $r, s \in R$ and $\gcd(r, s) = 1$. Then*

$$r \mid a_0, \quad s \mid a_n.$$

Proof. We know that $f(\beta) = 0$, and so

$$a_0 + a_1 \frac{r}{s} + a_2 \frac{r^2}{s^2} + \cdots + a_n \frac{r^n}{s^n} = 0$$
$$\Rightarrow a_0 s^n + a_1 r s^{n-1} + a_2 r^2 s^{n-2} + \cdots + a_n r^n = 0.$$

We rewrite the latter equation in two ways. First

$$-r(a_1 s^{n-1} + a_2 r s^{n-2} + \cdots + a_n r^{n-1}) = a_0 s^n,$$

and so $r \mid a_0 s^n$. Since $\gcd(r, s) = 1$, we conclude that $r \mid a_0$. Second,

$$-s(a_0 s^{n-1} + a_1 r s^{n-2} + \cdots + a_{n-1} r^{n-1}) = a_n r^n.$$

Hence $s \mid a_n r^n$, and again this results in $s \mid a_n$. $\qquad\square$

Example 19.34. Consider $p(x) = x^6 - 2x^5 + 3x^4 - 6x^3 + 9x^2 - 15x - 6 \in \mathbb{Z}[x]$. The field of fractions of $\mathbb{Z}$ is $\mathbb{Q}$, and so by the rational roots theorem, the only possible rational roots of p are ± 1, ± 2, ± 3, or ± 6. We check and see that ± 1 is not a root, but $x = 2$ is a root, and so $x - 2$ is a factor:

$$p(x) = x^6 - 2x^5 + 3x^4 - 6x^3 + 9x^2 - 15x - 6 = (x - 2)(x^5 + 3x^3 + 9x + 3).$$

The only possible rational roots of $q(x) = x^5 + 3x^3 + 9x + 3$ are ± 1, or ± 3. We already have seen that ± 1 is not a root of p so it cannot be a root of q either. So we only need to check ± 3. $q(3)$ is clearly positive, while $q(-3)$ is clearly negative and not zero. So the only rational root for p is 2.

Corollary 19.35. *Let $f(x) \in \mathbb{Z}[x]$, and let $f(x)$ be monic. If f has a root in $\mathbb{Q}$, then that root is in $\mathbb{Z}$.*

This corollary to the rational root theorem generalizes the fact, known since antiquity, that $\sqrt{2}$ is irrational!

Corollary 19.36. *Let $n \in \mathbb{Z}$ and $k \in \mathbb{Z}^{>1}$. Then $\sqrt[k]{n}$ is either an integer or irrational.*

Proof. Consider the monic polynomial $x^k - n \in \mathbb{Z}[x]$. A root of this polynomial is $\sqrt[k]{n}$. So by Corollary 19.35 if $\sqrt[k]{n}$ is rational, it must be an integer. $\qquad\square$

The rational roots theorem tells us that under certain circumstances if a polynomial has a linear factor in a bigger ring, then it also has a linear factor in a smaller ring. However, this theorem does not say anything about more general factorizations. This will be remedied by Gauss's lemma which is a generalization of the rational roots theorem.

Now $x^2 + 1$ is irreducible in $\mathbb{R}[x]$ and yet factors as $(x + i)(x - i)$ in $\mathbb{C}[x]$. This is not surprising since by extending the field from $\mathbb{R}$ to $\mathbb{C}$ we have gained much flexibility in factoring polynomials. This intuition works most of the time although we have to take care of a few technical mishaps.

Let $R \subset S$ be two integral domains. Of course, there are polynomials in $S[x]$ that do not exist in $R[x]$. But $R[x] \subseteq S[x]$ and hence given $f \in R[x]$, we also have $f \in S[x]$, and we can ask if irreducibility (or reducibility) if one translates to irreducibility (or reducibility) in another. In accordance with our intuition, most of the time—but not always—if f is irreducible in $S[x]$, then it remains irreducible in $R[x]$—after all, there are fewer potential factors for f in $R[x]$. Also, it seems likely—this will be developed and explored thoroughly in the chapters on fields and Galois theory—that we could make S big enough so that f will be reducible in $S[x]$. A (somewhat annoying) simple example shows that we need to be a bit more careful.

Example 19.37. Consider $\mathbb{Z}[x]$. In this ring, $2x + 4$ is reducible, since it factors into $2(x + 2)$, and neither 2 nor $x + 2$ are units. On the other hand, the same element is irreducible in $\mathbb{Q}[x]$, since 2 is now a unit in $\mathbb{Q}[x]$. Hence, in the case of this example, contrary to the intuition presented above, making the ring larger resulted in a reducible polynomial becoming irreducible. The problem is the simple issue of factoring constants.

In this section, we will add a bit of technical language to be able to say precisely when reducibility of $f \in R[x]$ implies reducibility of $f \in S[x]$, where $R \subseteq S$. (This is equivalent to conditions that imply irreducibility of $f \in R[x]$ given irreducibility of $f \in S[x]$.) Even though we have to be careful, this result will not be that profound. What is important is that, in certain cases and contrary to our intuition, the converse holds. This result, known as Gauss's lemma, generalizes the rational roots theorem and will have many applications. As a byproduct we will also show that if R is UFD then so is $R[x]$.

Note that, in the previous section, we considered the irreducible polynomials in $K[x]$ where K is a field. Here we want to consider irreducibles in $R[x]$ where R is an integral domain. While some of our results will be true in a slightly more general setting, we often will assume that R is a UFD.

Definition 19.38. Let R be an integral domain, and let $f \in R[x]$ with $\deg(f) > 0$. The polynomial f is *primitive* if the only elements of R that divide f are units of R.

If R is a UFD and $f(x) = a_0 + \cdots + a_n x^n$, then we define the *content* of f to be $c(f) = \gcd(a_0, \ldots, a_n)$.

Then, if R is a UFD, f is primitive if $c(f)$ is a unit of R, and for any $f \in R[x]$, we have $f = c(f)f^*$ where $f^* \in R[x]$ is primitive.

Remark 19.39. Note that to define the content, we need to know that greatest common divisors exist, and this certainly is true for UFDs (see Theorem 18.62). In a polynomial like $2x+4$ in $\mathbb{Z}[x]$, the content is 2 and we can write $2x+4 = 2(x+2)$, and as claimed we wrote the polynomial as the product of its content—which is a ring element—and a primitive polynomial. By considering contents and primitive polynomials, we are able to account for the fact that some elements of R are not invertible and hence are going to be counted as irreducible in $R[x]$.

Remark 19.40. The greatest common divisor, if it exists, is unique only up to associates (see Lemma 18.61), and, in that sense, the content is not well defined. Hence, by $c(f)$ we mean an equivalence class consisting of an element of R and all its associates, and by $c(f) = c(g)$, we mean the equality of equivalence classes and that the greatest common divisor of coefficients of f is an associate of the greatest common divisor of coefficients of g.

Lemma 19.41. *Let R be a UFD, let $\alpha, \beta \in R$, and let f^* and g^* be primitive polynomials in $R[x]$. Assume*

$$\alpha f^* = \beta g^*.$$

Then α and β are associates in R, and f^* and g^* are associates in $R[x]$. In particular, for $0 \neq h \in R[x]$, we have $h = c(h)h^*$ where $c(h)$ is the content of h and h^* is primitive, and this factorization is unique up to associates.

Proof. Since R is a UFD, α and β have a greatest common divisor (by Theorem 18.62). Let $\delta = \gcd(\alpha, \beta) \in R$. We then have $\alpha = \delta\alpha'$ and $\beta = \delta\beta'$ with α' and

β' relatively prime elements of R. We also have $\alpha' f^* = \beta' g^*$. Let b be any of the coefficients of g^*, then $\alpha' \mid \beta' b$. It follows, since $\gcd(\alpha', \beta') = 1$, that $\alpha' \mid b$ (Lemma 18.67). Thus α' divides every coefficient of g^*, a primitive polynomial. It follows that α' is a unit of R. Likewise, β' is a unit of R, and α and β are associates. Hence f^* and g^* are associates in $R[x]$ as well.   □

Example 19.42. It may be instructive to note that in Lemma 19.41, the condition that R be a UFD is important. Let $R = \mathbb{Z}[\sqrt{-5}]$, and consider the polynomial $9 + (6 + 3\sqrt{-5})x \in R[x]$. There are at least two ways of factoring an element of R from this polynomial:

$$3[3 + (2 + \sqrt{-5})x] = 9 + (6 + 3\sqrt{-5})x = (2 + \sqrt{-5})[2 - \sqrt{-5} + 3x].$$

3 and $2 + \sqrt{-5}$ are both irreducible (and as can be seen using norms) and are not associates.

Lemma 19.43. *Let R be an integral domain, and let $f \in R[x]$ with $\deg(f) > 0$. Then f irreducible $\Leftrightarrow$ f is primitive and f cannot be factored into two polynomials, in $R[x]$, of positive degree.*

In addition, irreducible elements of R are still irreducible in $R[x]$.

Proof. ($\Rightarrow$) Assume f is irreducible. Then, by the definition of irreducibility, f cannot be factored into two polynomials of positive degree. Now assume f is not primitive. Then $f = ag$ where $a \in R$ and a is not a unit—in fact, if R is a UFD, a can be taken to be the content of f. The polynomial g is not a unit either since $\deg(g) > 0$. Hence, contrary to assumption, f is reducible. Hence, f must be primitive.

($\Leftarrow$) Now assume that f is primitive and cannot be factored into two polynomials of positive degree. For a proof by contradiction, assume f is reducible, and write $f = gh$ with neither g nor h a unit. Since f cannot be factored into two polynomials of positive degree, we can assume, without loss of generality, that g is a constant. This means that $g \in R$ is a non-unit that divides f, and so f is not primitive. The contradition proves that f is irreducible.   □

Remark 19.44. Proposition 19.30 of the previous section stated that if F is a *field*, then $f \in F[x]$ with positive degree is irreducible if and only if it cannot be factored into two polynomials of positive degree. Lemma 19.43 is a more general version of this proposition in the case when the coefficients form just an *integral domain*. Polynomials over integral domains—unlike the ones over fields—are not always primitive, and hence we need the extra condition of primitivity.

Given a ring homomorphism, we can always define a corresponding map on polynomials. The reader should check the following.

Lemma 19.45 (Extending the canonical homomorphism). *Let R be a ring, and let I be an ideal of R. Let*

$$\overline{} : R \to R/I$$

be the canonical homomorphism. We extend this map by defining

$$\pi : R[x] \to (R/I)[x]$$

by

$$\pi(a_0 + a_1 x + \cdots + a_n x^n) = \overline{a_0} + \overline{a_1} x + \cdots + \overline{a_n} x^n.$$

Then π is an onto ring homomorphism and the kernel of π is $I[x]$ (polynomials with coefficients from I or the ideal in $R[x]$ generated by I).

Proposition 19.46. *R is a UFD. $f, g \in R[x]$. Then*

(a) *f and g primitive $\Leftrightarrow$ fg primitive.*

(b) *$c(fg) = c(f)c(g)$.*

Proof. CLAIM 1: *f and g primitive $\Rightarrow$ fg primitive.*

PROOF. Assume fg is not primitive, and let p be an irreducible in R with $p \mid c(fg)$. Let $\pi : R[x] \to (R/\langle p \rangle)[x]$ be the extension of the canonical homomorphism as in Lemma 19.45. The kernel of this map is the ideal in $R[x]$ generated by p. f and g primitive means that $\pi(f)$ and $\pi(g)$ are not zero. $R/\langle p \rangle[x]$ an integral domain means that $\pi(f)\pi(g) \neq 0$. Thus $\pi(fg) \neq 0$ and so $p \nmid fg$ and so $p \nmid c(fg)$. The contradiction proves the claim.

CLAIM 2: *$c(fg) = c(f)c(g)$.*

PROOF. $f = c(f)f^*$ and $g = c(g)g^*$ where f^* and g^* are primitive. Hence $fg = c(f)c(g)f^*g^*$ with f^*g^* primitive. Thus, by Lemma 19.41, $c(fg) = c(f)c(g)$.

CLAIM 3: *fg primitive $\Rightarrow$ f and g primitive.*

PROOF. $c(fg) = c(f)c(g)$. fg primitive means that $c(fg)$ is a unit. Hence $c(f)$ and $c(g)$ have to be units, and so f and g are primitive. $\qquad\square$

Example 19.47. Again the condition that R is a UFD is important. Let $R = \mathbb{Z}[\sqrt{-5}]$, and let

$$f = 3 + (2 + \sqrt{-5})x, \qquad g = 3 + (2 - \sqrt{-5})x.$$

Now f and g are both primitive, and yet $fg = 9 + 12x + 9x^2$ is not.

Remark 19.48. If R is a ring, $f, g \in R[x]$, and we say $f \mid g$ in $R[x]$, then we mean that $g = fh$ with $h \in R[x]$. In other words, another way of saying that g reduces in $R[x]$ is to say that f, a non-unit, divides g in $R[x]$.

Let $f \in R[x]$ be irreducible, and assume $R \subseteq S$. It is quite possible that f reduces in $S[x]$. Gauss's lemma says that if R is an integral domain and S is the field of fractions of R, then this cannot happen. (Well, to be precise, we have to add the condition that f is not a constant. For example, 2 is irreducible in $\mathbb{Z}[x]$ but not in $\mathbb{Q}[x]$.)

We will first prove a technical-looking theorem from which a number of versions—each useful in particular situations—of Gauss's lemma will follow. The fact that, in a UFD, $c(fg) = c(f)c(g)$ will play a crucial role in the proof. This fact, itself, depended on the fact that, in a UFD, the product of two primitive polynomials is primitive. Recall that in a UFD, in addition to greatest common divisors, we also have least common multiples. (See Lemma 18.73.)

Theorem 19.49. *Let R be a UFD, let F be the field of fractions of R, and let $f \in R[x]$.*

Assume that $f = gh$ with $g, h \in F[x]$.

Let a and b, respectively, be the least common multiple in R of the denominators of coefficients of g and h, so that ag and bh are in $R[x]$. Write $ag = c(ag)\, g'$ and $bh = c(bh)\, h'$ where g' and h' are primitive polynomials in $R[x]$. Then

$$c(ag)\, c(bh) = ab\, c(f), \text{ in } R, \text{ and}$$

$$f = c(f)\, g'h' \text{ in } R[x].$$

Proof. We have $abf = (ag)\,(bh)$ and hence, by Proposition 19.46(b), $ab\, c(f) = c(abf) = c(ag)\, c(bh)$, proving the first assertion. Now

$$abf = (ag)(bh) = c(ag)\, g'\, c(bh)\, h' = c(ag)\, c(bh)\, g'h' = ab\, c(f)\, g'h'.$$

Cancelling ab, we get the second equality.   □

Remark 19.50. Theorem 19.49 may look too complicated, but, if you get past the notation, it is pretty straightforward. It tells you exactly how to factor a polynomial in $R[x]$ if you know how to factor it in $F[x]$, where F is the field of fractions of the unique factorization domain R. You have a polynomial $f \in R[x]$. First, factor f in $F[x]$. This means that you may introduce fractions (of elements of R) in the factorization. If $f = gh$ in $F[x]$, then g and h are polynomials in $F[x]$ and not necessarily of $R[x]$. Next, you find a and b, respectively, the least common multiple of the *denominators* of the coefficients in g and in h. Now, a and b are elements of R and, moreover, ag and bh are polynomials in $R[x]$. In other words, by multiplying g and h by a and b, you have cleared their denominators. Finally, factor out the content of ag and bh, and write $ag = c(ag)g'$ and $bh = c(bh)h'$. Theorem 19.49 says that the factorization of f in $R[x]$ is $f = c(f)g'h'$.

Theorem 19.51 (Gauss's lemma). *Let R be a UFD, let F be the field of fractions of R, and let $0 \neq f \in R[x]$. Then*

(a) *Assume $\deg f > 0$. Then f is irreducible in $R[x]$ if and only if f is primitive in $R[x]$ and irreducible in $F[x]$.*

(b) *Assume g is primitive in $R[x]$. If $g \mid f$ in $F[x]$, then $g \mid f$ in $R[x]$.*

Proof. (a) ($\Leftarrow$) Assume f is primitive in $R[x]$, irreducible in $F[x]$, and that $f = gh$ in $R[x]$. Then $f = gh$ in $F[x]$ as well. But f is assumed to be irreducible in $F[x]$, and, hence, either g or h are units in $F[x]$. Without loss of generality, assume g is a unit in $F[x]$. But the units of $F[x]$ are exactly the non-zero elements of F. Hence, $\deg g = 0$ and g is an element of F. It is also an element of $R[x]$. This means that $g \in R$. It follows that the content of g in R is g. But $f = gh$ primitive in $R[x]$ means—by Proposition 19.46—that g is primitive in $R[x]$, and, hence, its content must be a unit of R. Thus g is a unit in R, and the proof is complete.

($\Rightarrow$) Assuming f is irreducible in $R[x]$, we already know—by Lemma 19.43—that f is primitive in $R[x]$. We need to prove that it remains irreducible in $F[x]$. Hence assume $f = gh$ in $F[x]$. We have to show that either g or h is a unit in $F[x]$.

By Theorem 19.49, we have $f = c(f)g'h'$ in $R[x]$. But f is irreducible in $R[x]$ and hence without loss of generality g' is a unit in $R[x]$. This means that g' is a unit of R. Recalling the definition of g', this means that g is a constant in F and hence a unit.

(b) Since $g \mid f$ in $F[x]$, we have $f = gh$ in $F[x]$. Again by Theorem 19.49, we have $f = c(f) \, g'h'$ in $R[x]$. However, we know that g is in $\dot{R}[x]$ and hence a, the least common multiple of the denominators of coefficients of g, is equal to 1. We also know that g is primitive, and hence $g = c(ag)g' = g' | f$ in $R[x]$. □

Remark 19.52. In Theorem 19.51(a) we do need $\deg f > 0$ since, for example, 2 is irreducible in $\mathbb{Z}[x]$ while not primitive.

Even when applied to the integers, Gauss's lemma has content. Wading through the technical language, we can record the following concrete result:

Corollary 19.53 (Gauss's lemma in $\mathbb{Z}$). *If $f \in \mathbb{Z}[x]$ is irreducible in $\mathbb{Z}[x]$ and $\deg(f) > 0$, then f is irreducible in $\mathbb{Q}[x]$.*

We now use Gauss's lemma to prove that a polynomial ring over a unique factorization domain is itself a unique factorization domain.

Theorem 19.54.

$$R \ UFD \ \Rightarrow \ R[x] \ UFD.$$

Proof. We have to prove that elements in $R[x]$ factor into irreducibles and that this factorization is unique.

Let $0 \neq f$ be a non-unit of $R[x]$. We want to factor f into irreducibles. First note that $f = c(f) \, f^*$ where f^* is primitive. Now R is a UFD and hence $c(f) = p_1 \cdots p_k$ with p_i, for $1 \leq i \leq k$, an irreducible element of R. Irreducible elements of R are irreducible elements of $R[x]$ and hence, so far, we have factored $c(f)$ into a product of irreducible elements of $R[x]$. Turning into f^*, if f^* cannot be factored into two polynomials of positive degree, then f^* is irreducible and the proof is complete. Otherwise $f^* = gh$ in $R[x]$ with $0 < \deg g, \deg h < \deg f^*$. Now f^* primitive implies that both g and h are primitive, and we can continue factoring them into primitive polynomials of lower and lower degree. Since the degrees are decreasing, this process cannot go on indefinitely, and eventually we will not be able to factor any further. At that point, we have achieved a factorization of f^* into irreducibles.

To prove that the factorization into irreducibles is unique (up to rearranging and associates), we show that every irreducible element is prime. (See Theorem 18.41.) Assume $f \in R[x]$ is irreducible. This means that f is primitive in $R[x]$. Now, let F be the field of fractions of R. Since f is irreducible in $R[x]$, by Gauss's lemma, f is irreducible in $F[x]$. But F is a field and hence $F[x]$ is a Euclidean domain. Hence f is prime in $F[x]$. To show that f is prime in $R[x]$, assume $f \mid gh$ in $R[x]$. Hence $f \mid gh$ in $F[x]$, and, in $F[x]$, f is a prime. So $f \mid g$ or $f \mid h$ in $F[x]$. But we know f is primitive, and hence by Gauss's lemma, $f \mid g$ or $f \mid h$ in $R[x]$, and the proof is complete. □

Remark 19.55. The strategy used in the proof of Theorem 19.54 is a common one. Often when we have a polynomial with coefficients in an integral domain R, we consider the polynomial as a polynomial in $F[x]$, where F is the field of fractions of R. The big advantage is that F is a field and, hence, $F[x]$ is a Euclidean domain (and therefore also a PID and a UFD).

Corollary 19.56.
$$R \ UFD \ \Rightarrow \ R[x_1, \ldots, x_k] \ UFD.$$

Example 19.57. Consider the ring $\mathbb{Z}[x]$. Since $\mathbb{Z}$ is a UFD, then—by Theorem 19.54—so is $\mathbb{Z}[x]$. On the other hand, the ideal $\langle 2, x \rangle$ is not principal (see Problem *16.1.17*) and so $\mathbb{Z}[x]$ is not a PID (nor, hence, an ED).

Problems

19.4.1. In Problem 15.1.1 you were asked to find if any integer solutions to $x^2 - 3y^2 = 0$ exists and to generalize your conclusion. Revisit that problem now.

19.4.2. Let $a, b \in \mathbb{Z}$ and $f = x^3 + ax^2 + bx + 1 \in \mathbb{Z}[x]$. For which values of a and b is f reducible in $\mathbb{Z}[x]$? What about in $\mathbb{Q}[x]$?

19.4.3. Reconcile Lemma 19.43 and Proposition 19.30. In particular, derive the latter from the former.

19.4.4. We gave a direct proof of the rational roots theorem (Theorem 19.33) and claimed that Gauss's lemma is a generalization. Use Gauss's lemma (Theorem 19.51) to give a different proof of the rational roots theorem.

19.4.5. Find a polynomial in $\mathbb{Z}[x]$ that has $\sqrt{2} + \sqrt{3}$ as a root. Use it to prove that $\sqrt{2} + \sqrt{3}$ is irrational.

19.4.6. Let f be a polynomial of degree 3 in $\mathbb{Q}[x]$. Assume that f has no integer roots. Is f necessarily irreducible in $\mathbb{Q}[x]$?

19.4.7. Show $\sqrt{2} + \sqrt[3]{2}$ is irrational.

19.4.8. Let A be an $n \times n$ matrix with integer entries. Assume that $\lambda \in \mathbb{Q}$ is an eigenvalue of A. Prove that $\lambda \in \mathbb{Z}$.

19.4.9. Let R be a UFD, let F be the field of fractions of R, and let $0 \neq f \in R[x]$. Assume $g \in R[x]$ is primitive, and $g \mid f$ in $F[x]$. Show that, in $R[x]$, g divides f^* where f^* is a primitive polynomial with $f = c(f)f^*$.

19.4.10. **Another version of Gauss's lemma.** Let R be a UFD, let F be the field of fractions of R, and let $0 \neq f \in R[x]$. Prove that if $f = gh$ for $g, h \in F[x]$, then there exists non-zero elements α and β in F such that αg and βh are in $R[x]$, and $f = (\alpha g)(\beta h)$ is a factorization of f in $R[x]$.

19.4.11. Let R be a UFD, and let F be the field of fractions of R. Let $\alpha \in F$, and assume that $f \in R[x]$ is a monic polynomial with $f(\alpha) = 0$. Prove that $\alpha \in R$.

19.4.12. Let S denote the set of polynomials in x with integer coefficients that have no linear term. In other words,
$$S = \{a_n x^n + a_{n-1} x^{n-1} + \cdots + a_2 x^2 + a_0 \mid n \in \mathbb{Z}^{\geq 0}, \ a_i \in \mathbb{Z}\}.$$
(a) Show that S is an integral domain.
(b) Is x^2 an irreducible in this ring? What about x^3?
(c) Is x^2 a prime in this ring? What about x^3?

(d) In S, does x^6 factor uniquely into irreducibles?

(e) In S do x^5 and x^6 have a greatest common divisor?

(f) Is S a UFD?[1]

19.4.13. Let $R = \mathbb{Z}/6\mathbb{Z}$, and let $p(x) = x^2 + 4x + 1 \in R[x]$. Which of the following give a factorization of $p(x)$ in $R[x]$:

$$(x+5)^2; \quad (3x^2 + 2x + 1)(2x + 1); \quad (x+5)(2x+1)(3x+5)?$$

What are the roots of $p(x)$? Does your answer contradict Corollary 19.25 or Theorem 19.54?

19.4.14. Let $f = x^5 - ax - 1 \in \mathbb{Z}[x]$. Find all values of a for which f is reducible in $\mathbb{Z}[x]$.

19.4.15. Let K be a field and $R = K[x]$. Let E be the field of fractions of R, and $h = -xy^3 - (x-1)y^2 - x + 1 \in R[y]$. Is h irreducible in $E[y]$?

19.4.16. Let $R = \mathbb{R}[x, y]$, and let S be the subring of R defined by $S = \mathbb{R}[x^3, xy, y^3]$. In the ring S, can you find an irreducible element that is not prime? Is R a UFD? What about S?

19.4.17. Let F be a field, and assume $\operatorname{char}(F) \neq 2$. Let $F(x)$ denote the field of rational functions over F. In other words,

$$F(x) = \left\{ \frac{p}{q} \mid p, q \in F[x], q \neq 0 \right\}.$$

Prove that $\sqrt{1 - x^2} \notin F(x)$. In other words, there is no element in $F(x)$ such that its square is $1 - x^2$. Was the condition $\operatorname{char}(F) \neq 2$ necessary?

19.4.18. Let $R = \mathbb{Z}[2\sqrt{2}]$, and let F be the field of fractions of R.

(a) What is a typical element of R? What about F? In particular, is $1 + \sqrt{2}$ an element of R and/or F?

(b) Can you factor $x^2 - 2$ in $R[x]$?

(c) Can you factor $x^2 - 2$ in $F[x]$?

(d) Are your answers consistent with Theorem 19.51 (Gauss's Lemma)?

(e) Is $\mathbb{Z}[2\sqrt{2}]$ a UFD? Why?

19.4.19. Is $(\mathbb{Z}[\frac{1}{2}])[x]$ a UFD? Why?

19.4.20. Is the ring of Laurent polynomials $\mathbb{Z}[x, \frac{1}{x}] = \{p(x, \frac{1}{x}) \mid p \in \mathbb{Z}[x, y]\}$ a UFD? Why?

19.4.21. Let $R = \mathbb{Z}[x]$, and let $M = \{1, (x-1), (x-1)^2, \ldots\} = \{(x-1)^n \mid n \geq 0\}$. Let $S = R[M^{-1}]$ (see Definition 17.9).

(a) What are the elements of S? Is S an integral domain? What are the invertible elements of S? What is the field of fractions of S?

(b) Can you find a non-trivial proper ideal of S? How about a maximal ideal of S?

---

[1] Adapted from Wildenberg [**Wil07**].

19.5. Irreducibility Criteria

Given R a commutative ring with identity, how do we decide if $p \in R[x]$ is irreducible? There is no sure-fire method for answering this question. Rather, we have a number of theorems each covering certain types of situations. Gauss's lemma often provides part of a one–two punch. Basically it says that if R is a UFD, then we can answer the question over F, the field of fractions of R. Over a field, degree 1 polynomials are always irreducible, and polynomials of degree 2 and 3 are irreducible if and only if they do not have any roots (Corollary 19.31). For higher degree polynomials, the problem is more difficult. Here, we give two examples— one is for polynomials over $\mathbb{Z}$ and one is the Schönemann-Eisenstein irreducibility criterion—of results that give us criteria for irreducibility. For unique factorization domains, Gauss's lemma then strengthens the result(s) to imply irreducibility over the field of fractions.

Theorem 19.58. *Let $p(x) \in \mathbb{Z}[x]$ be primitive and of degree d. Let $n > 0$ be an integer such that n does not divide the coefficient of x^d. Let $\pi : \mathbb{Z}[x] \to (\mathbb{Z}/n\mathbb{Z})[x]$ be the extension of the canonical homomorphism as in Lemma 19.45. In other words, $\pi(p(x))$ is the polynomial you get from $p(x)$ when you consider its coefficients $\bmod\, n$. Now, if $\pi(p(x))$ is irreducible in $(\mathbb{Z}/n\mathbb{Z})[x]$, then $p(x)$ is irreducible in $\mathbb{Z}[x]$.*

Proof. Assume that $p(x)$ is reducible in $\mathbb{Z}[x]$. Then $p(x) = g(x)h(x)$ with neither g nor h a unit. The units in $\mathbb{Z}[x]$ are ± 1 (see Lemma 19.29), and so neither g nor h are ± 1. We have also assumed that p is primitive and hence neither g nor h can be a constant. Applying π to both sides, we get $\pi(p(x)) = \pi(g(x))\pi(h(x))$ in $(\mathbb{Z}/n\mathbb{Z})[x]$. Since n does not divide the coefficient of x^d, we have that $\deg(\pi(p(x)) = \deg(p(x))$. This means that the degrees of $\pi(g(x))$ and $\pi(h(x))$ are, respectively, the same as the degrees of $g(x)$ and $h(x)$ (since applying π can only reduce the degrees). Hence $\pi(g(x))$ and $\pi(h(x))$ are not constants, and so $\pi(p(x))$ is reducible. $\qquad\square$

Remark 19.59. Note that both the assumption that p is primitive and that n does not divide the coefficient of x^d are necessary. Consider $p(x) = 3x$. This is a reducible polynomial in $\mathbb{Z}[x]$ but irreducible mod 2. Likewise, $3x^2 + x$ is reducible in $\mathbb{Z}[x]$ but reducible mod 3.

Example 19.60. Let $f(x) = 21x^3 - 3x^2 + 2x + 9$, and let $p = 2$. Now mod 2, this polynomial becomes $x^3 + x^2 + 1$. This polynomial has no roots in $\mathbb{F}_2 = (\mathbb{Z}/2\mathbb{Z}, +, \cdot)$ and so can have no linear factors. This means that $x^3 + x^2 + 1$ is irreducible in $\mathbb{F}_2[x]$ and hence, by Theorem 19.58, $21x^3 - 3x^2 + 2x + 9$ is irreducible in $\mathbb{Z}[x]$. Now Gauss's lemma tells us that this polynomial is also irreducible in $\mathbb{Q}[x]$.

Example 19.61. Let $f(x) = 21x^3 - 3x^2 + 2x + 8$. Proceeding as in the previous example, we see that in $\mathbb{F}_2[x]$, the polynomial becomes $x^3 + x^2 = x^2(x+1)$ which is reducible. However mod 5, we get $x^3 + 2x^2 + 2x + 3$. After trying every element of $\mathbb{F}_5 = (\mathbb{Z}/5\mathbb{Z}, +, \cdot)$, we see that the latter polynomial is irreducible in $\mathbb{F}_5[x]$. Hence, by Theorem 19.58, $21x^3 - 3x^2 + 2x + 8$ is irreducible in $\mathbb{Z}[x]$. Now, Gauss's lemma tells us that this polynomial is also irreducible in $\mathbb{Q}[x]$.

Example 19.62. Let $p(x) = x^4 + 15x^3 + 7 \in \mathbb{Z}[x]$. Just looking at it, it is not obvious whether this polynomial is irreducible or not. We can apply Theorem 19.58

with $n = 5$ and get $\pi(p(x)) = x^4 + 2$. Is this an irreducible polynomial in $\mathbb{F}_5[x]$? By plugging in every element of $\mathbb{F}_5 = (\mathbb{Z}/5\mathbb{Z}, +, \cdot)$, we see that $\pi(p(x))$ has no roots in $\mathbb{F}_5$ and hence no linear factors in $\mathbb{F}_5[x]$. The only other possibility is if $\pi(p(x))$ is factored into two polynomials of degree 2. We now write, in $\mathbb{F}_5[x]$,

$$x^4 + 2 = (x^2 + ax + b)(x^2 + cx + d).$$

Multiplying the left hand side and equating the like coefficients we get

$$\begin{cases} a + c = 0, \\ ac + b + d = 0, \\ ad + bc = 0, \\ bd = 2. \end{cases}$$

Since $c = -a$, we have $0 = ad + bc = a(d - b)$. Since $\mathbb{Z}/5\mathbb{Z}$ is an integral domain, we have to have $a = 0$ or $d = b$.

If $a = 0$, then $c = 0$ and we have to have $b + d = 0$ while $bd = 2$. This means $d = -b$ which implies $-b^2 = 2$ and so $b^2 = 3$. But $0^2 = 0$, $1^2 = 1$, $2^2 = 4$, $3^2 = 4$, and $4^2 = 1$. Hence we have no element b with $b^2 = 3$.

So $b = d$ and hence $b^2 = 2$. This again is impossible since in $\mathbb{Z}/5\mathbb{Z}$, as we saw in the previous case, the only perfect squares are 0, 1, and 4.

We conclude that $x^4 + 2$ is irreducible in $\mathbb{F}_5[x]$ and hence, by Theorem 19.58, $x^4 + 15x^3 + 7$ is irreducible in $\mathbb{Z}[x]$. Now, Gauss's lemma tells us that this polynomial is also irreducible in $\mathbb{Q}[x]$.

Recall that if A and B are ideals of a ring R, then the ideal AB consists of elements of the form $\sum_{i=1}^{n} a_i b_i$ where, for $1 \leq i \leq n$, $a_i \in A$, and $b_i \in B$. (See Definition 16.56 and Lemma 16.58.) In addition, the ideal AA is denoted by A^2.

Theorem 19.63 (Generalized Schönemann-Eisenstein criterion). *Let R be an integral domain. Let $f(x) = a_0 + a_1 x + \cdots + a_n x^n \in R[x]$ be a primitive polynomial of degree $n \geq 1$.*

Assume P is a prime ideal of R satisfying the following conditions:

(a) $a_n \notin P$, *and*

(b) $a_0, a_1, \ldots, a_{n-1} \in P$, *and*

(c) $a_0 \notin P^2$.

Then f is irreducible in $R[x]$.

Proof. Assume $f = gh$ with $g, h \in R[x]$ and such that neither g nor h is a constant. Let $g(x) = b_0 + b_1 x + b_2 x^2 + \cdots + b_r x^r$ and $h(x) = c_0 + c_1 x + \cdots + c_s x^s$. Then $r + s = n$ and $r, s < n$. We have $a_0 = b_0 c_0 \in P$, and since P is a prime ideal we must have $b_0 \in P$ or $c_0 \in P$. On the other hand if both b_0 and c_0 are in P, then $a_0 = b_0 c_0 \in P^2$ contradicting the third condition. So, without loss of generality, assume $b_0 \in P$ and $c_0 \notin P$. If every coefficient of g was in P, then so would be every coefficient of f which contradicts the first condition. So let b_i be the first coefficient of g such that $b_i \notin P$. Now $a_i = b_i c_0 + b_{i-1} c_1 + \cdots + b_0 c_i$ and so $b_i c_0 = a_i - b_{i-1} c_1 - \cdots - b_0 c_i$. Now every element of the right hand side is in P

(for $j < i$, $b_j c_{i-j} \in P$ since P is an ideal and $b_j \in P$). Thus $b_i c_0 \in P$ which implies $c_0 \in P$ or $b_i \in P$. Either one gives a contradiction. $\qquad \square$

A special case of the above—strengthened using Gauss's lemma—is often used:

Corollary 19.64 (Schönemann-Eisenstein criterion).[2] *Let R be a UFD, and let K be the fraction field of R. Let $f(x) = a_0 + a_1 x + \cdots + a_n x^n \in R[x]$. Suppose there exists a prime $p \in R$ such that*

(a) $p \nmid a_n$, *and*

(b) $p \mid a_i$ *for* $i = 0, \ldots, n-1$, *and*

(c) $p^2 \nmid a_0$.

Then f is irreducible in $K[x]$.

Example 19.65. The polynomial $f(x) = 5 + 10x^3 + x^5$ is irreducible in $\mathbb{Q}[x]$.

Problems

19.5.1. Factor $27x^3 + 13x + 15$ into irreducible factors in $\mathbb{Q}[x]$.

19.5.2. **Proof of Corollary 19.64.** Write down the details to show that Corollary 19.64 follows from Theorem 19.63 and Gauss's lemma.

19.5.3. Let R be an integral domain. Show that $f(x) \in R[x]$ is irreducible if and only if $f(x+1)$ is irreducible.

19.5.4. Is $x^8 + x^7 + \cdots + x^2 + x + 1$ irreducible in $\mathbb{Z}[x]$?

19.5.5. Let $\Phi_p(x) = x^{p-1} + x^{p-2} + \cdots + x + 1$, where p is a prime. Show that $\Phi_p(x)$ is irreducible in $\mathbb{Q}[x]$.

19.5.6. For which n is $x^{n-1} + x^{n-2} + \cdots + x + 1$ irreducible in $\mathbb{Q}[x]$?

19.5.7. Show that $x^4 + 1$ is irreducible in $\mathbb{Q}[x]$.

19.5.8. Is $x^4 + 1$ irreducible in $(\mathbb{Z}/5\mathbb{Z})[x]$?

19.5.9. Is $x^3 + 3x + 2$ irreducible in $\mathbb{Q}[x]$?

19.5.10. Let $f(x) = x^7 + 15x^2 - 21x + a$. Show that there are infinitely many choices for a to make f an irreducible polynomial in $\mathbb{Q}[x]$.

19.5.11. Is $x^4 + x^2 + 1$ irreducible in $\mathbb{Q}[x]$? Does it have any roots?

19.5.12. Is $x^6 + 4x^3 + 1$ irreducible in $\mathbb{Q}[x]$?

19.5.13. Factor $x^4 + 2x^3 + 3x^2 + 2x + 2 \in \mathbb{Z}[x]$ into irreducible factors.

19.5.14. Let $M = \mathbb{Z} - \langle 5 \rangle$, and let $\mathbb{Z}_{(5)} = \mathbb{Z}[M^{-1}]$ be the localization of $\mathbb{Z}$ at M. Is $x^5 + 10x^4 + 35x - 15$ irreducible in $\mathbb{Z}_{(5)}[x]$?

19.5.15. Let $g(x, y) = y^5 + x^2 y^4 + x^3 y^2 + xy + x \in \mathbb{Z}[x, y]$. Show that $g(x, y)$ is irreducible in $\mathbb{Z}[x, y]$. Is g irreducible in $\mathbb{Q}[x, y]$?

---

[2]Theodor Schönemann (1812–1868) and Ferdinand Gotthold Eisenstein (1823–1852) independently proved versions of this criterion. Even though Schönemann published his results before Eisenstein, this useful criterion is most often called just Eisenstein's criterion. See Cox [**Cox11**] for the fascinating history.

19.5.16. Let $p(x, y) \in \mathbb{C}[x, y]$ be a polynomial in two variables, and let $R = \mathbb{C}[x]$. Then $p \in R[y]$ and can be written as a polynomial in y and with coefficients from R:

$$p(x, y) = a_n(x)y^n + a_{n-1}(x)y^{n-1} + \cdots + a_1(x)y + a_0(x).$$

Assume
- x does not divide $a_n(x)$,
- for $i = 0, \ldots, n-1$, x divides $a_i(x)$, and
- x^2 does not divide $a_0(x)$.

Let $q(x, y) = (x + 1)p(x, y)$.
- (a) If we write q as an element of $R[y]$, then will the coefficients of the polynomial q also satisfy the conditions above?
- (b) Can q be an irreducible element of $\mathbb{C}[x, y]$?
- (c) In addition to the conditions above, assume p, as an element of $R[y]$, is primitive. Prove that p is an irreducible element of $\mathbb{C}[x, y]$.

19.6. Hilbert Basis Theorem*

As a final example of a result that shows a property of the coefficient ring being reflected onto the polynomial ring, we will prove the Hilbert Basis Theorem. This theorem has a long and distinguished history. It was proved in 1888 by David Hilbert (1862–1943), and it is sometimes said that this somewhat small theorem had a major role in convincing mathematicians that abstract axiomatic mathematics is powerful and important. Before this theorem, many mathematicians worked to find what was called "finite complete systems of invariants for forms". We will not discuss the details, but this amounted to finding a finite set of generators for specific ideals in polynomial rings. Hilbert, surprising the mathematical world, proved in one sweep and in an abstract way that, under mild conditions, all such ideals are finitely generated. This was a revolutionary result and one reason for its newness was that the original proof was not constructive. It provided no guidance to the task of actually finding the generators, and hence it was not clear if this was a satisfactory solution to the problems of the time. By the time the dust settled, axiomatic development of mathematics and non-constructive proofs became a cornerstone of mathematics in the first half of the twentieth century.[3]

Theorem 19.66 (Hilbert Basis Theorem). *Let R be a commutative ring with identity.*

$$R \text{ noetherian} \Rightarrow R[x] \text{ noetherian.}$$

---

[3]One of the leading mathematicians in invariant theory was Paul Gordan (1837–1912). It is often said that upon seeing Hilbert's proof, Gordan disapprovingly said (in translation) "This is not Mathematics, it is Theology!" As is often the case, the actual history of this debate is more complex. Apparently, the first time this quote appeared was after Gordan's death in 1912, and some contemporaries read the sentiment in the quote as high praise for Hilbert (that the proof demonstrates supernatural and divine insight). Gordan actually encouraged Hilbert and used his results in his own work. In fact, Gordan's doctoral student, Emmy Noether (1882–1935)—where the word "noetherian" comes from— was instrumental in the development of abstract algebra as a coherent, abstract, and axiomatic branch of mathematics. See McLarty [**McL12**] for an interesting discussion of the history and the mathematics related to the Hilbert Basis Theorem. For a readable and insightful biography of Hilbert, see Reid [**Rei96**]. For more on Emmy Noether, see Brewer and Smith [**BS81**].

Proof. Let J be an ideal of $R[x]$. We have to show that J is finitely generated. For $n \geq 0$, let

$$I_n = \{\text{leading coefficients of polynomials of degree } n \text{ in } J\}.$$

It is straightforward to show that I_n is an ideal of R. We also claim that $I_n \subseteq I_{n+1}$. To see this, let $a \in I_n$, and let $f \in J$ with $\deg f = n$ and the leading coefficient of f equal to a. Now xf is a polynomial of degree $n+1$ in J with leading coefficient equal to a. Hence $a \in I_{n+1}$.

Thus we have $I_0 \subseteq I_1 \subseteq \cdots \subseteq I_n \subseteq \cdots$. The ring R is assumed to be noetherian and hence this ascending chain of ideals stabilizes and we have, for some integer N,

$$I_0 \subseteq I_1 \subseteq \cdots \subseteq I_N = I_{N+1} = \cdots.$$

For $0 \leq i \leq N$, each I_i is finitely generated (since R is noetherian). So, for $0 \leq i \leq N$, let

$$I_i = \langle a_{i1}, \ldots, a_{ik_i} \rangle.$$

For $0 \leq i \leq N$ and $1 \leq j \leq k_i$, let f_{ij} be a polynomial of degree i in J with leading coefficient a_{ij}, and let

$$S_i = \{f_{ij} \mid j = 1, \ldots, k_i\} \qquad 0 \leq i \leq n.$$

We claim that $J = \langle S_0 \cup S_2 \cup \cdots \cup S_N \rangle$.

Clearly, the right-hand side is a subset of the left-hand side. Assume that the opposite inclusion does not hold. So let $g \in J$ with $\deg(g) = m$ be a polynomial of minimal degree such that $g \notin \langle S_0 \cup S_2 \cup \cdots \cup S_N \rangle$.

Let i_0 be defined by

$$i_0 = \begin{cases} m & \text{if } m < N \\ N & \text{if } m \geq N. \end{cases}$$

Then, by the definition of I_{i_0}, the leading coefficient of g is of the form

$$r_1 a_{i_0 1} + \cdots + r_{k_{i_0}} a_{i_0 k_{i_0}}.$$

This, in turn, is the leading coefficient of $r_1 f_{i_0 1} + \cdots + r_{k_{i_0}} f_{i_0 k_{i_0}}$. Thus the polynomial

$$h = g - x^{m-i_0}(r_1 f_{i_0 1} + \cdots + r_{k_{i_0}} f_{i_0 k_{i_0}})$$

has degree less than m and so is in $\langle S_0 \cup \cdots \cup S_N \rangle$. But now

$$g = h + x^{m-i_0}(r_1 f_{i_0 1} + \cdots + r_{k_{i_0}} f_{i_0 k_{i_0}}) \in \langle S_0 \cup \cdots \cup S_N \rangle.$$

The latter is a contradiction that proves the theorem.   □

Corollary 19.67. *R noetherian $\Rightarrow$ $R[x_1, \ldots, x_n]$ noetherian.*

Problems

19.6.1. **Converse to the Hilbert Basis Theorem.** Let R be a commutative ring with identity. Assume that $R[x]$ is noetherian. Prove that R is noetherian.

19.6.2. Let $R \subseteq S$ be commutative rings with identity, and assume $1_S = 1_R$. (In other words, S is a unitary overring of R.) Assume $S = R[s_1, s_2, \ldots, s_k]$ for $s_1, s_2, \ldots, s_k \in S$. Prove that if R is noetherian, then so is S.

19.6.3. Let $M = \mathbb{Z} - \langle 5 \rangle$, and let $\mathbb{Z}_{(5)} = \mathbb{Z}[M^{-1}]$ be the localization of $\mathbb{Z}$ at M. Is $\mathbb{Z}_{(5)}[x, y]$ noetherian?

19.6.4. Let $I = \langle xw - yz \rangle$ be an ideal of the polynomial ring $\mathbb{C}[x, y, z, w]$, and let $R = \mathbb{C}[x, y, z, w]/I$. If $p \in \mathbb{C}[x, y, z, w]$, we write $\bar{p}$ for $p + I \in R$.
  (a) Prove that R is a noetherian integral domain.
  (b) Is $\bar{x}$ an irreducible element of R?
  (c) Let $r = \bar{x}\,\bar{w} \in R$. Does r have a unique (up to associates and rearranging) factorization into irreducibles?
  (d) Is R an example of a noetherian domain that is not a UFD?
  (e) Is $R/\langle \bar{x} \rangle$ an integral domain?
  (f) Is $\bar{x}$ a prime element of R?

19.7. More Problems and Projects

The Ring of Formal Power Series

Definition 19.68 (The ring of formal power series). Let R be a commutative ring with identity. We define the ring $R[[x]]$ of *formal power series* in the indeterminate x with coefficients from R to be all formal infinite sums

$$\sum_{n=0}^{\infty} a_n x^n = a_0 + a_1 x + a_2 x^2 + \cdots,$$

where $a_0, a_1, \ldots \in R$. Define addition and multiplication similar to those for $R[x]$. In other words, we define

$$\sum_{n=0}^{\infty} a_n x^n + \sum_{n=0}^{\infty} b_n x^n = \sum_{n=0}^{\infty} (a_n + b_n) x^n,$$

$$\sum_{n=0}^{\infty} a_n x^n \times \sum_{n=0}^{\infty} b_n x^n = \sum_{n=0}^{\infty} (a_0 b_n + a_1 b_{n-1} + \cdots + a_n b_0) x^n.$$

Note that the elements of $R[[x]]$ are *formal* power series. This means that the x's are place holders, and we are not usually going to plug in anything for them. For this reason, we do not need to be concerned with issues of convergence. Instead of defining elements of $R[[x]]$ as power series, we could, in fact, have defined them as infinite sequences $(a_0, a_1, \ldots, a_n, \ldots)$. However, the definition of the product of two elements looks more natural when the elements are written as power series.

Problems

19.7.1. Let R be a commutative ring with identity.
    (a) Prove that $R[[x]]$ is a ring with identity.
    (b) Is $1 - x$ a unit in $R[[x]]$? If the answer is yes, give the inverse of $1 - x$ explicitly.

19.7.2. Assume R is an integral domain. Prove that $R[[x]]$ is also an integral domain.

19.7.3. Let R be a commutative ring with identity. Prove that $\sum_{n=0}^{\infty} a_n x^n$ is a unit in $R[[x]]$ if and only if a_0 is a unit in R.

19.7.4. Is $x^2 + 2x - 3$ irreducible in $\mathbb{Z}[x]$? Is it irreducible in $\mathbb{Z}[[x]]$? Why?

19.7.5. Let R be a commutative ring with identity. Assume that a_0 is an irreducible element of R, then is $\sum_{n=0}^{\infty} a_n x^n$ necessarily irreducible in $R[[x]]$?

19.7.6. Let $p(x) = 6 + x + x^2$.
    (a) Is $p(x)$ irreducible in $\mathbb{Z}[x]$? What about in $\mathbb{Q}[x]$?
    (b) In trying to see if $p(x)$ is reducible in $\mathbb{Z}[[x]]$, you write

$$p(x) = (a_0 + a_1 x + a_2 x^2 + \cdots)(b_0 + b_1 x + b_2 x^2 + \cdots),$$

    and you want neither factor to be a unit in $\mathbb{Z}[[x]]$. What are your choices for a_0 and b_0? Make a choice for these two coefficients.
    (c) Given your choice for a_0 and b_0, what are the choices for a_1 and b_1. Make a choice for these two coefficients as well.
    (d) Continue and find possible values for $a_0, \ldots, a_4$ and $b_0, \ldots b_4$.
    (e) Do you think that you can continue and find acceptable values for all the coefficients?
    (f) Is $p(x)$ irreducible in $\mathbb{Z}[[x]]$?

19.7.7. Let R be an integral domain. Prove
    (a) the ideal $\langle x \rangle$ is a prime ideal of $R[[x]]$, and
    (b) the ideal $\langle x \rangle$ is a maximal ideal of $R[[x]]$ if and only if R is a field.

19.7.8. Let F be a field. Prove that $F[[x]]$ is a DVR (discrete valuation ring—see Definition 18.75).

19.7.9. Let R be a noetherian commutative ring with identity. Prove that the ring of formal power series $R[[x]]$ is noetherian.

19.7.10. Let K be a field of characteristic 0, and let $R = K[[x]]$. Define $S = \langle x \rangle$ to be the ideal generated by x in R. Let $T = S/\langle x^2 \rangle$.
    (a) If u and v are elements of T, then what can you say about uv?
    (b) Show that T, in addition to being a ring, is a vector space over $\mathbb{Q}$.
    (c) Show that $(T, +)$ is an abelian divisible group. (See Definition 11.49.)

19.7.11. Let K be a field of characteristic 0, and let $R = K[[x]]$. Define $S = \langle x \rangle$ to be the ideal generated by x in R. Show that S is a ring (without identity) that has no maximal ideals.
You may find the following steps useful:
    Step 1: Let $f \in R$. Show, using Problem *19.7.3*, that if $fx \notin \langle x^2 \rangle$, then f is a unit of R.

By way of contradiction, assume M is a maximal ideal of S.

Step 2: Show, using Step 1, that if, for some $f \in R$, $fx \in M$ and $fx \notin \langle x^2 \rangle$, then $\langle x^2 \rangle \subseteq M$.

Step 3: Show, using Step 2, that $\langle x^2 \rangle \subseteq M$.

Step 4: Show that $M/\langle x^2 \rangle$ is a maximal ideal of $S/\langle x^2 \rangle$.

Step 5: Use Step 4, Problems 19.7.10 and 11.7.11 to arrive at a contradiction.[4]

Partial Fractions. In your calculus class, you have seen and used partial fraction decomposition of a rational function. For example, given a rational function such as $\dfrac{2x - 5}{x^2 - 3x + 2}$ and wanting, for example, to find an anti-derivative for it, a common approach is to find its partial fraction decomposition. You would first factor the denominator into a product of irreducible polynomials of $\mathbb{R}[x]$. In this case, we would have $x^2 - 3x + 2 = (x - 1)(x - 2)$. You would then write

(19.1) $$\frac{2x - 5}{x^2 - 3x + 2} = \frac{A}{x - 1} + \frac{B}{x - 2}$$

and would try to find real numbers A and B that would make the identity true. If successful, then instead of finding the anti-derivative of the original rational function, you would find the much easier anti-derivatives of the two new fractions. For more complicated rational functions, the form of the resulting decomposition would be different. For example, if one of the irreducible factors of the denominator was quadratic, then you would put $Ax+B$ in the numerator of the resulting fraction. If one of the irreducible factors of the denominator was $(x - 1)^3$, then you would have three terms $\frac{A}{x-1}$, $\frac{B}{(x-1)^2}$, and $\frac{C}{(x-1)^3}$ in your decomposition. And so on.

Why do we use these arcane rules for writing the partial fraction decomposition? The main reason is that we want to make sure that an actual partial fraction decomposition exists. The complete rules, as stated in your calculus book, guarantee that a partial fraction decomposition exists. Calculus books usually do not prove this fact but do use it extensively. If we know that an A and a B must exist to make equation (19.1) true, then finding A and B will also be easier. We can just plug specific values for x in order to find A and B. If we did not know that a solution exists, then it is possible that if we had plugged in the "right" values for x, we would have found a contradiction.

In the Problems, as an example, you are asked to prove that, for a particular type of a rational function, the desired partial fraction decomposition exists.

---

[4]Adapted from Malcolmson and Okoh [**MO00**].

19.7.12. Let k be a field, let f_1, f_2, $h \in k[x]$, and let $g = \gcd(f_1, f_2)$. Consider the equation

(19.2) $$u_1 f_1 + u_2 f_2 = h,$$

where the unknowns, u_1 and u_2, are polynomials in $k[x]$.[5]
(a) Show that equation (19.2) is solvable if and only if g divides h.
(b) Assume g divides h. Show that there exists *unique* u_1 and u_2 that satisfy equation (19.2) and such that $u_1 = 0$ or $\deg(u_1) < \deg(f_2) - \deg(g)$. Moreover, if $\deg(h) < \deg(f_1) + \deg(f_2) - \deg(g)$, then $\deg(u_2) < \deg(f_1) - \deg(g)$ or $u_2 = 0$.

19.7.13. Let $k = \mathbb{Q}$, $f_1 = x^3 - 1$, $f_2 = x^2 + x - 2$, and $h = x^3 - 4x + 3$. Find u_1 and u_2 that satisfy the conditions of Problem 19.7.12(b).

19.7.14. Assume that f_1 and f_2 are relatively prime and $h \in k[x]$ with $\deg(h) < \deg(f_1) + \deg(f_2)$. Show that there exists v_1 and $v_2 \in k[x]$ such that

$$\frac{h}{f_1 f_2} = \frac{v_1}{f_1} + \frac{v_2}{f_2},$$

with $v_i = 0$ or $\deg(v_i) < \deg(f_i)$, for $i = 1, 2$.

19.7.15. (a) Find the partial fraction decomposition of

$$\frac{x - 3}{x^3 + 3x^2 + 3x + 2}.$$

(b) In Problem 19.7.14, you proved that, under certain conditions, the partial fraction decomposition of a rational function exists. Is this theoretical fact of any use in finding an actual partial fraction decomposition as in the problem in the previous part?

---

[5]Adapted from Adams and Loustaunau [**AL94**, Problem 1.3.11, page 16].

Chapter 20

Gaussian Integers and (a little) Number Theory*

...where we see glimpses of the close relation between ring theory concepts and algebraic number theory, and, in particular, we bring together a few facts about Gaussian integers and see applications of unique factorization to Diophantine equations.

Here are three questions from number theory and one from ring/field theory to motivate this chapter:

Question 20.1. If, by curiosity, we look at examples of sums of two squares (see Table 20.1), we may wonder what integers occur as a sum of two squares. For example, 10, 13, 16, 17, 18, and 20 are the only integers between 10 and 20 that are the sum of two squares.

Table 20.1. Some integers occur as a sum of two squares; others do not.

$1^2 + 0^2 = 1$	$1^2 + 1^2 = 2$	$2^2 + 0^2 = 4$
$1^2 + 2^2 = 5$	$2^2 + 2^2 = 8$	$3^2 + 0^2 = 9$
$1^2 + 3^2 = 10$	$2^2 + 3^2 = 13$	$4^2 + 0^2 = 16$
$1^2 + 4^2 = 17$	$3^2 + 3^2 = 18$	$2^2 + 4^2 = 20$
$3^2 + 4^2 = 25$	$1^2 + 5^2 = 26$	$2^2 + 5^2 = 29$
$4^2 + 4^2 = 32$	$3^2 + 5^2 = 34$	$6^2 + 0^2 = 36$
$1^2 + 6^2 = 37$	$2^2 + 6^2 = 40$	$4^2 + 5^2 = 41$
$3^2 + 6^2 = 45$	$7^2 + 0^2 = 49$	$5^2 + 5^2 = 50$

To understand the integers that occur as a sum of two squares, we may ask about their prime factors. Certainly, if $a^2 + b^2 = A$, then $(ap)^2 + (bp)^2 = Ap^2$. So every prime p can occur in the factorization of the resulting integers. It will be more interesting to ask about the prime factors of $a^2 + b^2$ when $\gcd(a,b) = 1$.

If we look at Table 20.1, the only prime factors of $a^2 + b^2$ when $\gcd(a, b) = 1$ are $2, 5, 13, 17, 29, 37, 41, \ldots$, which themselves are a sum of two squares. In particular, we do not see $3, 7, 11, 19, 23, 31, 43, 47, \ldots$ as a sum of two squares or as a factor of $a^2 + b^2$ when $\gcd(a, b) = 1$. Other than 2, the list of primes that occur as a sum of two squares seems to precisely consist of those primes that have remainder 1 when divided by 4. Does this pattern continue, and what about its converse? In other words, does the set $\{p \in \mathbb{Z} \mid p = a^2 + b^2, p \text{ a prime}\}$ consist exactly of 2 and primes of the form $4k + 1$?

Question 20.2. Can the square of an integer plus 4 ever be the cube of another integer? In other words, what are the solutions to $x^2 + 4 = y^3$ where x and y are integers?

Question 20.3. In the field $\mathbb{R}$ of real numbers, there is no number whose square is -1. In the field $\mathbb{C}$ of complex numbers, there are two such numbers namely i and $-i$.[1] What about in the field $\mathbb{F}_p = (\mathbb{Z}/p\mathbb{Z}, +, \cdot)$, where p is a prime number? We can rephrase the question: For a prime p, find the number of solutions to $x^2 = -1$ (mod p).

Question 20.4. The ring $\mathbb{R}[i] = \{a + bi \mid a, b \in \mathbb{R}, i^2 = -1\}$ is a field (namely $\mathbb{C}$). For a prime p, let $\mathbb{F}_p = (\mathbb{Z}/p\mathbb{Z}, +, \cdot)$ denote the field with p elements. For which prime p, is $\mathbb{F}_p[i] = \{a + bi \mid a, b \in \mathbb{F}_p, i^2 = -1\}$ a field?

In the opening section of the ring theory part of the text (Section 15.1), we argued, through a few examples, that attempts to solve Diophantine equations could lead to the study of factorization in rings. In fact, the examples showed that if we know that $\mathbb{Z}[\sqrt{d}]$ is a unique factorization domain, we can use standard number theory arguments to solve certain Diophantine equations. This approach leads to two projects. The first one is to ascertain for which square-free integers d is the ring $\mathbb{Z}[\sqrt{d}]$ (and if $d = 1$ (mod 4), the larger $\mathbb{Z}[\frac{1+\sqrt{d}}{2}]$) a UFD. The second is to figure out what to do if $\mathbb{Z}[\sqrt{d}]$ is not a UFD. These projects go back to the nineteenth century and the attempts to resolve Fermat's Last Theorem and to generalize the so-called *quadratic reciprocity theorem*. They form the genesis of algebraic number theory. Much progress has been made on the first project, and for the second project the crucial insight has been to focus on "factorization of ideals" rather than factorization of numbers. Discussing either of these projects in detail takes us off track. In this optional chapter, we will just see some examples of using our ring theory knowledge to answer number theoretic questions, and examples of using unique factorization in rings of numbers larger than the integers to solve Diophantine equations. The first section of this chapter focuses on $\mathbb{Z}[i]$ the ring of Gaussian integers and, in particular, answers Questions 20.1, 20.3, and 20.4. The second section goes back to Diophantine equations and answers Question 20.2. Throughout, you will see the usefulness of the language of ring theory in reformulating number theoretic questions.[2]

---

[1] In the ring of quaternions over $\mathbb{R}$, there are an infinite number of elements whose square is -1. See Problem 15.2.26.

[2] For a very readable account of algebraic number theory, at the undergraduate level, see Stewart and Tall [**ST79**].

20.1. Gaussian Integers

We have already defined the Gaussian integers $\mathbb{Z}[i] = \{a + bi \mid a, b \in \mathbb{Z}, i^2 = -1\}$ as a subring of the complex numbers (page 320). We also have the norm map $N \colon \mathbb{Z}[i] \to \mathbb{Z}$ defined by $N(a + bi) = (a + bi)(a - bi) = a^2 + b^2$. We know that $N(\alpha\beta) = N(\alpha)N(\beta)$ for $\alpha, \beta \in \mathbb{Z}[i]$, and that $N(\alpha) = 1$ if and only if α is a unit in $\mathbb{Z}[i]$ (Definition 18.23 and Theorem 18.24). The only way $a^2 + b^2 = 1$ for ordinary integers a and b is if one of them is zero and the other is ± 1. Hence, ± 1 and $\pm i$ are the only units of $\mathbb{Z}[i]$. This means that every non-zero element of $\mathbb{Z}[i]$ has four associates. For example, $-i + 3$, $i - 3$, $3i + 1$, and $-3i - 1$ are all associates.

What are the primes of $\mathbb{Z}[i]$? Before answering this question, we note that $\mathbb{Z}[i]$ is a Euclidean domain (ED) and hence it is a PID and a UFD (Corollary 18.51), and being a prime in $\mathbb{Z}[i]$ is equivalent to being irreducible in $\mathbb{Z}[i]$ (Theorem 18.40). The fact that $\mathbb{Z}[i]$ is an ED was Problem 18.4.7 but we give the proof here for completeness.

Theorem 20.5. $\mathbb{Z}[i]$ *is a Euclidean domain. In particular, $\mathbb{Z}[i]$ is a principal ideal domain as well as a unique factorization domain.*

Proof. We want to show that the norm map N is a degree function and $\mathbb{Z}[i]$ is an ED with N as its degree function.[3]

Before we proceed, we extend the function N to all of the complex plane. In other words, if a and b are real numbers, then we can define $N(a + bi) = a^2 + b^2$. We have $N \colon \mathbb{C} \to \mathbb{R}$, and our original norm function is the restriction of this function to $\mathbb{Z}[i]$ (and when restricted to $\mathbb{Z}[i]$, the codomain of N is $\mathbb{Z}$). The advantage of this new and improved N is that we can calculate Euclidean distances with it. If z_1 and z_2 are two complex numbers, then the (Euclidean) distance between them on the complex plane is given by $\sqrt{N(z_1 - z_2)}$. (Why?) Also note that, with the same calculation as for the norms, we have $N(z_1 z_2) = N(z_1)N(z_2)$ for all $z_1, z_2 \in \mathbb{C}$.

To show that N is a degree function, first note that, for $x, y \in \mathbb{Z}[i]$, $N(xy) = N(x)N(y) \geq N(x)$ since $N(y) \geq 1$ as long as $y \neq 0$. Now let x and y be arbitrary elements of $\mathbb{Z}[i]$. We have to find the appropriate quotient and remainder.

In the complex plane the elements of $\mathbb{Z}[i]$ are exactly the points with integer coefficients, and these divide up the complex plane into squares of side 1 (see Figure 20.1). Consider the complex number x/y. It lies somewhere in the complex plane.

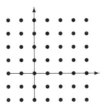

Figure 20.1. Elements of $\mathbb{Z}[i]$ divide up the complex plane into squares of side 1.

---

[3]If a ring $\mathbb{Z}[\sqrt{d}]$ is an ED with the absolute value of the norm map as its degree function (we do not need the absolute value if d is negative), then it is said to be *normed Euclidean*. There do exist rings $\mathbb{Z}[\sqrt{d}]$—for example, $\mathbb{Z}[\sqrt{14}]$—that are ED but not normed Euclidean. (See Harper [**Har04**].)

Let q be the closest (in Euclidean distance) element of $\mathbb{Z}[i]$ to x/y. Since the diagonals of the squares on the complex plane created by the elements of $\mathbb{Z}[i]$ have length $\sqrt{2}$, the distance of x/y from q is at most $\frac{\sqrt{2}}{2}$. Hence we have

$$\sqrt{N(x/y - q)} \le \frac{\sqrt{2}}{2} < 1 \quad \Rightarrow \quad N(x/y - q) < 1.$$

Let $r = x - yq$, then we have $x = yq + r$, and, more importantly, we have

$$N(r) = N(x - yq) = N(y(x/y - q)) = N(y)N(x/y - q) < N(y).$$

Hence we have a division algorithm in $\mathbb{Z}[i]$ as required, and $\mathbb{Z}[i]$ is a Euclidean domain. $\qquad\qquad\qquad\qquad\qquad\qquad\qquad\qquad\qquad\qquad\qquad\qquad\qquad\qquad\square$

We now know that primes and irreducibles are the same in $\mathbb{Z}[i]$, and that $\mathbb{Z}[i]$ is both a PID and a UFD. Our first question is whether the ordinary primes of $\mathbb{Z}$ continue to be primes in $\mathbb{Z}[i]$.

Example 20.6. We have $(1+i)(1-i) = 1^2 + 1^2 = 2$, $(2+i)(2-i) = 2^2 + 1^2 = 5$, and $(2+3i)(2-3i) = 2^2 + 3^2 = 13$. Since none of the factors are units, we conclude that 2, 5, and 13 are *not* primes in $\mathbb{Z}[i]$.

Looking at the factorizations of 2, 5, and 13, we suspect a close relationship with Question 20.1. One direction is very clear. If an ordinary prime p can be written as a sum of two squares, then it cannot be a prime in $\mathbb{Z}[i]$. For if $p = a^2 + b^2$, then $p = (a+bi)(a-bi)$. What about the converse? If an ordinary prime is not a prime in $\mathbb{Z}[i]$, then will it by necessity be a sum of two squares? If the answer is yes, then it would also mean that the only way to factor an ordinary prime in $\mathbb{Z}[i]$ is to possibly write it as a product of a Gaussian integer and its conjugate.

Example 20.7. Consider the ordinary prime 3. Can 3 be factored in $\mathbb{Z}[i]$? Assume $3 = (a+bi)(c+di)$ and take the norm of both sides. We get $9 = (a^2 + b^2)(c^2 + d^2)$. The left hand side is giving a factorization in $\mathbb{Z}$ of the ordinary integer 9. But 9 factors only in two ways: 9×1 or 3×3. If $a^2 + b^2$ (or $c^2 + d^2$) is equal to 1, then $a + bi$ (or $c + di$) will be a unit. On the other hand $a^2 + b^2$ cannot be equal to 3. Hence, we conclude that 3 is irreducible (and prime) in $\mathbb{Z}[i]$.

Again the connection to Question 20.1 is clear. The reason 3 could not be factored is that 3 could not be written as a sum of two squares.

Lemma 20.8. *Let $p \in \mathbb{Z}$ be a prime. Then p is a prime in $\mathbb{Z}[i]$ if and only if, in $\mathbb{Z}$, p cannot be written as a sum of two squares.*

Proof. Assume that p is a prime in $\mathbb{Z}[i]$. To arrive at a contradiction, assume that $p = a^2 + b^2$ with $a, b \in \mathbb{Z}$. Then $p = (a+bi)(a-bi)$, and neither $a + bi$ nor $a - bi$ is a unit in $\mathbb{Z}[i]$ since both of their norms equal p and not 1. Hence, p cannot be irreducible or prime. The contradiction proves one direction.

Now assume that p cannot be written as a sum of two squares in $\mathbb{Z}$. Again, by way of contradiction, assume that p is not a prime in $\mathbb{Z}[i]$. Then $p = (a+bi)(c+di)$ in $\mathbb{Z}[i]$ where neither $a + bi$ nor $c + di$ is a unit. Taking norms of both sides, we get $p^2 = (a^2 + b^2)(c^2 + d^2)$. Since p is a prime, factorization is unique in $\mathbb{Z}$, and neither $a^2 + b^2$ nor $c^2 + d^2$ is equal to one, we conclude that $a^2 + b^2 = p = c^2 + d^2$. The contradiction completes the proof. $\qquad\qquad\qquad\qquad\qquad\qquad\qquad\qquad\qquad\qquad\qquad\square$

A lot more is true. We found that ordinary primes that cannot be written as a sum of two squares remain prime in $\mathbb{Z}[i]$ and the rest of the ordinary primes factor as $\pi\overline{\pi}$ where $\pi \in \mathbb{Z}[i]$ and $\overline{\pi}$ is the complex conjugate of π. On the one hand, these factors π and $\overline{\pi}$ are the only other primes of $\mathbb{Z}[i]$ and, on the other hand, as we had suspected, we can completely characterize the ordinary primes that can (and cannot) be written as a sum of two squares.

Theorem 20.9. *Let $p \in \mathbb{Z}$ be a prime. Then the following are equivalent:*

(a) *p is not a prime in $\mathbb{Z}[i]$,*

(b) *$\exists\, a, b \in \mathbb{Z}$ with $p = a^2 + b^2$,*

(c) *$p = 2$ or $p = 1 \pmod 4$,*

(d) *$\mathbb{Z}/p\mathbb{Z}$ has an element whose square is -1,*

(e) *$x^2 + 1$ is not irreducible in $\mathbb{F}_p[x]$,*

(f) *$\mathbb{F}_p[i] = \{a + bi \mid a, b \in \mathbb{F}_p, i^2 = -1\}$ is not a field.*

Proof. (a) $\Leftrightarrow$ (b) This was Lemma 20.8.

(b) $\Rightarrow$ (c) Assuming p is odd, we have to show that $p = 1 \pmod 4$. Since $p = a^2 + b^2$ is odd, without loss of generality, we can assume a is odd and b is even. The integer a has remainder 1 or 3 when divided by 4, and so a^2 will have remainder 1 when divided by 4. On the other hand, b^2 will be divisible by 4. It follows that the remainder of p when divided by 4 is 1.

(c) $\Rightarrow$ (d) In $\mathbb{Z}/2\mathbb{Z}$, $1^2 = 1 = -1$, and so assume that $p - 1$ is divisible by 4. Let x be the product, in $\mathbb{Z}/p\mathbb{Z}$ of 1, 2, $\ldots$, $\dfrac{p-1}{2}$. To complete the proof, we show that $x^2 = -1$ in $\mathbb{Z}/p\mathbb{Z}$. Note that, in $\mathbb{Z}/p\mathbb{Z}$, we have $-1 = p - 1$, $-2 = p - 2$, $\ldots$, $-\dfrac{p-1}{2} = \dfrac{p+1}{2}$. Hence, in $\mathbb{Z}/p\mathbb{Z}$, $(p-1)! = (-1)^{\frac{p-1}{2}} x^2 = x^2$ since $(p-1)/2$ is even. The proof will be complete when we show that, in $\mathbb{Z}/p\mathbb{Z}$, $(p-1)! = -1$. This is actually Wilson's theorem and you were asked to prove it in Problem 1.3.15(c). For completeness, here is the argument: If k is an element of $((\mathbb{Z}/p\mathbb{Z})^{\times}, \cdot)$ that is its own inverse, then $k^2 - 1$ will be a multiple of p. This means that the prime p must divide $k - 1$ or $k + 1$. Since $0 \leq k \leq p - 1$, these two cases lead to $k = 1$ and $k = p - 1$, respectively. Now, in the product $1 \times 2 \times \cdots \times (p-1)$ every element cancels with its inverse except 1 and $p - 1$ which are their own inverses. As a result $(p-1)! = p - 1 = -1$ in $\mathbb{Z}/p\mathbb{Z}$. The proof is complete, since we found an element in $\mathbb{Z}/p\mathbb{Z}$ whose square is -1.

(d) $\Rightarrow$ (a) Assume that $a \in \mathbb{Z}/p\mathbb{Z}$ and $a^2 = -1$ in $\mathbb{Z}/p\mathbb{Z}$. This means that $a \neq 0$ and that p divides $a^2 + 1 = (a + i)(a - i)$. To arrive at a contradiction, assume that p is prime in $\mathbb{Z}[i]$. Then p divides $a + i$ or $a - i$. This means that there exists ordinary integers u and v such that $a + i = p(u + iv)$ or $a - i = p(u + iv)$. Multiplying these out and equating the real and imaginary parts, we get that either $pv = 1$ or $pv = -1$. In either case, this would mean that p is a unit which is a contradiction.

(d) $\Leftrightarrow$ (e) Note that $\mathbb{F}_p = (\mathbb{Z}/p\mathbb{Z}, +, \cdot)$ is a field. Hence, $x^2 + 1$ is reducible in $\mathbb{F}_p[x]$ if and only if it has a root in $\mathbb{F}_p$ (Corollary 19.31). In turn, this polynomial has a root in $\mathbb{F}_p$ if and only if there is an element of $\mathbb{F}_p$ whose square is -1.

(e) $\Leftrightarrow$ (f) The evaluation map $\theta : \mathbb{F}_p[x] \to \mathbb{F}_p[i]$ given by $\theta(p(x)) = p(i)$ is a ring homomorphism (Lemma 19.9) and its kernel is $\langle x^2 + 1 \rangle$. Hence,

$$\mathbb{F}_p[i] \cong \mathbb{F}_p[x]/\langle x^2 + 1 \rangle.$$

Now since $\mathbb{F}_p[x]$ is an ED and hence a PID, the ring $\mathbb{F}_p[i]$ is a field if and only if $x^2 + 1$ is irreducible in $\mathbb{F}_p[x]$ (see Corollary 18.13). $\square$

Remark 20.10. In the ring of ordinary integers, there is only one even prime, and the rest of the primes are odd. Theorem 20.9 says that the odd primes naturally fall into two types. The primes of the form $4k + 1$ are exactly those odd primes that can be written as a sum of two squares, and for which we can find an integer whose square has remainder -1 when divided by such a prime. In contrast, the primes p of the form $4k + 3$ are exactly those integer primes that continue to be a prime in the ring of Gaussian integers, and for these and only these primes $x^2 + 1$ remains irreducible in $\mathbb{F}_p[x]$ and $\mathbb{F}_p[i]$ is a field.

Remark 20.11. In Problems *16.2.7* and 16.2.8, you were asked to do special cases of Proposition 20.12. You may find it useful to do one of those problems before reading the proof of the next proposition.[4]

Proposition 20.12. *Let a and b be relatively prime ordinary integers, and define $n = a^2 + b^2$. Then the ring $\mathbb{Z}[i]/\langle a + bi \rangle$ is isomorphic to the ring $\mathbb{Z}/n\mathbb{Z}$.*

Proof. If either a or b are zero (and hence the other is 1), then the result is clear. So assume that neither a nor b are zero. Let $I = \langle a + bi \rangle$, $R = \mathbb{Z}[i]/I$, and denote $c + di + I$, a typical element of R, by $\overline{c + di}$.

Now define $\phi \colon \mathbb{Z} \to R$ by $\phi(m) = \overline{m} + I$. The proof will be complete when we show that ϕ is an onto ring homomorphism and $\ker(\phi) = n\mathbb{Z}$.

It is clear that ϕ is a ring homomorphism. To show that ϕ is onto, let $\overline{c + di}$ be an arbitrary element of R. We need to show that $\overline{c + di} = \overline{m}$ for some $m \in \mathbb{Z}$. Note that $\gcd(b, n) = 1$, and so b has a multiplicative inverse in $\mathbb{Z}/n\mathbb{Z}$. In other words, there exists $e \in \mathbb{Z}/n\mathbb{Z}$ with $be = 1$ (in $\mathbb{Z}/n\mathbb{Z}$). We have $n = (a - bi)(a + bi) \in I$ and so $\overline{eb} = \overline{1}$. Now, $\overline{a + bi} = \overline{0}$, and hence $\overline{i} = \overline{ebi} = \overline{-ae}$. Hence $\overline{c + di} = \overline{c - ade}$, and ϕ is onto.

To find the kernel of ϕ, first note that $n = (a + bi)(a - bi) \in I$ and so $n\mathbb{Z} \subseteq \ker(\phi)$. Now let $m \in \ker(\phi)$. The proof will be complete when we show that m is a multiple of n. From $m \in \ker(\phi)$, we conclude that $m \in I = \langle a + bi \rangle$. So $m = (a + bi)(c + di)$ for some $c, d \in \mathbb{Z}$. Multiplying out and equating the real and imaginary parts gives $m = ac - bd$ and $bc + ad = 0$. From $bc = -ad$ and the fact that $\gcd(a, b) = 1$, we get that $a \mid c$ and $b \mid d$. So $c = ak_1$ and $d = bk_2$. Moreover, $bc = -ad$ implies that $abk_1 = -abk_2$ and so $k_2 = -k_1$. Hence, $m = ac - bd = a^2 k_1 + b^2 k_1 = k_1 n$. $\square$

For the record, in the next corollary, we will explicitly list the primes of $\mathbb{Z}[i]$. If you start with a prime p of the integers $\mathbb{Z}$, then one of two things happen: either p stays a prime in $\mathbb{Z}[i]$, or it *splits* into the product of two primes of $\mathbb{Z}[i]$.

---

[4]For an expository account of the factor rings of Gaussian integers mixed with a little history, see Dresden and Dymàček [**DD05**].

Corollary 20.13. *Let $p \in \mathbb{Z}$ be a prime.*

(a) *If $p = 3 \pmod 4$, then p is also a prime in $\mathbb{Z}[i]$.*

(b) *If $p = 1 \pmod 4$ or $p = 2$, then $p = (a + bi)(a - bi)$ in $\mathbb{Z}[i]$ and both $a + bi$ and $a - bi$ are primes in $\mathbb{Z}[i]$.*

Conversely, if x is a prime in $\mathbb{Z}[i]$, then x is an associate of one of the primes of $\mathbb{Z}[i]$ listed in (a) or (b). Hence, in the ring $\mathbb{Z}[i]$, there are exactly two types of primes.

Proof. We have to prove that the listed elements are indeed primes in $\mathbb{Z}[i]$ and that there are no other primes in $\mathbb{Z}[i]$. By Theorem 20.9, a prime integer that is 3 (mod 4) remains a prime in $\mathbb{Z}[i]$. By the same theorem, 2 or a prime integer that is 1 (mod 4) is *not* a prime in $\mathbb{Z}[i]$ but it is a sum of two squares in $\mathbb{Z}$. Hence, in this case, $p = a^2 + b^2 = (a + bi)(a - bi)$. Now $N(a + bi) = N(a - bi) = a^2 + b^2 = p$ which is irreducible in $\mathbb{Z}$. Hence, by Theorem 18.24(c), $a + bi$ and $a - bi$ are both irreducibles—and hence primes—in $\mathbb{Z}[i]$.

Now, for the converse, assume that $x = a + bi$ is prime—and hence irreducible—in $\mathbb{Z}[i]$. Let $n = a^2 + b^2$. If $a = 0$ or $b = 0$, then x or one of its associates is a prime in $\mathbb{Z}$. It follows from Theorem 20.9 that x or one of its associates is 3 (mod 4). If neither a nor b are zero, then, in $\mathbb{Z}$, $\gcd(a, b) = 1$ since otherwise x would not be irreducible. Now, x irreducible implies—by Corollary 18.13—that $\mathbb{Z}[i]/\langle x \rangle$ is a field. On the other hand, by Proposition 20.12, we have $\mathbb{Z}[i]/\langle x \rangle \cong \mathbb{Z}/n\mathbb{Z}$. We conclude that $\mathbb{Z}/n\mathbb{Z}$ is a field and so $n = a^2 + b^2$ is a prime in $\mathbb{Z}$. This prime can evidently be written as a sum of two squares and so, by Theorem 20.9, n is either 2 or 1 (mod 4). Since $n = (a + bi)(a - bi)$, the proof is now complete. $\quad\square$

Sum of Two Squares. We are finally ready to answer the opening question of this chapter (Question 20.1). Which positive integers are the sum of two squares? Already, we know the answer for primes. The prime 2 is certainly a sum of two squares, and Theorem 20.9 tells us that an odd prime is a sum of two squares if and only if it is 1 (mod 4). We bootstrap this result in two steps:

Lemma 20.14. *Let n be a positive integer. If $n = a^2 + b^2$ with a and b relatively prime integers, then every odd prime that divides n is of the form 1 (mod 4).*

Proof. Let p be a prime number with $p \mid n$. Since a and b are relatively prime, p cannot divide both of them. Without loss of generality, assume that $\gcd(p, a) = 1$. Let $a = dp + r$ with d and r integers and $1 \le r < p$. Now, consider r as an element of the multiplicative group $(\mathbb{Z}/p\mathbb{Z}^\times, \cdot)$, and let $\bar{r}$ be its inverse. In other words, $r\bar{r} \equiv 1 \pmod p$. It follows that $a\bar{r} \equiv 1 \pmod p$. Let $x = \bar{r}b$. Then $x^2 = \bar{r}^2 n - (\bar{r}a)^2 \equiv -1 \pmod p$. Hence, in $\mathbb{Z}/p\mathbb{Z}$, there exists an element whose square is -1. By Theorem 20.9, p is either 2 or of the form 1 (mod 4), and the proof is complete. $\quad\square$

Finally, we consider numbers of the form $a^2 + b^2$ when a and b are not necessarily relatively prime.

Theorem 20.15. *Let n be a positive integer. Then n is a sum of two squares if and only if in the prime factorization of n, every prime of the form 3 (mod 4) appears an even number of times.*

Proof. ($\Leftarrow$) Assume that in the prime factorization of n, every prime of the form 3 (mod 4) appears an even number of times. We want to show that n is a sum of two squares. By the hypothesis, $n = r^2 s$ where r and s are integers and every prime in the prime factorization of s is either 2 or of the form 1 (mod 4). By Theorem 20.9, every prime factor of s is the sum of two squares. Using the identity

$$(a^2 + b^2)(c^2 + d^2) = (ac + bd)^2 + (ad - bc)^2,$$

we see that s is the sum of two squares. Finally, $r^2(\alpha^2 + \beta^2) = (r\alpha)^2 + (r\beta)^2$ and so n is the sum of two squares.

($\Rightarrow$) Let $n = a^2 + b^2$ where a and b are non-negative integers. Let $d = \gcd(a, b)$, and write $a = d\alpha$ and $b = d\beta$. Then $n = d^2(\alpha^2 + \beta^2)$, and $\gcd(\alpha, \beta) = 1$. By Lemma 20.14, the only prime divisors of $\alpha^2 + \beta^2$ can be 2 and primes of the form 1 (mod 4). Hence, if p is a prime of the form 3 (mod 4) and $p \mid n$, then $p \mid d$. If p^e is the highest power of p that divides d, then p^{2e} is the power of p that divides n. The proof is now complete. $\square$

Example 20.16. The theorem says that $1470 = 2 \times 3 \times 5 \times 7^2$ is not a sum of squares, since 3 appears an odd number of times in the prime factorization. However, $4410 = 2 \times 3^2 \times 5 \times 7^2$ is a sum of squares. Indeed, $4410 = 21^2 + 63^2$.

Problems

20.1.1. Write each of $3i$, $-5i$, $7i + 1$, and $7i + 2$ as a product of irreducibles in $\mathbb{Z}[i]$.

20.1.2. In Problem 19.3.4 you were asked to construct a field with nine elements. Use the results of this section to give a field with nine elements and another with 49 elements.

20.1.3. **Pythagorean triples from Gaussian integers.**[5] Three positive integers u, v, and w form a Pythagorean triple if $w^2 = u^2 + v^2$. Consider the Gaussian integer $z = 3 + 2i$, and let $N: \mathbb{Z}[i] \to \mathbb{Z}^{\geq 0}$ be the usual norm map. Now $N(z) = 13$, and $z^2 = 5 + 12i$. So

$$13^2 = N(z)^2 = N(z^2) = 5^2 + 12^2,$$

and we have found the Pythagorean triple 5, 12, and 13. Using this observation, start with the Gaussian integer $a + bi$, and find the corresponding Pythagorean triple.

20.1.4. Let r be a prime element of $\mathbb{Z}[i]$, and let $N: \mathbb{Z}[i] \to \mathbb{Z}^{\geq 0}$ be the usual norm map. Show that there exists a prime $p \in \mathbb{Z}$ such that $N(r) = p$ or $N(r) = p^2$.

20.1.5. Let $p \in \mathbb{Z}$ be a prime. Prove that the following are equivalent:
   (a) The prime p splits in $\mathbb{Z}[i]$ into the product of two primes (of $\mathbb{Z}[i]$).
   (b) $\mathbb{Z}[i]/\langle p \rangle$ is not an integral domain.
   (c) $\mathbb{F}_p[x]/\langle x^2 + 1 \rangle$ is not an integral domain.

---

[5] See Cuoco [**Cuo00**] for this and other elementary uses of norms.

20.1.6. Let d be a square-free integer, and let $p \in \mathbb{Z}$ be a prime number. Show that p is irreducible in $\mathbb{Z}[\sqrt{d}]$ if and only if $p \neq a^2 - db^2$ for any $a, b \in \mathbb{Z}$.

20.1.7. Let d be a square-free integer, and assume $\mathbb{Z}[\sqrt{d}]$ is a UFD. Let π be a prime of $\mathbb{Z}[\sqrt{d}]$. Prove that there exists a unique prime p of $\mathbb{Z}$ with $\pi \mid p$ (in $\mathbb{Z}[\sqrt{d}]$).

20.1.8. Let d be a square-free integer, and assume $\mathbb{Z}[\sqrt{d}]$ is a UFD. Let p be a prime of $\mathbb{Z}$. Prove that in $\mathbb{Z}[\sqrt{d}]$, either p is prime or a product of two primes.

20.1.9. Let n be a positive integer. How are the two rings $\mathbb{Z}[i]/\langle n \rangle$ and $(\mathbb{Z}/n\mathbb{Z})[i]$ related? Are they isomorphic? Prove your assertion.

20.1.10. Let a and b be integers, and let $d = \gcd(a, b) > 0$. What is the characteristic of the ring $\mathbb{Z}[i]/\langle a + bi \rangle$? Make a conjecture.

20.1.11. Strengthen Lemma 20.14, and, akin to Theorem 20.15, give a characterization of positive integers that can be written as a sum of two relatively prime integers.

20.1.12. For integers x and y, observe that $x + y\sqrt{2}$ is a unit of $\mathbb{Z}[\sqrt{2}]$ if and only if $x^2 - 2y^2 = \pm 1$. Is $1 + \sqrt{2}$ a unit of $\mathbb{Z}[\sqrt{2}]$? What about $(1 + \sqrt{2})^{47}$? Show that $\mathbb{Z}[\sqrt{2}]$ has an infinite number of units and that $x^2 - 2y^2 = \pm 1$ has an infinite number of solutions.

20.1.13. Prove that $\mathbb{Z}[\sqrt{2}]$ is a Euclidean domain with the absolute value of the norm as its degree function.

20.1.14. Consider the ring $\mathbb{Z}[2i] = \{a + 2bi \mid a, b \in \mathbb{Z}\}$. What are the units of $\mathbb{Z}[2i]$? Is $\mathbb{Z}[2i]$ a UFD?

20.1.15. Look at examples, and consider remainders when dividing by 20, in order to make a conjecture characterizing primes larger than 5 that can be written as $x^2 + 5y^2$ for integers x and y.

20.1.16. **Eisenstein integers.** Let $\omega = e^{\frac{2\pi i}{3}} = \frac{-1 + i\sqrt{3}}{2}$. Then $\omega^3 = 1$ and ω is called a *third root of unity*.[6] Recall from Definition 19.10, that $\mathbb{Z}[\omega] = \{g(\omega) \mid g(x) \in \mathbb{Z}[x]\}$. The elements of $\mathbb{Z}[\omega]$ are called *Eisenstein integers*.
  (a) What is ω^2? What is $1 + \omega + \omega^2$?
  (b) Show that $\mathbb{Z}[\omega] = \{a + b\omega \mid a, b \in \mathbb{Z}\}$.
  (c) Define $N \colon \mathbb{Z}[\omega] \to \mathbb{Z}$ by $N(a + b\omega) = (a + b\omega)(a + b\omega^2)$. Is this map well defined? In particular, is $(a + b\omega)(a + b\omega^2) \in \mathbb{Z}$?
  (d) Show that $N(\alpha\beta) = N(\alpha)N(\beta)$ for $\alpha, \beta \in \mathbb{Z}[\omega]$.
  (e) What are the units of $\mathbb{Z}[\omega]$?
  (f) Prove that $\mathbb{Z}[\omega]$ is a Euclidean domain with N as its degree function.

---

[6]See Section 27.2 for more on roots of unity.

20.2. Unique Factorization and Diophantine Equations

We are now ready to answer Question 20.2 and to rigorously discuss the examples of Diophantine equations from Section 15.1. We will not be attempting to systematically discuss the solution of Diophantine equations—a rich subfield of number theory—but rather provide a couple of examples to convince you that the language of ring theory and unique factorization of rings larger than the integers can be quite helpful in solving Diophantine equations. We begin with a lemma:

Lemma 20.17. *Let x and y be relatively prime integers in $\mathbb{Z}$.*

(a) *If x and y are of opposite parity (i.e., one is odd and the other even), then $x + iy$ and $x - iy$ are relatively prime in $\mathbb{Z}[i]$.*

(b) *If x and y are both odd, then $1 + i$ (or any of its associates) is the greatest common divisor of $x + iy$ and $x - iy$ in $\mathbb{Z}[i]$.*

Proof. Recall that the ring $\mathbb{Z}[i]$ is a unique factorization domain (Theorem 20.5), and we know all of its primes (Corollary 20.13). Let $\delta = a + bi$ be a prime in $\mathbb{Z}[i]$ that divides (in $\mathbb{Z}[i]$) both $x + iy$ and its conjugate $x - iy$. This prime δ must divide both $(x + iy) + (x - iy) = 2x$ and $(x + iy) - (x - iy) = 2iy$. Since i is a unit, we conclude that δ divides both $2x$ and $2y$. If δ does not divide 2, since δ is a prime, then δ must divide both x and y. Since x and y are relatively prime in $\mathbb{Z}$, we must have $\delta \notin \mathbb{Z}$, and so $\delta \neq \overline{\delta} = a - bi$. Now, if $x = \delta\gamma$ in $\mathbb{Z}[i]$, then conjugating both sides and remembering that $x \in \mathbb{Z}$, we have $x = \overline{\delta}\overline{\gamma}$ and so $\overline{\delta}$ also divides x. Hence, the real number $N(\delta) = \delta\overline{\delta}$ divides x. The same argument shows that $N(\delta)$ also divides y but $\gcd(x, y) = 1$, which gives a contradiction. The only other possibility is that $\delta \mid 2$ in $\mathbb{Z}[i]$, and so it is an associate of $1 + i$. Thus $\frac{x+iy}{1+i} = \frac{x+y}{2} - \frac{x-y}{2}i$ is an element of $\mathbb{Z}[i]$. This happens if and only if x and y have the same parity. We conclude that if x and y have opposite parity, then the greatest common divisor of $x + iy$ and $x - iy$ (in $\mathbb{Z}[i]$) is 1, and if both x and y are odd, then $1 + i$ (or any of its associates) is the only prime in $\mathbb{Z}[i]$ that divides both $x + iy$ and $x - iy$. In the latter case, since $\frac{x+y}{2}$ and $\frac{x-y}{2}$ are of opposite parity, we can also conclude that no higher power of $1 + i$ divides both $x + iy$ and $x - iy$, which completes the proof. $\square$

We can now solve one of the problems highlighted in the Preface.

Proposition 20.18. *Consider the Diophantine equation*

$$y^3 = x^2 + 4.$$

The only integer solutions to this equation are given by $(x = \pm 2, y = 2)$, and $(x = \pm 11, y = 5)$.

Proof. We operate in the ring $\mathbb{Z}[i]$ and write

$$y^3 = (x + 2i)(x - 2i).$$

Notice that if x is even, then we can further factor the right hand side. Hence, we consider the case x odd first. By Lemma 20.17, if x is odd, then the greatest common divisor of $x + 2i$ and $x - 2i$ in $\mathbb{Z}[i]$ is 1. Assuming x is odd, going back to $y^3 = (x + 2i)(x - 2i)$, and using unique factorization in $\mathbb{Z}[i]$, we get that if r is a prime divisor (in $\mathbb{Z}[i]$) of y, then r^3 divides either $x + 2i$ or $x - 2i$ (but not both).

The units in $\mathbb{Z}[i]$ all are cubes also. Hence, both $x + 2i$ and $x - 2i$ are perfect cubes (in $\mathbb{Z}[i]$). Write $x + 2i = (a + bi)^3$, expand, and equate real and imaginary parts to get

$$\begin{cases} a^3 - 3ab^2 & = x, \\ 3a^2b - b^3 & = 2. \end{cases}$$

The second integer equation is $b(3a^2 - b^2) = 2$. Hence either $b = \pm 1$ and $3a^2 - 1 = \pm 2$, or $b = \pm 2$ and $3a^2 - 4 = \pm 1$. Thus the only possibilities are $(a = \pm 1, b = 1)$ and $(a = \pm 1, b = -2)$. Then, from the first equation we get $x = \pm 2$ or $x = \pm 11$. Since x was assumed to be odd, we get that the only such possibility is $x = \pm 11$ and $y = 5$.

If x is even, write $x = 2z$, and note that y must be even also. Write $y = 2w$ and get $2w^3 = z^2 + 1$. We conclude that z must be odd, and we write $(1 + i)(1 - i)w^3 = (z + i)(z - i)$. By Lemma 20.17, we know that the greatest common divisor in $\mathbb{Z}[i]$ of $z + i$ and $z - i$ is $1 + i$ (or $1 - i = i(1 + i)$). Since $\mathbb{Z}[i]$ is a unique factorization domain and $(1 + i)(1 - i)w^3 = (z + i)(z - i)$, we see that in their factorization into primes each of $z + i$ and $z - i$ picks up one of $1 + i$ or $1 - i$ (these are associates and so it does not matter which) and then each picks up the cubes of some of the prime divisors of w. Hence,

$$z + i = (1 + i)(a + bi)^3.$$

Multiply out and equate real and imaginary parts to get

$$\begin{cases} a^3 - 3a^2b - 3ab^2 + b^3 & = z, \\ a^3 + 3a^2b - 3ab^2 - b^3 & = 1. \end{cases}$$

The second equation factors as $(a - b)(a^2 + 4ab + b^2) = 1$. These are just two integers whose product is 1, and so both factors must be ± 1. So $a = b \pm 1$ and $a^2 + 4ab + b^2 = 6b^2 \pm 6b + 1 = 6b(b \pm 1) + 1$. The only way the latter can be ± 1 is if $b = 0$ or $b = \mp 1$. These result in $(a = 1, b = 0)$ and $(a = 0, b = -1)$, and so the only possibilities for z are $z = \pm 1$. Hence, the only candidates for an even x are $x = \pm 2$ and $y = 2$ (which are indeed solutions).

We conclude that the only solutions to this Diophantine equation are $(x = \pm 2, y = 2)$ and $(x = \pm 11, y = 5)$ as claimed. □

As a prelude to a second example, we consider Pythagorean triples. Let $z = a + bi$ be an arbitrary element of $\mathbb{Z}[i]$. Then

$$(a^2 + b^2)^2 = N(z)^2 = N(z^2) = N(a^2 - b^2 + 2abi) = (a^2 - b^2)^2 + (2ab)^2.$$

Hence $x = a^2 - b^2$, $y = 2ab$, and $z = a^2 + b^2$ are a Pythagorean triple. As long as we look for relatively prime solutions—the so-called *primitive Pythagorean triples*—the converse is also true.

Lemma 20.19. *Let x, y, and z be integers with no common prime divisor. Then*

$$x^2 + y^2 = z^2$$

if and only if there exists integers a and b with

$$\begin{cases} x & = a^2 - b^2, \\ y & = 2ab, \\ z & = a^2 + b^2. \end{cases}$$

Proof. We have already shown—and it is easy to check directly—that if x, y, and z have the form prescribed above, then they form a Pythagorean triple.

For the converse, assume $x^2 + y^2 = z^2$. Note that if a prime divides two of the variables, it will have to divide the third, and so x, y, and z are pairwise relatively prime. Now, if both x and y are odd, then both x^2 and y^2 are 1 (mod 4). This would mean that z^2 is 2 (mod 4) which is impossible. Hence, x and y have opposite parity and z is odd. By Lemma 20.17, $x + iy$ and $x - iy$ are relatively prime in $\mathbb{Z}[i]$.

Now write $z^2 = (x + iy)(x - iy)$. Unique factorization in $\mathbb{Z}[i]$ implies that $x + iy$ (and $x - iy$) is a complete square (of an element in $\mathbb{Z}[i]$). Hence, $x + iy = (a + bi)^2$. Expand, and equate the real and imaginary parts to get $x = a^2 - b^2$ and $y = 2ab$. As a result $z = a^2 + b^2$, and the proof is complete.   □

For a final example, we consider a variation of the previous equation.

Proposition 20.20. *Let x, y, and z be integers with no common prime divisors. Then*

$$x^2 + y^2 = 2z^2$$

if and only if there exists integers a and b with

$$\begin{cases} x & = a^2 - b^2 - 2ab, \\ y & = a^2 - b^2 + 2ab, \\ z & = a^2 + b^2. \end{cases}$$

Proof. If p is a common prime divisor of x and y in $\mathbb{Z}$, then p^2 divides $x^2 + y^2$. This would mean that p divides z also which is a contradiction. So x and y are relatively prime in $\mathbb{Z}$. Now $x^2 + y^2$ is even (since it equals $2z^2$) and so x and y are both odd. Lemma 20.17 gives that $1 + i$ divides $x + iy$. Write $x + iy = (1 + i)(u + iv)$. From this we get that $x = u - v$ and $y = u + v$, and thus u and v are relatively prime in $\mathbb{Z}$. Moreover, taking norms, we get

$$2z^2 = x^2 + y^2 = 2(u^2 + v^2).$$

Thus $z^2 = u^2 + v^2$. Applying Lemma 20.19, we know that $z^2 = u^2 + v^2$ if and only if there exists integers a and b with $u = a^2 - b^2$, $v = 2ab$, and $z = a^2 + b^2$. The result now follows.   □

Problems

20.2.1. Does $x^2 + y^2 = 3z^2$ have any integer solutions?

20.2.2. Find and prove a version of Lemma 20.19 where you do not assume that the integers x, y, and z are relatively prime.

20.2.3. **Problem 15.1.6 redux.** Find all integer solutions to $y^3 = x^2 + 1$. Justify all your steps.

20.2.4. Show that $\mathbb{Z}[\sqrt{-2}]$ is a Euclidean domain with the norm as its degree function. Use this to justify the solution on page 312 for finding all integer solutions to $y^3 = x^2 + 2$.

Part 3

Fields and Galois Theory

Chapter 21

Introducing Field Theory and Galois Theory

... where prototypal problems of field theory are presented, a preliminary
example shows how the study of the roots of polynomials may be
related to fields and groups, and needed ring theory is briefly reviewed.

21.1. The Classical Problems of Field Theory

As is to be expected, field theory is the study of fields. A number of ancient
problems motivate our study of fields.

Question 21.1. Can you double a cube? More precisely, given one edge of a cube,
can you construct—using a straightedge and compass—another line segment such
that a cube with this new line segment as its side will have a volume twice as much
as the original cube?

This problem is an ancient one. It apparently was known to the Egyptians,
Greeks, and Indians. The Greek legend is the best known one. It is said that a
plague in Athens in 430 BCE compelled the citizens of Athens to consult the oracle
of Apollo at Delos. To defeat the plague, the oracle prescribed that the Athenians
double the size of their altar. They did so by doubling each side of the altar, but the
plague got worse. Apparently, the oracle meant that the *volume* of the altar must
be doubled. It is alleged—by Eratosthenes (276–194 BCE) as quoted by Theon
of Smyrna (circa 70–135 CE)—that Plato (427–347 BCE) had remarked that the
oracle really meant to "shame the Greeks for their neglect of mathematics and for
their contempt of geometry". The original legend did not specify the tools to be
used, and, in fact, solutions using a number of tools were found. However, in the
Greek/Alexandrian tradition of geometry, a magnitude was considered found if it
could be constructed by geometrical means and the elementary means for geometric
construction were constructions with a straightedge and compass.

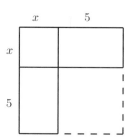

Figure 21.1. Completing the square to solve $x^2 + 10x = 39$. The area of the small square and the two rectangles is $x^2 + 10x$ and hence equal to 39. Thus the area of completed square is $39 + 25 = 64$. This means that $x + 5 = 8$, and hence $x = 3$.

Question 21.2. Can you trisect an angle? More precisely, given an arbitrary angle—and using only a straightedge and compass—can you draw an angle one third of the original angle?

Question 21.3. Can you square a circle? In other words, given a line segment, can you draw—using only a straightedge and compass—another line segment such that the area of a circle with radius equal to the old line segment is the same as the area of a square with the new line segment as its side?

Question 21.4. Given an arbitrary equation—in one variable—of degree ≥ 5, can you find its roots?

For the second degree equation, we have the quadratic formula possibly known to the Babylonians as early as 1800 BCE. (See Figure 21.1 for the geometric solution common in the Islamic period). An algebraic solution to arbitrary third and fourth degree equations was first published by Girolamo Cardano (1501–1576) based on the work of Niccolo Fontana Tartaglia (1500–1557), Scipione del Ferro (1465–1526), and Lodovico Ferrari (1525–1565). For the fifth and higher degree equations, the first question is whether we can find a formula—involving the four arithmetical operations and any combination of the coefficients—that gives the roots to any polynomial equation of degree 5. A weaker question is whether—regardless of the existence of a general formula—it is possible to write down any root of a polynomial equation using the four arithmetical operations and radicals.

Field theory provides the language and tools necessary to answer all of the above questions in the negative.

The Norwegian mathematician Niels Henrik Abel (1802–1829) was the first to give an acceptable proof of the impossibility of solving a quintic using radicals. Évariste Galois (1811–1832), however, took the theory to a new level by introducing different techniques and a different point of view. He clearly—although his writing was anything but clear—understood the relation of solutions of an algebraic equation and related groups of permutations. He introduced the idea of normal subgroups and was able to find precise conditions for solvability of a polynomial equation using radicals. While highly original and creative, the work of Galois on groups of permutations did have antecedents in the works of Gauss (1777–1855),

Lagrange (1736–1813), Cauchy (1789–1857), and Ruffini (1765–1822) among others.[1]

In the following chapters, we will see the solution to these classical problems.

Connection to High School Algebra. You may have wondered about the connection between the subject of this book and that of high school algebra. Clearly, the study of polynomial equations—as described above—gives one link. But there is more. Field theory allows us to step back and easily answer questions that look like somewhat complicated high school algebra problems. Here is a selection:

Question 21.5. Rationalize $\alpha = \frac{1}{1+5\sqrt[3]{2}+4\sqrt[3]{4}}$. In other words, give a fraction that is equal to α and has an integer in the denominator.

Question 21.6. Find a polynomial with rational coefficients that cannot be factored and has $2\sqrt[5]{17} + 5\sqrt[7]{19} - \sqrt[3]{2}$ as a root.

Question 21.7. Can you find one real number α such that rational linear combinations of $1, \alpha, \alpha^2, \dots$ include every one of $\sqrt[3]{47}$, $\sqrt[5]{17}$, and $18 - 2\sqrt[7]{19}$?

Question 21.8. Let $\alpha = \frac{\sqrt[5]{4}-\sqrt[5]{8}+\sqrt[5]{16}}{3+\sqrt[5]{2}}$. We know that α is not a rational number. Does there exist a polynomial $p(x)$ with rational coefficients such that $p(\alpha) = \sqrt[5]{2}$?

Question 21.9. If a polynomial with rational coefficients has $5 - 6\sqrt{14}$ as a root, then must it also have $5 + 6\sqrt{14}$ as a root?

Using tools of high school algebra, you can answer most—maybe all—of the above questions. On the other hand, field theory provides an easier but more

---

[1]Évariste Galois was born in 1811 in turbulent times near Paris, France, and died as a result of a duel in Paris only 21 years later. In his short life, he published five papers, none of which made a particularly big mark. He wrote, however, a number of manuscripts on solvability of equations—unpublished in his lifetime—that had a profound, lasting, and revolutionary impact on mathematics. Galois did submit three versions of a manuscript on solvability to prominent mathematicians for consideration. Cauchy returned Galois's first paper with suggestions for a revision. In fact, this is when Galois learned about the work of Abel, which was certainly relevant to his work. The second version of the paper was sent to Fourier to be considered for a prize. But Fourier died shortly thereafter, and the manuscript was lost. Poisson invited Galois to submit a third revision of his memoir to the Academy, and Galois did so, but the paper was rejected. Poisson wrote "His argument is neither sufficiently clear nor sufficiently developed to allow us to judge its rigor." While Poisson apparently failed to see the far-reaching importance of Galois's work—he did encourage Galois to rewrite and resubmit—many mathematicians agree with his judgement. Galois was a poor writer. After Galois's death, his brother and his friend Chevalier sent copies of his manuscript to many mathematicians. Galois had specifically asked—in a letter to Chevalier written the night before the duel—that his work be sent to Gauss and to Jacobi so that they could comment on the importance of the work. There is no record of any response from those two, but the mathematician Liouville worked on the paper, understood what it was all about, and presented it to the Academy in 1843, eleven years after Galois's death. The work was published in 1846. The work of Galois is not only the starting point of Galois theory, it is also the beginning of the deep study of groups. But Galois's short life was entangled in more than mathematics. He was politically active, an ardent republican, spent a few months in jail, and, in fact, did some of his mathematics while incarcerated. Galois had a rocky relationship with his teachers and the school system as well. He failed the entrance exam to the most prestigious college—the École Polytechnique—twice. Galois's father committed suicide three years before Galois's death, there is also a woman's name—Stephanie—scribbled on the margins of Galois's papers, and the list goes on. Not surprisingly, much has been written on Galois's life and many disputed stories and narratives have been circulating: that he didn't sleep the night before the duel because he was working out the details of his mathematics, that the whole duel was a right-wing conspiracy to eliminate a political activist, that he threw a chalk eraser at his examiner for not following Galois's mathematical argument, ... For a well-researched book on the life of Galois, see Rigatelli [**TR96**], and, for a fictionalized but an entertaining account see Infeld [**Inf78**]. Finally, to read Galois's own work in the original French with an English translation and with commentary, see Neumann [**Neu11**].

abstract approach that puts all of the above questions in a larger context. As you proceed in your study of fields, it may be instructive to occasionally come back to these questions to see if you have more to say about them.

21.2. Roots of Equations, Fields, and Groups—An Example

Galois basically invented the idea of a group to be able to tackle the problem of solutions of equations. But what do groups have to do with equations? In this short section, we look at an example in order to get an appreciation of the task ahead. Not everything will be proved—they will be in due time—but the reader will get some idea of the connections.

Galois theory relates three different worlds: equations, fields, and groups. Equations and their roots give rise to a chain of fields, and these in turn give rise to the so-called Galois groups. Galois himself went directly from equations to groups but the modern approach first translates the problems about equations to problems of *field extensions*.

We want to know whether it is possible to write the roots of a polynomial using the four arithmetical operations and radicals. We will see that if we know certain things about a chain of fields and a chain of groups, then we will be able to answer this question. We can actually do more and find ways of actually solving the equation, but we will not do that here.

Consider the polynomial $x^4 + x^3 - x^2 - 2x - 2$. Now in the case of this polynomial we can actually find the roots, and this will allow us to see possible connections with fields and groups and will guide our discussion. We are, of course, interested in the case when we do not know the roots. The polynomial is of fourth degree, and we have

$$x^4 + x^3 - x^2 - 2x - 2 = (x^2 - 2)(x^2 + x + 1).$$

Thus the four roots are

$$\alpha = \sqrt{2}, \ \beta = -\sqrt{2}, \ \gamma = \frac{-1 + \sqrt{3}i}{2}, \text{ and } \delta = \frac{-1 - \sqrt{3}i}{2}.$$

Note that these roots satisfy a number of equations. For example, $\alpha + \beta = 0$, $\gamma + \delta + 1 = 0$, $\alpha^2 - 2 = 0$, $\alpha^2\gamma^2 + \beta^2\delta^2 + 2\alpha\beta\delta\gamma - 2 = 0$, and so on.

Clearly, there are some symmetries in these roots. α and β are related, as are γ and δ. In all of the relations that they satisfy, you can switch the place of α and β or the place of δ and γ, and still get a valid relation. However, switching α and γ or some other permutation may not work.

Hence—thinking of the roots as the first, second, third, and the fourth roots— let $(1\ 2)$ denote the act of permuting (switching) α and β. Likewise, the permutation $(3\ 4)$ denotes permuting γ and δ. As usual, e stands for not permuting the roots, and $(1\ 2)(3\ 4)$ stands for permuting α and β as well as γ and δ.

So, related to the roots of this polynomial, we have the group

$$\{e, (1\ 2), (3\ 4), (1\ 2)(3\ 4)\} \cong \mathbb{Z}/2\mathbb{Z} \times \mathbb{Z}/2\mathbb{Z}.$$

Now let us look at the following chain of fields:

$$\mathbb{Q} \subset \mathbb{Q}(\gamma, \delta) \subset \mathbb{Q}(\alpha, \beta, \gamma, \delta).$$

$\mathbb{Q}(\gamma, \delta)$ is the field of rational expressions in γ and δ—e.g., $\frac{\gamma + 2\delta^2}{\gamma^3 - \delta}$—while $\mathbb{Q}(\alpha, \beta, \gamma, \delta)$ is the field of rational expressions in α, β, γ, and δ.

We also look at the following chain of subgroups:

$$G = \{e, (1\ 2), (3\ 4), (1\ 2)(3\ 4)\} \geq H = \{e, (1\ 2)\} \geq \{e\}.$$

Now G, and its subgroups, act on $\{\alpha, \beta, \gamma, \delta\}$ and, by extension, on $\mathbb{Q}(\alpha, \beta, \gamma, \delta)$. For example,

$$(1\ 2) \cdot \frac{\gamma + 2\alpha^2}{\beta^3 - \delta} = \frac{\gamma + 2\beta^2}{\alpha^3 - \delta}.$$

We now ask which elements of $\mathbb{Q}(\alpha, \beta, \gamma, \delta)$ are fixed by every element of H. In other words, which rational functions do not change at all when elements of H act on them? Clearly, every element of $\mathbb{Q}(\gamma, \delta)$ is fixed by this action. But you may think that there are more. For example, $\alpha^2 + \beta^2$ is also fixed by the action of H. But this element is also in $\mathbb{Q}(\gamma, \delta)$. Why is that? We have $\alpha^2 + \beta^2 = (\alpha + \beta)^2 - 2\alpha\beta = 0^2 - 2(-2) = 4 \in \mathbb{Q}$.

In fact, we can prove that $\text{Fix}(H)$, the elements of $\mathbb{Q}(\alpha, \beta, \gamma, \delta)$ that are fixed by every element of H, is $\mathbb{Q}(\gamma, \delta)$, and the only rational functions in $\mathbb{Q}(\alpha, \beta, \gamma, \delta)$ that are fixed by every element of G are the elements of $\mathbb{Q}$.

Now going back to the general case of a fourth degree equation—but keeping the above example in mind— assume that we have a fourth degree equation with roots α, β, γ, δ, and that we somehow know that the group $G = \{e, (1\ 2), (3\ 4), (1\ 2)(3\ 4)\}$ acts as permutations of the roots. Let $H = \{e, (1\ 2)\}$ be the subgroup of G, and assume that we know the following:

- the elements of $\mathbb{Q}(\alpha, \beta, \gamma, \delta)$ fixed by every element of G are exactly the elements of $\mathbb{Q}$; and
- the elements of $\mathbb{Q}(\alpha, \beta, \gamma, \delta)$ fixed by every element of H are exactly the elements of $\mathbb{Q}(\gamma, \delta)$;
- $\alpha + \beta$ and $\alpha\beta$ are fixed by every element of H;
- $\gamma + \delta$ and $\gamma\delta$ are fixed by every element of G.

Without knowing the roots of the polynomial and by just knowing the corresponding group and the above facts about fixed fields, we can argue as follows:

Just by the assumptions, we know that $\gamma + \delta$ and $\gamma\delta$ are elements of $\mathbb{Q}$ and $\alpha + \beta$ and $\alpha\beta$ are elements of $\mathbb{Q}(\gamma, \delta)$.

Now $(x - \gamma)(x - \delta) = x^2 - (\gamma + \delta)x + \gamma\delta$ is an equation with rational coefficients. Hence its roots—that is γ and δ—can be found using the quartic formula and written using radicals.

Now what about α and δ? Now these are the roots of $(x - \alpha)(x - \beta) = x^2 - (\alpha + \beta)x + \alpha\beta$ which is a polynomial with coefficients in $\mathbb{Q}(\gamma, \delta)$. Using the quadratic formula, we can find the roots of this quadratic, and write the roots using the four arithmetical operations and radicals of elements of $\mathbb{Q}(\gamma, \delta)$. But the latter themselves can be written using radicals of elements of $\mathbb{Q}$. Hence α and β can be written using $+$, $-$, $\times$, $\div$, and radicals of radicals. So, we can write the solutions to the quartic using rational numbers, the four arithmetical operations, and complicated radicals.

The moral of the story is that by knowing enough information about groups of permutations of roots, their subgroups, and the fields fixed by them, we can solve polynomials by first reducing them to lower degree polynomials. In the coming chapters, we will build tools and techniques to allow us to rigorously carry this task out.

21.3. A Quick Review of Ring Theory

To fruitfully engage field theory, we can either use many of the concepts introduced in the ring theory part of the text or reprove, in less generality, those aspects that we need. In fact, many books on Galois theory do not depend on a general analysis of rings since the actual facts needed can be proved directly, and the proofs are more straightforward when we limit ourselves to the specific situations actually needed. The more general theory of rings, however, puts many of the needed facts in a much more general context, and it allows you to *not* "miss the forest for the trees". The purpose of this section is to provide you with an opportunity to review some of the needed ring theory.

You can use the following two sample problems to guide your review of ring theory. Use the previous chapters and the "proofs" outlined here to arrive at a complete proof. When you are comfortable with these two proofs and the material in Chapter 19, then you are ready to proceed.

Problem 21.10. *Consider $\alpha = \sqrt[3]{2} + 6\sqrt{5}$. Consider the set of polynomials $p(x)$ with rational coefficients that have α as a root (i.e., $p(\alpha) = 0$). In other words, consider*

$$S = \{p \in \mathbb{Q}[x] \mid p(\alpha) = 0\}.$$

Then show that there exists one polynomial $m(x)$ such that

$$S = \{q(x)m(x) \mid q \in \mathbb{Q}[x]\}.$$

Solution. $R = \mathbb{Q}[x]$ is a ring, and S is an (non-trivial) ideal of R. Now, in fact, R is a Euclidean domain, and all Euclidean domains are principal ideal domains and hence S is generated by one element. That element is $m(x)$. □

Notation 21.11. We have been using the notation $\mathbb{F}_p$ for the field $(\mathbb{Z}/p\mathbb{Z}, +, \cdot)$. We now extend this notation and denote by $\mathbb{F}_q$ a finite field with q elements. This notation will be justified and refined later when we show that the only possible q's are powers of a prime (Theorem 22.32), and for every power of a prime q, there is a *unique* field of order q (Theorem 27.2).

Problem 21.12. *Construct a field with 625 elements.*

Construction. Consider $\mathbb{F}_5$. This is a field with five elements, and $R = \mathbb{F}_5[x]$ is a ring (with an infinite number of elements). $x^4 + 2$ has no roots in $\mathbb{F}_5$. This means that it has no linear factors. We can show that it cannot be factored into two polynomials of degree 2 either. Hence, it is an irreducible polynomial in R. Now $\mathbb{F}_5$ is a field, which means that R is an ED which, in turn, implies that R is a PID. In a PID, the ideal generated by an irreducible element is maximal. Hence $I = \langle x^4 + 2 \rangle$ is a maximal ideal of R. Now, in any commutative ring with 1, if I is a maximal ideal of R, then R/I is a field. Hence, $\mathbb{F}_5[x]/\langle x^4 + 2 \rangle$ is a field. Its elements are $\{a + bx + cx^2 + dx^3 + I \mid I = \langle x^4 + 2 \rangle, a, b, c, d \in \mathbb{F}_5\}$, and there are $5^4 = 625$ of these. ☐

Some of the facts and results that we need—many were used in the proofs above—are listed here.

- *Corollary 19.22.* If F is a field, then $F[x]$ is a Euclidean domain.

- *Corollary 18.51.* A Euclidean domain (ED) is a principal ideal domain (PID), and a PID is a unique factorization domain (UFD).

- *Field of fractions, Theorem 17.1.* Every integral domain D has a field of fractions F. The latter is a field containing an isomorphic copy D_0 of D, and each of its elements are of the form rs^{-1} where $r, s \in D_0$ and $s \neq 0$. In addition, this field of fractions is unique in the sense that if another field E contains a subring D_1 isomorphic to D, then E also has a subfield E_0 that is isomorphic to F and contains D_1.

- *Theorem 18.21.* In a PID—and, in particular, in $F[x]$ where F is a field—the ideal generated by an irreducible element is maximal.

- *Theorem 18.14.* If R is a commutative ring with 1 and M is a maximal ideal of R, then R/M is a field.

- *Prime subfield, Corollary 16.55.* If F is field, then the intersection of all subfields of F—called the prime subfield of F—is either isomorphic to $\mathbb{Q}$ or $\mathbb{F}_p$ where p is a prime number. The former are fields of characteristic 0 while the latter are fields of characteristic p.

- *Gauss's lemma, Theorem 19.51.* If R is a UFD, F its field of fractions, and f an irreducible polynomial in $R[x]$ of positive degree, then f is irreducible in $F[x]$.

- *Corollary 19.56.* If R is a UFD, then so is $R[x_1, \ldots, x_k]$.

- *Schönemann-Eisenstein's criterion, Corollary 19.64.* If R is a UFD, K its fraction field, $f(x) = a_0 + a_1 x + \cdots + a_n x^n \in R[x]$, p a prime, $p \nmid a_n$, $p \mid a_i$ for $0 \leq i \leq n - 1$, and $p^2 \nmid a_0$, then f is irreducible in $K[x]$.

Problems

21.3.1. By completing the details, write a complete solution to Problem 21.10.

21.3.2. By completing the details, write a complete solution to Problem 21.12.

21.3.3. Construct a field with 25 elements. Explicitly list the elements, and give an addition and a multiplication table.

21.3.4. Let A be an $n \times n$ matrix with real entries, and let $p(x) = a_0 + a_1 x + \cdots + a_k x^k$ be a polynomial with real coefficients. As is customary, we define

$$p(A) = a_0 I_n + a_1 A + a_2 A^2 + \cdots + a_k A^k.$$

Let $\mathbf{0}_n$ denote the $n \times n$ zero matrix, and define

$$S = \{p(x) \in \mathbb{R}[x] \mid p(A) = \mathbf{0}_n\}.$$

(a) Can I_n, A, A^2, ..., A^{n^2} be a linearly independent set of matrices? Can S consist only of the zero polynomial?

(b) Prove that there exists a polynomial $m(x)$ in $\mathbb{R}[x]$—called the *minimal polynomial* of A—such that $m(A) = \mathbf{0}_n$ and

$$S = \{m(x)q(x) \mid q(x) \in \mathbb{R}[x]\}.$$

(c) The Cayley–Hamilton theorem states that if $p(x)$ is the characteristic polynomial of A, then $p(A) = \mathbf{0}_n$. Assuming the Cayley–Hamilton theorem, state the relationship between the characteristic and the minimal polynomial of A.

(d) Give an example of a matrix where the minimal polynomial is different from the characteristic polynomial. Give another example where the two polynomials are the same.

Chapter 22

Field Extensions

...where we consider fields and bigger fields containing them, define algebraic and simple extensions, and make a distinction between $F[\alpha]$ and $F(\alpha)$ and, in addition, using vector spaces, we define the degree of a field extension and explore its relation with degrees of minimal polynomials.

22.1. Simple and Algebraic Extensions

Fields are a class of rings, and a field isomorphism is just a ring isomorphism (applied to fields). For convenience, we (re)record this definition.

Definition 22.1 (Field isomorphism). Let F and F' be fields. A *field isomorphism* from F to F' is the same as a ring isomorphism—i.e., a 1-1, onto, ring homomorphism—from F to F'. Such an isomorphism sends 0_F and 1_F to $0_{F'}$ and $1_{F'}$ respectively, and preserves all the four arithmetical operations. (See Problem 16.1.8). A field isomorphism from F to F is called a *field automorphism*.

Much of the focus of our study will be on field extensions.

Definition 22.2 (Extension field). A field E is an *extension field* of a field F if $F \subseteq E$. If E is an extension field of F, then we say that $F \subseteq E$ is a *field extension* (or just an *extension*).[1]

Remark 22.3. Recall—from Lemma 16.6—that, since fields are integral domains, if F and E are fields with $F \subseteq E$, then $1_F = 1_E$.

Remark 22.4. Often, a field E will contain an isomorphic copy of a field F. In other words, $K \subseteq E$ and K is a field isomorphic to F. Using a field isomorphism $\phi : F \to K$ and its inverse, we can translate back and forth between F and K. Technically, in such situations, E is *not* an extension field of F but rather it is an

---

[1] An alternative is to use the expression "E/F is a field extension" to mean "$F \subseteq E$ is a field extension." The symbol E/F is read as E over F.

extension field of $K = \phi(F)$. However (just as in the case of rings—see Remark 17.7), to avoid clutter, often in such situations, we identify F and its isomorphic copy K, and say that E is an extension field of F.

Definition 22.5. Let $F \subseteq E$ be fields, and $\alpha \in E$. Define $F(\alpha)$ to be the smallest subfield of E containing both F and α. In other words, $F(\alpha)$ is the intersection of all subfields of E that contain F and α. More generally, if $Y \subseteq E$, then $F(Y)$ is the smallest subfield of E containing F and Y.

To say that $F(\alpha)$ is the smallest subfield of E containing both F and α means that $F(\alpha)$ consists of elements of E that can be written using elements of F, α, and the four arithmetical operations $+$, $-$, $\times$, and $\div$. For fields $F \subseteq E$ and $\alpha \in E$, recall—from Definition 19.10—that, $F[\alpha] = \{g(\alpha) \mid g \in F[x]\}$ is an integral domain inside E containing F and α. In contrast to $F(\alpha)$, the ring $F[\alpha]$ is the collection of elements of E that can be written using elements of F, α, and the three arithmetical operations $+$, $-$, and $\times$.

Lemma 22.6. *Let $F \subseteq E$ be fields, and let $\alpha \in E$. Then $F(\alpha)$ is the field of fractions of $F[\alpha]$.*

Proof. The integral domain $F[\alpha]$ is contained in E, but may not be a field since some of its elements may not have an inverse in $F[\alpha]$. But since $F[\alpha]$ is an integral domain, it has a field of fractions K, and any field containing $F[\alpha]$ will contain a copy of K. Now, $F(\alpha)$ is a field inside E and it also contains $F[\alpha]$. Hence, it must contain a subfield isomorphic to K. But $F(\alpha)$ is also the smallest subfield of E containing both F and α. Hence $F(\alpha)$ must be the field of fractions of $F[\alpha]$.   □

$$
\begin{array}{c}
E \\
| \\
F(\alpha) \\
| \\
K \\
| \\
F[\alpha] \\
| \\
F
\end{array}
$$

Example 22.7. Consider $\mathbb{Q} \subseteq \mathbb{C}$. Now both $i \in \mathbb{C}$ and $\sqrt[3]{2}$ are in $\mathbb{C}$. In both cases, we have

$$\mathbb{Q}[i] = \{a + bi \mid a, b \in \mathbb{Q}\} = \mathbb{Q}(i),$$

$$\mathbb{Q}[\sqrt[3]{2}] = \{a + b\sqrt[3]{2} + c\sqrt[3]{4} \mid a, b, c \in \mathbb{Q}\} = \mathbb{Q}(\sqrt[3]{2}).$$

In other words, both $\mathbb{Q}[i]$ and $\mathbb{Q}[\sqrt[3]{2}]$ are already fields themselves. This is easy to see directly in the case of $\mathbb{Q}[i]$, since

$$\frac{1}{a + bi} = \frac{a - bi}{a^2 + b^2} = \frac{a}{a^2 + b^2} - \frac{b}{a^2 + b^2}i.$$

In the case of $\mathbb{Q}[\sqrt[3]{2}]$, while it is also possible to directly find the inverse of each element, it is easier to deduce the existence of inverses in $\mathbb{Q}[\sqrt[3]{2}]$ from general theorems that will follow.

In analogy with Lemma 22.6, we define $F(x)$:

Definition 22.8. Let F be a field, and let x be an indeterminate. Denote by $F(x)$ the field of fractions of the polynomial ring $F[x]$. The elements of $F(x)$ are *rational* *functions* in x. In other words, $F(x)$ consists of elements of the form

$$\frac{a_0 + a_1 x + \cdots + a_n x^n}{b_0 + b_1 x + \cdots + b_m x^m},$$

where $n, m \in \mathbb{Z}^{\geq 0}$, $a_0, \ldots, a_n, b_0, \ldots, b_m \in F$, and at least one of the b's is non-zero.

Remark 22.9. Note that there is a difference between Definitions 22.5 and 22.8 of $F(\alpha)$ and $F(x)$. The element α is a member of an extension field E of F. Hence, we can add, multiply, and divide α and elements of F using the operations of the field E. Hence, elements of $F(\alpha)$ are just elements of the field E. On the other hand, x is an indeterminate, and elements of $F(x)$ are ratios of polynomials in x with coefficients in F. Given $\alpha \in E$ with $F \subseteq E$, Lemma 22.6 can be translated as saying that

$$F(\alpha) = \{f(\alpha) \mid f \in F(x), \alpha \text{ is not a root of the denominator of } f\}.$$

Definition 22.10 (Primitive element and simple extensions). Let $F \subseteq E$ be fields. The field E is called a *simple extension* of F if $E = F(\alpha)$ for some $\alpha \in E$. An element $\alpha \in E$ such that $E = F(\alpha)$ is called a *primitive element* for the extension.

Example 22.11. Let $L = \mathbb{Q}(\sqrt{2}, \sqrt{3})$. Then L is the smallest subfield of the real numbers that contains all the rational numbers as well as $\sqrt{2}$ and $\sqrt{3}$. So, for example, $5\sqrt{2} + \frac{3}{2}\sqrt{3}$ and $\frac{2\sqrt{3}}{4\sqrt{2}-3\sqrt{3}}$ are elements of L. From the definition of L, it is not clear whether L is a simple extension of $\mathbb{Q}$. However,

CLAIM: $L = \mathbb{Q}(\sqrt{2} + \sqrt{3})$, and hence L is a simple extension of $\mathbb{Q}$.

PROOF OF CLAIM: Let $E = \mathbb{Q}(\sqrt{2} + \sqrt{3})$. Clearly $E \subseteq L$. We want to show that $L \subseteq E$. We know $\sqrt{2} + \sqrt{3} \in E$, hence $5 + 2\sqrt{6} = (\sqrt{2} + \sqrt{3})^2 \in E$. This means that $\sqrt{6} \in E$ and hence so is $\sqrt{6}(\sqrt{2} + \sqrt{3}) = 2\sqrt{3} + 3\sqrt{2}$. We now have $\sqrt{2} = (2\sqrt{3} + 3\sqrt{2}) - (2\sqrt{3} + 2\sqrt{2}) \in E$. It follows that $\sqrt{3} = (\sqrt{2} + \sqrt{3}) - \sqrt{2}$ is also in E. But if $\mathbb{Q}$, $\sqrt{2}$, and $\sqrt{3}$ are in E, then $L \subseteq E$ and the proof is complete.

Definition 22.12 (Algebraic over F). Let $F \subseteq E$ be fields, and let $\alpha \in E$. Then α is *algebraic* over F if there exists $0 \neq f \in F[x]$ with $f(\alpha) = 0$. If no such f exists then α is *transcendental* over F.

Example 22.13. Consider the extension $\mathbb{Q} \subset \mathbb{C}$, then $7, \sqrt{2}, i, \frac{3+i\sqrt[5]{7}}{9} \in \mathbb{C}$ are algebraic over $\mathbb{Q}$. On the other hand $\pi, e, 3 + \pi, e/7 \in \mathbb{C}$ are transcendental over $\mathbb{Q}$. The former are easy to establish, while the latter are difficult to prove.

Example 22.14. Let F be a field, and let x be an indeterminate. Consider the extension $F \subset F(x)$. The field $F(x)$ consists of ratios of polynomials in x, and we have $F \subset F[x] \subset F(x)$. Now $x \in F(x)$ and is transcendental over F since for $f \in F[x]$, $f(x) = f \neq 0$.

Definition 22.15 (Algebraic extension). Let $F \subseteq E$ be fields. E is *algebraic* over F or an *algebraic extension* of F if every $\alpha \in E$ is algebraic over F.

We are now ready to cast Problem 21.10 into the language of field theory and prove a stronger version.

Theorem 22.16. *Let $F \subseteq E$ be fields. Let $\alpha \in E$ be algebraic over F. Then there exists a unique $f \in F[x]$ with the following three properties:*

(a) *f is monic, and*

(b) *f is irreducible, and*

(c) *$f(\alpha) = 0$.*

Furthermore, $\{g \in F[x] \mid g(\alpha) = 0\} = \langle f \rangle$ in the ring $F[x]$.

Proof. Let $\epsilon_\alpha : F[x] \to E$ be the evaluation map at α. This map is defined by $\epsilon_\alpha(p(x)) = p(\alpha)$ and is a ring homomorphism. The kernel of this map is $\{g \in F[x] \mid g(\alpha) = 0\}$. The kernel is an ideal of $F[x]$, and $F[x]$ is a PID—in fact, since F is a field, $F[x]$ is an ED—hence, the kernel is generated by one element $f \in F[x]$. Since α is algebraic, the kernel is not just $\{0\}$, and so $f \neq 0$. If you multiply f by a unit, then it will still generate the same ideal—two associates generate the same ideal—and so we may choose f to be monic.

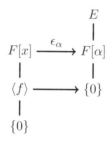

Figure 22.1. The evaluation homomorphism $\epsilon_\alpha : F[x] \to E$

We claim that f has to be irreducible. To prove the claim, assume $f = gh$ for $g, h \in F[x]$. Then $g(\alpha)h(\alpha) = f(\alpha) = 0$, and so $g(\alpha)$ or $h(\alpha)$ is zero. Without loss of generality, assume $g(\alpha) = 0$. Then $g \in \ker(\epsilon_\alpha) = \langle f \rangle$, and hence $f \mid g$. This means that $\deg(f) \leq \deg(g)$. But $f = gh$, and hence we must have $\deg(f) = \deg(g)$ and $\deg(h) = 0$. So, h is a unit. This proves that f is irreducible.

It remains to show that f, with the given properties, is unique. Hence assume that $f^* \in F[x]$ is monic and irreducible, and, furthermore, $f^*(\alpha) = 0$.

Since $f^* \in \ker(\epsilon_\alpha) = \langle f \rangle$, we have $f \mid f^*$. In other words, $f^* = fh$ which means—since both f^* and f are irreducible—that h is a constant.

But f and f^* are both monic, which means that $h = 1$, and $f^* = f$. $\square$

Definition 22.17 (Minimal polynomial). Let $F \subseteq E$ be fields. Let $\alpha \in E$ be algebraic over F. Then the unique monic irreducible polynomial in $F[x]$ that has α as a root is called the *minimal polynomial* of α over F, and denoted by $\min_F(\alpha)$.[2] If $n = \deg(\min_F(\alpha))$ then we say that α is algebraic of degree n over F.

Example 22.18. Let $\alpha = \sqrt{1 + \sqrt{3}} \in \mathbb{C}$. Then $\alpha^2 = 1 + \sqrt{3}$ and hence $\alpha^4 - 2\alpha^2 + 1 = 3$. Thus if we let $f = x^4 - 2x^2 - 2$, then α is a root of f. The monic polynomial

---

[2]Other notations for the minimal polynomial of α over F include $m_{\alpha,F}(x)$, $m_\alpha(x)$, or $\min(\alpha, F)$.

f is irreducible by the Schönemann-Eisenstein criterion (Corollary 19.64) and hence $f = \min_{\mathbb{Q}}(\alpha)$. The polynomial f has four roots and these are $\pm\sqrt{1 \pm \sqrt{3}}$.

Now, if $g \in \mathbb{Q}[x]$ is any polynomial that has α as a root, then, by Theorem 22.16, f will divide g. We conclude that if $g \in \mathbb{Q}[x]$ has $\sqrt{1 + \sqrt{3}}$ as a root, then, by necessity, it will have all four of $\pm\sqrt{1 \pm \sqrt{3}}$ as roots!

Problems

22.1.1. Prove that $\sqrt{3} \notin \mathbb{Q}(\sqrt{2})$.

22.1.2. Let p_1, p_2, ..., p_n, and let q be an $n+1$ distinct positive (integer) primes. Show that
$$\sqrt{q} \notin \mathbb{Q}[\sqrt{p_1}, \sqrt{p_2}, \ldots, \sqrt{p_n}].$$

22.1.3. Find the minimal polynomial of $4\sqrt{7} - 3$ over $\mathbb{Q}$.

22.1.4. Find the minimal polynomial of $\sqrt{-1 + \sqrt{2}}$ over $\mathbb{Q}$.

22.1.5. Let a and b be rational numbers. Find the minimal polynomial over $\mathbb{Q}$ of $a + bi \in \mathbb{C}$.

22.1.6. Let a and b be real numbers, and let $f \in \mathbb{R}[x]$. Prove that if $a + bi$ is a root of f, then so is $a - bi$.

22.1.7. Let $f \in \mathbb{Q}[x]$. Prove that if $5 + 6\sqrt{14}$ is a root of f, then so is $5 - 6\sqrt{14}$.

22.1.8. Find the minimal polynomial of $\sqrt{2} + 4\sqrt{7}$ over $\mathbb{Q}$.

22.1.9. Let $\zeta_7 = e^{2\pi i/7} = \cos(\frac{2\pi}{7}) + i\sin(\frac{2\pi}{7}) \in \mathbb{C}$. The complex number ζ_7 is called a seventh root of unity since $(\zeta_7)^7 = 1$. Find the minimal polynomial of ζ_7 over $\mathbb{Q}$.

22.1.10. Let $\zeta_7 = e^{2\pi i/7} = \cos(\frac{2\pi}{7}) + i\sin(\frac{2\pi}{7}) \in \mathbb{C}$. Find the minimal polynomial of ζ_7 over $\mathbb{Q}(\cos(\frac{2\pi}{7}))$.

22.1.11. Is $\mathbb{Q}[\sqrt{3}, i]$ a simple extension of $\mathbb{Q}$? If so, find a primitive element for the extension.

22.1.12. Is $\mathbb{Q}[\sqrt{4}, \sqrt{7}]$ a simple extension of $\mathbb{Q}$? If so, find a primitive element for the extension.

22.1.13. Let
$$\alpha = \sqrt{4 + \sqrt{7}} + \sqrt{4 - \sqrt{7}}.$$
Find $\min_{\mathbb{Q}}(\alpha)$ and all its roots. Is α one of the roots of $\min_{\mathbb{Q}}(\alpha)$? Which one?

22.1.14. Let $E = \mathbb{Q}[\sqrt{3}, \sqrt[3]{2}]$, and let $\alpha = \sqrt{3} + \sqrt[3]{2}$.
  (a) Show that $E = \mathbb{Q}[\alpha]$.
  (b) What is $|E : \mathbb{Q}|$?
  (c) Find the minimal polynomial of α over $\mathbb{Q}$ (and prove that it is the minimal polynomial).

$a^3 - 4 \rightarrow \quad x^3 - 4$

22.1.15. Let $f \in \mathbb{R}[x]$ and $\deg(f) > 2$. Can f be irreducible? Either prove that it cannot be or give an example where it is.

22.1.16. Let $f = x^4 + 2 \in \mathbb{R}[x]$. Is f irreducible in $\mathbb{R}[x]$? Either prove that it is or give a factorization of f into irreducible factors.

22.1.17. Let $F \subseteq E$ be fields, and let $\alpha \in E$ be algebraic over F. Let $f = \min_F(\alpha)$. Show that
$$F[\alpha] \cong F[x]/\langle f \rangle.$$
Conclude that $F[\alpha]$ is a field.

22.1.18. (a) Let $A = \dfrac{1}{1 + \sqrt[3]{2} + \sqrt[3]{4}}$.

   (i) Using Problem **22.1.17** and without any calculation, argue that there must exist rational numbers α, β, and γ so that
   $$A = \alpha + \beta \sqrt[3]{2} + \gamma \sqrt[3]{4}.$$

   (ii) Actually find the rational numbers α, β, and γ predicted in the previous part.

   (b) Find rational numbers α, β, and γ so that
   $$\frac{1}{1 + 2\sqrt[3]{2} + 5\sqrt[3]{4}} = \alpha + \beta \sqrt[3]{2} + \gamma \sqrt[3]{4}.$$

22.1.19. Let $\mathbb{F}_5 = (\mathbb{Z}/5\mathbb{Z}, +, \cdot)$ be the field with five elements, and let E be a field with $\mathbb{F}_5 \subseteq E$. Let $\alpha \in E$, and assume $\min_{\mathbb{F}_5}(\alpha) = x^4 + 4x^2 + 2$ is the minimal polynomial of α over $\mathbb{F}_5$. Can we express $\dfrac{1}{\alpha^2 + \alpha + 1}$ as a polynomial of degree no more than 3 in α?

22.1.20. Let R be an integral domain, and let F be its field of fractions. Prove that the field of fractions of $R[x]$ is $F(x)$.

22.1.21. Let $F \subseteq E$ fields, and let $\alpha \in E$. Let $\epsilon_\alpha : F[x] \to E$ be the evaluation homomorphism. Show that ϵ_α is 1-1 if and only if α is transcendental.

22.1.22. Let $F \subseteq E$ be a field extension, and let $\alpha, \beta \in E$. Show that α is algebraic over $F(\beta)$ if and only if β is algebraic over $F(\alpha)$.

22.1.23. Let $E = \mathbb{C}(x)$, let $F = \mathbb{Q}(x^6) \subset E$, and let $f = y^6 - x^6 \in F[y]$. What are the roots of f in E?

22.1.24. Let $\alpha = x^6$, and consider the fields $\mathbb{Q} \subseteq K \subseteq \mathbb{Q}(x)$, where $K = \mathbb{Q}(\alpha)$.
   (a) Find a polynomial in $K[y]$ that has x as a root.
   (b) What is $\min_K(x)$?

22.1.25. Let $\alpha = \frac{x}{x^3+1} \in \mathbb{Q}(x)$, and let $K = \mathbb{Q}(\alpha) \subseteq \mathbb{Q}(x)$. Is x algebraic over K? If so what is $\min_K(x)$?

22.1.26. Let $K = \mathbb{Q}(x)$ be the field of rational functions over $\mathbb{Q}$. Is $y^4 - 6y^2 + 3$ an irreducible element of $K[y]$? Prove your assertion.

22.1.27. It is known that both e and π are transcendental, but it is not known whether $e + \pi$ or $e\pi$ are irrational. Assuming the transcendence of e and π, prove that at most one of $e + \pi$ or $e\pi$ can be rational.

22.2. A Quick Review of Vector Spaces

We assume that the reader is familiar with vector spaces, but we repeat the necessary definitions here for completeness.

Definition 22.19 (Vector space). Let V be a non-empty set, and let F be a field. Assume that we have two operations: addition and scalar multiplication—-that is, for $v, w \in V$ and $\alpha \in F$, we have $v + w, \alpha v \in V$—such that

(a) $(V, +)$ is an abelian group. In other words, $+$ is a closed, associative, commutative binary operation on V that has a (additive) zero, and such that every element has an (additive) inverse.

(b) Scalar multiplication defines an action of the abelian group $(F - \{0\}, \cdot)$ on V. In other words for $\alpha \in F - \{0\}$ and $v \in V$, we have $1v = v$, and $\alpha(\beta v) = (\alpha \beta)v$.

(c) Addition and scalar multiplication satisfy the distributive laws. In other words for $\alpha, \beta \in F$ and $v, w \in V$, we have $\alpha(v+w) = \alpha v + \alpha w$, and $(\alpha+\beta)v = \alpha v + \beta v$.

Then V is called a *vector space over F*.

Remark 22.20. If V is a vector space over F, then both V and F have an additive zero. Usually, from the context, it is clear which zero we are talking about, and we can denote both of them by 0. If needed, we will denote the zero of V by 0_V, and the zero of F by 0_F.

Definition 22.21 (Subspaces, linear independence, span, basis, dimension). Let V be a vector space over the field F, and let $v_1, \ldots, v_k$ be a set of vectors in V.

A non-empty subset of V that is closed under addition and scalar multiplication is called a *subspace* of V. In other words, a subspace is a subset that—with the same operations as in V—is itself a vector space.

Any element of v of the form $\alpha_1 v_1 + \cdots + \alpha_k v_k$, where $\alpha_1, \ldots, \alpha_k \in \mathbb{R}$, is called a *linear combination* of $v_1, \ldots, v_k$. If S is a (possibly infinite) set of vectors in a vector space, then the collection of all linear combinations of a finite number of elements of S is called the *span* of S, and denoted by $\mathrm{Span}(S)$.

The set of elements $v_1, \ldots, v_k \in V$ is called *linearly independent over F* if the *only* linear combination of these elements equal to 0_V is $0_F v_1 + \cdots + 0_F v_k$. An infinite set of vectors is *linearly independent* if every finite subset of the vectors is linearly independent.

A linearly independent set of vectors of V whose span is all of V is called a *basis* for V. If V has a finite basis, then it is called *finite dimensional*; otherwise, it is called *infinite dimensional*.

The main fact from linear algebra that we require is the fact that *dimension* is well defined:

Theorem 22.22. *Let F be a field, and let V be a vector space over F. Assume that S and T are bases for V over F. Then $|S| = |T|$.*

Proof. Consult a text on linear algebra. □

Definition 22.23. Let F be a field, let V be a vector space over F, and let S be a basis for V over F. Then $|S|$ is called the *dimension* of V over F and is denoted by $\dim_F(V)$.

Remark 22.24. The statement that "all vector spaces have a basis" is equivalent to the axiom of choice. Hence, without some version of the axiom of choice, and assuming only the usual axioms of Zermelo–Fraenkel set theory, we can neither prove that all vector spaces have a basis nor find a counterexample! You were asked, in Problem 16.1.30, to use the Kuratowski–Zorn lemma (Axiom 16.35 which is equivalent to the axiom of choice) to prove that every vector space has a basis.

We will also occasionally use linear transformations between vector spaces, but most of what we use are general facts that apply to all homomorphisms between algebraic objects.

Definition 22.25. Let V and W be vector spaces over a field F, and let $T\colon V \to W$ be a map. T is called a *linear transformation* or an *F-linear* map if, for all $u, v \in V$ and $\alpha \in F$, we have

$$T(u + v) = T(u) + T(v), \text{ and } T(\alpha u) = \alpha T(u).$$

Given a linear transformation $T\colon V \to W$, we define the kernel and the image as usual, and one can prove the usual homomorphism theorems. For example, $V/\ker(T) \cong \operatorname{Im}(T) = T(V)$, and as a result we have:

Theorem 22.26. *Let $T\colon V \to W$ be a linear transformation, and assume that $\dim(V) < \infty$. Then*

$$\dim(\ker(T)) + \dim(\operatorname{Im}(T)) = \dim(V).$$

Problems

22.2.1. Let $V = \{(a+c, 2a+b+c, a+b, b-c, a+2b-c) \mid a,b,c \in \mathbb{R}\} \subseteq \mathbb{R}^5$. Is V a vector space over $\mathbb{R}$ (with the usual addition and scalar multiplication of vectors)? If so, find a basis for V over $\mathbb{R}$. What is $\dim_\mathbb{R}(V)$?

22.2.2. The field of complex numbers $\mathbb{C}$ is a vector space over $\mathbb{R}$ and over $\mathbb{C}$. What is $\dim_\mathbb{R}(\mathbb{C})$? What is $\dim_\mathbb{C}(\mathbb{C})$?

22.2.3. Let $V = \mathbb{R}[\sqrt{3}] = \{a + b\sqrt{3} \mid a,b \in \mathbb{R}\}$. Is V a vector space over $\mathbb{R}$ (with the usual addition and multiplication of numbers)? If so, what is $\dim_\mathbb{R}(V)$? Is V also a field?

22.2.4. Let $V = \{a + b\sqrt[3]{3} \mid a,b \in \mathbb{R}\}$. With the usual addition and multiplication of numbers, is V a vector space over $\mathbb{R}$? If so, what is $\dim_\mathbb{R}(V)$? Is V also a field? Is it a ring?

22.2.5. Let $V = \mathbb{R}[x] = \{a_0 + a_1 x + \cdots + a_n x^n \mid n \in \mathbb{Z}^{\geq 0}, a_0, a_1, \ldots, a_n \in \mathbb{R}\}$ be the ring of all polynomials in x over $\mathbb{R}$. With the usual addition and scalar multiplication for polynomials, is V a vector space over $\mathbb{R}$? Can you find a basis for V? What is $\dim_\mathbb{R}(V)$?

22.2.6. Let V be a vector space over a field F. Let B be a basis for V over F, and let $v \in V$. Prove that v can be written as a linear combination of elements of B in exactly one way.

22.2.7. Let X be an arbitrary set, and let F be a field. Let $\mathcal{F}(X, F) = \{f \colon X \to F\}$ be the set of functions from X to F. If $f, g \in \mathcal{F}(X, F)$, then $f + g$ is defined, as usual, by $(f + g)(x) = f(x) + g(x)$ for all $x \in X$. If $\alpha \in F$ and $f \in \mathcal{F}(X, F)$, then αf, the scalar multiplication of α and f, is defined, as usual, by $(\alpha f)(x) = \alpha f(x)$ for all $x \in X$.
  (a) Is $\mathcal{F}(X, F)$ always a vector space over F?
  (b) If $X = \{1, 2\}$ and $F = \mathbb{R}$, then what is a basis for $\mathcal{F}(X, F)$? What is $\dim_F(\mathcal{F}(X, F))$?
  (c) If $X = F = \mathbb{R}$, give five well known elements of $\mathcal{F}(X, F)$ that form a linearly independent set.

22.2.8. Let V be a finite dimensional vector space, and let $T \colon V \to V$ be a linear transformation. Prove that T is 1-1 if and only if T is onto.

22.2.9. Let $V = \mathbb{R}^\infty = \{(a_1, a_2, \ldots) \mid a_1, a_2, \ldots \in \mathbb{R}\}$, the vector spaces of infinite sequences with real entries. Let e_i denote the element in V with a 1 in the ith entry and 0's elsewhere. Let $A = \{e_i \mid 1 \leq i < \infty\}$.
  (a) Is A linearly independent?
  (b) Let $\mathbf{j} = (1, 1, 1, \ldots) \in V$ be the vector of all 1's. Is $A \cup \{\mathbf{j}\}$ linearly independent?
  (c) Is the vector $(1, 2, 3, 4, \ldots)$ in the span of $A \cup \{\mathbf{j}\}$?
  (d) Comment.

22.3. The Degree of an Extension

Lemma 22.27. *Let R be a commutative ring with identity, and let $F \subseteq R$ be a field. Assume that $1_F = 1_R$. Then R—using the ring addition and multiplication—is a vector space over F.*

Proof. We know that $(R, +)$ is an abelian group, and the scalar multiplication follows the axioms of a vector space. $\qquad\square$

Remark 22.28. Note that if F is a field, R a commutative ring with identity, and $F \subseteq R$, then $0_F = 0_R$. Hence, in the case of this vector space, the zero of the vector space is the same as the zero of the field.

Definition 22.29 (Degree of an extension). Let R be a commutative ring with identity, and let $F \subseteq R$ be a field. Consider R a vector space over F. The dimension of this vector space is denoted by $|R : F|$ or $\dim_F(R)$. If $F \subseteq E$ are fields, then $|E : F|$ is called the *degree* of the field extension.

Example 22.30. The set of elements $\{1, i\}$ forms a basis for the vector space $\mathbb{C}$ over the field $\mathbb{R}$ as well as for the vector space $\mathbb{Q}[i]$ over the field $\mathbb{Q}$. Hence, $|\mathbb{C} : \mathbb{R}| = 2 = |\mathbb{Q}[i] : \mathbb{Q}|$. Every element of $\mathbb{Q}[\sqrt[3]{2}]$ can be written as $\alpha + \beta\sqrt[3]{2} + \gamma\sqrt[3]{4}$, $\{1, \sqrt[3]{2}, \sqrt[3]{4}\}$ is a basis for $\mathbb{Q}[\sqrt[3]{2}]$ over $\mathbb{Q}$, and so $|\mathbb{Q}[\sqrt[3]{2}] : \mathbb{Q}| = 3$. Finally, $\{1, x, x^2, \ldots\}$ is a basis for $\mathbb{Q}[x]$ over $\mathbb{Q}$, and so $|\mathbb{Q}[x] : \mathbb{Q}| = \infty$.

Remark 22.31. Let F be a field, let R be a commutative ring with identity, and let $F \subset R$. If R is an infinite dimensional vector space over F, then we write $|R : F| = \infty$. However, when we write $|R : F| = |R' : F'|$—regardless of whether these are finite dimensional or infinite dimensional extensions—we mean that there is a 1-1 correspondence between a basis of R over F and a basis of R' over F'. In other words, if $F \subset R$ and $F' \subset R'$ are infinite dimensional extensions, then we only write $|R : F| = |R' : F'|$ if a basis for R over F has the same cardinality (see Definition 1.23) as the basis for R' over F'.

Just realizing the vector space structure of field extensions allows us to find the possible sizes of finite fields.

Theorem 22.32. *Assume that E is a field with a finite number of elements, then $|E| = p^n$ where p is a prime number.*

Proof. Every field has a subfield—the prime subfield—isomorphic to $\mathbb{Q}$ or $\mathbb{F}_p = (\mathbb{Z}/p\mathbb{Z}, +, \cdot)$ for some prime p (Corollary 16.55). Since E is finite, the prime subfield cannot be $\mathbb{Q}$, and so it is $\mathbb{F}_p$ for some p. Hence, we have $\mathbb{F}_p \subseteq E$, and E is a vector space over $\mathbb{F}_p$. Now E has a finite number of elements, and so the dimension of E over $\mathbb{F}_p$ must be finite. Assume $|E : \mathbb{F}_p| = n$ for some positive integer n. Let $\{e_1, \ldots, e_n\}$ be a basis for E over $\mathbb{F}_p$. Then each element of E can be written uniquely (see Problem 22.2.6) as

$$\alpha_1 e_1 + \cdots + \alpha_n e_n, \quad \text{with } \alpha_1, \ldots, \alpha_n \in \mathbb{F}_p.$$

Now, for each α_i, we have p choices, and so the total number of elements of E is p^n. $\qquad\square$

We next show that, in the case of finite dimensional extensions, nothing is lost if we limit ourselves to *field* extensions.

Lemma 22.33. *Let R be an integral domain, and let F be a field with $F \subseteq R$. Assume $|R : F| < \infty$, then R is a field.*

Proof. To show R is a field, let $0 \neq \beta \in R$. We need to find $\beta^{-1} \in R$.

Consider $1, \beta, \beta^2, \ldots \in R$. Since R is a finite dimensional vector space over F, eventually one of the powers of β will be linearly dependent on the ones before. Let m be the smallest positive integer such that β^m is linearly dependent on $1, \beta, \ldots, \beta^{m-1}$.

Hence, $c_m \beta^m + \cdots + c_1 \beta + c_0 1 = 0$, where $c_m \neq 0$.

Note that $c_0 \neq 0$, since otherwise $\beta(c_m \beta^{m-1} + \cdots + c_1) = 0$. This would mean—since R is an integral domain and $\beta \neq 0$—that $c_m \beta^{m-1} + \cdots + c_1 = 0$, contradicting the choice of m. We thus have

$$-c_0^{-1} \beta (c_m \beta^{m-1} + \cdots + c_1) = 1.$$

The proof is now complete since

$$\beta^{-1} = -c_0^{-1}(c_m \beta^{m-1} + \cdots + c_1) \in R. \qquad\square$$

Lemma 22.34. *Let $F \subseteq E$ be fields, and let $\alpha \in E$. Assume α is algebraic over F, and $\deg(\min_F(\alpha)) = n$. Then $\{1, \alpha, \alpha^2, \ldots, \alpha^{n-1}\}$ is a basis for $F[\alpha]$ over F. In particular, $|F[\alpha] : F| = \deg(\min_F(\alpha))$.*

Proof. Let $f(x) = \min_F(\alpha) = x^n + a_{n-1}x^{n-1} + \cdots + a_0$. We then have $f(\alpha) = 0$ which means that $\alpha^n = -a_{n-1}\alpha^{n-1} - \cdots - a_0$. By repeatedly using this, we can write every α^m as a linear combination, over F, of $1, \alpha, \alpha^2, \ldots, \alpha^{n-1}$. Now $F[\alpha]$ consists of linear combinations of powers of α and so it is spanned by $\{1, \alpha, \alpha^2, \ldots, \alpha^{n-1}\}$ over F.

To show that $\{1, \alpha, \alpha^2, \ldots, \alpha^{n-1}\}$ is linearly independent over F, assume

$$b_{n-1}\alpha^{n-1} + \cdots + b_0 1 = 0.$$

Define $g(x) = b_{n-1}x^{n-1} + \cdots + b_0 \in F[x]$. Then $g(\alpha) = 0$ and $\deg(g(x)) < n$. But $\deg(\min_F(\alpha)) = n$, and among the non-zero polynomials that have α as a root, the minimal polynomial has the least degree. Hence g is the zero polynomial, which means $b_i = 0$, for $0 \le i \le n-1$. This implies that $\{1, \alpha, \ldots, \alpha^{n-1}\}$ is linearly independent over F. $\qquad\square$

Theorem 22.35. *Let $F \subseteq E$ be fields, and let $\alpha \in E$. Then the following are equivalent:*

(a) *α is algebraic over F.*

(b) *$|F[\alpha] : F| < \infty$.*

(c) *$F[\alpha]$ is a field.*

(d) *$F[\alpha] = F(\alpha)$.*

(e) *$|F(\alpha) : F| < \infty$.*

(f) *There exists a field K such that $F \subseteq K \subseteq E$, $\alpha \in K$, and $|K : F| < \infty$.*

Proof. (a) $\Rightarrow$ (b) Lemma 22.34 gives a finite basis for $F[\alpha]$ over F, and hence $|F[\alpha] : F| < \infty$.

(b) $\Rightarrow$ (c) This is a special case of Lemma 22.33.

(c) $\Rightarrow$ (d) $F(\alpha)$ is the *smallest* field containing $F[\alpha]$. Now if $F[\alpha]$ is a field itself, then it is the smallest field containing itself.

(d) $\Rightarrow$ (a) We know $F[\alpha] = F(\alpha)$ is a field, and we want to show that α is algebraic over F. There is nothing to show if $\alpha = 0$, and so assume that $\alpha \ne 0$. Since $F[\alpha]$ is a field, $\alpha^{-1} \in F[\alpha]$, which means that $\alpha^{-1} = g(\alpha)$ some $g \in F[x]$. It follows that $\alpha g(\alpha) - 1 = 0$.

Define $f(x) = xg(x) - 1 \in F[x]$. Note that $f(\alpha) = 0$ and $f \ne 0$, which means that α is algebraic over F.

(a)$-$(d) $\Rightarrow$ (e) If $F[\alpha] = F(\alpha)$ and $|F[\alpha] : F| < \infty$, then clearly $|F(\alpha) : F| < \infty$.

(e) $\Rightarrow$ (b) We have $F \subseteq F[\alpha] \subseteq F(\alpha)$. Both of $F[\alpha]$ and $F(\alpha)$ are vector spaces over F, and so $F[\alpha]$ is a subspace of $F(\alpha)$. If $F(\alpha)$ is finite dimensional over F, then so is $F[\alpha]$.

(e) $\Rightarrow$ (f) Take $K = F(\alpha)$.

(f) $\Rightarrow$ (e) If K is a field that contains both F and α, then it must contain $F(\alpha)$. We have $F \subseteq F(\alpha) \subseteq K$, and so $F(\alpha)$ is a subspace of K. Since K is finite dimensional, so is $F(\alpha)$. $\qquad\square$

Example 22.36. It can be shown that π is transcendental over $\mathbb{Q}$. Hence, $\mathbb{Q}[\pi] \neq \mathbb{Q}(\pi)$. On the other hand, now we know $\mathbb{Q}[\sqrt[3]{2}]$ is a field, and so $\frac{a+b\sqrt[3]{2}+c\sqrt[3]{4}}{d+e\sqrt[3]{2}+f\sqrt[3]{4}}$ can be written as $\alpha + \beta\sqrt[3]{2} + \gamma\sqrt[3]{4}$. While we have proved that this can be done in theory, unless you are good with your (high school) algebra, it is not obvious how to do this practically. The above *proof* provides one possible road map—actually not the best possible—for the necessary calculation.

Remark 22.37. Let $F \subseteq E$ be a field extension, and let $\alpha \in E$ be algebraic over $\mathbb{F}$. See Problem *22.1.17* for a different proof that $F[\alpha]$ is a field.

Proposition 22.38. *Let F, K, and E be fields with $F \subseteq K \subseteq E$. Assume that $|K : F|$ and $|E : K|$ are finite. Then $|E : F| = |E : K| \, |K : F|$.*

Proof. Let $\{k_1, \ldots, k_n\}$ be a basis for K over F, and let $\{e_1, \ldots, e_m\}$ be a basis for E over K.

CLAIM: $\{k_i e_j \mid 1 \leq i \leq n, 1 \leq j \leq m\}$ is a basis for E over F.

PROOF OF CLAIM: *Span.* Let $x \in E$, then $x = y_1 e_1 + \cdots + y_m e_m$, where $y_1, \ldots, y_m \in K$. Now write each y_i in terms of $k_1, \ldots, k_n$, substitute, and get an expression for x in terms of $\{k_i e_j\}$.

Linear independence. Assume that, for $\alpha_{1,1}, \ldots, \alpha_{n,m} \in F$, we have

$$\alpha_{1,1} k_1 e_1 + \alpha_{1,2} k_1 e_2 + \alpha_{2,1} k_2 e_1 + \cdots + \alpha_{n,m} k_n e_m = 0.$$

We can collect terms and write:

$$(\alpha_{1,1} k_1 + \alpha_{2,1} k_2 + \cdots + \alpha_{n,1} k_n) e_1 + \cdots + (\alpha_{1,m} k_1 + \alpha_{2,m} k_2 + \cdots + \alpha_{n,m} k_n) e_m = 0.$$

Now since $\{e_1, \ldots, e_m\}$ is linearly independent over K, we conclude that each of the coefficients—that is, $\alpha_{1,j} k_1 + \alpha_{2,j} k_2 + \cdots + \alpha_{n,j} k_n$, for $1 \leq j \leq m$—must be zero. But then again $\{k_1, \ldots, k_n\}$ is linearly independent over F, and so all $\alpha_{i,j}$, for $0 \leq i \leq n$, $0 \leq j \leq m$, must be zero. □

Remark 22.39. Let $F \subseteq K \subseteq E$ be fields. Let S be a basis for K over F, and let T be a basis for E over K. Then basically the same proof as that of Proposition 22.38 shows—regardless of whether these bases are finite—that $R = \{st \mid s \in S, t \in T\}$ is a basis for E over F. Hence, it follows that a basis for E over F has the same cardinality—finite or infinite—as $S \times T$. In particular, E is finite dimensional over F if and only if both of the vector spaces E over K and K over F are finite dimensional.

Example 22.40. Let $E = \mathbb{Q}(\sqrt[3]{5}, \sqrt{3})$. Then E is a vector space over $\mathbb{Q}$. What is a basis for this vector space?

We first note that, by the Schönemann-Eisenstein criterion (Corollary 19.64), both $x^2 - 3$ and $x^3 - 5$ are irreducible over $\mathbb{Q}$, and hence they are, respectively, the minimal polynomials of $\sqrt{3}$ and $\sqrt[3]{5}$ over $\mathbb{Q}$. This means that $|\mathbb{Q}(\sqrt{3}) : \mathbb{Q}| = 2$ and $|\mathbb{Q}(\sqrt[3]{5}) : \mathbb{Q}| = 3$. Now

$$\mathbb{Q} \subseteq \mathbb{Q}(\sqrt{3}) \subseteq \mathbb{Q}(\sqrt{3}, \sqrt[3]{5}) \quad \text{and} \quad \mathbb{Q} \subseteq \mathbb{Q}(\sqrt[3]{5}) \subseteq \mathbb{Q}(\sqrt[3]{5}, \sqrt{3}).$$

Hence, by Proposition 22.38, both $|\mathbb{Q}(\sqrt{3}) : \mathbb{Q}|$ and $|\mathbb{Q}(\sqrt[3]{5}) : \mathbb{Q}|$ divide $|\mathbb{Q}(\sqrt{3}, \sqrt[3]{5}) : \mathbb{Q}|$. Since 2 and 3 are relatively prime, the latter dimension must be a multiple of 6.

But $x^2 - 3$ can be thought of as a polynomial in $\mathbb{Q}(\sqrt[3]{5})[x]$, hence $|\mathbb{Q}(\sqrt[3]{5}, \sqrt{3}) : \mathbb{Q}(\sqrt[3]{5})| \leq 2$. (Note that, unless we prove that $x^2 - 3$ is irreducible in $\mathbb{Q}(\sqrt[3]{5})[x]$, we cannot claim that this degree is actually 2.) So

$$\left|\mathbb{Q}(\sqrt{3}, \sqrt[3]{5}) : \mathbb{Q}\right| = \left|\mathbb{Q}(\sqrt{3}, \sqrt[3]{5}) : \mathbb{Q}(\sqrt[3]{5})\right| \left|\mathbb{Q}(\sqrt[3]{5}) : \mathbb{Q}\right| \leq 6.$$

Since $\left|\mathbb{Q}(\sqrt{3}, \sqrt[3]{5}) : \mathbb{Q}\right|$ was also a multiple of 6, we conclude that is exactly equal to 6. Moreover, we now conclude that $\left|\mathbb{Q}(\sqrt[3]{5}, \sqrt{3}) : \mathbb{Q}(\sqrt[3]{5})\right| = 2$.

Now that we know the degree of the extension, using Lemma 22.34, we have that $\{1, \sqrt[3]{5}, \sqrt[3]{25}\}$ and $\{1, \sqrt{3}\}$ are bases, respectively, for $\mathbb{Q}(\sqrt[3]{5})$ over $\mathbb{Q}$ and $\mathbb{Q}(\sqrt[3]{5}, \sqrt{3})$ over $\mathbb{Q}(\sqrt[3]{5})$. Per Remark 22.39, a basis for $\mathbb{Q}(\sqrt[3]{5}, \sqrt{3})$ over $\mathbb{Q}$ is

$$\{1, \sqrt[3]{5}, \sqrt[3]{25}, \sqrt{3}, \sqrt{3}\sqrt[3]{5}, \sqrt{3}\sqrt[3]{25}\}.$$

Partial Lattice Diagrams of Fields. If E is a field, then the collection of all subfields of E ordered by inclusion is a poset—in fact, a lattice—and hence we can draw its Hasse diagram. Just as for the poset of subgroups of a group and the poset of ideals of a ring, we often draw a *partial* lattice diagram and include only the subfields relevant to the work at hand. Proposition 22.38 suggests a labeling of the edges akin to what we did in the case of groups. If $F \subseteq E$ are fields, then we label the edge between F and E with $|E : F|$. Proposition 22.38 now says that consecutive edge lengths multiply. The drawing on the left of Figure 22.2 gives a partial lattice diagram for the fields of Example 22.40.

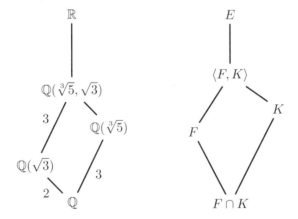

Figure 22.2. Partial lattice diagram of fields for Example 22.40 and for two fields in general position

If F and K are subfields of a field E, then, just as in group theory, we usually draw a partial lattice diagram that contains $F \cap K$ and $\langle F, K \rangle$. (See Problem 22.3.23 for guidance on the edge lengths for fields in general position as in the drawing on the right of Figure 22.2.) In field theory, $\langle F, K \rangle$, the subfield of E generated by F and K, is called the *compositum* of F and K. We record the definition here.

Definition 22.41. Let F and K be subfields of a field E, then the *compositum* of F and K in E, written $\langle F, K \rangle$ is the intersection of all subfields of E that contain both F and K. It is the smallest subfield of E containing both F and K. (An oft used notation for the compositum of F and K is FK.)

We now prove a straightforward but illuminating corollary of Proposition 22.38.

Corollary 22.42. *Let $F \subseteq E$ be fields, and assume $|E : F| < \infty$. Let $\alpha \in E$. Then $\deg(\min_F(\alpha))$ divides $|E : F|$, and in particular E is an algebraic extension of F.*

Proof. We have $F \subseteq F[\alpha] \subseteq E$, and hence

$$|E : F| = |E : F[\alpha]| \underbrace{|F[\alpha] : F|}_{\deg(\min_F(\alpha))}.$$

We conclude that $\deg(\min_F(\alpha))$ divides $|E : F|$ and is finite. $\square$

If $F \subseteq E$, and F and E are both fields, then, as we have seen, some elements of E are algebraic over F while others are transcendental. In many ways, the algebraic elements are easier to deal with, and it is natural—especially because of the next theorem—to consider the whole collection of algebraic numbers in E.

Theorem 22.43. *Let $F \subseteq E$ be fields. Let*

$$U = \{\alpha \in E \mid \alpha \text{ algebraic over } F\}.$$

Then U is a field.

Proof. Let $\alpha, \beta \in U$. We need to show that $\alpha + \beta \in U$ and, as long as $\beta \neq 0$, $\alpha/\beta \in U$. We do this indirectly and by using Theorem 22.35(f). Note that

$$F \subseteq F[\alpha] \subseteq (F[\alpha])[\beta] \subseteq E.$$

By Theorem 22.35, α algebraic over F means that $|F[\alpha] : F| < \infty$ and that $F[\alpha]$ is a field.

Now β is algebraic over F which certainly means that it is also algebraic over $F[\alpha]$ (whatever polynomial over F that had β as a root is also a polynomial over $F[\alpha]$). Hence, again by Theorem 22.35, $|(F[\alpha])[\beta] : F[\alpha]| < \infty$, and $(F[\alpha])[\beta]$ is a field.

Now, by Proposition 22.38, $|(F[\alpha])[\beta] : F| < \infty$. This means that $\alpha + \beta$ and, for $\beta \neq 0$, α/β are elements of a field—namely, $(F[\alpha])[\beta]$—that is a finite degree extension of F. It follows from Theorem 22.35(f) that they are both algebraic! $\square$

Remark 22.44. The above proof is non-constructive and should have come as a surprise. To prove that $\alpha + \beta$ is algebraic over F, we would expect to produce a polynomial over F that has $\alpha + \beta$ as a root. We did not. Instead we just showed that such a polynomial must exist. While this approach is sleek and somewhat appealing, it does not really help if you really want to find the minimal polynomial of $\alpha + \beta$.

Specializing to the complexes, we have the following definitions:

Definition 22.45 (Field of algebraic numbers). Let

$$\mathbb{A} = \{\alpha \in \mathbb{C} \mid \alpha \text{ is algebraic over } \mathbb{Q}\}.$$

Then $\mathbb{A}$ is called the *field of algebraic numbers*. In addition, any finite degree extension of $\mathbb{Q}$ is called an *algebraic number field*.

Algebraic number fields are the main object of study in algebraic number theory. Now, while every element of the field of algebraic numbers gives a finite degree extension of $\mathbb{Q}$, the whole field of algebraic numbers is not a finite degree extension.

Theorem 22.46. *Let $\mathbb{A} \subset \mathbb{C}$ be the field of algebraic numbers. Then $|\mathbb{A} : \mathbb{Q}| = \infty$.*

Proof outline. If you assume that $|\mathbb{A} : \mathbb{Q}| = m$, then you get a contradiction because, by the Schönemann-Eisenstein criterion (Corollary 19.64), $x^n - 2$ is irreducible in $\mathbb{Q}[x]$ for all $n \in \mathbb{Z}^{>0}$. You are asked to provide the details in Problem **22.3.32**. $\square$

There is more to say about algebraic numbers. Here is a useful proposition:

Proposition 22.47. *Let $F \subseteq E \subseteq L$ be fields. Assume E is algebraic over F. Let $\alpha \in L$, and assume α is algebraic over E. Then α is algebraic over F.*

Proof. This is Problem *22.3.31*. $\square$

Definition 22.48 (Algebraic integer). An *algebraic integer* is a complex number which is a root of a *monic* polynomial with integer coefficients. In other forms, algebraic integers are roots of polynomials of the form

$$x^n + a_{n-1}x^{n-1} + \cdots + a_0,$$

where $a_i \in \mathbb{Z}$ for $0 \le i \le n - 1$.

Problems

22.3.1. In Lemma 22.27, it is assumed that $1_R = 1_F$. Is that assumption necessary? Either give a proof that it is not, or give an example to show that, without this assumption, the conclusion does not hold.

22.3.2. Find a basis for $\mathbb{Q}(\sqrt[5]{47}, \sqrt{47})$ over $\mathbb{Q}$.

22.3.3. Let $\alpha = \sqrt[3]{2} + \sqrt[3]{4}$. Find the minimal polynomial of α over $\mathbb{Q}$. Is $\frac{1}{\alpha} + \frac{1}{\alpha^2} \in \mathbb{Q}[\alpha]$?

22.3.4. Assume that $F \subseteq K \subseteq E$ with F, K, and E all fields. Assume $|E : F| < \infty$. Show that $|E : K|$ and $|K : F|$ are both finite.

22.3.5. Let $K = \mathbb{Q}[\sqrt{2}]$ and $E = K[\sqrt{3}] = \mathbb{Q}[\sqrt{2}, \sqrt{3}]$. What is $\min_K(\sqrt{3})$? What is $|E : K|$? What is $|E : Q|$? What is a basis for E as a vector space over $\mathbb{Q}$?

22.3.6. Let $E = \mathbb{Q}(\sqrt{2} + \sqrt{3})$. What is $|E : \mathbb{Q}|$? What is the minimal polynomial of $\sqrt{2} + \sqrt{3}$ over $\mathbb{Q}$? Make sure that you give an argument for why your suggested polynomial is irreducible.

22.3.7. Let $E = \mathbb{Q}(\sqrt{2}, \sqrt{3})$, and let $V = \mathcal{F}(E, E)$ be the vector space of functions from E to E.
   (a) Let $\alpha \in E$. Show that $\alpha = a + b\sqrt{2} + c\sqrt{3} + d\sqrt{6}$.

(b) Define four functions id, $\sigma, \tau, \delta \in V$ by

$$\mathrm{id}(a + b\sqrt{2} + c\sqrt{3} + d\sqrt{6}) = a + b\sqrt{2} + c\sqrt{3} + d\sqrt{6},$$
$$\sigma(a + b\sqrt{2} + c\sqrt{3} + d\sqrt{6}) = a - b\sqrt{2} + c\sqrt{3} - d\sqrt{6},$$
$$\tau(a + b\sqrt{2} + c\sqrt{3} + d\sqrt{6}) = a + b\sqrt{2} - c\sqrt{3} - d\sqrt{6},$$
$$\delta(a + b\sqrt{2} + c\sqrt{3} + d\sqrt{6}) = a - b\sqrt{2} - c\sqrt{3} + d\sqrt{6}.$$

Show that σ is an automorphism of E. (The functions id, τ, and δ are also automorphisms of E.)

(c) Show that, as a subset of V, the set $\{\mathrm{id}, \sigma, \tau, \delta\}$ is linearly independent.

(d) Let $\circ$ denote function composition, and let $G = \{\mathrm{id}, \sigma, \tau, \delta\}$. Show that $(G, \circ)$ is a group.

(e) Find a familiar group that is isomorphic to $(G, \circ)$.

22.3.8. Can a field with 243 elements have a subfield with nine elements?

22.3.9. Let $\mathbb{F}_{p^m}$ and $\mathbb{F}_{p^n}$ denote fields with p^m and p^n elements, respectively. Show that if $\mathbb{F}_{p^m} \subseteq \mathbb{F}_{p^n}$, then m divides n.

22.3.10. Let $\alpha = \sqrt[3]{2} \in \mathbb{R}$. Write $\dfrac{\alpha - 1}{\alpha^2 - 1}$ as a linear combination of powers of α.

22.3.11. Let $\alpha \in \mathbb{C}$ be a root of $x^4 + x^2 + 5$. Let $E = \mathbb{Q}[\alpha]$, and let $\beta = \alpha^2 + \alpha + 1 \in E$. Write the inverse of β as a polynomial in α.

22.3.12. Let $f = x^4 - 10x^2 + 20 \in \mathbb{Q}[x]$, $\alpha \in \mathbb{C}$ a root of f, and $\beta = -4\alpha^2 + 20$.
(a) Is f irreducible over $\mathbb{Q}$?
(b) What is $|\mathbb{Q}[\alpha] : \mathbb{Q}|$?
(c) What is $|\mathbb{Q}[\beta] : \mathbb{Q}|$? What is $|\mathbb{Q}[\alpha] : \mathbb{Q}[\beta]|$?

22.3.13. Let $F \subseteq E$ be a field extension, let $\alpha \in E$, and let $f(x) = \min_F(\alpha)$. Let $\beta = a\alpha + b$ where $a, b \in F$ and $a \neq 0$.
(a) How is $F[\alpha]$ related to $F[\beta]$?
(b) How is $\min_F(\beta)$ related to $f(\frac{x-b}{a})$?

22.3.14. Let $p(x) = x^5 + \sqrt{2}x^3 + \sqrt{5}x^2 + \sqrt{7}x + \sqrt{11}$. The complex number α is a root of p. Is α algebraic over $\mathbb{Q}$? If the answer is no, give a proof, and if the answer is yes, give an upper bound for $\deg(\min_\mathbb{Q}(\alpha))$.

22.3.15. (a) What is $|\mathbb{Q}(\sqrt[5]{2}) : \mathbb{Q}|$?
(b) Let $\alpha = \dfrac{\sqrt[5]{4} - \sqrt[5]{8} + \sqrt[5]{16}}{3 + \sqrt[5]{2}}$. We know that α is not a rational number. Does there exist a polynomial $p(x)$ with rational coefficients such that $p(\alpha) = \sqrt[5]{2}$?

22.3.16. Let $F = \mathbb{Q}(\sqrt{14})$ and $E = \mathbb{Q}(\sqrt[8]{14})$, and consider the field extensions $\mathbb{Q} \subseteq F \subseteq E$. Find $|E : F|$ and $\min_F(\sqrt[8]{14})$.

22.3.17. Let $F \subseteq E$ be fields, and assume that $\alpha, \beta \in E$ are algebraic over F. Further assume that $\deg(\min_F(\alpha))$ and $\deg(\min_F(\beta))$ are relatively prime. Does $\min_F(\beta)$ have to be irreducible in $(F(\alpha))[x]$?

22.3.18. Let $F \subseteq E$ be a field extension, and let $f \in F[x]$ be an irreducible polynomial of degree $n > 1$. Assume that $|E : F| = m$ and $\gcd(m, n) = 1$.

Could f have a root in E? Either prove that it cannot or give an example to show that it may.

22.3.19. Can you find an infinite number of pairwise non-isomorphic fields E_1, E_2, ... such that, for all i, $|E_i : \mathbb{Q}| = 2$?

22.3.20. Let $F \subseteq E$ be a field extension, and assume that $\alpha_1, \alpha_2, \ldots, \alpha_k \in E$ are algebraic over F. Show that $F(\alpha_1, \alpha_2, \ldots, \alpha_k)$ is both a finite degree and an algebraic extension of F.

22.3.21. Let
$$\alpha = \sqrt{4 + \sqrt{11}} + \sqrt{4 - \sqrt{11}}.$$
Find $\min_{\mathbb{Q}}(\alpha)$ and all its roots. What is $|\mathbb{Q}[\alpha] : \mathbb{Q}|$? Can you find a field F with $\mathbb{Q} \subsetneq F \subsetneq \mathbb{Q}[\alpha]$? Either find one explicitly or prove that there are no such fields.

22.3.22. Let E be a field. Show that the poset of subfields of E ordered by inclusion is a lattice.

22.3.23. Let $F \subseteq E$ be fields, and let α and β be elements of E. Assume that α and β are algebraic over F.
(a) Show that
$$|F(\alpha, \beta) : F(\alpha)| \le |F(\beta) : F|.$$
(b) Draw a partial lattice diagram that includes F, $F(\alpha)$, $F(\beta)$, $F(\alpha, \beta)$, and E. What does the previous part say about the edge lengths? How does this compare with the corresponding result, Theorem 9.27, about the poset of subgroups of a group?

22.3.24. $K \subseteq L$ are fields. Assume that $|L : K|$ is a prime number. Show that L is a simple extension of K.

22.3.25. Assume $F \subseteq K \subseteq L$ and $F \subseteq E \subseteq L$ are field extensions. Assume that $|K : F|$ and $|E : F|$ are relatively prime integers. Show that $K \cap E = F$.

22.3.26. Let $F \subseteq E$ be fields, and let α and β be elements of E. Assume that α and β are algebraic over F. Assume that $|F(\alpha, \beta) : F|$ is equal to the product of $|F(\alpha) : F|$ and $|F(\beta) : F|$. Show that $F(\alpha) \cap F(\beta) = F$.

22.3.27. (a) Is π transcendental over $\mathbb{Q}(\pi^2)$?
(b) Assume that π is transcendental over $\mathbb{Q}$. Prove that $\pi^2 + \sqrt{5}$ is also transcendental over $\mathbb{Q}$.

22.3.28. Assume $\alpha \in \mathbb{C}$ and consider the field extension $\mathbb{Q} \subseteq \mathbb{Q}(\alpha)$.
(a) If α is algebraic over $\mathbb{Q}$, then could $\pi \in \mathbb{Q}(\alpha)$?
(b) If α is transcendental over $\mathbb{Q}$, then could $i \in \mathbb{Q}(\alpha)$?
(c) Is the field extension $\mathbb{Q} \subseteq \mathbb{Q}(\pi, i\pi)$ a simple extension?

22.3.29. Let $F \subseteq E$ with $\alpha \in E$ transcendental over F. Assume $E = F(\alpha)$. Let $\beta \in E - F$. Show
(a) α is algebraic over $F(\beta)$.
(b) β is transcendental over F.

22.3.30. Let $\alpha_1, \ldots, \alpha_n \in \mathbb{C}$ with the property that, for $1 \le i \le n$, $\alpha_i^2 \in \mathbb{Q}$. Let $E = \mathbb{Q}(\alpha_1, \alpha_2, \ldots, \alpha_n)$. Prove that $\sqrt[3]{2} \notin E$.

22.3.31. Let $F \subseteq E \subseteq L$ be fields. Assume that E is algebraic over F and that $\alpha \in L$ is algebraic over E. Show that α is algebraic over F.

22.3.32. Field of algebraic numbers. Complete the proof of Theorem 22.46. In other words, let $\mathbb{A} \subset \mathbb{C}$ be the field of algebraic numbers. Then show that $|\mathbb{A} : \mathbb{Q}| = \infty$.

22.3.33. The field $\mathbb{R}$ contains the square roots of all the positive prime integers. Let E be the smallest subfield of $\mathbb{R}$ that contains $\mathbb{Q}$ and all the square roots of all the positive prime integers. Is E a finite degree extension of $\mathbb{Q}$? Is E an algebraic extension of $\mathbb{Q}$? Why?

22.3.34. (a) Give examples of three algebraic integers: one in $\mathbb{Q}$; one in $\mathbb{R}$ but not in $\mathbb{Q}$; and one in $\mathbb{C}$ but not in $\mathbb{R}$.
(b) Show that the rational algebraic integers are precisely the elements of $\mathbb{Z}$.

22.3.35. Let $\alpha, \beta \in \mathbb{C}$. Assume that α and β are roots of polynomials with rational coefficients of degrees m and n, respectively. Show that $\alpha + \beta$ is a root of a polynomial with rational coefficients of degree mn.

22.3.36. An alternate proof of Lemma 22.33. Using the rank-nullity theorem of linear algebra, give an alternate proof for Lemma 22.33. (You should not use the lemma or any of its consequences.)

Let R be an integral domain, and let F be a field. Assume $F \subseteq R$ and $|R : F| < \infty$. To prove that R is a field, you may find the following steps useful:

STEP 1: Fix $0 \neq \beta \in R$. Define $T \colon R \to R$ by $T(\alpha) = \beta\alpha$ for all $\alpha \in R$. Show that T is a linear transformation.

STEP 2: Show that T is 1-1, and then, using the rank-nullity theorem of linear algebra, conclude that T is onto.

STEP 3: By choosing $\alpha \in R$ with $T(\alpha) = 1$, show that β has an inverse.

STEP 4: Rewrite the previous steps to get a short (and sleek) proof of Lemma 22.33.

22.3.37. Let F be any field, let $a \in F$, and let m and n be relatively prime positive integers. Show that $x^{mn} - a$ is irreducible over F if and only if $x^m - a$ and $x^n - a$ are irreducible over F.

22.3.38. Let $F \subseteq E$ be a field extension, and let $\alpha \in E$ be algebraic over F. Assume that the degree of $\min_F(\alpha)$ is odd. Show that $F(\alpha) = F(\alpha^2)$.

22.3.39. Let K be a field, and let $f = \frac{x^2+1}{x^3+x^2+1} \in K(x)$. Let $E = K(f)$, and consider the fields $K \subseteq E \subseteq K(x)$. What is $|K(x) : E|$?

22.3.40. Consider $\mathbb{Q}(x)$ the field of rational functions over $\mathbb{Q}$. What can you say about $\left| \mathbb{Q}(x) \colon \mathbb{Q}(\frac{x^2+1}{x+1}) \right|$?

22.3.41. Let F be a field, and let $p, q \in F[x]$ with $q \neq 0$ and $\gcd(p, q) = 1$. Define $\alpha = \frac{p}{q} \in F(x)$. Prove that

$$|F(x) \colon F(\alpha)| = \max\{\deg(p), \deg(q)\}.$$

Chapter 23

Straightedge and Compass Constructions

> ...where, using fields, we consider geometric constructions with a straight-edge and compass, and, in particular, we show that one cannot trisect an angle, double a cube, or square a circle.

In this chapter, we see how the basic theory of field extensions can be used to tackle ancient straightedge and compass constructions. In Euclid's *Elements* there are hardly any numbers and, while a number of geometric constructions have been interpreted as being algebraic, in the sense that they allow us to solve for an unknown, the primary point of view is geometric. In this tradition, a magnitude becomes known if we can construct a line segment whose length is the desired quantity. But, in such a construction, what are we allowed to use? The power and endurance of Euclid's *Elements* is in the fact that it starts with only five common notions, five "self-evident" postulates, and a number of definitions—Book I has 23 definitions and there are more definitions at the beginning of each book—to build the rest of the theory. His five postulates are as follows.

(a) A straight line segment can be drawn joining any two points.

(b) Any straight line segment can be extended indefinitely in a straight line.

(c) Given any straight line segment, a circle can be drawn having the segment as radius and one endpoint as center.

(d) All right angles are congruent.

(e) If two lines are drawn which intersect a third in such a way that the sum of the inner angles on one side is less than two right angles, then the two lines inevitably must intersect each other on that side if extended far enough.

The fifth postulate can be stated in a multitude of different ways, is the only one that could be argued is not self-evident, and is equivalent to the famous parallel postulate. The common notions—a category distinct from postulates in Euclid—are more general. They are as follows:

1. Things which equal the same thing also equal one another.

2. If equals are added to equals, then the wholes are equal.

3. If equals are subtracted from equals, then the remainders are equal.

4. Things which coincide with one another equal one another.

5. The whole is greater than the part.

Euclid's point is that he is starting with the most elementary assumptions and building an impressive edifice. In this tradition, if you wanted to go beyond the elementary methods, then you may, for example, assume that given the right data you can also construct conic sections—a subject that was treated extensively by Apollonius of Perga (circa 262–190 BCE)—but clearly it was preferred if you stuck to the elementary methods.

The postulates are what allow us to do geometry. One of them—somewhat curiously—states that all right angles are congruent, and one is the parallel postulate. The other three are all about geometric constructions. They tell you how to create lines and circles and, in the process, find new points—as points of intersection—from old points. These three postulates do what they do by using a ruler and compass. You are allowed to draw straight lines—through two points or to extend a given segment—and to draw circles given a center and its radius. Since all of Euclid's geometry is based on these five postulates, then, it can be argued, that, in this tradition, geometry means straightedge and compass constructions (plus the parallel postulate). Hence, it is no surprise that the ancients wanted to answer geometric questions using a straightedge and compass.

Note that the word "ruler" may conjure the image of a straightedge with markings, and this is *not* what Euclid intended. To make this clear, we use the word "straightedge" instead.

We are interested in questions such as this: Starting with a line segment of length 1, can we construct—using a straightedge and compass—a line segment of length $\sqrt[3]{2}$? This particular question is easily seen to be equivalent to the question of "doubling the cube" of Chapter 21.

23.1. The Field of Constructible Numbers

Definition 23.1 (Permissible constructions)**.** To *construct* a line segment, we may use the following *permissible* or *fundamental constructions*.

(a) Given two points, we can draw the line segment connecting them, and we also may draw a line through them extending indefinitely in each direction.

(b) Given a point and a line segment, we may draw a circle with center at the point and radius equal to the length of the line segment.

In addition, given two intersecting lines or circles, we are also given the point at their intersection.

Hence, starting with a number of given points, and, using the permissible constructions, we get new points. By repeating the process, we can generate a large number of "constuctible" points.

Definition 23.2 (Constructible number). A real number a is *constructible* if given a segment of length 1 it is possible to construct—using permissible constructions—a segment of length $|a|$.

We begin with a geometrical fact, followed by a few fundamental constructions.

Lemma 23.3. *If a quadrilateral has all sides equal, then the two diagonals are each other's perpendicular bisectors.*

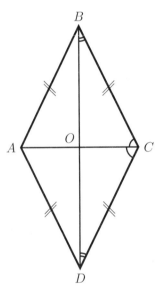

Figure 23.1. If all sides are equal, then the diagonals are perpendicular bisectors.

Proof. Consider Figure 23.1. The triangle ABC is congruent to the triangle ADC (they share one side, and the other two sides are all equal), and so $\angle ACB = \angle ACD$. The triangle BCD is isosceles and so $\angle CBD = \angle CDB$. We conclude that $\angle ACB + \angle CBD = \angle ACD + \angle CDB$, and each of these is one half of the sum of the interior angles of the triangle BCD. The sum of the interior angles of any triangle is 180 degrees, and so $\angle OCB + \angle CBO = 90$ degrees. This means that $\angle BOC$ is 90 degrees and BD is perpendicular to AC. Now, the triangles AOB and COB are congruent right triangles, and so $AO = OC$. Similarly, $BO = OD$, and so the two diagonals are perpendicular bisectors of each other. □

Lemma 23.4. *Given a line segment AB, we can draw—using a straightedge and compass—its perpendicular bisector.*

Proof. Place the compass at A, and draw a circle of arbitrary radius (as long as the radius is more than half of the length of the line segment; for example, the radius

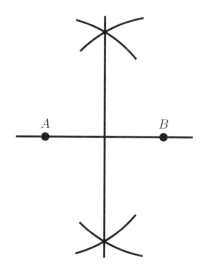

Figure 23.2. Perpendicular bisector of a line segment

could be the length of the segment). Using the same opening of the compass, draw a circle of the same radius but centered at B. The two circles have two intersections, and these two points of intersection together with A and B give a quadrilateral with equal sides. One diagonal of this quadrilateral is AB and hence—by Lemma 23.3—the other diagonal bisects it. See Figure 23.2.   □

Remark 23.5. Using similar and congruent triangles, it is straightforward to show that the perpendicular bisector of a line segment AB is the locus of all points that are equidistant from A and B. In other words, all points on the perpendicular bisector are the same distance from A and B, and conversely, any point that has the same distance from A and B lies on the perpendicular bisector.

Lemma 23.6. *Given a point P and a line ℓ, we can draw—using a straightedge and compass—a line through P and perpendicular to ℓ. If P is not on ℓ, we can also draw a line through P and parallel to ℓ.*

Proof. With the center of the compass at P—regardless of whether P is on ℓ or not—draw a circle large enough that intersects ℓ in two points A and B. Note that P is equidistant from A and B, since its distance to either A or B is equal to the radius of the circle. Now find the perpendicular bisector of AB. This will be perpendicular to ℓ and will go through P. See Figure 23.3.

To draw a line through P and parallel to ℓ, first draw a line k that is perpendicular to ℓ and goes through P. Then draw a line through P that is perpendicular to k. The latter will be parallel to ℓ. See Figure 23.4.   □

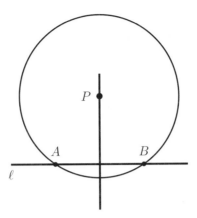

Figure 23.3. Constructing a perpendicular to ℓ through P

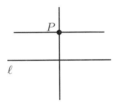

Figure 23.4. Constructing a line through P and parallel to ℓ

We now prove one of the main results of this chapter.

Proposition 23.7. *Constructible real numbers form a subfield of the reals.*

Proof. If α and β are constructible real numbers, then it is straightforward to see that so are $\alpha + \beta$ and $\alpha - \beta$. To show that the set of constructible real numbers form a field, we also have to show that $\alpha\beta$ and α/β (as long as $\beta \neq 0$) are also constructible.

Given the lengths α and β, Figures 23.5 and 23.6 give the constructions for $\alpha\beta$ and α/β, respectively. □

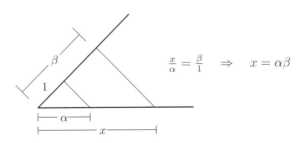

Figure 23.5. Constructing $\alpha\beta$ given α and β

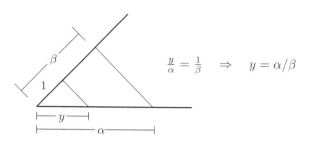

Figure 23.6. Constructing α/β given α and β

Corollary 23.8. *All rational numbers are constructible.*

Problems

23.1.1. Using the permissible constructions, give explicit constructions for bisecting an angle and trisecting a line segment.

23.1.2. We have a ruler that has two marks on it, and the distance between the two marks is r. Using this marked ruler and a compass, show that the method described below will trisect an arbitrary angle.

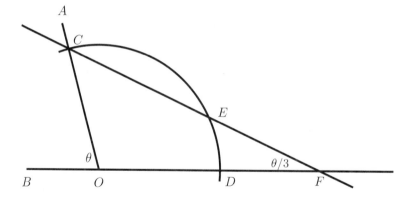

Figure 23.7. Trisecting an angle with a marked ruler

Given $\angle AOB = \theta$—see Figure 23.7—draw a circle centered at O with radius r, cutting OA at C, and (the extended) BO at D. Place the ruler with its edge through C and one of its marks on some point F on the line OD. Slide the ruler (making sure that its edge is always going through C and its mark at F stays on the line OD) until the other marked point is on the circle at a point that we call E. Then the angle $\angle EFO = \theta/3$.

23.1.3. Show that we can use a Tomahawk to trisect an angle.

A *Tomahawk* is a shape that is cut out of cardboard in the following manner: First draw a semicircle of radius 1. Extend the diameter of the

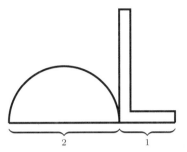

Figure 23.8. A Tomahawk

semicircle by one unit, and draw a line perpendicular to this diameter and tangent to the circle. See Figure 23.8.

To use a Tomahawk, place it on the given angle as follows (see Figure 23.9): Make sure that the semicircle is tangent to one of the sides of the angle, the end of the extended diameter is on the other side of the angle, and the perpendicular line goes through the angle. Then $\angle AOB$ (in Figure 23.9) is one third of the original angle.

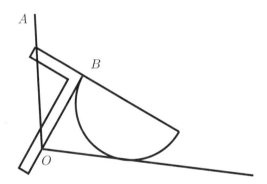

Figure 23.9. Trisecting an angle using a Tomahawk

23.1.4. Doubling a cube using a marked ruler. Assume that a unit is marked off on your ruler. Show that the following procedure—illustrated in Figure 23.10—allows you to construct $\sqrt[3]{2}$ using the marked ruler and a compass.

Construct an equilateral triangle ABC with unit length sides. Extend the segment AB and mark off the point D so that BD is of unit length. Draw the line DF through D and C. Extend the segment BC to get the line BE. Using the marked ruler (see next page) draw a line through A such that it intersects DF in G, intersects BE in H, and the length of GH is a unit. The length of AG is then $\sqrt[3]{2}$. (See Figure 23.10.)

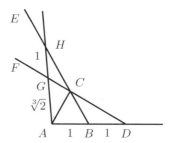

Figure 23.10. If ABC is an equilateral triangle with unit sides, and BD and GH are of unit length, then the length of AG will be $\sqrt[3]{2}$.

23.1.5. Constructions with a Rusty Compass. In his *Book on Geometrical Constructions Necessary for the Artisan*, the author Abū al-Wafā Būzjānī (940–997 CE), argues that constructions using a compass with one fixed opening are more reliable, and he goes on to give many such constructions (including ones for regular pentagons, octagons, and decagons). Assume that you have a compass—often called a *rusty* compass—that can only draw circles of radius 1. Given a line segment, show how, using the rusty compass and a straightedge, you can draw the perpendicular bisector of a given line segment. Make sure that you consider all cases (including the case when the line segment has a length of 2).[1]

23.2. Characterizing Constructible Numbers

We know that the constructible real numbers form a field and that the rationals are included in this field. Can we construct any irrational number? Or more generally, if we use lengths only in a field F, can we construct something outside of F? The following lemma shows that the answer is yes. 1

Lemma 23.9. *Given segments of length 1 and a, we can construct a segment of length $\sqrt{a}$.*

Proof. On a straight line, mark off two line segments AP and PB of lengths a and 1, respectively, so that AB is a line segment of length $a + 1$. By drawing its perpendicular bisector, find the midpoint O of AB. Draw a circle with center at O and radius of OA. Draw a perpendicular to AB at the point P, and call its point of intersection with the circle C. The triangles APC and BPC are similar. (Why?) Hence, if we let z be the length of PC, then we have

$$\frac{z}{a} = \frac{1}{z} \quad \Rightarrow \quad z = \sqrt{a}.$$

Hence, the length of PC is $\sqrt{a}$. See Figure 23.11.   □

---

[1]A remarkable conjecture by Jean Victor Poncelet (1788–1867) in 1822 that was proven by Jakob Steiner (1796–1863) in 1833 states that all constructions using a straightedge and compass can be made with a straightedge and only one use of the compass. As a result, Abū al-Wafā's project—constructions using a rusty compass—is entirely doable! An earlier—also remarkable—result proven independently by Jørgen Mohr (1640–1697) and Lorenzo Mascheroni (1750–1800) states that all straightedge and compass constructions can be constructed by compass alone.

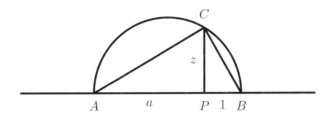

Figure 23.11. Constructing $\sqrt{a}$ from 1 and a

Example 23.10. Lemma 23.9 implies that given a unit length and using a straightedge and compass, we can construct $\sqrt{2}$. This, in turn, means that we can construct $3 + \sqrt{2}$, $\sqrt{3 + \sqrt{2}}$, and

$$\sqrt{\sqrt{3 + \sqrt{2}} + 47}.$$

Since the constructible real numbers form a field and by the fact that we can construct the square roots of constructible numbers, we have:

Corollary 23.11. *Let α be a real number, and assume that there is a sequence of fields*

$$\mathbb{Q} = F_0 \subseteq F_1 \subseteq F_2 \subseteq \cdots \subseteq F_n \subseteq \mathbb{R},$$

with $\alpha \in F_n$, and, for $0 \le j \le n - 1$, $F_{j+1} = F_j(\sqrt{k})$ for some $k \in F_j$. Then α is constructible.

Question 23.12. Are there any other constructible numbers?

Definition 23.13. Let $F \subseteq \mathbb{R}$ be a field. Recall that F^2, the *plane of F*, is defined by

$$F^2 = \{(x, y) \in \mathbb{R}^2 \mid x, y \in F\}.$$

A line passing through two points in F^2, will be called a *line in F^2*. Likewise, a circle with its center and some point in the circumference in F^2 is called a *circle in F^2*.

Lemma 23.14. *Using the points in F^2 and the permissible constructions, we can only get lines and circles in F^2. Moreover, a line in F^2 has an equation $ax + by + c = 0$ with $a, b, c \in F$. Likewise, a circle in F^2 has an equation of the form $x^2 + y^2 + ax + by + c = 0$, with $a, b, c \in F$.*

Proof. The first assertion follows directly from the definitions. In the equations for lines and circles in F^2, the coefficients are in F, since, to find them, we only use rational operations on coordinates of points in F^2. □

Lemma 23.15. *If ℓ_1 and ℓ_2 are two intersecting lines in F^2, then the point $\ell_1 \cap \ell_2$ is in F^2. If C_1 and C_2 are circles in F^2, then $\ell_1 \cap C_1$ and $C_1 \cap C_2$ are either in F^2 or in $(F(\sqrt{k}))^2$ for some $k \in F$.*

Proof. Write the equations and solve. When finding the point of intersection of two lines, we only use rational operations of the coefficients, and, hence, the answer will be a point in F^2. When we equate the equation of a line with that of a circle, we get a quadratic equation. This we can solve by using the quadratic formula, and the result will use at most one square root. Similarly, when you solve the equations of two circles, by equating them, you get the equation of the line that goes through their points of intersection, and then we solve this line together with one of the circles to get the points of intersection. Again, we use at most one square root. The result now follows. □

Theorem 23.16. *The following are equivalent:*

(a) $a \in \mathbb{R}$ *is constructible.*

(b) *There exists a finite sequence of fields*

$$\mathbb{Q} = F_0 \subset F_1 \subset \cdots \subset F_N \subset \mathbb{R}$$

with $a \in F_N$, and for every $0 \le j < N$, $F_{j+1} = F_j[\sqrt{k_j}]$ for some $k_j \in F_j$.

Proof. We have already shown—in Corollary 23.11—that (b) implies (a). To prove the converse, assume that a is a constructible real number. This means that using a segment of length 1, we can construct—by the permissible constructions—a segment of length a. Thus, in $\mathbb{R}^2$, the real plane, we are given the points $(0,0)$ and $(1,0)$, and, through a finite sequence of permissible constructions, we construct the point $P = (a, 0)$. In this sequence of constructions and before constructing P, many new points may have been constructed:

$$P_0 = (0,0), P_1 = (1,0), P_2, \ldots, P_M = P.$$

Each P_i is an intersection of already constructed lines and circles, and these lines and circles were constructed using the points $P_0, \ldots, P_{i-1}$. Now P_0 and P_1 are in $\mathbb{Q}^2$. A number of subsequent points in the sequence may also be in $\mathbb{Q}^2$. Let j_1 be the smallest positive integer such that $P_{j_1} \notin \mathbb{Q}^2$. Then, P_{j_1} is the intersection of two figures (lines or circles) in $\mathbb{Q}^2$, and, by Lemma 23.15, $P_{j_1} \in (\mathbb{Q}(\sqrt{k_0}))^2$ where $k_0 \in \mathbb{Q}$. Define $F_1 = \mathbb{Q}(\sqrt{k_0})$. Again, a number of points after P_{j_1} may be in the plane of F_1. Let j_2 be the smallest positive integer such that $P_{j_2} \notin F_1^2$. The point P_{j_2} is the intersection of two lines or circles in F_1^2, and, hence, again by Lemma 23.15, $P_{j_2} \in (F_1(\sqrt{k_1}))^2$ where $k_1 \in F_1$. Continue this process. We know it has to end since after a finite number of points, we will have constructed P. At that point, we have a finite sequence of fields

$$\mathbb{Q} = F_0 \subset F_1 \subset \cdots \subset F_N \subset \mathbb{R}$$

with $a \in F_N$, and for every $0 \le j < N$, $F_{j+1} = F_j[\sqrt{k_j}]$ for some $k_j \in F_j$. The proof is now complete. □

Corollary 23.17. *If a is a constructible real number, then a is algebraic over $\mathbb{Q}$ and $\deg(\min_{\mathbb{Q}}(a))$ is a power of 2.*

Proof. Since a is constructible, then by Theorem 23.16, a is in a field $F_N \in \mathbb{R}$, and we have a sequence of fields

$$\mathbb{Q} = F_0 \subset F_1 \subset \cdots \subset F_N \subset \mathbb{R},$$

where, for $0 \le j < N$, $|F_{j+1} : F_j| = 2$. Now

$$|F_N : \mathbb{Q}| = |F_N : F_{N-1}| \, |F_{N-1} : F_{N-2}| \, \cdots |F_1 : \mathbb{Q}| = 2^N.$$

On the other hand, $\mathbb{Q} \subseteq \mathbb{Q}(a) \subseteq F_N$, and so $\deg(\min_{\mathbb{Q}}(a)) = |\mathbb{Q}(a) : \mathbb{Q}|$ which divides $|F_N : \mathbb{Q}| = 2^N$. Hence, $\deg(\min_{\mathbb{Q}}(a))$ is a power of 2. In particular, a is algebraic over $\mathbb{Q}$. $\qquad\square$

Doubling a Cube. We can now answer one of the questions highlighted in the Preface.

Corollary 23.18. *It is impossible to double the cube.*[2]

Proof. Given a cube with a side of length 1, doubling it means constructing the side of a cube whose volume would be twice the given cube. This is the same as constructing $\sqrt[3]{2}$. But the minimal polynomial for $\sqrt[3]{2}$ is $x^3 - 2$ and 3 is not a power of 2. $\qquad\square$

Trisecting an Angle. Is it possible to trisect an angle? Sometimes the answer is yes.

Lemma 23.19. *It is possible—with a straightedge and compass—to trisect a right angle.*

Proof. Given a unit length, we know that we can construct a segment of length $\sqrt{3}$. Hence we can construct a 30-60-90 triangle. Having constructed a 30 degree angle, we have trisected a 90 degree angle. See Figure 23.12. $\qquad\square$

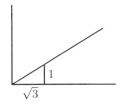

Figure 23.12. Trisecting a 90 degree angle

Corollary 23.20. *It is impossible to trisect a* 60 *degree angle using a straightedge and a compass.*

Proof. Assume that we can trisect a 60 degree angle. In Figure 23.12, we saw that we can construct a 60 degree angle starting with just a unit length. Hence, given our assumption that we can trisect a 60 degree angle, we can construct a 20 degree angle. This means that we can construct $\cos 20°$ and in turn $\alpha = 2\cos(20°)$. But what is the $\min_{\mathbb{Q}}(\alpha)$? We calculate

$$\cos 3\theta = \cos(2\theta + \theta)$$
$$= \cos 2\theta \cos \theta - \sin 2\theta \sin \theta$$
$$= (\cos^2 \theta - \sin^2 \theta) \cos \theta - 2 \sin \theta \cos \theta \sin \theta$$
$$= \cos^3 \theta - 3 \sin^2 \theta \cos \theta$$
$$= 4 \cos^3 \theta - 3 \cos \theta.$$

---

[2]In 1837, Pierre Wantzel (1814–1848) gave the first complete proof of the impossibility of doubling a cube.

Letting $\theta = 20°$ and using $\cos 3\theta = \frac{1}{2}$, we get that $\cos 20°$ is a root of $8u^3 - 6u - 1 = (2u)^3 - 3(2u) - 1$. Hence, $\alpha = 2\cos 20°$ is a root of $f(x) = x^3 - 3x - 1$.

We note that $f(x)$ is irreducible over $\mathbb{Z}/2\mathbb{Z}$ and so it is irreducible over $\mathbb{Q}$. Hence f is the minimal polynomial of α and since its degree is not a power of 2, we conclude that α is not constructible. The contradiction shows that it is not possible to trisect $60°$. $\square$

Squaring a Circle.

Theorem 23.21 (Lindemann 1882). *The number π is transcendental.*

Proof. The proof, first done by Ferdinand Lindemann (1852–1939), is somewhat involved, and it is not given here. $\square$

Corollary 23.22. *It is impossible to square a circle.*

Proof. Given a circle of radius 1, its area is π. Hence, to square this circle, we have to construct $\sqrt{\pi}$. If we can do that, then we can construct π. But π is transcendental, and all constructible numbers are algebraic. $\square$

Problems

23.2.1. True or false:
  (a) Every algebraic number is constructible.
  (b) Every constructible number is algebraic.
  (c) Every simple extension is algebraic.
  (d) Every extension of a finite field is a finite ring.
  (e) If α and β are transcendental over $\mathbb{Q}$, then so is $\alpha + \beta$.

23.2.2. Which of the following numbers are constructible?
  (a) $\sqrt{4} + \sqrt{7}$.
  (b) $\sqrt[5]{2} - 1$.
  (c) $\sqrt[4]{\sqrt{4} + \sqrt{7}}$.

23.2.3. Pick an arbitrary length as the unit length, and use an actual straightedge and an actual compass to construct $\dfrac{\sqrt{5}}{2}$.

23.2.4. Pick an arbitrary length as the unit length, and use a straightedge and compass to construct $\sqrt[4]{7}$.

23.2.5. Can you "square" a triangle? In other words, given the three vertices of a triangle, can you construct—using permissible constructions—a square with the same area as the triangle? Either show how or prove that it cannot be done.

23.2.6. Can you "rectangle" a circle? In other words, given a circle of radius 1, can you construct—using permissible constructions—a rectangle with the same area as the circle? Either show how or prove that it cannot be done.

23.2.7. Can you triple a cube? In other words, given a line segment of length 1 (one of the sides of a cube of volume 1), can you construct—using a straightedge and compass—the side of a cube of volume 3?

23.2.8. Can you trisect an angle measuring $54°$ using a straightedge and compass?

23.2.9. Show that the angle θ can be trisected by straightedge and compass if and only if the polynomial $4t^3 - 3t - \cos(\theta)$ is reducible over $\mathbb{Q}(\cos(\theta))$.

23.2.10. Let a and b be two real numbers, and let $a + bi \in \mathbb{C}$. Assume that a and b are constructible. Prove that $\deg(\min_{\mathbb{Q}}(a + bi))$ is a power of 2.

23.2.11. Can a regular 7-sided polygon be constructed using only a straightedge and compass?

Chapter 24

Splitting Fields
and Galois Groups

... where we show that, for any given polynomial, we can enlarge the field
so that the polynomial has a full set of roots, and where we define the
Galois group of a field extension—a fundamental object—and begin
exploiting the action of this group on roots of polynomials.

24.1. Roots of Polynomials, Field Extensions, and F-isomorphisms

We begin by repeating the familiar construction of moding out by the ideal generated by an irreducible polynomial. We used this construction to construct fields before. Now, we want to notice that the same construction basically gives a bigger field in which the irreducible polynomial has a root.

Theorem 24.1. *Let F be any field, and let $f \in F[x]$ with $\deg(f) > 0$. Then there exists a field E such that*

(a) *E has a subfield isomorphic to F (which we identify with F), and*

(b) *f has a root in E.*

Proof. It is enough to show that there exists a field containing an isomorphic copy of F in which an irreducible factor of f has a root. Hence, without loss of generality, we can assume that f is irreducible in $F[x]$. The ring $F[x]$ is a Euclidean domain, and hence $I = \langle f \rangle$ is a maximal ideal, and $E = F[x]/I$ is a field.

We want to show that E has an isomorphic copy of F and that f has a root in E. Note that

$$E = \{p + I \mid p \in F[x]\}.$$

Define $F_0 = \{c + I \mid c \in F\}$. The map $\phi : F \to F_0$ defined by $\phi(c) = c + I$ (the restriction of the canonical homomorphism $\pi : F[x] \to E$) is an isomorphism. Hence $F \cong F_0 \subseteq E$.

Identifying F and F_0—that is thinking of c and $c+I$ as two names for the same element—we can consider $f \in F_0[x]$. More precisely, if $f(x) = a_0 + a_1 x + \cdots + a_n x^n \in F[x]$, then $\hat{f}(x) = (a_0 + I) + (a_1 + I)x + \cdots + (a_n + I)x^n \in F_0[x]$ is the isomorphic copy of f. We claim that $\hat{f}(x)$ has a root in E. In fact, let $\alpha = x + I \in E$. We see that

$$
\begin{aligned}
\hat{f}(\alpha) &= (a_0 + I) + (a_1 + I)(x + I) + \cdots + (a_n + I)(x + I)^n \\
&= a_0 + a_1 x + \cdots + a_n x^n + I \\
&= f(x) + I = I.
\end{aligned}
$$

Now since we identify F and F_0, we also identify f and $\hat{f}$, and we say that f has a root in E.   □

Remark 24.2. Give a field F and a polynomial $f \in F[x]$. The polynomial may not have any roots in F. Hence, we cannot really talk about "a root of f" since one does not exist in our world. What the theorem shows is that we can always expand our world—i.e., construct a bigger field E that contains (a copy of) F—in which f has a root. Because of this, we can actually talk of the roots of f, even if f has no roots in $F[x]$.

Given a field F and a monic irreducible polynomial $f \in F[x]$, Theorem 24.1 gives us a bigger field in which f has a root. A little tweaking of what we know allows us to focus on a particularly useful extension field in which f has a root, and such that f is the minimal polynomial of that root.

Corollary 24.3. *Let F be any field, and let $f \in F[x]$. Assume that f is monic and irreducible. Then there exists a field K such that*

(a) *K has a subfield isomorphic to F (which we identify with F), and*

(b) *$K = F(\alpha) = F[\alpha]$ for some $\alpha \in K$, and*

(c) *$f = \min_F(\alpha)$.*

Proof. By Theorem 24.1, there exists a field E with $F \subseteq E$ such that $f(\alpha) = 0$ for some $\alpha \in E$. Now, let $K = F(\alpha) \subseteq E$. Now α is a root of $f \in F[x]$ and f is both monic and irreducible. Hence $f = \min_F(\alpha)$. Now since α is algebraic over F, we have $F(\alpha) = F[\alpha]$.   □

We continue to look at the relationship between polynomials and field extensions. If a polynomial $f \in F[x]$ is given, then f has many roots. What is the difference between extending F to $F(\alpha)$ or to $F(\beta)$ where α and β are two different roots of f? A useful heuristic is to recognize that as far as field extensions are concerned, **different roots of an irreducible polynomial are indistinguishable**. Theorem 24.6—the next substantial result—will make this statement precise by showing that two roots of the same irreducible polynomial not only give isomorphic extensions, but there is an isomorphism of the extensions that fixes the ground field and exchanges the role of the two roots. Hence, in a sense, a simple extension $F[\alpha]$ is characterized by the minimal polynomial $\min_F(\alpha)$. Theorem 24.6 is about isomorphisms and their interactions with polynomials, and before we state and prove it, we need to introduce a bit more notation and prove a lemma.

Extending a Homomorphism Between Rings to One Between Polynomial Rings. Assume K and L are rings and $\phi : K \to L$ is a map. Define $\widehat{\phi} : K[x] \to L[x]$ by $\widehat{\phi}(a_0 + a_1 x + \cdots + a_n x^n) = \phi(a_0) + \phi(a_1)x + \cdots + \phi(a_n)x^n$. The map $\widehat{\phi}$ is an *extension* of ϕ. In other words, the *restriction* of $\widehat{\phi}$ to K is just the map ϕ. Now, ϕ is a homomorphism if and only if $\widehat{\phi}$ is, and ϕ is an isomorphism if and only if $\widehat{\phi}$ is. We identify $\widehat{\phi}$ with ϕ and write $\widehat{\phi} = \phi$.

Remark 24.4. In Lemma 19.45, we had already defined a special case of this extension in the special case of the canonical homomorphism.

The following proposition may seem complicated at first, but—after deciphering what it says—it is really (re)stating the obvious.

Proposition 24.5. *Let K and L be rings, and let $\phi : K \to L$ a ring homomorphism. ϕ extends to a homomorphism $\phi : K[x] \to L[x]$. Let $\alpha \in K$ and $f \in K[x]$. Then*

$$\phi(f(\alpha)) = (\phi(f))(\phi(\alpha)).$$

Proof. Let $f(x) = a_0 + a_1 x + \cdots + a_n x^n$. Then $f(\alpha) = a_0 + a_1 \alpha + \cdots + a_n \alpha^n$, and

$$
\begin{aligned}
\phi(f(\alpha)) &= \phi(a_0 + a_1 \alpha + \cdots + a_n \alpha^n) \\
&= \phi(a_0) + \phi(a_1)\phi(\alpha) + \cdots + \phi(a_n)\phi(\alpha)^n.
\end{aligned}
$$

On the other hand,

$$(\phi(f))(x) = \phi(a_0) + \phi(a_1)x + \cdots + \phi(a_n)x^n,$$

and so

$$(\phi(f))(\phi(\alpha)) = \phi(a_0) + \phi(a_1)\phi(\alpha) + \cdots + \phi(a_n)\phi(\alpha)^n.$$

We conclude that $\phi(f(\alpha)) = (\phi(f))(\phi(\alpha))$. (See Figure 24.1.) $\square$

We are now ready to prove that different roots of irreducible polynomials give essentially indistinguishable extensions.

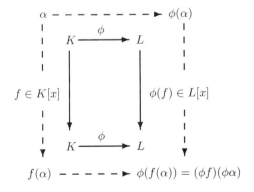

Figure 24.1. $\phi(f(\alpha)) = (\phi f)(\phi \alpha)$

Theorem 24.6. *Let F, E_1, and E_2 be fields with $F \subseteq E_1$ and $F \subseteq E_2$. Let $f \in F[x]$ be irreducible. Assume $f(\alpha) = 0$ for $\alpha \in E_1$ and $f(\beta) = 0$ for $\beta \in E_2$. Then there exists a unique isomorphism*

$$\phi : F[\alpha] \to F[\beta]$$

such that

(a) *$\phi(\alpha) = \beta$, and*
(b) *ϕ fixes every element of F.*

$$
\begin{array}{ccc}
\alpha\ E_1 & & \\
& & E_2\ \beta \\
\Big| & & \Big| \\
F[\alpha] & \overset{\phi}{\dashrightarrow} & F[\beta] \\
\Big| & \text{identity} & \Big| \\
f\ F & \longrightarrow & F\ f
\end{array}
$$

Proof. First note that, since f is an irreducible polynomial that has α and β as a root, f is a constant multiple of both $\min_F(\alpha)$ and $\min_F(\beta)$.

Every element of $F[\alpha]$ is of the form $g(\alpha)$ where $g \in F[x]$. We now define

$$\phi : F[\alpha] \to F[\beta],$$

by $\phi(g(\alpha)) = g(\beta)$, for all $g \in F[x]$.

Well defined. We first have to show that ϕ is well defined. In other words, an element of $F[\alpha]$ may have two different representations $g(\alpha) = h(\alpha)$ with both $g, h \in F[x]$, and we have to show that ϕ will produce the same result, regardless of which representation we use. Hence, assume $g(\alpha) = h(\alpha)$, and we have to show that $g(\beta) = h(\beta)$.

Now $g(\alpha) = h(\alpha)$ implies that $(g - h)(\alpha) = 0$. This means that $g - h$ is a multiple of the minimal polynomial of α and hence of f. Thus $g - h = fk$ for $k \in F[x]$. This means that $(g - h)(\beta) = f(\beta)k(\beta) = 0$, and hence $g(\beta) = h(\beta)$.

Homomorphism. We have

$$\phi(g(\alpha) + h(\alpha)) = \phi((g + h)(\alpha)) = (g + h)(\beta) = g(\beta) + h(\beta) = \phi(g(\alpha)) + \phi(g(\beta)),$$

$$\phi(g(\alpha)h(\alpha)) = \phi(gh(\alpha)) = gh(\beta) = g(\beta)h(\beta) = \phi(g(\alpha))\phi(g(\beta)).$$

Onto. Let $h(\beta)$ be an arbitrary element of $F[\beta]$. Then $h \in F[x]$, $h(\alpha) \in F[\alpha]$, and

$$\phi(h(\alpha)) = h(\beta).$$

One-to-one. If $\phi(g(\alpha)) = 0$, then $g(\beta) = 0$ which means that g is a multiple of f. If $g = fh$, then $g(\alpha) = f(\alpha)h(\alpha) = 0$. Hence $\ker(\phi) = \{0\}$ and ϕ is 1-1.

The image of α. Let $g(x) = x \in F[x]$. Then $\alpha = g(\alpha)$ and

$$\phi(\alpha) = \phi(g(\alpha)) = g(\beta) = \beta.$$

The image of elements of F. For $a \in F$, let $g(x) = a \in F[x]$ be the constant polynomial. Then

$$\phi(a) = \phi(g(\alpha)) = g(\beta) = a.$$

Hence, elements of F are fixed.

Uniqueness of ϕ. Assume $\psi : F[\alpha] \to F[\beta]$ is another F-isomorphism with $\psi(\alpha) = \beta$. Then ψ can be extended to a map sending $(F(\alpha))[x]$ to $(F(\beta))[x]$. Since ψ fixes elements of F, we have $\psi(g) = g$ for all $g \in F[x]$. Using Proposition 24.5, we now have

$$\psi(g(\alpha)) = (\psi(g))(\psi(\alpha)) = g(\beta).$$

This proves that $\psi = \phi$ defined above, and hence ϕ is unique. $\qquad\square$

Theorem 24.6 says more than $F[\alpha]$ is isomorphic to $F[\beta]$. The isomorphism is required to send α to β and to fix every element of F. In other words $F[\alpha]$ and $F[\beta]$ are not only isomorphic fields but α and β play essentially the same role in both fields. Prompted by this result, we define a stronger form of isomorphisms of two fields.

Definition 24.7 (*F-isomorphism*). Let F, E_1, and E_2 be fields with $F \subseteq E_1$ and $F \subseteq E_2$. Let $\phi : E_1 \to E_2$ be a field isomorphism. If ϕ fixes every element of F (in other words the restriction of ϕ to F, $\phi|_F$, is the identity map), then ϕ is called an *F-isomorphism*. If there exists an *F*-isomorphism from E_1 to E_2, then we write $E_1 \cong_F E_2$.

Remark 24.8. Let F, E_1, and E_2 be fields with $F \subseteq E_1$ and $F \subseteq E_2$. Let $\phi : E_1 \to E_2$ be an isomorphism. Then the following statements have exactly the same meaning:

(a) ϕ fixes every element of F.

(b) ϕ is an *F*-isomorphism.

(c) $\phi|_F$, the restriction of ϕ to F, is the identity map.

(d) ϕ is an extension of the identity map on F.

Example 24.9. There is no field isomorphism between $\mathbb{Q}(\sqrt{2})$ and $\mathbb{Q}(\sqrt{3})$ even though these are both degree 2 extensions over the rational numbers. In fact, if $f : \mathbb{Q}(\sqrt{2}) \to \mathbb{Q}(\sqrt{3})$ is a ring homomorphism, then it must be the zero homomorphism. This was Problem 16.1.10, but we repeat the argument here for completeness.

To show this, assume $f(x) \neq 0$ for $x \in \mathbb{Q}(\sqrt{2})$. Then $f(1x) = f(1)f(x)$ which means $f(x)[f(1) - 1] = 0$, and this, in turn, implies that $f(1) = 1$. Now assume that $f(\sqrt{2}) = a + b\sqrt{3}$, with $a, b \in \mathbb{Q}$. We have

$$2 = f(1 + 1) = f((\sqrt{2})^2) = (a + b\sqrt{3})^2 = a^2 + 3b^2 + 2ab\sqrt{3}.$$

Now, if a and b are both non-zero, we get that $\sqrt{3}$ is a rational number which is not true. Other possibilities, namely $b = 0$ or $a = 0$ result in contradictions as well. Hence, $f(x) = 0$ for all $x \in \mathbb{Q}(\sqrt{2})$.

Example 24.10. Let $f : \mathbb{Q}(\sqrt{3}) \to \mathbb{Q}(\sqrt{3})$ be defined by $f(a + b\sqrt{3}) = a - b\sqrt{3}$. It is straightforward to check that f is a $\mathbb{Q}$-isomorphism from $\mathbb{Q}(\sqrt{3})$ to itself. (This also follows directly and with no calculation from Theorem 24.6.) It comes as no surprise that we call f a $\mathbb{Q}$-*automorphism* of $\mathbb{Q}(\sqrt{3})$. The identity map is another $\mathbb{Q}$-automorphism of $\mathbb{Q}(\sqrt{3})$.

Theorem 24.6 can also be seen as a tool for creating *F*-isomorphisms. Example 24.9 already showed that these are less common than you may first think. In fact, Example 24.10 exhibited the only two possible $\mathbb{Q}$-automorphisms of $\mathbb{Q}(\sqrt{3})$. We could prove now that $\mathbb{Q}(\sqrt{3})$ has no other $\mathbb{Q}$-automorphisms, but we will wait until later when this fact will fall out of the theory with no additional (computational) cost.

Example 24.11. Consider the extension $\mathbb{Q} \subseteq \mathbb{C}$ and the polynomial $x^3 - 2 \in \mathbb{Q}[x]$. This polynomial is irreducible, and both $\sqrt[3]{2}$ and $\sqrt[3]{2}e^{2\pi i/3}$ are roots of this polynomial. Hence, by Theorem 24.6,

$$\mathbb{Q}[\sqrt[3]{2}] \cong_{\mathbb{Q}} \mathbb{Q}[\sqrt[3]{2}e^{2\pi i/3}],$$

even though one of these fields consists entirely of real numbers while the other has many complex numbers.

Example 24.12. Note that $\mathbb{Q}[\sqrt{2}] = \mathbb{Q}[2 + \sqrt{2}]$, and hence these two fields are isomorphic. In fact, the identity map is a $\mathbb{Q}$-isomorphism between them. However, the minimal polynomials of $\sqrt{2}$ and $2 + \sqrt{2}$ are not the same. For this reason, Theorem 24.6 does not apply. In fact, it can be shown—see the next proposition—that there does *not* exist a $\mathbb{Q}$-isomorphism from one to the other that sends $\sqrt{2}$ to $2 + \sqrt{2}$.

The appearance of F-isomorphisms in Theorem 24.6 was no coincidence, as is shown by the next proposition.

Proposition 24.13. *Let $F \subseteq E$ be fields, and let α and β be algebraic elements of E. Then the following are equivalent*

(a) $\min_F(\alpha) = \min_F(\beta)$.

(b) *There exists an F-isomorphism $\phi : F[\alpha] \to F[\beta]$ with $\phi(\alpha) = \beta$.*

Proof. One direction is an immediate corollary of Theorem 24.6. You are asked to prove the other direction in Problem *24.1.15*.   □

By modifying the proof of Theorem 24.6 slightly—there are no new ideas, just a need to keep track of the extra notation—we can prove the following more general version:

Theorem 24.14 (Theorem 24.6 generalized)**.** *Let $F_1 \subseteq E_1$ and $F_2 \subseteq E_2$ be fields, and assume $\theta : F_1 \to F_2$ is an isomorphism. Let $f_1 \in F_1[x]$ and $f_2 \in F_2[x]$ be irreducible, and assume that $\theta(f_1) = f_2$. Assume $f_1(\alpha) = 0$ for $\alpha \in E_1$ and $f_2(\beta) = 0$ for $\beta \in E_2$.*

Then there exists a unique isomorphism

$$\phi : F_1[\alpha] \to F_2[\beta]$$

such that

(a) $\phi|_{F_1} = \theta$, *and*

(b) $\phi(\alpha) = \beta$.

$$
\begin{array}{ccc}
E_1 & & \\
\vert & & E_2 \\
 & & \vert \\
F_1[\alpha] & \xrightarrow{\phi} & F_2[\beta] \\
\vert & & \vert \\
F_1 & \xrightarrow{\theta} & F_2 \\
f_1 & \dashleftarrow\dashrightarrow & f_2
\end{array}
$$

Proof. In Problem 24.1.20 you are asked to give a proof by mimicking the proof of Theorem 24.6.   □

Problems

24.1.1. Let $\sigma : \mathbb{C} \longrightarrow \mathbb{C}$ be a ring isomorphism that fixes every rational number. Could we have $\sigma(\sqrt{2}) = \sqrt{5}$?

24.1.2. Let $A = \mathbb{Q}[\sqrt{2}]$ and $B = \mathbb{Q}[\sqrt{3}]$. Not only A and B are fields, but they are also vector spaces over $\mathbb{Q}$. Are A and B isomorphic as fields? Are A and B isomorphic as vector spaces? What is the difference between these two questions?

24.1.3. Let $f = x^4 - 2 \in \mathbb{Q}[x]$. Describe a field K with $\mathbb{Q} \subseteq K$, $|K : \mathbb{Q}|$ as small as possible, and with f having a root in K. How many roots does f have in K?

24.1.4. Let $\mathbb{F}_3 = (\mathbb{Z}/3\mathbb{Z}, +, \cdot)$, and let $f = x^2 + 2x + 2 \in \mathbb{F}_3[x]$. Is f irreducible over $\mathbb{F}_3$? Construct a field K with $\mathbb{F}_3 \subseteq K$, $|K : \mathbb{F}_3|$ as small as possible, and with f have a root in K. How many roots does f have in K?

24.1.5. Let F be a field, and let $f \in F[x]$ be a monic irreducible polynomial of degree n. Define $I = \langle f \rangle \subseteq F[x]$, and let $E = F[x]/I$.
   (a) Show that E is a field with a subfield F' isomorphic to F.
   (b) Let $\alpha = x + I \in E$. Show that $E = F'[\alpha]$.
   (c) Show that $\min_{F'}(\alpha) = f$.
   (d) Show that $|E : F'| = n$.

24.1.6. Let $\mathbb{F}_2 = (\mathbb{Z}/2\mathbb{Z}, +, \cdot)$, and let $f(x) = x^4 + x + 1 \in \mathbb{F}_2[x]$.
   (a) Is $f(x)$ irreducible over $\mathbb{F}_2$?
   (b) Assume E is a field containing $\mathbb{F}_2$ such that $f(\alpha) = 0$ for some $\alpha \in E$. Find a basis for $\mathbb{F}_2(\alpha)$ over $\mathbb{F}_2$. How many elements does $\mathbb{F}_2(\alpha)$ have?
   (c) In $\mathbb{F}_2(\alpha)$ write $(1+\alpha)^{-1}$ as a linear combination of the basis elements.

24.1.7. The minimal polynomial of α over $\mathbb{Q}$ is $t^2 - 3$, and the minimal polynomial of β over $\mathbb{Q}$ is $t^2 - t - \frac{1}{2}$.
   (a) Is there a $\mathbb{Q}$-isomorphism $f : \mathbb{Q}(\alpha) \to \mathbb{Q}(\beta)$?
   (b) Is there a $\mathbb{Q}$-isomorphism $f : \mathbb{Q}(\alpha) \to \mathbb{Q}(\beta)$ with $f(\alpha) = \beta$?

24.1.8. Let $f = x^4 - 10x^2 - 20 \in \mathbb{Q}[x]$. Find all the roots of f in $\mathbb{C}$. For which pairs $\{\alpha, \beta\}$ of the roots of f, do we have $\mathbb{Q}[\alpha] \cong_{\mathbb{Q}} \mathbb{Q}[\beta]$?

24.1.9. Let F be a field, and let f_1, f_2, and f_3 be polynomials in $F[x]$. Show that there exists a field K with $F \subseteq K$, such that each of f_1, f_2, and f_3 have a root in K.

24.1.10. Let $E = \mathbb{Q}[\sqrt[3]{2}, e^{2\pi i/3}]$. Show that there exists $\sigma : E \to E$ such that σ is an $\mathbb{Q}$-automorphism of E, $\sigma(\sqrt[3]{2}) = \sqrt[3]{2}e^{2\pi i/3}$ and $\sigma(e^{2\pi i/3}) = e^{4\pi i/3}$.

24.1.11. Let $E = \mathbb{Q}(\sqrt[4]{47}, i)$.
   (a) What is $|E : \mathbb{Q}|$?
   (b) Show that there exists a $\mathbb{Q}$-automorphism of E that maps $\sqrt[4]{47}$ to $-i\sqrt[4]{47}$ and i to $-i$. Call this map σ.
   (c) What is $\sigma(i\sqrt[4]{47} + 5i)$?

24.1.12. Let $E = \mathbb{Q}(\sqrt{2}, \sqrt{3}, \sqrt{5}, \sqrt{47})$. Let $\sigma: E \to E$ be defined by

$$\sigma(a\sqrt{2} + b\sqrt{3} + c\sqrt{5} + d\sqrt{47} + e) = a\sqrt{2} - b\sqrt{3} + c\sqrt{5} - d\sqrt{47} + e.$$

With very little calculation and by repeated use of Theorem 24.14 (or Theorem 24.6), show that σ is a $\mathbb{Q}$-automorphism of E.

24.1.13. Let $\alpha = \sqrt{3 + \sqrt{6}}\, i$ and $\beta = \sqrt{3 - \sqrt{6}}\, i$.
   (a) Find two distinct polynomials $f_1, f_2 \in \mathbb{Q}[x]$ such that $f_1(\alpha) = f_2(\alpha) \in \mathbb{C}$. Can the degrees of both f_1 and f_2 be less than or equal to 3?
   (b) Can you find an example of $f_1, f_2 \in \mathbb{Q}[x]$ such that $f_1(\alpha) = f_2(\alpha) \in \mathbb{C}$ and $f_1(\beta) \neq f_2(\beta)$?

24.1.14. Let $F \subseteq E$ be a field extension, and let $\alpha \in E$ be transcendental over F. Show that $F(\alpha)$ is F-isomorphic to $F(x)$, where x is an indeterminate over the field F. Conclude that all simple transcendental extensions of F are F-isomorphic.

24.1.15. **Proof of Proposition 24.13.** Let $F \subseteq E_1$ and $F \subseteq E_2$ be field extensions. Let $\alpha \in E_1$ and $\beta \in E_2$. Assume that α is algebraic over F, and assume that $\phi : F[\alpha] \to F[\beta]$ is an F-isomorphism with the property that $\phi(\alpha) = \beta$. Prove that $\min_F(\alpha) = \min_F(\beta)$.

24.1.16. Let F be a field, and let $f, g \in F[x]$. Show that the greatest common divisor of f and g in $F[x]$ is 1 if and only if there is no field extension $F \subseteq E$ such that f and g have a common root in E.

24.1.17. Let $F \subseteq E$ be a field extension. Assume that $f, g \in F[x]$ are distinct, monic, and irreducible. Show that f and g cannot have a common root in E.

24.1.18. Let $F \subseteq E_1$ and $F \subseteq E_2$ be field extensions. A map $\phi: E_1 \to E_2$ is called an F-homomorphism from E_1 to E_2 if it is a ring homomorphism that fixes every element of F. Assume ϕ is an F-homomorphism from E_1 to E_2.
   (a) The fields E_1 and E_2 are vector space over F. Show that $\phi: E_1 \to E_2$ is a linear transformation of vector spaces.
   (b) Assume that $f \in F[x]$ and $\alpha \in E_1$ is a root of f. Show that $\phi(\alpha) \in E_2$ is also a root of f.

24.1.19. Let $F \subseteq E$ with $|E : F| < \infty$. Suppose $f \in F[x]$ is irreducible and $\deg(f) = p$, a prime. If f reduces in $E[x]$, show that p divides $|E : F|$.

24.1.20. **Proof of Theorem 24.14.** By mimicking the proof of Theorem 24.6, write down a complete proof of Theorem 24.14.

24.2. Splitting Fields

In the previous section, we studied—in some depth—field extensions that allow our favorite polynomial to have *one* root. What about the other roots? Can we enlarge the field enough so that we get *all* the roots of a given polynomial? As you may expect by repeatedly applying Theorem 24.1 (in the actual proof we will use induction), we can get to such a field. First we make two definitions.

Definition 24.15. Let F be a field, and let $f \in F[x]$. We say f *splits* in $F[x]$ if $f = c(x - \alpha_1) \cdots (x - \alpha_k)$ where $c, \alpha_1, \alpha_2, \ldots, \alpha_k$ are all in F.

Definition 24.16 (Splitting field). Let $F \subseteq E$ be fields, and let $f \in F[x]$. E is a *splitting field* for f over F if f splits in $E[x]$ and f does not split in $K[x]$ for any K with $F \subseteq K \subsetneq E$.

In other words, a splitting field is the *smallest* field in which a polynomial splits.

Remark 24.17. Let F be a field with $f \in F[x]$. Assume that $F' \subseteq E$ is a field extension, and that $\phi : F \to F'$ is an isomorphism of fields. Further assume that E is the splitting field for $\phi(f) \in F'[x]$. Continuing with our promise of identifying F and F', and f and $\phi(f)$, we say that E is the splitting field of f over F.

We next show that splitting fields exist.

Theorem 24.18. *Let F be a field, and let $f \in F[x]$. Then there exists a splitting field for f over F.*

Proof. We will induct on $\deg(f)$. For the base case if $\deg(f) = 1$, then f is linear, and hence it already splits over F, making F its splitting field.

Now, assume $\deg(f) > 1$ and that the theorem is true for all polynomials with degree less than degree of f.

By Theorem 24.1, we know that there exists a field K containing (an isomorphic copy of) F such that $f(\beta) = 0$ for some $\beta \in K$. Thus, in $K[x]$, we have $f = (x - \beta)g$. Since $\deg(g) < \deg(f)$, by the inductive hypothesis, g has a splitting field over K. Let L be the splitting field of g over K.

We have $F \subseteq K \subseteq L$ and f splits in L. Now let E be the intersection of all subfields of L that contain F and such that f splits in them. Then E is a splitting field for f over F. $\square$

Example 24.19. Consider the polynomial $x^3 - 2 \in \mathbb{Q}[x]$. One of the roots of this polynomial is $\sqrt[3]{2}$, and hence $\mathbb{Q}[\sqrt[3]{2}]$ is a field containing $\mathbb{Q}$ in which $x^3 - 2$ has a root. However, this is not a splitting field since $x^3 - 2$ also has complex numbers as roots. In fact, the roots of $x^3 - 2$ in $\mathbb{C}$ are

$$\sqrt[3]{2}, \ \sqrt[3]{2}e^{2\pi i/3}, \ \sqrt[3]{2}e^{4\pi i/3},$$

and so $\mathbb{Q}(\sqrt[3]{2}, e^{2\pi i/3})$ contains all roots of the polynomial. On the other hand, if a field contains all three roots, then it must contain both $\sqrt[3]{2}$ and $e^{2\pi i/3}$. Hence, $\mathbb{Q}(\sqrt[3]{2}, e^{2\pi i/3})$ is a splitting field for $x^3 - 2$ over $\mathbb{Q}$. What is the degree of this extension? We have

$$\mathbb{Q} \subseteq \mathbb{Q}(\sqrt[3]{2}) \subseteq \mathbb{Q}(\sqrt[3]{2}, e^{2\pi i/3}).$$

Now, $x^3 - 2$ is irreducible by the Schönemann-Eisenstein criterion (Corollary 19.64), and so it is the minimal polynomial of $\sqrt[3]{2}$ over the rationals. Hence, $|\mathbb{Q}(\sqrt[3]{2}) : \mathbb{Q}| = \deg \min_{\mathbb{Q}}(\sqrt[3]{2}) = 3$. The complex number $e^{2\pi i/3}$ is a root of $x^3 - 1 = (x - 1)(x^2 + x + 1)$, and so it is also a root of $x^2 + x + 1$. The field $\mathbb{Q}(\sqrt[3]{2})$ consists of only real numbers, and both roots of $x^2 + x + 1$ are not real. Hence, $x^2 + x + 1$ is irreducible over $\mathbb{Q}(\sqrt[3]{2})$ and so it is the minimal polynomial of $e^{2\pi i/3}$ over $\mathbb{Q}(\sqrt[3]{2})$. As a result $\left|\mathbb{Q}(\sqrt[3]{2}, e^{2\pi i/3}) : \mathbb{Q}(\sqrt[3]{2})\right| = 2$. We conclude that

$$\left|\mathbb{Q}(\sqrt[3]{2}, e^{2\pi i/3}) : \mathbb{Q}\right| = \left|\mathbb{Q}(\sqrt[3]{2}, e^{2\pi i/3}) : \mathbb{Q}(\sqrt[3]{2})\right| \, \left|\mathbb{Q}(\sqrt[3]{2}) : \mathbb{Q}\right| = 2 \times 3 = 6.$$

The next lemma is quite a straightforward fact that will be used often.

Lemma 24.20. *Let F be a field, and let $f \in F[x]$. Assume E is a splitting field for f over F. Assume K is a field with $F \subseteq K \subseteq E$. Then E is a splitting field for f over K.*

Proof. Consider f as a polynomial in $K[x]$. Then we know that f splits in E and that f cannot split in a smaller field containing K since otherwise E would not have been the splitting field for f over F. $\square$

We now prove that splitting fields are unique (up to isomorphism).

Theorem 24.21. *Let F_1 and F_2 be fields, and assume $\theta : F_1 \to F_2$ is an isomorphism. Let $f_1 \in F_1[x]$ and $f_2 \in F_2[x]$, and assume that $\theta(f_1) = f_2$. Assume E_1 is a splitting field for f_1 over F_1 and E_2 splitting field for f_2 over F_2. Then there exists an isomorphism $\phi : E_1 \longrightarrow E_2$ such that the restriction of ϕ to F_1, $\phi|_{F_1}$, is θ. (See Figure 24.2.)*

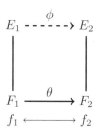

Figure 24.2. ϕ is an isomorphism that when restricted to F_1 is θ.

Proof. We will use induction on $|E_1 : F_1|$. For the base case note that if $|E_1 : F_1| = 1$, then $E_1 = F_1$, and f_1 splits over F_1. This means that f_2 splits over F_2, and hence $E_2 = F_2$ and we can use $\phi = \theta$.

Now assume $|E_1 : F_1| > 1$ and that the theorem has already been proved for smaller degree extensions.

Since $E_1 \neq F_1$, we know that f_1 does not split in F_1. Hence, f_1 has an irreducible factor g_1 of degree larger than 1. The polynomial g_1 must split in E_1, and so let $\alpha \in E_1$ be a root of g_1. Now let $g_2 = \theta(g_1)$. We have that g_2 is a factor of f_2, and so, since E_2 is a splitting field of f_2, there exists $\beta \in E_2$ such that β is a root of g_2.

By Theorem 24.14, there exists an isomorphism $\lambda : F_1[\alpha] \to F_2[\alpha]$ such that $\lambda|_{F_1} = \theta$ and $\lambda(\alpha) = \beta$. (See Figure 24.3.)

Now $|F_1[\alpha] : F_1| = \deg \min_{F_1}(\alpha) = \deg(g_1) > 1$, and so $F_1[\alpha] = F_1(\alpha)$ is a field, $f_1 \in (F_1[\alpha])[x]$, $f_2 \in (F_2[\beta])[x]$, and $\lambda(f_1) = \theta(f_1) = f_2$. In addition, by Lemma

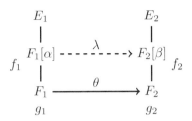

Figure 24.3. First we extend θ to $\lambda : F_1[\alpha] \to F_2[\beta]$.

24.20, E_1 and E_2 are, respectively, the splitting field of f_1 and f_2 over $F_1[\alpha]$ and $F_2[\beta]$.

Hence, we exactly have the hypothesis of the theorem—with F_1, F_2, and θ replaced by $F_1[\alpha]$, $F_2[\beta]$, and λ—except that $|E_1 : F_1[\alpha]| < |E_1 : F_1|$ since $|F_1[\alpha] : F_1| > 1$. Thus using the inductive hypothesis, we get that there exists $\phi : E_1 \to E_2$ such that $\phi|_{F_1[\alpha]} = \lambda$. Now,

$$\phi|_{F_1} = \lambda|_{F_1} = \theta,$$

and, the proof is complete. $\qquad\qquad\qquad\qquad\qquad\qquad\qquad\qquad\qquad\qquad\qquad\square$

Corollary 24.22. *Let F be a field, and let $f \in F[x]$. If E_1 and E_2 are splitting fields for f over F, then $E_1 \cong_F E_2$.*

Proof. In Theorem 24.21, let $F_1 = F_2 = F$ and $f_1 = f_2 = f$. The result follows. $\qquad\qquad\qquad\qquad\qquad\qquad\qquad\qquad\qquad\qquad\qquad\qquad\qquad\qquad\qquad\square$

Example 24.23. As usual, let $\mathbb{F}_2 = (\mathbb{Z}/2\mathbb{Z}, +, \cdot)$ be the field with two elements, and let $f = x^3 + x^2 + 1 \in \mathbb{F}_2[x]$. What is the splitting field of f over $\mathbb{F}_2$? When constructing splitting fields over $\mathbb{Q}$, we can proceed by finding the roots of the polynomial in $\mathbb{C}$ and then joining them to $\mathbb{Q}$. (See Example 24.19, for example.) Over $\mathbb{F}_2$, we cannot quite do the same, because we do not have a familiar bigger field that is guaranteed to contain all the roots. We can proceed in two (very related) ways. First, a polynomial of degree 3 is reducible over a field if and only if it has a root in that field. In our example, $f(0) = 1 = f(1)$, and so f is irreducible over $\mathbb{F}_2$.

METHOD 1: We first construct a field in which the polynomial has one root. We factor the polynomial over the new bigger field, and, if necessary, build an even bigger field that has an additional root. We proceed as in the proof of Theorem 24.1. The polynomial ring $\mathbb{F}_2[x]$ is a Euclidean domain, and f is an irreducible element. Hence $I = \langle f \rangle$ is a maximal ideal, $E = \mathbb{F}_2[x]/I$ is a field, and E contains a copy of $\mathbb{F}_2$, namely $\{I, 1 + I\}$. Identify this copy of $\mathbb{F}_2$ with $\mathbb{F}_2$ (i.e., $0 = I$, $1 = 1 + I$), and notice that $\alpha = x + I$ is a root of f in E, because

$$f(\alpha) = \alpha^3 + \alpha^2 + 1 = (x^3 + I) + (x^2 + I) + (1 + I) = f + I = I = 0.$$

In $E[x]$, f has a root and so it reduces. We use long division, and (remember that $-1 = +1$ in $\mathbb{F}_2$) we get

$$f(x) = (x + \alpha)\left[x^2 + (\alpha + 1)x + (\alpha^2 + \alpha)\right].$$

Let $g = x^2 + (\alpha + 1)x + (\alpha^2 + \alpha)$. The field $E = \{a + b\alpha + c\alpha^2 \mid a, b, c \in \mathbb{F}_2\}$ has eight elements. Using the quadratic formula or just by substituting elements of E

(and remembering that $\alpha^3 = \alpha^2 + 1$), we can see that

$$g(\alpha^2) = \alpha^4 + (\alpha + 1)\alpha^2 + \alpha^2 + \alpha = 0.$$

Hence, $g = (x + \alpha^2)(x + \alpha^2 + \alpha + 1)$. We conclude that in $E[x]$,

$$f = (x + \alpha)(x + \alpha^2)(x + 1 + \alpha + \alpha^2).$$

So E is the splitting field of f over $\mathbb{F}_2$, and the roots of f in $E[x]$ are α, α^2, and $1 + \alpha + \alpha^2$.

METHOD 2: We know that f has a unique splitting field over $\mathbb{F}_2$. Call this field E, and let α be a root of f in E. Then $\mathbb{F}_2 \subseteq E$, E has characteristic 2, and in E, we have $\alpha^3 + \alpha^2 + 1 = 0$. We can now proceed just as in the previous method, and factor f as $(x+\alpha)(x+\alpha^2)(x+\alpha^2+\alpha+1)$ in $E[x]$. We now note that $\mathbb{F}_2 \subseteq \mathbb{F}_2(\alpha) \subseteq E$ and that all the roots of f are in $\mathbb{F}_2(\alpha)$. Thus the splitting field of f over $\mathbb{F}_2$ is actually $\mathbb{F}_2(\alpha)$ (and $E = \mathbb{F}_2(\alpha) = \mathbb{F}_2[\alpha]$). Moreover, $\min_{\mathbb{F}_2}(\alpha) = f$, and so $|\mathbb{F}_2(\alpha) : F| = 3$. In fact, we now know that $\{1, \alpha, \alpha^2\}$ is a basis for the vector space $\mathbb{F}_2[\alpha]$ over $\mathbb{F}_2$, and so

$$E = \mathbb{F}_2[\alpha] = \{a + b\alpha + c\alpha^2 \mid a, b, c \in \mathbb{F}_2, \alpha^3 = \alpha^2 + 1\}$$

is a field of order 8 and the splitting field of f over $\mathbb{F}_2$.

In the second method, taking advantage of the existence (and uniqueness) of splitting fields, we constructed the splitting field (and a field of order 8) without recourse to quotient rings of polynomial rings. The two constructions, of course, are two different ways of saying the same thing, but now that we know splitting fields exist and are unique, we can construct fields as splitting fields of polynomials. In the particular case of finite fields, we will have more to say later and especially in Section 27.1.

Finally, if $F \subseteq E$ is a field extension, we show that there is a way to construct F-automorphisms of E with some specific properties as long as we know that E is a splitting field over F.

Theorem 24.24. *Let F be a field, and assume E is the splitting field of some polynomial over F. Assume L_1 and L_2 are two subfields of E that contain F. Let $\theta : L_1 \to L_2$ be an F-isomorphism. Then there exists an F-automorphism ϕ of E such that $\phi|_{L_1} = \theta$.*

Proof. Let $f \in F[x]$ be a polynomial such that E is a splitting field of f over F. By Lemma 24.20, E is a splitting field for f over both L_1 and L_2. Also, since θ is an F-isomorphism and $f \in F[x]$, we have $\theta(f) = f$. By Theorem 24.21, there exists an isomorphism $\phi : E \to E$ such that $\phi|_{L_1} = \theta$. Now $\phi|_F = \theta|_F$ is the identity map, and hence ϕ is an F-isomorphism. (See Figure 24.4.)   □

Remark 24.25. What is remarkable about Theorem 24.24 is that the hypothesis requires that E be a splitting field, but it does not specify the polynomial. In other words, the theorem does not really care for what polynomial E is a splitting field of. As long as E is a splitting field for some polynomial, then F-isomorphisms can be extended to E. Later, it will become even more clear that being a splitting field of *any* polynomial makes E special.

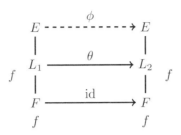

Figure 24.4. We can extend F-isomorphisms up to splitting fields.

It is also worthwhile to compare Theorem 24.24 with Theorem 24.14. Both theorems allow for extending an isomorphism between two distinct fields to an extension field. For Theorem 24.24, the only restriction is that the extension field be a splitting field. Theorem 24.14, on the other hand, requires that we adjoin two roots of an irreducible polynomial (or roots of two related polynomials) to get the extension field. In return, the conclusion of Theorem 24.14 gives more control over the newly constructed map. In summary, Theorem 24.24 has fewer hypotheses but Theorem 24.14 has a stronger conclusion. Both are useful.

Example 24.26. Let $g(t)$ be an arbitrary polynomial in $\mathbb{Q}[t]$, and consider the polynomial $f(t) = (t^2 - 3)g(t) \in \mathbb{Q}[t]$. Let E be the splitting field of f over $\mathbb{Q}$. In Example 24.10, we found two $\mathbb{Q}$-isomorphisms of $\mathbb{Q}(\sqrt{3})$—the identity and one that sends $\sqrt{3}$ to $-\sqrt{3}$. Now, by Theorem 24.24, both of these extend to $\mathbb{Q}$-isomorphisms of E.

Algebraic Closure and Algebraically Closed Fields.[*] If F is a field and $f \in F[x]$, then we have shown that we can safely talk about the splitting field of f over F. In other words, even if we do not a priori know of any fields containing F, we are assured that there is some such field that contains all the roots of f. Hence, if all we have is $f \in F[x]$, we can still talk about the roots of f, even if f has no roots in F. Then by a root of $f \in F[x]$, we mean a root of f in some splitting field for f over F. The fact that "splitting fields for f over F are F-isomorphic" means that there is no measurable difference—as far as field theory is concerned—between the choices for a splitting field. We can go one step further. Starting with a field F, we can find a field $\overline{F}$, called the *algebraic closure* of F, in which *every* polynomial in $F[x]$ splits. In other words, $\overline{F}$ contains the roots of *all* polynomials over F. In fact, even if $f \in \overline{F}[x]$—as opposed to $f \in F[x]$—then f will split over $\overline{F}$. As such, $\overline{F}$, will be an example of an *algebraically closed* field.

Definition 24.27. A field E is called *algebraically closed* if every non-zero polynomial in $E[x]$ splits in $E[x]$. If $F \subseteq E$ is a field extension, then E is called the *algebraic closure* of F, if E is an algebraic extension of F, and if every non-zero polynomial in $F[x]$ splits in $E[x]$. An algebraic closure of F is often denoted by $\overline{F}$.

Example 24.28. The field $\mathbb{C}$ of complex numbers is algebraically closed. The fact that every polynomial in $\mathbb{C}[x]$ splits in $\mathbb{C}[x]$ is the content of the fundamental theorem of algebra, Theorem 26.11.

Example 24.29. The field $\mathbb{Q}$ of rational numbers is not algebraically closed. Now $\mathbb{Q} \subseteq \mathbb{C}$ and $\mathbb{C}$ is algebraically closed. However, $\mathbb{C}$ is not an algebraic extension of $\mathbb{Q}$ and so it is *not* the algebraic closure of $\mathbb{Q}$. It turns out that the field $\mathbb{A}$ consisting of complex numbers that are algebraic over $\mathbb{Q}$ (Definition 22.45) is the algebraic closure of $\mathbb{Q}$. Hence, we can write $\mathbb{A} = \overline{\mathbb{Q}}$.

We have implied several claims—given a field F, the algebraic closure of F exists and is unique up to isomorphisms, $\mathbb{C}$ is algebraically closed, $\mathbb{A} = \overline{\mathbb{Q}}$—but we have provided no proofs. Some of the proofs are straightforward, some are not. That $\mathbb{C}$ is algebraically closed will be proved using Galois theory (see Theorem 26.11), and from this the fact that $\mathbb{A}$ is an algebraic closure for $\mathbb{Q}$ follows (see Corollary 24.33). We will construct the algebraic closure of a finite field in Section 27.4. The existence and uniqueness (up to isomorphism) of algebraic closures is a bit trickier and will be proved here using the axiom of choice (see Remark 16.34).[1] While knowing that algebraic closures exist is convenient—you do not have to worry about where roots of polynomials live, and instead you can always work inside the algebraic closure—often, and especially for our purposes, is not absolutely necessary. Hence, this section is optional, can be skipped in a first reading, and is only scantly referred to in what follows. We begin with two lemmas whose proofs are left to you in the problems.

Lemma 24.30. *A field E is algebraically closed if and only if every non-constant polynomial in $E[x]$ has a root in E.*

Lemma 24.31. *Let $F \subseteq E$ be an algebraic extension. Then the following are equivalent.*

(a) *E is algebraically closed.*

(b) *E is an algebraic closure of F.*

(c) *If $E \subsetneq L$ is a field extension, then L is not an algebraic extension of F.*

(d) *If $E \subsetneq L$ is a field extension, then L is not an algebraic extension of E.*

Proof. You are asked to provide the arguments in Problem 24.2.15. To prove (c) implies (d) use Proposition 22.47.   □

Proposition 24.32. *Assume $F \subseteq E$ is a field extension, and assume that E is algebraically closed. Let U be the set of elements of E that are algebraic over F, i.e.,*

$$U = \{\alpha \in E \mid \alpha \text{ is algebraic over } F\}.$$

Then U is an algebraic closure of F.

Proof. By Theorem 22.43, U is a field. Let $f \in F[x]$. To show that U is an algebraic closure of F, we need to show that f splits in U. We do know that f splits in E. Now if α is a root of f in E, then α is algebraic over F and hence $\alpha \in U$. Hence, all roots of f are in U and f splits over U.   □

---

[1] These results actually do not need the full force of the axiom of choice. See Banaschewski [**Ban92**].

In fact, we shall show that algebraic closures are unique (up to F-isomorphisms), and so, in Proposition 24.32, we could/should have said that U is *the* algebraic closure of F.

Corollary 24.33. *Let $\mathbb{A}$ be the field of algebraic numbers (in $\mathbb{C}$) over $\mathbb{Q}$. Then $\mathbb{A}$ is the algebraic closure of $\mathbb{Q}$.*

Proof. We already have the field extension $\mathbb{Q} \subset \mathbb{C}$. By the fundamental theorem of algebra, Theorem 26.11, $\mathbb{C}$ is algebraically closed. Hence, $\overline{\mathbb{Q}} = \mathbb{A}$ by Proposition 24.32. $\quad\square$

Theorem 24.34. *Let F be an arbitrary field. Then there exists an algebraic field extension $F \subseteq \overline{F}$ with $\overline{F}$ algebraically closed. In other words, an algebraic closure for F exists.*

Proof. This proof[2] uses a polynomial ring with an *infinite* number of variables. In such a ring, a particular element is a polynomial (not an infinite series) and, by necessity, involves only a finite number of the variables. Addition and multiplication is defined as in usual polynomial rings. We actually need a lot of variables, in fact, one variable for each non-constant polynomial in $F[x]$. For $f \in F[x]$ with $\deg(f) \geq 1$, let x_f be a variable, and let $\mathcal{J} = \{x_f \mid f \in F[x]\}$ be the set of variables. Let $R = F[\mathcal{J}]$ be the polynomial ring consisting of polynomials with variables from $\mathcal{J}$ and coefficients from F. Note that if $f \in F[x]$, then $f(x_f)$, the result of replacing the variable x with the variable x_f, is a polynomial in R (albeit one with only one variable). Let I be the ideal in R generated by all polynomials of the form $f(x_f) \in R$.

CLAIM: $I \neq R$.

PROOF OF CLAIM: By way of contradiction, assume $I = R$, then there would be polynomials $g_1, \ldots, g_n \in R$ and $f_1, \ldots, f_n \in F[x]$ such that

$$(24.1) \qquad g_1 f_1(x_{f_1}) + g_2 f_2(x_{f_2}) + \cdots + g_n f_n(x_{f_n}) = 1.$$

In this linear combination, only a finite number of polynomials, $f_1, f_2, \ldots, f_n$, from $F[x]$ are involved. We know we can find an extension $F \subseteq K_1$ where f_1 has a root (see Theorem 24.1). We can then find an extension $K_1 \subseteq K_2$ such that f_2 has a root in K_2. Repeating this, we find $F \subseteq K$, such that each of $f_1, \ldots, f_n$ have a root in K (this was Problem 24.1.9). Let $\alpha_1, \ldots, \alpha_n$ be roots in K of $f_1, f_2, \ldots, f_n$, respectively. Equation (24.1) is a polynomial identity and will remain true if we plug in any values for the variables. In this identity both sides are polynomials in $R = F[\mathcal{J}]$ but since $F \subseteq K$, this is also an identity in $K[\mathcal{J}]$. Hence, plug in α_1 for x_{f_1}, α_2 for x_{f_2}, $\ldots$, and α_n for x_{f_n}. We have $f_1(\alpha_1) = f_2(\alpha_2) = \cdots = f_n(\alpha_n) = 0$, and so the identity becomes $0 = 1$. The contradiction proves the claim.

Since I is a proper ideal of R, by Corollary 16.36—this is where we are using the axiom of choice—I is contained in M, a maximal ideal of R. Now $E = R/M$ is a field (Theorem 18.14), and the subset $\{a + M \mid a \in F\}$ is a subfield of E isomorphic to F. As usual, we identify F and its isomorphic copy, and hence $F \subseteq E$ is a field extension. If a was an element of the original F, then $a + M$ is the corresponding element in the new isomorphic copy of F. Hence to plug a into $f \in F[x]$, we find

---

[2]This particular way of proving the existence of algebraic closures is due to Emil Artin (1898–1962).

$f(a + M) = f(a) + M$. Now, $f(x_f + M) = f(x_f) + M = M$ since $f(x_f) \in I \subseteq M$. As a result, each $f \in F[x]$ with $\deg(f) \geq 1$, has a root in E.

Now let $E_0 = F$, $E_1 = E$, and repeat the process to get a field E_2, an extension field of E_1, such that *every* polynomial in $E_1[x]$ has a root in E_2. Continue this process and get fields , E_1, E_2, ..., such that

$$F = E_0 \subseteq E_1 \subseteq E_2 \subseteq \cdots \subseteq E_m \subseteq \cdots ,$$

and, for each $k \geq 1$, every polynomial in E_k has a root in E_{k+1}.

Let $\widetilde{E} = \bigcup_{i=1}^{\infty} E_i$. Then $\widetilde{E}$ is a field (the sum, difference, product, and quotient—by non-zero elements—of elements of $\widetilde{E}$ continue to be in $\widetilde{E}$, see Problem **15.2.24**). If $f \in \widetilde{E}[x]$, then the coefficients of f will lie in E_ℓ for some ℓ and so f has a root in $\widetilde{E}$ and cannot be irreducible in $\widetilde{E}[x]$. Hence, $\widetilde{E}$ is algebraically closed.

The field $\widetilde{E}$ contains F and is algebraically closed, but it may not be an algebraic extension of F. So let

$$\overline{F} = \{\alpha \in \widetilde{E} \mid \alpha \text{ is algebraic over } F\}.$$

Then by Proposition 24.32, $\overline{F}$ is an algebraic closure of F.   □

It remains to show that algebraic closures are unique up to isomorphism. For our proof we use the Kuratowski–Zorn lemma (Axiom 16.35) which is equivalent to the axiom of choice.

Theorem 24.35. *Let $F \subseteq E_1$ and $F \subseteq E_2$ be field extensions with both E_1 and E_2 algebraic closures of F. Then $E_1 \cong_F E_2$.*

Proof. Define a set of pairs of intermediate fields and 1-1 homomorphisms. More precisely, let

$$\mathcal{S} = \{(K, \theta) \mid F \subseteq K \subseteq E_1,\ \theta \colon K \to E_2 \text{ a 1-1 homomorphism with } \theta\big|_F = \text{ identity}\}.$$

For (K_1, θ_1), $(K_2, \theta_2) \in \mathcal{S}$, we say $(K_1, \theta_1) \leq (K_2, \theta_2)$ if $K_1 \subseteq K_2$ and $\theta_2\big|_{K_1} = \theta_1$. $(\mathcal{S}, \leq)$ is a non-empty partially ordered set and we claim that it satisfies the hypothesis of the Kuratowski–Zorn lemma. Given a chain of elements $(K_1, \theta_1) \leq (K_2, \theta_2) \leq \cdots$, we define $K = \bigcup_{i=1}^{\infty} K_i$. It is straightforward to show that this is a field (see Problem **15.2.24**). We define $\theta \colon K \to E_2$ as follows. If $\alpha \in K$, then $\alpha \in K_i$ for some i, and we define $\theta(\alpha) = \theta_i(\alpha)$. This map is well defined since if α is in both K_i and K_j and $i > j$, then $\theta_i\big|_{K_j} = \theta_j$. The pair $(K, \theta) \in \mathcal{S}$ is an upper bound for the chain of elements (the details are straightforward), and so applying Axiom 16.35, we get that the whole poset $\mathcal{S}$ has a maximal element. Let (M, ρ) be the maximal element of $\mathcal{S}$.

We have $F \subseteq M \subseteq E_1$, and $\rho \colon M \to E_2$ is a 1-1 homomorphism and $\rho\big|_F$ is the identity map. (See Figure 24.5.) Let $N = \rho(M)$, and note that ρ is an *isomorphism* from M to N.

We first claim that $M = E_1$. If not, let $\alpha \in E_1 - M$. The field E_1 is an algebraic extension of F, and hence α is algebraic over M. Let $f_1 = \min_M(\alpha) \in M[x]$ and $f_2 = \rho(f_1) \in N[x]$. Both f_1 and f_2 are irreducible. Since E_2 is algebraically closed and $f_2 \in E_2[x]$, f_2 has some root β in E_2. We now apply Theorem 24.14 and extend ρ to an isomorphism from $M(\alpha)$ to $N(\beta)$. However, (M, ρ) was a maximal

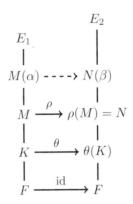

Figure 24.5. Finding an F-isomorphism between algebraic closures.

element of $\mathcal{S}$, and so $M(\alpha) = M$ and $\alpha \in M$, a contradiction. We conclude that $M = E_1$. Next we claim that $N = E_2$. The map ρ now maps E_1 isomorphically onto N while fixing F. Hence, N is an isomorphic copy of E_1 containing F. As a result N is an algebraic closure of F. But by Lemma 24.31, algebraic closures do not have proper algebraic extensions. Thus $E_2 = N$. Hence $\rho\colon E_1 \to E_2$ is an F-isomorphism between the two algebraic closures. $\qquad\square$

Problems

24.2.1. Assume $f \in \mathbb{Q}[x]$ is irreducible. Let $E \subseteq \mathbb{C}$ be the splitting field of f over $\mathbb{Q}$, and let $\alpha_1, \dots, \alpha_n$ be the roots of f in E. For each of the following questions, if the answer is yes, give a proof; otherwise, give a counterexample.
  (a) Is $E = \mathbb{Q}[\alpha_1, \alpha_2, \dots, \alpha_n]$?
  (b) Is $\{\alpha_1, \dots, \alpha_n\}$ a basis for E over $\mathbb{Q}$?
  (c) Is $|E : \mathbb{Q}| = n$?

24.2.2. Let $E = \mathbb{Q}[\sqrt[4]{5}]$, and let $f = x^4 - 25 \in \mathbb{Q}[x] \subseteq E[x]$.
  (a) Find the splitting field of f over $\mathbb{Q}$.
  (b) Find the splitting field of f over E.

24.2.3. Let $E = \mathbb{Q}[\sqrt[4]{5}]$, and let $\alpha = \sqrt[4]{5} + 1 \in E$.
  (a) What is $|E : \mathbb{Q}|$?
  (b) What is $\min_{\mathbb{Q}}(\alpha)$?
  (c) What is the splitting field of $\min_{\mathbb{Q}}(\alpha)$ over $\mathbb{Q}$?
  (d) What is the splitting field of $\min_{\mathbb{Q}}(\alpha)$ over E?

24.2.4. Let $f(x) = x^4 - 2x^2 - 2 \in \mathbb{Q}[x]$. What is the splitting field E of f over $\mathbb{Q}$? What is $|E : \mathbb{Q}|$?

24.2.5. Let $f(x) = x^4 - 4x^2 - 1$. Show that $f(x)$ is irreducible over $\mathbb{Q}$. Let $E \subseteq \mathbb{C}$ be a splitting field for f over $\mathbb{Q}$. Describe E and find $|E : \mathbb{Q}|$.

24.2.6. Construct a subfield of $\mathbb{C}$ which is a splitting field over $\mathbb{Q}$ for the polynomial $x^3 - 1$. Find the degree of this field over $\mathbb{Q}$. Do the same for $x^7 - 1$.

24.2.7. Construct a subfield of $\mathbb{C}$ which is a splitting field over $\mathbb{Q}$ for the polynomial $x^6 - 8$. Find the degree of this field over $\mathbb{Q}$.

24.2.8. Construct a splitting field E for $x^3 + x + 1$ over $\mathbb{F}_2 = (\mathbb{Z}/2\mathbb{Z}, +, \cdot)$. How many elements does E have? Find a familiar group that is isomorphic to $E^\times$, the group of units of E.

24.2.9. Construct a splitting field E for $x^3 + 2x + 1$ over $\mathbb{F}_3 = (\mathbb{Z}/3\mathbb{Z}, +, \cdot)$. What is $|E|$?

24.2.10. Let $\mathbb{Q} \subseteq E \subseteq \mathbb{C}$. Assume E is the (unique) splitting field for $x^p - 2$ over $\mathbb{Q}$ inside $\mathbb{C}$, and assume that p is a prime number. Find $|E : \mathbb{Q}|$.

24.2.11. Let $F \subseteq E$ and $f \in F[x]$ with $\deg(f) = n$. Assume f splits over E. Show that there exists a field L with $F \subseteq L \subseteq E$ such that f splits over L and $|L : F| \leq n!$.

24.2.12. Let $F \subseteq K$ be a field extension. Let $f \in F[x]$. Assume that M and L are the splitting fields of f over F and K, respectively. Show that $L = \langle K, M \rangle$, the compositum of K and M.

24.2.13. What is the algebraic closure of $\mathbb{Q}(\sqrt{2}, \sqrt[3]{2}, \sqrt[4]{2})$?

24.2.14. **Proof of Lemma 24.30.** Show that E is algebraically closed if and only if every non-constant polynomial in $E[x]$ has a root in E.

24.2.15. **Proof of Lemma 24.31.** Assume $F \subseteq E$ is an algebraic extension. Show that the conditions given in Lemma 24.31 are equivalent to E being an algebraic closure of F.

24.3. Galois Groups and Their Actions on Roots

Let $F \subseteq E$ be a field extension. We have already seen—in Proposition 24.13—that F-isomorphisms can play a role in identifying pair of elements of E that have the same minimal polynomial over F. In the course of our study of Galois theory in the coming chapters, we see the remarkable fact that the collection of *all* F-automorphisms of E contains much information about the extension $F \subseteq E$. We begin by defining the main object of our study:

Definition 24.36 (Galois group). Let $F \subseteq E$ be fields. Then $\mathrm{Aut}(E)$ is the group of automorphisms of E, and

$$\mathrm{Gal}(E/F) = \{\sigma \in \mathrm{Aut}(E) \mid \sigma \text{ is an } F\text{-automorphism of } E\}.$$

$\mathrm{Gal}(E/F)$ is called the *Galois group* of E over F.[3]

The group operation in $\mathrm{Aut}(E)$ is function composition (see Problem 16.1.5), and it is straightforward to show that $\mathrm{Gal}(E/F)$ is a subgroup of $\mathrm{Aut}(E)$.

---

[3]Some authors use the notation $\mathrm{Aut}(E/F)$ or $\mathrm{Aut}_F(E)$ instead of $\mathrm{Gal}(E/F)$ and reserve the expression "Galois group" for a more restricted class of field extensions.

For a field extension $F \subseteq E$, we want to understand $\mathrm{Gal}(E/F)$ and its relation to the field extension. Recalling our study of groups in Part 1, groups reveal their properties when they act on some set. The elements of Galois groups are automorphisms of a field E, and hence $\mathrm{Gal}(E/F)$ automatically acts on the set of elements of E. However, E is usually too big for this action—as an action—to provide very detailed information. There is another, much more productive, way that Galois groups act. If f is a non-zero polynomial in $F[x]$, then, as the next lemma shows, $\mathrm{Gal}(E/F)$ acts on the *roots* in E of f.

We already saw in Theorem 24.24 that field extensions that are splitting fields behave better than other extensions (see Remark 24.25). If $F \subseteq E$ is a splitting field and f is some irreducible polynomial in $F[x]$, then the action of $\mathrm{Gal}(E/F)$ will be transitive (that is, there is only one orbit).

Theorem 24.37. *Let $F \subseteq E$ be fields, and let $G = \mathrm{Gal}(E/F)$. Let $f \in F[x]$, and let $\Omega = \{\alpha \in E \mid f(\alpha) = 0\}$. Then G acts on Ω.*

Furthermore, if E is a splitting field of some polynomial over F, and if f is irreducible, then G is transitive on Ω.

Remark 24.38. In the second part of this theorem, E is assumed to be a splitting field of *some* polynomial over F and not necessarily the splitting field of f.

Proof. Let $\sigma \in G$. We first have to show that σ permutes the elements of Ω. In other words, if $\alpha \in \Omega$, then we have to show that $\sigma(\alpha) \in \Omega$. This will be true if $\sigma(\alpha)$ is a root of f. Note that $\sigma(f) = f$ since $f \in F[x]$, and σ fixes elements of F. Using this fact and Proposition 24.5, we have

$$0 = \sigma(0) = \sigma(f(\alpha)) = (\sigma(f))(\sigma(\alpha)) = f(\sigma(\alpha)).$$

Hence, $\sigma(\alpha) \in \Omega$.

Now, for $\sigma \in G$ and $\alpha \in \Omega$, we define $\sigma \cdot \alpha = \sigma(\alpha)$. We just saw that $\sigma \cdot \alpha \in \Omega$. Since elements of G are automorphisms, we have $1 \cdot \alpha = \alpha$ and $(\sigma\tau) \cdot \alpha = \sigma(\tau(\alpha)) = \sigma \cdot (\tau \cdot \alpha)$. Hence, $\cdot$ defines an action of G on Ω.

Next, we show that, given the two extra assumptions that E is a splitting field and that f is irreducible, G is transitive on Ω. So let $\alpha, \beta \in \Omega$. We need $\sigma \in G$ with $\sigma(\alpha) = \beta$.
Let $L = F[\alpha] \subseteq E$ and $M = F[\beta] \subseteq E$. By Proposition 24.13, there exists an F-isomorphism $\theta : L \to M$ such that $\theta(\alpha) = \beta$. Now applying Theorem 24.24, we get $\sigma \in G$ such that $\sigma|_L = \theta$ and hence $\sigma(\alpha) = \theta(\alpha) = \beta$.

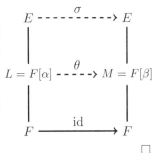

$\square$

Example 24.39. Consider the field extension $\mathbb{R} \subseteq \mathbb{C}$. What is $G = \mathrm{Gal}(\mathbb{C}/\mathbb{R})$? We know that $\mathbb{C} = \mathbb{R}[i]$ and $\{1, i\}$ is a basis for $\mathbb{C}$ over $\mathbb{R}$. Every $\sigma \in G$ is an $\mathbb{R}$-automorphism of $\mathbb{C}$. Hence, $\sigma(r) = r$ for all $r \in \mathbb{R}$. If we only knew $\sigma(i)$, then we would have $\sigma(a + bi) = a + b\sigma(i)$ and the map σ would be determined. So, what are the possibilities for $\sigma(i)$?

The polynomial $x^2 + 1 \in \mathbb{R}[x]$ has two roots i and $-i$. By Theorem 24.37, G acts on $\{i, -i\}$. Hence, $\sigma(i)$ is either i or $-i$. These are the only choices.

We can directly check that both of these choices work and each give us an element of G. Alternatively, we can appeal to Proposition 24.13. The minimal polynomial over $\mathbb{R}$ of both i and $-i$ is $x^2 + 1$ and hence there is a $\mathbb{R}$-isomorphism that takes i to i and another one that takes i to $-i$.

We conclude that $G = \{e, \sigma\}$ where e is the identity automorphism and $\sigma(a + bi) = a - bi$. Hence, $G \cong \mathbb{Z}/2\mathbb{Z}$.

Note that, since the Galois group acts on the roots of polynomials, it now follows immediately that, for $a, b \in \mathbb{R}$, if $a + bi$ is the root of a polynomial in $\mathbb{R}[x]$, then so is $a - bi$. (Earlier, in Problem 22.1.6, you gave a different proof of this fact.)

Example 24.40. Consider the field extension $\mathbb{Q} \subseteq \mathbb{Q}(\sqrt[3]{2})$. What is $G = \mathrm{Gal}(\mathbb{Q}(\sqrt[3]{2})/\mathbb{Q})$?

By Lemma 22.34, $\{1, \sqrt[3]{2}, \sqrt[3]{4}\}$ is a basis for $\mathbb{Q}(\sqrt[3]{2})$ over $\mathbb{Q}$. For $\sigma \in G$, if we know $\sigma(\sqrt[3]{2})$, then we know σ since $\sigma(a + b\sqrt[3]{2} + c\sqrt[3]{4}) = a + b\sigma(\sqrt[3]{2}) + c\sigma(\sqrt[3]{2})^2$. What are the possibilities for $\sigma(\sqrt[3]{2})$?

The polynomial $x^3 - 2$ has only one root in $\mathbb{Q}(\sqrt[3]{2})$, namely $\sqrt[3]{2}$—the other two roots of this polynomial are complex numbers—and, by Theorem 24.37, G acts on the sets of roots. Hence, $\sigma(\sqrt[3]{2}) = \sqrt[3]{2}$, and σ is the identity map.

We conclude that $G = \{e\}$ is a group with just one element.

Example 24.41. Consider $E = \mathbb{Q}[\sqrt{2}, \sqrt{3}]$. We have a chain of field extensions $\mathbb{Q} \subseteq \mathbb{Q}[\sqrt{2}] \subseteq E = \mathbb{Q}[\sqrt{2}, \sqrt{3}]$. We know that $x^2 - 2$ is the minimal polynomial of $\sqrt{2}$ over $\mathbb{Q}$ and so $|\mathbb{Q}[\sqrt{2}] : \mathbb{Q}| = 2$. We also know $\sqrt{3} \notin \mathbb{Q}[\sqrt{2}]$. (This was Problem 22.1.1 but the argument goes as follows: if $\sqrt{3} = a + b\sqrt{2}$, then squaring both sides and rearranging would give that $\sqrt{2}$ is rational.) Hence, $x^2 - 3$ is the minimal polynomial of $\sqrt{3}$ over not only $\mathbb{Q}$ but also $\mathbb{Q}[\sqrt{2}]$. We conclude that $|E : \mathbb{Q}[\sqrt{2}]| = 2$ and $|E : \mathbb{Q}| = |E : \mathbb{Q}[\sqrt{2}]|\ |\mathbb{Q}[\sqrt{2}] : \mathbb{Q}| = 2 \times 2 = 4$. So, as a vector space, E has a basis of four elements over $\mathbb{Q}$. We can show that $\sqrt{6}$ is not a $\mathbb{Q}$-linear combination of 1, $\sqrt{2}$, and $\sqrt{3}$, and so $E = \{a + b\sqrt{2} + c\sqrt{3} + d\sqrt{6} \mid a, b, c, d \in \mathbb{Q}\}$. (See also Problem 22.3.7.) We also now know that E is the splitting field of $(x^2 - 2)(x^2 - 3)$ over $\mathbb{Q}$ and both Theorems 24.14 and 24.37 apply. Every element of the $\mathrm{Gal}(E/\mathbb{Q})$ is determined by its action on $\sqrt{2}$ and $\sqrt{3}$. (This is because if $\sigma \in \mathrm{Gal}(E/\mathbb{Q})$, then σ would fix rational numbers, and $\sigma(\sqrt{6}) = \sigma(\sqrt{2})\sigma(\sqrt{3})$.)

The polynomial $x^2 - 2$ is irreducible in $\mathbb{Q}$ and so $\mathrm{Gal}(\mathbb{Q}(\sqrt{2})/\mathbb{Q})$ acts transitively on $\{\sqrt{2}, -\sqrt{2}\}$. So there are two Q-automorphisms of $\mathbb{Q}(\sqrt{2})$. One fixes $\sqrt{2}$ and the other sends $\sqrt{2}$ to $-\sqrt{2}$. Both of these fix the polynomial $x^2 - 3$, $\pm\sqrt{3}$ are the roots of this polynomial, and $x^2 - 3$ is irreducible over $\mathbb{Q}[\sqrt{2}]$. We now apply Theorem 24.14 (with $F_1 = F_2 = \mathbb{Q}[\sqrt{2}]$, $f_1 = f_2 = x^2 - 3$, and $\alpha = \sqrt{3}$, and $\beta = \pm\sqrt{3}$), and we conclude that each of the two $\mathbb{Q}$-automorphisms of $\mathbb{Q}[\sqrt{2}]$ can, in turn, be extended to two $\mathbb{Q}$ auto-morphisms of $\mathbb{Q}[\sqrt{2}, \sqrt{3}]$.

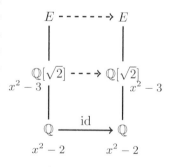

For example, we had one $\mathbb{Q}$-automorphism of $\mathbb{Q}[\sqrt{2}]$ that sent $\sqrt{2}$ to $-\sqrt{2}$. We can extend this to an automorphism of E that fixes $\sqrt{3}$. As a result, we have a

$\mathbb{Q}$-automorphism of E that sends $\sqrt{2}$ to $-\sqrt{2}$ and fixes $\sqrt{3}$. This automorphism maps $a+b\sqrt{2}+c\sqrt{3}+d\sqrt{6}$ to $a-b\sqrt{2}+c\sqrt{3}-d\sqrt{6}$. Hence, we have constructed four $\mathbb{Q}$-automorphisms of E. (Also see Problem 22.3.7.) There cannot be any others, since each automorphism is determined by its action of $\sqrt{2}$ and $\sqrt{3}$, and we already have every possible action. Hence, $\mathrm{Gal}(E/\mathbb{Q}) = \{e, \sigma, \tau, \sigma\tau\}$ where

$$e: a + b\sqrt{2} + c\sqrt{3} + d\sqrt{6} \mapsto a + b\sqrt{2} + c\sqrt{3} + d\sqrt{6},$$
$$\sigma: a + b\sqrt{2} + c\sqrt{3} + d\sqrt{6} \mapsto a - b\sqrt{2} + c\sqrt{3} - d\sqrt{6},$$
$$\tau: a + b\sqrt{2} + c\sqrt{3} + d\sqrt{6} \mapsto a + b\sqrt{2} - c\sqrt{3} - d\sqrt{6},$$
$$\sigma\tau: a + b\sqrt{2} + c\sqrt{3} + d\sqrt{6} \mapsto a - b\sqrt{2} - c\sqrt{3} + d\sqrt{6}.$$

Each element of the Galois group is of order 2 (if you repeat it twice, you get the identity), and so

$$\mathrm{Gal}(E/\mathbb{Q}) \cong \mathbb{Z}/2\mathbb{Z} \times \mathbb{Z}/2\mathbb{Z}.$$

Remark 24.42. If a group G acts on a set Ω of size n, then every element of G gives a permutation of Ω, and we can think of this permutation as an element of S_n, the symmetric group of order n. Hence, we have a map $\theta : G \to S_n$. We have actually proved—see Section 11.4 and Theorem 11.28—that θ is a group homomorphism (see Figure 24.6), and the kernel of this homomorphism consists of those elements of G that fix every element of Ω. We argued in Section 11.4, that this is a win-win situation. Either the kernel of this homomorphism is non-trivial, in which case we have found a normal subgroup of G, or else the kernel is trivial, and hence by the homomorphism theorems,

$$G \cong G/\ker(\theta) \cong \theta(G) \leq S_n.$$

In other words, G is isomorphic to a subgroup of S_n.

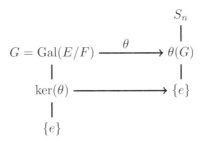

Figure 24.6. The action of the Galois group on the roots of a polynomial gives a homomorphism into a symmetric group.

In our case, we have $G = \mathrm{Gal}(E/F)$ acting on the roots of polynomials in $F[x]$. We apply the general principle above to get two facts about these Galois groups.

Theorem 24.43. *Let $F \subseteq E$ be fields. Assume $|E : F| < \infty$. Then $|\mathrm{Gal}(E/F)| < \infty$.*

Proof. We are assuming $|E : F| < \infty$, and hence the vector space E has a finite basis over F. Let $\alpha_1, \ldots, \alpha_k$ be a basis for E over F. Now let

$$f = \prod_{i=1}^{k} \min_F(\alpha_i)$$

and let

$$\Omega = \{\beta \in E \mid f(\beta) = 0\}.$$

Let $|\Omega| = n$. Now $G = \mathrm{Gal}(E/F)$ acts on Ω, and this gives a homomorphism $\theta : G \to S_n$. (See Figure 24.6.) Moreover, if an element of G fixes every element of Ω, then that element is an F-automorphism of E that fixes every element of the basis for E over F. Since every element of E can be written as a linear combination of the basis elements, this element of G fixes every element of E and must be the identity automorphism. Hence, the kernel of θ is $\{e_G\}$. We conclude that $G \cong G/\ker(\theta) \cong \theta(G) \leq S_n$. Hence $|G| \leq n! < \infty$. $\square$

A variation on the same argument, yields the following.

Theorem 24.44. *Let F be a field, and let $f \in F[x]$. Let E be a splitting field for f over F. Let $\Omega = \{\alpha \in E \mid f(\alpha) = 0\}$. Let $n = |\Omega|$. Then $\mathrm{Gal}(E/F)$ is isomorphic to a subgroup of S_n, the symmetric group of degree n.*

Proof. If $\Omega = \{\alpha_1, \ldots, \alpha_n\}$, then E is the smallest field that contains the roots of f, and hence $E = F(\alpha_1, \ldots, \alpha_n)$. Each α_i—by virtue of being a root of f—is algebraic over F, and hence $E = F[\alpha_1, \ldots, \alpha_n]$. In other words, every element of E is a polynomial in the α's. Hence, if an automorphism of E fixes F and $\alpha_1, \ldots, \alpha_n$, then that automorphism fixes every element of E and is the identity.

By Theorem 24.37, $G = \mathrm{Gal}(E/F)$ acts on Ω. This gives a group homomorphism $\theta : G \to S_n$. (See Figure 24.6.) Since the only F-automorphism that fixes every element of Ω is identity, this map is 1-1, and we have

$$G \cong G/\ker(\theta) \cong \theta(G) \leq S_n.$$ $\square$

Definition 24.45 (Gal(f)). Let F be a field, and let $f \in F[x]$. Let E be a splitting field for f over F. Then $\mathrm{Gal}(E/F)$ is called the *Galois group of f* and is denoted by $\mathrm{Gal}(f)$.

Remark 24.46. If f is a polynomial of degree n, then in Theorem 24.44 we proved that $|\mathrm{Gal}(f)| \leq n!$ In Theorem 24.43, we proved that the size of a Galois group of any finite degree extension is a finite number. We will have more to say about the size of Galois groups of finite degree extensions. In particular, in Problem *24.3.5*, you are asked to show that $|\mathrm{Gal}(E/F)| \leq |E : F|$ as long as E is a simple extension of F. Later, we will remove the condition that E is a simple extension, and we will also investigate situations where $|\mathrm{Gal}(E/F)| = |E : F|$. (See Theorems 25.50 and 25.51 and Problems *24.3.5* and *25.2.7*.)

Problems

24.3.1. Find $\mathrm{Gal}(\mathbb{Q}[\sqrt{7}]/\mathbb{Q})$.

24.3.2. Find $G = \mathrm{Gal}(\mathbb{Q}(\sqrt{7}, \sqrt{11})/\mathbb{Q})$. Give a complete argument for all your assertions. Choose a set S of generators for G, and let Ω be the set of roots, in $\mathbb{Q}(\sqrt{7}, \sqrt{11})$, of the polynomial $(x^2 - 7)(x^2 - 11)$. Draw the Cayley digraph of the action of G on Ω.

24.3.3. Let $E = \mathbb{Q}(\sqrt{2}, \sqrt{3}, i)$. Find $\mathrm{Gal}(E/\mathbb{Q})$.

24.3.4. Let $F \subseteq K \subseteq E$ be fields. Show that $\mathrm{Gal}(E/K)$ is a subgroup of $\mathrm{Gal}(E/F)$.

24.3.5. Let $F \subseteq E$ be fields with $|E : F| < \infty$. Assume $E = F[\alpha]$. Show

$$|\mathrm{Gal}(E/F)| \le |E : F|.$$

24.3.6. Let $\mathbb{Q}$ be the field of rational numbers, and let $\mathbb{R}$ be the field of real numbers. Find $\mathrm{Gal}(\mathbb{R}/\mathbb{Q})$.

24.3.7. Let $\alpha = \sqrt{2 + \sqrt{2}} \in \mathbb{C}$.
   (a) Compute $f = \min_{\mathbb{Q}}(\alpha)$.
   (b) Find $E \subseteq \mathbb{C}$ such that E is the splitting field for f over $\mathbb{Q}$. Compute $|E : \mathbb{Q}|$.
   (c) Show that $\mathrm{Gal}(E/\mathbb{Q})$ contains an element of order 4.

24.3.8. Find (up to isomorphism) the Galois group of $x^4 - 3x^2 + 4$ over $\mathbb{Q}$.

24.3.9. Let $f_1(x) = x^4 - x^2 + 1$ and $f_2 = x^4 + x^2 + 1$. Let E_1 and E_2, respectively, be the splitting fields of f_1 and f_2 over $\mathbb{Q}$.
   (a) Explicitly describe E_1 and E_2.
   (b) Find $|E_1 : \mathbb{Q}|$ and $|E_2 : \mathbb{Q}|$.
   (c) What is $\mathrm{Gal}(E_2/\mathbb{Q})$? Explicitly find its elements.
   (d) What is $\mathrm{Gal}(E_1/\mathbb{Q})$?

24.3.10. Find a familiar group that is isomorphic to the Galois group of $x^3 - 2$ over $\mathbb{Q}$.

24.3.11. What is $|\mathbb{Q}[\sqrt[4]{2}] : \mathbb{Q}|$? What is $\left|\mathrm{Gal}(\mathbb{Q}[\sqrt[4]{2}]/\mathbb{Q})\right|$? Find a familiar group that is isomorphic to $\mathrm{Gal}(\mathbb{Q}[\sqrt[4]{2}]/\mathbb{Q})$.

24.3.12. Let $f \in F[x]$ be an irreducible polynomial of degree 4, and let E be the splitting field of f over F. According to Theorem 24.44, $\mathrm{Gal}(E/F) \le S_4$. Can $\mathrm{Gal}(E/F) = \langle (1\ 2\ 3), (1\ 2) \rangle \cong S_3$?

24.3.13. **Conjugate Fields.** Let $F \subseteq K \subseteq E$ be fields, and let $\sigma \in \mathrm{Gal}(E/F)$. Then $\sigma(K)$ is called a (*Galois*) *conjugate* of K (under the action of $\mathrm{Gal}(E/F)$).
   (a) Show that $\sigma(K)$ is also a field containing F and contained in E.
   (b) Assume that E is the splitting field of some polynomial over F, and let $\alpha \in E$. Show that a field K is conjugate to $F(\alpha)$ (under the action of $\mathrm{Gal}(E/F)$) if and only if $K = F(\beta)$, where β is a root of $\min_F(\alpha)$.

24.3.14. Let $K = \mathbb{Q}(x)$ and let $F = \mathbb{Q}(x^6) \subset K$. What is the $\mathrm{Gal}(K/F)$?

24.3.15. Let $F \subseteq E$ be an algebraic extension of fields, and assume that $\phi \colon E \to E$ is a non-zero F-homomorphism (i.e., ϕ is a non-zero ring homomorphism that fixes every element of F; see Problem 24.1.18). Show that $\phi \in \mathrm{Gal}(E/F)$.

Chapter 25

Galois, Normal, and Separable Extensions

... where we go back and forth between subgroups of a Galois group and intermediate fields of an extension, define and study normal, separable, and Galois extensions, find some nice properties of Galois extensions, and use Galois groups to find minimal polynomials and to see that simple extensions are more common than you may have expected.

25.1. Subgroups of the Galois Group and Intermediate Fields

Given a field extension $F \subseteq E$, we have defined a group $\mathrm{Gal}(E/F)$. At least sometimes—see Example 24.40—the group carries little information about the field extension, and we have yet to see any significant use for the Galois group. To study the relationship between the group $\mathrm{Gal}(E/F)$ and the field extension $F \subseteq E$, we first see how to go from a subgroup of $\mathrm{Gal}(E/F)$ to a subfield of E containing F and, vice versa, how to go from such an intermediate field to a subgroup of the Galois group.

Definition 25.1 (Fixed field)**.** Let $F \subseteq E$ be fields, and let $G = \mathrm{Gal}(E/F)$. Let $H \leq G$ be a subgroup. Then the *fixed field* of H is denoted by $\mathrm{Fix}(H)$ and defined by

$$\mathrm{Fix}(H) = \{\alpha \in E \mid \text{ where } \sigma(\alpha) = \alpha \text{ for all } \sigma \in H\}.$$

The following lemma follows directly from the definitions.

Lemma 25.2. *Let $F \subseteq E$ be fields, and let $G = \mathrm{Gal}(E/F)$. Given $H \leq G$, $\mathrm{Fix}(H)$ is a field with*

$$F \subseteq \mathrm{Fix}(H) \subseteq E.$$

Similarly, given a field K with $F \subseteq K \subseteq E$, we have that $\mathrm{Gal}(E/K)$ is a subgroup of G. (See Figure 25.1.)

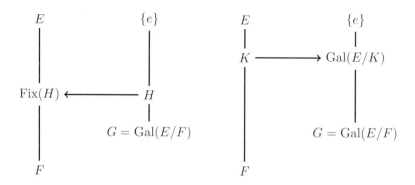

Figure 25.1. Subgroups of the Galois group and intermediate fields. Note that the group lattice diagram is drawn upside down.

It is helpful to depict the relationship between subgroups of the Galois group and the intermediate fields of a field extension using diagrams such as those in Figure 25.1. In such a figure, on one side we draw the (Hasse diagram of the) poset of fields K with $F \subseteq K \subseteq E$ ordered by inclusion. On the other side we draw (an upside down Hasse diagram of the) subgroups of the Galois group. Often we do not draw every subgroup and every intermediate field, but rather we draw those of interest based on the issue at hand. As in Figure 25.1, it is convenient to draw the group "upside-down", with the full group at the bottom and the identity subgroup at the top. The following straightforward lemma explains why:

Lemma 25.3. *Let $F \subseteq E$ be fields, and let $G = \mathrm{Gal}(E/F)$.*

(a) *If $H_1 \leq H_2 \leq G$, then $F \subseteq \mathrm{Fix}(H_2) \subseteq \mathrm{Fix}(H_1) \subseteq E$, and*

(b) *If $F \subseteq K_1 \subseteq K_2 \subseteq E$ are fields, then $\{e\} \leq \mathrm{Gal}(E/K_2) \leq \mathrm{Gal}(E/K_1) \leq G$.*

In other words, both $\mathrm{Fix}(\cdot)$ and $\mathrm{Gal}(E/\cdot)$ are order-reversing maps.

Lemma 25.2 has given us a way to go back and forth between the two structures (subgroups of $\mathrm{Gal}(E/F)$ and fields K with $F \subseteq K \subseteq E$). To see the relation of the two operations, we ask what happens if we compose the two maps.

Proposition 25.4. *Let $F \subseteq E$ be a field extension. Let K be a subfield of E containing F, and let H be a subgroup of $\mathrm{Gal}(E/F)$. Then (see Figure 25.2)*

$$
\begin{aligned}
K &\subseteq \mathrm{Fix}(\mathrm{Gal}(E/K)), \\
H &\leq \mathrm{Gal}(E/\mathrm{Fix}(H)).
\end{aligned}
$$

Proof. These follow directly from the definitions. $\mathrm{Gal}(E/K)$ is the collection of K-automorphisms of E, hence the fix field of these automorphisms must include K.

Likewise, $\mathrm{Fix}(H)$ is the collection of elements of E that are fixed by every element of H. Hence every element of H is an automorphism of E that fixes every element of $\mathrm{Fix}(H)$. □

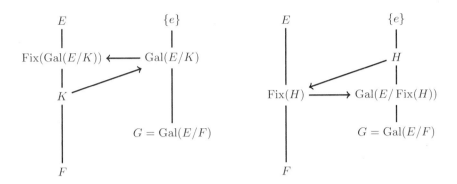

Figure 25.2. Fix(Gal(E/K)) and Gal($E/$ Fix(H))

Using Proposition 25.4 and Lemma 25.3, it is straightforward to see what happens if you compose the two maps more than once. You are asked to do this in Problem 25.1.5.

Problems

25.1.1. Find Gal($\mathbb{Q}(\sqrt[5]{7})/\mathbb{Q}$) and Fix(Gal($\mathbb{Q}(\sqrt[5]{7})/\mathbb{Q}$)).

25.1.2. Let $E = \mathbb{Q}[\sqrt[3]{2}, e^{2\pi i/3}]$. Let id $\in$ Gal($E/\mathbb{Q}$) denote the identity automorphism.
  (a) What is $|E : \mathbb{Q}|$? What is a basis for E over $\mathbb{Q}$?
  (b) Show that there exists $\sigma \in$ Gal($E/\mathbb{Q}$) with $\sigma(\sqrt[3]{2}) = \sqrt[3]{2}e^{2\pi i/3}$ and $\sigma(e^{2\pi i/3}) = e^{4\pi i/3}$.
  (c) What is $\sigma(\sqrt[3]{2}e^{2\pi i/3})$?
  (d) Is $H = \{$id$, \sigma\}$ a subgroup of Gal($E/\mathbb{Q}$)?
  (e) What is Fix(H)?

25.1.3. Let $F \subseteq E$ be fields, and let $K = $ Fix(Gal(E/F)). How are Gal(E/F) and Gal(E/K) related?

25.1.4. Let $F \subseteq E$ be fields, and let $K = $ Fix(Gal(E/F)).
  Show that Fix(Gal(E/K)) = K.

25.1.5. Let $F \subseteq E$ be fields. In Proposition 25.4 and Figure 25.2 we composed the two maps Fix($\cdot$) and Gal($E/\cdot$). What happens if you compose the maps one more time? Show that if $H \leq$ Gal(E/F) and K an intermediate field between F and E, then

$$\text{Fix}(\text{Gal}(E/\text{Fix}(H))) = \text{Fix}(H), \text{ and}$$
$$\text{Gal}(E/\text{Fix}(\text{Gal}(E/K))) = \text{Gal}(E/K).$$

25.2. Galois, Normal, and Separable Extensions

Example 24.40—where we showed that $\mathrm{Gal}(\mathbb{Q}[\sqrt[3]{2}]/\mathbb{Q}) = \{e\}$—was disappointing. For a reasonable looking field extension, the Galois group carried no information. In some sense, the problem was that $\mathbb{Q}[\sqrt[3]{2}]$ was not the right kind of extension. We had not extended the base field $\mathbb{Q}$ enough, and as a result we did not get any significant elements for the Galois group. In this case, we could remedy the situation by insisting that we extend the field all the way to a splitting field. In addition to being a splitting field, in more general situations and to get much of the Galois group, we need our extension to also be separable (defined later). Our ultimate goal will be to show that for these certain kinds of field extensions $F \subseteq E$, there will be a 1-1 correspondence between the subgroups of the Galois group and the intermediate fields L with $F \subseteq L \subseteq E$.

As we shall see, the kind of field extension that will be the most promising is where the two maps defined in the previous section are inverses of each other and we get equality in the relations of Proposition 25.4. Focusing on one such case gives us the important definition of a Galois extension.

Definition 25.5 (Galois extensions). Let $F \subseteq E$ be fields with $|E : F| < \infty$. We say that E is a *Galois extension* of F if $\mathrm{Fix}(\mathrm{Gal}(E/F)) = F$.

Example 25.6. Consider the extension $\mathbb{R} \subseteq \mathbb{C}$. We know—see Example 24.39— that $\sigma : \mathbb{C} \to \mathbb{C}$ defined by $\sigma(a + bi) = a - bi$ is an element of $\mathrm{Gal}(\mathbb{C}/\mathbb{R})$. Now, if a complex number $a + bi$ is fixed by σ, we must have $a + bi = a - bi$ which implies $b = 0$. Hence, the real numbers are the only complex numbers fixed by σ. We conclude that $\mathrm{Fix}(\mathrm{Gal}(\mathbb{C}/\mathbb{R})) = \mathbb{R}$, and $\mathbb{C}$ is a Galois extension of $\mathbb{R}$.

Example 25.7. Consider the extension $\mathbb{Q} \subseteq \mathbb{Q}[\sqrt[3]{2}]$. We saw in Example 24.40 that the Galois group of this extension is just the identity element. Now the identity element fixes everything, and so $\mathrm{Fix}(\mathrm{Gal}(\mathbb{Q}[\sqrt[3]{2}]/\mathbb{Q})) = \mathbb{Q}[\sqrt[3]{2}]$. Hence, this is not a Galois extension.

Which extensions are Galois extensions? What good are these extensions? As a warm up, we will state an oft-used lemma that identifies some Galois extensions.

Lemma 25.8. *Let $F \subseteq E$ be a field extension, and let $K = \mathrm{Fix}(\mathrm{Gal}(E/F))$. Then*

$$\mathrm{Gal}(E/F) = \mathrm{Gal}(E/K) \ and \ \mathrm{Fix}(\mathrm{Gal}(E/K)) = K.$$

In particular, if $|E : K| < \infty$, then E is Galois over K.

Proof. See Problems 25.1.3 and 25.1.4, but here is the argument.

Since $F \subseteq K$, we have $\mathrm{Gal}(E/K) \subseteq \mathrm{Gal}(E/F)$. But every F-automorphism of E also fixes K and so $\mathrm{Gal}(E/F) \subseteq \mathrm{Gal}(E/K)$. We conclude that $\mathrm{Gal}(E/F) = \mathrm{Gal}(E/K)$ and $K = \mathrm{Fix}(\mathrm{Gal}(E/F)) = \mathrm{Fix}(\mathrm{Gal}(E/K))$. To be a Galois extension, the extension must be a finite degree extension, and so if $|E : K| < \infty$, then E is Galois over K.

$$
\begin{array}{ccc}
E & & \{e\} \\
| & & | \\
K & \longleftarrow & \mathrm{Gal}(E/F) \\
| & & \\
F & &
\end{array}
$$

□

Turning to the advantages of a Galois extension, we begin with a theorem that, in the case of Galois extensions, allows us to find minimal polynomials of elements.

Theorem 25.9. *Assume E is a Galois extension of F, and let $G = \mathrm{Gal}(E/F)$. Let $\alpha \in E$, and let*

$$\Omega = \{\alpha_1 = \alpha, \alpha_2, \ldots, \alpha_r\}$$

be the orbit of α under the action of G. In other words, $\Omega = \{\beta \in E \mid \exists \sigma \in G \text{ with } \sigma(\alpha) = \beta\}$. Then

$$\min{}_F(\alpha) = (x - \alpha_1)(x - \alpha_2) \cdots (x - \alpha_r).$$

In particular, $\min_F(\alpha)$ splits and has distinct roots in E.

Proof. Let $f = \min_F(\alpha)$, and let $g = (x - \alpha_1)(x - \alpha_2) \cdots (x - \alpha_r)$. We want to show that $f = g$. Note that, to start with, we know that $f \in F[x]$ but we only know that $g \in E[x]$.

Since $f \in F[x]$, G acts on the roots of f. This means, since α is a root of f, that every element of Ω is also a root of f. Hence, $\deg(g) \leq \deg(f)$. (In fact, $g \mid f$.)

On the other hand, let $\sigma \in G$. Note that since σ permutes elements of Ω, $\{\sigma(\alpha_1), \sigma(\alpha_2), \ldots, \sigma(\alpha_r)\} = \Omega$. Now, σ is an F-automorphism of E and can be extended to an automorphism on $E[x]$ (see page 493). We apply the homomorphism σ to the polynomial $g \in E[x]$.

$$\sigma(g) = \sigma(x - \alpha_1)\sigma(x - \alpha_2) \cdots \sigma(x - \alpha_r) = (x - \sigma(\alpha_1)) \cdots (x - \sigma(\alpha_r)) = g.$$

We conclude that g is fixed by $\sigma \in G$. Hence, every coefficient of g is fixed by every element of G, and this means that every coefficient of g is in $\mathrm{Fix}(G)$. But, since E is a Galois extension of F, $\mathrm{Fix}(G) = F$. Hence, $g \in F[x]$. So g is a polynomial in $F[x]$ that has α as a root. This means that $f = \min_F(\alpha) \mid g$.

We conclude that $g = f$, and the proof is complete. $\square$

Example 25.10. Let $E = \mathbb{Q}[\sqrt{2}, \sqrt{3}]$. In Example 24.41, we found $\mathrm{Gal}(E/\mathbb{Q})$. In fact, $\mathrm{Gal}(E/\mathbb{Q}) = \{e, \sigma, \tau, \sigma\tau\}$, where σ fixes every element of $\mathbb{Q}[\sqrt{3}]$ and maps $\sqrt{2}$ to $-\sqrt{2}$, and, likewise, τ fixes every element of $\mathbb{Q}[\sqrt{2}]$ and maps $\sqrt{3}$ to $-\sqrt{3}$. The only elements of E that are fixed by every element of $\mathrm{Gal}(E/\mathbb{Q})$ are the rational numbers. Hence $\mathrm{Fix}(\mathrm{Gal}(E/\mathbb{Q})) = \mathbb{Q}$ and E is a Galois extension of $\mathbb{Q}$. Armed with this information and Theorem 25.9, we can easily find the minimal polynomial of any element of E. For example, let $\alpha = 2\sqrt{2} - 3\sqrt{3}$, then the roots of the minimal polynomial of α are the elements of E in the orbit of α under the action of $\mathrm{Gal}(E/\mathbb{Q})$. These are

$$\{2\sqrt{2} - 3\sqrt{3}, -2\sqrt{2} - 3\sqrt{3}, 2\sqrt{2} + 3\sqrt{3}, -2\sqrt{2} + 3\sqrt{3}\}.$$

Hence,

$$\min{}_{\mathbb{Q}}(\alpha) = (x - 2\sqrt{2} + 3\sqrt{3})(x + 2\sqrt{2} + 3\sqrt{3})(x - 2\sqrt{2} - 3\sqrt{3})(x + 2\sqrt{2} - 3\sqrt{3})$$
$$= x^4 - 70\,x^2 + 361.$$

In addition, we now know that $|\mathbb{Q}(\alpha) : \mathbb{Q}| = 4$, but, since $\mathbb{Q}(\alpha) \subseteq E$, we conclude that $E = \mathbb{Q}(\alpha)$ and α is a primitive element for the field extension $\mathbb{Q} \subseteq E$.

Theorem 25.9 is amazing both computationally—as the previous example showed—and theoretically. It says that if E is a Galois extension of F, then the minimal polynomial of *every* $\alpha \in E$ splits in E and has distinct roots. This surely should at least mean that a Galois extension is a splitting field, but it seems to imply even more. We will first name the two properties.

Definition 25.11 (Normal extension). Let $F \subseteq E$ be an algebraic extension of fields. Assume for all $\alpha \in E$ that $\min_F(\alpha)$ splits in $E[x]$. Then we call E a *normal extension* of F.

In other words, assuming that $F \subseteq E$ is a normal extension, if an irreducible polynomial in $F[x]$ has one root in E, then it splits over E.

Example 25.12. $\mathbb{Q} \subseteq \mathbb{Q}(\sqrt[3]{2})$ is not a normal extension since only one root of $x^3 - 2$ is in $\mathbb{Q}(\sqrt[3]{2})$.

Definition 25.13 (Separable polynomials). Let F be a field.

An irreducible polynomial $f \in F[x]$ is *separable* if f has $\deg(f)$ distinct roots in some splitting field over F.

A (not necessarily irreducible) polynomial $g \in F[x]$ is *separable* over F if each irreducible factor of g is separable over F.

A polynomial that is not separable is called *inseparable* over F.

Definition 25.14 (Separable elements and extensions). Let $F \subseteq E$ be fields, and let $\alpha \in E$. We say that α is *separable* over F if α is algebraic over F and if $\min_F(\alpha)$ is separable over F. The field E is a *separable extension* of F if every $\alpha \in E$ is separable over F.

The following is now an immediate corollary of Theorem 25.9:

Corollary 25.15. *Let E be a Galois extension of F. Then*

(a) *E is a normal extension of F, and*

(b) *E is a separable extension of F.*

Proof. The conclusion of Theorem 25.9 was that, since E is a Galois extension of F, the minimal polynomial of every element of E splits and has distinct roots in E. Hence, E is both a normal and a separable extension of F. □

Proposition 25.16. *Let $F \subseteq E$ be a field extension with $|E : F| < \infty$. Let $\{\alpha_1, \ldots, \alpha_n\}$ be a basis for E over F, and let*

$$f = \prod_{i=1}^{n} \min_F(\alpha_i).$$

Assume E is a normal extension of F. Then E is a splitting field for f over F.

Proof. A splitting field for f will include all its roots and hence will include $\alpha_1, \ldots, \alpha_n$. In other words, E will have to be a subfield of the splitting field. On the other hand, since E is a normal extension, f splits in E. Hence, E is the splitting field of f over F. □

Recall that a Galois extension, by definition, is a finite degree extension, and hence Proposition 25.16 and Corollary 25.15 together imply that a Galois extension is also a splitting field. In fact we can prove more:

Theorem 25.17. *Let $F \subseteq E$ be fields, and assume $|E : F| < \infty$. Then the following are equivalent:*

(a) *E is Galois over F.*

(b) *E is separable and normal over F.*

(c) *E is a splitting field for some separable polynomial over F.*

Proof. (a $\Rightarrow$ b) This is Corollary 25.15.

(b $\Rightarrow$ c) This follows from Proposition 25.16 since if E is a separable extension over F, then the polynomial f—defined in the statement of Proposition 25.16—is also separable since its irreducible factors will be separable.

(c $\Rightarrow$ a) We are assuming that E is the splitting field for a separable polynomial f over F. Let $K = \mathrm{Fix}(\mathrm{Gal}(E/F))$. We want to show that $K = F$.

We use induction on $|E : F|$.

For the base case, if $|E : F| = 1$, then $E = F$, and $K = F$ trivially. Hence, assume that $|E : F| > 1$, and further assume that the theorem has been proved for all field extensions where the degree of the extension is less than $|E : F|$.

The polynomial f may not be irreducible. Hence, let $g \in F[x]$ be a monic irreducible polynomial of degree > 1 and with $g \mid f$.

Let $\alpha \in E$ be a root of g. We have
$$F \subseteq F[\alpha] \subseteq E,$$
and $g = \min_F(\alpha)$. Hence, $|F[\alpha] : F| = \deg(g) > 1$.

We will show $K = F$ in two stages. In the first stage we show that $K \subseteq F[\alpha]$.

CLAIM. $K \subseteq F[\alpha]$.

PROOF OF CLAIM. By Lemma 24.20, E is a splitting field for f over $F[\alpha]$, f is separable, and $|E : F[\alpha]| < |E : F|$. Hence, we can apply the inductive hypothesis to the extension $F[\alpha] \subseteq E$ and conclude that E is a Galois extension of $F[\alpha]$.

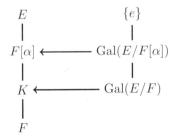

Figure 25.3. $\mathrm{Gal}(E/F[\alpha]) \subseteq \mathrm{Gal}(E/F)$ implies that $K \subseteq F[\alpha]$.

Now $\text{Gal}(E/F[\alpha]) \subseteq \text{Gal}(E/F)$ and (see Figure 25.3)

$$K = \text{Fix}(\text{Gal}(E/F)) \subseteq \text{Fix}(\text{Gal}(E/F[\alpha])) = F[\alpha].$$

So far, we have

$$F \subseteq K \subseteq F[\alpha]$$

and we want to show $K = F$. We know that $g = \min_F(\alpha)$, and we let $h = \min_K(\alpha)$. Since $|F[\alpha] : F| = \deg g$ and $|F[\alpha] : K| = \deg h$, the proof will be complete if we show that $\deg(g) = \deg(h)$.

Since g can be considered as a polynomial over K with α as a root, and h is the minimal polynomial of α over K, we have $h \mid g$ and so $\deg(h) \leq \deg(g)$.

Now g is separable—this is the one point in the proof where separability is used—and so it has distinct roots. Hence, if we show that every root of g is a root of h, then we have shown that $\deg(g) \leq \deg(h)$, and the proof will be complete.

Let β be a root of g. Since E is a splitting field over F and g is irreducible, $\text{Gal}(E/F)$ acts transitively on the roots of g. Hence, there is a $\sigma \in \text{Gal}(E/F)$ with $\sigma(\alpha) = \beta$. Since the coefficients of h are in $K = \text{Fix}(\text{Gal}(E/F))$, we note that $\sigma(h) = h$. We now have

$$h(\beta) = \sigma(h)(\beta) = \sigma(h)(\sigma(\alpha)) = \sigma(h(\alpha)) = \sigma(0) = 0.$$

Hence, every root of g is a root of h, and so $\deg(g) = \deg(h)$. It follows that $F = K$, and the proof is complete. $\square$

Corollary 25.18. *Let $F \subseteq K \subseteq E$ be fields. Suppose E is Galois over F. Then E is Galois over K.*

Proof. Since E is Galois over F, then E is the splitting field over F for some separable polynomial $f \in F[x]$. But you can consider $f \in K[x]$. E is the splitting field of f over K and f continues to be separable. Hence, E is Galois over K. $\square$

Example 25.19. Let $E = \mathbb{Q}(i, \sqrt{2})$, and consider the extension

$$\mathbb{Q} \subseteq E = \mathbb{Q}(i, \sqrt{2}).$$

First of all, E is the splitting field for $f = (x^2+1)(x^2-2)$ since f splits in E and any splitting field would have to contain both i and $\sqrt{2}$. Now f is a separable polynomial since it has distinct roots, and so we conclude that E is a Galois extension of $\mathbb{Q}$.

We have both $\mathbb{Q} \subseteq \mathbb{Q}(i) \subseteq E$ and $\mathbb{Q} \subseteq \mathbb{Q}(\sqrt{2}) \subseteq E$. Since $i \notin \mathbb{Q}(\sqrt{2})$, $x^2 + 1$ is irreducible over $\mathbb{Q}(\sqrt{2})$ and hence $x^2 + 1$ is the minimal polynomial of i over $\mathbb{Q}(\sqrt{2})$. Hence $|E : \mathbb{Q}(\sqrt{2})| = 2$. We conclude that $|E : \mathbb{Q}| = 2 \times 2 = 4$. It follows that $|E : \mathbb{Q}(i)| = 2$ as well. (See Figure 25.4.) In addition, E is Galois over $\mathbb{Q}(\sqrt{2})$ and over $\mathbb{Q}(i)$.

Now let $L = \mathbb{Q}(\sqrt{2})$, then $E = L(i)$ and $x^2 + 1$ is an irreducible polynomial in $L[x]$ with two roots i and $-i$ in E. Hence, by Theorem 24.6, there exists a unique isomorphism $\sigma : E = L(i) \to E = L(-i)$ such that $\sigma(i) = -i$ and $\sigma|_L$ is the identity map. Hence, we have $\sigma \in \text{Gal}(E/\mathbb{Q})$ with $\sigma(i) = -i$ and $\sigma(\sqrt{2}) = \sqrt{2}$.

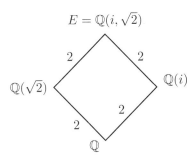

Figure 25.4. A partial Hasse diagram of subfields of $\mathbb{Q}(i, \sqrt{2})$ ordered by inclusion

Likewise, we have $\tau \in \text{Gal}(E/\mathbb{Q})$ with $\tau(\sqrt{2}) = -\sqrt{2}$ and $\tau(i) = i$.

Since elements of $\text{Gal}(E/\mathbb{Q})$ permute roots of $x^2 + 1$ and $x^2 - 2$, and since an isomorphism is determined by its action on these roots, we conclude that

$$\text{Gal}(E/\mathbb{Q}) = \{e, \sigma, \tau, \sigma\tau\} \cong \mathbb{Z}/2\mathbb{Z} \times \mathbb{Z}/2\mathbb{Z}.$$

Now, let $\alpha = \sqrt{2} + i$. Then, by Theorem 25.9, we have

$$\begin{aligned}
\min_{\mathbb{Q}}(\alpha) &= (x - \sqrt{2} - i)(x - \sqrt{2} + i)(x + \sqrt{2} - i)(x + \sqrt{2} + i) \\
&= ((x - \sqrt{2})^2 + 1)((x + \sqrt{2})^2 + 1) \\
&= x^4 - 2x^2 + 9.
\end{aligned}$$

It follows that $|\mathbb{Q}(\alpha) : \mathbb{Q}| = 4$. But $\alpha \in E$, and hence we conclude that

$$E = \mathbb{Q}(\sqrt{2} + i).$$

If we let $\beta = \sqrt{2}i$, then the orbit of β under the action of the Galois group will be $\{\beta, -\beta\}$. Hence, as expected,

$$\min_{\mathbb{Q}}(\beta) = (x - \beta)(x + \beta) = x^2 + 2.$$

We remark that we got most of the information about this extension without that much calculation. In particular, we found the minimal polynomial of α and the fact that $E = \mathbb{Q}(\alpha)$ with little or no calculation!

Problems

25.2.1. Is $\mathbb{R}$ a normal extension of $\mathbb{Q}$? Is $\mathbb{Q}(i)$ a normal extension of $\mathbb{Q}$? Is $\mathbb{Q}(\sqrt{47}, i)$ a normal extension of $\mathbb{Q}(i)$?

25.2.2. Let $f = x^3 - x - 1 \in \mathbb{Q}[x]$. Use calculus to draw an approximate graph of f. How many real roots does f have? Let α be one of the real roots, and let $K = \mathbb{Q}(\alpha)$. Is K a normal extension of $\mathbb{Q}$?

25.2.3. Let $F \subseteq E$ be fields, and assume that $|E : F| = 2$. Show that E is a normal extension of F.

25.2.4. Let E be a normal extension of F, and let $f \in F[x]$ be irreducible of degree greater than 1. Is it possible to find an example where f is irreducible in $E[x]$?

25.2.5. Let E be a Galois extension of F, and let $f \in F[x]$ be irreducible. If $\alpha \in E$ is a root of f, then show that $(x - \alpha)^2$ does not divide f.

25.2.6. Let $F \subseteq E$ be fields, and assume that K_1 and K_2 are two fields that contain F and are contained in E. Assume K_1 and K_2 are normal extensions of F. Is $K_1 \cap K_2$ necessarily a normal extension of F?

25.2.7. Let $F \subseteq E$ be fields with $|E : F| < \infty$. Assume $E = F[\alpha]$ and $|\mathrm{Gal}(E/F)| = |E : F|$. Show that E is a Galois extension of F.

25.2.8. Let $E = \mathbb{Q}(\sqrt[4]{2})$.
 (a) What is $|E : \mathbb{Q}|$? Is E a normal extension of $\mathbb{Q}$?
 (b) What are the elements of $\mathrm{Gal}(E/\mathbb{Q})$? Give a familiar group that is isomorphic to $\mathrm{Gal}(E/\mathbb{Q})$.
 (c) What is $\mathrm{Fix}(\mathrm{Gal}(E/\mathbb{Q}))$?
 (d) Let $K = \mathrm{Fix}(\mathrm{Gal}(E/\mathbb{Q}))$. Is E a Galois extension of K? Is K a Galois extension of $\mathbb{Q}$?
 (e) Let $L = E(i)$. Is L a Galois extension of $\mathbb{Q}$?

25.2.9. Use Theorem 25.9 to find $\min_{\mathbb{Q}}(\sqrt{2} + i\sqrt{3})$.

25.2.10. Let $E = \mathbb{Q}(\sqrt{3}, \sqrt{5}, \sqrt{7}, \sqrt{47})$, and let $\alpha = \sqrt{3} + \sqrt{5} + \sqrt{7} + \sqrt{47} \in E$.
 (a) What is $|E : \mathbb{Q}|$?
 (b) What is the size of the orbit of α under the action of $\mathrm{Gal}(E/\mathbb{Q})$?
 (c) What is the degree of the minimal polynomial of α?
 (d) Is α a primitive element for the field extension $\mathbb{Q} \subseteq E$?

25.2.11. Let $\zeta_5 = e^{2k\pi i/5}$. So ζ_5 is a non-real complex number with $(\zeta_5)^5 = 1$.
 (a) Find $|\mathbb{Q}[\zeta_5] : \mathbb{Q}|$.
 (b) Is $\mathbb{Q}[\zeta_5]$ Galois over $\mathbb{Q}$?
 Let $G = \mathrm{Gal}(\mathbb{Q}[\zeta_5]/\mathbb{Q})$.
 (c) Show that there exists $\sigma_1, \sigma_2 \in G$ with $\sigma_1(\zeta_5) = \zeta_5^4$ and $\sigma_2(\zeta_5) = \zeta_5^2$.
 (d) What is $\sigma_1(\zeta_5 + \zeta_5^4)$ and $\sigma_2(\zeta_5 + \zeta_5^4)$?
 (e) What is $\langle \sigma_1 \rangle$, the subgroup of G generated by σ_1?
 (f) What is $\mathrm{Fix}(G)$? Prove your assertion.
 (g) Is $\zeta_5 + \zeta_5^4$ a rational number? Prove your assertion.
 (h) What is $\mathrm{Fix}(\langle \sigma_1 \rangle)$? Prove your assertion.
 (i) What is G? Prove your assertion.

25.2.12. Let ζ_7 be a non-real complex number with $\zeta_7^7 = 1$. Let $\phi \in \mathrm{Gal}(\mathbb{Q}[\zeta_7]/\mathbb{Q})$ with $\phi(\zeta_7) = \zeta_7^3$.
 (a) What is the order of ϕ?
 (b) What is $|\mathrm{Gal}(\mathbb{Q}[\zeta_7]/\mathbb{Q})|$?
 (c) Find a well known group that is isomorphic to $\mathrm{Gal}(\mathbb{Q}[\zeta_7]/\mathbb{Q})$.
 (d) Find one irrational element of $\mathbb{Q}[\zeta_7]$ that is fixed by ϕ^3.

25.3. More on Normal Extensions

Theorem 25.20. *Let $F \subseteq E$ be fields with $|E : F| < \infty$. Then the following are equivalent:*

(a) *E is normal over F.*

(b) *E is a splitting field for some polynomial over F.*

(c) *If L and K are fields with $E \subseteq L$ and $F \subseteq K \subseteq L$ and if $\theta : E \to K$ is an F-isomorphism, then $K = E$.*

Proof. (a $\Rightarrow$ b) This is Proposition 25.16.

(b $\Rightarrow$ c) Let $F \subseteq E \subseteq L$ and $F \subseteq K \subseteq L$. Assume $\theta : E \to K$ is an F-isomorphism. (See Figure 25.5.) We know that E is the splitting field for some $g \in F[x]$. Since θ is an F-isomorphism, K is also a splitting field for g over F. Now E and K are both subfields of L and they are both splitting fields for g over F. But a splitting field is the *smallest* field in which g splits. We conclude that $K = E$.

(c $\Rightarrow$ a) Let $\alpha \in E$, and let $f = \min_F(\alpha)$. We want to show that f splits in E.

Let $M \supseteq E$ be a splitting field for f over E, and we want to show that $M = E$. We will construct an even bigger field L. Note that by Problem **24.2.11**, $|M : E| < \infty$, and, by Proposition 22.38, $|M : F| < \infty$. Let $\alpha_1, \ldots, \alpha_m$ be a basis for M over F. Let $g = \prod_{i=1}^{m} \min_F(\alpha_i)$, and let L be a splitting field for g over M.

Now $g \in F[x]$ and any splitting field of g over F would have to include $\alpha_1, \ldots, \alpha_m$, and hence would have to include M. Thus L—by virtue of being the splitting field of g over M—is also the splitting field of g over F.

Now, L is a splitting field (of g) over F, and f splits in L (since it already split in M). Let $\beta \in L$ with $f(\beta) = 0$. We want to show that $\beta \in E$. (This would show that f already splits in E completing the proof.)

We have $\alpha \in E \subseteq L$ and $f(\alpha) = 0$. We also know that f is irreducible and L is a splitting field over F. Hence, by Theorem 24.37, there exists $\sigma \in \mathrm{Gal}(L/F)$ with $\sigma(\alpha) = \beta$.

Now restricting σ to E, we have $\sigma|_E : E \to \sigma(E) \subseteq L$ is an F-isomorphism. By our hypothesis, we have $\sigma(E) = E$.

Now $\beta = \sigma(\alpha) \in \sigma(E) = E$. Hence, $\beta \in E$, and the proof is complete. $\quad\square$

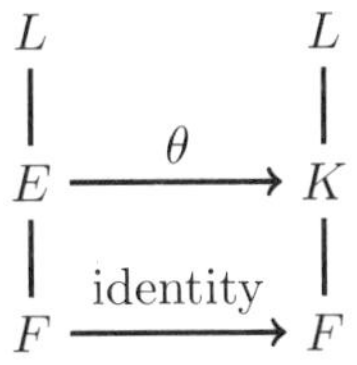

Figure 25.5. θ is an F-isomorphism.

Remark 25.21. At first it appeared that the concept of a normal extension was stronger than that of a splitting field. After all if E is a normal extension of F, then *every* irreducible in $F[x]$ that has a root in E splits in E, while to say that E is a splitting field over F means that there exists *one* polynomial in $F[x]$ for which E is a splitting field. But, we have shown that for finite degree extensions the two concepts are equivalent. In other words, if E is the splitting field of one polynomial over F, then every irreducible polynomial with a root in E, splits in E!

Definition 25.22 (Normal closure). Let $F \subseteq E$ be a field extension. A field L is called a *normal closure* for E over F if

(a) $E \subseteq L$.

(b) L is normal over F.

(c) If $E \subseteq K \subseteq L$ and K is normal over F, then $K = L$.

We can strengthen Proposition 25.16.

Proposition 25.23. *Let $F \subseteq E$ be a field extension with $|E : F| < \infty$. Let $\{\alpha_1, \ldots, \alpha_n\}$ be a basis for E over F, and let*

$$f = \prod_{i=1}^{n} \min_F(\alpha_i).$$

Then L is a normal closure for E over F if and only if L is a splitting field for F over E.

In particular, all normal closures of E over F are isomorphic, and E is a normal extension of F if and only if E is a splitting field for f over F.

Proof. In Problem *25.3.11*, you are asked to write up the proof. It basically follows from Proposition 25.16, Theorem 25.20, and the fact that the two splitting fields of the same polynomial are F-isomorphic (see Corollary 24.22). □

Remark 25.24. In Proposition 25.23, we showed that if E is a finite degree extension of F, then we can construct a normal closure L for E over F. In fact, L is the splitting field of $f = \prod_{i=1}^{n} \min_F(\alpha_i)$, where $\{\alpha_1, \ldots, \alpha_n\}$ is a basis for E over F. By Problem **24.2.11**, $|L : F|$ is no more than $\deg(f)!$, and it is finite. Now if, in addition to $|E : F| < \infty$, we knew that E is a separable extension of F, then L is a finite degree extension of F that is the splitting field of a separable polynomial over F. As a result, L will be a Galois extension of F (Theorem 25.17). Moreover, no proper subfield of L (that contains E) can be a Galois extension of F (since such a subfield cannot even be a normal extension of F). Such a field L is sometimes called the *Galois closure* of E over F. We record the most useful part of this observation.

Proposition 25.25. *Let $F \subseteq E$ be fields. Assume $|E : F| < \infty$, and that E is separable over F. Then there exists a finite degree field extension $L \supseteq E$ such that L is Galois over F.*

Problems

25.3.1. Which of the following extensions are normal?
   (a) $\mathbb{Q} \subseteq \mathbb{Q}(x)$.
   (b) $\mathbb{Q} \subseteq \mathbb{Q}(\sqrt{5}i)$.
   (c) $\mathbb{Q} \subseteq \mathbb{Q}(\alpha)$, where $\alpha = \sqrt[7]{5}$ is the real 7th root of 5.
   (d) $\mathbb{Q}(\alpha) \subseteq \mathbb{Q}(\sqrt{5}, \alpha)$, where $\alpha = \sqrt[7]{5}$.
   (e) $\mathbb{R} \subseteq \mathbb{R}(\sqrt{7}i)$.

25.3.2. Let $K = \mathbb{Q}(\sqrt{7})$ and $E = \mathbb{Q}(\sqrt[4]{7})$, and consider the field extensions $\mathbb{Q} \subseteq K \subseteq E$. Is K a normal extension of $\mathbb{Q}$? Is E a normal extension of K? Is E a normal extension of $\mathbb{Q}$? Which one of these extensions is a Galois extension?

25.3.3. Let $\alpha = \sqrt[4]{5} \in \mathbb{R}$, let $\beta = \alpha + i\alpha \in \mathbb{C}$, and let $\gamma = i\alpha^2 \in \mathbb{C}$. Is $\gamma \in \mathbb{Q}(\beta)$? If so, is $\mathbb{Q}(\beta)$ a normal extension of $\mathbb{Q}(\gamma)$? Is $\mathbb{Q}(\gamma)$ a normal extension of $\mathbb{Q}$? Is $\mathbb{Q}(\beta)$ a normal extension of $\mathbb{Q}$?

25.3.4. Let $F \subseteq E$ be fields, and let K and L be intermediate fields containing F and be contained in E. Assume that both K and L are normal extensions of F. Does $\langle K, L \rangle$, the compositum of K and L (see Definition 22.41), have to be a normal extension of F as well? Why?

25.3.5. Let $F \subseteq K \subseteq E$ be fields. Assume that E is a normal extension of F. Is E necessarily a normal extension of K? Is K necessarily a normal extension of F? In both cases, either prove that the answer is yes or give a counterexample.

25.3.6. Can you find three fields $F \subseteq K \subseteq E$ and a map $\phi : K \to \phi(K) \subseteq E$ such that E is Galois over F, ϕ is an F-isomorphism, but $\phi(K) \neq K$? Either give an explicit example or prove that this is impossible.

25.3.7. Let $F \subseteq K \subseteq E$ be fields. Assume $|E : F| < \infty$ and K is a Galois extension of F. Let $\sigma \in \text{Gal}(E/F)$. Show $\sigma\big|_K$, the restriction of σ to K, is an element of $\text{Gal}(K/F)$.

25.3.8. Let $F \subseteq K \subseteq E$ be fields and assume that E is a Galois extension of F. Show that K is a Galois extension of F if and only if $\sigma(K) = K$ for all $\sigma \in \text{Gal}(E/F)$.

25.3.9. Let $F = \mathbb{F}_5(x)$ be the field of rational functions over $\mathbb{F}_5$, and let $f = y^5 - x \in F[y]$. Let α be a root of f in some splitting field of f over F. Is $F(\alpha)$ a normal extension of F?

25.3.10. Let $f = x^6 + x^3 + 1 \in \mathbb{Q}[x]$, and let α be one of the complex roots of f. Let $\phi : \mathbb{Q}(\alpha) \to \mathbb{C}$ be a field homomorphism. Assume that ϕ is not the zero homomorphism.
   (a) Is $\mathbb{Q}(\alpha)$ the splitting field for f over $\mathbb{Q}$?
   (b) Is $\phi \in \text{Gal}(\mathbb{Q}(\alpha)/\mathbb{Q})$?
   (c) Explicitly find all possible ϕ.
   (d) Find a familiar group that is isomorphic to $\text{Gal}(f)$.

25.3.11. Let $F \subseteq E$ with $|E : F| < \infty$. By proving Proposition 25.23, show that a normal closure for E over F exists, and is a finite degree extension of F.

25.3.12. Let $E = \mathbb{Q}(\sqrt[4]{7})$. Find a normal closure for E over $\mathbb{Q}$.

25.3.13. Let $E = \mathbb{Q}(\sqrt{2}, \sqrt[3]{2}, \sqrt[4]{2})$. Find a normal closure for E over $\mathbb{Q}$.

25.3.14. Let $F \subseteq E$ be an algebraic extension, and let $\overline{E}$ denote the algebraic closure of E (see Definition 24.27). Let

$$\Omega = \{K \text{ a field} \mid E \subseteq K \subseteq \overline{E} \text{ and } K \text{ a normal extension of } F\}.$$

Now define $L = \bigcap_{K \in \Omega} K$. Show that L is a normal closure of E over F.

25.3.15. Let $f \in \mathbb{Q}[x]$ be a polynomial of degree 3, and assume that $\mathrm{Gal}(f)$ is a group of order 3. Prove that all the roots of f are real.

25.3.16. Let $F \subseteq K \subseteq E$ be fields, and let $\theta \colon K \to E$ be a non-zero F-homomorphism (i.e., ϕ is a non-zero ring homomorphism that fixes every element of F; see Problem 24.1.18). Assume that E is a normal extension of F. Show that there exists $\phi \in \mathrm{Gal}(E/F)$ such that the restriction of ϕ to K is θ. Is the assumption that E be a normal extension necessary? If so, give an example to show that the condition is necessary.

25.4. More on Separable Extensions

Recall that a field extension $F \subseteq E$ is separable if the minimal polynomial of every $\alpha \in E$ has distinct roots in some splitting field. We have seen that every Galois extension must be separable, and conversely every normal and separable extension is Galois. What we have not seen is an *example* of an inseparable polynomial. In other words, which irreducible polynomial has repeated roots? Such examples are not that common. We will first give an example and then proceed to develop the theory, which among other things, will explain the dearth of examples.

Example 25.26. Let $\mathbb{F}_3 = (\mathbb{Z}/3\mathbb{Z}, +, \cdot)$ be the field of size 3, and let $K = \mathbb{F}_3(Y)$. Elements of the field K are rational functions in Y—such as $\frac{2Y^2 + Y + 1}{Y + 1}$ and $\frac{1}{Y}$—and all the coefficients are in $\mathbb{F}_3$. Now consider $K[x]$ the ring of polynomials in x with coefficients in K. Let $f = x^3 - Y \in K[x]$.

Let $A = \mathbb{F}_3[Y]$ be the ring of polynomials in Y over $\mathbb{F}_3$. Then A is a Euclidean domain, the field of fractions of A is K, and Y is a prime in A. Hence, by the Schönemann-Eisenstein criterion, f is irreducible in $K[x]$.

Let E be a splitting field of f over K, and let $\beta \in E$ with $f(\beta) = 0$. Plugging β for x in the definition of f, we get that $\beta^3 = Y \in K$. We now have

$$f = x^3 - Y = x^3 - \beta^3 = (x - \beta)^3.$$

Thus f has only one root in E. Since f was irreducible, we conclude that f is inseparable and E is not a separable extension of F.

In addition, $E = K[\beta]$ is a normal extension—since it is a splitting field of a polynomial—of K and is not a Galois extension—since it is not separable. In fact, $\mathrm{Gal}(E/K)$ permutes the roots of f in E. But f has only one root in E, namely β.

Hence, every element of the Galois group fixes β. Since $E = K[\beta]$, this means that $\mathrm{Gal}(E/K) = \{e\}$ and hence $\mathrm{Fix}(\mathrm{Gal}(E/K)) = E$.

The derivative of a polynomial—which we define formally and without recourse to limits—will prove useful in studying separable and inseparable polynomials.

Definition 25.27 (Derivative). Let F be a field. For

$$f(x) = a_0 + a_1 x + \cdots + a_n x^n \in F[x],$$

define the *derivative* of f by

$$f'(x) = a_1 + 2a_2 x + \cdots + n a_n x^{n-1}.$$

The derivative follows the familiar rules:

(a) $(f + g)' = f' + g'$ for $f, g \in F[x]$,

(b) $(fg)' = f'g + fg'$ for $f, g \in F[x]$,

(c) $c' = 0$ for $c \in F$,

(d) $(cf)' = cf'$ for $c \in F$ and $f \in F[x]$.

Lemma 25.28. *Let E be a field, and let $0 \neq f \in E[x]$. Let $\alpha \in E$. Then $(x - \alpha)^2 \mid f$ if and only if $f(\alpha) = f'(\alpha) = 0$.*

Proof. You are asked to prove this in Problem **25.4.1**. □

Corollary 25.29. *Let F be a field, and let $f \neq 0$ a polynomial of degree n in $F[x]$. Let $E \supseteq F$ be a field in which both f and f' split. Assume that f and f' have no common zeros in E. Then f has n distinct roots in E.*

Proof. In E, f splits into n linear factors. If f did not have distinct roots, then, for some $\alpha \in E$, $(x - \alpha)^2$ would divide f. Then, by Lemma 25.28, f and f' would have a common root. The contradiction proves the claim. □

Theorem 25.30. *Let F be a field, and let $f \in F[x]$ be an irreducible polynomial. The polynomial f is inseparable if and only if f' is the zero polynomial.*

Proof. ($\Rightarrow$) Let $E \supseteq F$ be a splitting field for f over F. Assume that f is not separable. By definition, there exists $\alpha \in E$ with $(x - \alpha)^2 \mid f(x)$, and, hence, by Lemma 25.28 $f(\alpha) = f'(\alpha) = 0$. Since f is irreducible and α is a root of f, we have $f = c \min_F(\alpha)$. Now $f'(\alpha) = 0$ which means that $\min_F(\alpha) \mid f'$, and so $f \mid f'$. However, $\deg(f') < \deg(f)$ and so f' must be the zero polynomial.

($\Leftarrow$) Assume $f' = 0$. Let α be a root of f in a splitting field, then $f(\alpha) = f'(\alpha) = 0$ which, by Lemma 25.28, implies that $(x - \alpha)^2 \mid f(x)$. This means that f is not separable. □

Theorem 25.31. *Let F be a field with $\mathrm{char}(F) = 0$. Then all polynomials $f \in F[x]$ are separable.*

Furthermore, if $\mathrm{char}(F) = p > 0$, and $f \in F[x]$ is inseparable, then

$$f = a_0 + a_1 x^p + a_2 x^{2p} + \cdots + a_n x^{np}.$$

In other words, $f(x) = g(x^p)$ for some $g \in F[x]$.

Proof. Without loss of generality, we can assume f is irreducible. $\deg(f) \geq 1$ which in the case of a field of characteristic zero means that $\deg(f') \geq 0$. In other words f' is not the zero polynomial. Thus, by Theorem 25.30, f is separable.

If $\mathrm{char}(F) = p$, to show that $f(x) = g(x^p)$, without loss of generality, we can assume that f is irreducible since if $f_1(x) = g_1(x^p)$, $f_2(x) = g_2(x^p)$, ..., $f_k(x) = g_k(x^p)$, then $f_1 f_2 \ldots f_k(x) = g_1 g_2 \cdots g_k(x^p)$. Now if $f = a_0 + a_1 x + \cdots + a_n x^n$, then $f' = a_1 + \cdots + n a_n x^{n-1}$. This is the zero polynomial since f is irreducible and inseparable. So, $k a_k = 0$ for all $1 \leq k \leq n$. Hence, for $1 \leq k \leq n$, if $a_k \neq 0 \mod p$, then $k = 0 \mod p$. The result now follows. $\qquad\square$

Definition 25.32. Let F be a field of characteristic $p > 0$, and define $\Phi : F \to F$ by $\Phi(a) = a^p$. Then Φ is called the *Frobenius map*. The image of the Frobenius map is denoted by F^p.

Lemma 25.33. *Let F be a field of characteristic $p > 0$, and let Φ be the Frobenius map. Then Φ is an endomorphism (i.e., a 1-1 ring homomorphism).*

Furthermore, if F is finite, then Φ is onto (i.e., $F^p = F$), and if $F = \mathbb{Z}/p\mathbb{Z}$, then Φ is the identity map.

Proof. Let $a, b \in F$. We have $(ab)^p = a^p b^p$ since F is commutative. By the binomial theorem and since $\mathrm{char}(F) = p$, we also have $(a + b)^p = a^p + b^p$. (All the other terms in the expansion of $(a+b)^p$ have a coefficient divisible by p and become zero; see Lemma 7.4.) These show that Φ is a ring homomorphism. Since F is a field, it has no non-trivial ideals, $\Phi(1) = 1$ and so the kernel of the map cannot be F, and hence $\ker(\Phi) = \{0\}$, and the map is 1-1.

If F is finite, a 1-1 map from F to F will also be onto, and hence $F^p = F$. For $F = \mathbb{Z}/p\mathbb{Z}$, Fermat's Little Theorem, Corollary 5.35, gives that $a^p = a$ for all $a \in F$, and hence Φ is the identity map. $\qquad\square$

Proposition 25.34. *Let F be a finite field, then all $f \in F[x]$ are separable.*

Proof. Let $\mathrm{char}(F) = p$, and, without loss of generality, assume that f is irreducible and inseparable. By Theorem 25.31, we have

$$f = a_0 + a_1 x^p + a_2 x^{2p} + \cdots + a_n x^{np}.$$

Now since F is finite, the Frobenius map is onto and hence, for each i with $0 \leq i \leq n$, there exists $b_i \in F$ with $b_i^p = a_i$. Hence,

$$f = b_0^p + b_1^p x^p + b_2^p x^{2p} + \cdots + b_n^p x^{np} = (b_0 + b_1 x + b_2 x^2 + \cdots + b_n x^n)^p.$$

Now let $g = b_0 + b_1 x + b_2 x^2 + \cdots + b_n x^n \in F[x]$. We have $f = g^p$ is not irreducible. The contradiction proves the proposition. $\qquad\square$

We conclude that if $f \in F[x]$ is inseparable, then F is an infinite field of characteristic p. This explains the scarcity of examples over fields that we are most familiar with. Recall that a field extension $F \subseteq E$ is separable if every element of E is separable over F (Definition 25.14). We now bring together a few facts about separable extensions. First we record the abundance of separable extensions.

Corollary 25.35. *Let $F \subseteq E$ be a field extension with $|E : F| < \infty$. If $\mathrm{char}(F) = 0$ or if F is a finite field, then E is a separable extension of F.*

Proof. For $\alpha \in E$, the condition $|E : F| < \infty$ means that α is algebraic over F. By Theorem 25.31 and Proposition 25.34, $\min_F(\alpha)$ is separable over F, and so α is separable over F. Since α was arbitrary, E is a separable extension of F. □

Lemma 25.36. *Let $F \subseteq K \subseteq E$ be fields. Let E be a separable algebraic extension of F. Then E is separable over K and K is separable over F.*

Proof. Let $\alpha \in E$, then $\min_F(\alpha)$ is separable. But $\min_K(\alpha)$ divides $\min_F(\alpha)$ and hence it must be separable as well. We conclude that α is separable over K, and so E is separable over K.

Let $\beta \in K$. Then $\beta \in E$ also, and hence $\min_F(\beta)$ is separable since E is separable over F. We conclude that K is separable over F. □

Proposition 25.37. *Let $F \subseteq E$ be fields. Assume $E = F[\alpha]$ and that α is separable over F. Then E is separable over F.*

Proof. By definition, a separable element is algebraic, and so $|E : F| < \infty$. Now, let $f = \min_F(\alpha)$, and let L be the splitting field of f over E. Then L is the splitting field of f over F as well, and f is separable. Hence, L is Galois over F. But this means that L is separable over F, and hence E is separable over F. □

Proposition 25.37 can be generalized—using a similar proof—to show that if you join a finite number of separable elements to the ground field, then you get a separable extension. In Problem **25.4.13** you are asked to prove this generalization.

Problems

25.4.1. **Proof of Lemma 25.28.** Let E be a field, and let $0 \neq f \in E[x]$. Let $\alpha \in E$. Show that

$$(x - \alpha)^2 \mid f \text{ if and only if } f(\alpha) = f'(\alpha) = 0.$$

25.4.2. Let p be a prime, and let $\mathbb{F}_p = (\mathbb{Z}/p\mathbb{Z}, +, \cdot)$. For which n does $x^n - 1 \in \mathbb{F}_p[x]$ have distinct roots?

25.4.3. The integer 10006429 is a prime number. Let F be a finite field of order 10006429^3, let $\beta \in F$, and let $f(x) = x^{10006429} - \beta$. Can f be irreducible in $F[x]$? If E is the splitting field for f over F, then what is $|E : F|$?

25.4.4. Let $F = \mathbb{Q}(y)$ be the field of rational functions in the indeterminate y. Let $f \in F[x]$ be defined by

$$f(x) = 2x^3 + (y^2 - 4y + 1)x^2 - 2y(y^2 - y + 1)x + y^2(y^2 + 1).$$

(a) Does $(x - y) \mid f$ in $F[x]$?

(b) Does $(x - y)^2 \mid f$ in $F[x]$?

Perfect Fields.

Definition 25.38. Perfect fields. Let $p \in \mathbb{Z}^{>0}$ be a prime, and let F be a field of characteristic p. F is called *perfect* if every element of F has a pth root in F.

25.4.5. Let F be a field of characteristic $p > 0$.
    (a) Show that F is perfect if and only if $F^p = F$.
    (b) Show that if F is finite, then it is perfect.
    (c) Show that $F(x)$, the field of rational functions over F, is not perfect.

25.4.6. Let K be a field of characteristic $p > 0$, and assume that K is not perfect. Prove that there exists an irreducible inseparable polynomial in $K[x]$.

25.4.7. Let F be a field of characteristic $p > 0$, $E = F[\alpha]$ with $\alpha^p \in F$. Assume E is a separable extension of F. What are the possibilities for $|E : F|$?

25.4.8. Let F be a field of characteristic $p > 0$, and let E be an extension of F. Assume $\alpha \in E - F$. Show that $\alpha^p \notin F^p$.

25.4.9. $F \subseteq E$ are fields of characteristic 0. Let $f, g \in F[x]$. Assume that f splits in E. Further suppose that there exists $\alpha \in E - F$ with $f(\alpha) = g(\alpha) = 0$. Show that there exists $\beta \in E - F$ with $\beta \neq \alpha$ and $f(\beta) = g(\beta) = 0$.

25.4.10. Let F be a field, and let $f \in F[x]$ be an irreducible polynomial. Let E be a splitting field for f over F. Show that all roots of f in E have the same multiplicity. (See Definition 19.26.)

25.4.11. Let F be a field of characteristic $p > 0$, and let $f(x) \in F[x]$ be a polynomial of degree n. Assume that $f'(x)$ is the zero polynomial. Show that $p \mid n$ and that f has at most n/p distinct roots.

25.4.12. Assume F is a field of characteristic $p > 0$, $F \subseteq E$ a finite degree field extension, and $p \nmid |E : F|$. Show that E is a separable extension of F.

25.4.13. Let $E = F[\alpha_1, \alpha_2, \ldots, \alpha_n]$ with each α_i separable over F. Show E is separable over F.

25.4.14. Let $F \subseteq E$ be a field extension, and let $\alpha, \beta \in E$. Assume that both α and β are separable over F. Does $\alpha + \beta$ have to be separable over F? Why?

25.5. Simple Extensions

Let $F \subseteq E$ be fields. Recall—from Definition 22.10—that E is a *simple* extension of F if $E = F(\alpha)$ for some $\alpha \in E$, and such an α, if it exists, is called a *primitive element* for the extension. We already saw in Example 25.19 that if an extension is simple, then the Galois group can help in finding the primitive element. In this section, we seek conditions that guarantee that an extension is simple. We treat finite and infinite fields separately.

Finite Fields. Before turning to simple extensions, we prove the already interesting fact that all finite fields have cyclic multiplicative groups. Recall Definition 1.45 that, for n a positive integer, the Euler ϕ-function (or totient function) is defined by

$$\phi(n) = |\{a \in \mathbb{Z} \mid 1 \le a \le n, \text{ and } \gcd(a, n) = 1\}|.$$

Lemma 25.39.

$$\sum_{d \mid n} \phi(d) = n.$$

Proof. Let H be a cyclic group of order n. For every $d \mid n$, H has a unique subgroup of order d which has $\phi(d)$ generators. Thus the total number of generators of subgroups of H is $\sum_{d \mid n} \phi(d)$. On the other hand every element of H is the generator of one and only one subgroup of H. So, $\sum_{d \mid n} \phi(d) = n$. $\qquad\square$

Lemma 25.40. *Let G be a finite group with the property that for each positive integer k, there are at most k elements $x \in G$ such that $x^k = 1$. Then G is cyclic.*

Proof. If G contains no element of order d, then we certainly have

$$|\{x \in G \mid o(x) = d\}| = 0 \le \phi(d).$$

Suppose G *does* contain an element a of order d. Then $\langle a \rangle$ is cyclic of order d and every element $x \in \langle a \rangle$ satisfies $x^d = 1$. Thus $\langle a \rangle$ contains d solutions to the equation $x^d = 1$, and, thus, by the hypothesis, contains all the solutions in G of this equation. How many elements of order d are there in the cyclic group $\langle a \rangle$ of order d? These are exactly the generators of $\langle a \rangle$, and thus there are $\phi(d)$ of them. Thus in this case we have

$$|\{x \in G \mid o(x) = d\}| = |\{x \in \langle a \rangle \mid o(x) = d\}| = \phi(d);$$

and so in all cases we have

$$|\{x \in G \mid o(x) = d\}| \le \phi(d).$$

Now every one of the n elements of G has order d for some d dividing n, and thus

$$n = \sum_{d \mid n} |\{x \in G \mid o(x) = d\}| \le \sum_{d \mid n} \phi(d) = n.$$

We thus have equality throughout, and $|\{x \in G \mid o(x) = d\}| = \phi(d)$ for all $d \mid n$. In particular taking $d = n$, there exists $x \in G$ with $o(x) = n$. Then $|\langle x \rangle| = n$, and so $G = \langle x \rangle$ is cyclic. $\qquad\square$

If F is a field, then every non-zero element of F is invertible, and so $F^{\times}$, the group of units of F, is $(F - \{0\}, \times)$. For finite fields, this group is always cyclic! In fact, we prove that, even if F is infinite, every finite subgroup of $F^{\times}$ is cyclic.

Theorem 25.41. *Let $(F, +, \times)$ be a field, and let G be a finite subgroup of $F^{\times} = (F - \{0\}, \times)$. Then G is cyclic. In particular, if F is a finite field, then $F^{\times}$ is a cyclic group.*

Proof. The group G is finite, and the polynomial $x^k - 1$ has at most k roots in any field (Corollary 19.27). Hence, $x^k - 1$ has at most k roots in G which is a subset of $F^{\times}$. Thus G is a cyclic group by Lemma 25.40. $\qquad\square$

Corollary 25.42. *Let F be a finite field, and let $F \subseteq E$. Assume E is a field with $|E : F| < \infty$. Then E is a simple extension of F. In other words, $E = F[\alpha]$ for some $\alpha \in E$.*

Proof. If $|F| = q$ and $|E : F| = n$, then $|E| = q^n < \infty$. Hence $E^\times$ is cyclic. Assume that $E^\times = \langle \alpha \rangle$ for $\alpha \in E$. That means that every non-zero element of E is a power of α. Hence, in particular, $E = F[\alpha]$ is a simple extension of F. $\square$

Primitive Element Theorem. Having considered finite fields, we now turn to general fields.

Theorem 25.43. *Let $F \subseteq E$ be fields, and assume $|E : F| < \infty$. Assume that there exists only finitely many fields K with $F \subseteq K \subseteq E$. Then E is a simple extension over F.*

Remark 25.44. In fact, a stronger theorem than the one above is true. Let $F \subseteq E$ be an algebraic extension. Then E is a simple extension of F if and only if there are only finitely many intermediate fields K with $F \subseteq K \subseteq E$. (See Problem 25.5.12.)

Proof. In Corollary 25.42, we proved the case when $|F|$ is finite. So assume that $|F| = \infty$, and induct on $|E : F|$.

For the base case, it is clear that if $|E : F| = 1$, then $E = F$, and E is a simple extension of F.

Now assume $E > F$ and that the theorem has been proved for smaller indices. Choose $\alpha \in E - F$, then $F \subsetneq F[\alpha] \subseteq E$. We have that $|E : F[\alpha]| < |E : F| < \infty$ and that there are only finitely many intermediate fields between $F[\alpha]$ and E (since any of these would be also an intermediate field between F and E). Hence, by induction, there exists $\beta \in E$ with $E = F[\alpha][\beta] = F[\alpha, \beta]$.

For each $u \in F$, define $K_u = F[\alpha + u\beta]$. Each K_u is an intermediate field, and we can construct one for each $u \in F$. The field F is infinite, but we are assuming that there are only a finite number of intermediate fields between F and E. As a result, there exists at least two distinct elements u and v in F with $K_u = K_v$. Let $K = K_u = K_v$. We have $\alpha + u\beta$ and $\alpha + v\beta$ are in K. Hence, $(u - v)\beta \in K$. Now $u - v$ is a non-zero element of $F \subseteq K$, and hence $\beta \in K$. Since $\alpha + u\beta$ as well as u and β are elements of K, we also have that $\alpha \in K$. Thus

$$E = F[\alpha, \beta] \subseteq K \subseteq E.$$

Hence, we have equality throughout, and so $E = K = F[\alpha + u\beta]$ is a simple extension of F. $\square$

Theorem 25.45 (Primitive element). *Let $F \subseteq E$ be a separable field extension. Further assume $|E : F| < \infty$. Then E is a simple extension of F. In other words, $E = F[\alpha]$ for some $\alpha \in E$.*

Proof. To show that E is a simple extension of F, we need to show that there are only finitely many intermediate fields containing F and contained in E (Theorem 25.43). Since E is a separable extension of F, by Proposition 25.25 there exists $L \supseteq E$ such that L is Galois extension of F and $|L : F| < \infty$. We will show that there are finitely many intermediate fields K such that $F \subseteq K \subseteq L$. This will then force the number of intermediate fields between F and E to be finite as well. Define a map

$\Lambda \colon \{\text{fields } K \mid F \subseteq K \subseteq L\} \to \{\text{Subgroups of } \mathrm{Gal}(L/F)\}$

by $\Lambda(K) = \mathrm{Gal}(L/K)$.

Now $|L : F| < \infty$ implies—see Theorem 24.43—that $\mathrm{Gal}(L/F)$ is a finite group, and hence has a finite number of subgroups. We prove that Λ is 1-1, and hence there are a finite number of intermediate fields as well.

$$
\begin{array}{ccc}
L & & \{e\} \\
| & & | \\
K & \xrightarrow{\;\Lambda\;} & \mathrm{Gal}(L/K) \\
| & & | \\
F & & \mathrm{Gal}(L/F)
\end{array}
$$

If $F \subseteq K \subseteq L$, then L is Galois over K (Corollary 25.18) and $K = \mathrm{Fix}(\mathrm{Gal}(L/K))$. Hence, for two intermediate fields K_1 and K_2, if $\Lambda(K_1) = \Lambda(K_2)$, then $\mathrm{Gal}(L/K_1) = \mathrm{Gal}(L/K_2)$, and we have

$$K_1 = \mathrm{Fix}(\mathrm{Gal}(L/K_1)) = \mathrm{Fix}(\mathrm{Gal}(L/K_2)) = K_2.$$

Hence, Λ is 1-1, there are only a finite number of intermediate fields, and E—as well as L—is a simple extension of F. $\qquad\square$

Remark 25.46. In the proof of the primitive element theorem, Theorem 25.45, for a Galois extension $F \subseteq L$, we used the fact that the map $\mathrm{Gal}(L/\cdot) \colon \{\text{fields } K \mid F \subseteq K \subseteq L\} \to \{\text{Subgroups of } \mathrm{Gal}(L/F)\}$—that maps K to $\mathrm{Gal}(L/K)$—is 1-1. This is a part of the "Galois correspondence" that will be explored more fully in the fundamental theorem of Galois theory, Theorem 26.9. As you may expect, this map is also onto, and its inverse is the map $\mathrm{Fix}(\cdot)$, that sends a subgroup of the Galois group to its fixed field. We have chosen to prove this bit of the correspondence here, in order to have the very powerful primitive element theorem at our disposal. Later we will use this theorem to prove that, for Galois extensions, the degree of the extension is equal to the size of the Galois group. This, in turn, will facilitate proving the other parts of the fundamental theorem.

Corollary 25.47. *Let $F \subseteq E$ be a field extension. Then E is a simple extension of F if any of the following conditions hold:*

(a) *$F \subseteq E$ is a Galois extension.*

(b) *$|E : F| < \infty$ and $\mathrm{char}(F) = 0$.*

(c) *$|E : F| < \infty$ and F is a finite field.*

Proof. In all three cases, $|E : F| < \infty$ and E is a separable extension of F (Corollaries 25.15 and 25.35). Hence the primitive element theorem, Theorem 25.45, applies. $\qquad\square$

Corollary 25.47 is strong and unexpected. It tells us that simple extensions abound, and it shows the power of the concept of a Galois extension. We are now ready to answer one of the questions posed in the Preface.

Question 25.48. Does there exist a real number α such that rational linear combinations of 1, α, α^2, ... include every one of $\sqrt[3]{47}$, $\sqrt[5]{17}$, and $18 - 2\sqrt[7]{19}$?

Answer. While it will take some work to actually find such an α, we can now easily prove that such an α exists. Consider the field $E = \mathbb{Q}[\sqrt[3]{47}, \sqrt[5]{17}, \sqrt[7]{19}]$. This is a finite degree separable extension of $\mathbb{Q}$ (it is separable since $\mathbb{Q}$ is field of characteristic 0), and so it is a simple extension by Theorem 25.45. We conclude that $E = \mathbb{Q}[\alpha]$ for some $\alpha \in E \subseteq \mathbb{R}$. This means that each of $\sqrt[3]{47}$, $\sqrt[5]{17}$, and $18 - 2\sqrt[7]{19}$ are rational linear combinations of $1, \alpha, \alpha^2, \ldots$. $\qquad\square$

Remark 25.49. Let $F \subseteq E$ be fields. This extension is very well behaved if $|E : F| < \infty$ and E is a separable extension of F. Two facts stand out: By Proposition 25.25, we can embed this extension in a Galois extension. In other words, we can find $L \supseteq E$ such that L is a Galois extension of F. Theorem 25.45 proved that E is a simple extension of F. In other words, $E = F[\alpha]$ for some $\alpha \in E$. We can bring these two facts together, by letting L be the splitting field of the separable polynomial $\min_F(\alpha)$ over E. Then L will automatically be the splitting field of $\min_F(\alpha)$ over F, and a Galois extension of F. (See Remark 25.24.)

Size of $\mathrm{Gal}(E/F)$. Using the primitive element theorem, Theorem 25.45, we next consider $|\mathrm{Gal}(E/F)|$. In Theorem 24.43, we had already shown that for a finite degree extension, the size of the Galois group is finite. Here, using the primitive element theorem, Theorem 25.45, we strengthen that result and prove that, for a finite degree extension, the size of the Galois group is bounded by the degree of the extension. (In fact, the size of the Galois group *divides* the degree of the extension; see Problem 25.5.9.) Following this theorem, we will focus on Galois extensions and prove an important result. We show that, for a *Galois* extension, the size of the Galois group is actually *equal* to the degree of the extension. Knowing the size of the Galois group will indeed be very useful.

Theorem 25.50. *Let $F \subseteq E$ be fields, and assume that $|E : F| < \infty$. Then $|\mathrm{Gal}(E/F)| \leq |E : F|$.*

Proof. The strategy is to first prove the result for simple extensions and then to reduce the general case to that of simple extensions.

SPECIAL CASE. Assume that $E = F[\alpha]$ is a simple finite degree extension of F.

PROOF OF THEOREM FOR THE SPECIAL CASE. You were asked to prove this in Problem *24.3.5*. The basic argument is straightforward. If $E = F[\alpha]$, then every F-automorphism of E is determined by its effect on α. Moreover, an F-automorphism can send α only to one of the other roots of $\min_F(\alpha)$. The result follows since $\min_F(\alpha)$ has at most $|E : F|$ roots.

For the general case, let $K = \mathrm{Fix}(\mathrm{Gal}(E/F))$. Then by Lemma 25.8, $\mathrm{Gal}(E/F) = \mathrm{Gal}(E/K)$ and E is Galois over K. By Corollary 25.47 E is a simple extension of K, and we have $E = K[\alpha]$ for some $\alpha \in E$. Now, by the special case proved above, $|\mathrm{Gal}(E/K)| \leq |E : K|$. We conclude

$$|\mathrm{Gal}(E/F)| = |\mathrm{Gal}(E/K)| \leq |E : K| \leq |E : F|.$$

$$
\begin{array}{ccc}
E & & \{e\} \\
| & & | \\
K & \longleftarrow & \mathrm{Gal}(E/F) \\
| & & \\
F & &
\end{array}
$$

$\qquad\square$

We now turn to the case of Galois extensions and prove that, for such extensions (and only for such extensions) the size of the Galois group is the same as the degree of the extension.

Theorem 25.51. *Let $F \subseteq E$ be fields. Then E is Galois over F if and only if $|\mathrm{Gal}(E/F)| = |E : F| < \infty$.*

Proof. ($\Rightarrow$) Assume E is Galois over F. Then, by Corollary 25.47, $E = F[\alpha]$ for some $\alpha \in E$, and let $f = \min_F(\alpha)$. Let

$$\Omega = \{\beta \in E \mid f(\beta) = 0\}.$$

We have $\deg(f) = |E : F|$, and since E is both normal and separable, f has $\deg(f)$ roots in E. Hence $|\Omega| = |E : F|$.

The polynomial f is irreducible and E is a splitting field (since it is normal) and hence $G = \mathrm{Gal}(E/F)$ acts transitively on Ω (see Theorem 24.37). If any element of G fixes α, then this element fixes every element of $E = F[\alpha]$, and so it is the identity element of G. Hence, by the fundamental counting principle, Theorem 6.1,

$$|G| = \frac{|G|}{|\mathrm{Stab}_G(\alpha)|} = |\mathcal{O}_\Omega(\alpha)| = |\Omega| = \deg(f) = |E : F|.$$

($\Leftarrow$) Assume that $|\mathrm{Gal}(E/F)| = n = |E : F|$.
Let $K = \mathrm{Fix}(\mathrm{Gal}(E/F))$. By Lemma 25.8, $\mathrm{Gal}(E/F) = \mathrm{Gal}(E/K)$ and E is Galois over K. For a Galois extension, we just proved that the size of the Galois group is equal to the degree of the extension, and so $|\mathrm{Gal}(E/K)| = |E : K|$. Hence,

$$n = |\mathrm{Gal}(E/F)| = |\mathrm{Gal}(E/K)| = |E : K|.$$

Now $F \subseteq K \subseteq E$ and $|E : F| = n = |E : K|$. This implies that $|K : F| = 1$, and so $K = F$. Hence $F = \mathrm{Fix}(\mathrm{Gal}(E/F))$, proving that E is a Galois extension of F.

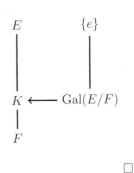

$\square$

Remark 25.52. If $F \subseteq E$ are fields with $|E : F| < \infty$, there are two tricks/techniques that could be helpful in taking advantage of what we know of Galois extensions. If we know that E is a separable extension of F—for example, if $\mathrm{char}(F) = 0$—then we consider a field L with $F \subseteq E \subseteq L$ and with L a Galois extension of F (Proposition 25.25 and Remark 25.49). This technique was used, for example, in the proof of the Primitive Element Theorem 25.45. The second trick is to consider $K = \mathrm{Fix}(\mathrm{Gal}(E/F))$. We know (Lemma 25.8) that $F \subseteq K \subseteq E$, E is a Galois extension of F and $\mathrm{Gal}(E/K) = \mathrm{Gal}(E/F)$. This approach could be utterly futile—for example, in the case when K is all of E—but the combination of Theorems 25.50 and 25.51 say that E is a Galois extension of F when $\mathrm{Gal}(E/F)$ is as large as it could be. There are contexts when smaller Galois groups are actually easier to deal with, and this trick can allow us to reduce the general case of a result to the case when the extension is a Galois extension. The proofs of Theorems 25.50 and 25.51 provide examples of such a situation.

Problems

25.5.1. True or False:
   (a) If E is the splitting field of a polynomial over $\mathbb{Q}$, then E is Galois over $\mathbb{Q}$.
   (b) Every finite degree extension of $\mathbb{Q}$ is a simple extension.
   For the remaining statements, assume $F \subseteq L \subseteq E$, and E is a Galois extension of F.
   (c) E has to be a Galois extension of L.
   (d) L has to be a Galois extension of F.
   (e) E has to be a separable extension of L.
   (f) L has to be a separable extension of F.
   (g) E has to be a simple extension of F.

25.5.2. In the proof of Theorem 25.41, we invoked Corollary 19.27 to get that the polynomial $x^k - 1$ has at most k roots in a field. But Corollary 19.27 applies to *any* integral domain (and not just fields). Is it possible that one could strengthen Theorem 25.41? If the answer is yes, show how, and if the answer is no, explain why.

25.5.3. Let $f = (x^3 - 2)(x^2 + 1) \in \mathbb{Q}[x]$. Let E be the splitting field of f over $\mathbb{Q}$, let $\alpha = 5\sqrt[3]{2} + 7i\sqrt{3}$, and let $g = \min_{\mathbb{Q}}(\alpha)$.
   (a) Is $\alpha \in E$? Why?
   (b) What is $|\mathrm{Gal}(E/\mathbb{Q})|$? Prove your assertion.
   (c) Does g split in E? Why?
   (d) What is the orbit of α under the action of $\mathrm{Gal}(E/\mathbb{Q})$?
   (e) What is the degree of g? Give a factorization of g in E.
   (f) Use a symbolic algebra software (e.g., Maple, Mathematica, or Sage) to multiply out your factorization of the previous part to find g (and its coefficients) explicitly.
   (g) Explicitly write down an element $\beta \in E$ such that the orbit of β under the action of $\mathrm{Gal}(E/\mathbb{Q})$ has $|E : \mathbb{Q}|$ elements?
   (h) Is $E = \mathbb{Q}[\alpha]$? Is $E = \mathbb{Q}[\beta]$? Prove your assertions.

25.5.4. Let $\mathbb{F}_{13} = (\mathbb{Z}/13\mathbb{Z}, +, \cdot)$ be the field with 13 elements. Find all the generators of $\mathbb{F}_{13}^{\times}$, the group of units of $\mathbb{F}_{13}$.

25.5.5. Find all the generators of $\mathbb{F}_{17}^{\times}$, the group of units of $\mathbb{F}_{17}$.

25.5.6. Let $E = \mathbb{F}_7[i] = \{a + bi \mid a, b \in \mathbb{F}_7, \ i^2 = -1\}$. What is $|E|$? Is E a field? If so, find a generator for $E^{\times}$, the group of units of E.

25.5.7. Let $f(x) = x^4 - 4x^2 - 1 \in \mathbb{Q}[x]$. Find the Galois group of $f(x)$.

25.5.8. Let E be a field of order p^n, and let $\alpha \in E$ with $E^{\times} = \langle \alpha \rangle$. Prove that the minimal polynomial of α over $\mathbb{F}_p$ has degree n.

25.5.9. **Strengthening Theorem 25.50.** Let $F \subseteq E$ be fields with $|E : F| < \infty$. Prove that $|\mathrm{Gal}(E/F)|$ divides $|E : F|$.

25.5.10. Let $F \subseteq K \subseteq E$ be fields, and assume that E is a Galois extension of F. Recall that the fields $\sigma(K)$ for $\sigma \in \mathrm{Gal}(E/F)$ are called the conjugates of

K (Problem 24.3.13). Let L be the compositum of the conjugates of K. Show that L is a normal closure for K over F.

25.5.11. Let $K \subseteq E$ be infinite fields. Assume that u and v are elements of E that are algebraic and separable over K. Must there exist $a \in K$ with $K(u, v) = K(u + av)$? Either prove that the answer is always yes, or provide a counterexample.

25.5.12. **Converse of Theorem 25.43.** Let $F \subseteq E$ be fields with $|E : F| < \infty$. Assume that $E = F(\alpha)$ for some $\alpha \in E$. Show that there are only a finite number of intermediate fields containing F that are contained in E. You may find the following steps helpful.
STEP 1: Let $f(x) = \min_F(\alpha)$. Let $\mathcal{A} = \{K \text{ a field } \mid F \subseteq K \subseteq E\}$ be the set of intermediate fields between F and E. Let $\mathcal{B} = \{g(x) \in E[x] \mid g \text{ divides } f \text{ in } E[x]\}$ be the set of factors of f in $E[x]$.
STEP 2: If $K \in \mathcal{A}$, define $\theta(K) = \min_K(\alpha)$. Show that θ defines a map from $\mathcal{A}$ to $\mathcal{B}$.
STEP 3: If $K \in \mathcal{A}$, define L to be the subfield of K containing F and the coefficients of $\theta(K)$. Show that $L = K$.
STEP 4: Show the map θ is 1-1, and complete the proof.

25.5.13. Let $F \subseteq K \subseteq E$ be fields. Assume that $E = F(\alpha)$ where $\alpha \in E$ is algebraic over F. Prove that $K = F(\beta)$ for some $\beta \in K$.
REMARK: The condition that α be algebraic over F is actually not necessary, but the stronger result is (much) harder to prove. The case when α is not algebraic is known as Lüroth's theorem.

25.5.14. Let K be an infinite field of characteristic $p > 0$, and let $L = K[u, v]$ where $u^p, v^p \in K$ and $|L : K| = p^2$. Show that there are an infinite number of intermediate fields contained in L and containing K. Conclude that L is not a simple extension of K.

25.6. More Problems and Projects

Polynomials with a Specified Root. Let $F \subseteq E$ be a field extension, let $\alpha \in E$, and assume that α is algebraic over F. How do we find a non-zero polynomial in $F[x]$ that has α as its root? Since α is algebraic over F, by definition, such a polynomial exists. In fact, the minimal polynomial of α over F is such a polynomial. But how do we actually find $\min_F(\alpha)$? We usually start with the definition of α and somewhat haphazardly find a polynomial that has α as a root. As an example, let $\mathbb{Q} \subseteq \mathbb{C}$ and $\alpha = \sqrt{2} + i\sqrt{3}$. We find $\alpha^2 = -1 + 2i\sqrt{6}$, and so $\alpha^2 + 1 = 2i\sqrt{6}$. Squaring both sides gives $\alpha^4 + 2\alpha^2 + 1 = -24$. We conclude that α is a root of $x^4 + 2x^2 + 25$. At this point, if we want the *minimal* polynomial of α, we either have to prove that the polynomial that we found is irreducible or factor the polynomial into irreducible factors. One of these irreducible factors will be the minimal polynomial. An alternate—more elegant—way is to use Theorem 25.9. We first make sure that $\alpha \in E$ where E is a Galois extension of F—in our case, we could take $E = \mathbb{Q}(\sqrt{2}, \sqrt{3}, i)$—and then use $\mathrm{Gal}(E/F)$ to find the other roots of the minimal polynomial. This latter method is pleasing—and unlike our haphazard method, it does not require us to factor the resulting polynomial—but it does require finding the exact Galois group of an extension that contains α. Here, in the Problems, we present a method that does not rely on finding the Galois group and yet systematically finds a polynomial— not necessarily irreducible—that has α as a root. Just like our original haphazard method, factoring the resulting polynomial into irreducible factors will give the minimal polynomial of α. This method requires a good bit of calculation but is well suited for machines. The main idea is to use the action of the full symmetric group S_n instead of the Galois group, since, after all, the Galois group is a *subgroup* of an appropriate symmetric group.

Problems

25.6.1. Let $f \in \mathbb{Q}[x]$ and $E = \mathbb{Q}[r_1, \ldots, r_n]$, where $r_1, \ldots, r_n$ are all the roots of f. Let $G = \mathrm{Gal}(E/\mathbb{Q})$.
  (a) The group G acts on $\{r_1, \ldots, r_n\}$. Show that, based on this action, for every $\tau \in G$, there is a unique element $\phi(\tau) \in S_n$ such that $\{\phi(\tau) \mid \tau \in G\}$ is a subgroup of S_n isomorphic to G.
  (b) Identify every $\sigma \in G$ with $\phi(\sigma) \in S_n$ from the previous part. In other words, every $\sigma \in G$ is, simultaneously, a $\mathbb{Q}$-automorphism of E and a permutation of $\{1, \ldots, n\}$. Let $p \in \mathbb{Q}[x_1, \ldots, x_n]$ be a polynomial in n variables. Prove that, for every $\sigma \in G$,

$$\sigma(p(r_1, \ldots, r_n)) = p(r_{\sigma(1)}, r_{\sigma(2)}, \ldots, r_{\sigma(n)}).$$

25.6.2. Let $a = \sqrt[3]{2}$, $\omega = e^{2\pi i/3}$, and $\alpha = a + \omega$. Our goal is to find a polynomial in $\mathbb{Q}[x]$ that has α as a root. For now, answer the following questions:
  (a) Let E be the splitting field of $x^3 - 2$ over $\mathbb{Q}$. Find the roots of $x^3 - 2$ in E and call them r_1, r_2, and r_3.
  (b) Is $E = \mathbb{Q}[r_1, r_2, r_3]$? Is $\alpha \in E$?
  (c) Find a polynomial in three variables $p(x_1, x_2, x_3) \in \mathbb{Q}[x_1, x_2, x_3]$ such that $p(r_1, r_2, r_3) = \alpha$.
  (d) In p, if you plug in r_1 for x_1, r_2 for x_2, and r_3 for x_3, you get α. Now permute these substitutions and get other elements of E. For example, plug in r_2 for x_1, r_1 for x_2 and r_3 for x_3. This permutation corresponds to the element $(1\ 2) \in S_3$, switching the places of 1 and 2. Do this for every element of S_3 and get six elements of E:

$$\alpha = \alpha_1, \alpha_2, \ldots, \alpha_6.$$

  (e) By multiplying out (possibly using a symbolic algebra software), explicitly find the polynomial

$$h(x) = (x - \alpha_1)(x - \alpha_2) \cdots (x - \alpha_6).$$

  (f) Is $h(x) \in \mathbb{Q}[x]$? Is $h(\alpha) = 0$?
  (g) Is $\mathrm{Gal}(E/\mathbb{Q})$ a subgroup of S_3?
  (h) Make a guess about the relationship between the polynomial $h(x)$ found here and the polynomial produced by Theorem 25.9.

25.6.3. Based on Problem 25.6.2, we have the following conjecture:
CONJECTURE: Let $f \in \mathbb{Q}[x]$, and let E be the splitting field of f over $\mathbb{Q}$. Let r_1, r_2, ..., r_n be the roots of f in E. It follows that $E = \mathbb{Q}[r_1, r_2, \ldots, r_n]$. Let $\alpha \in E$, and let g be the minimal polynomial of α over $\mathbb{Q}$. There exists $p(x_1, x_2, \ldots, x_n) \in \mathbb{Q}[x_1, \ldots, x_n]$ such that $\alpha = p(r_1, r_2, \ldots, r_n)$. Let $\sigma \in S_n$ be the symmetric group of degree n. Define $\sigma(\alpha) = p(r_{\sigma(1)}, r_{\sigma(2)}, \ldots, r_{\sigma(n)})$. Let $T = \{\sigma(\alpha) \mid \sigma \in S_n\}$. Then *all* the roots of g are among the elements of T, and furthermore $h(x) = \prod_{\sigma \in S_n}(x - \sigma(\alpha))$ has α as a root and has coefficients in $\mathbb{Q}$.
Starting with the setup and the conclusion of Problem 25.6.1, prove the conjecture.

25.6.4. Using the scheme of Problem 25.6.3, find a polynomial in $\mathbb{Q}[x]$ that has $\sqrt[3]{2} - 2e^{2\pi i/3}$ as a root.

Roots in $\mathbb{F}_p$ but not in $\mathbb{Z}$. If a polynomial $f \in \mathbb{Z}[x]$ has a root $\alpha \in \mathbb{Z}$, then $f(\alpha) = 0$. This remains true if we consider both the polynomial and α mod p. Hence if a polynomial has a root in $\mathbb{Z}$, it also has a root in $\mathbb{F}_p$ for every prime p. The converse is not true, as Problem 25.6.6 shows.

25.6.5. Let $p \in \mathbb{Z}^{>0}$ be a prime. If $f \in \mathbb{Z}[x]$ has a root $\alpha \in \mathbb{Q}$, then does f necessarily have a root in $\mathbb{F}_p$?

25.6.6. Let $f = (x^2 - 2)(x^2 - 3)(x^2 - 6)$. Show that f has no roots in $\mathbb{Z}$ but, for every positive prime p, f has a root in $\mathbb{F}_p$. Can you replace 2, 3, and 6 with another set of integers? You may find the following steps helpful.
STEP 1: Show that f does have a root in $\mathbb{F}_2$, $\mathbb{F}_3$, and $\mathbb{F}_5$, so assume p is a prime greater than 5. (Step 1 is not really necessary.)
STEP 2: Define $\phi \colon \mathbb{F}_p^\times \to \mathbb{F}_p^\times$ by $\phi(a) = a^2$. Is ϕ a group homomorphism?
STEP 3: Assume that for a particular p, at least one of 2 or 3 or 6 is in the image of ϕ. Show that for such a p, the polynomial f has a root in $\mathbb{F}_p$.
STEP 4: What is the $|\ker(\phi)|$? Draw a homomorphism diagram, and determine $|\mathbb{F}_p^\times : \mathrm{Im}(\phi)|$.
STEP 5: Let G be the group $\mathbb{F}_p^\times$, and let $H = \mathrm{Im}(\phi)$ a subgroup of G. Show that if 2 and 3 are not in the image of ϕ, then they belong to the same coset of H in G, and as a result 2×3 belongs to H.
STEP 6: Write a complete and coherent proof for the claim, and answer the accompanying question.

Trace and Norm. Let $F \subseteq E$ be a Galois extension, and let $\alpha \in E$. Define $T_{E/F}(\alpha)$, the *trace* of α in E over F, and define $N_{E/F}(\alpha)$, the *norm* of α in E over

F by

$$T_{E/F}(\alpha) = \sum_{\sigma \in \mathrm{Gal}(E/F)} \sigma(\alpha), \quad N_{E/F}(\alpha) = \prod_{\sigma \in \mathrm{Gal}(E/F)} \sigma(\alpha).$$

If $\sigma \in \mathrm{Gal}(E/F)$ and $\alpha \in E$, then the element $\sigma(\alpha)$ is called a *Galois conjugate* of α. The norm as defined here is a generalization of the norm for the quadratic integer rings that was so heavily used in the ring theory sections.

25.6.7. Let $F \subseteq E$ be a Galois extension.
    (a) Show that $T_{E/F}(\alpha)$ and $N_{E/F}(\alpha)$ are elements of F.
    (b) Show $T_{E/F}\colon E \to E$ is a linear transformation of the vector space E over F.
    (c) Show $N_{E/F}\colon E^{\times} \to F^{\times}$ is a group homomorphism.

25.6.8. Let d be a square-free integer, and consider the field extension $\mathbb{Q} \subseteq \mathbb{Q}(\sqrt{d})$. Let $\alpha = a + b\sqrt{d} \in \mathbb{Q}(\sqrt{d})$.
    (a) Find $N_{\mathbb{Q}(\sqrt{d})/\mathbb{Q}}(\alpha)$ in terms of a, b, and d. Have you seen this function before?
    (b) Find $T_{\mathbb{Q}(\sqrt{d})/\mathbb{Q}}(\alpha)$ in terms of a, b, and d.

25.6.9. Let $F \subseteq E$ be a field extension. Fix $\alpha \in E$, and define $f_{\alpha}\colon E \to E$ by $f_{\alpha}(\beta) = \alpha\beta$.
    (a) The field E is a vector space over F. Show that f_{α} is a linear transformation of vector spaces.
    (b) Is f_{α} 1-1?
    (c) If $|E : F| < \infty$, is f_{α} onto?
    (d) Let $\mathbb{Q} \subseteq \mathbb{Q}(i\sqrt{3})$ be a field extension, and let $B = \{1, i\sqrt{3}\}$ be a basis for the vector space $\mathbb{Q}(i\sqrt{3})$ over $\mathbb{Q}$. Let $\alpha = a + bi\sqrt{3} \in \mathbb{Q}(i\sqrt{3})$, and find the matrix of f_{α} with respect to B. Call this matrix M.
    (e) What is the determinant of M? Had you seen this determinant before?
    (f) What is the trace of M? Had you seen this trace before?

25.6.10. Let $E = \mathbb{Q}[\sqrt{2}, \sqrt{3}]$ and $\sigma \in \mathrm{Gal}(E/\mathbb{Q})$. Then E is a vector space over $\mathbb{Q}$, $\mathcal{B} = \{1, \sqrt{2}, \sqrt{3}, \sqrt{6}\}$ is a basis for E over $\mathbb{Q}$, and $\sigma\colon E \to E$.
    (a) Let $\alpha = 4 - 7\sqrt{2} + 3\sqrt{3} - 5\sqrt{6} \in E$. What is $[\alpha]_{\mathcal{B}}$, the coordinate vector of α with respect to the basis $\mathcal{B}$?
    (b) Is σ a linear transformation of vector spaces?
    (c) For each $\sigma \in \mathrm{Gal}(E/\mathbb{Q})$, find the matrix of σ with respect to the basis $\mathcal{B}$. In other words, find a 4×4 matrix M_{σ} such that, for all $\beta \in E$, $M_{\sigma}[\beta]_{\mathcal{B}}$ is equal to $[\sigma(\beta)]_{\mathcal{B}}$, the coordinate vector of $\sigma(\beta)$ with respect to $\mathcal{B}$.
    (d) Is $\{M_{\sigma} \mid \sigma \in \mathrm{Gal}(E/\mathbb{Q})\}$ a group? If so, how is it related to $\mathrm{Gal}(E/\mathbb{Q})$?

(e) Find

$$\sum_{\sigma \in \mathrm{Gal}(E/\mathbb{Q})} M_\sigma \begin{bmatrix} 4 \\ -7 \\ 3 \\ -5 \end{bmatrix}.$$

Why does the answer have so many zero entries? Base your answer on Galois theory.

(f) Let $\alpha = 4 - 7\sqrt{2} + 3\sqrt{3} - 5\sqrt{6} \in E$. Find the trace $T_{E/\mathbb{Q}}(\alpha)$ of α in E over $\mathbb{Q}$.

(g) How much of your answers depend on this particular field? Can you generalize?

25.6.11. Let $E = \mathbb{Q}[\sqrt{2}, \sqrt{3}]$, and let $\alpha = 4 - 7\sqrt{2} + 3\sqrt{3} - 5\sqrt{6} \in E$.

(a) Find $N_{E/\mathbb{Q}}(\sqrt{3})$, the norm of $\sqrt{3}$ in E over $\mathbb{Q}$.

(b) Find $N_{E/\mathbb{Q}}(\alpha)$.

Cubic Equations. The solution to cubic equations over $\mathbb{Q}$ has been known since the sixteenth century when it was solved in the Italian peninsula. In the problems here, you learn how to solve a cubic equation as well as how to find its Galois group.

25.6.12. The reduction.

(a) Consider the cubic equation $y^3 + 6y^2 + 7y - 11 = 0$. Change variables by letting $x = y - 2$. What is the resulting new cubic equation?

(b) Starting with $y^3 + ay^2 + by + c = 0$, find α such the substitution $x = y - \alpha$ will result in an equation of the form $x^3 + px + q = 0$. Conclude that to solve a general cubic equation over $\mathbb{Q}$, we need only to solve equations of the form $x^3 + px + q = 0$.

25.6.13. An example. Consider the cubic equation $x^3 + 3x + 6 = 0$.

(a) Let $f = x^3 + 3x + 6 \in \mathbb{Z}[x]$. Is f irreducible over $\mathbb{Q}$? What is the derivative of f? Is the function f always increasing? How many real zeros does f have?

(b) Change the variable by letting $x = u - v$, expand, and get an equation in two variables. Find a linear polynomial $p(u, v)$ such that the new equation can be written in the form

$$(u^3 - v^3) + 6 + (u - v)p(u, v) = 0.$$

(c) To find some solution to our cubic, we try to find solutions for u and v in the system of equations

$$\begin{cases} u^3 - v^3 + 6 = 0, \\ p(u, v) = 0. \end{cases}$$

In the linear equation $p(u, v) = 0$, solve for v in terms of u, substitute the result for v in $u^3 - v^3 + 6 = 0$, and by clearing denominators get a sixth degree equation for u.

(d) Substitute $z = u^3$ in the sixth degree equation of the previous part and get a quadratic equation for z.

(e) Solve the quadratic, and choose one of the two solutions for z. (It turns out the other will end up giving the same final result.) Call it a.

(f) Solve the equation $u^3 - a = 0$, and find three solutions for u. Among the three solutions, choose your favorite one and call it α. Write the other ones in terms of α and appropriate roots of unity.

(g) Go back and, for each solution for u, find the corresponding v.

(h) What are the three roots of the original equation? Did you find all the roots?

25.6.14. The solution. Let p and q be arbitrary complex numbers, and consider the cubic equation $x^3 + px + q = 0$. Mimic your solution to Problem 25.6.13 and find the solutions to this cubic. In the parlance of Problem 25.6.13, your solutions will be in terms of one cube root of a and appropriate roots of unity.

25.6.15. The reducible cubics. Let $\mathbb{Q} \subseteq F \subseteq \mathbb{C}$ be fields, and let $f = x^3 + px + q \in F[x]$. Let $E \subseteq \mathbb{C}$ be the splitting field for f over F, and let $G = \text{Gal}(E/F)$. The polynomial f has three not-necessarily-distinct roots in $\mathbb{C}$.

(a) Show that either none, one, or three of the roots of f are in F.

(b) If all three of the roots of f are in F, then what is E? What is G?

(c) If only one of the roots of f is in F, then what is $|E : F|$? What is G? Is $E = F(\alpha)$ where α is one of the roots of f?

(d) If none of the roots of f are in F, then argue that f is irreducible over F.

(e) If f is irreducible over F, then argue that f cannot have repeated roots. In this case, what are the possibilities for $|E : F|$ and $\text{Gal}(E/F)$?

25.6.16. The discriminant. Let $\mathbb{Q} \subseteq F \subseteq \mathbb{C}$ be fields, and let $f = x^3 + px + q \in F[x]$. Let $E \subseteq \mathbb{C}$ be the splitting field for f over F, and let $G = \text{Gal}(E/F)$. Let α_1, α_2, and α_3 be the roots of f in E. Define

$$\delta(f) = (\alpha_2 - \alpha_1)(\alpha_3 - \alpha_1)(\alpha_3 - \alpha_2) = \prod_{1 \leq i < j \leq n} (\alpha_j - \alpha_i),$$

$$\Delta(f) = \delta(f)^2 = \prod_{1 \leq i < j \leq n} (\alpha_j - \alpha_i)^2.$$

The quantity $\Delta(f)$ is called the *discriminant* of f in E.

(a) Let $\sigma \in G$. Prove that $\sigma(\delta(f)) = \pm\delta(f)$.

(b) Prove that $\Delta(f) \in F$.

(c) If you and I had found the same roots for f but in a different order, then could your and my values for $\delta(f)$ be different? What about our values for $\Delta(f)$?

(d) Show that $\Delta(f) \neq 0$ if and only if f has three distinct roots.

25.6.17. The Galois group. Let $\mathbb{Q} \subseteq F \subseteq \mathbb{C}$ be fields, and let $f = x^3 + px + q \in F[x]$. Let $E \subseteq \mathbb{C}$ be the splitting field for f over F, and let $G = \text{Gal}(E/F)$.

Assume that f is irreducible in $F[x]$, and let α_1, α_2, and α_3 be the three distinct roots of f in E.

(a) Show that G is isomorphic to a subgroup of S_3 and that 3 divides $|G|$.

(b) What are the subgroups of S_3 whose order is divisible by 3?

(c) Show that $\delta(f) \notin F$ if and only if G has an element of order 2.

(d) Assume $\delta(f) \notin F$. Show that $G \cong S_3$, $|E:F| = 6$ and $F \subseteq F(\delta(f)) \subseteq E$.

(e) Assume $\delta(f) \in F$. Show that $G \cong A_3 \cong \mathbb{Z}/3\mathbb{Z}$, $|E:F| = 3$, $E = F(\alpha_1)$, and all the roots of f are real.

25.6.18. Let $\mathbb{Q} \subseteq F \subseteq \mathbb{C}$ be fields, and let $f = x^3 + px + q \in F[x]$. Let $E \subseteq \mathbb{C}$ be the splitting field for f over F, and let α_1, α_2, and α_3 be the three roots of f in E.

(a) By writing $f = (x - \alpha_1)(x - \alpha_2)(x - \alpha_3) \in E[x]$, show that

$$\alpha_1 + \alpha_2 + \alpha_3 = 0,$$

$$\alpha_1 \alpha_2 + \alpha_1 \alpha_3 + \alpha_2 \alpha_3 = p,$$

$$\alpha_1 \alpha_2 \alpha_3 = -q.$$

(b) Show that $\alpha_1^2 + \alpha_2^2 + \alpha_3^2 = -2p$.

(c) Find $\alpha_1^3 + \alpha_2^3 + \alpha_3^3$ in terms of p and q.

(d) Start with $0 = (\alpha_1 + \alpha_2 + \alpha_3)^4$, expand the right hand side, and use the identities already found to give an identity for $\alpha_1^4 + \alpha_2^4 + \alpha_3^4$ in terms of p and q.

25.6.19. **Formula for the discriminant.** Let $\mathbb{Q} \subseteq F \subseteq \mathbb{C}$ be fields, and let $f = x^3 + px + q \in F[x]$. Let $E \subseteq \mathbb{C}$ be the splitting field for f over F, and let α_1, α_2, and α_3 be the three roots of f in E. Furthermore, let $\delta(f)$ and $\Delta(f)$, the discriminant of f in E, be defined as in Problem 25.6.16.

(a) The matrix

$$V = \begin{bmatrix} 1 & \alpha_1 & \alpha_1^2 \\ 1 & \alpha_2 & \alpha_2^2 \\ 1 & \alpha_3 & \alpha_3^2 \end{bmatrix}$$

is called a *Vandermonde matrix*. Show that $\delta(f) = \det(V)$.

(b) Show that

$$\Delta(f) = \det(V^t V) = \det \begin{bmatrix} 3 & 0 & \sum \alpha_i^2 \\ 0 & \sum \alpha_i^2 & \sum \alpha_i^3 \\ \sum \alpha_i^2 & \sum \alpha_i^3 & \sum \alpha_i^4 \end{bmatrix}.$$

(c) By evaluating the determinant in the previous part and by using the results in Problem 25.6.18, show that

$$\Delta(f) = -4p^3 - 27q^2.$$

25.6.20. Determine the Galois group over $\mathbb{Q}$ of $f = x^3 - 3x + 1$ and of $g = x^3 + 3x + 1$ with very little calculation and by relying on the results of the previous problems. Does either polynomial have three real roots?

25.6.21. **Omar Khayyam's solution to the cubic.** The mathematician, philosopher, and poet Omar Khayyam (1048–1131) used conic sections to solve cubic equations. Using our modern symbolic notation and the parlance of analytic geometry, consider the cubic equation $x^3 + 2x - 7 = 0$. Draw a circle of radius $7/4$ with center at $(7/4, 0)$, and the parabola $y = x^2/\sqrt{2}$. Other than the origin, the two curves have one other point of intersection. Show that the x-coordinate of this point is a solution to the cubic. (See Figure 25.6.) Generalize to any cubic of the form $x^3 + cx = d$ where c and d are positive real numbers.

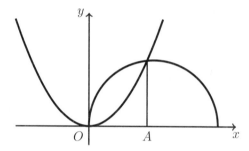

Figure 25.6. Given a circle of radius $7/4$ with center at $(7/4, 0)$ and the parabola $y = x^2/\sqrt{2}$, the length of OA is a solution to $x^3 + 2x - 7 = 0$.

Chapter 26

Fundamental Theorem of Galois Theory

...where we prove, for Galois extensions, the fundamental correspondence between subgroups of the Galois group and the intermediate fields in the field extension, and we then use the correspondence to prove the fundamental theorem of algebra and to study examples of Galois groups.

26.1. Galois Groups and Fixed Fields

Let $F \subseteq E$ be a finite degree field extension. Recall that such an extension is a Galois extension if $\mathrm{Fix}(\mathrm{Gal}(E/F)) = F$ (Definition 25.5). If E is a Galois extension of F, we have shown, among other things, that $E = F[\alpha]$ for some $\alpha \in E$ (Corollary 25.47) and that $|\mathrm{Gal}(E/F)| = |E:F|$ (Theorem 25.51). As a step in the proof of the Galois correspondence, in this section we show that the map $\mathrm{Fix}(\cdot)$ that sends subgroups of $\mathrm{Gal}(E/F)$ to subfields of E that contain F has a left inverse—namely the map $\mathrm{Gal}(E/\cdot)$—and is therefore 1-1. This will be true as long as $|E:F| < \infty$, and even if E is *not* a Galois extension of F. Our starting point will be Lemma 25.8. In that lemma, we started with a finite degree field extension $F \subseteq E$, and considered $K = \mathrm{Fix}(\mathrm{Gal}(E/F))$. We showed that $\mathrm{Gal}(E/F) = \mathrm{Gal}(E/K)$, and while E may not be a Galois extension of F, it is a Galois extension of K. It follows that $|\mathrm{Gal}(E/F)| = |\mathrm{Gal}(E/K)| = |E:K|$. Here, we want to prove a similar result but replace $\mathrm{Gal}(E/F)$ with an arbitrary subgroup of $\mathrm{Aut}(E)$. We can easily prove the following:

Lemma 26.1. *Let E be a field, and let H be a subgroup of $\mathrm{Aut}(E)$. Let $K = \mathrm{Fix}(H) \subseteq E$. Assume $|E:K| < \infty$. Then E is Galois over K.*

In particular, $|H|$ is finite, and $|H| \leq |E:K|$.

Proof. Since every element of H fixes every element of K, we have $H \leq \mathrm{Gal}(E/K)$. Hence,

$$K \subseteq \mathrm{Fix}(\mathrm{Gal}(E/K)) \subseteq \mathrm{Fix}(H) = K.$$

We conclude $\mathrm{Fix}(\mathrm{Gal}(E/K)) = K$ and so E is Galois over K. We know that H is a subgroup of $\mathrm{Gal}(E/K)$ and by Theorem 25.51 we now have $|\mathrm{Gal}(E/K)| = |E : K|$. Hence, H is a finite group with no more than $|E : K|$ elements. $\qquad \square$

We actually can prove more. Under the hypotheses of the above lemma, H *is the Galois group of* E *over* K and, as a result, has exactly $|E : K|$ elements. (Compare with Lemma 25.8.) The proof will actually follow if we show that $|E : K| \leq |H|$ (see proof of Theorem 26.3 below). To prove this, we need to somehow limit the number of elements of E that are linearly independent over K. A technical lemma will accomplish this. Given a field extension $K \subseteq E$, the field E is a vector space over K and a vector space over E. Assume you have a set of two or more elements of E. When you consider E as a vector space over E, then this set of elements will certainly be linearly dependent. On the other hand, as elements of a vector space over K, your set of elements may or may not be linearly dependent. The lemma says that if the set of elements satisfy a strong form of linear dependence over E, then they will also be linearly dependent over K.

Lemma 26.2. *Let E be a field, and let $H \leq \mathrm{Aut}(E)$. Let $K = \mathrm{Fix}(H) \subseteq E$. Let $\alpha_1, \alpha_2, \ldots, \alpha_m \in E$ and $c_1, \ldots, c_m \in E$. Assume that not all of the c's are zero. Suppose, for every $\sigma \in H$,*

$$\sum_{i=1}^{m} \sigma(c_i)\alpha_i = 0,$$

then $\{\alpha_1, \ldots, \alpha_m\}$ is linearly dependent over K.

Proof. We induct on m.

For the base case, if $m = 1$, then $c_1\alpha_1 = 0$ and $c_1 \neq 0$. Hence $\alpha_1 = 0$ and $1\alpha_1 = 0$. So α_1 is dependent over K.

Now let $m > 1$, and assume that the result is true for any collection of fewer than m elements of E. Hence, if any of the c_i is zero, then we are done by induction. So we can assume, for $1 \leq i \leq m$, that $c_i \neq 0$, and let $b_i = c_i/c_m$. Note that $b_m = 1$.

Letting σ be the identity of H, we get $\sum c_i\alpha_i = 0$. and dividing through by c_m results in $\sum b_i\alpha_i = 0$. The proof will be complete if we show that $b_i \in K$ for all i. We do this by showing that $b_i \in \mathrm{Fix}(H) = K$. For all $\sigma \in H$, we have

$$(26.1) \qquad \sum_i \sigma(b_i)\alpha_i = \sum_i \sigma\left(\frac{c_i}{c_m}\right)\alpha_i = \frac{1}{\sigma(c_m)}\sum_i \sigma(c_i)\alpha_i = 0.$$

Now fix $\tau \in H$, then, for all $\sigma \in H$, $\sigma\tau \in H$. So we have, from equation (26.1), that $\sum_i \sigma\tau(b_i)\alpha_i = 0$ as well as $\sum_i \sigma(b_i)\alpha_i = 0$. Hence,

$$(26.2) \qquad \sum_i \sigma(\tau(b_i) - b_i)\alpha_i = 0 \quad \forall \, \sigma \in H.$$

Note that, $\tau(1) = 1$, since τ is an automorphism in H. So $\tau(b_m) - b_m = \tau(1) - 1 = 0$. This means that, in equation (26.2), we have a linear combination of fewer than m elements equal to zero. This means that we are done by induction—i.e., a

K-linear combination of some of the α's is zero—unless all the coefficients in equation (26.2) are zero.

So, for all $1 \leq i \leq m$ and $\tau \in H$, we have $\tau(b_i) - b_i = 0$. Hence $b_i \in \text{Fix}(H) = K$, and $\sum b_i \alpha_i = 0$. Thus α_i's are dependent over K, and we are done. □

We can now strengthen Lemma 26.1.

Theorem 26.3. *Let E be a field, and let $H \leq \text{Aut}(E)$ with $|H| < \infty$. Let $K = \text{Fix}(H) \subseteq E$. Then E is Galois over K, $H = \text{Gal}(E/K)$, and $|H| = |E : K|$.*

Proof. CLAIM: The theorem will be proved if we show that $|E : K| \leq |H|$.

PROOF OF CLAIM: If $|E : K| \leq |H|$, then we would know that $|E : K| < \infty$, and hence Lemma 26.1 applies. That lemma showed that E is Galois over K and $|H| \leq |E : K|$. Since E is Galois over K, we know, by Theorem 25.51, that $|E : K| = |\text{Gal}(E/K)|$. If we can show $|E : K| \leq |H|$, then we would get that $|H| = |E : K| = |\text{Gal}(E/K)|$. It is also clear that $H \leq \text{Gal}(E/K)$, and hence $H = \text{Gal}(E/K)$. The proof would be complete.

CLAIM: $|E : K| \leq |H|$.

PROOF OF CLAIM: Assume the claim is not true. Let $H = \{\sigma_1, \sigma_2, \ldots, \sigma_n\}$, and choose $\alpha_1, \ldots, \alpha_m \in E$ linearly independent over K and with $m > n = |H|$.

Consider the $n \times m$ matrix A with entries in E:

$$A = \begin{bmatrix} \sigma_1(\alpha_1) & \sigma_1(\alpha_2) & \ldots & \sigma_1(\alpha_m) \\ \sigma_2(\alpha_1) & \sigma_2(\alpha_2) & \ldots & \sigma_2(\alpha_m) \\ \vdots & \vdots & \ddots & \vdots \\ \sigma_n(\alpha_1) & \sigma_n(\alpha_2) & \ldots & \sigma_n(\alpha_m) \end{bmatrix}.$$

The matrix A has more columns than rows and hence it has a non-trivial nullspace (the rank of A is at most n and the nullity of A is at least $m - n$). Let $\begin{bmatrix} c_1 \\ c_2 \\ \vdots \\ c_m \end{bmatrix}$ be a non-zero vector in the nullspace of A. Thus $c_1, \ldots, c_m \in E$ are not all zero, and we have

$$\sum_i c_i \sigma(\alpha_i) = 0 \quad \forall\, \sigma \in H.$$

Applying σ^{-1} to each of these equations, we get

$$\sum_i \sigma^{-1}(c_i)\alpha_i = 0 \quad \forall\, \sigma \in H.$$

The set $\{\sigma^{-1} \mid \sigma \in H\}$ is the same as H. Hence, since the above equation is true for *all* $\sigma \in H$, we have

$$\sum_i \sigma(c_i)\alpha_i = 0 \quad \forall\, \sigma \in H.$$

Now by Lemma 26.2, $\alpha_1, \ldots, \alpha_m$ are linearly dependent over K. The contradiction proves the claim. □

We now strengthen one part of Proposition 25.4. Note that for the next corollary, we are not assuming that E is a Galois extension of F. We only need that $F \subseteq E$ is a finite degree extension.

Corollary 26.4. *Let $F \subseteq E$ be fields, and assume $|E : F| < \infty$. Let $H \leq \mathrm{Gal}(E/F)$. Then*

$$\mathrm{Gal}(E/\mathrm{Fix}(H)) = H.$$

In particular, the map $\mathrm{Fix}(\cdot)\colon \{\text{Subgroups of } \mathrm{Gal}(E/F)\} \to \{\text{fields } K \text{ with } F \subseteq K \subseteq E\}$ has a left inverse and is 1-1.

Proof. Since $|E : F| < \infty$, we have $|H| \leq |E : F| < \infty$, and Theorem 26.3 applies. As a result, $\mathrm{Gal}(E/\mathrm{Fix}(H)) = H$. This means that the map $\mathrm{Gal}(E/\cdot)$ from $\{\text{fields } K \text{ with } F \subseteq K \subseteq E\}$ to $\{\text{Subgroups of } \mathrm{Gal}(E/F)\}$ is a left inverse for $\mathrm{Fix}(\cdot)$, and the latter is 1-1. (If $\mathrm{Fix}(H_1) = \mathrm{Fix}(H_2)$ for subgroups H_1 and H_2, then $H_1 = \mathrm{Gal}(E/\mathrm{Fix}(H_1)) = \mathrm{Gal}(E/\mathrm{Fix}(H_2)) = H_2$.) $\qquad\square$

Problems

26.1.1. Let E be a field, let $H \leq \mathrm{Aut}(E)$, and let $K = \mathrm{Fix}(H) \subseteq E$. Show that $|E : K| < \infty$ if and only if $|H| < \infty$.

26.1.2. Let E be a field, let $H \leq \mathrm{Aut}(E)$, and let $K = \mathrm{Fix}(H) \subseteq E$. Assume $|E : K| < \infty$. Then prove that $H = \mathrm{Gal}(E/K)$ and $|H| = |E : K|$. Explain the difference between the statement of this problem and the statement of Theorem 26.3.

26.1.3. Let $F \subseteq E$ be fields with $|E : F| < \infty$. Let $G = \mathrm{Gal}(E/F)$ and $K = \mathrm{Fix}(G)$. Let $F \subseteq L \subseteq K \subseteq E$. Is $\mathrm{Gal}(E/L) = G$?

26.1.4. Let $F \subseteq E$ be fields with $|E : F| < \infty$. Give an example to show that the map $\mathrm{Gal}(E/\cdot)$ from $\{\text{fields } K \text{ with } F \subseteq K \subseteq E\}$ to $\{\text{subgroups of } \mathrm{Gal}(E/F)\}$ is not always 1-1. Show that this map is 1-1 if we assume that E is a Galois extension of F.

26.1.5. Let $F \subseteq E$ be fields. Figure 25.2 showed the effects of composing the maps $\mathrm{Fix}(\cdot)$ and $\mathrm{Gal}(E/\cdot)$. If you know that $|E : F| < \infty$, you can modify one of the diagrams. Which one and how?

26.1.6. Let $F \subseteq E$ be a Galois extension with $|E : F| = 180$. Show that there are intermediate fields K_1, K_2, and K_3 containing F and contained in E such that $|K_1 : F| = 45$, $|K_2 : F| = 20$, and $|K_3 : F| = 36$.

26.1.7. Let $\mathbb{Q} \subseteq L$ be fields. Assume that L is a normal extension of $\mathbb{Q}$ and $|L : \mathbb{Q}| = 81$. Show that there exist intermediate fields K_1, K_2, and K_3 with

$$K_0 = \mathbb{Q} \subseteq K_1 \subseteq K_2 \subseteq K_3 \subseteq K_4 = L,$$

where, for $i = 1, \ldots, 4$, $|K_i : K_{i-1}| = 3$.

26.2. Fundamental Theorem of Galois Theory

In this section, we state and prove the Galois correspondence between intermediate fields in a Galois extension $F \subseteq E$ and the subgroups of the Galois group $\mathrm{Gal}(E/F)$. In fact, most parts of the theorem have already been proved and used. We already have maps, $\mathrm{Gal}(E/\cdot)$ and $\mathrm{Fix}(\cdot)$, from the intermediate fields to subgroups of the Galois group and vice versa. In this section—using the results of Section 26.1—we will note that, for a Galois extension, these two maps are inverses of each other and, as a result, both are bijections and give a 1-1 correspondence between subgroups of the Galois group and the intermediate fields in the extension. However, one remarkable feature of this correspondence that we have yet to discuss is the role of normal subgroups, and that is where we begin.

If $F \subseteq K \subseteq E$ are fields, then $\mathrm{Gal}(E/K) \leq \mathrm{Gal}(E/F)$. But when is this subgroup a normal subgroup? Recall that a normal subgroup is a subgroup that is fixed under the conjugation action. What is the corresponding feature of intermediate fields? In fact, $\mathrm{Gal}(E/F)$ acts on the set of intermediate fields, since if $\sigma \in \mathrm{Gal}(E/F)$ and K is a field with $F \subseteq K \subseteq E$, then $\sigma(K)$ is also a field contained in E and containing F (and the conditions of an action are easily satisfied). The field $\sigma(K)$ is called a (Galois) conjugate of K (see Problem 24.3.13), and the next lemma clarifies the relationship between conjugate fields and conjugate subgroups.

Lemma 26.5. *Let $F \subseteq K \subseteq E$ be fields, and let $\sigma \in \mathrm{Gal}(E/F)$. Then*

$$\mathrm{Gal}(E/\sigma(K)) = \sigma \, \mathrm{Gal}(E/K)\sigma^{-1}.$$

In particular, if $\sigma(K) = K$ for all $\sigma \in \mathrm{Gal}(E/F)$, then $\mathrm{Gal}(E/K) \lhd \mathrm{Gal}(E/F)$.

Proof. Let $\tau \in \sigma \, \mathrm{Gal}(E/K)\sigma^{-1}$, then $\tau = \sigma\delta\sigma^{-1}$ for $\delta \in \mathrm{Gal}(E/K)$. Now τ is the composition of three automorphisms of E and hence $\tau \in \mathrm{Aut}(E)$. We claim that τ fixes every element of the field $\sigma(K)$. A typical element of $\sigma(K)$ is $\sigma(a)$ with $a \in K$, and

$$\tau(\sigma(a)) = (\sigma\delta\sigma^{-1})(\sigma(a)) = \sigma(\delta(a)) = \sigma(a).$$

Hence $\tau \in \mathrm{Gal}(E/\sigma(K))$, and we have proved that $\sigma \, \mathrm{Gal}(E/K)\sigma^{-1} \subseteq \mathrm{Gal}(E/\sigma(K))$.

Now let $\tau \in \mathrm{Gal}(E/\sigma(K))$. Consider $\sigma^{-1}\tau\sigma \in \mathrm{Aut}(E)$. If $a \in K$, then $\tau(\sigma(k)) = \sigma(k)$ since τ fixes every element of $\sigma(K)$. Hence,

$$(\sigma^{-1}\tau\sigma)(k) = \sigma^{-1}(\tau(\sigma(k))) = \sigma^{-1}(\sigma(k)) = k.$$

Hence, $\sigma^{-1}\tau\sigma$ is a K-automorphism of E and an element of $\mathrm{Gal}(E/K)$. As a result, $\tau \in \sigma \, \mathrm{Gal}(E/K)\sigma^{-1}$. So $\mathrm{Gal}(E/\sigma(K)) \subseteq \sigma \, \mathrm{Gal}(E/K)\sigma^{-1}$, and the proof is complete. $\square$

In the case of Galois extensions, we expect and get a more precise relationship.

Lemma 26.6. *Let $F \subseteq K \subseteq E$, and assume that E is a Galois extension of F. Let $\sigma \in \mathrm{Gal}(E/F)$. Then*

$$\mathrm{Gal}(E/\sigma(K)) = \mathrm{Gal}(E/K) \quad \text{if and only if} \quad \sigma(K) = K.$$

Proof. The if direction is obvious. Assume that $\mathrm{Gal}(E/\sigma(K)) = \mathrm{Gal}(E/K)$. Since E is a Galois extension of F, E is also a Galois extension of both K and $\sigma(K)$

(Corollary 25.18). Hence,

$$K = \operatorname{Fix}(\operatorname{Gal}(E/K)) = \operatorname{Fix}(\operatorname{Gal}(E/\sigma(K)) = \sigma(K). \qquad \square$$

We are now ready to see the role of normal subgroups in a Galois extension.

Proposition 26.7. *Let $F \subseteq K \subseteq E$ be fields, and assume that E is a Galois extension of F. Then the following are equivalent:*

(a) *K is a Galois extension of F.*

(b) *K is a normal extension of F.*

(c) *$\sigma(K) = K$ for all $\sigma \in \operatorname{Gal}(E/F)$.*

(d) *$\operatorname{Gal}(E/K) \lhd \operatorname{Gal}(E/F)$.*

Furthermore, if any of these conditions are satisfied, then

$$\operatorname{Gal}(E/F)/\operatorname{Gal}(E/K) \cong \operatorname{Gal}(K/F).$$

Proof. (a $\Rightarrow$ b) Every Galois extension is both normal and separable.

(b $\Rightarrow$ a) Since E is a finite degree separable extension of F, the intermediate field K is also a finite degree separable extension of F (Lemma 25.36)). Now K is also assumed to be normal and, as a result, it is a Galois extension (Theorem 25.17).

(b $\Rightarrow$ c) Fix $\sigma \in \operatorname{Gal}(E/F)$. Then K is a finite degree extension of F and $\sigma \colon K \to \sigma(K)$ is an F-isomorphism. Now by Theorem 25.20 and because K is assumed to be a normal extension of F, we have $K = \sigma(K)$.

(c $\Rightarrow$ b) Let $\alpha \in K$ and let $f = \min_F(\alpha)$. We have to show that f splits in K. We already know that f splits in E, and so we let $\beta \in E$ be a root of f. We want to show that $\beta \in K$. Now E is Galois over F, and so it is the splitting field of a polynomial. Hence, $\operatorname{Gal}(E/F)$ acts transitively on the roots of the irreducible polynomial f (Theorem 24.37). So there exists $\sigma \in \operatorname{Gal}(E/F)$ such that $\sigma(\alpha) = \beta$. But we know $\sigma(K) = K$ and so $\beta \in \sigma(K) = K$.

(c $\Rightarrow$ d) This is Lemma 26.5.

(d $\Rightarrow$ c) Let $\sigma \in \operatorname{Gal}(E/F)$. By Lemma 26.5, since $\operatorname{Gal}(E/K) \lhd \operatorname{Gal}(E/F)$, we have $\operatorname{Gal}(E/\sigma(K)) = \operatorname{Gal}(E/K)$. But then Lemma 26.6 implies that $\sigma(K) = K$.

Now assume that each of the equivalent conditions is satisfied, and define

$$\theta \colon \operatorname{Gal}(E/F) \to \operatorname{Gal}(K/F)$$

by $\theta(\sigma) = \sigma|_K$, where $\sigma|_K$ is the restriction of the F-automorphism σ to K. Since, by assumption, $\sigma(K) = K$, the map $\sigma|_K$ is indeed an F-automorphism of K. Clearly, θ is a group homomorphism with $\ker(\theta) = \operatorname{Gal}(E/K)$. Now, if $\tau \in \operatorname{Gal}(K/F)$, then since E is a splitting field over F, we can (Theorem 24.24) extend τ to $\sigma \in \operatorname{Gal}(E/F)$. Now $\theta(\sigma) = \tau$ proving that θ is onto. (See Figure 26.1.) Hence,

$$\operatorname{Gal}(E/F)/\operatorname{Gal}(E/K) = \operatorname{Gal}(E/F)/\ker(\theta) \cong \operatorname{Im}(\theta) = \operatorname{Gal}(K/F). \qquad \square$$

Remark 26.8. We have proved that, in the case of a Galois extension, a subgroup of the Galois group is a normal subgroup if and only if its fixed field is a normal

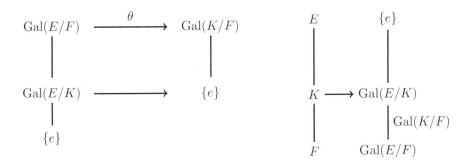

Figure 26.1. If E and K are Galois extensions of F, then $\mathrm{Gal}(K/F) \cong \mathrm{Gal}(E/F)/\mathrm{Gal}(E/K)$.

extension of the base field. In fact, this is why normal subgroups are called *normal* subgroups: they are the subgroups that correspond to normal field extensions!

We are finally ready to state the Galois correspondence.

Theorem 26.9 (Fundamental Theorem of Galois Theory). *Let $F \subseteq E$ be a Galois extension of fields. Let $G = \mathrm{Gal}(E/F)$. Let*

$$\mathcal{F} = \{K \mid K \text{ is a field and } F \subseteq K \subseteq E\},$$
$$\mathcal{G} = \{H \mid H \leq G\}.$$

Then

(a) *The maps*

$$\mathrm{Gal}(E/\cdot) : \mathcal{F} \longrightarrow \mathcal{G}$$

and

$$\mathrm{Fix}(\cdot) : \mathcal{G} \longrightarrow \mathcal{F}$$

are bijections between $\mathcal{F}$ and $\mathcal{G}$. The two maps are the inverse of each other, and both maps reverse containments.

(b) *If $K \in \mathcal{F}$ and $H \in \mathcal{G}$ with $K \longleftrightarrow H$—i.e., $\mathrm{Fix}(H) = K$ and $\mathrm{Gal}(E/K) = H$—then $|H| = |E : K|$ and $|G : H| = |K : F|$.*

(c) *If $K \in \mathcal{F}$ and $H \in \mathcal{G}$ with $K \longleftrightarrow H$, then $H \lhd G$ if and only if K is Galois over F, and in this case $G/H \cong \mathrm{Gal}(K/F)$.*

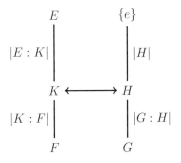

Proof. (a) We already know that the two maps reverse containments (Lemma 25.3). To show that the two maps are inverses of each other—and, hence both are bijections—we verify that each followed by the other is the identity map.

If $K \in \mathcal{F}$, then E is Galois over K (see Corollary 25.18), and so $\mathrm{Fix}(\mathrm{Gal}(E/K)) = K$. On the other hand, if $H \in \mathcal{G}$ by Corollary 26.4, $\mathrm{Gal}(E/\mathrm{Fix}(H)) = H$.

(b) $|H| = |\text{Gal}(E/K)| = |E : K|$. In addition,

$$|G : H| = |G| / |H| = |E : F| / |E : K| = |K : F|.$$

(c) This was Proposition 26.7.

$\square$

Remark 26.10. Theorem 26.9 gives a correspondence when E is a *Galois* extension of F. However, by now (see Remark 25.52) we know how to bootstrap this result for field extensions $F \subseteq E$ when it is only known that $|E : F| < \infty$. In such a case, let $K = \text{Fix}(\text{Gal}(E/F))$. We then know that E is Galois over K and $\text{Gal}(E/K) = \text{Gal}(E/F)$ (Lemma 25.8). Hence, Theorem 26.9 applies, and we get a 1-1 correspondence between subgroups of $\text{Gal}(E/F)$ and subfields of E that contain K.

The Fundamental Theorem of Algebra. The fundamental theorem of algebra says that every non-constant polynomial in $\mathbb{C}[x]$ can be factored into linear factors.[1] In other words, the only irreducible polynomials in $\mathbb{C}[x]$ are of degree 1, and there are no algebraic extensions of $\mathbb{C}$. More succinctly, it says that $\mathbb{C}$ is algebraically closed (Definition 24.27). It could be argued that the fundamental theorem of algebra is neither "fundamental" to algebra nor really a theorem of "algebra". This theorem is specifically about complex numbers, and the latter are constructed from real numbers. The construction of real numbers is *not* an algebraic construction, and it requires analytical tools (such as limits, convergence, Cauchy sequences, or Dedekind cuts). As a result, it is not possible to give a purely algebraic proof of the fundamental theorem of algebra. Each of the many proofs of this theorem incorporate some analytical facts or tools. Here, we use the Galois correspondence to minimize what is needed from analysis. In fact, we need two facts only: (1) a polynomial $p(x)$ in $\mathbb{R}[x]$ of odd degree has a real root (this follows from the intermediate value theorem since $\lim_{x \to -\infty} p(x)$ and $\lim_{x \to \infty} p(x)$ have different signs), and (2) if $\alpha \in \mathbb{C}$, then there exists $\beta \in \mathbb{C}$ with $\beta^2 = \alpha$ (i.e., complex numbers have square roots—if $z = re^{i\theta}$ with $r \geq 0$, then $\sqrt{z} = \sqrt{r}e^{i\theta/2}$).

Theorem 26.11 (The Fundamental Theorem of Algebra). *If $f \in \mathbb{C}[x]$, then f factors into linear factors in $\mathbb{C}[x]$. In other words, $\mathbb{C}$ is algebraically closed.*

Proof. CLAIM: It is enough to prove the theorem for $f \in \mathbb{R}[x]$.

PROOF OF CLAIM: Given a polynomial $f \in \mathbb{C}[x]$, let $\overline{f}$ be the polynomial gotten by replacing each coefficient of f with its conjugate (i.e., $\overline{f}$ is the result of applying to f the conjugation automorphism that sends $a + bi$ to $a - bi$). Now the polynomial $g = f\overline{f}$ has real coefficients since g is fixed under conjugation. If we know the theorem to be true for polynomials in $\mathbb{R}[x]$, then g factors into linear factors. If $g = f\overline{f}$ has such a factorization, then f must too (since $\mathbb{C}[x]$ is a UFD).

CLAIM: There are no fields L with $\mathbb{R} \subseteq L$ and $|L : \mathbb{R}|$ an odd number greater than 1.

---

[1]Statements asserting something like the fundamental theorem of algebra began appearing in the early seventeenth century, and in the eighteenth century many—including d'Alembert, Euler, Lagrange, Laplace, and Gauss—attempted to prove the theorem, but all of their proofs had gaps (the gap in Gauss's original proof was topological and subtle and was not filled until 1920). The first mostly rigorous proof was published in 1806 by Jean-Robert Argand (1768–1822), an amateur mathematician. Carl Friedrich Gauss (1777–1855) produced a number of other mostly rigorous proofs some years later.

PROOF OF CLAIM: Assume such a field exists. Since $\text{char}(\mathbb{R}) = 0$, the field L is a finite degree separable extension of $\mathbb{R}$, and so, by the primitive element theorem, Theorem 25.45, $L = \mathbb{R}(\alpha)$ is a simple extension. The degree of $\min_{\mathbb{R}}(\alpha)$ is equal to $|L : \mathbb{R}|$ and hence odd. Every odd degree polynomial over the reals has a real root, and yet $\min_{\mathbb{R}}(\alpha)$ is supposed to be irreducible. The contradiction proves the claim.

CLAIM: There are no fields M with $\mathbb{C} \subseteq M$ and $|M : \mathbb{C}| = 2$.

PROOF OF CLAIM: Assume such a field exists. Again, M is a finite degree separable extension of $\mathbb{C}$, and so, by the primitive element theorem, Theorem 25.45, $M = \mathbb{C}(\beta)$ for some $\beta \in M$. This time the degree of $\min_{\mathbb{C}}(\beta)$ is 2. Without loss of generality, assume $\min_{\mathbb{C}}(\beta) = x^2 + 2bx + c$. Then the roots of this polynomial, by the quadratic equation, are $-b \pm \sqrt{b^2 - ac}$. But every complex number has a square root in $\mathbb{C}$, and so $\min_{\mathbb{C}}(\beta)$ is not irreducible. The contradiction proves the claim.

We now turn to the proof of the theorem. Given $f \in \mathbb{R}[x]$, we want to show that f already splits in $\mathbb{C}$. Let E be the splitting field of $(x^2 + 1)f$ over $\mathbb{R}$. Since E contains both i and $\mathbb{R}$, it must contain $\mathbb{C} = \mathbb{R}(i)$. The field E is the splitting field of a polynomial and the characteristic of $\mathbb{R}$ is zero, and so E is a Galois extension of $\mathbb{R}$. Assume that $|\text{Gal}(E/\mathbb{R})| = |E : R| = 2^a m$, where m is an odd number. We now employ the Galois correspondence.

$$E$$
$$|$$
$$\mathbb{C}$$
$$|$$
$$\mathbb{R}$$

Let $G = \text{Gal}(E/\mathbb{R})$, and let H be a Sylow 2-subgroup of G. Then $|H| = 2^a$ and $|G : H| = m$. Let L be the intermediate field containing $\mathbb{R}$ and contained in E that corresponds—in the Galois correspondence—to H. Now $|L : \mathbb{R}| = |G : H| = m$, an odd number. We have proved that the only odd degree extension of $\mathbb{R}$ is $\mathbb{R}$ itself. So $L = \mathbb{R}$. As a result $m = 1$ and $G = H$ is a 2-group.

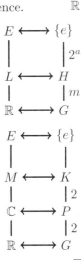

Now let P be the subgroup of G corresponding—in the Galois correspondence—to $\mathbb{C}$. Then $|G : P| = 2$. If $a > 1$, then since P is a 2-group, by Theorem 12.1, P has a subgroup K of index 2. Let M be the intermediate field containing $\mathbb{C}$ and contained in E corresponding—in the Galois correspondence—to K. Then M is an extension of $\mathbb{C}$ with $|M : \mathbb{C}| = 2$. We proved that such an extension does not exist. The contradiction proves that $a = 1$, $P = \{e\}$, $G \cong \mathbb{Z}/2\mathbb{Z}$, and $E = \mathbb{C}$. The proof is complete.

$\square$

Problems

26.2.1. In Problem 22.1.1, you were asked to show that $\sqrt{3} \notin \mathbb{Q}(\sqrt{2})$. It also follows from Problem 22.1.2 or by direct calculation that $\sqrt{5} \notin \mathbb{Q}(\sqrt{2}, \sqrt{3})$. Here, instead, use the Galois correspondence to prove, with little calculation, that $\sqrt{5} \notin \mathbb{Q}(\sqrt{2}, \sqrt{3})$. You may find the following steps helpful.
STEP 1: Let $K = \mathbb{Q}(\sqrt{2}, \sqrt{3})$, and find the number of intermediate fields L with $\mathbb{Q} \subsetneq L \subsetneq K$. Example 24.41 may be relevant.
STEP 2: Assume $\sqrt{5} \in K$, produce more intermediate fields than promised in Step 1, and arrive at a contradiction.

26.2.2. Let $E = \mathbb{Q}(\sqrt{2}, \sqrt{3}, \sqrt{5}) \subset \mathbb{C}$.
(a) Find $\mathrm{Gal}(E/\mathbb{Q})$.
(b) Find all the intermediate fields L, such that $\mathbb{Q} \subset L \subset E$.
(c) Is E a simple extension of $\mathbb{Q}$? Can you identify an element $\alpha \in E$ with $E = \mathbb{Q}[\alpha]$?

26.2.3. What is the minimal polynomial of $\sqrt{2} + 2\sqrt{3} - \sqrt{6} + 3\sqrt{5} - \sqrt{10}$? If you need to multiply algebraic expressions, you can use a symbolic math software (such as Maple, Mathematica, or Sage).

26.2.4. Let $F \subset L \subset E$.
(a) Assume that L is a normal extension of F and that E is a normal extension of L. Must E be a normal extension of F?
(b) Assume that L is a Galois extension of F and that E is a Galois extension of L. Further assume that every F-automorphism of L can be extended to an automorphism of E. Must E be a Galois extension of F?

26.2.5. In the proof of Proposition 26.7, we used Lemma 26.5 to show that if $\sigma(K) = K$, then $\mathrm{Gal}(E/K) \lhd \mathrm{Gal}(E/F)$. Give an alternate argument by using the map $\theta \colon \mathrm{Gal}(E/F) \to \mathrm{Gal}(K/F)$ defined by $\theta(\sigma) = \sigma|_K$.

26.2.6. In the final argument in the proof of Proposition 26.7, we used Theorem 24.24 to show that the map θ is onto. Give an alternate argument by first showing that $|\mathrm{Im}(\theta)| = |\mathrm{Gal}(K/F)|$.

26.2.7. Let E be a Galois extension of F, and let p be a prime number. Assume that p divides $|E : F|$. Show that there exists a field L, such that $F \subseteq L \subseteq E$ and $|E : L| = p$.

26.2.8. Let $f(x) = p(x)q(x)$, where $p(x)$ and $q(x)$ are irreducible polynomials in $\mathbb{Q}[x]$, and $\deg p = \deg q = 2$. What are the possible Galois groups for $f(x)$?

26.2.9. Let $F \subseteq E$ be a Galois extension with $G = \mathrm{Gal}(E/F)$. Let L and M be intermediate fields containing F and contained in E, and assume that, in the Galois correspondence, they correspond to subgroups H and K, respectively.
(a) Show that, in the Galois correspondence, $L \cap M$ corresponds to $\langle H, K \rangle$.

(b) Show that, in the Galois correspondence, $\langle L, M \rangle$, the compositum of L and M (see Definition 22.41), corresponds to $H \cap K$.

26.2.10. Let $F \subseteq E$ be a Galois extension, and assume that K_1 and K_2 are two intermediate fields containing F and contained in E. Assume K_1 and K_2 are Galois extensions of F. Is $K_1 \cap K_2$ necessarily a Galois extension of F?

26.2.11. Let $F \subseteq E$ be a Galois extension with $|E : F| = 12$. Show that there exists an intermediate field K with $F \subseteq K \subseteq E$ and $|K : F| = 4$. Furthermore, show that K is a Galois extension of F if and only if K is the only intermediate field containing F and contained in E with $|K : F| = 4$.

26.2.12. Let $F \subseteq E$ be a Galois extension of fields. Assume that $|E : F| = 15$.
   (a) Show that there exists two intermediate fields K and L with $K \cap L = F$ and $\langle K, L \rangle = E$.
   (b) Do the fields K and L in the previous part have to be Galois extensions of F?
   (c) What is the number of intermediate fields K with $F \subseteq K \subseteq E$?

26.2.13. Let $F \subseteq E$ be a Galois extension of fields. Assume that $|E : F| = 8{,}225$.
   (a) Show that there exists a field K such that K is a Galois extension of F and $|K : F| = 175$.
   (b) Show that there exists a field L with $F \subseteq L \subseteq E$ and $|L : F| = 35$.

26.2.14. Let $F \subseteq E$ be a Galois extension, and assume $|E : F| = 5{,}145$. Let K be an intermediate field with $F \subseteq K \subseteq E$ and $|K : F| = 5$. Prove that K is a Galois extension of F.

26.2.15. Let $E = \mathbb{Q}(\sqrt[4]{2}, \sqrt[4]{3})$.
   (a) Is E Galois over $\mathbb{Q}$? Find all the elements of $G = \mathrm{Gal}(E/\mathbb{Q})$.
   (b) What is $K = \mathrm{Fix}(G)$? Find all the elements of $H = \mathrm{Gal}(K/\mathbb{Q})$.
   (c) The map $\mathrm{Fix}(\cdot)$ is a bijection between subgroups of G and which subfields of E? Exhibit the correspondence.
   (d) The map $\mathrm{Fix}(\cdot)$ is a bijection between subgroups of H and which subfields of E? Exhibit the correspondence.

26.2.16. Let $E = \mathbb{Q}(\sqrt[4]{2}, \sqrt{3})$.
   (a) What is $|E : \mathbb{Q}|$?
   (b) Is E Galois over $\mathbb{Q}$? Find all the elements of $G = \mathrm{Gal}(E/\mathbb{Q})$ and identify a familiar group isomorphic to G.
   (c) For each subgroup $H \leq G$, find $\mathrm{Fix}(H)$.
   (d) Let $L = \mathbb{Q}(\sqrt{3})$. What are the elements of $\mathrm{Gal}(E/L)$?
   (e) Is $\mathrm{Gal}(E/L) \leq \mathrm{Gal}(E/\mathbb{Q})$? Is $\mathrm{Gal}(E/L) \lhd \mathrm{Gal}(E/\mathbb{Q})$? What is $\mathrm{Fix}(\mathrm{Gal}(E/L))$?

26.2.17. In Figure 25.1, we used two diagrams to show the back and forth between subgroups H of $\mathrm{Gal}(E/F)$ and the intermediate fields K with $F \subseteq K \subseteq E$.
   (a) Could we have saved some ink and used one diagram as in Figure 26.2? Either prove that $\mathrm{Fix}(H) \subseteq K \Leftrightarrow \mathrm{Gal}(E/K) \subseteq H$ or give a counterexample.
   (b) Would you give a different answer if H and $\mathrm{Fix}(H)$ were drawn above K and $\mathrm{Gal}(E/K)$?

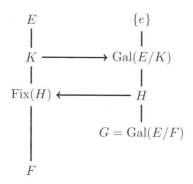

Figure 26.2. Does $\mathrm{Fix}(H) \subseteq K$ always imply that $\mathrm{Gal}(E/K) \subseteq H$?

26.2.18. Let $\mathbb{R} \subseteq E$ be a field extension with $|E : \mathbb{R}| < \infty$. Prove that either $E = \mathbb{R}$ or E is a splitting field of $x^2 + 1$ over $\mathbb{R}$.

26.2.19. Let $F \subseteq E$ be fields with $\mathrm{char}(F) = 0$ and $|E : F| = 2$. Assume
    (a) if $p \in F[x]$ with $\deg(p)$ odd, then p has a root in F, and
    (b) if $q \in E[x]$ with $\deg(q) = 2$, then q has a root in E.
Prove that E is the splitting field over E of every polynomial $f \in E[x]$.

26.3. Examples of Galois Groups

Example 26.12. Let $f = (t^2 - 3)(t^2 + 1)$. In Example 24.26, we saw that the splitting field of f is $E = \mathbb{Q}[\sqrt{3}, i]$, and $|E : \mathbb{Q}| = 4$. Since E is the splitting field of a separable polynomial over $\mathbb{Q}$, we know that E is a Galois extension of $\mathbb{Q}$ and $|\mathrm{Gal}(E/\mathbb{Q})| = |E : \mathbb{Q}| = 4$. If $\sigma \in \mathrm{Gal}(E/\mathbb{Q})$, then σ acts on the roots of $t^2 - 3$, and so $\sigma(\sqrt{3}) = \pm\sqrt{3}$. Likewise, $\sigma(i) = \pm i$. Since, every element of $\mathrm{Gal}(E/\mathbb{Q})$ is determined by its actions on $\sqrt{3}$ and on i, we have at most four possibilities for σ. But we know that $\mathrm{Gal}(E/\mathbb{Q})$ has four elements. Hence, each of the four possibilities must occur. So

$$\mathrm{Gal}(E/\mathbb{Q}) = \{e, \sigma, \tau, \sigma\tau\},$$

where e fixes every element of E, σ fixes every element of $\mathbb{Q}(i)$, and $\sigma(\sqrt{3}) = -\sqrt{3}$. On the other hand τ fixes every element of $\mathbb{Q}(\sqrt{3})$ and $\tau(i) = -i$.

The group $G = \mathrm{Gal}(E/\mathbb{Q})$ is a group of order 4 where each non-identity element is of order 2. Thus $G \cong \mathbb{Z}/2\mathbb{Z} \times \mathbb{Z}/2\mathbb{Z}$. This group has three subgroups of order 2:

$$\langle\sigma\rangle = \{e, \sigma\}, \ \langle\tau\rangle = \{e, \tau\}, \ \langle\sigma\tau\rangle = \{e, \sigma\tau\}.$$

We conclude, from the fundamental theorem of Galois theory, that—in addition to $\mathbb{Q}$ and E—there are exactly three intermediate fields between $\mathbb{Q}$ and E (see Figure 26.3):

$$\mathrm{Fix}(\langle\sigma\rangle) = \mathbb{Q}(i), \ \mathrm{Fix}(\langle\tau\rangle) = \mathbb{Q}(\sqrt{3}), \ \mathrm{Fix}(\langle\sigma\tau\rangle) = \mathbb{Q}(i\sqrt{3}).$$

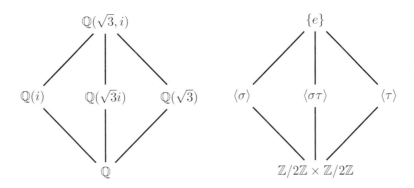

Figure 26.3. Subgroups of $\mathbb{Z}/2\mathbb{Z} \times \mathbb{Z}/2\mathbb{Z}$ correspond to the intermediate fields K with $\mathbb{Q} \subseteq K \subseteq \mathbb{Q}(\sqrt{3}, i)$.

Example 26.13. Let $f = x^2 + x + 1 \in \mathbb{F}_2[x]$. Neither 0 nor 1 are the roots of f, and so f is irreducible in $\mathbb{F}_2[x]$. Let E be the splitting field of f, and let $\alpha \in E$ be a root of f. Hence, $\alpha^2 + \alpha + 1 = 0$ and so $\alpha^2 = \alpha + 1$ (recall that the characteristic is 2 and so $-1 = 1$). Now $|\mathbb{F}_2(\alpha) : \mathbb{F}_2| = \deg f = 2$ and so $\mathbb{F}_2(\alpha)$ has $2^2 = 4$ elements. These are $0, 1, \alpha, \alpha^2 = \alpha + 1$. We see

$$f(\alpha + 1) = (\alpha + 1)^2 + \alpha + 1 + 1 = \alpha^2 + 1 + \alpha = 0.$$

Hence, $f = (x - \alpha)(x - \alpha - 1) = (x + \alpha)(x + \alpha + 1) \in \mathbb{F}_2(\alpha)[x]$. We conclude that $E = \mathbb{F}_2(\alpha)$. Since E is the splitting field of a separable polynomial over $\mathbb{F}_2$, we have that E is a Galois extension of $\mathbb{F}_2$ and $|\mathrm{Gal}(E/\mathbb{F}_2)| = |E : \mathbb{F}_2| = 2$. This Galois group acts transitively on the roots of f. Hence there exists $\sigma \in \mathrm{Gal}(E/\mathbb{F}_2)$ with $\sigma(\alpha) = \alpha + 1$, and

$$\mathrm{Gal}(E/\mathbb{F}_2) = \{e, \sigma\} \cong \mathbb{Z}/2\mathbb{Z}.$$

Example 26.14. Let $f = x^4 - 2 \in \mathbb{Q}[x]$, and let E be the splitting field of f over $\mathbb{Q}$. We seek $G = \mathrm{Gal}(f) = \mathrm{Gal}(E/\mathbb{Q})$.

Let $\alpha = \sqrt[4]{2}$. Then the roots of f are $\pm\alpha$ and $\pm i\alpha$, and in $E[x]$ we have

$$f = (x - \alpha)(x + \alpha)(x - i\alpha)(x + i\alpha).$$

Both α and $i = i\alpha/\alpha$ must be in E, and so $E = \mathbb{Q}(\alpha, i)$. By the Schönemann-Eisenstein criterion f is irreducible over $\mathbb{Q}$ and so $f = \min_\mathbb{Q}(\alpha)$. Hence $|\mathbb{Q}(\alpha) : \mathbb{Q}| = \deg f = 4$. On the other hand, $i \notin \mathbb{Q}(\alpha)$ since the latter consists of only real numbers. Thus $\min_{\mathbb{Q}(\alpha)} i$ has degree larger than 1. Since i is a root of $x^2 + 1$, the latter must be $\min_{\mathbb{Q}(\alpha)} i$. Hence,

$$|E : \mathbb{Q}| = |E : \mathbb{Q}(\alpha)| \; |\mathbb{Q}(\alpha) : \mathbb{Q}| = 2 \times 4 = 8.$$

Since E is the splitting field of a separable polynomial over $\mathbb{Q}$, we have that E is a Galois extension of $\mathbb{Q}$. So $|G| = |E : \mathbb{Q}| = 8$.

By Theorem 24.44, we know that $G \leq S_4$. Hence, G is a subgroup of order 8 of the symmetric group of degree 4. But $|S_4| = 24 = 8 \times 3$ and hence G must be a Sylow 2-subgroup of S_4. All Sylow subgroups of a group are isomorphic to each other, and hence, up to isomorphism, G has only one type of Sylow 2-subgroup.

One such Sylow subgroup is $\langle (1\ 2\ 3\ 4), (1\ 2) \rangle$ which is the familiar group D_8 (see Example 1.32). Hence,

$$G \cong D_8.$$

We want to find the elements of G explicitly and to observe the Galois correspondence. Let $K = \mathbb{Q}(i)$. We know that f is irreducible over $\mathbb{Q}$, but is it irreducible over K? Since, $|K : \mathbb{Q}| = 2$ and $|E : \mathbb{Q}| = 8$, we must have $|E : K| = 4$. But $E = K(\alpha)$ and so $\deg \min_K(\alpha) = |E : K| = 4$. Since α is a root of f and f has degree 4, we must have $f = \min_K(\alpha)$, and so f is irreducible over K.

The advantage of knowing that f is irreducible over K is that we now know that $\mathrm{Gal}(E/K)$—which is a subgroup of $G = \mathrm{Gal}(E/\mathbb{Q})$—acts transitively on the roots of f. Hence, there is an element $\sigma \in \mathrm{Gal}(E/K)$ with $\sigma(\alpha) = i\alpha$. Every element of $\mathrm{Gal}(E/K)$ fixes every element of K, and hence $\sigma(i) = i$. In addition, σ is an automorphism of E and fixes every element of $\mathbb{Q}$. So we have $\sigma \in \mathrm{Gal}(E/\mathbb{Q}) = G$ with $\sigma(\alpha) = i\alpha$ and $\sigma(i) = i$. Repeatedly applying σ, we get that

$$\sigma : \alpha \mapsto i\alpha \mapsto -\alpha \mapsto -i\alpha \mapsto \alpha.$$

We conclude that σ is an element of order 4 in G. Incidentally, since E is the splitting field of f over K, we have that E is a Galois extension of K, and $|\mathrm{Gal}(E/K)| = |E : K| = 4$, and so

$$\mathrm{Gal}(E/K) = \langle \sigma \rangle \cong \mathbb{Z}/4\mathbb{Z}.$$

Likewise, let $F = \mathbb{Q}(\alpha)$. We already know that $x^2 + 1$ is irreducible over F, and hence $\mathrm{Gal}(E/F)$—which is another subgroup of G—acts transitively on its roots. So we have $\tau \in G$ with $\tau(\alpha) = \alpha$ and $\tau(i) = -i$. Now combining σ and τ, we get all the elements of G. In fact, by checking their effects on α and i, we see that, as expected, we have $\tau\sigma = \sigma^3\tau$. Hence,

$$G = \mathrm{Gal}(E/\mathbb{Q}) = \langle \sigma, \tau \mid \sigma^4 = \tau^2 = e, \tau\sigma = \sigma^3\tau \rangle \cong D_8.$$

To illustrate the Galois correspondence, first note that

$$\sigma(\sqrt{2}) = \sigma(\alpha^2) = \sigma(\alpha)^2 = -\sqrt{2}.$$

Hence, for example, $\sigma^2(\sqrt{2}) = \sigma(-\sqrt{2}) = \sqrt{2}$, and so $\sqrt{2} \in \mathrm{Fix}(\langle \sigma^2 \rangle)$.

We know the subgroups of D_8, and for each we can find the corresponding fixed fields. We are aided by the fact that we know, from the Galois correspondence, the degree of the extensions. For example, the three subgroups of order 4 of D_8 correspond to extensions of degree 2 of $\mathbb{Q}$, and we have

$$\mathrm{Fix}(\langle \sigma \rangle) = \mathbb{Q}(i) = K,$$
$$\mathrm{Fix}(\langle \sigma^2, \tau \rangle) = \mathbb{Q}(\sqrt{2}),$$
$$\mathrm{Fix}(\langle \sigma^2, \sigma\tau \rangle) = \mathbb{Q}(i\sqrt{2}).$$

Also note that $e^{\pi i/4} = \frac{\sqrt{2}}{2} + i\frac{\sqrt{2}}{2}$. So $\sigma(e^{\pi i/4}) = -e^{\pi i/4}$, $\sigma^2(e^{\pi i/4}) = e^{\pi i/4}$, and $\tau(e^{\pi i/4}) = \frac{\sqrt{2}}{2} - i\frac{\sqrt{2}}{2}$. We leave it to the reader to verify the details of the Galois correspondence illustrated in Figure 26.4.

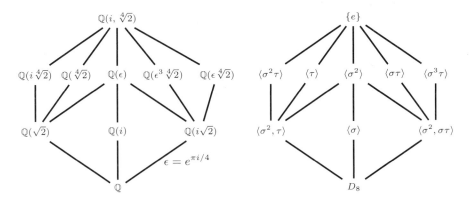

Figure 26.4. Galois correspondence between subgroups of D_8 and intermediate fields L with $\mathbb{Q} \subseteq L \subseteq \mathbb{Q}(i, \sqrt[4]{2})$

Problems

26.3.1. Let $f(x) = x^3 - 2 \in \mathbb{Q}[x]$. Let E be the splitting field of $f(x)$ over $\mathbb{Q}$, and let G be the Galois group of $f(x)$. Exhibit the Galois correspondence between the subgroups of G and the fields K with $\mathbb{Q} \subseteq K \subseteq E$. Indicate which extensions are normal.

26.3.2. Find explicitly a polynomial $f \in \mathbb{Q}[x]$ with splitting field of degree 3 over $\mathbb{Q}$.

26.3.3. Let $\mathbb{Q} \subseteq K$ be a field extension with $|K : \mathbb{Q}| = 2$. Show that $K = \mathbb{Q}[\sqrt{m}]$ for some $m \in \mathbb{Q}$.

26.3.4. Let $\beta = \sqrt{3 + 2\sqrt{2}}$, $g = \min_{\mathbb{Q}}(\beta)$, and E be the splitting field for g over $\mathbb{Q}$. Find g, E, $|E : \mathbb{Q}|$, and $\mathrm{Gal}(g)$. Can you write β without any nested radicals?

26.3.5. Let $\alpha = \sqrt{5 + \sqrt{21}}$, and let $f = \min_{\mathbb{Q}}(\alpha)$.
   (a) Find f.
   (b) Find the splitting field E for f over $\mathbb{Q}$. What is $|E : \mathbb{Q}|$?
   (c) Find $\mathrm{Gal}(f)$.
   (d) Find all intermediate fields K with $\mathbb{Q} \subseteq K \subseteq E$. If for any of these fields K we have $|K : \mathbb{Q}| = 2$, then write K as $K = \mathbb{Q}[\sqrt{m}]$ for some $m \in \mathbb{Q}$.
   (e) Can you (explicitly) find two positive integers m_1 and m_2 such that $E = \mathbb{Q}[\sqrt{m_1}, \sqrt{m_2}]$?
   (f) Can you (explicitly) write α without any nested radicals?

26.3.6. Let E be the splitting field of $x^4 - 10x^2 - 20$ over $\mathbb{Q}$.
   (a) Find $\mathrm{Gal}(E/\mathbb{Q})$ explicitly and identify a familiar group that is isomorphic to $\mathrm{Gal}(E/\mathbb{Q})$.
   (b) Find all the subgroups of $\mathrm{Gal}(E/\mathbb{Q})$ and the intermediate fields between $\mathbb{Q}$ and E. Which subgroup corresponds to which field in the Galois correspondence?

26.3.7. Let E be the splitting field of $x^4 - 10x^2 + 36$ over $\mathbb{Q}$.
  (a) Find $\mathrm{Gal}(E/\mathbb{Q})$ explicitly and identify a familiar group that is isomorphic to $\mathrm{Gal}(E/\mathbb{Q})$.
  (b) Find all the subgroups of $\mathrm{Gal}(E/\mathbb{Q})$ and the intermediate fields between $\mathbb{Q}$ and E. Which subgroup corresponds to which field in the Galois correspondence?

26.3.8. Let $E = \mathbb{Q}(\sqrt[8]{14}, i)$ and $F = \mathbb{Q}(\sqrt{14})$. Explicitly, find the elements of $\mathrm{Gal}(E/F)$, and give a familiar group that is isomorphic to $\mathrm{Gal}(E/F)$. Exhibit the Galois correspondence between subgroups of $\mathrm{Gal}(E/F)$ and intermediate fields K with $F \subseteq K \subseteq E$.

26.3.9. Let $K = \mathbb{Q}(x)$, $F = \mathbb{Q}(x^6) \subset K$, and $f = \min_F(x)$. Find $\mathrm{Gal}(f)$.

26.3.10. **Galois group of a simple transcendental extension.** Let F be a field, and let $F(x)$ be the field of rational functions in x over F. Our aim in the next two problems is to find $\mathrm{Gal}(F(x)/F)$. For $A = \begin{bmatrix} a & b \\ c & d \end{bmatrix} \in \mathrm{GL}(2, F)$, define $\phi_A \colon F(x) \to F(x)$ by

$$\phi_A(f(x)) = f\left(\frac{ax+b}{cx+d}\right).$$

In other words, to find $\phi_A(f)$, we replace, in the rational function $f(x)$, every occurrence of x with $\frac{ax+b}{cx+d}$. (Such a substitution is called a *fractional linear transformation*.)
  (a) Show that ϕ_A is a ring homomorphism that fixes every element of F.
  (b) What can $\ker(\phi_A)$ be? Show that ϕ_A is 1-1.
  (c) Show that $F(x) = F\left(\frac{ax+b}{cx+d}\right)$, and as a result ϕ_A is onto. Conclude that $\phi_A \in \mathrm{Gal}(F(x)/F)$.
  (d) Prove that, if $A, B \in \mathrm{GL}(2, F)$, then $\phi_{AB} = \phi_B \phi_A$. (Note that the operation in $\phi_B \phi_A$ is function composition.)
  (e) Define a map $\Theta \colon \mathrm{GL}(2, F) \to \mathrm{Gal}(F(x)/F)$ by

$$\Theta(A) = \phi_{A^{-1}}.$$

Show that Θ is a group homomorphism.

26.3.11. Continuing with the notation and assumptions of Problem 26.3.10, let $\psi \in \mathrm{Gal}(F(x)/F)$. Then $\psi \colon F(x) \to F(x)$ and $\psi \colon x \longmapsto \frac{p(x)}{q(x)}$ where $p, q \in F[x]$, $q \neq 0$, and $\gcd(p, q) = 1$.
  (a) Show that, for $f \in F(x)$, $\psi(f(x)) = f\left(\frac{p(x)}{q(x)}\right)$. Conclude that the image of ψ is $F\left(\frac{p(x)}{q(x)}\right)$.
  (b) Using Problem 22.3.41, show that p and q are linear polynomials in x.
  (c) Show that $\psi = \phi_A$ for some $A \in \mathrm{GL}(2, F)$. Conclude that the map Θ is onto. (The maps ϕ_A and Θ are defined in Problem 26.3.10.)
  (d) Show that $\ker(\Theta) = \{ \begin{bmatrix} \lambda & 0 \\ 0 & \lambda \end{bmatrix} \mid \lambda \in F \}$.
  (e) Conclude that $\mathrm{Gal}(F(x)/F) \cong \mathrm{GL}(2, F)/\mathbf{Z}(\mathrm{GL}(2, F))$. The latter is called a *projective linear group* and is denoted by $\mathrm{PGL}(2, F)$.

26.3.12. Let $A = \begin{bmatrix} i & 0 \\ 0 & 1 \end{bmatrix}$ and $B = \begin{bmatrix} 0 & 1 \\ 1 & 0 \end{bmatrix}$ be elements of $\mathrm{GL}(2, \mathbb{C})$. Let $G = \mathrm{Gal}(\mathbb{C}(x)/\mathbb{C})$, and define $\phi_A, \phi_B \in G$ as in Problems 26.3.10 and 26.3.11. Let $H = \langle \phi_A, \phi_B \rangle \leq G$.

  (a) Find a familiar group that is isomorphic to H.

  (b) Show that $\mathbb{C}(x^4 + x^{-4}) \subseteq \mathrm{Fix}(H)$.

  (c) What is $|\mathbb{C}(x) : \mathbb{C}(x^4 + x^{-4})|$? Show that $\mathbb{C}(x^4 + x^{-4}) = \mathrm{Fix}(H)$.

  (d) What is $\mathrm{Gal}(\mathbb{C}(x)/\mathbb{C}(x^4 + x^{-4}))$?

26.3.13. Let F be a field, and let $A = \begin{bmatrix} 0 & 1 \\ -1 & 1 \end{bmatrix}$ and $B = \begin{bmatrix} 0 & 1 \\ 1 & 0 \end{bmatrix}$ elements of $\mathrm{GL}(2, F)$. Let $G = \mathrm{Gal}(F(x)/F)$, and define $\phi_A, \phi_B \in G$ as in Problems 26.3.10 and 26.3.11. Let $H = \langle \phi_A, \phi_B \rangle \leq G$.

  (a) Find a familiar group that is isomorphic to H.

  (b) Let $K = F\left(\frac{(x^2 - x + 1)^3}{x^2(x-1)^2}\right)$. Show that $K \subseteq \mathrm{Fix}(H)$.

  (c) What is $|F(x) : K|$? Show that $K = \mathrm{Fix}(H)$.

  (d) What is $\mathrm{Gal}(F(x)/K)$?

Chapter 27

Finite Fields
and Cyclotomic Extensions

... where we give a fuller treatment of finite fields, study the splitting field and the Galois group of $x^n - 1$ and more generally $x^n - a$ and, in the process, prove the irreducibility of the so-called cyclotomic polynomials.

27.1. Finite Fields

In the previous chapters, we have proved a number of results on finite fields and their extensions. In this section, we complete our treatment of finite fields.

Let E be a finite field. What can we say about E? We know that a finite field must have characteristic p for some prime p (see Theorem 16.52), and this means that F, the prime subfield of E, is isomorphic to $\mathbb{F}_p = (\mathbb{Z}/p\mathbb{Z}, +, \cdot)$ (Corollary 16.55). So we have $F \subseteq E$, and $|E : F| = n$ for some integer n. This results in $|E| = |F|^n = p^n$ (Theorem 22.32). Thus the only possible sizes for finite fields are powers of primes. But are there fields for each of these possible sizes? In other words, if p is an arbitrary prime and n an arbitrary positive integer, then does there exist a field of order p^n? The answer is yes, and, in fact, there is a *unique* field of order p^n, as we shall show in Theorem 27.2.

For the special case, when $n = 2$ and $p \cong 3 \pmod 4$, we have already proved— see Theorem 20.9—that $\mathbb{F}_p[i]$ is a field of order p^2. More generally, our original method (see pages 374 and 411) for constructing fields was to use the fact that if f is an irreducible polynomial of degree n in $\mathbb{F}_p[x]$, then $\mathbb{F}_p[x]/\langle f \rangle$ is a field of order p^n (see Proposition 19.28). In fact, we carried this method out in Section 19.3 and produced a field with four elements. The problem with this method is that while we have argued that there are irreducible polynomials of arbitrary large degree (see Problem **19.3.24**), we have not—at least so far—produced irreducible polynomials of *every* possible degree n. Here, we will take a different route, and first construct a field of order p^n as a splitting field of a polynomial. We will then use this field to prove that, in $\mathbb{F}_p[x]$, there does exist irreducible polynomials of every possible degree n (see Corollary 27.5).

Of course, many of the results about fields in the previous chapters also apply to finite fields. We highlight three results more specific to finite fields. Let F be a finite field. In Proposition 25.34, we showed that *every* $f \in F[x]$ is separable, and Theorem 25.41 proved that $F^\times$, the group of units of F, is a finite cyclic group. Finally, if E is finite degree extension of F, then we proved (Corollary 25.42) that $E = F[\alpha]$ for some $\alpha \in E$.

We will begin by constructing and characterizing a finite field of prime power order.

Proposition 27.1. *Let p be a prime, let n be a positive integer, let $q = p^n$, and let K be a set containing $\mathbb{F}_p = (\mathbb{Z}/p\mathbb{Z}, +, \cdot)$. Let $f = x^q - x \in \mathbb{F}_p[x]$. Then the following are equivalent:*

(a) *K is a field with q elements,*

(b) *K is the set of all the roots of f (in some splitting field),*

(c) *K is a splitting field of f over $\mathbb{F}_p$.*

Proof. (a $\Rightarrow$ b) Suppose that K is a field with q elements. Then $K^\times$ is a (cyclic) group of order $q - 1$, and so, for every $u \in K$, we have $u^{q-1} = 1$. Hence, for every $u \in K$ (including $u = 0$), we have $u^q = u$. Thus f has its full quota of q distinct roots in K, and K *is* the set of all of the roots of f.

(b $\Rightarrow$ c) Now, assume that K is the set of all the roots of f in some splitting field for f. That splitting field has characteristic p, and so, for $u, v \in K$, we have $(u + v)^q = u^q + v^q = u + v$, $(uv)^q = u^q v^q = uv$, and $(u^{-1})^q = (u^q)^{-1} = u^{-1}$. Thus if u and v are roots of f, then so are $u + v$, uv, u^{-1}. Hence, the latter are also in K, and K is closed under addition, multiplication, and taking inverses. We conclude that K is a field, and since it contains all the roots, K is a splitting field of f over $\mathbb{F}_p$.

(c $\Rightarrow$ a) Finally, suppose that K is the splitting field of f over $\mathbb{F}_p$. We saw in the previous part of the proof that the set of all roots of f forms a field. Hence, K is exactly the set of roots of f. Now the derivative of f is -1, and so there are no common roots between the polynomial and its derivative. Hence $(x - \alpha)^2 \nmid x^q - x$ for all $\alpha \in K$, and so the q roots of the polynomial are distinct. Thus K is a field with q elements. $\qquad\square$

Theorem 27.2. *Let p be a prime, let n be a positive integer, and let $q = p^n$. Then there exists a unique (up to isomorphism) field with q elements. Furthermore, if m is not a power of a prime, then there are no fields of order m.*

Proof. We have already proved that the order of a finite field must be a power of a prime (Theorem 22.32). Now let $q = p^n$, where p is a prime and n a positive integer. In the previous proposition, we saw that the splitting field of $x^q - x$ over $\mathbb{F}_p$ has q elements, and thus there exists a field of order q. Also in the previous proposition, we showed that *every* field of order q is a splitting field of $x^q - x$ over

$\mathbb{F}_p$. We know that two splitting fields for the same polynomial are isomorphic and hence all fields of order q are isomorphic. □

We are finally ready to justify and formalize our use of the notation $\mathbb{F}_q$ for *the* field of order q for every prime power q.

Definition 27.3. Let n be a positive integer, and let $p \in \mathbb{Z}^{>0}$ a prime. Let $q = p^n$ be a power of a prime. Then the unique field of order q is called the *Galois field* of order q and is denoted by $\mathbb{F}_q$ or $\mathrm{GF}(q)$.

Remark 27.4. If p is a prime, then $\mathbb{F}_p = \mathrm{GF}(p) \cong (\mathbb{Z}/p\mathbb{Z}, +, \cdot)$. However, if q is not a prime (but a power of a prime), then $\mathbb{F}_q$ is definitely *not* isomorphic to $(\mathbb{Z}/q\mathbb{Z}, +, \cdot)$ since the latter is not even an integral domain.

We can now prove:

Corollary 27.5. *Let q be a power of a prime, and let n be a positive integer. Then there exists an irreducible polynomial of degree n in $\mathbb{F}_q[x]$.*

Proof. Let E be a field of order q^n. Then $\mathbb{F}_q \subseteq E$ and $|E : \mathbb{F}_q| = n$. Hence, by Corollary 25.42, $E = \mathbb{F}_q[\alpha]$ for some $\alpha \in E$. Let $f = \min_{\mathbb{F}_q}(\alpha)$. Then f is irreducible, and its degree is equal to $|E : \mathbb{F}_q| = n$. □

We now turn our attention to determining the Galois group of finite degree extensions of $\mathbb{F}_p$.

Question 27.6. Let $q = p^n$, let E be a field of order q, and let F be the prime subfield of E. Then what is $\mathrm{Gal}(E/F)$?

Recall that for any field E of characteristic p, we defined (see Definition 25.32) the Frobenius ring homomorphism

$$\Phi : E \to E$$

by $\Phi(a) = a^p$. In addition to being a ring homomorphism, Φ is always 1-1, and if E is finite, then Φ is also onto. In addition, Φ is the identity map if $E \cong \mathbb{F}_p$ (see Lemma 25.33). This means that if E is a field of order $q = p^n$ and F the prime subfield of E, then $\Phi \in \mathrm{Gal}(E/F)$. So, for finite fields, the Frobenius map always gives one element of the Galois group. In fact, much more is true.

Theorem 27.7. *Let p be a prime, and let n be a positive integer. Let E be a field of order p^n, and let F be the prime subfield of E. Then E is Galois over F, and $\mathrm{Gal}(E/F)$ is a cyclic group of order n generated by the Frobenius map Φ.*

Proof. Let $q = p^n$, then, by Proposition 27.1, E is splitting field of $f = x^q - x$ over F. Over a finite field, all polynomials are separable—this is Proposition 25.34 but we also know directly that f has q distinct roots—and so E is the splitting field of a separable polynomial over F. Hence, E is Galois over F and $|\mathrm{Gal}(E/F)| = |E : F| = n$.

We know that $\Phi \in \mathrm{Gal}(E/F)$ and hence $\langle \Phi \rangle = \{e, \Phi, \Phi^2, \ldots\} \subseteq \mathrm{Gal}(E/F)$. The group $\mathrm{Gal}(E/F)$ is a finite group of order n, and so the order of every element divides n. Assume that $o(\Phi) = j$. Thus, for all $\alpha \in E$, we have $\Phi^j(\alpha) = \alpha$. However,

$$\Phi^j(\alpha) = \Phi^{j-1}(\alpha^p) = \Phi^{j-2}(\alpha^{(p^2)}) = \cdots = \alpha^{(p^j)}.$$

Hence,

$$\alpha^{(p^j)} = \alpha, \quad \text{for all } \alpha \in E.$$

As a result, the equation $x^{(p^j)} - x$ has p^n roots—that is, every element of E—in E. The number of roots can be no more than the degree of the polynomial, and so $p^j \geq p^n$ which means $j \geq n$. Since $j \mid n$, we have $j = n$, and Φ is a generator for $\mathrm{Gal}(E/F)$. $\qquad\Box$

Example 27.8. Let $\mathbb{F}_8$ be the field of order 8. We know that such a field exists, and we are justified in calling it *the* field of order 8 since all such fields are isomorphic (Theorem 27.2). We also know that $\mathbb{F}_8^\times$ is a cyclic group of order 7 (Theorem 25.41). Because of the latter, we can immediately write

$$\mathbb{F}_8 = \{0, 1, \alpha, \alpha^2, \alpha^3, \alpha^4, \alpha^5, \alpha^6\}, \text{ where } \alpha^7 = 1.$$

With this representation of $\mathbb{F}_8$, it is clear how to multiply any two elements, and, for instance, $\alpha^3 \alpha^5 = \alpha$. But what about addition? Since $\mathbb{F}_8$ is unique, does that mean that there is only one way that we can define addition? Is it possible a priori and without having made any other choices to decide what $\alpha^3 + 1$ is? One thing we do know is that $\mathrm{char}(\mathbb{F}_8) = 2$, and so $x + x = 0$ for all $x \in \mathbb{F}_8$. Thus $(\mathbb{F}_8, +)$ is an abelian group where every element has order 2. It follows that $(\mathbb{F}_8, +) \cong \mathbb{Z}/2\mathbb{Z} \times \mathbb{Z}/2\mathbb{Z} \times \mathbb{Z}/2\mathbb{Z}$.

Before answering the question about addition, we ask a different—but as we shall see a very related—question: What are the irreducible polynomials of degree 3 in $\mathbb{F}_2[x]$? A polynomial of degree 3 is irreducible in $\mathbb{F}_2[x]$ if and only if it does not have a root in $\mathbb{F}_2$. By plugging in 0 and 1, we see that there are exactly two irreducible polynomials of degree 3 over $\mathbb{F}_2$, namely $x^3 + x^2 + 1$ and $x^3 + x + 1$. If δ is a root of either polynomial in a splitting field, then $|\mathbb{F}_2(\delta) : \mathbb{F}_2| = \deg(\min_{\mathbb{F}_2}(\delta)) = 3$, and hence the roots of these polynomials live in a field of order 8. The field $E_1 = \mathbb{F}_2[x]/\langle x^3 + x^2 + 1\rangle$ is of order 8, and $x^3 + x^2 + 1$ has a root in E_1, namely $x + I$. In fact, $x^3 + x^2 + 1$ splits in E_1 (see Example 24.23). Similarly, $x + I$ is a root of $x^2 + x + 1$ in the field $E_2 = \mathbb{F}_2[x]/\langle x^3 + x + 1\rangle$.

Now, E_1 and E_2 are both fields of order 8, and so they are both isomorphic to $\mathbb{F}_8$ above. This shows that the rules for addition for $\mathbb{F}_8$, while giving isomorphic fields, are not unique. In E_1, if we let $\alpha = x + I$, then $E_1 = \{0, 1, \alpha, \ldots, \alpha^6\}$ and $\alpha^3 + \alpha^2 + 1 = 0$ (and, as a result, $\alpha^3 + 1 = \alpha^2$). While in E_2, if we let $\beta = x + I$, we have $E_2 = \{0, 1, \beta, \ldots, \beta^6\}$ with $\beta^3 + \beta + 1 = 0$ (and as a result $\beta^3 + 1 = \beta$).

Identifying $\mathbb{F}_8$ with E_1, consider the field extension $\mathbb{F}_2 \subseteq E_1 = \mathbb{F}_8$. This means that α now satisfies $\alpha^3 + \alpha^2 + 1 = 0$. The Galois group $\mathrm{Gal}(E_1/\mathbb{F}_8)$ is a cyclic group of order 3 generated by the Frobenius map Φ. The Galois group acts on the roots of polynomials, and so $\Phi(\alpha) = \alpha^2$ as well as $\Phi^2(\alpha) = \alpha^4$ are also—in addition to α—roots of $x^3 + x^2 + 1$. We conclude that α, α^2, and α^4 are the roots of $x^3 + x^2 + 1$ in $\mathbb{F}_8$, and in $\mathbb{F}_8[x]$ we have (remember we are in characteristic 2 and so $-1 = 1$)

$$x^3 + x^2 + 1 = (x + \alpha)(x + \alpha^2)(x + \alpha^4).$$

But $x^3 + x + 1$ also has roots in $\mathbb{F}_8$. Where are those? As you may expect, α^3, $\Phi(\alpha^3) = \alpha^6$, and $\Phi^2(\alpha) = \alpha^5$ are the roots of $x^3 + x + 1$. We can plug in α^3 to see that it is a root of $x^3 + x + 1$, but we can also argue that two monic irreducible polynomials over a field F have no common roots in any extension field (Problem 24.1.17). Hence, in $\mathbb{F}_8$, we have

$$x^3 + x + 1 = (x + \alpha^3)(x + \alpha^5)(x + \alpha^6).$$

If we let $\beta = \alpha^3$, then $\mathbb{F}_8 = \{0, 1, \beta, \ldots, \beta^6\}$, and this time $\beta^3 = \beta + 1$.

To summarize, we can write $\mathbb{F}_8 = \{0, 1, \alpha, \ldots, \alpha^6\}$ with $\alpha^7 = 1$. However, the addition depends on a choice of an irreducible polynomial of degree 3 over $\mathbb{F}_2$.

Knowing that the Galois group of a finite field extension is cyclic, we can use the Galois correspondence to get information about subfields of a finite field.

Corollary 27.9. *Let p be a prime, let n be a positive integer, and let E be a field of order p^n. Then*

(a) *if $m \mid n$, then E has a unique subfield of order p^m;*

(b) *if $m \nmid n$, then E does not have a subfield of order p^m; and*

(c) *if $K \subseteq E$ is a subfield of order p^m, then E is Galois over K and $\mathrm{Gal}(E/K)$ is a cyclic group of order n/m generated by Φ^m (where Φ is the Frobenius map).*

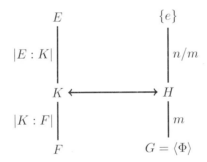

Figure 27.1. Finite field extensions and the Galois correspondence

Proof. Let F be the prime subfield of E. We know that $|F| = p$, that E is a Galois extension of F, and that $G = \mathrm{Gal}(E/F)$ is a cyclic group of order n generated by the Frobenius map Φ. Note that all subfields of E are intermediate fields containing F. See Figure 27.1.

(a) If $m \mid n$, then the cyclic group G has a unique subgroup H of order n/m. By the Galois correspondence, there exists a unique field $K = \mathrm{Fix}(H)$ with $F \subseteq K \subseteq E$ and with $|K : F| = |G : H| = m$. The latter implies that $|K| = p^m$.

(b) If E has a subfield K of order p^m, then $|K : F| = m$, and by the Galois correspondence, we get that $H = \mathrm{Gal}(E/K)$ is a subgroup of G of order n/m. But this means that $m \mid n$, which is a contradiction.

(c) Since $G = \langle \Phi \rangle$ is cyclic, then, for each m dividing n, G has a unique subgroup of order n/m generated by Φ^m. The conclusion now follows from the Galois correspondence. $\qquad\qquad\square$

Problems

27.1.1. Let $P = \{\mathbb{F}_{3^i} \mid 1 \leq i \leq 10\}$ be a collection of fields. The set P ordered by inclusion is a partial order. Draw the Hasse diagram of this poset.

27.1.2. Draw the Hasse diagram of all the subfields of $\mathbb{F}_{5^{30}}$ ordered by inclusion.

27.1.3. Let $F \subseteq E$ be a field extension. Assume $|F|$ and $|E : F|$ are both finite. Prove that E is a Galois extension of F.

27.1.4. A "proof" of Proposition 27.1 contains the following argument:
> All finite degree extensions of finite fields are separable, and so the polynomial $x^q - x$ has q distinct roots in its splitting field over $\mathbb{F}_p$.

Comment.

27.1.5. Let $\mathbb{F}_9$ be the field of order 9, and let $\mathbb{F}_9^\times = \langle \alpha \mid \alpha^8 = 1 \rangle$.
   (a) List all irreducible polynomials of degree 2 over $\mathbb{F}_3$.
   (b) In $\mathbb{F}_9$, what are the possibilities for $1 + \alpha$? In other words, what are the possibilities for i so that $1 + \alpha = \alpha^i$?
   (c) Let $I = \langle x^2 + 1 \rangle$ in $\mathbb{F}_3[x]$, and identify $\mathbb{F}_9$ with $E = \mathbb{F}_3[x]/I = \{a + bx + I \mid a, b \in \mathbb{F}_3\}$. Which elements of E have multiplicative order equal to 8?
   (d) What are the roots of $x^2 + 1$ in E?
   (e) For each irreducible polynomial of degree 2 over $\mathbb{F}_3$, find its roots in E.

27.1.6. Let $F \subseteq E$ be fields with $|F| = 3^3$ and $|E| = 3^{15}$. Describe $\mathrm{Gal}(E/F)$ as explicitly as possible.

27.1.7. With as little calculation as possible, show that there exists $\alpha \in \mathbb{F}_5$ such that
$$x^5 - x = x(x-1)(x-\alpha)(x-\alpha^2)(x-\alpha^3).$$

27.1.8. Let $f = x^{25} - x$, and let E be a splitting field for f over $\mathbb{F}_5$. Is there an element $\alpha \in E$ such that
$$f = x(x-1)(x-\alpha)(x-\alpha^2)\cdots(x-\alpha^{23})?$$
If so, how many such α's exist? If not, why not?

27.1.9. If $x^{25} - x \in \mathbb{F}_5[x]$ is factored into a product of irreducible polynomials, then what are the degrees of those irreducible polynomials?

27.1.10. Let f be a monic irreducible polynomial of degree 2 in $\mathbb{F}_5[x]$. Show that f divides $x^{25} - x$.

27.1.11. Find the number of monic irreducible polynomials of degrees 1 and 2 in $\mathbb{F}_5[x]$ without actually listing them.

27.1.12. Let p be a prime, let n be a positive integer, and let $q = p^n$. Let $f \in \mathbb{F}_p[x]$ be an irreducible polynomial of degree m. Prove that f divides $x^q - x$ if and only if m divides n.

27.1.13. **Factoring** $x^q - x$. Consider the field $\mathbb{F}_5$, and let $q = 5^{30}$. Prove that the product of all monic irreducible polynomials in $\mathbb{F}_5[x]$ whose degree divides 30 is $x^q - x$. Is there anything special about the numbers 5 and 30? Generalize.

27.1.14. **Counting irreducible polynomials.** Let p and r be primes. Find the number of monic irreducible polynomials of degree r in $\mathbb{F}_p[x]$.

27.1.15. Let $\mathbb{F}_4 = \{a + b\alpha \mid a, b \in \mathbb{F}_2, \alpha^2 = \alpha + 1\}$ be the field with four elements. Find two irreducible polynomials, one of degree 2 and one of degree 3, in $\mathbb{F}_4[x]$.

27.1.16. Let $\mathbb{F}_4$ be the field with four elements. Find an irreducible polynomial of degree 4 in $\mathbb{F}_4[x]$.

27.1.17. Let p be a prime, let $f = x^3 - 1 \in \mathbb{F}_p[x]$, and let E be the splitting field for f over $\mathbb{F}_p$. For each of $p = 3$, 5, and 7,
 (a) find $|E|$; and
 (b) if $E^\times = \langle \alpha \rangle$, then find the powers of α that are the roots of f.

27.1.18. Let $f(x) = x^{30} - 1 \in \mathbb{F}_7[x]$, and let E be the splitting field of f over $\mathbb{F}_7$. What are $|E : \mathbb{F}_7|$ and $|E|$?

27.1.19. Let $f = x^7 - x - 3 \in \mathbb{F}_7[x]$. Let E be a splitting field for f over $\mathbb{F}_7$, and let $\alpha \in E$ be a root of f. Write F for $\mathbb{F}_7$.
 (a) Show that the orbit of α under the action of $\mathrm{Gal}(E/F)$ is

$$\{\alpha, \alpha + 1, \dots, \alpha + 6\}.$$

 (b) What are the roots of $\min_F(\alpha)$? What is the degree of the $\min_F(\alpha)$?
 (c) Is f irreducible in $F[x]$? Why?
 (d) Is $E = F[\alpha]$?
 (e) What are $|E : F|$ and $|E|$?

27.1.20. Let p be a positive prime, let $0 \neq a \in \mathbb{F}_p$, let $f(x) = x^p - x - a \in \mathbb{F}_p[x]$, and let E be a splitting field for f over $\mathbb{F}_p$. Show that f is irreducible in $\mathbb{F}_p[x]$, and find $|E|$.

27.1.21. Let p be a prime, and let n and m be positive integers. Prove that $p^m - 1$ divides $p^n - 1$ if and only if m divides n.

27.1.22. Let F be a field with $\mathrm{char}(F) = p > 0$. Let $a \in F$ and write $f(x) = x^p - a$. Show that either f splits over F or f is irreducible in $F[x]$.

27.1.23. Let p be a positive prime with $p \neq \pm 1 \pmod 5$ and $p \neq 5$. Let $f = x^4 + x^3 + x^2 + x + 1 \in \mathbb{F}_p[x]$. Let E be a splitting field of f and $\alpha \in E$ a root of f.
 (a) Show $\alpha^5 = 1$ and $o(\alpha) = 5$ in the group $E^\times$.
 (b) Show $\alpha \notin \mathbb{F}_p$. Conclude that f has no linear factors.
 (c) Can $\alpha \in K$ where $|K : F| = 2$? Can f have any quadratic factors?
 (d) Prove that f is irreducible in $\mathbb{F}_p[x]$.

27.1.24. Let p be a positive prime with $p = 3 \pmod 7$ or $p = 5 \pmod 7$. Let $f = \frac{x^7 - 1}{x - 1} \in \mathbb{F}_p[x]$. Prove that f is irreducible.

27.2. Cyclotomic Extensions

In this section, we study roots of unity as well as the splitting field and the Galois group of $x^n - 1$ over a field F. After a (relatively brief) discussion of the general case, we turn to the important case where $F = \mathbb{Q}$, the rationals.

Definition 27.10. Let F be a field and n a positive integer. We call $\zeta \in F$ an *nth root of unity* if $\zeta^n = 1$. We call ζ a *primitive nth root of unity* if $\zeta^n = 1$ and $\zeta^m \ne 1$, for $1 \le m < n$.

Example 27.11. In $\mathbb{Q}$, the field of rational numbers, the only roots of unity are ± 1. The multiplicative identity 1 is, of course, an nth root of unity for all positive integers n, and it is a primitive nth root of unity only for $n = 1$. Its negative, -1 is a primitive 2nd root of unity.

Example 27.12. In $\mathbb{C}$, the field of complex numbers, there are nth roots of unity and primitive nth roots of unity for *every* positive integer n. For example, let U be the set of 8th roots of unity in $\mathbb{C}$. To find elements of U, we solve $x^8 = 1$. Using Euler's formula (see page 415), we can write $1 = \cos(2k\pi) + i\sin(2k\pi) = e^{2k\pi i}$, where k can be any integer. As a result, $x = e^{\frac{k\pi i}{4}}$. Plugging in $k = 0, \ldots, 7$ (other values of k will cycle back and give the same numbers), we get that the set of 8th roots of unity in $\mathbb{C}$ is

$$U = \{1, e^{\pi i/4}, e^{\pi i/2}, e^{3\pi i/4}, e^{\pi i}, e^{5\pi i/4}, e^{3\pi i/2}, e^{7\pi i/4}\}.$$

Using Euler's formula, we can write these numbers in the form $a + bi$ and get $U = \{\pm 1, \pm i, \pm\frac{\sqrt{2}}{2} \pm \frac{\sqrt{2}}{2}i\}$. The set U together with multiplication (of complex numbers) is a cyclic group of order 8. (This is not a coincidence as we shall see momentarily.) In fact, $U = \langle e^{\pi i/4} \rangle$, and the primitive 8th roots of unity are the generators of this cyclic group. If we let $\zeta = e^{\pi i/4}$, we know from our early work on cyclic groups (Problem *2.3.19*) that the generators of $\langle \zeta \rangle$ are all the elements of the form ζ^ℓ where $1 \le \ell \le 8$ with $\gcd(\ell, 8) = 1$. Hence, the primitive 8th roots of unity in $\mathbb{C}$ are

$$e^{\pi i/4}, e^{3\pi i/4}, e^{5\pi i/4}, e^{7\pi i/4}.$$

Example 27.13. Let $\mathbb{F}_7 = (\mathbb{Z}/7\mathbb{Z}, +, \cdot)$ be the (unique) field with seven elements. Then we know that $\mathbb{F}_7^\times$ is a cyclic group (Theorem 25.41) of order 6. In a cyclic group of order 6, there are $\phi(6) = 2$ elements of order 6, $\phi(3) = 2$ elements of order 3 (the generators of the unique subgroup of order 3), $\phi(2) = 1$ element of order 2, and one element of order 1. In fact, $\mathbb{F}_7^\times = \{3, 3^2 = 2, 3^3 = 6, 3^4 = 4, 3^5 = 5, 3^6 = 1\} = \langle 3 \rangle$. Hence, 3 and $3^{-1} = 5$ are primitive 6th roots of unity, 2 and $2^{-1} = 4$ are primitive 3rd roots of unity, 6 is a primitive 2nd root of unity, and 1 is a primitive 1st root of unity. As an example, every element of $\mathbb{F}_7$ is also a 12th root of unity, but $\mathbb{F}_7$ has no primitive 12th roots of unity.

Lemma 27.14. *Let F be a field, and let n be a positive integer. Let U denote the set of roots in F of the polynomial $x^n - 1$. Then U is a finite cyclic subgroup of $F^\times = (F - \{0\}, \cdot)$. The order of U divides n, elements of U are those elements of $F^\times$ whose order divide n, and they are precisely the nth roots of unity in F. The field F contains primitive nth roots of unity if and only if $|U| = n$. If $|U| = n$, then the generators of U—that is elements of order n in $F^\times$—are precisely the primitive nth roots of unity in F.*

Proof. If α and β are two roots of $x^n - 1$, then $(\alpha\beta)^n = \alpha^n\beta^n = 1$ and $(\alpha^{-1})^n = (\alpha^n)^{-1} = 1$. So, U is closed under multiplication and taking inverses. Hence, U is a subgroup of $F^\times$. The number of roots of $x^n - 1$ in a field F is at most n (Corollary 19.27), and so U is a finite subgroup. Such a subgroup is cyclic by Theorem 25.41. If ζ is a generator of U, the $\zeta^n = e$ and so $|U| = o(\zeta) \mid n$. A primitive root of unity is an element of order n in U, a cyclic group. Hence, F—and therefore U—has a primitive root of unity if and only if $|U| = n$. The rest of the claims about elements of U and its generators are straightforward. $\qquad\square$

Lemma 27.15. *Let F be a field, let n be a positive integer, and let $f = x^n - 1 \in F[x]$. Let E be the splitting field of f over F, and let U be the set of roots of f in E. Then, the following are equivalent:*

(a) $\operatorname{char}(F) \nmid n$ *(this includes the case when $\operatorname{char}(F) = 0$).*

(b) $|U| = n$.

(c) *E contains a primitive root of unity.*

(d) $E = F[\zeta]$, *where ζ is a primitive root of unity.*

Proof. (a $\Rightarrow$ b) $|U| = n$ if and only if the roots of $x^n - 1$ are distinct. The derivative of $x^n - 1$ is nx^{n-1}. If $\operatorname{char}(F) \nmid n$, then zero is the only root of the derivative. Zero is not the root of $x^n - 1$, and hence, by Corollary 25.29, unless $\operatorname{char}(F) \mid n$, $x^n - 1$ has n distinct roots in an splitting field.

(b $\Rightarrow$ a) If $\operatorname{char}(F) = p$ and $p \mid n$, then $n = pm$. Since p is a prime, we have $x^n - 1 = (x^m - 1)^p$ has at most m distinct roots in E.

(b $\Leftrightarrow$ c) This was proved in Lemma 27.14.

(d $\Rightarrow$ c) Obvious.

(c $\Rightarrow$ d) Let $\zeta \in E$ be a primitive root of unity. Then $F \subseteq F(\zeta) \subseteq E$. But $\{\zeta^i \mid 1 \le i \le n\}$ are the n distinct roots of f in $F(\zeta)$. We conclude that $F(\zeta) = F[\zeta]$ is the splitting field of f. $\qquad\square$

Turning to the Galois group of $x^n - 1$, first recall (see Definition 1.51) that $(\mathbb{Z}/n\mathbb{Z})^\times$ is the group of units of $(\mathbb{Z}/n\mathbb{Z}, +, \cdot)$. In fact (see Theorem 1.55) $(\mathbb{Z}/n\mathbb{Z})^\times = \{\, a \in \mathbb{Z}/n\mathbb{Z} \mid \gcd(a, n) = 1 \,\}$. In particular, $|(\mathbb{Z}/n\mathbb{Z})^\times| = \phi(n)$.

Proposition 27.16. *Let F be a field with $\operatorname{char}(F) \nmid n$ (this includes the case when $\operatorname{char}(F) = 0$), and let E be the splitting field of $x^n - 1$ over F. Then E is Galois over F and $\operatorname{Gal}(E/F)$ is isomorphic to a subgroup of $(\mathbb{Z}/n\mathbb{Z})^\times$. In particular, this group is abelian, and if n is a prime, then $\operatorname{Gal}(E/F)$ is cyclic and its order divides n.*

Proof. By Lemma 27.15, $x^n - 1$ has n distinct roots in E and so it is a separable polynomial. The field E is the splitting field of a separable polynomial over F, and hence E is Galois over F.

Again by Lemma 27.15, $E = F[\zeta]$ where ζ is a primitive nth root of unity, and, by Lemma 27.14, the roots of $x^n - 1$ form a cyclic subgroup of $E^\times$ of order n. The primitive nth roots of unity in E are exactly the generators of this cyclic subgroup, and these are of the form ζ^i where $\gcd(n, i) = 1$. Now, if $\sigma \in \text{Gal}(E/F)$, then $\sigma \colon E \to E$ is an automorphism. As a result, σ sends ζ to another primitive root of unity. ($\sigma : E^\times \to E^\times$ is a group automorphism and so $o(\sigma(\zeta)) = o(\zeta)$.) So $\sigma(\zeta) = \zeta^i$ where $\gcd(n, i) = 1$. Define

$$\Theta \colon \text{Gal}(E/F) \to (\mathbb{Z}/n\mathbb{Z})^\times, \text{ by } \Theta(\sigma) = i \text{ where } \sigma(\zeta) = \zeta^i.$$

If σ and τ are arbitrary elements of $\text{Gal}(E/F)$, then, for some positive integers i and j, we have $\sigma(\zeta) = \zeta^i$, and $\tau(\zeta) = \zeta^j$. Hence, $\sigma(\tau(\zeta)) = \sigma(\zeta^j) = \zeta^{ij}$, and so

$$\Theta(\sigma\tau) = ij = \Theta(\sigma)\Theta(\tau),$$

proving that Θ is a group homomorphism. If $\Theta(\sigma) = 1$, then $\sigma \in \text{Gal}(F(\zeta)/F)$ fixes both ζ and F. As a result σ is the identity automorphism and Θ is 1-1. We conclude that $\text{Gal}(E/F) = \text{Gal}(E/F)/\ker(\Theta) \cong \Theta(\text{Gal}(E/F)) \leq (\mathbb{Z}/n\mathbb{Z})^\times$ as promised. $\qquad\square$

In the case of finite fields, we already know a fair amount about possible extension fields, and, as an example, we can prove:

Theorem 27.17. *Let p be a prime, and assume $p \nmid n$. Let E be the splitting field of $x^n - 1$ over $\mathbb{F}_p$. Then $|E : \mathbb{F}_p|$ is the least positive integer k such that $n \mid p^k - 1$.*

Proof. Let U be the set of roots of $f = x^n - 1$ in E. By Lemma 27.15, $|U| = n$, and by Lemma 27.14, U is a cyclic group of order n. Assume that $|E : \mathbb{F}_p| = k$, then E is a field with p^k elements and $E^\times$ is a cyclic group of order $p^k - 1$. So U is a (cyclic) subgroup of order n of a cyclic group of order $p^k - 1$. Hence, $n \mid p^k - 1$. Conversely, if $n \mid p^k - 1$, then $E^\times$ will have a subgroup of order n and every element of this subgroup will satisfy the equation $x^n = 1$. Hence, $x^n - 1$ will have n roots in E and split. The conclusion now follows. $\qquad\square$

We now focus on the extensions of $\mathbb{Q}$. The following proposition is straightforward.

Proposition 27.18. *Let $\mathbb{C}$ denote the field of complex numbers, and let n be a positive integer. Let*

$$C = \{e^{2\pi k i/n} \mid 0 \leq k < n\}.$$

Then C is a cyclic group of order n, and its elements are the nth roots of unity in $\mathbb{C}$. The generators of C are the primitive nth roots of unity in $\mathbb{C}$. They are

$$\{e^{2\pi k i/n} \mid 0 \leq k < n, \ \gcd(k, n) = 1\}.$$

The number of primitive roots of unity in $\mathbb{C}$ is $\phi(n)$ where ϕ is the Euler ϕ function.

Proof. Let $\zeta = e^{2\pi i/n}$. Then $\zeta^n = 1$, and $C = \{\zeta^k \mid 0 \leq k < n\} = \langle \zeta \rangle$ is clearly cyclic of order n. Each of the n elements of C are a root of $x^n - 1$, and this polynomial has at most n roots. Hence C is exactly the set of nth roots of unity. The generators of a cyclic group such as C are ζ^k where $\gcd(k, n) = 1$, and they number $\phi(n)$. By definition, these generators are the primitive nth roots of unity. $\qquad\square$

Definition 27.19. If ζ is a primitive nth root of unity in $\mathbb{C}$, then $\mathbb{Q}(\zeta)$—which is the splitting field of $x^n - 1$ over $\mathbb{Q}$—is called the nth *cyclotomic field*.[1] More generally, if F is a field and $E = F(\zeta)$ where ζ is a root of unity, then E is called a *cyclotomic extension* of F.

Remark 27.20. Recall (see Problem **19.5.5**) that if p is a prime, then $x^{p-1} + x^{p-2} + \cdots + x + 1$ is irreducible over $\mathbb{Q}$, and, hence, it is the minimal polynomial of $e^{\frac{2\pi i}{p}}$, a primitive pth root of unity. It follows that, for p a prime,

$$|\mathbb{Q}(\zeta) : \mathbb{Q}| = p - 1, \quad \text{if } \zeta \text{ is a primitive } p\text{th root of unity.}$$

Definition 27.21. Let $n > 0$, and let U denote the set of primitive nth roots of unity in $\mathbb{C}$. Define

$$\Phi_n(x) = \prod_{\zeta \in U} (x - \zeta) = \prod_{\substack{0 \le k < n \\ \gcd(k,n)=1}} (x - e^{2k\pi i/n}).$$

$\Phi_n(x)$ is a monic polynomial of degree $\phi(n)$ and is called the nth *cyclotomic polynomial*.

Example 27.22.

$$\Phi_1(x) = x - 1, \qquad\qquad\qquad \Phi_2(x) = x + 1,$$
$$\Phi_3(x) = x^2 + x + 1, \qquad\qquad \Phi_4(x) = x^2 + 1,$$
$$\Phi_5(x) = x^4 + x^3 + x^2 + x + 1, \qquad \Phi_6(x) = x^2 - x + 1,$$
$$\Phi_7(x) = x^6 + x^5 + x^4 + x^3 + x^2 + x + 1, \qquad \Phi_8(x) = x^4 + 1.$$

Lemma 27.23. *Let n be a positive integer, then*

$$x^n - 1 = \prod_{\substack{d > 0 \\ d \mid n}} \Phi_d(x).$$

Proof. Let C be the set consisting of all nth roots of unity. Each $\zeta \in C$ is a *primitive* dth root of unity for exactly one $d \mid n$ with $d > 0$. Hence, C is the disjoint union of $U_{d_1}, \ldots, U_{d_k}$ where $d_1, \ldots, d_k$ are all the positive divisors of n and U_{d_i} is the set of primitive d_ith roots of unity. Now the roots of $x^n - 1$ are distinct (the derivative and the polynomial have no common roots) and are exactly the elements of C. Hence,

$$x^n - 1 = \prod_{\zeta \in C} (x - \zeta) = \prod_{i=1}^{k} \prod_{\zeta \in U_{d_i}} (x - \zeta)$$

$$= \prod_{i=1}^{k} \Phi_{d_i}(x) = \prod_{\substack{d > 0 \\ d \mid n}} \Phi_d(x). \qquad\square$$

---

[1] The nth roots of unity in $\mathbb{C}$ all lie on the unit circle, and divide the circle into equal arcs. This is the origin for the use of the word "cyclotomy" which means "circle dividing".

The following lemma will be used at least twice.

Lemma 27.24. *Let R and S be integral domains with $R \subseteq S$, and let $f, g \in R[x]$. Assume that the leading coefficient of f is a unit of R and that there exists $h \in S[x]$ with $g = fh$. Then $h \in R[x]$.*

Proof. This is Problem 19.2.10, but we give the argument here as well. Since the leading coefficient of f is a unit, by Theorem 19.21, there exists $q, r \in R[x]$ with $g = fq + r$ and with $r = 0$ or $\deg r < \deg f$.

We now have $fh = fq + r$ which means $f(h - q) = r$. If $h - q \neq 0$, then $\deg r \geq \deg f$ which is a contradiction. Hence, $h = q \in R[x]$. □

Corollary 27.25. *Let n be a positive integer, then $\Phi_n \in \mathbb{Z}[x]$.*

Proof. First let $f = \prod_{\substack{d|n \\ 0 < d < n}} \Phi_d(x)$. Then, by Lemma 27.23, we have, in $\mathbb{C}[x]$,

$$x^n - 1 = f(x)\Phi_n(x).$$

We now prove $\Phi_n \in \mathbb{Z}[x]$ by induction on n. For the base case, we know the statement to be true for small values of n. Assuming that $\Phi_d \in \mathbb{Z}[x]$ for all $d < n$, we get that $f \in \mathbb{Z}[x]$. By Lemma 27.24, it follows that $\Phi_n(x) \in \mathbb{Z}[x]$. □

Remark 27.26. By looking at examples for small values of n, one may conjecture that the coefficients of Φ_n are not only integers but are always from the set $\{0, \pm 1\}$. If you are interested in seeing whether this is true or not, you should try Problem *27.2.23*.

We now prove an important and non-trivial result attributed to Gauss.

Theorem 27.27 (Gauss). *Let n be a positive integer, then Φ_n is irreducible in $\mathbb{Q}[x]$.*

Proof. To prove that Φ_n is irreducible in $\mathbb{Q}[x]$, it is enough, by Gauss's lemma (see Corollary 19.53), to show that Φ_n is irreducible in $\mathbb{Z}[x]$.

By way of proof by contradiction, assume that Φ_n is reducible in $\mathbb{Z}[x]$. Since Φ_n is monic, it is primitive, and we conclude, by Lemma 19.43, that $\Phi_n(x) = f(x)g(x)$ with $f, g \in \mathbb{Z}[x]$ and $\deg f$ and $\deg g$ less than $\deg \Phi_n$. Since Φ_n is monic, the leading coefficients of f and g are invertible, and, hence, we can choose f and g to be monic. Further, without loss of generality, we can assume that f is irreducible in $\mathbb{Z}[x]$. (If not, then factor f further, and continue until you get an irreducible factor.) Since f is irreducible in $\mathbb{Z}[x]$, it also irreducible in $\mathbb{Q}[x]$ by Gauss's lemma, Corollary 19.53.

We have $\Phi_n(x) = f(x)g(x)$ and the roots of Φ_n are precisely all the primitive nth roots of unity. Hence, the primitive nth roots of unity are partitioned into two non-empty sets: roots of f and roots of g. Let ζ be a root of f, and let δ be a root of g. Since ζ is a *primitive* root of unity, all roots of unity—including all primitive roots of unity—are powers of ζ. Hence, for some k with $\gcd(k, n) = 1$, $\delta = \zeta^k$. Now among all such choices of pairs ζ and δ, pick one for which k is the smallest possible positive integer. In other words, ζ and δ are both primitive roots of unity, ζ is a root of f, δ is a root of g, k is a positive integer with $\gcd(k, n) = 1$, $\delta = \zeta^k$, and we cannot find another pair of roots of f and g with a smaller k.

CLAIM: $k = p$ a prime number with $p \nmid n$.

PROOF OF CLAIM: Let p be a prime divisor of k. Then $(\zeta^p)^{k/p} = \delta$. If ζ^p is a root of f, then ζ^p and δ give a pair of roots with a smaller k. This contradiction proves that ζ^p is a root of g. But then ζ and ζ^p are a pair of admissible roots of f and g. Since k is the smallest possible exponent for such pairs, we conclude that $p \geq k$. But $p \mid k$, and hence $k = p$. Since $\gcd(k, n) = 1$, we have $p \nmid n$.

Note that since f is irreducible, we have $f = \min_{\mathbb{Q}} \zeta$. Let $\ell(x) = g(x^p) \in \mathbb{Z}[x]$. Then $\ell(\zeta) = g(\zeta^p) = g(\delta) = 0$. So $f \mid \ell$ in $\mathbb{Q}[x]$. So $\ell(x) = f(x)h(x)$ in $\mathbb{Q}[x]$. But f monic implies, by Lemma 27.24, that $h \in \mathbb{Z}[x]$. Thus

$$g(x^p) = \ell(x) = f(x)h(x) \in \mathbb{Z}[x].$$

Let $\mathbb{F}_p = (\mathbb{Z}/p\mathbb{Z}, +, \cdot)$, and let $\pi : \mathbb{Z} \to \mathbb{F}_p$ be the canonical ring homomorphism. Extend this homomorphism to $\pi : \mathbb{Z}[x] \to \mathbb{F}_p[x]$. We have

$$\pi(\ell) = \pi(f)\pi(h).$$

Applying π to a polynomial is the same as just considering the polynomial as a polynomial in $\mathbb{F}_p[x]$ and hence reducing its coefficients mod p. Now what is $\pi(g(x)^p)$? In $\mathbb{F}_p[x]$, $(a_0 + a_1 x + \cdots + a_n x^n)^p = a_0^p + a_1^p x^p + \cdots + a_n^p x^{np}$, and the latter is just $a_0 + a_1 x^p + \cdots + a_n x^{np}$, since $a^p = a \mod p$ by Fermat's Little Theorem (Corollary 5.33). Hence, $\pi(g(x)^p) = \pi(g(x^p))$. So, since π is a homomorphism,

$$\pi(g(x))^p = \pi(g(x)^p) = \pi(g(x^p)) = \pi(\ell) = \pi(f)\pi(h).$$

It follows that $\pi(f)$ divides $\pi(g(x))^p$ in $\mathbb{F}_p[x]$ which is a UFD. Hence, $\pi(f)$ and $\pi(g(x))$ have a common irreducible factor $d(x)$.

But $fg = \Phi_n(x) \mid x^n - 1$ in $\mathbb{Z}[x]$, and hence $\pi(f)\pi(g) \mid x^n - 1$ in $\mathbb{F}_p[x]$. This means that, in $\mathbb{F}_p[x]$, $d^2(x)$ divides $x^n - 1$. We conclude that $x^n - 1$ does not have distinct roots in $\mathbb{F}_p[x]$. But, since $p \nmid n$, by Lemma 27.14, the roots of $x^n - 1$ *are* distinct. The contradiction proves that $\Phi_n(x)$ is irreducible in $\mathbb{Q}[x]$. $\square$

Corollary 27.28. *Let n be a positive integer, and let ζ be a primitive nth root of unity in $\mathbb{C}$. Then $\mathbb{Q}(\zeta)$ is the splitting field for both $x^n - 1$ and $\Phi_n(x)$, and*

$$|\mathbb{Q}(\zeta) : \mathbb{Q}| = \phi(n).$$

Proof. All the roots of $\Phi_n(x)$ and $x^n - 1$ are powers of ζ and so $\mathbb{Q}(\zeta)$ is the splitting field of both polynomials. It follows from Theorem 27.27 that $\Phi_n(x)$ is the minimal polynomial of ζ over $\mathbb{Q}$, and hence $|\mathbb{Q}(\zeta) : \mathbb{Q}| = \deg(\Phi_n(x)) = \phi(n)$. $\square$

Let ζ be a primitive nth root of unity. Then we know that the nth cyclotomic field $\mathbb{Q}(\zeta)$ is the splitting field over $\mathbb{Q}$ of $x^n - 1$ or $\Phi_n(x)$. Hence $\mathbb{Q}(\zeta)$ is Galois over $\mathbb{Q}$. What is the Galois group? We basically know the result already.

Theorem 27.29. *Let n be a positive integer, and let ζ be a primitive nth root of unity in $\mathbb{C}$. Then*

$$\mathrm{Gal}(\mathbb{Q}(\zeta)/\mathbb{Q}) \cong (\mathbb{Z}/n\mathbb{Z})^{\times}.$$

In particular, this group is abelian.

Proof. By Proposition 27.16, we know that $\mathbb{Q}(\zeta)$ is a Galois extension of $\mathbb{Q}$ and $\mathrm{Gal}(\mathbb{Q}(\zeta)/\mathbb{Q})$ is isomorphic to a subgroup of $(\mathbb{Z}/n\mathbb{Z})^\times$, a group of order $\phi(n)$. By Corollary 27.28, we know that $|\mathrm{Gal}(\mathbb{Q}(\zeta)/\mathbb{Q})| = |\mathbb{Q}(\zeta) : \mathbb{Q}| = \phi(n)$. We conclude that $\mathrm{Gal}(\mathbb{Q}(\zeta)/\mathbb{Q}) \cong (\mathbb{Z}/n\mathbb{Z})^\times$. $\qquad\square$

Problems

27.2.1. For a field F, let A be the set of roots of $x^3 - 1$ in F. Find a familiar group that is isomorphic to A if F is the field $\mathbb{F}_3$. Do the same if F is $\mathbb{F}_5$, $\mathbb{F}_9$, $\mathbb{Q}$, $\mathbb{Q}(i\sqrt{3})$, or $\mathbb{C}$.

27.2.2. For a field F, let A be the set of roots of $x^{48} - 1$ in F. Find a familiar group that is isomorphic to A if F is the field $\mathbb{F}_7$. Do the same if F is $\mathbb{F}_3$, $\mathbb{F}_{49}$, or $\mathbb{C}$.

27.2.3. Find a field E with $\mathbb{Q} \subseteq E$ and with $\mathrm{Gal}(E/\mathbb{Q}) \cong \mathbb{Z}/46\mathbb{Z}$.

27.2.4. Find a field E with $\mathbb{Q} \subseteq E$ and $\mathrm{Gal}(E/\mathbb{Q}) \cong \mathbb{Z}/4\mathbb{Z} \times \mathbb{Z}/2\mathbb{Z}$.

27.2.5. In Problem 24.3.9, you found the Galois group of $f = x^4 + x^2 + 1$ over $\mathbb{Q}$. Redo this problem, and find a familiar group isomorphic to $\mathrm{Gal}(f)$ without any calculations other than noticing that $(x^2 - 1)(x^4 + x^2 + 1) = x^6 - 1$.

27.2.6. Let p be a prime, and let ζ be a primitive pth root of unity in $\mathbb{C}$. Both $(\mathbb{C}, +)$ and $(\mathbb{C}^\times, \cdot)$ are groups. What is the subgroup generated by ζ in each of them? Find a familiar group that is isomorphic to each of these subgroups and a familiar group that is isomorphic to $\mathrm{Gal}(\mathbb{Q}(\zeta)/\mathbb{Q})$. Are any of these three groups isomorphic?

27.2.7. Let ζ be a primitive 8th root of unity in $\mathbb{C}$, and let $G = \mathrm{Gal}(\mathbb{Q}(\zeta)/\mathbb{Q})$. Does there exists $\sigma \in G$ with $\sigma(\zeta) = \zeta^2$? What about with $\sigma(\zeta) = \zeta^3$?

27.2.8. Let ζ be a primitive 8th root of unity in $\mathbb{C}$. Find $\mathrm{Gal}(\mathbb{Q}(\zeta)/\mathbb{Q}(i))$.

27.2.9. Let $f = x^{13} - 1 \in \mathbb{Q}[x]$, and let E be the splitting field of f over $\mathbb{Q}$. Find $\mathrm{Gal}(E/\mathbb{Q})$ and all the intermediate fields K with $\mathbb{Q} \subseteq K \subseteq E$.

27.2.10. Let p be a prime and let $f = x^6 - 1 \in \mathbb{F}_p[x]$. Let E be a splitting field for f over $\mathbb{F}_p$.
  (a) If $p = 11$, then how many roots does f have in $\mathbb{F}_{11}$? What is $|E|$? What is $\mathrm{Gal}(E/\mathbb{F}_{11})$?
  (b) If $p = 13$, then how many roots does f have in $\mathbb{F}_{13}$? What is $|E|$? What is $\mathrm{Gal}(E/\mathbb{F}_{13})$?

27.2.11. Let $f = x^{13} - 1 \in \mathbb{F}_5[x]$. Let E be the splitting field of f over $\mathbb{F}_5$. What is $|E|$? What is $\mathrm{Gal}(E/\mathbb{F}_5)$? If you factor $x^{13} - 1$ in $\mathbb{F}_5[x]$, what are the degrees of the irreducible factors?

27.2.12. Let $f = x^{16} - 1 \in \mathbb{F}_3[x]$, and let E be the splitting field of f over $\mathbb{F}_3$.
  (a) What is $|E|$? How many roots does f have in E?
  (b) How many roots does f have in $\mathbb{F}_3$?
  (c) Consider the field extension $\mathbb{F}_3 \subseteq \mathbb{F}_9$. How many roots does f have in $\mathbb{F}_9$?

(d) Consider the field extension $\mathbb{F}_3 \subseteq \mathbb{F}_{27}$. How many roots does f have in $\mathbb{F}_{27}$?

(e) If $\alpha \in \mathbb{F}_9$ is a root of f and $\alpha \notin \mathbb{F}_3$, then what is the degree of $\min_{\mathbb{F}_3}(\alpha)$? Does this minimal polynomial divide f in $\mathbb{F}_3[x]$?

(f) We want to factor f in $\mathbb{F}_3[x]$. Without doing so, can you predict the degrees of the irreducible factors?

27.2.13. Let $f = x^{12} - 1 \in \mathbb{F}_3[x]$, and let E be the splitting field of f over $\mathbb{F}_3$.

(a) Does f have 12 distinct roots? Is f separable?

(b) Factor f into irreducible factors in $\mathbb{F}_3[x]$.

(c) What is $|E|$?

(d) Does E contain a primitive 12th root of unity?

(e) Is $E = \mathbb{F}_3[\zeta]$ where ζ is a primitive root of unity?

(f) What is $\mathrm{Gal}(E/\mathbb{F}_3)$?

27.2.14. Let $F \subseteq E$ be a field extension (with no assumption on the characteristic of F). Assume $E = F[\zeta]$ where ζ is a root of unity. Prove that E is a Galois extension of F and that $\mathrm{Gal}(E/F)$ is isomorphic to a subgroup of $(\mathbb{Z}/n\mathbb{Z})^\times$ for some positive integer n.

27.2.15. Let ζ be a primitive 11th root of unity in $\mathbb{C}$, and let $\alpha = \zeta + \zeta^{-1}$. Let f be the minimal polynomial of α over $\mathbb{Q}$, and let E be the splitting field of f over $\mathbb{Q}$. Finally, let $G = \mathrm{Gal}(\mathbb{Q}(\zeta)/\mathbb{Q})$.

(a) Find the elements of the set $\{\sigma(\alpha) \mid \sigma \in G\}$.

(b) Is $\mathbb{Q}(\alpha)$ a Galois extension of $\mathbb{Q}$? Why?

(c) Find a familiar group that is isomorphic to $\mathrm{Gal}(E/\mathbb{Q})$.

27.2.16. Let ζ be a primitive 11th root of unity in $\mathbb{C}$, and let $G = \mathrm{Gal}(\mathbb{Q}(\zeta)/\mathbb{Q})$.

(a) Is G cyclic? If so, explicitly give a generator for G. In other words, give a $\mathbb{Q}$-automorphism of $\mathbb{Q}(\zeta)$ that generates G.

(b) List all of the subgroups of G.

(c) For each subgroup $H \leq G$, give the corresponding fixed field, and exhibit the Galois correspondence.

(d) Is every intermediate field K—containing $\mathbb{Q}$ and contained in $\mathbb{Q}(\zeta)$—a simple extension of $\mathbb{Q}$? If so, for each intermediate field K, find $\alpha \in \mathbb{Q}(\zeta)$ with $K = \mathbb{Q}(\alpha)$.

(e) Which of the intermediate fields K are real fields (i.e., are contained in $\mathbb{R}$)? What is $\mathbb{Q}(\zeta) \cap \mathbb{R}$?

27.2.17. Let ζ be a primitive nth root of unity, and let $\alpha = \zeta + \zeta^{-1}$.

(a) Show that $\alpha \in \mathbb{R}$.

(b) Consider the extension $\mathbb{Q}(\alpha) \subseteq \mathbb{Q}(\zeta)$. Argue that complex conjugation is an element of $\mathrm{Gal}(\mathbb{Q}(\zeta)/\mathbb{Q}(\alpha))$.

(c) If $f \in \mathbb{Q}(\alpha)$ is an irreducible polynomial that has ζ as a root, then can you identify another root of f?

(d) Find the minimal polynomial of ζ over $\mathbb{Q}(\alpha)$.

(e) What is $|\mathbb{Q}(\zeta) : \mathbb{Q}(\alpha)|$?

(f) Show that $\mathbb{Q}(\alpha) = \mathbb{Q}(\zeta) \cap \mathbb{R}$. In other words, show that $\mathbb{Q}(\zeta + \zeta^{-1})$ is the largest real field contained in $\mathbb{Q}(\zeta)$.

(g) What are the elements of $\mathrm{Gal}(\mathbb{Q}(\zeta)/\mathbb{Q}(\alpha))$?

27.2.18. Let ζ_3 and ζ_9 be a primitive 3rd and 9th root of unity in $\mathbb{C}$, respectively. Then $\mathbb{Q}[\zeta_3] \subseteq \mathbb{Q}[\zeta_9]$. Is this a Galois extension? What is $\mathrm{Gal}(\mathbb{Q}[\zeta_9]/\mathbb{Q}[\zeta_3])$?

27.2.19. Let p be a prime number, and let n be a positive integer with $p \mid n$. Let ζ be a primitive nth root of unity in $\mathbb{C}$, and let $m = np$. Prove that $x^p - \zeta$ has p distinct roots in $\mathbb{C}$ and that each is a primitive mth root of unity. Conversely, show that if α is a primitive mth root of unity, then $\zeta = \alpha^p$ is a primitive nth root of unity.

27.2.20. Let p be a prime number, and let n be a positive integer with $p \mid n$. Prove

$$\Phi_{pn}(x) = \Phi_n(x^p).$$

27.2.21. Let p be a prime number.
  (a) Verify that $\Phi_1(x^p) = \Phi_p(x)\Phi_1(x)$.
  (b) Let n be a positive integer. Verify each of the following:

$$x^{pn} - 1 = \prod_{d \mid n} \Phi_d(x^p),$$

$$x^{pn} - 1 = \prod_{d \mid n} \Phi_d(x) \prod_{d \mid n} \Phi_{pd}(x) \quad \text{if } p \nmid n.$$

  (c) Let n be a positive integer not divisible by p. Prove

$$\Phi_{pn}(x) = \frac{\Phi_n(x^p)}{\Phi_n(x)}.$$

27.2.22. Use Problems 27.2.20 and 27.2.21 to find $\Phi_n(x)$ for $1 \leq n \leq 16$.

27.2.23. (a) Use Problem 27.2.21 to find $\Phi_{21}(x)$.
  (b) Use Problem 27.2.21 and a symbolic algebra software (such as Maple, Mathematica, or Sage) to find $\Phi_{105}(x)$. Anything interesting? (The necessary calculation for Φ_{105} can be carried out by hand but it is tedious.)

27.2.24. Let n be an odd integer greater than 1. How is $\Phi_{2n}(x)$ related to $\Phi_n(x)$? Prove your assertion.

27.2.25. Let $n = 2^3 \times 3^5 \times 5^2$, let ζ be a primitive nth root of unity, and let $G = \mathrm{Gal}(\mathbb{Q}(\zeta)/\mathbb{Q})$. Let $n_1 = 8$, $n_2 = 243$, and $n_3 = 25$, and, for $1 \leq i \leq 3$, let ζ_i be a primitive n_ith root of unity, and let $P_i = \mathrm{Gal}(\mathbb{Q}(\zeta_i)/\mathbb{Q})$.
  (a) Show that G is isomorphic to $P_1 \times P_2 \times P_3$.
  (b) Show that $\mathbb{Q}(\zeta_1) \cap \mathbb{Q}(\zeta_2) \cap \mathbb{Q}(\zeta_3) = \mathbb{Q}$.
  (c) Show that $\langle \mathbb{Q}(\zeta_1), \mathbb{Q}(\zeta_2), \mathbb{Q}(\zeta_3) \rangle = \mathbb{Q}(\zeta)$, where $\langle \mathbb{Q}(\zeta_1), \mathbb{Q}(\zeta_2), \mathbb{Q}(\zeta_3) \rangle$ is the compositum of $\mathbb{Q}(\zeta_1)$, $\mathbb{Q}(\zeta_2)$, and $\mathbb{Q}(\zeta_3)$. (See Definition 22.41.)
  (d) What is special about n? Generalize to other n.

27.3. The Polynomial $x^n - a$

In the previous section we considered the splitting field of $x^n - 1$ over a field F. In this section we turn to the polynomial $x^n - a$ where $a \in F$. Consider the already interesting case when $F = \mathbb{Q}$. The roots of $x^n - a$ in $\mathbb{C}$ are $\sqrt[n]{a}e^{2k\pi i/n}$ where

$0 \leq k < n$. Hence, a splitting field for $x^n - a$ will have to include both $\sqrt[n]{a}$ and $e^{2\pi i/n}$. We conclude that $E = \mathbb{Q}(\sqrt[n]{a}, e^{2\pi i/n})$ is the splitting field of $x^n - a$ over $\mathbb{Q}$. Hence, to get to E, we can first adjoin a primitive root of unity to $\mathbb{Q}$ to get a field K and then adjoin one root of $x^n - a$ to K. As a result, we want to study splitting fields of $x^n - a$ over fields that contain a primitive root of unity. Ultimately, in the next chapter, we want to turn to solvability of polynomials, and we will repeatedly consider the splitting fields of polynomials of the form $x^n - a$.

Remark 27.30. Let $n > 1$ be a positive integer, and let $a \in \mathbb{R}$. Even in subfields of $\mathbb{C}$, the notation $\sqrt[n]{a}$ can be ambiguous. If n is odd and $a \geq 0$, then there exists a unique real number α with $\alpha^n = a$. Hence, in this case, $\sqrt[n]{a}$ is defined, unambiguously, to be α. If n is even and $a \geq 0$, there are two real numbers whose nth power is a (if $a = 0$, then the two roots coincide). In this case also, $\sqrt[n]{a}$ is unambiguously defined. We define $\sqrt[n]{a}$ to be the unique *positive* real number α with $\alpha^n = a$. (The other real root of $x^n - a$, in this case, is $-\sqrt[n]{a}$.) If $a < 0$ and $n = 2$, then by $\sqrt{a}$ we mean $i\sqrt{|a|}$, and the other root of $x^2 = a$ is $-i\sqrt{|a|}$. For bigger values of n, we could do something similar and use one specific primitive root of unity to define $\sqrt[n]{a}$. However, this is not standard. For the case $a < 0$, $n > 2$, by $\sqrt[n]{a}$, we just mean *some* root, in $\mathbb{C}$, of $x^n - a$. While convenient as a shorthand, when stating formal theorems and definitions, we will usually refrain from using the notation $\sqrt[n]{a}$ if a can be negative. If we are in a field other than $\mathbb{C}$, then $\sqrt[n]{a}$ again just means *some* nth root of a in the field, if such a root exists.

Theorem 27.31. *Let n be a positive integer, and assume that the field K contains a primitive nth root of unity. Let $a \in K$, and let E be the splitting field for $x^n - a$ over K. Then*

(a) $\mathrm{char}(K) \nmid n$.

(b) $E = K(\alpha)$ *where α is one root of $x^n - a$.*

(c) E *is Galois over K.*

(d) $\mathrm{Gal}(E/K)$ *is cyclic and its order divides n.*

Proof. By Lemma 27.15, since K contains a primitive nth root of unity, $\mathrm{char}(K) \nmid n$. Let $\zeta \in K$ be a primitive root of unity. Then $\{\alpha\zeta^k \mid 0 \leq k < n\}$ is the set of n roots of $x^n - a$, and these roots all belong to $K(\alpha)$. We conclude that $x^n - a$ is a separable polynomial over K and that its splitting field over K is $E = K(\alpha)$. Being the splitting field of a separable polynomial, E is a Galois extension of K (Theorem 25.17). It remains to show that $\mathrm{Gal}(E/K)$ is cyclic and its order divides n.

Let $\sigma \in \mathrm{Gal}(E/K)$. Then σ is determined by its action on α. Consider $\frac{\sigma(\alpha)}{\alpha}$. (Both $\sigma(\alpha)$ and α are elements of $E^{\times}$, and hence the division is in the field E.) Since $\sigma(\alpha)$ is another root of $x^n - a$, we have $\sigma(\alpha) = \alpha\zeta^k$ for some $1 \leq k < n$. Hence,

$$\frac{\sigma(\alpha)}{\alpha} = \zeta^k.$$

Hence, we define a map

$$\Theta : \mathrm{Gal}(E/K) \to \langle \zeta \rangle,$$

by $\Theta(\sigma) = \sigma(\alpha)/\alpha$.

We show that Θ is a 1-1 group homomorphism which means that $\text{Gal}(E/K) \cong \Theta(\text{Gal}(E/K)) \leq \langle \zeta \rangle$. Hence, $\text{Gal}(E/K)$ is isomorphic to a subgroup of a cyclic group of order n. As a result $\text{Gal}(E/K)$ is cyclic itself and its order divides n.

To show that Θ is a homomorphism, let $\sigma, \tau \in \text{Gal}(E/K)$ with $\sigma(\alpha) = \alpha\zeta^i$ and $\tau(\alpha) = \alpha\zeta^j$. Note that both σ and τ fix K and hence fix roots of unity. We have

$$\Theta(\sigma\tau) = \frac{(\sigma\tau)(\alpha)}{\alpha} = \frac{\sigma(\alpha\zeta^j)}{\alpha} = \frac{\alpha\zeta^i\zeta^j}{\alpha} = \zeta^i\zeta^j$$

and

$$\Theta(\sigma)\Theta(\tau) = \frac{\sigma(\alpha)}{\alpha}\frac{\tau(\alpha)}{\alpha} = \zeta^i\zeta^j.$$

So, $\Theta(\sigma\tau) = \Theta(\sigma)\Theta(\tau)$ and Θ is a group homomorphism. If $\sigma \in \ker(\Theta)$, then $\sigma(\alpha)/\alpha = 1$, and so $\sigma(\alpha) = \alpha$. Hence σ fixes all of $E = F[\alpha]$ which means that $\sigma = \{e\}$. This means that Θ is 1-1, and the proof is complete. $\square$

Remark 27.32. Theorem 27.31 says that if the field K contains a primitive nth root of unity, and if E is the splitting field of $x^n - a$ over K, then $\text{Gal}(E/K)$ is cyclic and its order divides n. The converse of this is also true. To prove that a polynomial solvable by radicals has a solvable Galois group, we only need Theorem 27.31. After all, in such a situation we already know that solving the polynomial comes down to repeatedly finding the roots of polynomials of the form $x^n - a$, and we would like to control the structure of the Galois group. On the other hand, to prove that a solvable Galois group implies that the polynomial is solvable by radicals, we need the converse which we will prove as Theorem 28.24.

Corollary 27.33. *Let n be a positive integer, and let F be a field with $\text{char}(F) \nmid n$. Let $a \in F$, let $f = x^n - a \in F[x]$, let E be a splitting field for f over F, and let $G = \text{Gal}(E/F)$. Then E is Galois over F and there exists an $N \lhd G$ with*

(a) *N cyclic, and*

(b) *G/N abelian.*

Proof. If $n = 1$ or if $a = 0$, then $E = F$ and the result follows trivially. So assume $n > 1$ and $a \neq 0$. Since $a \neq 0$, 0 is not a root of f, and since $\text{char}(F) \nmid n$, the only root of $f' = nx^{n-1}$ is 0. Hence, f and its derivative have no roots in common, and so, by Corollary 25.29, $x^n - a$ has n distinct roots. Let

$$\alpha = \alpha_1, \alpha_2, \ldots, \alpha_n$$

be the roots of f in E. Then

$$\frac{\alpha_1}{\alpha}, \frac{\alpha_2}{\alpha}, \ldots, \frac{\alpha_n}{\alpha}$$

are n distinct nth roots of 1 in E. They are nth roots of 1 since $(\alpha_i/\alpha)^n = a/a = 1$, and they are distinct since $\alpha_i/\alpha = \alpha_j/\alpha$ implies that $\alpha_i = \alpha_j$. Thus E contains the full set of n nth roots of unity, and so it contains ζ a primitive nth root of unity.

Now, E is the splitting field of a separable polynomial over F and hence it is Galois over F. We have $F \subseteq K = F(\zeta) \subseteq E$. Let $N = \text{Gal}(E/K)$. By Theorem 27.31, E is Galois over K and N is cyclic. The field K is the splitting field of $x^n - 1$, a separable polynomial, over F, and so K is Galois over F. By the Galois correspondence, we have $N \lhd G$ and $G/N \cong \text{Gal}(K/F)$. Now, the latter is a

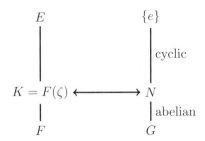

Figure 27.2. As long as $\mathrm{char}(F) \nmid n$, the Galois group of $x^n - a$ is cyclic by abelian.

cyclotomic extension and, hence, by Proposition 27.16, it is abelian. See Figure 27.2. $\qquad\qquad\qquad\qquad\qquad\qquad\qquad\qquad\qquad\qquad\qquad\qquad\qquad\qquad$ $\square$

Problems

27.3.1. Let $K = \mathbb{Q}(i)$, and let E be the splitting field of $x^4 - 4$ over K. Find a familiar group that is isomorphic to $\mathrm{Gal}(E/K)$.

27.3.2. Let E be the splitting field of $x^8 - 1$ over $\mathbb{Q}$, and let K be the splitting field of $x^4 - 4$ over $\mathbb{Q}$. How are E and K related? Is either contained in the other?

27.3.3. Let $f = x^8 - 3 \in \mathbb{Q}[x]$. Let E be the splitting field of f over $\mathbb{Q}$, and let $G = \mathrm{Gal}(E/\mathbb{Q})$. Find $|G|$. Does G have an element σ of order 8? If so, what is the fixed field of $\langle \sigma \rangle$?

27.3.4. Let $f = x^{11} - 2 \in \mathbb{Q}[x]$, and let E be the splitting field of f over $\mathbb{Q}$. Find $\mathrm{Gal}(E/\mathbb{Q})$ and all the intermediate fields K with $\mathbb{Q} \subseteq K \subseteq E$.

27.3.5. Let $f = (x^3 - 2)(x^3 - 7) \in \mathbb{Q}[x]$, let $\omega = e^{2\pi i/3}$ be a primitive third root of unity. Let E be the splitting field of f over $\mathbb{Q}(\epsilon)$. What is $|E \colon \mathbb{Q}(\omega)|$? What is $\mathrm{Gal}(E/\mathbb{Q}(\omega))$? Find a splitting field for f over $\mathbb{Q}$.

27.3.6. Let $f = x^3 + 1 \in \mathbb{F}_5[x]$. Let E be the splitting field of f over $\mathbb{F}_5$. What is $\mathrm{Gal}(E/\mathbb{F}_5)$?

27.3.7. Let $f = x^6 - 2 \in \mathbb{F}_{11}[x]$.
    (a) What is the (multiplicative) order of 2 in $\mathbb{F}_{11}^\times$?
    (b) If $\mathbb{F}_{11} \subseteq E$ is a field extension, and $\alpha \in E$ with $\alpha^6 = 2$, then what is $o(\alpha)$ in $E^\times$?
    (c) Does f have any roots in $\mathbb{F}_{11}$?
    (d) Consider the field extension $\mathbb{F}_{11} \subseteq \mathbb{F}_{11^2}$. Does f have any roots in $\mathbb{F}_{11^2}$?
    (e) If E is the splitting field of f over $\mathbb{F}_{11}$, then what is $|E|$?
    (f) What is $\mathrm{Gal}(E/\mathbb{F}_{11})$?

27.3.8. Let $E = \mathbb{Q}(i, \sqrt[8]{2})$, $F = \mathbb{Q}(i)$, $K = \mathbb{Q}(\sqrt{2})$, and $L = \mathbb{Q}(i\sqrt{2})$.

(a) Is E the splitting field of a polynomial over $\mathbb{Q}$? Is E a Galois extension of $\mathbb{Q}$?

(b) What is $|E:\mathbb{Q}|$?

(c) Find a familiar group isomorphic to $\mathrm{Gal}(E/F)$.

(d) Find a familiar group isomorphic to $\mathrm{Gal}(E/K)$.

(e) Find a familiar group isomorphic to $\mathrm{Gal}(E/L)$.

27.3.9. Let p be a positive prime, let F be a field, and let $f = x^p - b \in F[x]$.

(a) Assume $\mathrm{char}(F) = p$. Show that f is reducible in $F[x]$ if and only if F is the splitting field for f over F.

(b) Assume $\mathrm{char}(F) = 0$. Show that f is reducible in $F[x]$ if and only if the splitting field for f over F is $F(\epsilon)$, where ϵ is a primitive pth root of unity.

27.3.10. Let p be a prime, let $f = x^p - 2 \in \mathbb{Q}[x]$, and let E be the splitting field of f over $\mathbb{Q}$. Let ζ be a primitive pth root of unity, and let $\alpha = \sqrt[p]{2}$ be the real pth root of 2. Finally, let $G = \mathrm{Gal}(E/\mathbb{Q})$.

(a) What is $|G|$?

(b) Let a and b be positive integers with $1 \le a \le p-1$ and $0 \le b \le p-1$. Does there exists $\sigma \in G$ with $\sigma(\zeta) = \zeta^a$ and $\sigma(\alpha) = \alpha\zeta^b$?

(c) Define $H = \left\{ \begin{bmatrix} a & b \\ 0 & 1 \end{bmatrix} \mid a \in \mathbb{F}_p^\times, b \in \mathbb{F}_p \right\}$. Show that $H \le \mathrm{GL}(2,p)$.

(d) Show that $G \cong H$. Make sure that you give the group isomorphism and that you prove that it is an isomorphism.

(e) What is the order of the element $\begin{bmatrix} 1 & 1 \\ 0 & 1 \end{bmatrix} \in H$? For p odd, what is the order of the element $\begin{bmatrix} 2 & 0 \\ 0 & 1 \end{bmatrix} \in H$?

(f) For p odd, is G abelian?

27.4. More Problems and Projects

Problems

27.4.1. **Constructibility of a regular n-gon.** Let $\alpha = a + bi \in \mathbb{C}$. We say that α is *constructible* (by a straightedge and compass) if and only if a and b are constructible real numbers (see Definition 23.2). Let $n \geq 3$, and let ϵ_n be a primitive nth root of unity.

  (a) Show that it is possible to construct a regular n-gon with a straightedge and compass if and only if ϵ_n is a constructible complex number.

  (b) Show that it is possible to construct a regular n-gon with a straightedge and compass if and only if $\phi(n)$ is a power of 2. (See Definition 1.45, for the definition of the Euler ϕ-function.)

  (c) Is it possible to construct a regular 9-gon? What about a regular 17-gon?

Chinese Remainder Theorem. One modern form of the Chinese Remainder Theorem, whose roots go to the Chinese classic mathematical manual *Sunzi suanjing* written most likely in the fifth century CE, is the following:

Theorem 27.34 (Chinese Remainder Theorem). *Let n be a positive integer, let $p_1, \ldots, p_m$ be distinct primes, and assume that $n = p_1^{k_1} p_2^{k_2} \cdots p_m^{k_m}$ is a unique factorization of n into a product of prime powers. Then*

$$(\mathbb{Z}/n\mathbb{Z})^\times \cong (\mathbb{Z}/p_1^{k_1}\mathbb{Z})^\times \times \cdots \times (\mathbb{Z}/p_m^{k_m}\mathbb{Z})^\times.$$

One can prove this theorem directly and with no recourse to field theory (see Problem 16.1.25). In the problems, however, you are asked to deduce it as a consequence of results on cyclotomic extensions.

27.4.2. Let m and n be relatively prime positive integers, and let ζ_m and ζ_n be a primitive mth and nth root of unity, respectively.

  (a) Let $\zeta_{mn} = \zeta_m \zeta_n$. Show that ζ_{mn} is a primitive mnth root of unity.

  (b) Recall that if K and L are fields, then $\langle K, L \rangle$ denotes the compositum of K and L (see Definition 22.41). Show that $\mathbb{Q}(\zeta_{mn}) \subseteq \langle \mathbb{Q}(\zeta_m), \mathbb{Q}(\zeta_n) \rangle$.

  (c) Show that $\langle \mathbb{Q}(\zeta_m), \mathbb{Q}(\zeta_n) \rangle \subseteq \mathbb{Q}(\zeta_{mn})$. Conclude that
  $$\mathbb{Q}(\zeta_{mn}) = \langle \mathbb{Q}(\zeta_m), \mathbb{Q}(\zeta_n) \rangle.$$

27.4.3. We continue with the notation and assumptions of Problem 27.4.2. In particular, $\gcd(n, m) = 1$.

  (a) Without making any assumptions about $\mathbb{Q}(\zeta_m) \cap \mathbb{Q}(\zeta_n)$, draw a (provisional) partial lattice diagram of fields that includes $\mathbb{Q}(\zeta_m)$, $\mathbb{Q}(\zeta_n)$, $\mathbb{Q}(\zeta_{mn})$, $\mathbb{Q}(\zeta_m) \cap \mathbb{Q}(\zeta_n)$, and $\mathbb{Q}$.

   (b) Using your diagram, show that

$$|\mathbb{Q}(\zeta_{mn}) : \mathbb{Q}| \leq |\mathbb{Q}(\zeta_n) : \mathbb{Q}(\zeta_m) \cap \mathbb{Q}(\zeta_n)| \; |\mathbb{Q}(\zeta_m) : \mathbb{Q}| .$$

   (c) Prove that the above inequality is indeed an equality. Conclude that $\mathbb{Q}(\zeta_m) \cap \mathbb{Q}(\zeta_n) = \mathbb{Q}$.

   (d) Modify your partial lattice diagram appropriately.

27.4.4. We continue with the notation and results of Problems 27.4.2 and 27.4.3. Let $G = \mathrm{Gal}(\mathbb{Q}(\zeta_{mn})/\mathbb{Q})$, $N = \mathrm{Gal}(\mathbb{Q}(\zeta_m)/\mathbb{Q})$, and $M = \mathrm{Gal}(\mathbb{Q}(\zeta_n)/\mathbb{Q})$.

   (a) Using Galois's correspondence, draw the partial lattice diagram of groups that corresponds to the partial lattice diagram of fields of Problem 27.4.3.

   (b) Use the correspondence to argue that G has normal subgroups isomorphic with N and M.

   (c) Show that $G \cong N \times M$.

   (d) Prove the Chinese Remainder Theorem 27.34.

Algebraic Closure of $\mathbb{F}_q$. In Theorems 24.34 and 24.35, we showed that every field F has a *unique algebraic closure* $\overline{F}$. The field $\overline{F}$ is an algebraic extension of F and is *algebraically closed*. The latter means that every polynomial in $\overline{F}[x]$—including polynomials in $F[x]$—split over $\overline{F}$ (Definition 24.27). In Theorem 26.11, we proved that $\mathbb{C}$ is algebraically closed, which means that $\overline{R} = \mathbb{C}$. We also know (Corollary 24.33) that $\overline{Q} = \mathbb{A}$, the field of algebraic numbers over $\mathbb{Q}$. In the problems here you are asked to find the algebraic closure of finite fields.

27.4.5. Let $\mathbb{F}_3$ be the field of order 3. Consider the infinite sequence of field extensions:

$$\mathbb{F}_3 \subseteq \mathbb{F}_{3^2} \subseteq \mathbb{F}_{3^6} \subseteq \cdots \subseteq \mathbb{F}_{3^{n!}} \subseteq \cdots .$$

Let $\mathrm{GF}(3^\infty) = \bigcup_{n=1}^\infty \mathbb{F}_{3^{n!}}$.

   (a) Is $\mathrm{GF}(3^\infty)$ a field? What is the characteristic of $\mathrm{GF}(3^\infty)$?

   (b) Is $\mathrm{GF}(3^\infty)$ an algebraic extension of $\mathbb{F}_3$?

   (c) What is $\overline{\mathbb{F}_3}$, the algebraic closure of $\mathbb{F}_3$?

   (d) What is $|\overline{\mathbb{F}_3} : \mathbb{F}_3|$?

27.4.6. Let $\mathrm{GF}(3^\infty)$ be defined as in Problem 27.4.5.

   (a) Is $\mathrm{GF}(3^\infty)$ a perfect field? (See Definition 25.38.)

   (b) Does $\mathrm{GF}(3^\infty)$ have a subfield isomorphic to $\mathbb{F}_{27}$?

   (c) What is $\overline{\mathbb{F}_{27}}$, the algebraic closure of $\mathbb{F}_{27}$?

Wedderburn's Theorem on Finite Division Rings. Early in our study of ring theory, we stated, without proof, Wedderburn's Theorem 15.25, that every finite division ring is a field. We are now ready to prove this theorem, and you will be guided to do so in the Problems.

27.4.7. Let $n > 1$ be an integer, let $\zeta \neq 1$ be an nth root of unity, and let $r \geq 2$ a real number. Recall that the absolute value $|a + bi|$ of a complex number is $\sqrt{a^2 + b^2}$, its Euclidean distance from the origin. The nth cyclotomic polynomial is denoted by $\Phi_n(x)$, as usual.
   (a) Show that $|r - \zeta| > r - 1$.
   (b) Show that $|\Phi_n(r)| > r - 1$.

27.4.8. Let $m < n$ be positive integers, let q be a power of a prime, and assume $\frac{q^n - 1}{q^m - 1}$ is an integer. Show that $q^n - 1$ and $\dfrac{q^n - 1}{q^m - 1}$ are both multiples of $\Phi_n(q)$.

27.4.9. Let D be a finite division ring, and let $\mathbf{Z}(D)$ be the center of D. Let $a \in D$ and $\mathbf{C}_D(a)$ be the centralizer of a in D. Recall from Problem 16.1.27 that $\mathbf{Z}(D)$ is a field, and, as a result, $\mathbf{Z}(D) \cong \mathbb{F}_q$ where q is a power of a prime. Identify $\mathbf{Z}(D)$ with $\mathbb{F}_q$.
   (a) Show that $D^{\times}$ is a finite group.
   (b) Show that $\mathbf{C}_D(a)$ is a vector space over $\mathbb{F}_q$. Denote $\dim_{\mathbb{F}_q}(\mathbf{C}_D(a))$ by $\dim(a)$.
   (c) Show that $|\mathbf{C}_D(a)| = q^{\dim(a)}$. Let $n = \dim(1)$, and show $|D| = q^n$.
   (d) Let S be a set of representatives for the conjugacy classes with more than one element of the group $D^{\times}$. Using the class equation of Corollary 6.17 for the group $D^{\times}$, show

$$q^n - 1 = (q - 1) + \sum_{a \in S} \frac{q^n - 1}{q^{\dim(a)} - 1}.$$

27.4.10. **Proof of Wedderburn's Theorem 15.25.** Let D be a finite division ring, and let $\mathbf{Z}(D)$ be the center of D. By Problem 27.4.9, D is a vector space over the finite field $\mathbf{Z}(D)$. Let $q = |\mathbf{Z}(D)|$, and let n be the dimension of D over $\mathbf{Z}(D)$.
   (a) Use Problems 27.4.8 and 27.4.9 to show that $\Phi_n(q)$ divides $q - 1$.
   (b) Use Problem 27.4.7 to show that $n = 1$ and $D = \mathbf{Z}(D)$.
   (c) Rewrite your work on Problems 27.4.7–27.4.10 to write a coherent and concise proof of Wedderburn's Theorem 15.25.

Chapter 28

Radical Extensions, Solvable Groups, and the Quintic

... where we prove Galois's theorem that a polynomial has roots that can be expressed using radicals and the four arithmetical operations if and only if it has a solvable Galois group, and where we use this theorem to exhibit a specific quintic not solvable by radicals.

28.1. Solvability by Radicals

We now turn our attention to the classical problem of solvability of a general polynomial of degree n (over $\mathbb{C}$). There are formulas for solving second, third, and fourth degree equations, but what about higher degree polynomials? We can ask two questions:

Question 28.1. Is there a formula for finding the roots of every polynomial of degree n for specific $n \geq 5$?

Question 28.2. Regardless of the existence of a general formula, given a specific polynomial and a root of the polynomial, can we write down a reasonable expression for the root?

What would a reasonable expression be? It is not unreasonable to demand the ability to write down the roots of a polynomial using the arithmetical operations $(+, -, \times, \div)$ and the taking of roots. Hence, we would accept an expression of the form

$$\xi = \sqrt[3]{11}\sqrt[5]{13 + \sqrt[3]{17}} + \sqrt[3]{\frac{\sqrt{47 - \sqrt[3]{17} - \sqrt[5]{11}}}{\sqrt{3 + i\sqrt[11]{31}}}}.$$

In writing down an expression for a root, we are *not* allowing the use of functions such as ln or cos, and we are not allowing the use of limits. This should make intuitive sense. Consider a polynomial such as $x^5 - 10x + 5$. The root of this polynomial is a complex number: after raising it to the fifth power and subtracting ten times the number, we get -5. It is not unreasonable to expect that such a number be expressed without resort to limits or functions other than $\sqrt[n]{x}$.

As complicated as the expression for ξ looks, we can deconstruct it using field extensions. Which reasonable looking subfield of $\mathbb{C}$ contains ξ? We can construct such a field E by starting with $\mathbb{Q}$ and one by one adjoining new elements. To be precise, let $\alpha, \beta, \gamma, \ldots, \zeta$ be real numbers such that $\alpha^3 = 11$, $\beta^3 = 17$, $\gamma^5 = 13 + \beta$, $\delta^5 = 11$, $\epsilon^2 = 47 - \alpha - \delta$, and $\zeta^{11} = 31$, and let η and θ be complex numbers with $\eta^2 = 3 + i\zeta$ and $\theta^3 = \epsilon/\eta$. Now, if we define

$$E = \mathbb{Q}(\alpha, \beta, \gamma, \delta, \epsilon, \zeta, i, \eta, \theta),$$

then we have $\xi \in E$ and

$$\begin{aligned}
\mathbb{Q} &\subseteq \mathbb{Q}(\alpha) \subseteq \mathbb{Q}(\alpha, \beta) \subseteq \mathbb{Q}(\alpha, \beta, \gamma) \subseteq \mathbb{Q}(\alpha, \beta, \gamma, \delta) \subseteq \mathbb{Q}(\alpha, \beta, \gamma, \delta, \epsilon) \\
&\subseteq \mathbb{Q}(\alpha, \beta, \gamma, \delta, \epsilon, \zeta) \subseteq \mathbb{Q}(\alpha, \beta, \gamma, \delta, \epsilon, \zeta, i) \subseteq \mathbb{Q}(\alpha, \beta, \gamma, \delta, \epsilon, \zeta, i, \eta) \\
&\subseteq \mathbb{Q}(\alpha, \beta, \gamma, \delta, \epsilon, \zeta, i, \eta, \theta) = E.
\end{aligned}$$

Note that each of the extensions is a simple extension of the previous field, and some power of the element adjoined belongs to the previous field. In other words, $\alpha^3 \in \mathbb{Q}$, $\beta^3 \in \mathbb{Q}(\alpha)$, $\gamma^5 \in \mathbb{Q}(\alpha, \beta)$, $\ldots$, $\theta^3 \in \mathbb{Q}(\alpha, \beta, \gamma, \delta, \epsilon, \zeta, i, \eta)$. This latter condition—a power of the element adjoined be in the previous field—ensures that all of the elements in the extension can be written using the four arithmetical operations and the taking of roots. The example provides a template for the definition of a repeated radical extension.

Definition 28.3 (Repeated radical extension). Let $F \subseteq E$ be a field extension. Then E is a *repeated radical extension* (r.r.e., hereafter) of F if there exist fields

$$F = F_0 \subseteq F_1 \subseteq F_2 \subseteq \cdots \subseteq F_m = E,$$

where, for $1 \leq i \leq m$, $F_i = F_{i-1}[\alpha_i]$ with $\alpha_i^{n_i} \in F_{i-1}$ for some $n_i \in \mathbb{Z}^{>0}$.

Remark 28.4. We sometimes refer to

$$F = F_0 \subseteq F_1 \subseteq F_2 \subseteq \cdots \subseteq F_m = E,$$

where, for $1 \leq i \leq m$, $F_i = F_{i-1}[\alpha_i]$ with $\alpha_i^{n_i} \in F_{i-1}$ for some $n_i \in \mathbb{Z}^{>0}$, as a *radical tower*.

Also, note that if E is an r.r.e. of F, then $|E : F| < \infty$.

As in our example, a root that lives in a repeated radical extension of our base field F, in principle, can be written using the four arithmetical operations together with the taking of roots.

Definition 28.5 (Solvable by radicals). Let F be a field, and let $f \in F[x]$. Then f is *solvable by radicals* (over F) if there exists an r.r.e. of F over which f splits.

In other words, $f \in F[x]$ is solvable by radicals if there exist fields E and L with $F \subseteq E \subseteq L$ such that E is a splitting field of f and L is an r.r.e. of F.

Note that for $f \in F[x]$ to be solvable by radicals, the splitting field of F does not have to be an r.r.e. of F. It just has to be contained in such a field. There certainly are polynomials—for example, $x^3 - 4x + 2$—that are solvable by radicals but whose splitting field itself is not an r.r.e. However, it takes some non-trivial work to prove the details of such examples.[1]

We shall see that not all polynomials are solvable by radicals, and this means that the answer to both introductory questions is no. Not only there is not a formula for higher degree polynomials, there are polynomials for which we cannot even write down the roots in a reasonable fashion. Galois was able to give a characterization of those polynomials that are solvable by radicals.

Theorem 28.6 (Galois). *Let F be a field with* $\mathrm{char}(F) = 0$, *and let $f \in F[x]$. Let E be a splitting field for f over F. Then f is solvable by radicals if and only if* $\mathrm{Gal}(E/F)$ *is a solvable group.*

To show that there are polynomials that are *not* solvable by radicals, we need to prove the only if direction. For that, we actually do not need the characteristic zero assumption—or even that E is a splitting field. We will prove the more general fact that if $F \subseteq E \subseteq L$ are fields and L is an r.r.e. of F, then $\mathrm{Gal}(E/F)$ is solvable. The other direction—that $\mathrm{Gal}(E/F)$ implies that f is solvable by radicals—is not true without an assumption on $\mathrm{char}(F)$ (see Problem 28.3.13). We will prove that if $F \subseteq E$ is a finite-dimensional Galois extension, $\mathrm{char}(F)$ does not divide $|E : F|$, and $\mathrm{Gal}(E/F)$ solvable, then there exists L an r.r.e. of F with $F \subseteq E \subseteq L$. Galois's Theorem 28.6 then follows. The first-time reader can—and maybe should—assume $\mathrm{char}(F) = 0$—or even $F = \mathbb{Q}$—throughout. Even if you do not restrict yourself to the characteristic 0 case, it is a good exercise to figure out which proofs can be streamlined with this extra assumption. One advantage, of course, is that in characteristic 0, all finite degree extensions are separable, and as a result are both simple and can be extended to a Galois extension (see Remark 25.49).

For the rest of this section, we collect a few elementary facts about repeated radical extensions and prove one important—but admittedly technical—fact. If the field K is an r.r.e. of the field F, then the normal closure of K over F is also an r.r.e. of F.

Lemma 28.7. *Assume that $F \subseteq K$ are fields with $|K : F| < \infty$, and assume L is a field containing F that is F-isomorphic to K. If K is an r.r.e. of F, then so is L.*

Proof. Let $\sigma \colon K \to L$ be an F-isomorphism. Since K is an r.r.e. of F, we have $F = F_0 \subseteq F_1 \subseteq \cdots \subseteq F_m = K$ where, for $1 \leq i \leq m$, $F_i = F_{i-1}[\alpha_i]$ with $\alpha_i^{n_i} \in F_{i-1}$ for some $n_i \in \mathbb{Z}^{>0}$. Now we have $F = F_0 \subseteq \sigma(F_1) \subseteq \cdots \subseteq \sigma(F_m) = \sigma(K)$, and, for $1 \leq i \leq m$, $\sigma(F_i) = \sigma(F_{i-1}[\alpha_i]) = \sigma(F_{i-1})[\sigma(\alpha_i)]$. Furthermore, $(\sigma(\alpha_i))^{n_i} = \sigma(\alpha_i^{n_i}) \in \sigma(F_{i-1})$. Hence, $\sigma(K)$ is also an r.r.e. of F. $\square$

Recall that $\langle K, L \rangle$ denotes the compositum of K and L (Definition 22.41).

Lemma 28.8. *Let K and L be two intermediate fields that contain the field F and are contained in the field E.*

---

[1] See Isaacs [**Isa94**, pp. 350–354] or Isaacs and Moulton [**IM98**] for the details and much more.

(a)   *If K is an r.r.e. of F, then $\langle K, L \rangle$ is an r.r.e. of L.*

(b)   *If both K and L are r.r.e.s of F, then $\langle K, L \rangle$ is an r.r.e. of F.*

Proof. We have $F = F_0 \subseteq F_1 \subseteq \cdots \subseteq F_m = K$ where, for $1 \leq i \leq m$, $F_i = F_{i-1}[\alpha_i]$ with $\alpha_i^{n_i} \in F_{i-1}$ for some $n_i \in \mathbb{Z}^{>0}$. Now consider the following tower of fields starting with L: $L = L_0 \subseteq L_1 = L[\alpha_1] \subseteq L_2 = L_1[\alpha_2] \subseteq \cdots \subseteq L_m = L_{m-1}[\alpha_m]$. Note that $\alpha_1^{n_1} \in F \subseteq L$. Likewise $\alpha_2^{n_2} \in F_1 = F[\alpha_1] \subseteq L[\alpha_1] = L_1$. Continuing (by induction, for example), we get that, for $1 \leq i \leq m$, $\alpha_i^{n_i} \in F_{i-1} \subseteq L_{i-1}$. Finally, L_m is the smallest field that contains F, L, and $\alpha_1, \ldots, \alpha_m$. Hence, L_m is the smallest field containing L and K. We conclude that we have a radical tower starting from L leading to $\langle K, L \rangle$. Appending this tower to a radical tower from F to L—assuming one exists—we get a radical tower from F to $\langle K, L \rangle$. $\square$

Corollary 28.9. *Let $F \subseteq K \subseteq E$ be fields. Assume that E is an r.r.e. of F, then E is an r.r.e. of K.*

Proof. Since $E = \langle K, E \rangle$, the conclusion follows from Lemma 28.8(a). $\square$

We need the following characterization of normal closures.

Proposition 28.10. *Let $F \subseteq K$ be a finite degree field extension. Then the normal closure of K over F is the compositum of a finite number of fields, each isomorphic to K.*

Proof. Let $K = F[\alpha_1, \ldots, \alpha_n]$, let $f = \prod_{i=1}^{n} \min_F(\alpha_i)$, and let E be the splitting field of f over K. Then E is the normal closure of K over F (Proposition 25.23). Let $\beta \in E$ be a root of $\min_F(\alpha_i)$ for some $1 \leq i \leq n$. Then $F(\alpha_i)$ is F-isomorphic to $F(\beta) \subseteq E$ (Theorem 24.6). If $\theta_\beta \colon F(\alpha_i) \to F(\beta)$ is such an F-isomorphism, then, since E is a normal extension of F, this isomorphism can be extended to an F-isomorphism ϕ_β from E to E (Theorem 24.24). As a result $F(\beta) \subseteq \phi_\beta(K) \cong K$. Let Ω be the set of all the roots of f, then L, the compositum of the fields $\{\phi_\beta(K) \mid \beta \in \Omega\}$, is a subfield of E that contains all the roots of f, and so f splits over L. But E is the splitting field of f. As a result, $L = E$ and E is the compositum of fields isomorphic to K. $\square$

Corollary 28.11. *Let $F \subseteq K$ be a finite degree field extension. If K is an r.r.e. of F, then so is the normal closure of K over F.*

Proof. Let E be the normal closure of K over F. If K is an r.r.e. of F, then, by Lemma 28.7, so is every field that is F-isomorphic to K. By Lemma 28.10, E is the compositum of a finite number of fields that are F-isomorphic to K. Hence, E is the compositum of a finite number of fields that are r.r.e.s of F. By Lemma 28.8, this compositum is also an r.r.e. of F. $\square$

Problems

28.1.1. Let $F \subseteq E$ be a field extension with $|E : F| = 2$. Is E necessarily a repeated radical extension of F?

28.1.2. Let $f(x) = ax^2 + bx + c$ be a polynomial of degree 2 in $\mathbb{Q}[x]$. Using Definition 28.5, show that f is solvable by radicals.

28.1.3. Let $f = x^2 + 1 \in \mathbb{F}_2[x]$. What is a splitting field for f over $\mathbb{F}_2$? What are the roots of f in the splitting field? Can we use the usual quadratic formula to find the roots of f over $\mathbb{F}_2$?

28.1.4. Let $\alpha = \sqrt{2} + i\sqrt{3} \in \mathbb{C}$. Find a field $E \subseteq \mathbb{C}$ such that $\alpha \in E$ and E is an r.r.e. of $\mathbb{Q}$. Is $\mathbb{Q}(\alpha)$ an r.r.e. of $\mathbb{Q}$?

28.1.5. Find an r.r.e. of $\mathbb{Q}$ that contains $\sqrt[4]{3 + 2\sqrt{5}}$. What is the degree of the extension?

28.1.6. Let $E = \mathbb{Q}(\sqrt{2}i, \sqrt[4]{2}(1 - i))$.
(a) Is E an r.r.e. of $\mathbb{Q}$?
(b) Find $|E : \mathbb{Q}|$.
(c) Is E normal over $\mathbb{Q}$?
(d) Find the normal closure of E over $\mathbb{Q}$.

28.1.7. Let $L = \mathbb{Q}(\sqrt[3]{3}, \sqrt[3]{5})$.
(a) Is L a repeated radical extension of $\mathbb{Q}$?
(b) What is the normal closure of L over $\mathbb{Q}$? Call it $\widetilde{L}$.
(c) Is $\widetilde{L}$ a repeated radical extension of $\mathbb{Q}$? If so, explicitly produce a radical tower of fields from $\mathbb{Q}$ to $\widetilde{L}$.
(d) Is $\widetilde{L}$ a repeated radical extension of L?
(e) Is $\widetilde{L}$ a Galois extension of $\mathbb{Q}$? Is L a Galois extension of $\mathbb{Q}$? Is $\widetilde{L}$ a Galois extension of L?

28.1.8. Let $L = \mathbb{Q}(\sqrt[6]{3}, \zeta)$, where ζ is a primitive 8th root of unity. Can you find a radical tower

$$\mathbb{Q} = F_0 \subseteq F_1 \subseteq F_2 \subseteq \cdots \subseteq F_m = L$$

such that, for $1 \leq i \leq m$, $F_i = F_{i-1}[\alpha_i]$ with $\alpha_i^{p_i} \in F_{i-1}$ for some positive *prime* number p_i?

28.1.9. Assume that you have already proved the following:
Let $F \subseteq E \subseteq L$ be fields with $\mathrm{char}(F) = 0$, E a Galois extension of F, and L an r.r.e. of F. Then $\mathrm{Gal}(E/F)$ is solvable.
Show that the same statement is true without the assumption that E is a Galois extension of F.

28.1.10. Let F be a field, and let $f \in F[x]$ be an irreducible polynomial. Assume f has one root expressible by radicals (i.e., this root is in an r.r.e. of F). Show that f is solvable by radicals.

28.1.11. Let $\alpha = \sqrt[4]{\sqrt[3]{5} + \sqrt{13} + \sqrt[5]{47}}$, and let $f = \min_{\mathbb{Q}}(\alpha)$. Is f solvable by radicals?

28.1.12. Let $F \subseteq E$ be a field extension. Define $\mathcal{R}$ to be the union of all subfields of E that are an r.r.e. of F.

(a) Show that $\mathcal{R}$ is a field.

(b) If $|E : F| < \infty$, show that $\mathcal{R}$ is an r.r.e.

28.2. A Solvable Polynomial Has a Solvable Galois Group

In this section, we prove one direction of Galois's theorem, Theorem 28.6. We will prove that if a polynomial is solvable by radicals then its Galois group is a solvable group. To prove and use this theorem, we will be working with solvable groups, and, in fact, this would be a good time for the reader to review basic facts about solvable groups from Section 14.1. Recall that a group is *solvable* if there is a finite chain of subgroups starting with the identity and ending with the whole group such that each subgroup in the chain is normal in the next one and that the factor groups are abelian (Definition 14.1). There are a number of equivalent characterizations of solvable groups, and, for example, in the definition we can insist that the factor groups be cyclic (instead of abelian). Alternatively, we can insist that the subgroups in the chain be normal in the whole group (and not just normal in the next one). See Theorem 14.16 for precise statements. Finally, subgroups and factor groups of solvable groups are solvable, and if G is a group, $N \lhd G$, with both N and G/N solvable, then G is solvable as well (Theorem 14.18).

To begin, we need a straightforward lemma for the characteristic p case.

Lemma 28.12. *Let F be a field of characteristic p, $E = F[\alpha]$ with $\alpha^p \in F$. If E is a separable extension of F, then $F = E$.*

Proof. This was Problem **25.4.7** but here is the argument. Let $\beta = \alpha^p \in F$. Then $g = x^p - \beta \in F[x]$ has α as a root. Hence $\min_F(\alpha) \mid g$. But, in $E[x]$, we have $g = (x - \alpha)^p$, and hence $\min_F(\alpha)$ splits in E and α is its only root. However, E is a separable extension of F and so $\min_F(\alpha)$ has distinct roots. We conclude that $\min_F(\alpha) = x - \alpha$ and $\alpha \in F$. Hence, $E = F[\alpha] = F$. $\square$

Let F be a field, let $f \in F[x]$, and let E be the splitting field of f over F. We want to show that if f is solvable by radicals, then $\mathrm{Gal}(E/F)$ is solvable. By the definition of solvability by radicals, $F \subseteq E \subseteq L$ with L an r.r.e. of F. We will first prove the theorem in the special case when $E = L$ is a Galois extension of F.

Theorem 28.13. *Let $F \subseteq L$ be fields. Assume that L is a Galois extension and a repeated radical extension of F. Then $\mathrm{Gal}(L/F)$ is a solvable group.*

Proof. The field L is a repeated radical extension of F. Hence, we have a radical tower going from F to L:

$$F \subseteq F(\alpha_1) \subseteq F(\alpha_1, \alpha_2) \subseteq \cdots \subseteq F(\alpha_1, \alpha_2, \ldots, \alpha_r) = L,$$

where there exists positive integers $n_1, \ldots, n_r$ with $\alpha_1^{n_1} \in F$, and, for $2 \leq i \leq r$, $\alpha_i^{n_i} \in F(\alpha_1, \ldots, \alpha_{i-1})$.

Assume that for some $1 \leq i \leq r$, we have $n_i = hk$. Let $\beta = \alpha_i^h$. Then refine the radical tower and insert an additional intermediate field:

$$F \subseteq \cdots \subseteq F(\alpha_1, \ldots \alpha_{i-1}) \subseteq F(\alpha_1, \ldots, \alpha_{i-1}, \beta) \subseteq F(\alpha_1, \ldots, \alpha_{i-1}, \beta, \alpha_i) \subseteq \cdots L.$$

Note that $\beta^k = \alpha_i^{n_i} \in F(\alpha_1, \ldots \alpha_{i-1})$, and $\alpha_i^h = \beta \in F(\alpha_1, \ldots, \alpha_{i-1}, \beta)$. Repeating this process as many times as necessary, we can assume that each n_i is a prime number.

Since L is a Galois extension of F, it is also a separable extension of F. As a result $F(\alpha_1, \ldots, \alpha_i)$ is also a separable extension of $F(\alpha_1, \ldots, \alpha_{i-1})$. By Lemma 28.12, we can assume that $\mathrm{char}(F) \neq n_i$ for $1 \leq i \leq r$.

For $r > 1$, let $n_1 = p$, a prime, and induct on r. The base case $r = 1$ follows from Corollary 27.33. (Alternatively, one can use $r = 0$ as a trivial base case.) Let M_0 be the splitting field of $x^p - 1$ over L, and let M_1 be the splitting field for $x^p - 1$ over F in M_0. Since $\mathrm{char}(F) \nmid p$, by Proposition 27.16, M_1 is a Galois extension of F and $\mathrm{Gal}(M_1/F)$ is abelian (in fact, cyclic) and hence solvable. (M_1 is a cyclotomic extension of F.)

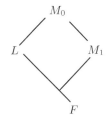

Since L is a Galois extension of F, L is the splitting field of some separable polynomial $g(x)$ in $F[x]$. By Lemma 27.15, since $\mathrm{char}(F) \nmid p$, $x^p - 1$ is a separable polynomial, M_0 is the splitting field of the separable polynomial $g(x)(x^p - 1) \in F[x]$. Hence, M_0 is a Galois extension of F. Consequently, M_0 is also Galois over L and over M_1. The field M_0 is the same as $\langle L, M_1 \rangle$, the compositum of L and M_1 (see Problem 24.2.12), and so by Lemma 28.8, M_0 is an r.r.e. of M_1. In fact, $M_1 \subseteq M_1(\alpha_1) \subseteq \cdots \subseteq M_1(\alpha_1, \ldots, \alpha_r) = M_0$ is a radical tower. We conclude that M_0 is a Galois extension and an r.r.e. of both M_1 and $M_1(\alpha_1)$. By the inductive hypothesis—recall we were proving the theorem by induction on r—$\mathrm{Gal}(M_0/M_1(\alpha_1))$ is solvable.

Since $n_1 = p$, we know that $\beta = \alpha_1^p \in F \subseteq M_1$, and α_1 is a root of $x^p - \beta \in M_1[x]$. Moreover, M_1 is the splitting field of $x^p - 1$ over F, and $\mathrm{char}(F) \nmid p$. So M_1 contains a primitive pth root of unity (Lemma 27.15). By Theorem 27.31, $M_1(\alpha_1)$ is the splitting field of $x^p - \beta$, $M_1(\alpha_1)$ is a Galois extension of M_1, and $\mathrm{Gal}(M_1(\alpha_1)/M_1)$ is cyclic and hence solvable.

We repeatedly apply Theorem 14.18 to complete the proof. Since both $\mathrm{Gal}(M_0/M_1(\alpha_1))$ and $\mathrm{Gal}(M_1(\alpha_1)/M_1)$ are solvable, then so is $\mathrm{Gal}(M_0/M_1)$. Now both $\mathrm{Gal}(M_0/M_1)$ and $\mathrm{Gal}(M_1/F)$ are solvable, and hence $\mathrm{Gal}(M_0/F)$ is solvable (see the left-hand diagram in Figure 28.1). As a result—again by Theorem 14.18—$\mathrm{Gal}(L/F) \cong \mathrm{Gal}(M_0/F)/\mathrm{Gal}(M_0/L)$ is solvable (see the right-hand diagram in Figure 28.1). $\qquad\square$

We now bootstrap the result of Theorem 28.13, weaken the hypothesis, and prove a much stronger version.

Theorem 28.14. *Let $F \subseteq E \subseteq L$ be fields. Assume that L is a repeated radical extension of F. Then $\mathrm{Gal}(E/F)$ is solvable.*

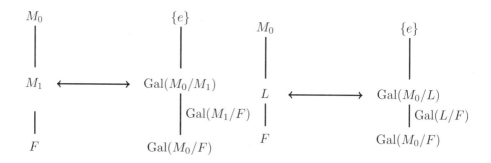

Figure 28.1. M_0 is a Galois extension of F, and we have two Galois correspondences.

Proof. The heavy lifting for this theorem has already been done in Theorem 28.13. We prove the theorem through several claims, each time with slightly weaker hypotheses.

CLAIM 1: If L is an r.r.e. of F, then $\mathrm{Gal}(L/F)$ is solvable.

PROOF OF CLAIM 1: Let $M = \mathrm{Fix}(\mathrm{Gal}(L/F))$, then $F \subseteq M \subseteq L$. By Lemma 25.8, L is a Galois extension of M and $\mathrm{Gal}(L/F) = \mathrm{Gal}(L/M)$. By Corollary 28.9, L is an r.r.e. of M. Hence, by Theorem 28.13, $\mathrm{Gal}(L/F) = \mathrm{Gal}(L/M)$ is solvable.

CLAIM 2: If $F \subseteq E \subseteq L$ are fields, E is a Galois extension of F, and L is both a normal extension and an r.r.e. of F, then $\mathrm{Gal}(E/F)$ is solvable.

PROOF OF CLAIM 2: (If we also knew that L is a separable extension of F, then L would be a Galois extension of F, and we could use Galois correspondence directly. Here we mimic the proof of Theorem 26.9(c).) Since E is a Galois extension of F, by Theorem 25.20 (see Problem *25.3.7* and proof of Theorem 26.9(c)), if $\sigma \in \mathrm{Gal}(L/F)$, then $\sigma\big|_E$, the restriction of σ to E, is an element of $\mathrm{Gal}(E/F)$. Hence, we can define a map $\theta \colon \mathrm{Gal}(L/F) \to \mathrm{Gal}(E/F)$ by $\theta(\sigma) = \sigma\big|_E$. The map θ is clearly a group homomorphism. We claim that θ is onto. If $\tau \in \mathrm{Gal}(E/F)$, then, since L is normal over F, we can—by Theorem 24.24—extend τ to an element $\sigma \in \mathrm{Gal}(L/F)$. Hence $\sigma\big|_E = \tau$ and θ is onto. We conclude that $\mathrm{Gal}(E/F) \cong \mathrm{Gal}(L/F)/\ker(\theta)$. By Claim 1, $\mathrm{Gal}(L/F)$ is solvable, and by Theorem 14.18—quotients of solvable groups are solvable—so is $\mathrm{Gal}(E/F)$.

CLAIM 3: If $F \subseteq E \subseteq L$ are fields, E is a Galois extension of F, and L is an r.r.e. of F, then $\mathrm{Gal}(E/F)$ is solvable.

PROOF OF CLAIM 3: Let $\widetilde{L}$ be the normal closure of L over F. So we have $F \subseteq E \subseteq \widetilde{L}$ with $\widetilde{L}$ normal over F. In addition, by Corollary 28.11, $\widetilde{L}$ is an r.r.e. of F. By Claim 2, $\mathrm{Gal}(E/F)$ is solvable.

CLAIM 4: If $F \subseteq E \subseteq L$ are fields and L is an r.r.e. of F, then $\mathrm{Gal}(E/F)$ is solvable.

PROOF OF CLAIM 4: Let $K = \mathrm{Fix}(\mathrm{Gal}(E/F))$. Then $F \subseteq K \subseteq E \subseteq L$. Now (by Lemma 25.8) E is a Galois extension of K, and, by Corollary 28.9, L is an r.r.e. of K. Hence, by Claim 3, $\mathrm{Gal}(E/K)$ is solvable. But (again by Lemma 25.8) $\mathrm{Gal}(E/F) = \mathrm{Gal}(E/K)$, and so $\mathrm{Gal}(E/F)$ is solvable as well. $\qquad\square$

Remark 28.15. Note that the statement of Theorem 28.14 did not mention Galois extensions, normal closures, separability, or even splitting fields. Even so, the proof relied heavily on much of what we know about normal and Galois extensions. This is remarkable. It is one thing to introduce specialized concepts and then proceed to prove things about them. It is quite another feat to introduce jargon, but then proceed to use that language and the accompanying machinery to prove results that we could have stated before any mention of Galois theory.

Corollary 28.16 (Galois). *Let F be a field, and let $f \in F[x]$. If f is solvable by radicals, then $\mathrm{Gal}(f)$ is a solvable group.*

Proof. Let E be the splitting field of f over F. Since f is assumed to be solvable by radicals, we know that there exists L, an r.r.e. of F, with $F \subseteq E \subseteq L$. By Theorem 28.14, $\mathrm{Gal}(f) = \mathrm{Gal}(E/F)$ is solvable. $\qquad\square$

An Insolvable Quintic. To give a specific example of an insolvable quintic, we first have to give a class of polynomials whose Galois groups are the full symmetric group. Let G be an arbitrary finite group. Does there exists a field extension $\mathbb{Q} \subseteq E$ with $\mathrm{Gal}(E/\mathbb{Q}) \cong G$? This is known as the inverse Galois problem and has been the object of intense study. Much is known, but the inverse Galois problem remains an open problem.

Lemma 28.17. *Let p be a prime, and let f be an irreducible polynomial of degree p over $\mathbb{Q}$. Assume that f has exactly two non-real roots in $\mathbb{C}$. Then*

$$\mathrm{Gal}(f) \cong S_p.$$

In particular, if $p \geq 5$, then f is not solvable by radicals.

Proof. Let E be the splitting field of f over $\mathbb{Q}$ in $\mathbb{C}$, let Ω be the set of roots of f in $\mathbb{C}$, and let $G = \mathrm{Gal}(E/\mathbb{Q})$.

All polynomials over a field of characteristic 0 are separable and f is irreducible. Hence $|\Omega| = \deg f = p$. The group G acts on Ω, and, by Theorem 24.44, $G \leq S_p$. We want to show that $G = S_p$.

Let $\sigma : E \to E$ denote complex conjugation. Then $\sigma \in \mathrm{Gal}(E/\mathbb{Q})$, and since f has exactly two non-real roots, σ switches these two non-real roots and stabilizes all the other roots. Hence, G contains a 2-cycle.

Since f is irreducible and if α is a root of f, then $p = \deg(f) = |\mathbb{Q}[\alpha] : \mathbb{Q}|$ divides $|E : \mathbb{Q}|$. Now $|G| = |E : \mathbb{Q}|$ and $p \mid |G|$. Hence, G has an element τ with $o(\tau) = p$. This means that G has a p-cycle. Together, σ and τ generate all of S_p (see Problem 3.1.15), and hence $G = S_p$.

By Corollary 14.21, G is not solvable for primes p greater or equal to 5. $\qquad\square$

We can now answer one of the problems highlighted in the Preface.

Example 28.18. The polynomial $x^5 - 10x + 5$ is not solvable by radicals.

Proof. This polynomial is irreducible by the Schönemann-Eisenstein criterion. We also see by graphing it that it has three real roots and hence exactly two non-real roots. Hence, by Lemma 28.17, the polynomial is not solvable by radicals. $\qquad\square$

Remark 28.19. Because of Lemma 28.17, after being told the Schönemann-Eisenstein criterion, any middle schooler armed with a graphing calculator can create polynomials not solvable by radicals—a problem that had vexed humanity for over 2000 years!

Problems

28.2.1. Show that $x^7 - 14x^5 + 21x + 7$ is not solvable by radicals.

28.2.2. Show that $x^5 - 4x + 2$ and $x^5 - 12x + 3$ are not solvable by radicals.

28.2.3. Give a polynomial $f \in \mathbb{Q}[x]$ of degree 11 with $\mathrm{Gal}(f) = S_{11}$.

28.2.4. Can you give a polynomial in $\mathbb{R}[x]$ that is not solvable by radicals?

28.2.5. Give a polynomial of degree 6 in $\mathbb{Q}[x]$ that is not solvable by radicals.

28.2.6. Is $p(x) = x^5 + 5x^4 + 10x^3 + 10^2 - x - 2 \in \mathbb{Q}[x]$ solvable by radicals?

28.2.7. Let $K = \mathbb{Q}(\sqrt{2} + \sqrt[3]{2} + \sqrt[7]{2})$. Is $\mathrm{Gal}(K/\mathbb{Q})$ solvable?

28.2.8. Let $K = \mathbb{F}_3(x)$ be the field of rational functions over $\mathbb{F}_3$. Let

$$E = K\left(\sqrt[6]{\frac{\sqrt{x}+1}{\sqrt{x}-1}} + \sqrt[3]{\frac{\sqrt{x}+1}{\sqrt{x}-1}} + \sqrt{\frac{\sqrt{x}+1}{\sqrt{x}-1}}\right).$$

Is $\mathrm{Gal}(E/K)$ solvable?

28.2.9. Assume that $f \in \mathbb{Q}[x]$ and that f is solvable by radicals. Corollary 28.16 asserts that $\mathrm{Gal}(f)$ is a solvable group. Streamline the proof of this result in the case when the ground field is the field of rational numbers. In particular, the proof of Theorem 28.14 can be shortened considerably.

28.3. A Solvable Galois Group Corresponds to a Solvable Polynomial*

In this final (optional) section, we prove the other direction of Galois's theorem, Theorem 28.6. We will prove that if a polynomial has a solvable Galois group and if the characteristic of the field does not divide the index of the splitting field of the polynomial, then the polynimial is indeed solvable by radicals. The main task for the proof is to give a partial converse to Theorem 27.31. In that theorem, we investigated the Galois group of the polynomial $x^n - a$. This was relevant since if $F_i \subseteq F_{i+1}$ is a step in a radical tower, then $F_{i+1} = F_i[\alpha]$ where α is a root of a polynomial of the form $x^n - a$. We now have to start with a nice Galois group and show that it must come from such an extension. To do so requires fine tuning a few of our tools. In fact, in this section, we prove two results—Dedekind's theorem, Theorem 28.20, and the so-called theorem on natural irrationalities, Theorem 28.25—that are independently interesting and are useful in deeper studies and applications of Galois theory. We will also briefly encounter cocycles. This is the tip of an iceberg and the first step into the area of group cohomology, which we will not explore.

We will begin with Dedekind's theorem, a general theorem asserting the linear independence of automorphisms of a field. (For a particular example, see Problem 22.3.7.) Recall that, if E is a field, the set $\mathcal{F}(E, E)$ of functions from E to E with the usual addition and scalar multiplication of functions is a vector space (see Problem 22.2.7).

Theorem 28.20 (Dedekind). *Let E be a field, and let $V = \mathcal{F}(E, E)$ be the vector space over E of functions from E to E. Let S be a finite subset of $\mathrm{Aut}(E)$. Then the set S is linearly independent in V.*

Proof. Induct on $|S|$. For $|S| = 1$, there is nothing to prove. So assume that $|S| = n > 1$ and the proposition has been proved for sets of automorphisms with fewer than n elements. Let $S = \{\sigma_1, \sigma_2, \ldots, \sigma_n\}$, and assume that, for $\lambda_1, \ldots, \lambda_n \in E$, we have

(28.1) $$\lambda_1 \sigma_1 + \lambda_2 \sigma_2 + \cdots + \lambda_n \sigma_n = 0.$$

Since $\sigma_1 \neq \sigma_n$, there exists $a \in E$ with $\sigma_1(a) \neq \sigma_n(a)$. On the one hand, multiplying equation (28.1) by the scalar $\sigma_1(a)$, we have

(28.2) $$\lambda_1 \sigma_1(a)\sigma_1 + \lambda_2 \sigma_1(a)\sigma_2 + \cdots + \lambda_n \sigma_1(a)\sigma_n = 0.$$

On the other hand, for any $b \in E$, we can plug in ab into equation (28.1).

$$0 = \lambda_1 \sigma_1(ab) + \lambda_2 \sigma_2(ab) + \cdots + \lambda_n \sigma_n(ab)$$
$$= \lambda_1 \sigma_1(a)\sigma_1(b) + \lambda_2 \sigma_2(a)\sigma_2(b) + \cdots + \lambda_n \sigma_n(a)\sigma_n(b).$$

Since $b \in E$ was arbitrary, we have

(28.3) $$\lambda_1 \sigma_1(a)\sigma_1 + \lambda_2 \sigma_2(a)\sigma_2 + \cdots + \lambda_n \sigma_n(a)\sigma_n = 0.$$

Subtracting equation (28.3) from equation (28.2), the σ_1 term is eliminated, and we get

$$\lambda_2(\sigma_1(a) - \sigma_2(a))\sigma_2 + \cdots + \lambda_n(\sigma_1(a) - \sigma_n(a))\sigma_n = 0.$$

By the inductive hypothesis $\{\sigma_2, \ldots, \sigma_n\}$ is linearly independent, and so we must have all the coefficients equal to zero. In particular $\lambda_n(\sigma_1(a) - \sigma_n(a)) = 0$. Since $\sigma_1(a) \neq \sigma_n(a)$, we have $\lambda_n = 0$. But this means that $\{\sigma_1, \ldots, \sigma_{n-1}\}$ is linearly dependent, and this contradicts the inductive hypothesis. $\square$

Remark 28.21. If S is an infinite set in a vector space V, then S is said to be linearly independent if every *finite* subset of S is linearly independent (Definition 22.21). Hence, Theorem 28.20 shows that, if E is a field, then $\mathrm{Aut}(E)$ is a linearly independent subset of $\mathcal{F}(E, E)$.

Cocycles. Let $F \subseteq E$ be a field extension. As usual, let $E^\times = E - \{0_E\}$ denote the invertible elements of E, and fix an element $\alpha \in E^\times$. Define a function $f_\alpha \colon \mathrm{Gal}(E/F) \to E^\times$ by

$$f_\alpha(\sigma) = \frac{\sigma(\alpha)}{\alpha}.$$

(Both $\sigma(\alpha)$ and α are in $E^\times$ and so the division is in the field E. We actually have already seen this function in the proof of Theorem 27.31.) The function f_α, in some sense, measures how much α is moved by elements of $\mathrm{Gal}(E/F)$. We also can recover $\sigma(\alpha)$ from f_α, since $\sigma(\alpha) = \alpha f_\alpha(\sigma)$.

Both $\mathrm{Gal}(E/F)$ and $E^{\times}$ are groups, but the function f_{α} is *not* a group homomorphism. It does however satisfy an interesting property. Let $\sigma, \tau \in \mathrm{Gal}(E/F)$, then, remembering that α, $\sigma(\alpha)$, and $\tau(\alpha)$ are elements of the field E and hence commute, we have

$$f_{\alpha}(\sigma\tau) = \frac{(\sigma\tau)(\alpha)}{\alpha} = \frac{\sigma(\tau(\alpha))}{\alpha}$$

$$= \frac{\sigma(\alpha\frac{\tau(\alpha)}{\alpha})}{\alpha} = \frac{\sigma(\alpha)\sigma\left(\frac{\tau(\alpha)}{\alpha}\right)}{\alpha}$$

$$= f_{\alpha}(\sigma)\sigma(f_{\alpha}(\tau)).$$

As an aside, functions (from $\mathrm{Gal}(E/F)$ to $E^{\times}$) with this property[2] are called a *1-cocycle* (or a *crossed homomorphism*) and are studied in group cohomology. We want to show the converse of the above observation.

Proposition 28.22. *Assume $F \subseteq E$ is a finite degree extension. Let $f\colon \mathrm{Gal}(E/F) \to E^{\times}$. Then the following are equivalent:*

(a) $f(\sigma\tau) = f(\sigma)\sigma(f(\tau))$ *for every $\sigma, \tau \in \mathrm{Gal}(E/F)$.*

(b) *There exists an element $\alpha \in E^{\times}$ with the property that $f(\sigma) = \dfrac{\sigma(\alpha)}{\alpha}$, for every $\sigma \in \mathrm{Gal}(E/F)$.*

Proof. We already proved one direction in the discussion before the proposition. For the other direction assume that for every $\sigma, \tau \in \mathrm{Gal}(E/F)$, we have $f(\sigma\tau) = f(\sigma)\sigma(f(\tau))$. Let $\mathrm{Gal}(E/F) = \{\sigma_1, \sigma_2, \ldots, \sigma_n\}$—$\mathrm{Gal}(E/F)$ is a finite group since $|E:F| < \infty$—and consider

$$f(\sigma_1)\sigma_1 + f(\sigma_2)\sigma_2 + \cdots + f(\sigma_n)\sigma_n.$$

This is a linear combination of F-automorphisms of E, and, by Dedekind's theorem, Theorem 28.20, a finite set of automorphisms of E is linearly independent. So this linear combination cannot be identically zero. Hence, there exists $b \in E^{\times}$ with

$$f(\sigma_1)\sigma_1(b) + f(\sigma_2)\sigma_2(b) + \cdots + f(\sigma_n)\sigma_n(b) \neq 0.$$

$f(\sigma_1)\sigma_1(b) + f(\sigma_2)\sigma_2(b) + \cdots + f(\sigma_n)\sigma_n(b)$ is a non-zero element of E and so it has an inverse in E. Call the inverse of this element α. We want to show that α is the element promised by the proposition. For σ an arbitrary element of $\mathrm{Gal}(E/F)$, we have $\sigma(f(\sigma_i)) = f(\sigma)^{-1}f(\sigma\sigma_i)$, and so

$$\sigma(\alpha^{-1}) = \sigma\left(f(\sigma_1)\sigma_1(b) + f(\sigma_2)\sigma_2(b) + \cdots + f(\sigma_n)\sigma_n(b)\right)$$

$$= \sigma(f(\sigma_1))\sigma\sigma_1(b) + \sigma(f(\sigma_2))\sigma\sigma_2(b) + \cdots + \sigma(f(\sigma_n))\sigma\sigma_n(b)$$

$$= f(\sigma)^{-1}f(\sigma\sigma_1)\sigma\sigma_1(b) + f(\sigma)^{-1}f(\sigma\sigma_2)\sigma\sigma_2(b) + \cdots + f(\sigma)^{-1}f(\sigma\sigma_n)\sigma\sigma_n(b)$$

$$= f(\sigma)^{-1}\left(f(\sigma\sigma_1)\sigma\sigma_1(b) + f(\sigma\sigma_2)\sigma\sigma_2(b) + \cdots + f(\sigma\sigma_n)\sigma\sigma_n(b)\right)$$

$$= f(\sigma)^{-1}\left(f(\sigma_1)\sigma_1(b) + f(\sigma_2)\sigma_2(b) + \cdots + f(\sigma_n)\sigma_n(b)\right)$$

$$= f(\sigma)^{-1}\alpha^{-1}.$$

The penultimate equality follows since, for a fixed σ, the set $\{\sigma\sigma_1, \sigma\sigma_2, \ldots, \sigma\sigma_n\} = \mathrm{Gal}(E/F)$. We conclude that $f(\sigma) = \sigma(\alpha)/\alpha$ for every $\sigma \in \mathrm{Gal}(E/F)$. $\square$

---

[2]Equations of the form $f_{\alpha}(\sigma\tau) = f_{\alpha}(\sigma)\sigma(f_{\alpha}(\tau))$ are sometimes called *Noether's equations*.

We actually just need a special case of the above proposition.

Corollary 28.23. *Let E be a field, let G be a finite group of automorphisms of E, and let $F = \mathrm{Fix}(G)$. Let $f: G \to F^\times$ be a group homomorphism. Then there exists $\alpha \in E^\times$ with $f(\sigma) = \sigma(\alpha)\alpha^{-1}$ for all $\sigma \in G$.*

Proof. We know (Theorem 26.3) that E is Galois over F (hence $|E : F| < \infty$) and $G = \mathrm{Gal}(E/F)$. Let $\sigma, \tau \in G$. We know $f(\tau) \in F$ and σ fixes elements of F, and so $\sigma(f(\tau)) = f(\tau)$. Thus

$$f(\sigma\tau) = f(\sigma)f(\tau) = f(\sigma)\sigma(f(\tau)).$$

By Proposition 28.22, there exists an element $\alpha \in E^\times$ such that $f(\sigma) = \sigma(\alpha)\alpha^{-1}$. $\square$

We now use Corollary 28.23 to give a converse to Theorem 27.31.

Theorem 28.24 (Kummer). *Let $K \subseteq E$, and assume that K contains a primitive nth root of unity. Then the following are equivalent:*

(a) *E is Galois over K, and $\mathrm{Gal}(E/K)$ is cyclic of order dividing n.*

(b) *$E = K[\alpha]$ for some element $\alpha \in E$ with $\alpha^n \in K$.*

Proof. Let $\epsilon \in K$ be a primitive nth root of unity.

(b $\Rightarrow$ a) Let $a = \alpha^n \in K$. Then, for $1 \leq k \leq n$, $\alpha\epsilon^k$ is a root of $x^n - a$. These are all the roots of $x^n - a$ and they are all elements of E. Hence, E is a splitting field for $x^n - a$ over K. Hence, by Theorem 27.31, E is Galois over K, $\mathrm{Gal}(E/K)$ is cyclic, and its order divides n.

(a $\Rightarrow$ b) Let $\mathrm{Gal}(E/K) = \langle \sigma \rangle$ with $o(\sigma) = m$ and $m \mid n$. The group $\langle \epsilon \rangle$ is a cyclic group of order n and the map $f: \mathrm{Gal}(E/K) \to \langle \epsilon \rangle \subseteq K^\times$ defined by $f(\sigma^i) = \epsilon^{ni/m}$ is a 1-1 group homomorphism. By Corollary 28.23, there exists $\alpha \in E^\times$ with $f(\tau) = \tau(\alpha)/\alpha$ for all $\tau \in \mathrm{Gal}(E/K)$. We first claim that $\alpha^n \in K$. Let $\delta = \epsilon^{n/m} = f(\sigma)$, then $\delta = \sigma(\alpha)/\alpha$. So $\sigma(\alpha) = \delta\alpha$ and $\sigma(\alpha^n) = \delta^n\alpha^n = \alpha^n$. Hence, $\alpha^n \in \mathrm{Fix}(\mathrm{Gal}(E/K)) = K$. It remains to show that $E = K[\alpha]$. We have $K \subseteq K[\alpha] \subseteq E$ and E is Galois over K. Hence E is Galois over $K[\alpha]$ and $|E : K[\alpha]| = |\mathrm{Gal}(E/K[\alpha])|$. Hence, the proof will be complete if we show that $\mathrm{Gal}(E/K[\alpha])$ consists of only the identity element. Let $\tau \in \mathrm{Gal}(E/K[\alpha]) \subseteq \mathrm{Gal}(E/K)$. Then $\tau(\alpha) = \alpha$ and $f(\tau) = \tau(\alpha)/\alpha = \alpha/\alpha = 1$. This implies τ is the identity since f is 1-1. The proof is now complete. $\square$

Before proceeding, we need to prove a general useful theorem—reminiscent of Theorem 11.43—about Galois extensions. Recall (Definition 22.41) that if F and K are subfields of E, then the *compositum* of F and K is the intersection of all subfields of E that contain both F and K.

Theorem 28.25 (Natural irrationalities). *Let F, K, L, and E be fields with K and L both containing F and contained in E. Let $M = K \cap L$ be the intersection, and let $N = \langle K, L \rangle$ be the compositum of K and L. (See Figure 28.2.) Assume that K is a Galois extension of F. Then N is Galois over L and $\mathrm{Gal}(N/L) \cong \mathrm{Gal}(K/M)$. In particular, $|N : L| = |K : M|$.*

Proof. The field K is a Galois extension of F. Hence K is the splitting field of a separable polynomial $f \in F[x]$ (Theorem 25.17). This polynomial is also a polynomial in $L[x]$, and the field N is the splitting field of f over L. This is because f splits in N—since $K \subseteq N$—and any splitting field of f over L will have to contain both K—since K is generated by F and the roots of f—and L. Hence, N is the splitting field of a separable polynomial—namely f—over L, and it is Galois over L.

To show $\mathrm{Gal}(N/L) \cong \mathrm{Gal}(K/M)$, we define a map $\Theta\colon \mathrm{Gal}(N/L) \to \mathrm{Gal}(K/M)$. Let $\sigma \in \mathrm{Gal}(N/L)$. Then σ is an L-automorphism of N, and in particular, it is also an F-automorphism or an M-automorphism of N. We ask what is $\sigma(K)$? Since K is a normal extension of F, by Theorem 25.20, $\sigma(K) = K$. Hence, $\sigma\big|_K$ is an M-automorphism of K. Hence, for $\sigma \in \mathrm{Gal}(N/L)$, we define $\Theta(\sigma) = \sigma\big|_K$. The map Θ is clearly a group homomorphism. It remains to show that Θ is 1-1 and onto.

Let $\sigma \in \ker(\Theta)$. Then σ is an L-automorphism of N which when restricted to K is the identity. Hence, σ fixes all of K and all of L, and as a result it must fix $\langle K, L \rangle = N$. We conclude that σ is the identity and Θ is 1-1.

To show that Θ is onto, we want to show that $\Theta(\mathrm{Gal}(N/L)) = \mathrm{Gal}(K/M)$. We know that K is Galois over M and both $\Theta(\mathrm{Gal}(N/L))$ and $\mathrm{Gal}(K/M)$ are subgroups of $\mathrm{Gal}(K/M)$. Hence, we can show that these subgroups are equal by using the Galois correspondence from the fundamental theorem of Galois theory (Theorem 26.9) and by showing that $\mathrm{Fix}(\Theta(\mathrm{Gal}(N/L))) = \mathrm{Fix}(\mathrm{Gal}(K/M))$. Let $M_0 = \mathrm{Fix}(\Theta(\mathrm{Gal}(N/L))) \subseteq K$. Since $\Theta(\mathrm{Gal}(N/L)) \subseteq \mathrm{Gal}(K/M)$, we have that $M_0 \supseteq M$. On the other hand, $M_0 \subseteq K$ and every element of M_0 is fixed by every element of $\mathrm{Gal}(N/L)$ when restricted to K. Hence, $M_0 \subseteq \mathrm{Fix}(\mathrm{Gal}(N/L)) = L$.

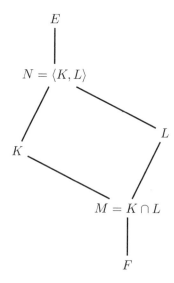

Figure 28.2. If K is Galois over F, then N is Galois over L.

We conclude that $M_0 \subseteq K \cap L = M$. Thus $\text{Fix}(\Theta(\text{Gal}(N/L))) = M_0 = M = \text{Fix}(\text{Gal}(K/M))$, and Θ is onto.

Finally, $|N:L| = |\text{Gal}(N/L)| = |\text{Gal}(K/M)| = |K:M|$.   □

Now, we are ready to prove that Galois extensions with solvable Galois groups are contained in repeated radical extensions.

Theorem 28.26. *Assume $F \subset E$ is a Galois field extension with $\text{Gal}(E/F)$ solvable. Further, assume that $\text{char}(F)$ does not divide $|E:F|$. Then there exists a repeated radical extension L of F with $F \subseteq E \subseteq L$.*

Proof. Let $n = |E:F|$. By Lemma 27.15, the splitting field of $x^n - 1$ over E, can be written as $E[\epsilon]$ with ϵ a primitive nth root of unity. The field $E[\epsilon]$ is the smallest field containing E and $F[\epsilon]$, and so $E[\epsilon] = \langle F[\epsilon], E \rangle$ is the compositum of the two. See Figure 28.3.

By the theorem of natural irrationalities, Theorem 28.25, since E is Galois over F, we have that $E[\epsilon]$ is Galois over $F[\epsilon]$, and $\text{Gal}(E[\epsilon]/F[\epsilon]) \cong \text{Gal}(E/F[\epsilon] \cap E) \leq \text{Gal}(E/F)$. Hence, by Theorem 14.18, $\text{Gal}(E[\epsilon]/F[\epsilon])$ is a solvable group. Now we have to turn the solvability of this group into an r.r.e. of $F[\epsilon]$.

Let $G = \text{Gal}(E[\epsilon]/F[\epsilon])$. By Theorem 14.16, since G is solvable, there exists a finite chain of subgroups

$$\{e\} = H_0 \leq H_1 \leq \cdots \leq H_{n-1} \leq H_n = G,$$

such that, for $0 \leq i < n$, $H_i \lhd H_{i+1}$ and H_{i+1}/H_i is a cyclic group of prime order. Note that the order of H_{i+1}/H_i divides $|G|$ which in turn divides $|\text{Gal}(E/F)| = |E:F| = n$.

Let $F_i = \text{Fix}(H_i)$. We then have

$$F[\epsilon] = F_n \subseteq F_{n-1} \subseteq \cdots \subseteq F_1 \subseteq F_0 = E[\epsilon].$$

By the fundamental theorem of Galois theory, Theorem 26.9, F_i is a Galois extension of F_{i+1} and $\text{Gal}(F_i/F_{i+1}) \cong H_{i+1}/H_i$ is a cyclic group of order dividing n. We now apply Kummer's theorem, Theorem 28.24, to the extension $F_{i+1} \subseteq F_i$—here is where we need F_{i+1} to contain a primitive nth root of unity and why we adjoined ϵ to F and E—and we conclude that $F_i = F_{i+1}[\alpha]$ where $\alpha^n \in F_i$.

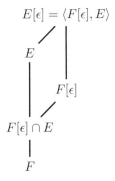

Figure 28.3. Adjoining a root of unity to F and E

Hence, $E[\epsilon]$ is an r.r.e. of $F[\epsilon]$. Since $F[\epsilon]$ itself is a radical extension of F, we conclude that $E[\epsilon]$ is an r.r.e. of F. Letting $L = E[\epsilon]$, the proof is now complete. $\square$

We are finally ready to prove—in fact, we have already proved—the if direction of Galois's theorem, Theorem 28.6.

Corollary 28.27. *Let F be a field, let $f \in F[x]$, and let E be the splitting field of f over F. Assume that $\mathrm{char}(F)$ does not divide $|E : F|$ (automatically satisfied if $\mathrm{char}(F) = 0$). If $\mathrm{Gal}(E/F)$ is solvable, then f is solvable by radicals.*

Proof. Let g be an irreducible factor of f, then $\deg(g) \mid |E : F|$ (since $\deg(g) = \deg(\min_F(\alpha)) = |F[\alpha] : F|$ where α is a root of g). Thus $\mathrm{char}(F) \nmid \deg(g)$. As a result, the derivative of g cannot be identically zero, and so g is separable (Theorem 25.30—also see Problem **25.4.12**). Hence, f is separable and E is the splitting field of a separable polynomial over F. So E is a Galois extension of F with a solvable Galois group. To show that f is solvable by radicals, we need a repeated radical extension L of F with $F \subseteq E \subseteq L$. This is exactly what Theorem 28.26 promises. $\square$

Problems

28.3.1. Let $E = \mathbb{Q}(\sqrt{2}, \sqrt{3})$.
  (a) What is a basis for E as a vector space over $\mathbb{Q}$?
  (b) Explicitly describe the elements of $\mathrm{Gal}(E/\mathbb{Q})$.
  (c) Let $V = \mathcal{F}(E, E)$ be the vector space of functions from E to E. Directly, and without appealing to Dedekind's theorem, Theorem 28.20, show that $S = \mathrm{Gal}(E/\mathbb{Q})$ is a linearly independent subset of V.

28.3.2. Let $K \subseteq E$, and assume that K contains a primitive nth root of unity. Prove that the following are equivalent:
  (a) E is Galois over K, and $\mathrm{Gal}(E/K)$ is a cyclic group of order n.
  (b) E is the splitting field of an irreducible polynomial $x^n - a \in K[x]$.

28.3.3. Given the situation and notation of the theorem of natural irrationalities, Theorem 28.25, assume $|N : F| < \infty$. Show that $|N : K| = |L : M|$.

28.3.4. Let $F \subseteq E$ be a field extension. Let K and L be intermediate fields containing F and contained in E. Further assume that $K \cap L = F$ and that K is a Galois extension of F. Let $\alpha \in L$ be algebraic over F. Show that $\min_F(\alpha)$ is irreducible in $K[x]$.

28.3.5. Is $p(x) = x^4 + 3x^3 - 2x^2 + 7x - 47 \in \mathbb{Q}[x]$ solvable by radicals? Make sure that your reasoning is complete.

28.3.6. Let F be a field of characteristic 0, and let $f \in F[x]$ be irreducible. Let E be the splitting field of f over F. Assume $|E : F| = pq$ where p and q are primes. Show that f is solvable by radicals.

28.3.7. Let F be a field of characteristic 0, and let $f \in F[x]$ be irreducible. Let E be the splitting field of f over F. Assume $|E \colon F| = pqr$ where p, q, and r are distinct primes. Show that f is solvable by radicals.

28.3.8. We know that a group of order 15 is cyclic, and cyclic groups are solvable. Does this mean that every irreducible polynomial of degree 15 is solvable by radicals? Either give a complete proof or point out the flaw in the argument.

28.3.9. Let $f \in \mathbb{Q}[x]$ with $\deg(f) \leq 4$. Prove that f is solvable by radicals.

28.3.10. Let F be a field of characteristic 0, and let $f \in F[x]$ be of degree 5. Let E be the splitting field of f over F. If $|E \colon F| < 60$, then must f be solvable by radicals? Is the converse true as well?

28.3.11. Let n be a positive integer, let F be a field, and let $f \in F[x]$ be a polynomial of degree n. Assume that $\operatorname{char}(F)$ does not divide $n!$ and that $\operatorname{Gal}(f)$ is solvable. Prove that f is solvable by radicals.

28.3.12. Let p be a prime, and let $K = \mathbb{F}_p(t)$ be the field of rational functions in t (t is a transcendental over $\mathbb{F}_p$). Let $f = x^p - x - t \in K[x]$. Show that f has no roots in K. You may find the following steps helpful.
Step 1: Let $\alpha = g/h \in K$ where $g, h \in \mathbb{F}_p[t]$ with $\gcd(g, h) = 1$. Assume α is a root of f, and show that h has to be a constant and g is either a constant or a constant multiple of t.
Step 2: Show that if α is a constant multiple of t, then t cannot be transcendental over $\mathbb{F}_p$.
Step 3: Show that α cannot be a constant either. Complete the proof.

28.3.13. **Cyclic Galois group yet not solvable by radicals.** Let p be a prime, and let $K = \mathbb{F}_p(t)$ where t is a transcendental over $\mathbb{F}_p$. Let $f = x^p - x - t \in K[x]$. Show that $\operatorname{Gal}(f)$ is cyclic and yet f is not solvable by radicals. You may find the following steps helpful.
Step 1: Let E be a splitting field for f over K, and let $\alpha \in E$ be a root of f. Show that, for $0 \leq i \leq p - 1$, $\alpha + i$ is also a root of f.
Step 2: Show that $E = K[\alpha]$ and that f is a separable polynomial.
Step 3: Assume that $g \in K[x]$ was a polynomial of degree $0 < m < p$ and $g \mid f$. By considering the sum of the roots of g, show that $m\alpha$ must be in K. Appeal to Problem 28.3.12, and conclude that f is irreducible in $K[x]$.
Step 4: Prove that $\operatorname{Gal}(f)$ is cyclic of order p.
Step 5: Assume that f is solvable by radicals, then show that there exist fields $K \subseteq L \subseteq M$ with $M = L[u]$, $\alpha \in M$, $\alpha, u \notin L$, and $u^q \in L$ where q is a prime.
Step 6: Show that $x^q - u^q \in L[x]$ is irreducible, and conclude that $|M \colon L| = q$.
Step 7: Show that $L \subseteq E \subseteq M$, and conclude that $E = M$ and $q = p$.
Step 8: Use Proposition 25.37 to show that E is a separable extension of L. Conclude that $u \in M$ is separable over L.
Step 9: Arrive at a contradiction, and complete the proof.

28.4. More Problems and Projects

The Inverse Galois Problem. Let G be a finite group. Does there exist a Galois extension $\mathbb{Q} \subseteq E$ such that $\mathrm{Gal}(E/\mathbb{Q}) \cong G$? This is an important problem, and, as of this writing, it is unresolved. The answer is known to be yes if G is a solvable group, or a variety of other groups, but the answer is not known in general. In the problems, you are first asked to show—assuming Dirichlet's theorem—that the answer is yes if G is a cyclic group (you can take this a step further and replace "cyclic" with "abelian"). Then you show that, if p is a prime, you can have the Galois group be isomorphic to S_p where p is any prime. You are then asked to use this result and show that, for an arbitrary finite group G, there exists fields $\mathbb{Q} \subseteq L \subseteq E$ with $\mathrm{Gal}(E/L) = G$.[3]

Problems

28.4.1. Find a field extension $\mathbb{Q} \subseteq E$ such that $\mathrm{Gal}(E/\mathbb{Q}) = \mathbb{Z}/5\mathbb{Z}$.

28.4.2. Let n be a fixed positive integer, and consider the arithmetic sequence

$$1, n+1, 2n+1, \ldots, kn+1, \ldots.$$

A special case of Dirichlet's theorem on primes in arithmetic sequences says that this sequence contains an infinite number of primes. Assume that p is one such prime. Use p to find a field extension $\mathbb{Q} \subseteq E$ with $\mathrm{Gal}(E/\mathbb{Q}) \cong \mathbb{Z}/n\mathbb{Z}$.

28.4.3. Let $f = x^2(x-2)(x-4)(x-6) \in \mathbb{Q}[x]$, and let $g(x) = f(x) - 2$.
 (a) How many real roots (counting multiplicity) does f have? How many complex roots does f have?
 (b) Without using graphing software, give an approximate graph of f and g.
 (c) How many real roots and how many complex roots does g have?
 (d) What is $\mathrm{Gal}(g)$? Is g solvable by radicals?
 (e) Using f and g as a template, find a polynomial of degree 7 not solvable by radicals and with S_7 as its Galois group over $\mathbb{Q}$.

28.4.4. Let p be a prime. Give an explicit polynomial $g \in \mathbb{Q}[x]$ such that if E is the splitting field of g over $\mathbb{Q}$, then $\mathrm{Gal}(E/\mathbb{Q}) = S_p$.

28.4.5. Let G be a finite group. Show that there are fields L and E with $\mathbb{Q} \subseteq L \subseteq E$ such that $\mathrm{Gal}(E/L) \cong G$. You may find the following steps helpful.
 STEP 1: Let p be a prime larger than $|G|$. Is G isomorphic to a subgroup of S_p? (Is Cayley's theorem, Theorem 11.35—or its proof—relevant?)
 STEP 2: Find an irreducible polynomial $g \in \mathbb{Q}[x]$ with $\mathrm{Gal}(g) \cong S_p$. (Is Problem 28.4.4 relevant?)
 STEP 3: Let E be the splitting field of g over $\mathbb{Q}$. Use Galois correspondence to construct L.

---

[3]Problems 28.4.3–28.4.5 adapted from Osofsky [**Oso99**].

Hilbert's Theorem 90.[4] In the problems below, you are asked to prove and explore the following:

Theorem 28.28 (Hilbert's Theorem 90). *Let* $F \subseteq E$ *be a finite degree field extension. Assume* $\mathrm{Gal}(E/F) = \langle \sigma \rangle$ *is a cyclic group. Let* $\beta \in E$. *Then*

$$\prod_{\tau \in \mathrm{Gal}(E/F)} \tau(\beta) = 1$$

if and only if there exists $\alpha \in E^\times$ *with* $\dfrac{\sigma(\alpha)}{\alpha} = \beta$.

28.4.6. Let ϵ be a primitive fifth root of unity in $\mathbb{C}$, and let $G = \mathrm{Gal}(\mathbb{Q}(\epsilon)/\mathbb{Q})$.
  (a) Show that there exists an element $\sigma \in G$ that sends ϵ to ϵ^2.
  (b) What is $\langle \sigma \rangle$ (in G)?
  (c) Let $\beta = \frac{\sigma(\epsilon^3)}{\epsilon^3}$. Find

$$\prod_{\tau \in G} \tau(\beta).$$

28.4.7. Let $F \subseteq E$ be a finite degree field extension. Assume $\mathrm{Gal}(E/F) = \langle \sigma \rangle$ is a cyclic group. Let $\alpha \in E^\times$ and define $\beta = \dfrac{\sigma(\alpha)}{\alpha}$. Show that

$$\prod_{\tau \in \mathrm{Gal}(E/F)} \tau(\beta) = 1.$$

28.4.8. **Hilbert's Theorem 90.** Prove Theorem 28.28.
     You may find the following steps helpful.
     STEP 1: One (easier) direction is Problem 28.4.7.
     STEP 2: Define a map $f \colon \mathrm{Gal}(E/F) \to E^\times$ by

$$f(e) = 1, \ f(\sigma) = \beta, \ f(\sigma^2) = \sigma(\beta)\beta, \ \ldots, f(\sigma^i) = \sigma^{i-1}(\beta)f(\sigma^{i-1}), \ \ldots.$$

     STEP 3: Let $n = |\mathrm{Gal}(E/F)|$, and let $i + j$ be a multiple of n. Show that

$$f(\sigma^i \sigma^j) = 1 = f(\sigma^i)\sigma^i(f(\sigma^j)).$$

     STEP 4: Assume $i + j$ has remainder r when divided by n, and $0 < r < n$. Show that

$$f(\sigma^i \sigma^j) = \sigma^{r-1}(\beta) \cdots \sigma^2(\beta)\sigma(\beta)\beta = f(\sigma^i)\sigma^i(f(\sigma^j)).$$

     STEP 5: Conclude that f is a crossed homomorphism and complete the proof using Proposition 28.22.

28.4.9. Consider the field extension $\mathbb{Q} \subseteq \mathbb{Q}(i)$. Recall (similar to Example 24.39) that $\mathrm{Gal}(\mathbb{Q}(i)/\mathbb{Q}) \cong \mathbb{Z}/2\mathbb{Z}$. Denote by σ the non-identity element of the Galois group that maps $c + di \in \mathbb{Q}(i)$ to its conjugate $c - di$. Let $\beta = a + bi \in \mathbb{Q}(i)$, and assume that $\prod_{\tau \in \mathrm{Gal}(\mathbb{Q}(i)/\mathbb{Q})} \tau(\beta) = 1$.
  (a) Show that (a, b) is a rational point (i.e., a point with rational coordinators) on the unit circle.

---

[4]A version of this theorem was "Satz 90" in Hilbert's influential *Zahlbericht*. For a new English translation of the book, see Hilbert [**Hil98**] where "Satz 90" is still "Theorem 90".

(b) Hilbert's Theorem 90 says that there exists $\alpha \in \mathbb{Q}(i)^{\times}$ with $\dfrac{\sigma(\alpha)}{\alpha} = \beta$.

Plug $\alpha = c + di$ in $\dfrac{\sigma(\alpha)}{\alpha}$, and show that we can, in fact, choose an appropriate $\alpha \in \mathbb{Z}[i]$ with the same property.

(c) Show that (a, b) is a rational point on the unit circle if and only if there are integers c and d with

$$a = \frac{c^2 - d^2}{c^2 + d^2} \quad \text{and} \quad b = \frac{-2cd}{c^2 + d^2}.$$

28.4.10. (a) Akin to Problem 28.4.9, use Hilbert's Theorem 90 to find a parametrization for all the rational points on the hyperbola $x^2 - 3y^2 = 1$.

(b) We are looking for integers X, Y, and Z that satisfy the equation $X^2 - 3Y^2 = Z^2$. Let $x = X/Z$ and $y = Y/Z$. Using your answer to the previous part, find a parametrization for integer solutions to $X^2 - 3Y^2 = Z^2$. (This was one of the problems highlighted in the Preface.)

Abel and Abelian Groups.[5] The Norwegian mathematician Niels Henrik Abel (1802–1829) gave the first accepted proof[6] that a general quintic cannot be solved by radicals.[7] Abel was close to completing a solution to the general question of solvability by radicals, but he died before he could complete the project.[8] A few years after Abel's death, the problem was solved by Galois. In a paper *Mémoire sur une classe particulière d'équations résolubes algébriquement*[9] published in 1829, Abel states the following theorem:

> If the roots of an equation of arbitrary degree are related among themselves in such a way that *all* the roots can be expressed rationally by means of one of them, which we denote by x; if in addition whenever one denotes by θx, $\theta_1 x$ two other arbitrary roots, one has
>
> $$\theta\theta_1 x = \theta_1\theta x,$$
>
> then the equation to which they belong will always be solvable algebraically.

We want to investigate (a restatement in modern parlance of) this theorem.

Definition 28.29. Let $f \in \mathbb{Q}[x]$ be of positive degree, and let E be a splitting field for f over $\mathbb{Q}$. Assume that $E = \mathbb{Q}(\alpha)$ where α is a root of f. Let $\alpha_1 = \alpha$, α_2, ...,

---

[5] Adapted from Cox [**Cox12**, pp. 143–145].

[6] An earlier proof with some gaps was given by the Italian mathematician Ruffini (1765-1822).

[7] Abel sent a copy of his paper to Gauss (1777-1855). The unopened letter was found after Gauss' death.

[8] While in poor health, he traveled by sled to visit his fiancé for Christmas of 1828, got very sick in the process, and died on April 6, at the age of 27.

[9] Abel, Niels Henrik *Œuvres complètes. Tome I (French) [Complete works. Vol. I]* Edited and with a preface by L. Sylow and S. Lie. Reprint of the second (1881) edition. Éditions Jacques Gabay, Sceaux, 1992, pp. 478-507.

α_n be the roots of f in E. For $1 \leq i \leq n$, choose $g_i \in \mathbb{Q}(x)$ such that $\alpha_i = g_i(\alpha)$. Assume that

$$g_i(g_j(\alpha)) = g_j(g_i(\alpha)), \quad 1 \leq i, j \leq n.$$

We then call f an *abelian polynomial*.

Abel's theorem then translates to:

Theorem 28.30. *Let $f \in \mathbb{Q}[x]$ be an abelian polynomial. Then f is solvable by radicals.*

This theorem follows from the following—which is the reason why commutative groups are called abelian—and Galois's theorem, Theorem 28.6, on polynomials solvable by radicals:

Theorem 28.31. *Let $f \in \mathbb{Q}[x]$ be an abelian polynomial. Then $\mathrm{Gal}(f)$ is an abelian group.*

In the problems, you are asked to investigate these theorems.

28.4.11. Let $f \in \mathbb{Q}[x]$ be of positive degree, and let E be a splitting field for f over $\mathbb{Q}$. Let $\alpha_1 = \alpha$, α_2, ..., α_n be the roots of f in E.
  (a) Show that $E = \mathbb{Q}(\alpha)$ if and only if, for $1 \leq i \leq n$, we can find $g_i \in \mathbb{Q}(x)$ such that $\alpha_i = g_i(\alpha)$. Can we insist that $g_i \in \mathbb{Q}[x]$?
  (b) Assume σ and τ are elements of $\mathrm{Gal}(E/\mathbb{Q})$ with the property that $\sigma(\alpha) = \alpha_2$ and $\tau(\alpha) = \alpha_3$. Further, assume that g_2 and g_3 are polynomials in $\mathbb{Q}[x]$ with $g_2(\alpha) = \alpha_2$ and $g_3(\alpha) = \alpha_3$. Show that

$$\sigma(\tau(\alpha)) = g_3(g_2(\alpha)).$$

28.4.12. (a) Prove Theorem 28.31.
  (b) Explain how Theorem 28.30 follows from Theorems 28.31 and 28.6.

28.4.13. Is $x^3 - 2$ an abelian polynomial?

28.4.14. Let $\alpha = i + 2\sqrt{2}$, and let $f = \min_{\mathbb{Q}}(\alpha)$. Let E be the splitting field for f over $\mathbb{Q}$.
  (a) Give a factorization in $E[x]$ of f into linear factors.
  (b) Is $E = \mathbb{Q}[\alpha]$? Why?
  (c) Find α^3, and use it to write i and $\sqrt{2}$ as polynomials in α.
  (d) If $\alpha = \alpha_1$, α_2, ..., α_n are the roots of f, for $1 \leq i \leq n$, find $g_i \in \mathbb{Q}[x]$ with $g_i(\alpha) = \alpha_i$.
  (e) Is f an abelian polynomial? Convince yourself of the answer, explain your reasoning, and reproduce $g_i(g_j(\alpha))$ and $g_j(g_i(\alpha))$ for a few illustrative cases.
  (f) What is $\mathrm{Gal}(f)$?

Hints for Selected Problems

1.3.3. You may want to do Problem 1.3.2 first.

1.3.8. Be careful about the case $a = b = 0$.

1.3.13. Use Theorem 1.49 to find u and v with $ua + vb = 1$. Multiply both sides by c. Does a have to divide one side of the resulting equation?

1.3.15. Consider the product of all of the elements of G.

1.3.18. Problems 1.3.16 and 1.3.17 may be relevant.

1.3.19. Problem 1.3.18 may be relevant.

1.4.4. How many possible first rows are there? Given the first row, how many possibilities for the second row? Recall that a 2×2 matrix is invertible if and only if the first row is not all zeros and the second row is not a multiple of the first. Problem 1.4.3 may be relevant.

1.4.5. Problem 1.4.3 is relevant. You may also want to do Problem 1.4.4 first.

1.4.7. It may be a good idea to do Problem 1.4.6 first. Pair an invertible $n \times n$ matrix A with the matrix whose first row is -1 times the first row of A and its other rows are identical to those of A. How does the determinant of A compare with that of the matrix that A is paired with?

1.4.8. First do Problems 1.4.4 and 1.4.7.

1.4.9. You may want to do Problems 1.4.7 and 1.4.8 first.

1.5.8(c). Use induction. Split your board into four parts. You may have to judiciously cut out a few more squares.

2.2.5. Write $G = \{e, a, b\}$, and see what choices you have for the multiplication table.

2.3.7. Problem 2.2.4a may be relevant.

2.3.10. Remember that $G = \langle a \rangle$ for some $a \in G$. For the second question, Proposition 2.45 may be relevant.

2.3.14. You could do this problem directly, or you could use the previous problem.

2.3.15. For the second example, consider the symmetries of an appropriate regular polygon, and recall that the product of two reflections is a rotation.

2.3.16. If you suspect that $o(xy) = k$—where k depends on $o(x)$ and $o(y)$—first show that $(xy)^k = e$. Proposition 2.45 (Problem **2.3.8**) implies that $o(xy) \mid k$. Then show that $o(xy)$ cannot be a proper factor of k.

2.3.17. Begin by applying Theorem 1.49 to $o(a)$ and ℓ.

2.3.18. Let $o(x) = b \gcd(i, o(x))$. After showing $o(x^i) \mid b$, Problem 1.3.11 may come in handy in showing $b \mid o(x^i)$.

2.3.19(b). If g^m is a generator for G, then $g = (g^m)^k$ for some positive integer k. Show that $mk - 1 = n\ell$, for some integer ℓ. From this derive the value of $\gcd(n, m)$. Conversely, show that, for every m with this value for $\gcd(n, m)$, the element g^m is a generator for G. In making the above arguments, Proposition 2.45, Problem 1.3.12, and Theorem 1.49 may be relevant.

2.3.20. Problem *2.3.19* is relevant.

2.3.24(a). Problems 2.3.9 and 2.3.17 may be relevant.

2.3.24(b). Let $x \in G$ with $x^3 = a$ and consider $(b^{-1}xb)^3$.

2.4.5. As a warmup, do Problem 2.4.4.

2.4.6. Use your answer in part (a) as a guide for the other parts.

2.4.10. Consider a regular hexagon centered at the origin. The group D_{12} is the group of symmetries of this hexagon. Each of these symmetries can be the result of a linear transformation on $\mathbb{R}^2$. You need to find a basis for $\mathbb{R}^2$ such that each symmetry of D_{12} can be realized as a 2×2 matrix with integer entries. Use two adjacent "radii" of the hexagon.

2.4.17. One of them divides the other.

2.5.7(b). Use your conjecture in the first part.

2.5.10. Problems 2.3.20, 2.5.8, and *2.5.9* may be relevant.

2.6.21. Make sure that you understand the proof of Theorem 2.75, and use the division algorithm.

2.6.22. Problem 2.6.21 may be relevant.

2.6.25. Using the division algorithm, write $n = qs + r$. Argue that $a^r \in H$, and conclude that $r = 0$.

2.6.26. Problem **2.3.16** is relevant.

2.6.35. If $x \in G$, what can you say about $\mathbf{C}_G(x)$? Use Problem 2.6.34.

2.7.10(d). The group G is itself the centralizer of e. Problems 2.7.10(b) and 2.6.34 may be useful.

2.7.11(b). Problem 1.3.14(a) maybe relevant.

2.7.11(c). Problem 1.3.14(b) maybe relevant.

2.7.15. Use Problem 2.7.14.

2.7.17. Use Problem 2.7.15 and 2.7.16.

3.1.10. Do Problem *3.1.9* first.

3.1.14. Problem 3.1.13 may be relevant. Relabel 1, ..., n so that σ' and τ' become σ and τ.

3.1.15. Problem 3.1.14 may be relevant.

3.2.10. Write $(a\ b)(a\ c)$ and $(a\ b)(c\ d)$ as the product of three cycles.

3.2.12(d). Can you write every transposition as a product of $3k$ transpositions?

3.3.3. First decide which of $1, \ldots, 100$ are going to be in the long cycle, then create the cycle, and then decide how the remaining elements are going to be permuted.

3.3.4. You may want to do Problem 3.3.3 first.

4.2.4. If you have done Problem 4.1.7, then you can use those calculations here.

4.3.7. You may first want to try Problem 4.3.6.

4.4.16. Problem **4.4.15** may be relevant.

4.5.4(c). The previous two parts may be relevant.

5.1.9. Mimic the proof of Lemma 5.9.

5.1.15. Let Hg^t be an arbitrary right coset. Use the division algorithm to show that $Hg^t = Hg^r$ where $0 \le r \le m - 1$.

5.1.17(b). Consider H, Hx, and Hx^2. Could they all be distinct?

5.1.17(c). The previous part is relevant. Consider g^4.

5.1.19. You may want to use Problems 3.2.6 and **5.1.17**.

5.2.11. Problem **5.1.19** could be useful.

5.2.13. Let A be the set of right cosets of $V \cap U$ in V, and let B be the set of right cosets of U in G. Define an appropriate map $\theta : A \to B$, and show that it is well defined and 1-1. Use Lemma 5.6 whenever necessary.

5.2.14. In the proof of Problem **5.2.13** we used a map θ. We have the equality if the map θ is onto.

5.2.15. Problems **5.2.13** and **5.2.14** are very relevant.

5.3.6. Use Fermat's little theorem.

5.3.7. What if we lived in $\mathbb{Z}/4\mathbb{Z}$?

5.4.1(d). Problem **2.3.16** may be relevant.

5.4.1(e). Problem 2.3.7, Problem **5.1.17**, and the previous part of this problem could be relevant.

5.4.4. Problem **5.1.6** or Theorem 5.24(b) can be used. Note that because of Problem 5.4.3, the element ba has order 2 and is its own inverse.

5.4.7. Corollary 5.23 is relevant. What is the smallest possible value for $|G : \mathbf{Z}(G)|$?

5.4.8. Problem *5.4.7* is relevant.

5.4.11. First do Problem 5.4.10. Use Problem 2.6.5 and/or Proposition 5.23.

5.4.12. Use the result in Problem 5.4.11.

5.4.13. Define an appropriate action of $H \times K$ on $\Omega = G$ so that the double cosets are the orbits.

6.2.5. Problem *2.3.13* maybe relevant.

6.2.8. To show $\subseteq$, let $u, v \in G$ with $uv = a$ and $vu = b$. Note that it is enough to show that $u = cg$ for some $c \in C$. Define $c = ug^{-1}$, and show that it commutes with a. Problem 6.2.7 may help a bit.

6.2.14. Problem 4.4.16 may be relevant.

6.2.15. Use Problem 6.2.13 and the discussion preceding it.

6.4.6. From Problem 6.4.5, if $B_G(a) \neq \emptyset$ and $a^2 \neq e$, then $E_G(a)$ is a subgroup of even order.

7.2.9. In one of the cases, you may want to use Cauchy's theorem.

7.2.10. Problem *2.3.16* may be relevant.

7.3.2. See Problems *7.2.7* and *7.3.1*.

8.1.4. What is the average of $\mathrm{fix}(g)$ where g ranges over all elements of G? Is there any $g \in G$ with $\mathrm{fix}(g)$ above average?

8.1.5. For the first part, note that if $gxH \neq xH$, then $x^{-1}gx \notin H$. For the second part, let $H = \langle S \rangle$.

8.3.4. Mimic the steps for Problem 8.3.1.

9.2.1. Do Problem 5.2.11 first. Problem **5.1.19** could be useful.

9.2.2. Note that 3-cycles are elements of A_4, and use the lattice diagram of Figure 9.11.

10.1.12. Consider the $\mathbf{Z}(G)$.

10.1.13(b). Draw a partial lattice diagram as in Figure 10.1.

10.1.14. Mimic the proof of Proposition 10.13.

10.1.16. For one direction, look at Example 10.7. For the other direction, draw a partial lattice diagram as in Figure 10.1.

10.1.17. Draw a partial lattice diagram of S_4 that includes A_4 and a second subgroup of order 12. Proposition 10.6 and Problem **5.1.19** may be relevant.

10.1.18. Draw a partial lattice diagram that includes H and A_n.

10.2.3. Problem **5.1.19** could be useful.

10.2.6. You may want to use Proposition 10.18(e).

10.2.14. Draw a partial lattice diagram for $\mathbf{N}_G(P)$ that includes P and Q.

10.2.17. As a warmup do Problem 10.2.16 first.

10.2.21. Where does element $mnm^{-1}n^{-1}$ live?

10.2.23(a). Let G act on the set of its subgroups by conjugation. Count the number of elements in the subgroups that are in the orbit of H.

10.2.23(b). Look at Figure 10.3, and use the previous part.

10.3.6. Proposition 5.36 can be used to simplify your work on the last part.

10.3.9. Consider the group G/N and one of its elements.

10.3.12. First do Problem *10.3.11*.

10.3.13. Consider an appropriate quotient group. Problem 6.3.3 may be relevant.

10.3.14. Problem 6.3.3 may be relevant.

10.3.15. Use Problem 10.1.12, and form a quotient group. Problem *10.3.11* may be relevant.

10.4.4. Let G be the group, and let $H \leq G$. Let $N_0, N_1, \ldots, N_k$ be the series of subgroups of G that demonstrate the solvability of G. Draw a partial lattice diagram of the group N_i and show that $|H \cap N_i : H \cap N_{i-1}|$ is either one or the same as $|N_i : N_{i-1}|$. Proposition 9.28, Remark 10.12), and Lemma 10.21 may be relevant.

10.4.5.   Problems 2.6.25, 5.1.5, and **10.3.9** may be relevant. To prove Problem 10.4.5(d) $\Rightarrow$ 10.4.5(a) first show that $H, aH, a^2H, \ldots, a^{k-1}H$ are the distinct left cosets of H in G.

10.4.8. Take $n(a) = |G|$.

10.4.9. Use Cauchy's theorem, Corollary 7.11, and Problem 10.4.8 to find a normal subgroup of order 47. Then apply Problem **10.2.21**.

10.4.10(b). Let $K = aHa^{-1}$. Assuming that $K \neq H$, draw a partial lattice diagram, argue that $G = KH$, and get $a = kh$ with $k \in K$ and $h \in H$.

11.1.12. Problem **10.2.21** may be relevant.

11.3.4. Problem 11.1.5 may be relevant.

11.3.13. Problem 11.3.12 may be relevant.

11.3.14. Problem 11.3.13 may be relevant.

11.4.8. Consider the action of this group on the faces of the tetrahedron. Problem *10.1.17* may be relevant.

11.5.8. Problem **2.6.24** may be relevant.

11.6.6. For every $u \in (\mathbb{Z}/n\mathbb{Z})^\times$ show that the map $\phi_u : C \to C$ defined by $\phi_u(c) = uc$ is an automorphism of C. As a warm-up, do Problem *11.6.5* first.

11.6.11(b). What are the possibilities for the image of a and for the image of b?

11.6.12. What are the choices for the images of $(1,0)$ and $(0,1)$?

11.6.13. Draw the partial lattice diagram in Figure 11.26, and modify it using the given assumptions. Use Theorem 11.47 and Problem 11.6.11.

11.7.4. Draw pictures. Is either α or β 1-1 or onto?

11.7.17. Before finding $\mathrm{Aut}_c(D_8)$, you may want to do Problem 11.6.11.

12.1.1. Follow the strategy in the proof of Theorem 12.1.

12.1.4. Do Problem *12.1.3* first.

12.1.5. Do Problems *12.1.3* and 12.1.4 first.

12.1.6. Do Problem 12.1.5 first.

12.1.7. Show that P acts on N by conjugation. Then, similar to the proof of Corollary

6.19, consider the orbit sizes and the orbits of size 1.

12.3.3. Problem 12.3.2 may be relevant.

12.3.11. First do Problem 12.3.10.

12.3.13. First do Problem 12.3.12. Show that the Sylow subgroups of the subgroup of order 15 are normal in G.

12.3.17. Theorem 12.15 may be relevant.

12.3.21. First do Problem 12.3.20.

12.4.1.   What are the possibilities for $|\mathrm{Syl}_3(G)|$? Proceed as in the proof of Lemma 12.23.

12.4.3. Problem 12.4.2 may be relevant.

12.4.5. Let G act on $\mathrm{Syl}_7(G)$ (or on the cosets of $\mathbf{N}_G(P)$ where $P \in \mathrm{Syl}_7(G)$). Get a homomorphism and draw a partial lattice diagram including S_8, A_8, and the image of G.

12.4.6. As a warm up do Problem 12.3.13. Then show that G cannot have subgroups of order 15, 20, or 30.

13.1.7. First do Problem 13.1.1.

14.1.2. You could use Corollary 6.19.

14.1.4. Problems 14.1.2 and 14.1.3 are relevant.

14.1.20.   The steps outlined in Problem 14.1.19 are relevant.

14.2.13. Draw a diagram and note that H must be contained in a maximal subgroup of G.

14.2.15. To show that all Sylow p-subgroups of $\Phi(G)$ are normal, let $P \in \mathrm{Syl}_p(\Phi(G))$. The Frattini argument and Problem 14.2.13 can be used to show $P \lhd G$.

14.2.16.   To show (a) $\Rightarrow$ (b), let M be a maximal subgroup of G. Use Problem 14.2.8 and Corollary 14.11 to show $G' \leq M$. To show (d) $\Rightarrow$ (a), show all maximal subgroups of G are normal. Use homomorphism theorems to go back and forth between G and $G/\Phi(G)$.

14.2.17. To show that all Sylow p-subgroups of K are normal, let $P \in \mathrm{Syl}_p(K)$. Draw a diagram and argue that $P\Phi(G)/\Phi(G) \in \mathrm{Syl}_p(K/\Phi(G))$. Then apply the Frattini argument to $P\Phi(G)$ and use Problem 14.2.13 to show $P \lhd G$.

14.2.18. Let M be a maximal subgroup of G. To show that $\Phi(N) \leq M$, draw a partial lattice diagram with G, M, $\Phi(N)$, and $M \cap \Phi(N)$, then using Problem 9.2.9 insert N and $M \cap N$, and then use Problem 14.2.13.

14.2.19. For one direction, Problems 14.2.16 and 14.2.18 may be relevant. For the other, draw a partial lattice diagram with N, $\Phi(G)$, $N\Phi(G)$, $N \cap \Phi(G)$ and N', and use the direct diamond theorem (Theorem 11.43), Corollary 14.11, Problem 14.2.17, and Corollary 14.28.

14.2.20. Draw a diagram, apply Problem 14.2.17 to $K\Phi(G)$ with the help of the direct diamond theorem (Theorem 11.43), and then use Corollary 14.28.

14.3.13. Problem 14.1.19 may be relevant.

15.3.6. Try $u = 1 - a + b$.

16.1.10. After assuming that $f(r) \neq 0$ for some $r \in R$, first prove that $f(1) = 1$, and then consider $f((\sqrt{2})^2)$.

16.1.26(b). If (a, b) is in K, is $(a, 0)$ in K as well?

16.1.29. Problem 16.1.2 may be relevant.

16.1.30. Let P be the poset of linearly independent subsets of V.

16.2.14. Problem 16.2.13 is relevant.

16.3.8. If the characteristic of R is zero, let $S = \mathbb{Z}$ be the ring of integers. If the characteristic of R is n, then let $S = \mathbb{Z}/n\mathbb{Z}$ be the ring of integers mod n. Let $T = R \times S = \{(a, z) \mid a \in R, z \in S\}$. Define addition and multiplication on T as follows: $(a, z) + (b, w) = (a + b, z + w)$ and $(a, z)(b, w) = (ab + wa + zb, zw)$. Prove that T has the desired properties.

16.4.17. Find an appropriate homomorphism $\phi : S \rightarrow (S + I)/I$ and determine its kernel.

17.2.8. Mimic the proof of Theorem 17.21. You only have to make a few adjustments.

17.2.10. The previous problem is relevant.

18.1.18. Norms may be helpful.

18.1.19. Norms may be helpful.

18.1.20. You may want to use a spreadsheet or a simple computer program.

18.1.30. Do Problem 18.1.29 first.

18.1.31. Problems 16.1.26 and 16.2.14 may be relevant.

18.2.9. By way of contradiction, let $\mathcal{F}$ be the family of ideals of R that do not satisfy the claim. Use Problem 18.2.3 to find a maximal element of $\mathcal{F}$ and use the fact that this ideal is not prime.

18.3.15. Problem 18.3.14 and Theorem 18.41 may be relevant.

18.4.5. First show that the value of the function d is never less than $d(1)$.

18.4.7. Show that the norm is the degree function. First show that given a and b in $\mathbb{Z}[i]$, we can find an element q of $\mathbb{Z}[i]$ such that the distance in the complex plane of b/a and q is less than $\frac{\sqrt{2}}{2} < 1$.

18.4.8. Is Problem 18.3.10 relevant?

18.4.9. Problem 18.3.11 may be relevant.

18.5.5. Problem 18.5.4 may be relevant.

18.6.1. Corollary 16.36 may be relevant.

18.6.12. Use the representation given in Problem 18.6.11 to define an appropriate function $d: D - \{0\} \rightarrow \mathbb{Z}^{\geq 0}$.

18.6.14(g). Problem 18.6.1(c) may be relevant.

18.6.24(b). Theorem 18.63 may be relevant.

18.6.25. Mimic the proof of Theorem 18.50.

18.6.26. First show that the first three conditions are equivalent. For the final condition, Problems 18.6.24 and 18.6.25 are relevant.

18.6.27. Use Zorn's Lemma.

18.6.28. Problems 18.3.12 and 18.6.27 may be relevant.

19.2.6. Problem *16.1.17* maybe relevant.

19.2.7. Problem 18.1.13 maybe relevant.

19.2.11. Consider the Euclidean algorithm for finding the greatest common divisors.

19.3.15. Write $x^4 = -1 = e^{(2k+1)i\pi}$ where k is an integer.

19.3.16. Do Problem *19.3.15* first.

19.3.22. If $\alpha \in \mathbb{F}_p$ is a root of f, then what is the order of α as an element of the group $\mathbb{F}_p^\times$?

). As a warmup do Problem 19.3.22. If $\alpha \in \mathbb{F}_p$ is a root of f, then α is also a root of $x^3 - 1$.

19.3.24. Assume that the answer is no and proceed as in Euclid's proof of the infinitude of primes.

19.4.7. Corollary 19.35 may be relevant.

19.4.15. Is h irreducible in $(K[y])[x]$? Is Gauss's Lemma relevant?

19.4.19. Problem 18.3.15 may be relevant.

19.4.20. Problems 18.3.15 and 19.1.7 may be relevant.

19.5.4. Write $1 + x + \cdots + x^8 = \frac{x^9 - 1}{x - 1}$. Do the same for $1 + x + x^2$ and $1 + x^3 + x^6$.

19.5.5. First find $(x - 1)\Phi_p(x)$, then use Problem **19.5.3** and Schönemann-Eisenstein criterion.

19.5.6. Problems 19.5.4 and **19.5.5** are relevant.

19.5.7. Problem **19.5.3** may be relevant.

19.5.13. Write the polynomial as $(x^2 + x + 1)^2 + 1$, then find all of its complex roots.

19.5.15. Let $R = \mathbb{Z}[x]$ so that $\mathbb{Z}[x, y] = R[y]$.

19.6.2. Corollary 19.67 and Problem 18.2.1 may be relevant.

19.7.5. Problem *19.7.3* may be relevant.

19.7.8. First show that $\langle x \rangle$ is the unique maximal ideal, then show that $F[[x]]$ is a PID.

19.7.14. Problem 19.7.12(b) may be relevant.

20.1.13. Mimic the proof of Theorem 20.5.

20.2.1. Any solution over $\mathbb{Z}/4\mathbb{Z}$?

20.2.4. Modify the proof of Theorem 20.5.

22.1.2. Induct on n, and let $F = \mathbb{Q}[\sqrt{p_1}, \ldots, \sqrt{p_{n-1}}]$.

22.1.6. What is the minimal polynomial of $a + bi$? What about $a - bi$?

22.1.9. Problem **19.5.5** may be helpful.

22.1.10. Consider $\left(\zeta_7 - \cos\left(\frac{2\pi}{7}\right)\right)^2$.

22.1.15. Problem 22.1.6 may be relevant.

22.1.17. The evaluation homomorphism maybe helpful.

22.1.18(a). What is $(1 + x + x^2)(1 - x)$?

22.1.24(b). Do Problem 19.4.15 first.

22.1.25. Do Problems 19.4.15 and 22.1.24 first.

22.1.27. Consider $(x - \pi)(x - e)$.

22.3.5. Problem 22.1.1 may be relevant.

22.3.6. Example 22.11 and Problem 22.3.5 may be relevant.

22.3.15(b). This question can be resolved with hardly any calculations.

22.3.19. See Example 24.19.

22.3.23(a). Consider the logic of Example 22.40.

22.3.31. Let $K = F(Y)$ where Y is the set of coefficients of $\min_F(\alpha)$. Show α is algebraic over K, and use the fact that $F \subseteq F[\alpha] \subseteq K[\alpha]$.

22.3.33. See Example 24.19.

22.3.37. Let γ be a root of $x^{nm} - a$, $\alpha = \gamma^n$, and $\beta = \gamma^m$. Considering the fields $F(\gamma)$, $F(\alpha)$, $F(\beta)$, and $F(\alpha, \beta)$ may prove useful.

22.3.39. Problems 22.1.25 and 19.4.15 may be relevant.

22.3.41. Do Problems *22.3.39* and 22.3.40 first.

23.1.4. Draw a perpendicular HJ from H to the line through A and C. Triangles ACG and AJH are similar.

23.2.8. $5 \times 18 = 90$.

23.2.11. What is the minimal polynomial of $\cos(2\pi/7) + i\sin(2\pi/7)$? Problem 23.2.10 may be relevant.

24.1.10. Use Theorem 24.14.

24.1.12. Problem 22.1.2 may be relevant.

24.1.17. Problem 24.1.16 could be relevant.

24.1.18(b). Proposition 24.5 may be relevant.

24.1.19. Consider a field $L \supseteq E$ in which f has a root.

24.2.11. After choosing L, possibly use induction on n.

24.3.5. Consider the action of the Galois group on the roots of $\min_F(\alpha)$.

24.3.6. First show that any automorphism of the reals will have to preserve order.

24.3.9. Do Problem 19.3.17 first.

24.3.10. See Example 24.19.

24.3.12. Is the action of the Galois group on the roots transitive?

24.3.14. Problems 22.1.24 and 22.1.23 may be relevant.

24.3.15. To show that ϕ is onto, let $\alpha \in E$, and consider the restriction of ϕ to the smallest subfield of E that contains F and all the roots of $\min_F(\alpha)$.

25.2.7. Lemma 25.8 and Problem *24.3.5* may be relevant.

25.2.10. Problems 22.1.2 and 24.1.12 may be relevant.

25.3.15. Consider conjugation as a possible element of the Galois group.

25.3.16. Theorem 24.24 may be relevant.

25.4.10. Theorem 24.6 may be relevant.

25.4.14. Problem *25.4.13* may be relevant.

25.5.6. Theorem 20.9 may be relevant.

25.5.13. Problem 25.5.12 and Theorem 25.43 may be relevant.

25.5.14. Consider $K[u + \beta v]$ where $\beta \in K$.

25.6.1. Theorem 24.44 and Proposition 24.5 may be relevant.

25.6.18(b). What is $(\alpha_1 + \alpha_2 + \alpha_3)^2$?

26.1.7. Corollary 12.7 can be relevant.

26.2.3. Do Problem 26.2.2 first.

26.2.12. Use your knowledge of groups.

26.2.13. Parts of Problem *12.3.15* may be relevant.

26.2.14. Problem *12.3.17* may be relevant.

26.2.17(a). Problem 26.2.16 may be relevant.

26.2.19. Mimic the proof of the fundamental theorem of algebra, Theorem 26.11.

26.3.2. First find a field $E = \mathbb{Q}(\tau)$ such that E is a Galois extension of $\mathbb{Q}$ and $|E : \mathbb{Q}| = 6$. Then use Galois theory.

26.3.8. Do Problem 22.3.16 first.

26.3.9. Problem 24.3.14 may be relevant.

26.3.10(c). Problem 22.3.41 may be relevant.

26.3.11(e). Problem 1.5.6 may be relevant.

26.3.12(c). Problem 22.3.41 may be relevant.

27.1.5. Read over Example 27.8 first.

27.1.13. Do Problem 27.1.12 first.

27.1.14. Think over Example 27.8 and do Problem 24.1.17 first. Alternatively, use Problem 27.1.13.

27.1.18. Use Theorem 25.41.

27.1.19. Theorems 27.7 and 25.9 may be relevant.

27.1.20. Do Problem *27.1.19* first.

27.1.21. Consider a field of order p^n and a possible subfield of order p^m. Lagrange's theorem may be relevant.

27.1.23. Lagrange's theorem from group theory may be relevant.

27.1.24. Do Problem 27.1.23 first.

27.2.4. Theorem 27.29 and Problem 2.5.5 may be relevant.

27.2.5. Is the splitting field a cyclotomic extension?

27.2.11. You don't have to actually factor the polynomial.

27.2.14. What is n?

27.2.20. Problem 27.2.19 may be relevant.

27.2.21(c). Use induction on n and the previous parts.

27.2.24. Two monic polynomials of the same degree and with the same roots are equal.

27.3.9(b). If α is the constant term of a factor of degree r for f in $F[x]$, then show that, for some integers s and t, $\alpha^s b^t$ is in F and is a root of f.

27.4.1(b). Corollary 23.17 and Problem 23.2.10 may be relevant.

27.4.3(b). Problem 22.3.23 may be relevant.

27.4.3(c). Problem 2.5.10 may be relevant.

27.4.4(c). Problem 11.1.12 may be relevant.

27.4.5(a). Problem **15.2.24** may be relevant.

27.4.8. Problem *27.1.21* and Lemma 27.23 may be relevant.

28.2.6. Consider $p(x - 1)$.

28.3.4. Draw a diagram and use the theorem of natural irrationalities, Theorem 28.25.

28.4.2. Let ζ be a pth root of unity.

28.4.4. Do Problem 28.4.3 first.

28.4.7. As a warmup do Problem 28.4.6.

28.4.12(a). Problem 28.4.11 may be relevant.

28.4.13. Use Theorem 28.31.

Short Answers for Selected Problems

1.1.5. $\mathbf{Z}(D_8) = \{R_0, R_{180}\}$.

1.1.6. $\mathbf{Z}(D_6) = \{e\}$.

1.2.8. f (but not g) necessarily 1-1.

1.2.16. No right inverses, ∞ left inverses.

1.2.21. Yes.

1.3.6. Yes.

1.4.5(b). $p^n - p$.

1.4.11. Yes.

1.5.8. $2^{2n} - 1$. No. Yes. No.

2.1.4(a). No.

2.1.4(c). Yes.

2.1.5. Yes, yes.

2.1.6. $n = 1, 2$.

2.3.10. $o(b) \mid n$.

2.3.16. pq.

2.3.20. $\phi(n)$.

2.3.21. Yes.

2.4.7. 8. No.

2.4.8. No, yes.

2.4.14. Yes.

2.4.15. Yes. No.

2.5.3. Yes.

2.5.6. No.

2.5.9. $\mathrm{lcm}(o(h), o(k))$.

2.5.11. Yes.

2.5.13. No.

2.6.10. Yes.

2.6.13. 21. $\mathbb{Z}/21\mathbb{Z}$

2.6.15. Yes.

2.6.18. Yes.

2.6.22. 0 or 1.

2.6.30. G.

3.1.4. 8. 11.

3.1.5. Yes.

3.1.6. Yes. Yes. Yes.

3.1.7. 2,520.

3.2.2. No.

3.2.4. 15.

3.2.7. No.

3.2.8. No.

3.2.9. No.

4.3.7. $k!(n-k)!$.

4.4.4. No.

5.1.4(a). 4.

5.1.5. No.

5.1.14. No.

5.1.19. 8. No.

5.2.2. 27.

5.2.4(a). 14.

5.2.8. 143.

5.3.3. 0.

5.3.4. 12.

5.3.6. No.

5.3.7. No.

6.1.1. No.

6.1.3. 5.

6.2.3(e). 6.

6.2.4. Yes.

6.2.11. 3, 6.

6.2.15. 5 with sizes 1, 12, 12, 15, and 20.

6.2.17. $p(p + 1)$.

6.3.2. 3.

7.1.2. Even.

7.1.3. 1.

7.2.5. 1.

7.2.8. No.

7.2.9. 1, 21.

7.3.6. No. Yes. Yes. Yes.

8.2.2. 396.

8.2.3. 56.

8.2.7. 10,763,361.

8.2.10. 6, D_8.

8.2.11. 96.

9.2.5. 240. 15. Yes. No.

9.2.6. 27, 9.

9.2.7(d). 5.

10.1.2. No. Yes.

10.1.6. Yes.

10.2.10. Yes. 99.

10.3.1. 6. S_3.

10.3.3(c). $\mathbb{Z}/2\mathbb{Z} \times \mathbb{Z}/2\mathbb{Z}$.

11.1.2. Yes. $\{0, 4\}$. No.

11.1.10. Yes. No. Yes. No.

11.3.1. No.

11.3.6. S_3. $\{e\}$.

11.4.2. 5.

11.4.8. A_4.

11.4.9. $\frac{m^2(m^2+11)}{12}$

11.5.1. 60, 15

11.5.2. 120. 40. $\mathbb{Z}/8\mathbb{Z}$. Yes. $\mathbb{Z}/3\mathbb{Z}$.

11.5.3. 24. 12. $\mathbb{Z}/3\mathbb{Z}$. Yes. $\mathbb{Z}/2\mathbb{Z}$.

11.6.3. Yes.

11.7.1. Yes.

12.3.1. 13. 45. 13.

14.1.1. $A_5 \times \mathbb{Z}/2\mathbb{Z}$.

14.1.12. Yes.

14.3.5. No. Yes. Yes. Yes. Yes.

14.3.12. Yes. Yes. Yes.

15.1.6. $y = 1$, $x = 0$.

15.2.1. Yes.

15.2.4. No. No.

15.2.6. No. No. Yes.

15.2.8. No. Yes.

15.2.9. Yes. Yes. Yes. No.

15.2.16. Yes. No.

15.2.24. Yes.

16.1.1. No.

16.1.2. Yes.

16.1.7. 1.

16.1.9. Yes, no, yes.

16.1.13. No, yes.

16.1.16. Yes.

16.1.17. No.

16.2.3. Yes. $\mathbb{Z}/3\mathbb{Z}$.

16.2.8. $\mathbb{Z}/29\mathbb{Z}$. Yes.

16.3.1. Yes.

16.3.2. Yes.

16.3.6. No. Yes.

16.4.1. $\langle 30 \rangle$

16.4.5. $\langle 6 \rangle$

17.1.3. $\mathbb{Q}$.

18.1.2. Yes.

18.1.6. No.

18.1.8. Not necessarily.

18.1.16. Yes. P. $\mathbb{C}[y]$. Yes. No. No.

18.1.21. ± 1. Yes. No.

18.1.24. Yes. No. No.

18.1.25. No. No. Yes.

18.1.30. No. Yes.

18.2.2. No.

18.2.6. Yes. No.

18.3.2. No.

18.6.3. Yes. Yes. No.

18.6.24. No. Yes.

19.1.4. No.

19.2.10. Yes.

19.3.5. Yes. 8. Yes. 2. 7. 2.

19.3.8. Yes. No. Yes.

19.3.11. 6. No. Yes. 27.

19.3.17. $\pm\frac{\sqrt{3}}{2} \pm \frac{1}{2}i$.

19.3.19. 3. 3. 6.

19.3.24. Yes.

19.4.6. No.

19.4.12. Yes. Yes. No. No. No. No. No.

19.4.14. -1, 0, 2.

19.5.4. No.

19.5.8. No.

19.7.4. No. Yes.

20.1.14. ± 1. No.

22.1.19. Yes.

22.2.4. Yes. 2. No. No.

22.2.9. Yes. Yes. No.

22.3.5. $x^2 - 3$. 2. 4. $\{1, \sqrt{2}, \sqrt{3}, \sqrt{6}\}$.

22.3.8. No.

22.3.17. Yes.

22.3.19. Yes.

22.3.27(a). No.

22.3.28. No. No. No.

22.3.33. No. Yes.

23.2.5. Yes.

23.2.6. No.

23.2.7. No.

23.2.8. Yes.

23.2.11. No.

24.1.2. No. Yes.

24.2.5. $\mathbb{Q}[\sqrt{2+\sqrt{5}}, i]$. 8.

24.2.10. $p(p-1)$.

24.2.13. $\mathbb{A}$.

24.3.1. $\mathbb{Z}/2\mathbb{Z}$.

24.3.6. $\{e\}$.

24.3.8. $\mathbb{Z}/2\mathbb{Z} \times \mathbb{Z}/2\mathbb{Z}$.

24.3.10. S_3.

24.3.12. No.

25.2.1. No. Yes. Yes.

25.2.2. 1. No.

25.2.4. Yes.

25.2.6. Yes.

25.2.10. 16. 16. 16. Yes.

25.2.12(a). 6.

25.3.3. Yes. Yes. Yes. No.

25.3.4. Yes.

25.3.5. Yes. No.

25.3.6. Yes.

25.3.9. Yes.

25.4.3. No. 1.

25.4.7. 1.

25.4.14. Yes.

25.6.5. No.

25.6.12(a). $x^3 - 5x - 9 = 0$.

25.6.12(b). $\alpha = a/3$.

25.6.18(c). $-3q$.

25.6.18(d). $2p^2$.

26.2.4. No. Yes.

26.2.10. Yes.

26.2.12. Yes. 4.

26.3.12(d). D_8

26.3.13(d). S_3

27.1.8. Yes. 8.

27.1.9. 1 and 2.

27.1.14. $\dfrac{p^r - p}{r}$.

27.2.7. No. Yes.

27.2.11. 5^4. $\mathbb{Z}/4\mathbb{Z}$. $1, 4, 4, 4$.

27.2.24. $\Phi_{2n}(x) = \Phi_n(-x)$.

27.3.7. 10. 60. No. Yes. 121. $\mathbb{Z}/2\mathbb{Z}$.

27.4.5. Yes. 3. Yes. $\mathrm{GF}(3^\infty)$. ∞.

27.4.6. Yes. Yes. $\mathrm{GF}(3^\infty)$.

28.1.1. Yes.

28.1.6. Yes. 4. No.

28.3.10. Yes. Yes.

28.4.13. No.

Complete Solutions for Selected (Odd-Numbered) Problems

1.1.1. Multiplication table for D_8, the symmetries of a square (recall that $R_0 = e$, $R_{90} = a$, $R_{180} = a^2$, $R_{270} = a^3$, $H = b$, $D' = ab$, $V = a^2b$, $D = a^3b$) is as follows:

	e	a	a^2	a^3	b	ab	a^2b	a^3b
e	e	a	a^2	a^3	b	ab	a^2b	a^3b
a	a	a^2	a^3	e	ab	a^2b	a^3b	b
a^2	a^2	a^3	e	a	a^2b	a^3b	b	ab
a^3	a^3	e	a	a^2	a^3b	b	ab	a^2b
b	b	a^3b	a^2b	ab	e	a^3	a^2	a
ab	ab	b	a^3b	a^2b	a	e	a^3	a^2
a^2b	a^2b	ab	b	a^3b	a^2	a	e	a^3
a^3b	a^3b	a^2b	ab	b	a^3	a^2	a	e

There are many interesting patterns, and in some sense that is what our study of groups is all about. One pattern is that every row and every column is a permutation of all of the elements. In other words, no element is repeated in any row or in any column. An $n \times n$ table where the entries are from n distinct symbols and there is no repetition in any row or in any column is called a *latin square*. Sudoku puzzles give examples of a special kind of latin squares. Another pattern is the way the rotations and the reflections are separated in the table. $\{e, a, a^2, a^3\}$ are the rotations and $\{b, ab, a^2b, a^3b\}$ are the reflections. Looking at the table, a—somewhat straightforward—fact jumps at you:

	Rotation	Reflection
Rotation	Rotation	Reflection
Reflection	Reflection	Rotation

You can—and, in fact, later we will—think of this as a multiplication table of a group as well:

1.4.5. To count the number of invertible $n \times n$ matrices with entries in $\mathbb{Z}/p\mathbb{Z}$, we proceed one row at a time. There are p^n choices for the first row, and we have to exclude the row of all zeros. Hence, there are $p^n - 1$ choices for the first row. For the second row, we have to exclude all the multiples of the first row. There are p multiples of the first row—since there are exactly p scalars—and hence the number of possibilities for the second row is $p^n - p$. Now for the third row, we have to exclude all the linear combinations of the first two rows. To find a linear combination of the first two rows, we have to choose two scalars, one to multiply the first row by and one to multiply by the second row. Hence there are p^2 possible linear combinations of the first two rows. So the number of rows that we can use for the third row is $p^n - p^2$. Continuing in this way, we get

THEOREM.
$$|\text{GL}(n,p)| = (p^n - 1)(p^n - p)(p^n - p^2)\cdots(p^n - p^{n-1}).$$

1.4.9. If a matrix is in $\text{GL}(n,p)$, then it is invertible and so its determinant cannot be zero. Hence, the choices for its determinant are 1, 2, ..., $p - 1$. The point is that the matrices in $\text{GL}(n,p)$ are distributed evenly among these different possible determinants, and so the number of matrices in $\text{GL}(n,p)$ that have determinant equal to 1 is exactly

$$|\text{SL}(n,p)| = \frac{|\text{GL}(n,p)|}{p-1} = \frac{(p^n - 1)(p^n - p)(p^n - p^2)\cdots(p^n - p^{n-1})}{p-1}.$$

Let α and β be two integers between 1 and $p - 1$. The only thing that remains to be shown is that the number of matrices with determinant α is equal to the number of matrices with determinant β. Let $\mathcal{S}_1 = \{M \in \text{GL}(n,p) \mid \det(M) = \alpha\}$ and $\mathcal{S}_2 = \{M \in \text{GL}(n,p) \mid \det(M) = \beta\}$. Let A be the $n \times n$ matrix defined by $A = \begin{bmatrix} \beta/\alpha & 0 \\ 0 & I_{n-1} \end{bmatrix}$, where I_{n-1} denotes the $(n-1) \times (n-1)$ identity matrix. Note that $\det(A) = \beta/\alpha$, and A is invertible. Define $f : \mathcal{S}_1 \to \mathcal{S}_2$ by $f(N) = AN$.

First note that if $N \in \mathcal{S}_1$, then $\det(AN) = \det(A)\det(N) = \frac{\beta}{\alpha}\alpha = \beta$, and so AN is indeed an element of $\mathcal{S}_2$. Hence, f is an actual function. If $AN_1 = AN_2$, then multiplying by A^{-1} on the left, we get $N_1 = N_2$. Thus f is 1-1. If M is an arbitrary element of $\mathcal{S}_2$, then $\det(A^{-1}M) = \det(A^{-1})\det(M) = \frac{\alpha}{\beta}\beta = \alpha$ and so $A^{-1}M \in \mathcal{S}_1$. In addition, $f(A^{-1}M) = M$ and so f is onto. f is a 1-1, onto map and so $|\mathcal{S}_1| = |\mathcal{S}_2|$, and the proof is complete.

2.2.3. We have $a^2 = e$ for all elements of G. This means that every element of G is its own inverse. Now let x and y be two arbitrary elements of G. xy is also an element of G, and thus $xy = (xy)^{-1} = y^{-1}x^{-1} = yx$. We conclude that the group is abelian.

2.2.5. Let $|G| = 3$, then $G = \{e, a, b\}$. We know (because of the cancellation property) that no row and no column of the multiplication table can have repeats. Fill in the row and the column for e, and then see what are the other possibilities. As we shall see, there will be only one possibility.

	e	a	b
e	e	a	b
a	a		
b	b		

Consider the element aa. It can be either e, a, or b. If $aa = a$, then $a = e$, which is a contradiction. $aa = e$ would force $ab = b$, but this would also imply $a = e$. Thus $aa = b$, and the rest of the table follows since in each row we need exactly one e, one a, and one b.

	e	a	b
e	e	a	b
a	a	b	e
b	b	e	a

Thus, up to isomorphism, there is only one group of order 3. In this group $b = a^2$ and hence the group is $G = \{e, a, a^2\}$ with $a^3 = e$. We can also write

$$G = \langle a \mid a^3 = e \rangle.$$

The group is clearly abelian.

NOTE. We know that $(\mathbb{Z}/3\mathbb{Z}, +)$ is a group of order 3, and we now know that there is really only one group of order 3. This means that, after proper relabeling of elements, the multiplication table of every group of order 3 will be identical to the table for $(\mathbb{Z}/3\mathbb{Z}, +)$ which in turn is identical to the multiplication table found above.

2.3.13. Let $g = yxy^{-1}$, $n = o(x)$, and $m = o(g)$. We want to show that $n = m$. The order of an element h is the smallest positive integer ℓ such that $h^\ell = e$. Hence, if k is a positive integer and $h^k = e$, we can conclude that $o(h) \leq k$. We have

$$g^n = (yxy^{-1})^n = (yxy^{-1})(yxy^{-1}) \cdots (yxy^{-1}) = yx^n y^{-1} = yey^{-1} = e.$$

We conclude that $m = o(g) \leq n$. On the other hand, from $g = yxy^{-1}$ we get $x = y^{-1}gy$ and so

$$x^m = (y^{-1}gy)^m = (y^{-1}gy)(y^{-1}gy) \cdots (y^{-1}gy) = y^{-1}g^m y = e.$$

This time, we conclude that $n = o(x) \leq m$. Since we have proved that $m \leq n$ and $n \leq m$, we now have $m = n$ and the proof is complete.

2.3.19. THEOREM. *Let $G = \langle g \rangle$ be a cyclic group of order n. Then g^m is a generator of G if and only if $\gcd(n, m) = 1$.*

PROOF. Note that since g is a generator for the cyclic group G, we have that $o(g) = n$.

Now, if g^m is a generator, then $g \in G = \langle g^m \rangle$ and so $g = (g^m)^k$ for some integer k. Thus $g^1 = g^{mk}$ which implies that $g^{mk-1} = e$ and so $n \mid (mk - 1)$ by Problem **2.3.8** (remember that this problem says that if $g^s = 1$, then $o(g) \mid s$). It follows that $mk - 1 = n\ell$ and hence $mk - n\ell = 1$. This implies that $\gcd(n, m) = 1$, since any common divisor of n and m would have to divide 1.

Conversely, if $\gcd(n, m) = 1$, we know that there exist integers k and w such that $mk + nw = 1$—remember that the gcd of two integers can always be written as an integer linear combination of the two integers—and thus

$$g = g^1 = (g^m)^k (g^n)^w = (g^m)^k e^w = (g^m)^k,$$

and so $g \in \langle g^m \rangle$. If $g \in \langle g^m \rangle$, and since $\langle g^m \rangle$ is a subgroup, then all powers of g must be in $\langle g^m \rangle$ as well. Hence $G = \langle g \rangle \subseteq \langle g^m \rangle$. All the powers of g^m are clearly in G also, and hence we also have $\langle g^m \rangle \subseteq G$. It follows that $G = \langle g^m \rangle$.

The number of integers between 1 and n that are relatively prime to n is $\phi(n)$ (Definition 1.45), and so we have proved that a cyclic group of order n has $\phi(n)$ different generators.

2.4.7. To find the elements of H, we find all possible powers and products of the two generators, and we get that

$$H = \left\{ a = \begin{bmatrix} 0 & i \\ i & 0 \end{bmatrix}, a^2 = b^2 = \begin{bmatrix} -1 & 0 \\ 0 & -1 \end{bmatrix}, a^3 = \begin{bmatrix} 0 & -i \\ -i & 0 \end{bmatrix}, a^4 = b^4 = e = \begin{bmatrix} 1 & 0 \\ 0 & 1 \end{bmatrix}, \right.$$
$$\left. b = \begin{bmatrix} 0 & 1 \\ -1 & 0 \end{bmatrix}, ab = \begin{bmatrix} -i & 0 \\ 0 & i \end{bmatrix}, a^2 b = b^3 = \begin{bmatrix} 0 & -1 \\ 1 & 0 \end{bmatrix}, a^3 b = ba = \begin{bmatrix} i & 0 \\ 0 & -i \end{bmatrix} \right\}.$$

The relations are $a^4 = e$, $b^2 = a^2$, and $ba = a^3 b$. Hence,

$$H = \langle a, b \mid a^4 = e, a^2 = b^2, ba = a^3 b \rangle.$$

The group H has order 8 and is not isomorphic to D_8. For example, in D_8 there are only two elements of order 4, but in H there are six elements of order 4. The group H is called the quaternion group of order 8 and denoted by Q_8.

2.4.9. Let $|G| = 4$. If a is any element of G, then all of its powers must be in G also. Thus $\langle a \rangle \subseteq G$, and so $o(a) = |\langle a \rangle| \leq |G|$. Pick a in G such that $o(a)$ is as large as possible. $o(a)$ can be 4, 3, or 2.

If $o(a) = 4$, then $G = \{a, a^2, a^3, e\}$. This is a cyclic group, and all cyclic groups of order 4 are isomorphic.

If $o(a) = 3$, then $a^{-1} = a^3$ and $G = \{e, a, a^2, b\}$, where b is the fourth element of the group. Now ab must be in G. Which element is it? $ab = b$ implies $a = e$; $ab = a^2$ implies $b = a$; $ab = a$ implies $b = e$; and finally $ab = e$ implies that b is the inverse of a. All of these are not true, and thus such a group cannot exist.

If $o(a) = 2$, then the order of all non-identity elements must be 2, since otherwise a would not have the largest order. Thus in this case $G = \{e, a, b, c\}$, and $a^2 = b^2 = c^2 = e$. With an argument identical to the above, the elements ab and ba cannot be e, a, or b. Thus $ab = c = ba$. We can now write the complete multiplication table. This group is called the *Klein 4-group* and is isomorphic to $\mathbb{Z}/2\mathbb{Z} \times \mathbb{Z}/2\mathbb{Z}$.

Thus there are only two groups of order 4: The cyclic group of order 4, and the Klein 4-group. These two groups are not isomorphic since one is cyclic (it has an element of order 4) and the other is not cyclic. The multiplication table for these two groups follow.

	e	a	a^2	a^3
e	e	a	a^2	a^3
a	a	a^2	a^3	e
a^2	a^2	a^3	e	a
a^3	a^3	e	a	a^2

	e	a	b	ab
e	e	a	b	ab
a	a	e	ab	b
b	b	ab	e	a
ab	ab	b	a	e

2.4.17. CLAIM. $o(\theta(g))$ divides $o(g)$.

PROOF. Assume $o(g) = n$, and let $x = \theta(g)$. We have
$$x^n = \theta(g)^n = \theta(g^n) = \theta(e) = e.$$
Thus $o(x)$ divides n.

Note that $o(g)$ does not have to equal $o(\theta(g))$. For example, consider $\phi : D_8 \to \{e\}$ defined by $\phi(x) = e$ for all $x \in D_8$. Then ϕ is a homomorphism, and $o(\theta(g)) = 1$ for all $g \in D_8$.

QUESTION. Can you show that $o(g) = o(\theta(g)) \cdot |\langle g \rangle \cap \ker(\theta)|$?

2.5.9. CLAIM. $o((h, k)) = \operatorname{lcm}(o(h), o(k))$.

PROOF OF CLAIM. Let $m = \operatorname{lcm}(o(h), o(k))$, and let $n = o(h, k)$ (in the group $H \times K$). We want to show that $m = n$.

The positive integer m is the least common multiple of $o(h)$ and $o(k)$. This means that m is a multiple of these two positive integers and also that m divides any other integer that is a multiple of both integers. Hence we have
$$m = uo(h) \quad \text{where } u \in \mathbb{Z}^{>0},$$
$$m = vo(k) \quad \text{where } v \in \mathbb{Z}^{>0}.$$
We now have
$$(h, k)^m = (h^m, k^m) = (h^{uo(h)}, k^{vo(k)}) = ((h^{o(h)})^u, (k^{o(k)})^v) = (e_H, e_K) = e_{H \times K}.$$
It follows that $n \mid m$. (Recall that if $g^s = e$, then $o(g) \mid s$.) On the other hand,
$$(e_H, e_K) = e_{H \times K} = (h, k)^n = (h^n, k^n).$$

Hence $h^n = e_H$ and $k^n = e_K$. Now $h^n = e_H$ implies that $o(h) \mid n$, and similarly $k^n = e_K$ implies that $o(k) \mid n$. So n is a multiple of both $o(h)$ and $o(k)$, and by the definition of least common multiple, this means that $m \mid n$.

We have $n \mid m$ and $m \mid n$, and that both are positive integers, hence $n = m$.

2.6.33. ($\Leftarrow$) Assume $HK = KH$, to show that HK is a subgroup of G. We need to show closure and the existence of inverses. Note that $HK = KH$ implies that every element of the form kh, where $k \in K$ and $h \in H$, can be written in the form $h'k'$ where h' is an element of H and k' is an element of K. You have to be careful in that $HK = KH$ does *not* imply that $kh = hk$. Now assume that $x, y \in HK \Rightarrow x = h_1 k_1$, and $y = h_2 k_2$. Thus

$$xy = (h_1 k_1)(h_2 k_2) = h_1(k_1 h_2)k_2 = h_1(h'k')k_2 = (h_1 h')(k'k_2) \in HK.$$

Thus closure is proved, and we have to show the existence of inverses. Let $x \in HK$. Then $x = hk$ for $h \in H$, $k \in K$. This implies

$$x^{-1} = k^{-1}h^{-1} = h'k' \in HK.$$

In the above we again used the fact that any product of the form $k^{-1}h^{-1}$ can be written in the form $h'k'$ since $KH = HK$.

($\Rightarrow$) Assume that HK is a subgroup of G. To show $HK = KH$.

We will first show that $KH \subseteq HK$. Let $x \in KH \Rightarrow x = kh$ for $k \in K, h \in H$. To show that $x \in HK$, we instead show that x is the product of two elements in HK. This does prove that $x \in HK$ since HK is a subgroup and hence closed under multiplication. But why is x a product of two elements of HK? The reason is that

$$x = (ek)(he) \quad \text{and} \quad ek \in HK, \ he \in HK.$$

So every element of KH is in HK, and thus we have proved that $KH \subseteq HK$.

On the other hand, let $y \in HK$. We need to show that $y \in KH$. Since HK is a group, we must have $y^{-1} \in HK$. Thus $y = hk$ for some $h \in H$, $k \in K$, and $k^{-1}h^{-1} = y^{-1} = h'k' \in HK$. So what? Note that we have

$$y = (y^{-1})^{-1} = (h'k')^{-1} = (k')^{-1}(h')^{-1} \in KH.$$

Thus $HK \subseteq KH$ and hence $HK = KH$.

3.1.9. To understand what the problem says, let us look at an example. Let $\sigma = (1\ 4\ 3\ 8)(2\ 6\ 5) \in S_8$, and let $\tau = (1\ 6\ 3)(7\ 5\ 2) \in S_8$. To conjugate σ by τ, we have to find $\tau\sigma\tau^{-1}$. In a general group to do this you will have to multiply the three elements. This problem says that for the group S_n there is a shortcut for conjugating. All you have to do is to apply τ to the *entries* of σ. For example, τ sends 1 to 6 and so we replace 1 by 6 in σ. We get that

$$\tau\sigma\tau^{-1} = (6\ 4\ 1\ 8)(7\ 3\ 2).$$

Now for the proof, observe that if $\sigma(i) = j$, then

$$\delta(\tau(i)) = \tau\sigma\tau^{-1}(\tau(i)) = \tau\sigma(i) = \tau(j).$$

Thus, if the ordered pair i, j appears in the cycle decomposition of σ, then the ordered pair $\tau(i), \tau(j)$ appears in the cycle decomposition of $\tau\sigma\tau^{-1}$. This completes the proof.

4.1.5.

(a) Recall that in the conjugation action on subgroups, $g \cdot H = gHg^{-1}$ and that $gHg^{-1} = \{ghg^{-1} \mid h \in H\}$. To prove that this is an action, we first have to know that $g \cdot H$ is a subgroup of G. Problem **2.6.27** already stated that for $g \in G$ and $H \leq G$, gHg^{-1} is a subgroup of G, but here is the proof anyway. If gh_1g^{-1} and gh_2g^{-1} are two elements of gHg^{-1}, then so is their product since $(gh_1g^{-1})(gh_2g^{-1}) = g(h_1h_2)g^{-1}$.

Additionally, the inverse of a typical element of gHg^{-1} is also an element of gHg^{-1} since $(ghg^{-1})^{-1} = gh^{-1}g^{-1}$.

To show that we have an action, we have to show that the two conditions of an action hold. First, note that $e \cdot H = eHe^{-1} = H$. Second $x \cdot (y \cdot H) = x \cdot (yHy^{-1}) = xyHy^{-1}x^{-1} = xyH(xy)^{-1} = xy \cdot H$.

(b) We have that $H = \{e, a\} = \{e, (1\ 2)\}$. We calculate the following:

$$a \cdot H = aHa^{-1} = \{aea^{-1}, aaa^{-1}\} = \{e, a\} = H,$$

$$b \cdot H = bHb^{-1} = \{beb^{-1}, bab^{-1}\} = \{e, (1\ 3)(1\ 2)(1\ 3)\} = \{e, (2\ 3)\} = \{e, c\},$$

$$c \cdot H = cHc^{-1} = \{cec^{-1}, cac^{-1}\} = \{e, (2\ 3)(1\ 2)(2\ 3)\} = \{e, (1\ 3)\} = \{e, b\}.$$

4.2.5. Here the group is $G = D_8$, the set Ω is the set of subgroups of G, the set of generators for G is $\{a, b\}$, and the action of G on Ω is conjugation. The set Ω has ten elements. These are the subgroups of G and each is a vertex in the Cayley digraph. We have to have a and b act on each subgroup to see what other subgroup we get. The element b actually fixes eight of the ten subgroups and switches the remaining two. More specifically, $b \cdot \langle ab \rangle = \langle a^3 b \rangle$ and $b \cdot \langle a^3 b \rangle = \langle ab \rangle$. (I have omitted the actual calculations but, for example, $b \cdot \langle ab \rangle = b \langle ab \rangle b = \{beb, babb\} = \{e, a^3 b\} = \langle a^3 b \rangle$.) Likewise, a fixes six of the ten subgroups, switches $\langle b \rangle$ and $\langle a^2 b \rangle$, and switches $\langle ab \rangle$ and $\langle a^3 b \rangle$. Using solid color lines for a and dashed colored lines for b, we get Figure C.1 as the Cayley digraph of this action for this set of generators.

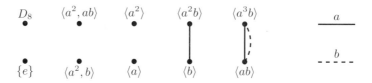

Figure C.1. Cayley digraph of the action of $D_8 = \langle a, b \rangle$ on its subgroups by conjugation

Two things can help the calculations: First, if $h \in H$, then hH is the product of h with every element of H, and the result is exactly the same as the elements in the row, corresponding to h, of the multiplication table of H. Hence $hH = H$. Likewise $Hh^{-1} = H$. We conclude that $hHh^{-1} = H$ for all $h \in H$. Second, the subgroups H and gHg^{-1} have the same size. Often, after finding only a few elements of gHg^{-1}, we can identify gHg^{-1}. Finally, note that six of the ten subgroups are completely fixed by the action. The subgroups of a group that are fixed under conjugation action will later be called the *normal* subgroups of the group and will play an important role in the theory.

4.3.3. Recall if G is a group, $x \in G$ and $H \leq G$, then $\mathbf{N}_G(H)$ is a subgroup, and $x \in \mathbf{N}_G(H)$ if and only if $xHx^{-1} = H$. First note that if $h \in H$, then $hHh^{-1} = H$. (See the solution to Problem *4.2.5* above.) As a result, $H \leq \mathbf{N}_G(H)$. Similarly if $z \in \mathbf{C}_G(H)$—the elements of G that commute with every element of H—then $zH = Hz$ and so $zHz^{-1} = Hzz^{-1} = H$.

Now for $H = \langle b \rangle$, since both H and $\mathbf{Z}(G)$ are in $\mathbf{N}_G(H)$, we have $\langle b, a^2 \rangle \leq \mathbf{N}_G(H)$. If x is any element not in $\langle b, a^2 \rangle$, then $\langle b, a^2, x \rangle$ is the whole group D_8. Hence—since $\mathbf{N}_G(H)$ is a subgroup—the normalizer of H in G is either $\langle b, a^2 \rangle$ or all of D_8. But $aHa^{-1} \neq H$ and so the normalizer cannot be all of D_8. We conclude that $\mathbf{N}_G(H) = \langle b, a^2 \rangle$.

Now $a^2 \in \mathbf{Z}(D_8)$ and so $gKg^{-1} = K$ for all $g \in G$. We conclude that $\mathbf{N}_G(K) = D_8$.

You can also use the Cayley digraph that was constructed in Problem *4.2.5* to find normalizers. The elements of G that fix a subgroup (in the conjugation action of G on subgroups) constitute the normalizer of that subgroup. So, for example, $K = \langle a^2 \rangle$ is an isolated vertex of that graph. This means that *every* element of G fixed K. Thus, $\mathbf{N}_G(K) = G$. On the other hand, $H = \langle b \rangle$ is fixed by b and moved to $\langle a^2 b \rangle$ by a. Going through the elements of G and, for each, using the Cayley digraph, we see that $\{e, b, a^2, a^2 b\}$ all fix H, and hence $\mathbf{N}_G(H) = \langle b, a^2 \rangle$.

5.2.15. We have $U \cap V \leq U \leq G$, and so $|G : U \cap V| = |G : U| \, |U : U \cap V|$. Thus $|G : U|$ divides $|G : U \cap V|$. However, we also have $U \cap V \leq U \leq G$, and so $|G : U \cap V| = |G : V| \, |V : U \cap V|$. Now, $|G : U|$ divides $|G : U \cap V|$ and $\gcd(|G : U|, |G : V|) = 1$. We conclude that $|G : U|$ divides $|V : U \cap V|$. In particular $|G : U| \leq |V : U \cap V|$.

In Problem **5.2.13** we had proved that we *always* have $|V : U \cap V| \leq |G : U|$. Thus we must have $|G : U| = |V : U \cap V|$, and so by Problem **5.2.14**, we have $G = VU$.

5.4.7. We know $|G : \mathbf{Z}(G)| \neq 1$ and, by Proposition 5.23, $|G : \mathbf{Z}(G)| \neq 2, 3$. Hence $|G : \mathbf{Z}(G)| \geq 4$ which means that $|\mathbf{Z}(G)| \leq \frac{1}{4}|G|$. The group D_8 has eight elements and its center—the subgroup generated by the 180 degree rotation—has two elements. Hence, exactly one fourth of the elements of D_8 are in the center.

6.2.5. Two elements in the same conjugacy class always have the same order. (See the solution to Problem *2.3.13*.) Since x is the unique element of order 2, we conclude that the conjugacy class of x consists of only x. But this means that x is in the center. The reason is that if $g \in G$ is arbitrary, then $gxg^{-1} = x$ implies that $gx = xg$ proving that x commutes with every element of the group. Alternatively, if $|\mathrm{cl}_G(x)| = 1$, then $|\mathbf{C}_G(x)| = |G| / |\mathrm{cl}_G(x)| = |G|$.

6.2.11. We want to find $|\mathbf{C}_G(g)|$ when $G = S_4$ or $G = S_5$ and $g = (1\ 2\ 3)$. We will first find $|\mathrm{cl}_G(g)|$, the size of the conjugacy class of g in G, and then, to compute $|\mathbf{C}_G(g)|$, we use the fact that

$$|G| = |\mathbf{C}_G(g)| \ |\mathrm{cl}_G(g)|.$$

Let $G = S_4$. In Problems *3.1.9* and *4.4.15* we saw that two elements of S_4 are conjugate if and only if they have the same cycle type. So how many elements have the same cycle type as g? For this cycle type, we need one element of the set $[4] = \{1, 2, 3, 4\}$ to be fixed, and then for the remaining three elements, there are two choices (e.g., $(1\ 2\ 4)$ and $(1\ 4\ 2)$). Hence, the total number of elements that are conjugate to g in S_4 are $4 \times 2 = 8$. We conclude that

$$|\mathbf{C}_{S_4}(g)| = 4!/8 = 3.$$

So three elements in S_4 commute with g. In fact, these are e, g, and g^2. We now know that nothing else in S_4 commutes with g.

Now let $G = S_5$. Proceeding as before, we see that $|\mathrm{cl}_G(g)| = \binom{5}{2} \times 2 = 20$, and

$$|\mathbf{C}_{S_5}(g)| = 5!/20 = 6.$$

This time, there are six elements that commute with $(1\ 2\ 3)$. It is easy to write them down:

$$\mathbf{C}_{S_5}(g) = \{e, (1\ 2\ 3), (1\ 3\ 2), (4\ 5), (1\ 2\ 3)(4\ 5), (1\ 3\ 2)(4\ 5)\}.$$

7.2.7. If H is a subgroup of the group G, we construct xHx^{-1} by taking elements of H one at a time, premultiplying them by x, and postmultiplying them by x^{-1}. The resulting set of elements is xHx^{-1}.

LEMMA. *Let $H \leq G$ be groups, and let $x \in G$. Then xHx^{-1} is a subgroup of G, and if H is a finite group, then $|xHx^{-1}| = |H|$.*

PROOF. This is Problem **2.6.27**, but we repeat the proof here for completeness:

(a) First, $e = geg^{-1} \in gHg^{-1}$. If $x, y \in gHg^{-1}$, then $x = ghg^{-1}$ and $y = gkg^{-1}$ for $h, k \in H$, and we have $xy = ghkg^{-1} \in gHg^{-1}$ since $hk \in H$. Also if $x \in gHg^{-1}$, then $x = ghg^{-1}$ for some $h \in H$, and thus $x^{-1} = (ghg^{-1})^{-1} = gh^{-1}g^{-1} \in gHg^{-1}$. Thus, gHg^{-1} is non-empty and closed under products and inverses, and, hence is a subgroup.

(b) Basically it follows from the cancellation laws that $|H| = |xHx^{-1}|$. To see this, let $H = \{h_1, h_2, \ldots, h_m\}$. Then $xHx^{-1} = \{xh_1x^{-1}, xh_2x^{-1}, \ldots, xh_mx^{-1}\}$. There are no repetitions in this list since $xh_ix^{-1} = xh_jx^{-1}$ implies—by cancellation laws—that $h_i = h_j$, and hence $i = j$. Thus, we showed that if H has m elements, then so does xHx^{-1}.

COROLLARY. *Let P be a Sylow p-subgroup of a finite group G. Let x be an arbitrary element of G. Then xPx^{-1} is also a Sylow p-subgroup of G.*

PROOF. A subgroup of G is a Sylow p-subgroup of G if and only if its order is the highest power of p dividing the order of G. By the lemma, xPx^{-1} is a subgroup of G and has the same size as P. Thus xPx^{-1} is a Sylow p-subgroup of G.   □

Note. The converse of the above corollary is also true, though harder to prove. In other words, every two Sylow p-subgroups (for the same prime p) are conjugate in G.

LEMMA. *If P is the unique Sylow p-subgroup of G, then $xP = Px$.*

PROOF. Let x be an arbitrary element of G. We need to show that $xP = Px$. This is equivalent to showing that $xPx^{-1} = P$. But xPx^{-1} is another Sylow p-subgroup and since P is the only Sylow p-subgroup, we must have that $xPx^{-1} = P$.   □

Note. The converse of this lemma is also true; it follows from the converse of the first corollary.

7.3.1. Let $\Omega = \{xP | x \in G\}$ be the set of left cosets of P in G. The group Q acts on Ω by left multiplication, i.e., $q \cdot xP = qxP$, and we know that the orbits partition Ω. By the FCP the size of orbits are divisors of $|Q|$, and so are powers of p.

CLAIM. There exists some orbit of size 1.

PROOF. If all orbits had size bigger than 1, then the size of each orbit would be divisible by p. But the size of Ω is the sum of the sizes of orbits (since orbits partition Ω) and if the size of each orbit is divisible by p, then the size of all of Ω would be divisible by p. Now the size of Ω is the same as the number of cosets of P in G and so it is $|G : P|$. But $P \in \mathrm{Syl}_p(G)$ which implies that p does not divide $|G : P| = |G| / |P|$. The contradiction implies that not all orbit sizes can be greater than 1. Thus there must be an orbit of size 1.

Let $\{gP\}$ be an orbit of size 1.

CLAIM. $Q \leq gPg^{-1}$.

PROOF. Let $x \in Q$. We need to show that $x \in gPg^{-1}$. Now gP is in an orbit of size 1 and is thus fixed by the action of elements of Q. We have

$$x \cdot gP = gP \Rightarrow xgP = gP \Rightarrow xgPg^{-1} = gPg^{-1}.$$

If we let $H = gPg^{-1}$, then we have shown

$$xH = H \Rightarrow x \in H = gPg^{-1}.$$

Thus every p-subgroup of G (a p-subgroup of G is a subgroup of G whose size is a power of p) is inside a conjugate of P. Now conjugates of P are also Sylow p-subgroups, and, hence, we have proved that every p-subgroup of G is inside some Sylow p-subgroup of G.

7.3.3. Let Ω be the set of subgroups of G. Then G acts on Ω by conjugation. The subgroup P is an element of Ω, and the stabilizer of P in G is $\mathbf{N}_G(P)$. The FCP says that the size of the orbit of P in G is $|G : \mathbf{N}_G(P)|$. Our only question: What are the elements of the orbit of P in this action?

Every element of the orbit of P is a subgroup of the form xPx^{-1}. But xPx^{-1} is itself a Sylow p-subgroup of G. In other words, every member of the orbit of P is a Sylow p-subgroup. So far it is possible that there are Sylow subgroups of G outside of the orbit of P.

However, in Corollary 7.14, we proved that if Q is another Sylow p-subgroup of G, then $Q = xPx^{-1}$ for some $x \in G$. In other words, Q is a conjugate of P and is in the orbit of P. Hence *all* Sylow p-subgroups of G are in the orbit of P. The argument used in that corollary was based on the Sylow D theorem, Theorem 7.13—proved in Problem **7.3.1**—and is straightforward: Since P is a Sylow p-subgroup and Q is a p-subgroup, then, by the Sylow D theorem, $Q \leq xPx^{-1}$. But Q is actually a Sylow p-subgroup and, hence, the same size as xPx^{-1}. We conclude that $Q = xPx^{-1}$.

It follows that $\mathrm{Syl}_p(G)$ is exactly the orbit of P in this action. We know that the size of this orbit is $|G : \mathbf{N}_G(P)|$, and hence this is exactly the number of Sylow p-subgroups (or the size of $\mathrm{Syl}_p(G)$ which is the set of Sylow p-subgroups in G). So we have $|\mathrm{Syl}_p(G)| = |G : \mathbf{N}_G(P)|$. It follows that the *number* of Sylow p-subgroups divides the order of G. $\square$

7.3.5.

(a) N is a subgroup of G, and so the order of N divides the order of G. This means that the highest power of the prime p dividing $|N|$ is no more than the highest power of p dividing $|G|$. On the other hand, every subgroup is inside its own normalizer and, hence, $P \leq N$. This means that $|P|$—which is the highest power of p dividing $|G|$—divides $|N|$. We conclude that the size of Sylow p-subgroups of G and N are the same. Now $P \leq N$ and hence $P \in \mathrm{Syl}_p(N)$.

(b) If $x \in N$, then, by definition of $\mathbf{N}_G(P)$, $xPx^{-1} = P$. Hence $x \in \mathbf{N}_N(P)$. We conclude that $\mathbf{N}_N(P)$ is all of N.

(c) The number of Sylow p-subgroups of N is equal to $|N : \mathbf{N}_N(P)| = |N : N| = 1$. Hence, there is only one Sylow p-subgroup in N and that is P.

7.3.7. Assume that $Q \in \mathrm{Syl}_p(G)$ is fixed by the conjugation action of P on $\mathrm{Syl}_p(G)$. Then $xQx^{-1} = Q$ for all $x \in P$. This means that for all $x \in P$, we have $x \in \mathbf{N}_G(Q)$. Thus $P \leq \mathbf{N}_G(Q)$. Now consider the group $N = \mathbf{N}_G(Q)$. Both P and Q are subgroups of N and both are Sylow p-subgroups of N. By Lemma *7.3.5*, N has only one Sylow p-subgroup, and hence $Q = P$.

7.3.9. Let $\mathcal{S} = \mathrm{Syl}_p(G) = \{P_1, P_2, \ldots, P_s\}$ be the set of Sylow p-subgroups of G. Let $P = P_1$. Now the group P acts on the set $\mathcal{S}$ by conjugation. In other words, for $x \in P$ and $Q \in \mathcal{S}$, we define $x \cdot Q = xQx^{-1}$. Note that we had already seen that the whole group G acts on subgroups of G and that one of the orbits of this action is the set of Sylow p-subgroups of G. (see Problem **7.3.3** or Corollary 7.15.) Because $\mathrm{Syl}_p(G)$ is one of the orbits, we can restrict the action of G to $\mathrm{Syl}_p(G)$. Here, we actually want to restrict the action further, and have P, a subgroup of G, act on $\mathrm{Syl}_p(G)$. But this time, we may not have one orbit any more (if we had the whole group G act, then there would be one orbit). We know the orbits of the action partition the set. So the orbits of the action of P on $\mathcal{S}$ partition $\mathcal{S}$. What are the sizes of the orbits?

The size of each orbit divides $|P|$, the size of the group acting. Now $|P|$ is a power of p, and hence all orbit sizes are powers of p. How many orbits of size 1? (Note that $1 = p^0$ is a power of p.) In Problem *7.3.7*, we showed that P is the only element of $\mathcal{S}$ that is fixed by this action.

Hence, there is exactly one orbit of size 1 and all the other orbits have sizes divisible by p.

The orbits partition the set, and so the sum of the orbit sizes is $|\mathcal{S}|$, the size of the set. So we have

$$|\mathcal{S}| = 1 + kp \quad \text{for some non-negative integer } k.$$

9.2.5. To guide your argument, draw a partial lattice diagram of G that includes H, N, $H \cap N$, and $\langle N, H \rangle$. Since $\mathbf{N}_G(N) = G$ (in Chapter 10, we say that N is normal in G), we know, by Proposition 9.32, that NH is a subgroup of G. This is crucial, since it implies, by Proposition 9.28, that $\langle N, H \rangle = NH$, we can draw a parallelogram, and $|NH : N| = |H : H \cap N|$.

We know that $|N| = 60$ and $|H| = 16$. Looking at the diagram, we let $a = |G : NH|$ and $b = |NH : N| = |H : H \cap N|$. The integer b divides $|H|$, and so it is a power of 2. On the other hand a divides $|G : H|$ which is 45. Since $ab = |G : N| = 12$, we conclude that $a = 3$ and $b = 4$.

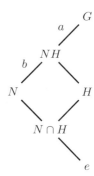

Now, we can calculate the other indices. Have $45 = |G : H| = a|NH : H|$ and so $|NH : N| = 15$. Thus $|NH| = |NH : H||H| = 15 \times 16 = 240$.

Finally, $|H| = |N \cap H| \, |H : N \cap H| = |N \cap H| b$ and so $|N \cap H| = 4$. This makes $N \cap H$ a Sylow 2-subgroup of N but not of H.

10.1.17. Assume H is a subgroup of S_4 of order 12 distinct from A_4. Then, since $|S_4 : A_4| = 2$, $A_4 \lhd S_4$ and $HA_4 = S_4$.

We draw a partial lattice diagram of subgroups. Since $HA_4 = S_4$, we can draw a parallelogram, and so we can conclude that $|A_4 : A_4 \cap H| = |S_4 : H| = 2$, which means that $A_4 \cap H$ is a subgroup of order 6 in A_4. However, we have proved in Problem **5.1.19** that A_4 does not have a subgroup of order 6. The contradiction proves that H does not exist.

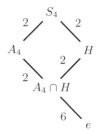

10.2.15. If N is a subgroup of G and C is a conjugacy class of elements of G, then it is quite possible that some of elements of C are in N while others are not in N. This problem says that this is not possible if $N \lhd G$. In this problem we are assuming that we already know that N is a subgroup. It certainly is possible to have a union of conjugacy classes not be a subgroup, but this problem says that if a union of conjugacy classes is a subgroup, then it must be a normal subgroup.

($\Rightarrow$) To show that N is the union of some of the conjugacy classes of G, we have to show that if $x \in N$ and if y is conjugate to x, then y is also in N. Now $y = gxg^{-1}$ for some $g \in G$. We also know that $gNg^{-1} = N$ since N is normal in G. Thus

$$y = gxg^{-1} \in gNg^{-1} = N.$$

($\Leftarrow$) Assume N is a subgroup and is the union of some of the conjugacy classes of G. To show that N is normal in G. We will show that $gNg^{-1} \subseteq N$ for all $g \in G$, and the proof will be complete by Lemma 10.17.

Let $x \in gNg^{-1}$. Then $x = gng^{-1}$ for some $n \in N$. This means that n and x are in the same conjugacy class, and thus since n is in N so must x. Thus $x \in N$ and we are done.

10.3.11. To show K is a subgroup of G, let x and y be two elements of K. Then $x \in Nx$ and $y \in Ny$ where Nx and Ny are elements of E. Since E is assumed to be a subgroup of G/N, we know that $(Nx)(Ny)$ and $(Nx)^{-1}$ are elements of E as well. Now, $xy \in Nxy = (Nx)(Ny) \in E$ and $x^{-1} \in Nx^{-1} = (Nx)^{-1} \in E$. Thus K is closed under multiplication and taking inverses and is thus a subgroup of G. Since N is the identity element of E, every element of N will be in K and so we must have $N \subseteq K$.

Now K is the union of $|E|$ cosets of N. These cosets are disjoint and each is size $|N|$. Hence
$$|K| = |E| \, |N| = |K : N| \, |N|.$$

10.3.13. Let $Z = \mathbf{Z}(G)$. We first determine $|Z|$. This was done in Problem 6.3.3 (or Problem 6.3.2), but we briefly recall the argument here: By Lagrange's theorem $|Z| = 1, 3, 9,$ or 27. Since G is a p-group, $|Z| > 1$. Since G is non-abelian, then $|Z| < 27$. In fact, if G is non-abelian, $|Z| \neq 9$ since the index of the center cannot be prime (Proposition 5.23). We conclude that $|Z| = 3$.

Now G/Z is a group of order $9 = 3^2$ and is therefore abelian. The cosets Zx and Zy are two elements of G/Z and, because they commute, we can write
$$Zxy = ZxZy = ZyZx = Zyx.$$
Thus, $yx \in Zyx = Zxy$. This implies that $yx = zxy$ for some $z \in Z$.

11.1.11. CLAIM. $\theta(\mathbf{Z}(G)) \leq \mathbf{Z}(\theta(G))$.

PROOF. Let $x \in \theta(\mathbf{Z}(G))$ and $h \in \theta(G)$. We want to show that $xh = hx$. Since both x and h are in the image of θ, we have $z \in \mathbf{Z}(G)$ and $g \in G$ with $\theta(z) = x$ and $\theta(g) = h$. Now we have
$$xh = \theta(z)\theta(g) = \theta(zg) = \theta(gz) = \theta(g)\theta(z) = hx.$$

It is very possible for the two groups not to be equal. For example, let $G = S_3$, and let $N = \langle (1\ 2\ 3) \rangle$. Let $\theta : G \longrightarrow G/N$ be the canonical homomorphism. In this example we have $\theta(\mathbf{Z}(G)) = \{e\} \neq G/N = \mathbf{Z}(\theta(G))$.

11.2.5. We first identify the groups $(\mathbb{Z}/9\mathbb{Z})^{\times}$ and $(\mathbb{Z}/11\mathbb{Z})^{\times}$.
$$(\mathbb{Z}/9\mathbb{Z})^{\times} = \{1, 2, 4, 5, 7, 8\} = \langle 2 \rangle \cong (\mathbb{Z}/6\mathbb{Z}, +),$$
$$(\mathbb{Z}/11\mathbb{Z})^{\times} = \{1, 2, 3, 4, 5, 6, 7, 8, 9, 10\} = \langle 2 \rangle \cong (\mathbb{Z}/10\mathbb{Z}, +).$$

The homomorphism $\theta \colon (\mathbb{Z}/9\mathbb{Z})^{\times} \to (\mathbb{Z}/11\mathbb{Z})^{\times}$ is non-trivial, and so the image is not just the identity.

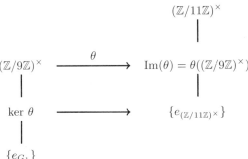

Since θ is non-trivial, $|\mathrm{Im}(\theta)| = \left|(\mathbb{Z}/9\mathbb{Z})^{\times} : \ker(\theta)\right| > 1$. Two is the only divisor that $(\mathbb{Z}/9\mathbb{Z}))^{\times}$ and $(\mathbb{Z}/11\mathbb{Z}))^{\times}$ have in common. So $|\mathrm{Im}(\theta)| = \left|(\mathbb{Z}/9\mathbb{Z}))^{\times} : \ker(\theta)\right| = 2$ and, as a result, $|\ker(\theta)| = 3$. Both groups are cyclic and thus have a unique subgroup of every order dividing the order of the group. Hence, $\ker(\theta) = \langle 4 \rangle = \{4, 7, 1\}$, and $\mathrm{Im}(\theta) = \langle 10 \rangle = \{10, 1\}$.

11.3.9. The canonical homomorphism $\pi\colon G \to G/N$ is defined by $\pi(g) = Ng$. In other words, every group element goes to the coset of N that contains that element. π is a homomorphism because $\pi(gh) = Ngh = NgNh = \pi(g)\pi(h)$.

Define $\psi\colon G/N \to H$ by $\psi(Nx) = \phi(x)$. On the face of it, we have defined a map from G/N to H, but we have to be careful since every element of G/N has many different representations. In other words, the same coset could be called Ng but also Nx for *different* elements g and x, and given our definition, we have sent the same coset to both $\phi(g)$ and $\phi(x)$. We have to make sure that the latter are equal, since otherwise we do not have a function.

To prove that ψ is well defined, we assume that $Nx = Ny$ and we need to show that $\phi(x) = \phi(y)$. If $Nx = Ny$, then $y = nx$ for some $n \in N$, and so $\phi(y) = \phi(nx) = \phi(n)\phi(x) = \phi(x)$ since $n \in N = \ker(\phi)$.

Next, we show that ψ is a homomorphism:

$$\psi(NxNy) = \psi(Nxy) = \phi(xy) = \phi(x)\phi(y) = \psi(Nx)\psi(Ny).$$

Finally to show that the diagram commutes, we let $x \in G$. We have $\psi(\pi(x)) = \psi(Nx) = \phi(x)$, and so the diagram commutes.

11.3.11. Write $D_8 = \langle a, b \mid a^4 = b^2 = e, ba = a^3b \rangle$. If two elements commute, then their commutator will be the identity. In fact, the commutators in D_8 are $e = aaa^{-1}a^{-1}$ and $a^2 = aba^{-1}b^{-1}$. They generate the group $\langle a^2 \rangle = \{e, a^2\}$ which is a normal subgroup of order 2 and the commutator subgroup of D_8. S_3 is the group consisting of the six permutations of $\{1, 2, 3\}$. Other than e, we have $(1\ 3\ 2) = (1\ 2)(2\ 3)(1\ 2)^{-1}(2\ 3)^{-1}$ and $(1\ 2\ 3) = (2\ 3)(1\ 2)(2\ 3)(1\ 2)$ as commutators, and they generate $\langle (1\ 2\ 3) \rangle$ which is a normal subgroup of order 3 and the commutator subgroup of S_3. Note that in both cases, the commutator subgroup ended up being equal to the set of commutators. This is *not* true in general, the commutator subgroup is the subgroup *generated* by the commutators.

11.4.3. We label each element of D_8 as follows: $1 = e$, $2 = a$, $3 = a^2$, $4 = a^3$, $5 = b$, $6 = ab$, $7 = a^2b$, and $8 = a^3b$. Now the action of b by left multiplication is

$$b \cdot 1 = 5, \quad b \cdot 2 = 8, \quad b \cdot 3 = 7, \quad b \cdot 4 = 6, \quad b \cdot 5 = 1, \quad b \cdot 6 = 4, \quad b \cdot 7 = 3, \quad b \cdot 8 = 2.$$

Now $\theta(b)$ is the element of S_8 corresponding to the action of b on D_8. Hence,

$$\theta(b) = (1\ 5)(2\ 8)(3\ 7)(4\ 6).$$

11.5.7. The group G acts on a set of size 3 and hence we have a homomorphism θ from G to S_3. The kernel of this homomorphism is the collection of elements of G that do not move any of the set elements. Thus we also know that $|\ker(\theta)| = 5$. We have the homomorphism diagram on the left of Figure C.2. Since $|S_3| = 6$, $\theta(G) \cong G/\ker(\theta)$, and $|G/\ker(\theta)| = 6$, we must have $S_3 = \theta(G)$ and the map must be onto. Thus, we modify the homomorphism diagram, and we get the diagram on the right of Figure C.2. Now, by the homomorphism theorems, every subgroup of S_3 corresponds to a subgroup of G containing the kernel. S_3 has non-trivial subgroups of orders 2 and 3. These translate to subgroups of orders 10 and 15 for G. See Figure C.3.

11.5.11.

(a) Clearly $L \le KL$ and thus $\phi(L) \le \phi(KL)$. Conversely, if $x \in \phi(KL)$, then $x = \phi(kl)$ where $k \in K$ and $l \in L$. Now, $x = \phi(kl) = \phi(k)\phi(l) = \phi(l) \in \phi(L)$.

(b) Now $K \lhd G$ and so KL is a group and thus $|L : K \cap L| = |KL : K|$. On the other hand $|\phi(L)| = |\phi(KL)| = |KL : \ker(\phi)| = |KL : K|$. So all three numbers are equal. See the homomorphism diagram in Figure C.4.

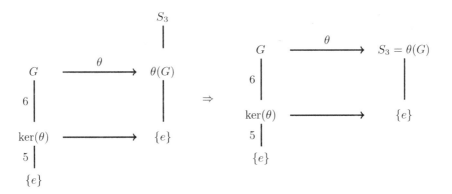

Figure C.2

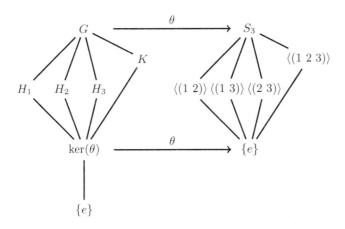

Figure C.3

11.6.5.

(a) PROPOSITION. *If $\theta\colon G \to G$ is an automorphism, then $o(g) = o(\theta(g))$ for all $g \in G$.*
PROOF. Let $n = o(g)$ and $m = o(\theta(g))$. We have $e = \theta(e) = \theta(g^n) = \theta(g)^n$ and this proves that $m \mid n$. On the other hand $e = \theta(g)^m = \theta(g^m)$ which means that $g^m \in \ker(\theta) = \{e\}$. Hence, $g^m = e$ and so $n \mid m$. We conclude that $n = m$.
    In the special case of this problem, we have $o(\sigma(1)) = o(1) = 12$.

(b) $(\mathbb{Z}/12\mathbb{Z})^{\times}$ consists of those elements $n \in \mathbb{Z}/12\mathbb{Z}$ with $\gcd(n, 12) = 1$. These are also exactly the generators of $(\mathbb{Z}/12\mathbb{Z}, +)$. Hence, every element of order 12 in $(\mathbb{Z}/12\mathbb{Z}, +)$ is in $(\mathbb{Z}/12\mathbb{Z})^{\times}$. We conclude that $\sigma(1) \in (\mathbb{Z}/12\mathbb{Z})^{\times}$.

(c) It follows from the previous part that the possibilities for $\sigma(1)$ are 1, 5, 7, and 11, which are precisely the elements of order 12 in $(\mathbb{Z}/12\mathbb{Z}, +)$ as well as the elements of $(\mathbb{Z}/12\mathbb{Z})^{\times}$. These are also exactly the numbers $1 \le n \le 12$ with $\gcd(n, 12) = 1$. We have shown that $\sigma(1)$ cannot be anything else other than one of these. However, we have not yet shown that each of these four are actually possible as $\sigma(1)$. We will show that in the final part of this problem.

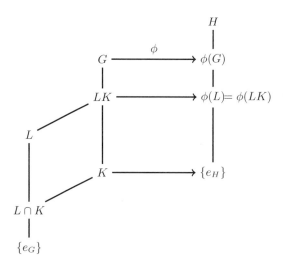

Figure C.4. If $K = \ker(\phi)$, then $\phi(L) = \phi)LK$).

(d) If $k \in \mathbb{Z}/12\mathbb{Z}$, then $\sigma(k) = \sigma(\underbrace{1 + 1 + \cdots + 1}_{k}) = \sigma(1) + \sigma(1) + \cdots + \sigma(1) = k\sigma(1)$.

The point of this part is that if we know where 1 is mapped to, then we know where every element of $\mathbb{Z}/12\mathbb{Z}$ is mapped to as well. Hence, by the previous part, there can be at most four automorphisms for $(\mathbb{Z}/12\mathbb{Z}, +)$.

(e) Let $a \in (\mathbb{Z}/12\mathbb{Z})^\times = \{1, 5, 7, 11\}$. Define $\phi_a \colon (\mathbb{Z}/12\mathbb{Z}, +) \to (\mathbb{Z}/12\mathbb{Z}, +)$ by $\phi_a(k) = ka$ (where ka means $\underbrace{a + a + \cdots + a}_{k}$). We want to show that ϕ_a is an automorphism.

Remember that the group is $(\mathbb{Z}/12\mathbb{Z}, +)$ and the operation is addition. We have

$$\phi_a(k + \ell) = (k + \ell)a = ka + \ell a = \phi_a(k) + \phi_a(\ell),$$

and so ϕ_a is a homomorphism. Next we show that ϕ_a is 1-1 by considering the kernel of ϕ_a. Assume that $\phi_a(k) = 0$ for some $0 \le k \le 11$. This means that $ka = a + a + \cdots + a = 0$. This means that $o(a) \mid k$. But $o(a) = 12$ and so $k = 0$. This proves that ϕ_a is 1-1. Since ϕ_a is a map from a finite set to another finite set of the same size, the fact that it is 1-1 means that it is onto as well. Hence, ϕ_a is an automorphism.

WHAT DOES THIS SAY ABOUT $\mathrm{Aut}(\mathbb{Z}/12\mathbb{Z}, +)$? Let $C = (\mathbb{Z}/12\mathbb{Z}, +)$. In the final part of the problem, we proved that each element of $(\mathbb{Z}/12\mathbb{Z})^\times$ gives an automorphism of C. In the first parts of the problem we proved that there can be no other automorphisms of C and that each automorphism of C does indeed correspond to multiplication by an element of $(\mathbb{Z}/12\mathbb{Z})^\times$. Hence, we have a 1-1, onto map

$$\theta \colon (\mathbb{Z}/12\mathbb{Z})^\times \to \mathrm{Aut}(C),$$

given by $\theta(a) = \phi_a$. We can show that the map θ is itself also a homomorphism. This is because $\phi_{ab} = \phi_a\phi_b$ (where the operation on the right-hand side is function composition), and so $\theta(ab) = \theta(a)\theta(b)$ (again on the right-hand side $\theta(a)\theta(b) = \theta(a) \circ \theta(b)$ is the composition of the two automorphisms). We conclude that θ is an isomorphism and

$$\mathrm{Aut}(\mathbb{Z}/12\mathbb{Z}, +) = (\mathbb{Z}/12\mathbb{Z})^\times.$$

There is, of course, nothing special about 12, and the same argument works for other cyclic groups as well.

11.6.13. This problem shows the power of the N/C Theorem 11.47. The group H is sitting inside G and this, by itself, puts no restriction on the size or structure of G. For example, the group $D_8 \times \mathbb{Z}/100\mathbb{Z}$ is a group of order 800 with D_8 as a normal subgroup. Somehow, knowing that there are no elements outside of H that commute with every element of H severely restricts $|G|$ (and, in fact, the structure of G).

It is best practice to draw the partial lattice diagram in Figure 11.26, and then modify it based on the information given in this problem. Ultimately however, our argument should not depend on a diagram, and so here we proceed without it. The subgroup H is normal in G and so $\mathbf{N}_G(H) = G$. We also know that $\mathbf{C}_G(H) = \mathbf{Z}(H)$ and the size of the latter is 2. The N/C theorem says that $\mathbf{C}_G(H) \lhd \mathbf{N}_G(H)$ and $\mathbf{N}_G(H)/\mathbf{C}_G(H)$ is isomorphic to a subgroup of $\mathrm{Aut}(H)$. Thus, in our case, $G/\mathbf{Z}(H)$ is isomorphic to a subgroup of $\mathrm{Aut}(H)$.

What is $\mathrm{Aut}(D_8)$? In Problem 11.6.11, you were led to show that $\mathrm{Aut}(D_8) \cong D_8$. For the current problem, we do not need the full force of that result. We just need to know that $|\mathrm{Aut}(D_8)| \leq 8$. To prove this, let $D_8 = \langle a, b \mid a^4 = b^2 = e, ba = a^3b \rangle$, and note that an automorphism of D_8 will have to send a to an element of order 4 and b to an element of order 2. Hence, there are two choices for where a can go—namely, a and a^3—and four choices for where b can go—namely, b, ab, a^2b, and a^3b. As soon as we have decided where a and b go, the map is determined on all of D_8 since $\{a, b\}$ is a set of generators for the group. We conclude that there are at most $2 \times 4 = 8$ choices for automorphisms of D_8. (To complete Problem 11.6.11, you have to show that each of these choices are indeed possible and that the resulting group of order 8 is isomorphic to D_8.)

Since $|\mathrm{Aut}(D_8)| \leq 8$, we conclude that $|G/\mathbf{Z}(H)| \leq 8$. But if this group had fewer than eight elements, then H would not be a *proper* subgroup of G (recall that $H/\mathbf{Z}(H) \leq G/\mathbf{Z}(H)$ is of size 4). Thus $|G/\mathbf{Z}(H)| = 8$ and

$$|G| = |G : \mathbf{Z}(H)| \ |\mathbf{Z}(H)| = 8 \times 2 = 16.$$

12.1.3. Let K be the normal subgroup of order 11. Let $P \in \mathrm{Syl}_{13}(G)$, then P is a subgroup of order 13^5. Since P is a p-group, it has subgroups of order 13, 13^2, 13^3, and 13^4. Let Q be a subgroup of order 13^i for $1 \leq i \leq 4$. Now $K \cap Q = \{e\}$ since $\gcd(|K|, |Q|) = 1$. On the other hand, KQ is a subgroup since $K \lhd G$. Hence, $|KQ| = |K| |Q| = 11 \times 13^i$. We conclude that G has subgroups of order 1, 13, 13^2, 13^3, 13^4, 13^5, 11, 11×13, 11×13^2, 11×13^3, 11×13^4, and 11×13^5. By Lagrange's theorem there can be no subgroups of some other size. Hence, for such a group G, the converse of Lagrange's theorem holds.

12.1.7. The group P certainly acts on P by conjugation. Now since N is normal in P, P also acts on $\Omega = N$ by conjugation. In other words, for $g \in P$ and $n \in N$, define $g \cdot n = gng^{-1}$ and note that $gng^{-1} \in N$ since $N \lhd P$. Clearly, $e \cdot n = n$ and $g \cdot (h \cdot n) = (gh) \cdot n$, and we have an action.

The orbit sizes divide the order of the group and hence must be a power of p. We know that e is in an orbit of size 1 since $geg^{-1} = e$ for all $g \in P$.

If we assume that $\{e\}$ is the only orbit of size 1, we get that the size of N is $1 + kp$ since orbits partition the set (and we have one orbit of size 1 and all other orbits have an order that is a multiple of p). But $\{e\} < N \leq P$ and so, by Lagrange's theorem, $|N|$ is a power of p (and is bigger than 1). But $1 + kp$ has remainder 1 when divided by p and cannot be a power of p. The contradiction shows that there must be other orbits of size 1.

Assume that $\{z\}$ is another orbit of size 1. Then $gzg^{-1} = z$ for every $g \in P$ which means that $gz = zg$ for all $g \in P$. We conclude that z is an element of the center of P. But $e \neq z \in N$ and so $N \cap \mathbf{Z}(P) \neq \{e\}$.

12.2.3. Name the group G. We know that G acts on a set with four elements. This immediately gives a homomorphism θ from G into S_4. By the homomorphism theorems, $G/\ker(\theta) \cong \theta(G) \leq S_4$. We are told that $\ker(\theta) \neq G$. Could $\ker(\theta) = \{e\}$? If this were the case, then G would be isomorphic to a subgroup of S_4. But G has 60 elements while S_4 has 24 elements. The impossibility shows that $\ker(\theta)$ is not trivial and G has a non-trivial normal subgroup.

12.2.5. Draw a diagram. The subgroup N is normal in G, and thus NK is a subgroup of G, and thus $|NK : N| = |K : N \cap K|$.

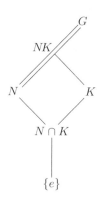

Note that we always first draw the diagram in general position and without any prior assumptions. As we make arguments, we might change the diagram. For example, here $|N \cap K|$ may be one or larger, and NK is a subgroup somewhere between N and G. We know that $|N \cap K|$ divides both $|N| = 5 \times 11$ and $|K| = 5 \times 7$. Thus $|N \cap K|$ is either 1 or 5. If $|N \cap K| = 1$, then $|NK : N| = |K : N \cap K| = 35$ which would imply that $|NK| = 35 \times 55 > 1155$. This is a contradiction and hence $|N \cap K| = 5$. Thus $|NK : N| = |K : N \cap K| = \frac{35}{5} = 7$. Thus $|NK| = 385$. Now $|G : NK| = 3$, which is the smallest prime divisor of $|G|$, and hence $NK \lhd G$.

12.3.3. If $p = 2$, then G is a group of order 8 and cannot be simple. If p is odd, let n_p denote the number of Sylow p-subgroups of G. This number has to divide $|G|$ and have a remainder 1 when divided by p. Hence, n_p will have to divide 4. This means that $n_p = 1, 2,$ or 4. But 2 is not 1 (mod p) for any prime p, and so $n_p \neq 2$. In addition if $n_p = 1$, then the unique Sylow p-subgroup of G is normal in G, and G is not simple. So the only case left is if $n_p = 4$. In this case, we need $4 = 1$ (mod p). This means that $p = 3$ and $|G| = 36 = 2^2 \times 3^2$.

Let P be a Sylow 3-subgroup of G. Then P has nine elements and $|G : P| = 4$. By Theorem 12.4, G is guaranteed to have a normal subgroup N such that $N \leq P$ and that $|G : N|$ divides $4! = 24$. Now $N \neq \{e\}$ since $36 = |G| = |G : \{e\}|$ does not divide 24. Hence, N is a non-trivial normal subgroup of G and G is not simple.

12.3.11. Let $n_p(G) = |\mathrm{Syl}_p(G)|$ be the number of Sylow p-subgroups of G. We know, by the Sylow theorems, that $n_p(G) = |G : N_G(P)|$ divides $|G|$ and that $n_p(G) = 1 + kp$ for some non-negative integer k. Now the divisors of 99 are 1, 3, 9, 11, 33, and 99. The only one that is of the form $1 + 3k$ is 1 and so $n_3(G) = 1$. Also the only one that is of the form $1 + 11k$ is 1 and hence $n_{11}(G) = 1$. So there is a unique Sylow 3-subgroup P, and a unique Sylow 11-subgroup Q.

By the Sylow theorems, being unique, each of these subgroups are normal in G. P and Q have relatively prime orders, and so $P \cap Q = \{e\}$. They also have relatively prime indices and so (by Theorem 5.24c) $PQ = G$.

Now $|P| = 9$ and $|Q| = 11$ and so both P and Q are abelian. Elements of P also commute with elements of Q by Problem **10.2.21**. Thus if g and h are elements of G, then $g = x_1 y_1$ and $h = x_2 y_2$ with $x_1, x_2 \in P$ and $y_1, y_2 \in Q$. Now $gh = x_1 y_1 x_2 y_2 = x_2 y_2 x_1 y_1 = hg$ and so G is abelian.

We conclude that every group of order 99 is abelian. In fact, there are only two groups of order 99:

$$\mathbb{Z}/11\mathbb{Z} \times \mathbb{Z}/9\mathbb{Z}, \quad \text{and} \quad \mathbb{Z}/11\mathbb{Z} \times \mathbb{Z}/3\mathbb{Z} \times \mathbb{Z}/3\mathbb{Z}.$$

12.3.15.

(a) The number of Sylow p-subgroups must divide the order of the group and must have remainder 1 when divided by p. Checking these two conditions for primes 5, 7 and 47, we get that $|\mathrm{Syl}_5(G)| = |\mathrm{Syl}_7(G)| = |\mathrm{Syl}_{47}(G)| = 1$. In other words, there is unique Sylow p-subgroup for $p = 5, 7$ and 47. Each of these subgroups will be a normal subgroup of G.

(b) Since both P and Q are normal subgroups, PQ is not only a subgroup but a normal subgroup of G ($xPQ = PxQ = PQx$ for all $x \in G$).

By Lagrange's theorem $P \cap Q = \{e\}$ and hence every element of P commutes with every element of Q. Since both P and Q are abelian, we conclude that PQ is abelian.

If $P = \langle x \rangle$ and $Q = \langle y \rangle$, then $o(x) = 47$ and $o(y) = 7$. Since these are relatively prime numbers and $xy = yx$, we have that $o(xy) = 7 \times 47 = |PQ|$. We conclude that $PQ = \langle xy \rangle$ is a cyclic group of order $7 \times 47 = 329$ and is a normal subgroup of G.

(c) By Cauchy's theorem and the Sylow theorems, we have subgroups of order 1, 5, 7, 25, 47 and 8225. Since each of the Sylow subgroups are normal, we also have that the products of these subgroups are also subgroups. Hence, we also have subgroups of order 35, 235, 175, 329, and 1175. PQ is also a normal subgroup of G and hence the product of this subgroup with a subgroup of order 5 is also a subgroup of order 1645. By Lagrange's theorem, there are no other subgroups. We conclude that if k divides the order of G, then there is a subgroup of order k in G.

(d) R is a normal subgroup of order 25 while PQ is a cyclic normal subgroup of order 329. Their indices are relatively prime and hence $PQR = G$.

(e) We have $PQR = G$ and $PQ \cap R = \{e\}$ (since their orders are relatively prime), and so elements of PQ commute with elements of R. PQ itself is a cyclic group and R is a group of order 25 and so is also abelian. We conclude that every element of R and every element of PQ commute with every element of G. This means that R and PQ and hence RPQ is inside $\mathbf{Z}(G)$. We conclude that G is abelian.

(f) As we have seen $PQ = \langle xy \rangle$ is a cyclic subgroup of order 329. Now let $S = \langle z \rangle$ be a subgroup of order 5 in G. Again, see G is abelian, $o(xyz) = 329 \times 5 = 1645$, and hence $\langle xyz \rangle$ is a cyclic subgroup of order 1645.

Consider the group $G = \mathbb{Z}/47\mathbb{Z} \times \mathbb{Z}/7\mathbb{Z} \times \mathbb{Z}/5\mathbb{Z} \times \mathbb{Z}/5\mathbb{Z}$. This example, shows that G does not have to be cyclic, and so the largest cyclic subgroup that G is guaranteed to have is one of order 1645.

In fact, using the classification of abelian groups (Chapter 13), we now know that there are only two groups of order 8225: $\mathbb{Z}/8255\mathbb{Z}$ and $\mathbb{Z}/1645\mathbb{Z} \times \mathbb{Z}/5\mathbb{Z}$.

12.3.17. Either H is normal in G or, by Theorem 12.4, H has a subgroup $N \lhd G$ with $|H : N| = 3$. But then G/N is a group of order 15 and, by Theorem 12.15, all such groups are cyclic. All subgroups of cyclic groups are normal and so $H/N \lhd G/N$ which implies, by the homomorphism theorems, that $H \lhd G$.

12.4.5. In Problem 12.4.4, we showed that G has eight Sylow 7-subgroups and hence $|G : \mathbf{N}_G(P)| = 8$. We now have an action of G on a set with eight elements. There are actually two such actions. We could let G act on $\mathrm{Syl}_7(G)$ by conjugation or G could act on the cosets of $H = \mathbf{N}_G(P)$, where $P \in \mathrm{Syl}_7(G)$, by $g \cdot xH = gxH$. In either case, we get a homomorphism $\theta : G \to S_8$, and, since G is simple, the kernel will have to be trivial.

Thus $G \cong \theta(G) \leq S_8$. So so far we have that G is isomorphic to a subgroup of S_8. But why would G be isomorphic to a subgroup of A_8?

Assume $\theta(G)$ is not contained in A_8, and draw a partial lattice diagram including S_8, A_8 and $\theta(G)$. We know that $|S_8 : A_8| = 2$ and hence we have a parallelogram. See Figure C.5.

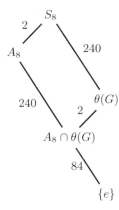

Figure C.5. If $\theta(G)$ is not contained in A_8, it will have a subgroup of index 2.

We now get that $A_8 \cap \theta(G)$ has index 2 in $\theta(G)$ and hence is a normal subgroup of $\theta(G)$. But $\theta(G)$ is isomorphic to G and hence is simple and has no normal subgroups. The contradiction shows that, contrary to our assumption, $\theta(G) \leq A_8$. Hence A_8 has a subgroup isomorphic to G.

15.2.1. Every ring has an addition table and a multiplication table. For example, the addition and multiplication tables for $\mathbb{Z}/3\mathbb{Z}$, which is a field, follow.

+	0	1	2
0	0	1	2
1	1	2	0
2	2	0	1

·	0	1	2
0	0	0	0
1	0	1	2
2	0	2	1

The multiplication table of every ring will have a row and a column of all zeros. If you throw this row and column out and if you have a field, then the rest is the multiplication table of a commutative group. The two tables are not independent of each other. The distributive laws connect the two together.

In this problem we want a three-element ring R that is not a field. Regardless of what the R is, $(R, +)$ will be a commutative group of order 3, and we know, from group theory, that there is only one such group, and that is $\mathbb{Z}/3\mathbb{Z}$. Hence our new ring R will have the same addition table as $\mathbb{Z}/3\mathbb{Z}$. Here is an example of an addition and a multipication table that gives a ring.

+	0	1	2
0	0	1	2
1	1	2	0
2	2	0	1

·	0	1	2
0	0	0	0
1	0	0	0
2	0	0	0

Note that this ring does not even have an identity. We could have come up with this ring in another way as well. Consider the ring $\mathbb{Z}/9\mathbb{Z}$. This ring has nine elements and with $+$ it is the cyclic group of order 9. It has a unique subgroup of order 3 which consists of multiples of 3: $\{0, 3, 6\}$. This is actually a subring of $\mathbb{Z}/9\mathbb{Z}$ (the operations are the same as in $\mathbb{Z}/9\mathbb{Z}$, i.e., modulo 9 arithmetic) and has the addition and multiplication tables isomorphic to the ones above:

$+$	0	3	6		$\cdot$	0	3	6
0	0	3	6		0	0	0	0
3	3	6	0		3	0	0	0
6	6	0	3		6	0	0	0

In ring theory, the set of multiples of 3 in $\mathbb{Z}/9\mathbb{Z}$ is called an *ideal* and is denoted by $\langle 3 \rangle$.

15.2.7. The only choices for a and b are 0 and 1, and so we have

$$E = \left\{ \mathbf{0} = \begin{bmatrix} 0 & 0 \\ 0 & 0 \end{bmatrix}, \; I = \begin{bmatrix} 1 & 0 \\ 0 & 1 \end{bmatrix}, \; A = \begin{bmatrix} 0 & 1 \\ 1 & 1 \end{bmatrix}, \; B = \begin{bmatrix} 1 & 1 \\ 1 & 0 \end{bmatrix} \right\}.$$

Hence, E does have four elements. Note that $B = I + A$ and so $E = \{\mathbf{0}, I, A, I + A\}$. Since the entries are in $\mathbb{Z}/2\mathbb{Z}$, every element added to itself is zero, and so E is both closed under addition and every element of E has an additive inverse (namely itself). Associativity is inherited from matrix addition, and so $(E, +)$ is a group and is isomorphic to the Klein 4-group, $\mathbb{Z}/2\mathbb{Z} \times \mathbb{Z}/2\mathbb{Z}$.

Note that $A^2 = \begin{bmatrix} 1 & 1 \\ 1 & 0 \end{bmatrix} = B$ and $A^3 = I$ and so $(E - \{\mathbf{0}\}, \cdot) = \{A, A^2, I\}$ is the cyclic group of order 3. Hence E is a field of order 4.

15.2.25. The ring of quaternions with coefficients in F is, a priori, a non-commutative ring with identity. In this problem we want to show that it is really a division ring. To do this we need to show that every non-zero element has an inverse.

The multiplicative inverse of $a + bi + cj + dk$ is $(a - bi - cj - dk)/(a^2 + b^2 + c^2 + d^2)$. To verify, just multiply the two to see that you get 1.

How did we come up with this element? Mimicking what we do with complex numbers, we started with $\frac{1}{a+bi+cj+dk}$ and multiplied top and bottom with the "conjugate" of the denominator, i.e., with $a - bi - cj - dk$.

16.1.1. We have defined an integral domain to be a commutative ring with identity that has no zero divisors. A subring of an integral domain will necessarily be commutative and will continue not to have zero divisors, but it may not have an identity. For an example, let $D = \mathbb{Z}$ be the ring of integers, and let $R = 2\mathbb{Z}$ be the subring of even integers. If we had required—many authors do—that a ring have an identity and a subring contain that same identity, then the answer to the question would be yes. With such a definition, $2\mathbb{Z}$ would *not* be a subring of $\mathbb{Z}$.

16.1.13.

(a) No. The ideal generated by 6 in $(\mathbb{Z}/12\mathbb{Z}, +, \cdot)$ is $\langle 6 \rangle = \{0, 6\}$ and $6 \times 6 = 0$. On the other hand in $(\mathbb{Z}/2\mathbb{Z}, +, \cdot) = \{0, 1\}$, we have $1 \cdot 1 = 1$.

(b) Yes. The ideal generated by 4 in $(\mathbb{Z}/12\mathbb{Z}, +, \cdot)$ is $\langle 4 \rangle = \{0, 4, 8\}$. Map $0 \to 0$, $4 \to 1$ and $8 \to 2$, and we get an isomorphism.

16.1.17. Consider the ideal $\langle 2, x \rangle$ generated by 2 and x. The elements of this ideal are polynomials of the form $2q(x) + xr(x)$ where q and r are arbitrary polynomials in $\mathbb{Z}[x]$. Hence, $\langle 2, x \rangle$ consists of polynomials with an even constant term. If this ideal was generated by one element $p(x) \in \mathbb{Z}[x]$, then the polynomial 2 would have to be a multiple of $p(x)$. This would mean that $p(x) = k \in \mathbb{Z}$ is a constant. Now we have to have k times a polynomial with integer coefficients equal to x. The only way this is possible is if $k = 1$. However, $p(x) = 1$ generates the whole ring and not just $\langle 2, x \rangle$. We conclude that $\langle 2, x \rangle$ is not a principal ideal.

16.1.19. For a commutative ring with identity, the ideal generated by d is $\{rd \mid r \in R\}$. But if R does not contain identity, then $\{rd \mid r \in R\}$ may not even contain d. Note that, in the expression $rd + nd$, the product rd is the ring product, while nd is a shorthand for $\underbrace{d + \cdots + d}_{n}$. So the claim is that if R does not have an identity and to find $\langle d \rangle$, we have to find all the sums of the form $rd + nd$ where $r \in R$ and $n \in \mathbb{Z}$.

We need to show two things: First, that $J = \{rd + nd \mid r \in R, n \in \mathbb{Z}\}$ is an ideal of R containing d, and second, that it is the *smallest* ideal of R with such a property. It is straightforward to verify that $(J, +)$ is an abelian group. Also if $s \in R$, then $s(ad + nd) = (as + ns)d + 0d \in J$. Thus, J is an ideal. Now, if an ideal of R contains d, it must also contain $2d, 3d, \ldots$, and it must also contain elements of the form rd. Hence, such an ideal contains every element in J. We conclude that J is the smallest ideal of R containing d, and hence it is the ideal generated by d.

16.1.23. To prove the statement by contradiction, assume a is a zero divisor of R. Then, by definition, a is not zero and there exists a non-zero element $b \in R$ with $ab = 0$. Since I is non-trivial, it has a non-zero element g. Since I is an ideal, we have $ga \in I$, and we also have $(ga)b = g(ab) = 0$. We conclude that ga is a zero divisor of R in I. But there were supposed to be no such elements. The contradiction completes the proof.

16.2.5. First make sure that you know what the elements of I are. R is a commutative ring with identity, and hence the ideal generated by $a \in R$ is $\langle a \rangle = Ra = \{ra \mid r \in R\}$. So the elements of I are products of elements of R with $x^2 - 2x$. Hence $I = \{p(x)(x^2 - 2x) \mid p(x) \in \mathbb{Z}[x]\}$. For example, $(x^3 - 5)(x^2 - 2x)$ is an element of I while $x^2 + 3$ is not.

Elements of R/I are of the form $r + I$ where $r \in R$. So for example, $x + I$ and $x - 2 + I$ are elements of R/I. We also know that the zero of R/I is I. Finally, $r + I = I$ if and only if $r \in I$. Hence, for example, $x^2 - 2x + I = I$.

(a) $x + I$ and $x - 2 + I$ are both zero divisors since neither is the zero element and yet $(x + I)(x - 2 + I) = x^2 - 2x + I = I$.

(b) Let $J = \langle x \rangle$. Then elements of J are all polynomials in R that have x as a factor. This includes every element of I since every element of I has $x(x - 2) = x^2 - 2x$ as a factor. The element x is in J but not I, and the polynomial $x + 1$ is not in J. Hence $I \subsetneq J \subsetneq R$.

(c) We have $x^2 - 2x + I = I$ which means that $x^2 + I = 2x + I$. (Also $x^2 + I = 2x + (x^2 - 2x) + I = 2x + I$.) Now

$$x^5 + I = (x^2 + I)(x^2 + I)(x + I) = (2x + I)(2x + I)(x + I) = (4 + I)(x^2 + I)(x + I)$$

$$= (4 + I)(2x + I)(x + I) = (4 + I)(2 + I)(x^2 + I) = (8 + I)(2x + I)$$

$$= 16x + I.$$

16.2.7. The ideal generated by $1 + 3i$ consists of all the elements of the form $h = (1 + 3i)(a + bi) = a - 3b + (3a + b)i$.

(a) Since $1 + 3i \in I$, we have $I = I + (1 + 3i) = (I + 1) + (I + 3i)$. This means that $I + 3i = -(I + 1) = I - 1$ and so

$$I + i = (I - i)(I - 1) = (I - i)(I + 3i) = I + 3.$$

There is an alternative notation—more streamlined but a bit confusing at first—that makes the above calculation more apparent. We have two different rings here: $\mathbb{Z}[i]$ and R. Elements such as $1 + 3i$ and $-1 + 2i$ are elements of $\mathbb{Z}[i]$, while elements such as $I + i$ and $I + 3 + 2i$ are elements of R. In R, the element I is the zero, and so if we denote it by 0 and remember that adding 0 does not change things, then we can write $3 + 2i$ for $I + 3 + 2i$. The only confusion is that with this notation we do not know if $3 + 2i$ is an element of $\mathbb{Z}[i]$ or an element of $\mathbb{R}$. We fix this by stating the ring in which the calculation is taking place. So, with this new streamlined notation, we could redo the above calculation as follows:

In R we have $1 + 3i = 0$. Thus in R we have $3i = -1$, and so multiplying on both sides by $-i$, we get $i = 3$ in R.

From $i = 3$, we get $-1 = i^2 = 9$, and so in R we have $9 = -1$ and hence $10 = 0$. In our original more cumbersome but more transparent notation, these statements are written as $I + 9 = I - 1$ and $I + 10 = I$.

(b) A typical element of R is $I + a + bi$ where $a, b \in \mathbb{Z}$. In the previous part we saw that in R, $i = 3$ and $10 = 0$. Thus in R, we have $a + bi = a + 3b$ and $a + 3b$ is equal to an integer between 0 and 9. Thus $I + a + bi = I + m$ where $0 \leq m \leq 9$. So R has at most ten elements. (It is possible that there are other relations that we have not uncovered that further reduce the size of R.)

(c) We will prove that $R \cong \mathbb{Z}/10\mathbb{Z}$, and hence establish that R has exactly ten elements.

Define $\phi : \mathbb{Z} \to R$ by $\phi(n) = I + n$. This is clearly a ring homomorphism. What is the kernel? What is the image?

CLAIM 1. ϕ is onto.

PROOF. In the previous part we saw that every element of R is of the form $I + m$ where m is some integer. Now $I + m = \phi(m)$ and hence ϕ is onto.

CLAIM 2. $\ker(\phi) = 10\mathbb{Z}$.

PROOF. If $m = 10k$ for some integer k, then $\phi(m) = I + 10k = (I + 10)(I + k) = I$ and hence $m \in \ker(\phi)$. (We could have just said that in R, we have $10 = 0$ and hence $10k = 0$.) So $10\mathbb{Z} \subseteq \ker(\phi)$.

Now assume that $n \in \ker(\phi)$. Thus $I + n = I$, and so $n \in I$. We have to show that n is a multiple of 10. $n \in I$ implies that $n = (a + bi)(1 + 3i) = (a - 3b) + (3a + b)i$ for some integers a and b. n is an integer, and so we must have $3a + b = 0$ and $n = a - 3b$. Thus $b = -3a$ and $n = a - 3(-3a) = 10a \in 10\mathbb{Z}$. So $\ker(\phi) \subseteq 10\mathbb{Z}$. Hence $\ker(\phi) = 10\mathbb{Z}$. Now by the homomorphism theorem, we have

$$\mathbb{Z}/10\mathbb{Z} = \mathbb{Z}/\ker(\phi) \cong \operatorname{Im}(\phi) = R.$$

18.1.11. ($\Leftarrow$) Assume that $\langle p \rangle$ is a non-zero prime ideal. We claim that p must be a prime element. First of all since $\langle p \rangle$ is non-zero, we have $p \neq 0$. Furthermore, a prime ideal cannot be the whole ring, and hence p cannot be a unit (since if p were a unit, then $\langle p \rangle$ would be the whole ring). Now assume $p \mid ab$, for $a, b \in R$. We have to show that $p \mid a$ or $p \mid b$. $p \mid ab$ means that $ab = pc$ for some $c \in R$. Thus $ab \in \langle p \rangle \Rightarrow$ (since $\langle p \rangle$ is a prime ideal) $a \in \langle p \rangle$ or $b \in \langle p \rangle$. Without loss of generality, assume that $a \in \langle p \rangle$. R is a commutative ring with identity, and hence $\langle p \rangle = \{ rp \mid r \in R \}$. So $a = rp$ for some $r \in R$, and so $p \mid a$. Thus p is a prime element.

($\Rightarrow$) Conversely, assume p is a prime element, and we have to show that $\langle p \rangle$ is a prime ideal. First note that since p is neither a zero nor a unit, $\langle p \rangle$ cannot be zero or the whole ring. Now assume $ab \in \langle p \rangle$. R is a commutative ring with identity, and hence $\langle p \rangle$ is all the multiples of p. So we have $ab = pc$, and so $p \mid ab$. p is a prime element, and so $p \mid a$ or $p \mid b$, and thus either $a \in \langle p \rangle$ or $b \in \langle p \rangle$.

18.1.15. Let $R = \mathbb{Z}[x]$. This is an integral domain and x is an irreducible element of R. Now $\langle 2, x \rangle$ is a bigger ideal than $\langle x \rangle$ and is not all of R. Hence, $\langle x \rangle$ is not maximal.

AN ALTERNATE ARGUMENT. The evaluation map at zero $\epsilon_0 \colon \mathbb{Z}[x] \to \mathbb{Z}$ defined by $\epsilon_0(p) = p(0)$ is an onto homomorphism and the kernel is $\langle x \rangle$. So $\mathbb{Z}[x]/\langle x \rangle \cong \mathbb{Z}$ which is an integral domain but not a field. Hence, $\langle x \rangle$ is a prime ideal but not a maximal ideal. Since $\langle x \rangle$ is a prime ideal, we have that x is a prime element, and in an integral domain all prime elements are irreducible.

18.1.23. Define a map $f \colon \mathbb{Z} \times \mathbb{Z} \to \mathbb{Z}/4\mathbb{Z}$ by $f(x, y) = x \bmod 4$. For example, $f(5, 11) = 1$, and $f(7, 11) = 3$. The map f is an onto homomorphism, as is easily checked.

Which elements go to 0? Exactly the elements of $\mathbb{Z} \times \mathbb{Z}$ that have an integer divisible by 4 in their first coordinate. Thus $\ker(f) = B$. So, B is an ideal of $\mathbb{Z} \times \mathbb{Z}$ and $\mathbb{Z} \times \mathbb{Z}/B$ is isomorphic to $\mathbb{Z}/4\mathbb{Z}$. Now $\mathbb{Z}/4\mathbb{Z}$ has 0-divisors ($2 \times 2 = 0$), and so $\mathbb{Z}/4\mathbb{Z}$ is neither an integral domain nor a field. We conclude B is neither prime nor maximal.

Now to find a maximal ideal containing B, we first find a maximal ideal of $\mathbb{Z}/4\mathbb{Z}$ and then find its inverse image. The ideal $\langle 2 \rangle = \{0, 2\}$ is maximal in $\mathbb{Z}/4\mathbb{Z}$, and $C = f^{-1}(\{0, 2\}) = \{(2x, y) \mid x, y \in \mathbb{Z}\}$ is a maximal ideal of $\mathbb{Z} \times \mathbb{Z}$ that contains B.

$$
\begin{array}{ccc}
\mathbb{Z} \times \mathbb{Z} & \xrightarrow{\ f\ } & \mathbb{Z}/4\mathbb{Z} \\
| & & | \\
C = f^{-1}(\langle 2 \rangle) & \longleftrightarrow & \langle 2 \rangle \\
| & & | \\
B & \longrightarrow & \{0\} \\
| & & \\
\{0\} & &
\end{array}
$$

18.3.13.

(a) R is closed under addition and multiplication and contains 1. Hence it is a subring of $\mathbb{Q}[x]$. Since the latter does not have any zero divisors, then neither does R. Hence, R is an integral domain.

(b) If $p, q \in R$ with $p(x)q(x) = 1$, then $\deg p + \deg q = 0$. This means that they are both constants, and the only invertible constants are ± 1. So ± 1 are the only units of R.

(c) We have $x = (\frac{1}{2}x)(2)$, neither factor is a unit and so x is not irreducible.

Assume $x = p_1 p_2 \cdots p_k$ where, for $1 \le i \le k$, p_i is irreducible. Then comparing the degrees of both sides, we see that precisely one of the factors has to be of degree 1 and all others are constants. Say $p_1(x) = ax + b$, with $a \ne 0$. If $b \ne 0$, then the product will have a constant term and not equal x. Hence $b = 0$ and $p_1(x) = ax$. But $ax = (\frac{1}{2})(2ax)$ and neither $\frac{1}{2}$ nor $2ax$ are units. Hence ax is not irreducible. But we had assumed p_1 is irreducible. The contradiction proves that x is not a product of irreducibles.

(d) In an integral domain x prime implies x irreducible. Since x is not irreducible, it is not prime either. R is not a UFD since we do not even have factorizations into irreducibles.

(e) Since R is a commutative ring with identity, the ideal generated by x is Rx. This means that elements of $\langle x \rangle$ are those polynomials that have no constant term *and* have a linear term with an integer coefficient. The derivative at 0 picks out the coefficient of the linear term, and so we can write

$$\langle x \rangle = \{p \in \mathbb{Q}[x] \mid p(0) = 0, p'(0) \in \mathbb{Z}\}.$$

(f) Let $I = \langle x \rangle$. By definition, $R/I = \{p(x) + I \mid p(x) \in R\}$. Every polynomial with no constant and no linear term is in I and hence $R/I = \{ax + b + I \mid a \in \mathbb{Q}, b \in \mathbb{Z}\}$. But we can simplify this more.

For $a \in \mathbb{Q}$, let $\lfloor a \rfloor$ denote the biggest integer that is smaller than a, and let $\mathrm{frac}(a) = a - \lfloor a \rfloor$ be the fractional part of a. So $a = \lfloor a \rfloor + \mathrm{frac}(a)$, and $0 \le \mathrm{frac}(a) < 1$. For example, $23.8 = 23 + .8$ and $-15.3 = -16 + .7$. Now, every *integer* multiple of x is in I, and, so $ax + b + I = \lfloor a \rfloor x + \mathrm{frac}(a)x + b + I = \mathrm{frac}(a)x + b + I$. Since $\mathrm{frac}(a)$

is a rational number and $0 \leq \mathrm{frac}(a) < 1$, we have

$$R/\langle x \rangle = \left\{ \frac{c}{d}x + b + \langle x \rangle \mid b, c, d \in \mathbb{Z}, 0 \leq c < d \right\}.$$

Since x is not a prime element, $\langle x \rangle$ is not a prime ideal, and hence $R/\langle x \rangle$ is not an integral domain. Hence it must have zero-divisors. One such zero-divisor is $\frac{1}{2}x + I$. This is not the zero element since $\frac{1}{2}x \notin I$. In addition, $2 \notin I$ and so $2 + I$ is also not the zero element. On the other hand,

$$(\frac{1}{2}x + I)(2 + I) = x + I = I.$$

(g) R is not noetherian. We will give two proofs.

First, if R was noetherian, then it would be a noetherian integral domain and these rings would satisfy $UF1$, meaning that every element is either zero or a unit or a product of irreducibles. We have seen that x is not a product of irreducibles, and hence R is not noetherian.

The second proof is a direct proof. We have the following infinitely ascending chain of ideals:

$$\langle x \rangle \subsetneq \langle \frac{1}{2}x \rangle \subsetneq \langle \frac{1}{4}x \rangle \subsetneq \cdots .$$

This proves that R is not noetherian.

Now can you find an ideal that is not finitely generated?

18.4.1. Assume u is a unit. So there exists $v \in R$ such that $uv = 1$. Then $d(1) \leq d(1u) = d(u)$ and $d(u) \leq d(uv) = d(1)$. Thus $d(u) = d(1)$.

Assume $d(u) = d(1)$. By the division algorithm there exist q and r such that $1 = uq + r$ with $r = 0$ or $d(r) < d(u)$. However, $d(u) = d(1) \leq d(1r) = d(r)$. Thus $r = 0$, and so $1 = uq$. In other words u is a unit.

18.6.3. In every commutative ring with 1, maximal ideals are prime, and so I—being a maximal ideal—is a prime ideal. Since $p \neq 0$, $I = \langle p \rangle$ is a non-zero prime ideal, and—again in any commutative ring with 1—this means that p is a prime element. Hence, p is prime.

Now, if q is an associate of p (i.e., if $q = up$ where u is a unit), we have $\langle q \rangle = \langle p \rangle$ is a non-zero prime ideal and hence q is also prime. Thus p and all of its associates are primes in S.

Assume that the element $x \in S$ was another prime. x, being prime, cannot be a unit (or zero), and hence $\langle x \rangle$ is a non-zero proper ideal of S. This ideal is either maximal or contained in a maximal ideal. But I is the sole maximal ideal in S and so $\langle x \rangle \subseteq \langle p \rangle$. This means that $p \mid x$ in S, or equivalently $x = py$ where $y \in S$.

Now S is an integral domain and, in an integral domain, all primes are irreducible. Hence x is irreducible, and so, in the factorization $x = py$, either p or y must be units of S. Now, p is not a unit (since it is a prime), and so y is a unit and x is an associate of p. We conclude that p and its associates in S are the only primes in S.

18.6.7. ($\Rightarrow$) A DVR is a PID with a unique irreducible element. Every PID is a UFD, and so it satisfies UF1.

($\Leftarrow$) Since D has a unique irreducible element (up to associates), to show that D is a DVR, we have to show that D is a PID. Let x be an irreducible element in D. (This element is essentially the unique irreducible element. All other irreducibles are of the form ux where u is a unit of D.) Since D satisfies UF1, elements of D are either 0, a unit, or can be written as a product of irreducibles. Hence, every non-zero element of D is of the form ux^i where u is a unit and $i \geq 0$. For the units, we have $i = 0$, and for non-zero, non-units we have $i > 0$. Let I be a proper ideal of D. Then none of the elements of I are units, and hence they all are of the form ux^i with $i > 0$. As a result $I \subseteq \langle x \rangle$. Let E be the set of exponents (the i in ux^i) of the elements of I. Then E is a non-empty subset of the positive integers, and hence (by the well ordering of the integers) has a least element j. Thus $x^j \in I$ and $I = \langle x^j \rangle$. We have shown that I is a principal ideal and D is a PID. We conclude that D is a discrete valuation ring.

19.2.9. We have $(3x + 2)(x + 4) = 3x^2 + 4x + 3 = (4x + 1)(2x + 3)$. Thus both $3x + 2$ and $4x + 1$ divide $p(x)$. Now $4(3x + 2) = 2x + 3$ and 4 is a unit, and so $3x + 2$ and $2x + 3$ are associates. Also $4(x + 4) = (4x + 1)$ and hence $x + 4$ and $4x + 1$ are associates. This, therefore, does not violate unique factorization. In fact $\mathbb{F}_5$ is a field and hence $\mathbb{F}_5[x]$ is a Euclidean domain, which means that it is both a PID and a UFD.

19.2.11. In $F[x]$, we can write $f = qg + r$ with $r = 0$ or $\deg(r) < \deg(g)$. Moreover, q and r are unique elements of $F[x]$ (as well as elements of $E[x]$). In $E[x]$, we can do the same, but uniqueness (of the quotient and remainder in $E[x]$) means that we will get $f = qg + r$ exactly as before. Hence, the Euclidean algorithm will result in the same greatest common divisor whether we perform it in $F[x]$ or in $E[x]$.

19.3.3. There are eight polynomials of degree 3 in $\mathbb{F}_2[x]$. Going through them, we find two that have no roots: $x^3 + x + 1$ and $x^3 + x^2 + 1$. A polynomial of degree 3 in $\mathbb{F}_2[x]$ is irreducible if and only if it has no roots, and so these two polynomials are all the irreducible polynomials of degree 3 in $\mathbb{F}_2[x]$.

Let $I = \langle x^3 + x + 1 \rangle$. Then I is a maximal ideal and $E = \mathbb{F}_2[x]/I$ is a field. In E, every occurrence of x^3 can be replaced by $x + 1$ since $x^3 + I = x + 1 + I$. Hence, in E every polynomial can be transformed to a polynomial of degree 2 or less. Hence, we have

$$E = \{a + bx + cx^2 + I \mid a, b, c \in \mathbb{F}_2, x^3 = x + 1\}.$$

There are two choices for each a, b, and c, and so E has eight elements. Every element of E when added to itself is zero (since the arithmetic is mod 2) and hence $(E, +)$ is an abelian group where every element is of order 2. This means that $E \cong \mathbb{Z}/2\mathbb{Z} \times \mathbb{Z}/2\mathbb{Z} \times \mathbb{Z}/2\mathbb{Z}$. We claim that $(E - \{0\}, \times)$ is a cyclic group of order 7 and, hence, isomorphic to $\mathbb{Z}/7\mathbb{Z}$. (In fact, if F is *any* finite field, then $(F - \{0\}, \times)$ is a cyclic group. See Theorem 25.41.) To prove this for E, we need to find a generator for the group. In fact, from group theory, we know that in $\mathbb{Z}/7\mathbb{Z}$ *every* non-identity element is a generator. So, if we find the consecutive powers of x, we should get every non-zero element of E. Using $x^3 = x + 1$, we have $x^4 = xx^3 = x^2 + x$, $x^5 = xx^4 = x^3 + x^2 = x^2 + x + 1$, and so on. We get the following table.

x	x^2	x^3	x^4	x^5	x^6	x^7
x	x^2	$x + 1$	$x^2 + x$	$x^2 + x + 1$	$x^2 + 1$	1

Hence, every non-zero element of E has two representations. Every element can be written as a linear combination of 1, x, and x^2, or as a power of x. For example, $x^5 = x^2 + x + 1$ are two ways of writing the same element. The representation as a linear combination makes addition easy, and the representation as powers makes multiplication easy. For example, it is not obvious what $x^3 + x^6$ is. But replacing x^3 and x^6 with $x + 1$ and $x^2 + 1$, respectively, we have

$$x^3 + x^6 = x + 1 + x^2 + 1 = x^2 + x = x^4.$$

On the other hand, it takes a little computation to find $(x^2 + x)(x^2 + x + 1)$. But if we write each factor as a power of x—using the table above and the fact that $x^7 = 1$—we get

$$(x^2 + x)(x^2 + x + 1) = x^4 x^5 = x^9 = x^2.$$

Given this double representation, it is easy to write the addition table and multiplication table of $E = \{0, 1, x, x^2, \ldots, x^6\}$. These are given in Figure C.6.

+	0	1	x	x^2	x^3	x^4	x^5	x^6
0	0	1	x	x^2	x^3	x^4	x^5	x^6
1	1	0	x^3	x^6	x	x^5	x^4	x^2
x	x	x^3	0	x^4	1	x^2	x^6	x^5
x^2	x^2	x^6	x^4	0	x^5	x	x^3	1
x^3	x^3	x	1	x^5	0	x^6	x^2	x^4
x^4	x^4	x^5	x^2	x	x^6	0	1	x^3
x^5	x^5	x^4	x^6	x^3	x^2	1	0	x
x^6	x^6	x^2	x^5	1	x^4	x^3	x	0

$\times$	0	1	x	x^2	x^3	x^4	x^5	x^6
0	0	0	0	0	0	0	0	0
1	0	1	x	x^2	x^3	x^4	x^5	x^6
x	0	x	x^2	x^3	x^4	x^5	x^6	1
x^2	0	x^2	x^3	x^4	x^5	x^6	1	x
x^3	0	x^3	x^4	x^5	x^6	1	x	x^2
x^4	0	x^4	x^5	x^6	1	x	x^2	x^3
x^5	0	x^5	x^6	1	x	x^2	x^3	x^4
x^6	0	x^6	1	x	x^2	x^3	x^4	x^5

Figure C.6. The addition table and multiplication table for $E = \{0, 1, x, x^2, \ldots, x^6 \mid x^7 = 1\}$

19.3.13.

(a) We can just check every element of $\mathbb{F}_7$, and we see that 0, 1, 2, 3, 4, 5, and 6 are all roots of $x^7 - x$. We could have predicted this in another way. By Fermat's Little Theorem, if p is a prime number, then a^{p-1} has remainder 1 when divided by p as long as $p \nmid a$. This implies that a^p and a have the same remainder when divided by p (regardless of whether $p \mid a$ or not). We write this as $a^p \equiv a \pmod{p}$. In other words, for every a, we have $a^p - a \equiv 0 \pmod{p}$. This means that, as long as p is a prime, every element of $\mathbb{F}_p$ is a root of $x^p - x \in \mathbb{F}_p[x]$.

(b) We know that if a is a root of a polynomial over a field, then $x - a$ divides the polynomial. Hence for each $a \in \mathbb{F}_7$, we have that $x - a$ divides f. So
$$f = x(x-1)(x-2)(x-3)(x-4)(x-5)(x-6).$$

19.3.15. We solve $x^4 = -1$ using Euler's formula: $e^{ix} = \cos(x) + i\sin(x)$. For $x = \pi, 3\pi, 5\pi, \ldots$, we have $e^{ix} = -1$. Hence, we write
$$x^4 = -1 = e^{(2k+1)\pi i} \quad \text{for } k \in \mathbb{Z}.$$

We take fourth roots of both sides to get:
$$x = e^{\frac{1}{4}(2k+1)\pi i} \quad \text{with } k \in \mathbb{Z}.$$

By plugging in various integers for k, we get all the solutions for x. For example, for $k = 0$, we get $x = e^{i\pi/4} = \cos(\pi/4) + i\sin(\pi/4) = \frac{\sqrt{2}}{2} + \frac{\sqrt{2}}{2}i$. Because of the periodicity of the exponential function, we only get four distinct solutions:
$$x = \pm\frac{\sqrt{2}}{2} \pm \frac{\sqrt{2}}{2}i.$$

Hence $x^4 + 1$ has four *complex* roots, and these come in pairs of conjugates. Let $\alpha = \frac{\sqrt{2}}{2} + \frac{\sqrt{2}}{2}i$ and $\beta = -\frac{\sqrt{2}}{2} + \frac{\sqrt{2}}{2}i$. Let $\overline{\alpha}$ and $\overline{\beta}$ denote the conjugates of α and β, respectively. Hence in $\mathbb{C}[x]$ we have
$$x^4 + 1 = (x - \alpha)(x - \overline{\alpha})(x - \beta)(x - \overline{\beta}).$$

Now $(x - \alpha)(x - \overline{\alpha}) = x^2 - (\alpha + \overline{\alpha})x + \alpha\overline{\alpha} = x^2 - \sqrt{2}x + 1$ and this has *real* coefficients! $(x - \beta)(x - \overline{\beta})$ also gives real coefficients, and we get
$$x^4 + 1 = (x^2 - \sqrt{2}x + 1)(x^2 + \sqrt{2}x + 1).$$

The quadratics do not have any real roots, and so this is the factorization of $x^4 + 1$ in $\mathbb{R}[x]$. Of course, now that we know the factorization, we can come up with a different and clever way of deriving it:
$$x^4 + 1 = x^4 + 2x^2 + 1 - 2x^2 = (x^2 + 1)^2 - 2x^2 = (x^2 + 1 + \sqrt{2}x)(x^2 + 1 - \sqrt{2}x).$$

⌣⌣ you could just use a symbolic mathematical software such as Sage, Maple, or Mathematica.

In fact, the method that we used for $x^4 + 1$—i.e., using the fact that complex roots come in conjugate pairs—can be generalized to show that in $\mathbb{R}[x]$ all irreducible polynomials are of degree 1 or 2.

If you originally believed that $x^4 + 1$ was irreducible in $\mathbb{R}[x]$, then you are in good company. In 1702, Leibniz thought that he had proved the fundamental theorem of algebra (Theorem 26.11) wrong by showing that $x^4 + 1$ could not be factored over the reals. Forty years later, Euler showed that Leibniz was mistaken.

19.4.5. If $\alpha = \sqrt{2} + \sqrt{3}$, then $\alpha^2 = 5 + 2\sqrt{6}$. Thus $(\alpha^2 - 5)^2 = 24$. Hence $\alpha^4 - 10\alpha^2 + 1 = 0$. Thus α is a root of $x^4 - 10x^2 + 1$. By the rational roots test, the only possible rational roots for this polynomial are ± 1. Inspection shows that neither 1 nor -1 are roots, and so we conclude that there are no rational roots. Thus α—which is a root of this polynomial—must be irrational.

19.4.17. We give two proofs. The first mimics the proof of the irrationality of $\sqrt{2}$, the second uses Gauss's lemma.

FIRST PROOF. Note that $F[x]$ is an ED, and hence every two elements have a greatest common divisor. Thus every element of $F(x)$ can be written as $\frac{p}{q}$ with $p, q \in F[x]$ and $\gcd(p, q) = 1$. Assume $\sqrt{1 - x^2} = p/q$ with $p, q \in F[x]$ and with $\gcd(p, q) = 1$. Then $(1 - x)(1 + x)q^2 = p^2$. $1 + x$ is irreducible in $F[x]$—all linear polynomials are—and hence it is a prime. We have $1 + x \mid p^2$ which implies that $1 + x \mid p$. This means that $(1 + x)^2 \mid p^2$, which in turn means that $1 + x \mid (1 - x)q^2$. The only way that $1 + x$ could divide $1 - x$ would be if $1 + x = 1 - x$ which would mean $2x = 0$. This only happens in a field of characteristic 2. Hence $1 + x$ a prime that does not divide $1 - x$. This means that $1 + x$ divides q^2. Again this means that $1 + x \mid q$ showing that $\gcd(p, q) \neq 1$. The contradiction proves the claim.

SECOND PROOF. There is no polynomial in $F[x]$—remember that $F[x]$ is not the same as $F(x)$—whose square is $1 - x^2$. This is because such a polynomial would have to be linear, and the square of $ax + b \in F[x]$—we must have $a \neq 0$—is $a^2x^2 + 2abx + b^2$. Since $\mathrm{char}(F) \neq 2$, we would have to have $ab = 0$, which means that b is zero, and we cannot get $1 - x^2$.

Now consider the polynomial $y^2 - (1 - x^2) \in F[x, y] = (F[x])[y]$. This polynomial has no root in $F[x]$ and so by Gauss's lemma—since $F(x)$ is the field of fractions of $F[x]$—it has no root in $F(x)$. This means that $\sqrt{1 - x^2} \notin F(x)$.

We certainly need $\mathrm{char}(F) \neq 2$, since in $\mathbb{F}_2[x]$ we have $\sqrt{1 - x^2} = 1 + x \in \mathbb{F}_2[x] \subseteq \mathbb{F}_2(x)$.

19.5.3. The key fact is that $f(x) = g(x)h(x) \Leftrightarrow f(x + 1) = g(x + 1)h(x + 1)$. Also, note that, since R is an integral domain, $\deg(k(x)) = \deg(k(x + 1))$ for all $k \in R[x]$.

Since R is an integral domain, the irreducibles in $R[x]$ are the irreducibles of R together with those polynomials of positive degree that cannot be factored into two polynomials, in $R[x]$, of positive degree. (See Lemma 19.43.)

If $\deg(f(x)) = \deg(f(x + 1)) = 0$, then $f(x) = f(x + 1)$, and so $f(x)$ is irreducible if and only if $f(x + 1)$ is irreducible.

If $\deg(f(x)) = \deg(f(x+1)) > 0$, then $f(x)$ is reducible if and only if $f(x) = g(x)h(x)$ with g and h both polynomials of positive degree. The latter happens if and only if $f(x + 1) = g(x + 1)h(x + 1)$ with $g(x + 1)$ and $h(x + 1)$ polynomials of positive degree which is equivalent to $f(x + 1)$ being reducible.

19.7.3. ($\Rightarrow$) We assume that $a_0 + a_1 x + a_2 x^2 + \cdots + a_n x^n + \cdots$ has an inverse. Thus for some $b_0, b_1, \ldots \in R$ we must have

$$(a_0 + a_1 x + a_2 x^2 + \cdots + a_n x^n + \cdots)(b_0 + b_1 x + b_2 x^2 + \cdots + b_n x^n + \cdots) = 1.$$

This means that $a_0 b_0 = 1$, and so a_0 is a unit in R.

($\Leftarrow$) Assume that a_0 is a unit. We want to know whether we can find $b_0, b_1, \ldots \in R$ such that

$$(a_0 + a_1 x + a_2 x^2 + \cdots + a_n x^n + \cdots)(b_0 + b_1 x + b_2 x^2 + \cdots + b_n x^n + \cdots) = 1.$$

Thus the question is whether we can solve the following system of equations where the unknowns are $b_0, b_1, \ldots, b_n, \ldots$:

$$\begin{cases} a_0 b_0 &= 1 \\ a_0 b_1 + a_1 b_0 &= 0 \\ a_0 b_2 + a_1 b_1 + a_2 b_0 &= 0 \\ \quad\vdots \\ a_0 b_n + a_1 b_{n-1} + \cdots + a_n b_0 &= 0 \\ \quad\vdots \end{cases} \Rightarrow \begin{cases} b_0 &= a_0^{-1} \\ b_1 &= a_0^{-1}[-a_1 b_0] \\ b_2 &= a_0^{-1}[-a_1 b_1 - a_2 b_0] \\ \quad\vdots \\ b_n &= a_0^{-1}[-a_1 b_{n-1} - \cdots - a_n b_0] \\ \quad\vdots \end{cases}$$

It is clear that if $(a_0)^{-1}$ exists, then one by one we can find $b_0, b_1, \ldots, b_n, \ldots$.

22.1.9. We know that $x^7 - 1 = (x - 1)(x^6 + x^5 + \cdots + x + 1)$. Now ζ_7 is a root of $x^7 - 1$ and hence—since $\mathbb{C}$ is an integral domain—it must be a root of $x - 1$ or of $x^6 + \cdots + 1$. Now $\zeta_7 = \cos(2\pi/7) + i \sin(2\pi/7)$ and $\sin(2\pi/7) \neq 0$. Hence ζ_7 is not a real number and so it is not equal to 1. We conclude that ζ_7 is a root of $\Phi_7(x) = x^6 + x^5 + \cdots + x + 1$. This polynomial is called the seventh *cyclotomic polynomial* and was shown to be irreducible in Problem **19.5.5**. Thus $\Phi_7(x)$ is the minimal polynomial of ζ_7 over $\mathbb{Q}$.

22.1.13. We have

$$\alpha^2 = 8 + 2\sqrt{9} = 14 \quad \Rightarrow \quad \alpha^2 - 14 = 0.$$

Now let $f(x) = x^2 - 14$. This polynomial is monic, irreducible by Schönemann-Eisenstein, and has α as a root. So it must be the minimal polynomial of α. Its roots are $\pm\sqrt{14}$. We know that α is positive and is one of these roots. Hence, we conclude—maybe surprisingly—that

$$\sqrt{4 + \sqrt{7}} + \sqrt{4 - \sqrt{7}} = \sqrt{14}.$$

22.1.17. Consider the evaluation map $\epsilon_\alpha : F[x] \to E$ defined by $\epsilon_\alpha(p) = p(\alpha)$. This map is a ring homomorphism, its image is $F[\alpha]$, and the kernel—by Theorem 22.16—is $\langle f \rangle$.

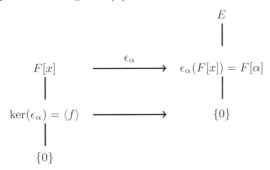

By the homomorphism theorem, Theorem 16.45, we now have $F[\alpha] \cong F[x]/\langle f \rangle$. Since f is irreducible and $F[x]$ is a PID (in fact, an ED), we have that $\langle f \rangle$ is maximal in $F[x]$ and $F[x]/\langle f \rangle$ is a field. We conclude that $F[\alpha]$ is a field and hence $F[\alpha] = F(\alpha)$.

22.3.3. We have

$$\alpha^3 = (\sqrt[3]{2} + \sqrt[3]{4})^3 = 2 + 3\sqrt[3]{4}\sqrt[3]{4} + 3\sqrt[3]{2}\sqrt[3]{16} + 4$$
$$= 6 + 6\sqrt[3]{2} + 6\sqrt[3]{4} = 6 + 6\alpha,$$

and, hence, α is a root of $x^3 - 6x - 6$. The latter is irreducible by the Schönemann-Eisenstein criterion. Thus, $\min_{\mathbb{Q}}(\alpha) = x^3 - 6x - 6$. α is algebraic over $\mathbb{Q}$ and thus $\mathbb{Q}[\alpha] = \mathbb{Q}(\alpha)$. Now $\frac{1}{\alpha} + \frac{1}{\alpha^2} \in \mathbb{Q}(\alpha) = \mathbb{Q}[\alpha]$.

22.3.15.

(a) $x^5 - 2$ is irreducible by the Schönemann-Eisenstein criterion and hence $x^5 - 2 = \min_{\mathbb{Q}}(\sqrt[5]{2})$. Now $|\mathbb{Q}(\sqrt[5]{2}) : \mathbb{Q}| = \deg \min_{\mathbb{Q}}(\sqrt[5]{2}) = 5$.

(b) We first note that $\alpha \in \mathbb{Q}(\sqrt[5]{2}) = \mathbb{Q}[\sqrt[5]{2}]$. Thus

$$\mathbb{Q} \subseteq \mathbb{Q}[\alpha] \subseteq \mathbb{Q}[\sqrt[5]{2}].$$

So, by Proposition 22.38, $|\mathbb{Q}[\alpha] : \mathbb{Q}|$ divides $|\mathbb{Q}[\sqrt[5]{2}] : \mathbb{Q}|$ which is 5. We conclude that $|\mathbb{Q}[\alpha] : \mathbb{Q}|$ is either 1 or 5. The degree of the extension cannot be 1 since $\alpha \notin \mathbb{Q}$. Thus $|\mathbb{Q}[\alpha] : \mathbb{Q}| = 5$. This means that $\mathbb{Q}[\alpha]$, which was a subspace of $\mathbb{Q}[\sqrt[5]{2}]$, has the same dimension as the latter. Thus these two vector spaces must be equal. Hence, $\sqrt[5]{2} \in \mathbb{Q}[\alpha]$. Thus there exists a polynomial $p \in \mathbb{Q}[x]$ with $p(\alpha) = \sqrt[5]{2}$.

$$\begin{array}{c} \mathbb{Q}[\sqrt[5]{2}] \\ | \\ 5 \quad \mathbb{Q}[\alpha] \\ | \\ \mathbb{Q} \end{array}$$

22.3.31. Let $f = \min_E(\alpha)$. Then $f(x) = x^n + a_{n-1}x^{n-1} + \cdots + a_0$ where $a_i \in E$. The latter are algebraic over F since E itself is algebraic over F. We have

$$F \subseteq F[a_0] \subseteq F[a_0, a_1] \subseteq \cdots \subseteq F[a_0, \ldots, a_{n-1}] = K.$$

Each a_i is algebraic over the preceding field, and so each of the field extensions is finite dimensional. Hence, $|K : F| < \infty$. In addition, α is algebraic over K since $f(\alpha) = 0$ and $f \in K[x]$.

Thus we have $F \subseteq K \subseteq K[\alpha]$ where both extensions are finite, and so $|K[\alpha] : F| < \infty$. Now $F \subseteq F[\alpha] \subseteq K[\alpha]$ which implies that $|F[\alpha] : F| \leq |K[\alpha] : F| < \infty$. As a result, α is algebraic over F.

22.3.37. Let γ be a root of $x^{nm} - a$, $\alpha = \gamma^n$, and $\beta = \gamma^m$. Then α is a root of $x^m - a$ and β is a root of $x^n - a$. Since the degree of a simple field extension $F(\delta)$ is given by the degree of the minimal polynomial of δ, we know that $|F(\gamma) : F| \leq nm$, $|F(\alpha) : F| \leq m$, and $|F(\beta) : F| \leq n$. Also, since $\min_F(\alpha)$ can be thought of as a polynomial with coefficients in $F(\beta)$, we have $|F(\alpha, \beta) : F(\beta)| \leq |F(\alpha) : F| \leq m$. Thus,

$$|F(\alpha, \beta) : F| = |F(\alpha, \beta) : F(\beta)| \, |F(\beta) : F| \leq mn.$$

Finally, note that $\alpha \in F(\gamma)$ and so $F(\alpha) \subseteq F(\alpha, \gamma) = F(\gamma)$. Furthermore, since $\gamma^n - \alpha = 0$, we have that γ is a root of $x^n - \alpha \in F(\alpha)[x]$, and

$$|F(\gamma) : F(\alpha)| \leq n.$$

($\Rightarrow$) Assume that $x^{mn} - a$ is irreducible over F. Then, this polynomial is the minimal polynomial of γ over F and so $|F(\gamma) : F| = nm$. But, if we assume $|F(\alpha) : F| < m$, we would have

$$|F(\gamma) : F| = |F(\gamma) : F(\alpha)| \, |F(\alpha) : F| < nm.$$

This is a contradiction proving that $|F(\alpha) : F| = m$. This means that $\deg(\min_F(\alpha)) = m$ and hence $x^m - a$ must be irreducible (since otherwise the minimal polynomial would be a factor of $x^m - a$ and would have smaller degree). An identical argument reversing the roles of α and β shows that $x^n - a$ is also irreducible.

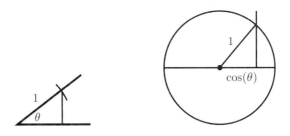

Figure C.7. Constructing $\cos(\theta)$ from θ (left) and vice versa (right)

($\Leftarrow$) Assume that $x^m - a$ and $x^n - a$ are irreducible over F. Then these polynomials are the minimal polynomials of α and β respectively, and $|F(\alpha) : F| = m$, and $|F(\beta) : F| = n$. Since $|F(\alpha, \beta) : F| = |F(\alpha, \beta) : F(\beta)| \; |F(\beta) : F|$, we have that n divides $|F(\alpha, \beta) : F|$. Similarly, m also divides $|F(\alpha, \beta) : F|$. Since n and m are relatively prime, we have nm divides $|F(\alpha, \beta) : F|$. But the latter was always $\leq nm$. Hence $|F(\alpha, \beta) : F| = nm$.

Both α and β are in $F(\gamma)$ and hence so is $F(\alpha, \beta)$. In other words,

$$F \subseteq F(\alpha, \beta) \subseteq F(\gamma).$$

We know that $|F(\gamma) : F| \leq mn$. If we assume that $|F(\gamma) : F| < mn$, then we would have

$$|F(\alpha, \beta) : F| \leq |F(\gamma) : F| < mn.$$

The contradiction proves that $|F(\gamma) : F| = mn$. Hence $x^{mn} - a$ *is* the minimal polynomial of γ over F and hence it must be irreducible.

22.3.39. Let $h = (y^2 + 1) - f(y^3 + y^2 + 1)$, and note that h is a polynomial in y and with coefficients in $K(f) = E$. So $h \in E[y]$, and, in fact, $h = -fy^3 - (f-1)y^2 - f + 1$ is a polynomial of degree 3 in $E[y]$. Since $f \in K(x)$, $E \subseteq K(x)$ is a field extension, and we can plug elements of the bigger field $K(x)$ into polynomials with coefficients in E. Thus, we can plug x into h. Now $h(x) = (x^2 + 1) - f(x^3 + x^2 + 1) = 0$, and so x is the root of a third-degree polynomial over E. Hence, x is algebraic over E and $E[x] = E(x)$. Now $E(x)$ is the smallest field that contains K, f, and x, and so $E(x) = K(x)$. Hence, $|K(x) \colon E| = |E[x] \colon E| = \deg(\min_E(x))$. If we knew that h was irreducible in $E[y]$, then h would be a constant multiple of the minimum polynomial of x, and $\deg(h) = \deg(\min_E(x))$. We would conclude that $|K(x) \colon E| = 3$. To prove that h is irreducible in $E[y]$ (this was Problem 19.4.15), note that $E = K(f)$ is the field of fractions of $K[f]$. Furthermore, $K[f]$ is a UFD. So by Gauss's lemma, h is irreducible in $E[y] = K(f)[y]$ if it is irreducible in $(K[f])[y]$. Hence the question is whether $h \in K[f, y] = (K[y])[f]$ is irreducible. But $h = (-y^3 - y^2 - 1)f + y^2 + 1$ is a primitive polynomial of degree 1 in $(K[y])[f]$ (since $x^2 + 1$ and $x^3 + x^2 + 1$ are relatively prime) and so, by Lemma 19.43, it must be irreducible in $K[y, f]$. As a result it is also irreducible in $K[f, y]$ and $K(f)[y] = E[y]$. We conclude that $|K(x) \colon E| = 3$.

23.2.9. Note that $\cos(\theta) = 4\cos^3(\frac{\theta}{3}) - 3\cos(\frac{\theta}{3})$, and so $\cos(\frac{\theta}{3})$ is a root of $4t^3 - 3t - \cos(\theta)$. We also claim that the angle θ is constructible if and only if the real number $\cos(\theta)$ is constructible. If you are given θ, then use the compass to mark a unit length on one side of the angle, and then draw a perpendicular to the other side to get $\cos(\theta)$ (see the drawing on the left of Figure C.7). If you are given $\cos(\theta)$, then mark it off on the radius of a unit circle, and draw a perpendicular to the radius. Drawing the radius to the intersection of the circle and the perpendicular gives the angle θ (see the drawing on the right of Figure C.7).

($\Rightarrow$) Assume that the angle θ can be trisected. Then if we are given $\cos(\theta)$, we can construct θ, trisect it, and then construct $\cos(\theta/3)$. Thus the ability to trisect θ implies that $\cos(\theta/3)$ can be constructed using permissible constructions from $\cos(\theta)$. This means that $|\mathbb{Q}(\cos(\theta), \cos(\theta/3)) : \mathbb{Q}(\cos(\theta))|$ is a power of 2. As a result the degree of the minimal polynomial of $\cos(\theta/3)$ over $\mathbb{Q}(\cos(\theta))$ is a power of 2. However, $\cos(\theta/3)$ is also a root of $4t^3 - 3t - \cos(\theta)$. We conclude that the latter cannot be the minimal polynomial and must be reducible.

($\Leftarrow$) Assume that $4t^3 - 3t - \cos(\theta)$ is reducible over $\mathbb{Q}(\cos(\theta))$. This means that the degree of the minimal polynomial of $\cos(\theta/3)$ over $\mathbb{Q}(\cos(\theta))$ is 1 or 2. We conclude that given $\cos(\theta)$, we can construct $\cos(\theta/3)$. Now, given θ, we first construct $\cos(\theta)$, then construct $\cos(\theta/3)$, and then construct $\theta/3$. Hence, we can trisect θ.

24.1.15. Let $f = \min_F(\alpha)$. The coefficients of f are in F and ϕ is an F-isomorphism, thus $\phi f = f$. Now we have

$$0 = \phi(0) = \phi(f(\alpha)) = (\phi f)(\phi(\alpha)) = f(\beta).$$

The polynomial f is monic and irreducible, and $f(\beta) = 0$. Hence $f = \min_F(\beta)$.

Note that Theorem 24.6 proved the converse. In other words, if α and β have the same minimal polynomials, then there is an F-isomorphism from $F[\alpha]$ to $F[\beta]$ that sends α to β. So in this sense, simple extensions are characterized by minimal polynomials.

24.1.19. Let $L \supseteq E$ be a field in which f has a root, and let $\alpha \in L$ with $f(\alpha) = 0$. f is irreducible and α is a root of f and so $\min_F(\alpha) = cf$, and so $|F[\alpha] : F| = \deg f = p$. Now,

$$|E[\alpha] : F| = |E[\alpha] : F[\alpha]|\ |F[\alpha] : F|,$$

and, hence, p divides $|E[\alpha] : F|$.

Let $g = \min_E(\alpha)$. Now $f \in E[x]$ and α is a root of f, and thus $g \mid f$ in $E[x]$. f is reducible in $E[x]$ and hence $\deg(g) < \deg(f) = p$ which implies that $p \nmid \deg(g) = |E[\alpha] : E|$. Now

$$|E[\alpha] : F| = |E[\alpha] : E|\ |E : F|.$$

We have that $|E[\alpha] : F|$ is divisible by p and $|E[\alpha] : E|$ is not divisible by p. Hence, p divides $|E : F|$.

24.2.3. $\sqrt[4]{5}$ is a root of $x^4 - 5$ which is irreducible by the Schönemann-Eisenstein criterion. Hence $\min_{\mathbb{Q}}[\sqrt[4]{5}] = x^4 - 5$ and $|E : \mathbb{Q}| = 4$. Now, since $\alpha \in E$, we have $\mathbb{Q}[\alpha] = E$, and hence $|\mathbb{Q}[\alpha] : \mathbb{Q}| = 4$. This means that the degree of the minimum polynomial of α is four. Now, since $\alpha - 1 = \sqrt[4]{5}$, we have $(\alpha - 1)^4 = 5$, and so α is a root of $(x-1)^4 - 5 = x^4 - 4x^3 + 6x^2 - 4x - 4$ which is monic and of degree 4. Since we know that the minimal polynomial of α is of degree 4, we conclude that this polynomial must be irreducible (since otherwise α would be the root of one of the irreducible factors and would have a minimal polynomial of degree less than 4). Hence

$$\min_{\mathbb{Q}}(\alpha) = x^4 - 4x^3 + 6x^2 - 4x - 4.$$

To find the splitting field of this polynomial, we need the roots of the polynomial in $\mathbb{C}$. From our construction, we know that $x^4 - 4x^3 + 6x^2 - 4x - 4 = (x-1)^4 - 5$, and so $\min_{\mathbb{Q}}(\alpha) = 0$ is equivalent to $(x-1)^4 = 5 = 5e^{2\pi k i}$, for any integer k. Hence, $x - 1 = \sqrt[4]{5}e^{\frac{\pi k}{2}i}$. The possibilities for $e^{\frac{\pi k}{2}i}$ are ± 1 and $\pm i$. Hence, the roots of $\min_{\mathbb{Q}}(\alpha)$ are $1 \pm \sqrt[4]{5}$ and $1 \pm i\sqrt[4]{5}$. Hence, in $\mathbb{C}$, this polynomial factors as $(x-1-\sqrt[4]{5})(x-1+\sqrt[4]{5})(x-1-i\sqrt[4]{5})(x-1+i\sqrt[4]{5})$. The splitting field of $\min_{\mathbb{Q}}(\alpha)$ must include all these four roots. Hence, it must also contain $\frac{1}{2}(1 + \sqrt[4]{5}) - \frac{1}{2}(1 - \sqrt[4]{5}) = \sqrt[4]{5}$. The splitting field must also include $\frac{(1+i\sqrt[4]{5})-1}{\sqrt[4]{5}} = i$. Hence, $\mathbb{Q}[\sqrt[4]{5}, i]$ must be contained in the splitting field. On the other hand, all four roots live in

$\mathbb{Q}[\sqrt[4]{5}, i]$. We conclude that $\mathbb{Q}[\sqrt[4]{5}, i] = E[i]$ is the splitting field of $\min_{\mathbb{Q}}(\alpha)$ over $\mathbb{Q}$ and over E.

24.2.7. $x^6 - 8 = 0$ implies that $x^6 = 8e^{2k\pi i}$ for $k \in \mathbb{Z}$, and hence $x = \sqrt{2}e^{\frac{k\pi i}{3}}$. Plugging in $k = 0, 1, \ldots, 5$, we get that the roots of $x^6 - 8$ are

$$\sqrt{2}, \sqrt{2}e^{\pi i/3}, \sqrt{2}e^{2\pi i/3}, -\sqrt{2}, \sqrt{2}e^{4\pi i/3}, \sqrt{2}e^{5\pi i/3}.$$

The splitting field for $x^6 - 8$ must include all these roots. A field that contains these roots will also include $\sqrt{2}$ and $e^{\pi i/3}$. Hence, $\mathbb{Q}(\sqrt{2}, e^{\pi i/3})$ is contained in the splitting field. But this field actually does contain all the roots, and so it is the splitting field. Now note that $e^{\pi i/3}$ is a root of $x^6 - 1$ and this polynomial factors as $(x-1)(x+1)(x^2+x+1)(x^2-x+1)$. By inspection $e^{\pi i/3}$ is a root of $x^2 - x + 1$. Now, we claim that this polynomial is irreducible in $\mathbb{Q}(\sqrt{2})$. This is true, since otherwise $e^{\pi i/3}$ would be in $\mathbb{Q}(\sqrt{2})$. This clearly is not the case, since $\mathbb{Q}(\sqrt{2})$ only contains real numbers. Thus $\min_{\mathbb{Q}(\sqrt{2})}(e^{\pi i/3}) = x^2 - x + 1$. So

$$|\mathbb{Q}(\sqrt{2}, e^{\pi i/3}) : \mathbb{Q}| = |\mathbb{Q}(\sqrt{2}, e^{\pi i/3}) : \mathbb{Q}(\sqrt{2})| \, |\mathbb{Q}(\sqrt{2}) : \mathbb{Q}| = 2 \times 2 = 4.$$

24.3.5. Let $\Omega = \{\text{roots of } \min_F(\alpha) \text{ in } E\}$. The Galois group $\mathrm{Gal}(E/F)$ acts on Ω. This gives a map $\phi \colon \mathrm{Gal}(E/F) \to \Omega$ defined by $\phi(\sigma) = \sigma(\alpha)$. To show that this map is 1-1, assume that $\phi(\sigma_1) = \phi(\sigma_2)$, for $\sigma_1, \sigma_2 \in \mathrm{Gal}(E/F)$. By the definition of the map, we have $\sigma_1(\alpha) = \sigma_2(\alpha)$. Hence, the automorphisms σ_1 and σ_2 agree on α and they both fix elements of F. But $E = F[\alpha]$ and if two automorphisms agree on F and on α, they must be identical. Hence, $\sigma_1 = \sigma_2$, proving that ϕ is 1-1. Since the map is 1-1 and Ω is a finite set, we have

$$|\mathrm{Gal}(E/F)| \leq |\Omega| \leq \deg(\min_F(\alpha)) = |E : F|.$$

25.2.7. Assume that $|\mathrm{Gal}(E/F)| = |E : F|$. Let $G = \mathrm{Gal}(E/F)$, and let $K = \mathrm{Fix}(G)$. We want to show that $K = F$.

How are $\mathrm{Gal}(E/F)$ and $\mathrm{Gal}(E/K)$ related? Every K-automorphism of E is certainly an F-automorphism since $F \subseteq K$, and hence $\mathrm{Gal}(E/K) \subseteq \mathrm{Gal}(E/F)$. On the other hand, every F-automorphism of E fixes K since K is the fixed field. Hence $\mathrm{Gal}(E/F) \subseteq \mathrm{Gal}(E/K)$. We conclude that $\mathrm{Gal}(E/K) = \mathrm{Gal}(E/F) = G$.

Now, by assumption, $E = K[\alpha]$ and hence, by Problem *24.3.5*, we have $|G| \leq |E : K|$. Hence, using the assumption that $|G| = |E : F|$, we have

$$|E : K| \geq |G| = |E : F| = |E : K||K : F| \Rightarrow |K : F| \leq 1 \Rightarrow |K : F| = 1 \Rightarrow K = F.$$

25.2.11.

(a) ζ_5 is a root of $x^5 - 1 = (x-1)(x^4 + x^3 + x^2 + x + 1)$ which means that it is a root of $f = x^4 + x^3 + x^2 + x + 1$. f is irreducible by Problem **19.5.5**, and hence $\min_{\mathbb{Q}}(\zeta_5) = f$ and $|\mathbb{Q}[\zeta_5] : \mathbb{Q}| = 4$.

(b) The roots of $f = \min_{\mathbb{Q}}(\zeta_5)$ are ζ_5, ζ_5^2, ζ_5^3, and ζ_5^4—they are all distinct and when raised to the fifth power give 1—and they are all in $\mathbb{Q}[\zeta_5]$. Thus $\mathbb{Q}[\zeta_5]$ is a splitting field of a separable polynomial, and hence it is a Galois extension of $\mathbb{Q}$.

(c) $\mathbb{Q}[\zeta_5]$ is a splitting field of an irreducible polynomial, and hence it acts transitively on $\{\zeta_5, \zeta_5^2, \zeta_5^3, \zeta_5^4\}$, the roots of an irreducible polynomial. The claim follows.

(d)
$$\sigma_1(\zeta_5 + \zeta_5^4) = \sigma_1(\zeta_5) + \sigma_1(\zeta_5)^4 = \zeta_5^4 + (\zeta_5^4)^4 = \zeta_5^4 + \zeta_5,$$
$$\sigma_2(\zeta_5 + \zeta_5^4) = \sigma_2(\zeta_5) + \sigma_2(\zeta_5)^4 = \zeta_5^2 + (\zeta_5^2)^4 = \zeta_5^2 + \zeta_5^3.$$

(e) Note that $\sigma_1^2(\zeta_5) = \sigma_1(\zeta_5^4) = (\zeta_5^4)^4 = \zeta_5$. Hence σ_1^2 fixes all of $\mathbb{Q}[\zeta_5]$ and thus must be the identity map. We conclude that $\langle \sigma_1 \rangle = \{e, \sigma_1\}$.

(f) $\mathrm{Fix}(G) = \mathbb{Q}$ since $\mathbb{Q}[\zeta_5]$ is a Galois extension of $\mathbb{Q}$.

(g) $\zeta_5 + \zeta_5^4$ is *not* fixed by σ_2 and hence it is not in the $\mathrm{Fix}(G) = \mathbb{Q}$.

(h) We know that $\zeta_5 + \zeta_5^4$ is an irrational number that is in the $\mathrm{Fix}(\langle \sigma_1 \rangle)$. In addition, σ_1 does not fix ζ_5 and hence $\mathrm{Fix}(\langle \sigma_1 \rangle) \neq \mathbb{Q}[\zeta_5]$. Thus

$$\mathbb{Q} \subsetneq \mathbb{Q}[\zeta_5 + \zeta_5^4] \subseteq \mathrm{Fix}(\langle \sigma_1 \rangle) \subsetneq \mathbb{Q}[\zeta_5].$$

Now $|\mathbb{Q}[\zeta_5] : \mathbb{Q}| = 4$ and thus $|\mathbb{Q}[\zeta_5 + \zeta_5^4] : \mathbb{Q}| = |\mathrm{Fix}(\langle \sigma_1 \rangle) : \mathbb{Q}| = 2$. We conclude that $\mathrm{Fix}(\langle \sigma_1 \rangle) = \mathbb{Q}[\zeta_5 + \zeta_5^4]$.

(i) Every element of G is determined by its action on ζ_5. Now if $\sigma \in G$, then $\sigma(\zeta)$ can be one of $\{\zeta_5, \zeta_5^2, \zeta_5^3, \zeta_5^4, \}$. Hence, G is a group of order 4 and we also can see—by calculating $\sigma_2^2(\zeta_5) = \sigma_2(\zeta_5^2) = \zeta_5^4 = \sigma_1(\zeta_5)$—that $\sigma_2^2 = \sigma_1$. Thus σ_2 is an element of order 4 and thus $G = \{e, \sigma_2, \sigma_2^2 = \sigma_1, \sigma_2^3\} \cong \mathbb{Z}/4\mathbb{Z}$.

25.3.7. σ is a 1-1, onto homomorphism from E to E that fixes every element of F. Thus as a map on K it will also be 1-1, a homomorphism, and will fix elements of F. However, it is not clear why $\sigma(K) = K$. In other words, we have to show that σ maps elements of K into K and also that $\sigma\big|_K$ is onto. These can be shown directly. However, we can also say that K is Galois over F and hence it is normal over F. Now a characterization of normal extensions was that if σ is an F-isomorphism from E to E and K is an intermediate field such that K is a normal extension of F, then $\sigma(K) = K$. The proof is thus complete.

25.3.11. Let $\alpha_1, \alpha_2, \ldots, \alpha_m$ be a basis for E over F. This means that $E = F[\alpha_1, \ldots, \alpha_m]$. Let $f = \prod \min_F(\alpha_i)$. Let L be the splitting field for f over E. Proposition 25.23 claims that L is the normal closure for E over F.

First note that any splitting field for f over F would have to include $\alpha_1 \ldots, \alpha_m$, and hence it would include E. This means that L is a splitting field for f over F—in addition to being the splitting field for f over E as originally defined—and thus L is normal over F.

Now assume that $E \subseteq K \subseteq L$ with K normal over F. Now K contains α_i, for $1 \leq i \leq m$, and so each $\min_F(\alpha_i)$ has a root in K. K is assumed to be normal and so $\min_F(\alpha_i)$ splits in K. We conclude that f splits in K. But L was the splitting field of f over E and $K \subseteq L$. Hence $K = L$, and L is a normal closure for E over F.

Note that if E itself is a splitting field for f over F, then E is a normal extension. We had proved the converse of this fact in Proposition 25.16. Hence, we conclude that a finite degree extension E is a normal extension of F if and only if it is the splitting field of f over F.

25.4.13. Let $f = \prod_{i=1}^{n} \min_F(\alpha_i)$. f is separable over F since all of its irreducible factors are separable over F. Let L be the splitting field for f over E. Then L is also a splitting field for f over F. (This needs an argument, but we have made similar arguments often—for example, in the solution to Problem *25.3.11*—so here it is left to you.) This means that L is Galois over F. Thus L is a separable extension of F. Now $F \subseteq E \subseteq L$ and hence E is a separable extension of F.

25.5.3. Let $\omega = e^{2\pi i/3} = -\frac{1}{2} + \frac{\sqrt{3}}{2}i$. Then $\omega^3 = 1$ and the roots of f in E are $\sqrt[3]{2}, \sqrt[3]{2}\omega$, $\sqrt[3]{2}\omega^2$, i, and $-i$. Hence, E contains $\sqrt[3]{2}$, ω, i, and $\sqrt{3}$. On the other hand, if a field contains $\{\sqrt[3]{2}, \sqrt{3}, i\}$ or $\{\sqrt[3]{2}, \omega, i\}$, then it will contain all the roots. Hence,

$$E = \mathbb{Q}[\sqrt[3]{2}, \sqrt{3}, i] = \mathbb{Q}[\sqrt[3]{2}, \omega, i].$$

(a) Since $\sqrt[3]{2}$, i, and $\sqrt{3}$ are in E, so is α.

(b) E is the splitting field of a separable polynomial over $\mathbb{Q}$, and hence E is a Galois extension of $\mathbb{Q}$. We conclude that $|\mathrm{Gal}(E/\mathbb{Q})| = |E : \mathbb{Q}|$. To find the latter, note that

$$\mathbb{Q} \subseteq \mathbb{Q}[\sqrt[3]{2}] \subseteq \mathbb{Q}[\sqrt[3]{2}, \sqrt{3}] \quad \text{and} \quad \mathbb{Q} \subseteq \mathbb{Q}[\sqrt{3}] \subseteq \mathbb{Q}[\sqrt[3]{2}, \sqrt{3}].$$

Now $\min_{\mathbb{Q}}(\sqrt[3]{2}) = x^3 - 2$ and $\min_{\mathbb{Q}}(\sqrt{3}) = x^2 - 3$. Hence, $|\mathbb{Q}[\sqrt[3]{2}] : \mathbb{Q}| = 3$ and $|\mathbb{Q}[\sqrt{3}] : \mathbb{Q}| = 2$. So both 2 and 3 divide $|\mathbb{Q}[\sqrt[3]{2}, \sqrt{3}] : \mathbb{Q}|$. Hence the latter is divisible by 6. Now $|\mathbb{Q}[\sqrt[3]{2}, \sqrt{3}] : \mathbb{Q}[\sqrt[3]{2}]|$ is at most 2, and hence

$$\left|\mathbb{Q}[\sqrt[3]{2}, \sqrt{3}] : \mathbb{Q}\right| = 6.$$

Now i is not a real number and hence not in $\mathbb{Q}[\sqrt[3]{2}, \sqrt{3}]$. It is a root of $x^2 + 1$ and so this must be its minimal polynomial over this field. Now, we have

$$\mathbb{Q} \subseteq \mathbb{Q}[\sqrt[3]{2}, \sqrt{3}] \subseteq \mathbb{Q}[\sqrt[3]{2}, \sqrt{3}, i] = E.$$

We conclude that

$$|\mathrm{Gal}(E/\mathbb{Q})| = |E : \mathbb{Q}| = 6 \times 2 = 12.$$

(c) E is a Galois extension and hence a normal extension of $\mathbb{Q}$. The irreducible polynomial g already has one root in E namely α, and so it splits in E.

(d) The Galois group has 12 elements. Each element is determined by its effect on $\sqrt[3]{2}$, $\sqrt{3}$, and i.

   The Galois group permutes transitively the roots of $x^3 - 2$, $x^2 - 3$, and $x^2 + 1$. Hence, there are three choices for the images of $\sqrt[3]{2}$, two choices for the images of $\sqrt{3}$, and two choices for the images of i. If we could do these choices independently, we would have exactly $3 \times 2 \times 2 = 12$ elements. Since we already know that there are exactly 12 different elements, then we must be able to have all combinations of these choices.

   Hence, for example, there is an element σ of the Galois group with $\sigma(i) = i$, $\sigma(\sqrt{3}) = \sqrt{3}$ and $\sigma(\sqrt[3]{2}) = \sqrt[3]{2}\omega$. Since $\omega = -\frac{1}{2} + \frac{\sqrt{3}}{2}i$ and σ fixes i and $\sqrt{3}$, we have $\sigma(\omega) = \omega$. We now know that $\sigma(\sqrt[3]{2}\omega) = \sqrt[3]{2}\omega^2$ and $\sigma(\sqrt[3]{2}\omega^2) = \sqrt[3]{2}$.

   Applying these 12 elements to α, we get six different elements in the orbit of α:

$$\{5\sqrt[3]{2} + 7\sqrt{3}i, \; 5\sqrt[3]{2}\omega + 7\sqrt{3}i, \; 5\sqrt[3]{2}\omega^2 + 7\sqrt{3}i, \; 5\sqrt[3]{2} - 7\sqrt{3}i, \; 5\sqrt[3]{2}\omega - 7\sqrt{3}i, \; 5\sqrt[3]{2}\omega^2 - 7\sqrt{3}i\}.$$

(e) In E, the roots of g are exactly the elements of the orbit of α under the action of the Galois group. Hence $\deg(g) = 6$, and g factors as

$$g(x) = (x - \alpha_1)(x - \alpha_2)(x - \alpha_3)(x - \alpha_4)(x - \alpha_5)(x - \alpha_6),$$

where $\alpha_1, \ldots, \alpha_6$ are the the elements of E in the orbit of α (identified above).

(f) Maple multiplies this out to give

$$g(x) = \min_{\mathbb{Q}}(\alpha) = 3239023 + 220500\,x + 64827\,x^2 - 500\,x^3 + 441\,x^4 + x^6.$$

We also now have a proof that this polynomial is irreducible.

(g) Arguing as in the case of α, we see that $\beta = i + \sqrt[3]{2} + \sqrt{3}$ or $\beta = \sqrt{3} + \sqrt[3]{2}i$ have orbits of size 12. (One other way to argue this is to say that no element of the Galois group fixes them.)

(h) $|\mathbb{Q}(\alpha) : \mathbb{Q}| = \deg \min_{\mathbb{Q}}(\alpha) = 6$. Hence $E \neq \mathbb{Q}(\alpha)$. On the other hand, $|\mathbb{Q}(\beta) : \mathbb{Q}| = \deg \min_{\mathbb{Q}}(\beta) = 12$. Hence $E = \mathbb{Q}(\beta)$.

   We already knew that E is a simple extension of $\mathbb{Q}$, but here we were able to find the primitive element using the Galois group.

25.5.7. The Galois group of $f(x)$ is $\mathrm{Gal}(E/\mathbb{Q})$ where E is the splitting field of f over $\mathbb{Q}$. Since E is a splitting field of f, then we know that $\mathrm{Gal}(E/\mathbb{Q})$ is a subgroup of S_n where n is the number of roots of f in E. (This was Theorem 24.44 but the argument is that the Galois group acts on the roots, and this gives a homomorphism from $\mathrm{Gal}(E/\mathbb{Q})$ into S_n. Since $\mathbb{Q}$-automorphisms of E are determined by their effect on the roots, the kernel of the action is trivial and so $\mathrm{Gal}(E/\mathbb{Q})$ is isomorphic to the image of the homomorphism which is a subgroup of S_n.) Hence, regardless of what n is, $\mathrm{Gal}(E/\mathbb{Q})$ is a subgroup of S_4. (For example, if f had just two roots in E, then $\mathrm{Gal}(E/\mathbb{Q})$ would be a subgroup of S_2, but S_2 itself is a subgroup of S_4.) We also know that E is the splitting field of a separable polynomial (all polynomials are separable since we are in characteristic 0), and so E is a Galois extension of $\mathbb{Q}$ and $|\mathrm{Gal}(E/\mathbb{Q})| = |E\colon \mathbb{Q}|$.

To find $|E\colon \mathbb{Q}|$ (this was Problem 24.2.5 but we provide the argument here), we first find the roots explicitly. The quadratic equation gives $x^2 = \frac{4 \pm \sqrt{16+4}}{2} = 2 \pm \sqrt{5}$, and so $x = \pm\sqrt{2 \pm \sqrt{5}}$. If we let $\alpha = \sqrt{2 + \sqrt{5}}$ be one of the roots of f. Then $\alpha\sqrt{2 - \sqrt{5}} = i$. Hence, $i \in E$, and the four roots of f can be written as $\pm\alpha$ and $\pm\frac{i}{\alpha}$. We conclude that $E = \mathbb{Q}[\alpha, i]$.

Let $K = \mathbb{Q}[\alpha]$, then $|E\colon \mathbb{Q}| = |E\colon K| \, |K\colon \mathbb{Q}|$. The element i is in E but it is not in K, and so $x^2 + 1$ is irreducible in $K[x]$. Hence, $x^2 + 1$ is the minimal polynomial of i over K, and, since $E = K[i]$, we have $|E\colon K| = 2$. We also know that $|K\colon \mathbb{Q}| = \deg(\min_{\mathbb{Q}}(\alpha))$. If we knew that f is irreducible over $\mathbb{Q}$, then we would know that f is the minimal polynomial of α over $\mathbb{Q}$. We would conclude that $|K\colon \mathbb{Q}| = 4$, and, as a result, $|E\colon \mathbb{Q}| = 8$. This would mean that $\mathrm{Gal}(E/\mathbb{Q})$ is a subgroup of S_4 of size 8. All subgroups of size 8 in S_4 are isomorphic to D_8 and hence we would conclude that $\mathrm{Gal}(E/\mathbb{Q}) \cong D_8$. (Any subgroup of order 8 in S_4 is a Sylow 2-subgroup of S_4, and all Sylow p-subgroups of a group are isomorphic to each other. Since S_4 does have a subgroup isomorphic to D_8, all of its subgroups of order 8 must be isomorphic to D_8.)

That argument depended on knowing that f is irreducible. To show that f is irreducible over $\mathbb{Q}$, we consider f as a polynomial in $\mathbb{F}_3[x]$. Mod 3, f becomes $x^4 + 2x^2 + 2$. This polynomial has no roots in $\mathbb{F}_3 = (\mathbb{Z}/3\mathbb{Z}, +, \cdot)$ and so the only way it could be reducible is if it was a product of two irreducible polynomials of degree 2. The only monic irreducibles of degree 2 over $\mathbb{F}_3$ are $x^2 + 1$, $x^2 + x + 2$, $x^2 + 2x + 2$, and it is straightforward to check that these cannot multiply to give us f. Hence f is irreducible mod 3 and hence irreducible in $\mathbb{Q}[x]$ (Theorem 19.58).

We conclude that the Galois group of f is isomorphic to D_8. Have we "found" the Galois group of $f(x)$? We have certainly identified this group as an abstract group. It is isomorphic to D_8. But usually we are interested in explicitly knowing the elements of the Galois group. In other words, what are the actual $\mathbb{Q}$-automorphisms of $E = \mathbb{Q}[\alpha, i]$? Every $\mathbb{Q}$-automorphism of E is determined by its action on α and i. To know an element of $\mathrm{Gal}(E/\mathbb{Q})$, we just need to know where it sends α and i.

Here is one way to proceed. Let $L = \mathbb{Q}[i]$, then K and L are both intermediate fields containing $\mathbb{Q}$ and contained in E. We know that $\mathrm{Gal}(E/K)$ and $\mathrm{Gal}(E/L)$ are subgroups of $\mathrm{Gal}(E/\mathbb{Q})$, and so, by finding their elements, we can find elements of $\mathrm{Gal}(E/\mathbb{Q})$. Moreover, we know, a priori, the size of these groups. Since E is Galois over $\mathbb{Q}$, it is also a Galois extension of both K and L. As a result $|\mathrm{Gal}(E/K)| = |E\colon K| = 2$ and $|\mathrm{Gal}(E/L)| = |E\colon L| = 4$. We can also argue that f is irreducible over not only over $\mathbb{Q}$ but also over L. From $|E : \mathbb{Q}| = |L[\alpha] : L| \, |L : \mathbb{Q}|$, we get $|L[\alpha] : L| = 4$, which means that f cannot be reducible over L.

The field E is the splitting field of the irreducible polynomial $x^2 + 1$ over K, and so $\mathrm{Gal}(E/K)$ acts transitively on the roots of $x^2 + 1$. Hence, there exists σ an automorphism of E that fixes every element of K (including every element of $\mathbb{Q}$ and α) and sends i to $-i$. Thus, we have $\sigma \in \mathrm{Gal}(E/\mathbb{Q})$ with

$$\sigma(i) = -i, \quad \sigma(\alpha) = \alpha.$$

On the other hand, E is the splitting field of the irreducible polynomial f over L, and so $\mathrm{Gal}(E/L)$ acts transitively on the roots of f. Hence, in addition to identity, there are elements $\tau_1, \tau_2, \tau_3 \in \mathrm{Gal}(E/L)$ with

$$\tau_1(\alpha) = -\alpha, \ \tau_2(\alpha) = \frac{i}{\alpha}, \ \tau_3(\alpha) = \frac{-i}{\alpha}, \quad \text{and } \tau_1(i) = \tau_2(i) = \tau_3(i) = i.$$

We conclude that $\mathrm{Gal}(E/K) = \{e, \sigma\}$ while $\mathrm{Gal}(E/L) = \{e, \tau_1, \tau_2, \tau_3\}$. Composing each of these elements with itself gives a map that fixes both i and α and so it is the identity. As a result, so far, every non-identity element that we have found is an element of order 2.

What other elements does $\mathrm{Gal}(E/\mathbb{Q})$ have? Since, so far, we have found five distinct elements, there must be three more elements. These are $\tau_1\sigma$, $\tau_2\sigma$, and $\tau_3\sigma$. We have

$$\tau_1\sigma: \begin{array}{l} i \mapsto -i, \\ \alpha \mapsto -\alpha; \end{array} \qquad \tau_2\sigma: \begin{array}{l} i \mapsto -i, \\ \alpha \mapsto \dfrac{i}{\alpha}; \end{array} \qquad \tau_3\sigma: \begin{array}{l} i \mapsto -i, \\ \alpha \mapsto \dfrac{-i}{\alpha}. \end{array}$$

The element $\tau_1\sigma$ is again an element of order 2, but $\tau_2\sigma$ and $\tau_3\sigma$ are elements of order 4. Hence $\mathrm{Gal}(E/\mathbb{Q}) = \{e, \sigma, \tau_1, \tau_2, \tau_3, \tau_1\sigma, \tau_2\sigma, \tau_3\sigma\}$. In fact, if we let $\rho = \tau_2\sigma$, then $\rho^2 = \tau_1\sigma$ and $\rho^3 = \tau_3\sigma$. As a result, $\rho\sigma = \tau_2\sigma^2 = \tau_2$, $\rho^2\sigma = \tau_1$, and $\rho^3\sigma = \tau_3$. Thus $\mathrm{Gal}(E/\mathbb{Q}) = \{e, \rho, \rho^2\rho^3, \sigma, \rho\sigma, \rho^2\sigma, \rho^3\sigma\}$ with $\rho^4 = e = \sigma^2$ and $\sigma\rho = \rho^3\sigma$. This is the familiar presentation for the group D_8.

26.2.7. We have that the prime p divides $|E : F| = |\mathrm{Gal}(E/F)|$. Thus by Cauchy's theorem, Corollary 7.11—if a prime p divides the order of a group, then the group has an element of order p—there exists $\sigma \in \mathrm{Gal}(E/F)$ with $o(\sigma) = p$. Thus $\langle \sigma \rangle$ is a subgroup of order p in $\mathrm{Gal}(E/F)$. Let $L = \mathrm{Fix}(\langle \sigma \rangle)$. By the Galois correspondence $F \subseteq L \subseteq E$ and $|E : L| = |\langle \sigma \rangle| = p$.

26.3.1. Let $\omega = e^{\frac{2\pi i}{3}} = -\frac{1}{2} + \frac{\sqrt{3}}{2}i$, then the roots of f are $\sqrt[3]{2}$, $\sqrt[3]{2}\omega$, and $\sqrt[3]{2}\omega^2$. Now let $E = \mathbb{Q}[\sqrt[3]{2}, \omega] = \mathbb{Q}[\sqrt[3]{2}, \sqrt{3}i]$. Clearly E is the splitting field of f over $\mathbb{Q}$ (by now this is clear, but also see Example 24.19). E is a Galois extension of $\mathbb{Q}$, and we have seen in Problem 24.3.10 that $\mathrm{Gal}(f) = \mathrm{Gal}(E/\mathbb{Q}) \cong S_3 \cong D_6$.

We repeat some of the solution to Problem 24.3.10 here. The elements of the Galois group are determined by their effects on $\sqrt[3]{2}$ and on ω (or on $\sqrt[3]{2}$ and on $\sqrt{3}i$). These group elements act on the roots of f and on the roots of $x^2 + x + 1$ (or on the roots of $x^2 + 3$), and there are six possible combinations. Since $|\mathrm{Gal}(E/\mathbb{Q})| = |E : \mathbb{Q}| = 6$, every one of these combinations must actually be possible.

Thus there exists $\sigma \in \mathrm{Gal}(E/\mathbb{Q})$ with $\sigma(\omega) = \omega^2$, and $\sigma(\sqrt[3]{2}) = \sqrt[3]{2}$. This is an element of order 2 and, in fact, $\mathrm{Gal}(E/\mathbb{Q}[\sqrt[3]{2}]) = \{e, \sigma\}$.

There also exists $\tau \in \mathrm{Gal}(E/\mathbb{Q})$ with $\tau(\sqrt[3]{2}) = \sqrt[3]{2}\omega$ and $\tau(\omega) = \omega$. Now since τ fixes ω it is easy to check that $\tau^3 = e$. In addition, we see that $\sigma\tau(\sqrt[3]{2}) = \tau^2\sigma(\sqrt[3]{2})$ and $\sigma\tau(\omega) = \tau^2\sigma(\omega)$. Hence $\sigma\tau(\alpha) = \tau^2\sigma(\alpha)$ for all $\alpha \in E$. We conclude that $\sigma\tau = \tau^2\sigma$. Thus $\mathrm{Gal}(E/\mathbb{Q}) = \{e, \tau, \tau^2, \sigma, \tau\sigma, \tau^2\sigma\}$ with relations $\sigma^2 = \tau^3 = e$ and $\sigma\tau = \tau^2\sigma$. As we mentioned before, this is D_6 (or S_3). We can see the Galois correspondence in Figure C.8.

To find the intermediate field corresponding to a subgroup H, we just find $\mathrm{Fix}(H)$. The calculation is simplified by the fact that we know the degree of this extension over $\mathbb{Q}$ since it is the same as $|D_6 : H|$.

The whole group, $\langle \tau \rangle$, and $\{e\}$, are the only normal subgroups of $\mathrm{Gal}(E/\mathbb{Q}) = D_6$, and hence $\mathbb{Q}$, $\mathbb{Q}(\omega)$, and E are the only normal extensions of $\mathbb{Q}$ in E.

27.1.9. Recall from Proposition 27.1 that the roots of $x^{25} - x$ all form the field $\mathbb{F}_{25}$ of order 25, and $|\mathbb{F}_{25} : \mathbb{F}_5| = 2$. Now let f be an irreducible factor of $x^{25} - x$ in $\mathbb{F}_5[x]$. The polynomial f either splits in $\mathbb{F}_5$ or in $\mathbb{F}_{25}$ since all the roots of f are in $\mathbb{F}_{25}$. Hence the degree of f is either 1 or 2.

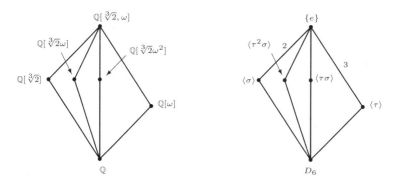

Figure C.8. The Galois correspondence between subgroups of D_6 and fields K with $\mathbb{Q} \subseteq K \subseteq \mathbb{Q}[\sqrt[3]{2}, \omega]$.

27.1.19. The field E is a finite field of characteristic 7 and F is its prime subfield. Regardless of $|E|$, let $\Phi : E \to E$ be the Frobenius map defined by $\Phi(a) = a^7$ for $a \in E$. We know that E is a Galois extension of F and $\mathrm{Gal}(E/F)$ is a cyclic group generated by Φ. (Note that we know the generator of the Galois group without knowing what $\mathrm{Gal}(E/F)$ or $|E|$ is.) Hence, we can find the orbit of α by applying Φ over and over. Since α is a root of f, we have $\alpha^7 = \alpha + 3$. Hence,

$$\Phi(\alpha) = \alpha^7 = \alpha + 3,$$
$$\Phi^2(\alpha) = \Phi(\alpha + 3) = \Phi(\alpha) + 3 = \alpha + 6,$$
$$\Phi^3(\alpha) = \Phi(\alpha + 6) = \alpha + 2,$$
$$\Phi^4(\alpha) = \Phi(\alpha + 2) = \alpha + 5,$$
$$\Phi^5(\alpha) = \Phi(\alpha + 5) = \alpha + 1,$$
$$\Phi^6(\alpha) = \Phi(\alpha + 1) = \alpha + 4,$$
$$\Phi^7(\alpha) = \Phi(\alpha + 4) = \alpha.$$

We conclude that the orbit of α under the action of Ω as well as the set of roots of $\min_F(\alpha)$ is

$$\{\alpha, \alpha + 1, \alpha + 2, \ldots, \alpha + 6\},$$

and so

$$\min_F(\alpha) = (x - \alpha)(x - \alpha - 1) \cdots (x - \alpha - 6).$$

In particular, $\min_F(\alpha)$ is of degree 7. Since α is a root of f, and f is monic of degree 7, we conclude that $f = \min_F(\alpha)$. As a result, f is irreducible in $F[x]$.

We have that $F[\alpha]$ is a field containing F and all the roots of f and hence it is a splitting field of f and $E = F[\alpha]$. As a result, $|E : F| = \deg \min_F(\alpha) = 7$, and so $|E| = 7^7$.

27.1.21. Let E be a field of order p^n. Then $E^\times = (E - \{0\}, \cdot)$ is a cyclic group of order $p^n - 1$.

Now assume that m divides n, then E has a subfield K of order p^m, then $K^\times \leq E^\times$. This implies by the Lagrange's theorem that $|K^\times| = p^m - 1$ divides $|E^\times| = p^n - 1$.

Conversely, assume that $p^m - 1$ divides $p^n - 1$. To show that m divides n, by Corollary 27.9 it is enough to show that E has a subfield of order p^m. In a cyclic group, for each order dividing the order of the group, there is exactly one subgroup of that order. Hence, $E^\times$ has one subgroup of order $p^m - 1$. Let $q = p^m$. Every element of this subgroup, if raised to $q-1$—which is the order of the group—is equal to 1. We have that there are $q-1$ non-zero elements of E that satisfy the equation $x^{q-1}=1$. These same elements as well as zero satisfy $x^q = x$, and hence there are q elements of E that are roots of $x^q - x$. Hence $x^q - x$ splits in E. This means that there is a splitting field L for $x^q - x$ in E. By Proposition 27.1, L has $q = p^m$ elements. (In fact, L is the subgroup of order $q-1$ in E together with zero.) Since E has a subfield of order p^m, we conclude that m divides n.

27.2.5. $x^4 + x^2 + 1 = 0$ implies that $(x^2-1)(x^4+x^2+1) = 0$ which means that $x^6 - 1 = 0$. Hence, the roots of f are 6th roots of unity. In fact, the only 6th roots of unity that are *not* a root of f are ± 1 (the roots of $x^2 - 1$). So among the roots are primitive 6th roots of unity (as well as primitive 3rd roots of unity). Hence the splitting field of f is $\mathbb{Q}(\zeta_6)$ where ζ_6 is a primitive 6th root of unity. We conclude that $\mathrm{Gal}(f) \cong (\mathbb{Z}/6\mathbb{Z})^\times \cong \mathbb{Z}/2\mathbb{Z}$.

You could also have noticed that $x^4 + x^2 + 1 = (x^2+x+1)(x^2-x+1) = \Phi_3(x)\Phi_6(x)$, and its splitting field is the same as that of $\Phi_6(x)$ and $x^6 - 1$.

27.2.15. There are several ways of doing this problem. Here is one requiring very little actual calculation.

Consider the cyclotomic extension $L = \mathbb{Q}(\zeta)$. We know that L is Galois over $\mathbb{Q}$ and $|\mathrm{Gal}(L/\mathbb{Q})| = |L : \mathbb{Q}| = 10$. The polynomial $\Phi_{11}(x)$ is irreducible and has ten roots all which are primitive 11th roots of unity. In fact, these roots are ζ, ..., ζ^{10}. Now $G = \mathrm{Gal}(L/\mathbb{Q})$ acts on these roots transitively, and every element of G is determined by its action on ζ. Hence,

$$G = \{\sigma_i \mid 1 \leq i \leq 10,\ \sigma_i(\zeta) = \zeta^i\}.$$

Now, $\alpha = \zeta + \zeta^{-1} \in L$ and, hence, the roots of its minimal polynomial f are exactly the elements of the orbit of α under the action of G. Since $\zeta^{-1} = \zeta^{10}$ and $\sigma(\zeta^{-1}) = \sigma(\zeta)^{-1}$, we have

$$\sigma_1(\alpha) = \zeta + \zeta^{10} = \sigma_{10}(\alpha),$$
$$\sigma_2(\alpha) = \zeta^2 + \zeta^9 = \sigma_9(\alpha),$$
$$\sigma_3(\alpha) = \zeta^3 + \zeta^8 = \sigma_8(\alpha),$$
$$\sigma_4(\alpha) = \zeta^4 + \zeta^7 = \sigma_7(\alpha),$$
$$\sigma_5(\alpha) = \zeta^5 + \zeta^6 = \sigma_6(\alpha).$$

Hence, $\deg(f) = 5$. We conclude that $|\mathbb{Q}(\alpha) : \mathbb{Q}| = \deg f = 5$. We also know that G is an abelian group (in fact cyclic and isomorphic to $((\mathbb{Z}/11\mathbb{Z})^\times, \cdot) \cong (\mathbb{Z}/10\mathbb{Z}, +)$). Hence, all of its subgroups are normal. By the Galois correspondence, every intermediate field K with $\mathbb{Q} \subseteq K \subseteq L$ is a normal extension of $\mathbb{Q}$. Hence, $\mathbb{Q}(\alpha)$ is a normal extension of $\mathbb{Q}$ and so it is the splitting field of f over $\mathbb{Q}$. Hence, $E = \mathbb{Q}(\alpha)$ is a Galois extension of $\mathbb{Q}$. We conclude that

$$|\mathrm{Gal}(E/\mathbb{Q})| = |E : \mathbb{Q}| = |\mathbb{Q}(\alpha) : \mathbb{Q}| = 5.$$

But, there is only one group of order 5 and so

$$\mathrm{Gal}(E/\mathbb{Q}) \cong \mathbb{Z}/5\mathbb{Z}.$$

Since we know the roots of f, we have a factorization of f into linear factors. By multiplying them out (preferably using a symbolic math software), we can actually find $f(x) = x^5 + x^4 - 4x^3 - 3x^2 + 3x + 1$.

27.2.23.

(a) We have

$$\Phi_{21}(x) = \Phi_3(x^7)/\Phi_3(x) = \frac{x^{14} + x^7 + 1}{x^2 + x + 1} = x^{12} - x^{11} + x^9 - x^8 + x^6 - x^4 + x^3 - x + 1.$$

(b) We use Maple to get

$$
\begin{aligned}
\Phi_{105}(x) &= \Phi_{21}(x^5)/\Phi_{21}(x) \\
&= 1 - x^9 - x^8 - x^6 + x^2 + x^{14} + x^{13} - 2\,x^7 - x^5 + x + x^{12} - x^{40} \\
&\quad - x^{20} + x^{15} + x^{48} + x^{46} + x^{47} - x^{43} - 2\,x^{41} - x^{39} + x^{36} + x^{35} + x^{34} \\
&\quad + x^{33} + x^{32} + x^{31} - x^{28} - x^{26} - x^{24} - x^{22} + x^{17} - x^{42} + x^{16}.
\end{aligned}
$$

Note that the coefficients of x^7 and x^{41} are 2. This is the first instance of the cyclotomic polynomial where the coefficients are anything but 0 and ± 1.

28.1.9. Let $F \subseteq E \subseteq L$ be fields with $\mathrm{char}(F) = 0$, and let L be an r.r.e. of F. Then we want to show that $\mathrm{Gal}(E/F)$ is solvable.

Let $K = \mathrm{Fix}(\mathrm{Gal}(E/F))$. Then $F \subseteq K \subseteq E$ and $\mathrm{Gal}(E/K) = \mathrm{Gal}(E/F)$. Hence, $K = \mathrm{Fix}(\mathrm{Gal}(E/F)) = \mathrm{Fix}(\mathrm{Gal}(E/K))$, and so E is Galois over K. Using the assumption, with F replaced with K, we get that $\mathrm{Gal}(E/K)$ is solvable. But $\mathrm{Gal}(E/F) = \mathrm{Gal}(E/K)$, and hence it must be solvable.

Bibliography

[AL94]   William W. Adams and Philippe Loustaunau, *An introduction to Gröbner bases*, Graduate Studies in Mathematics, vol. 3, American Mathematical Society, Providence, RI, 1994.

[AL00]   H. Azad and A. Laradji, *On a theorem of Clay*, College Math. J. **31** (2000), no. 5, 405–406.

[And]   Lars Døvling Andersen, *Combinatorics: ancient and modern* (Robin J. Wilson and John. J. Watkins, editors), Oxford University Press, Oxford, 2013, pp. 251–284.

[Ban92] Bernhard Banaschewski, *Algebraic closure without choice*, Z. Math. Logik Grundlag. Math. **38** (1992), no. 4, 383–385.

[Ber86] J. L. Berggren, *Episodes in the mathematics of medieval islam*, Springer Verlag, New York-Berlin-Heidelberg-London-Paris-Tokyo, 1986.

[Ber05] Ruth I. Berger, *Hidden group structure*, Math. Mag. **78** (2005), no. 1, 45–48.

[BJN94] P. B. Bhattacharya, S. K. Jain, and S. R. Nagpaul, *Basic abstract algebra*, second ed., Cambridge University Press, Cambridge, 1994.

[BM00]   Michael Brennan and Desmond MacHale, *Variations on a theme: A_4 definitely has no subgroup of order six!*, Math. Mag. **73** (2000), 36–40.

[BS81]   James W. Brewer and Martha K. Smith (eds.), *Emmy Noether*, Monographs and Textbooks in Pure and Applied Mathematics, vol. 69, Marcel Dekker, Inc., New York, 1981, A tribute to her life and work.

[BS94]   Sarah Marie Belcastro and Gary J. Sherman, *Counting centralizers in finite groups*, Math. Mag. **67** (1994), no. 5, 366–374.

[Bur55] W. Burnside, *Theory of groups of finite order*, second ed., Dover Publications Inc., New York, 1955, Reprint by photo-offset of the 2d edition [Cambridge, 1911].

[BW93]   Thomas Becker and Volker Weispfenning, *Gröbner bases*, Graduate Texts in Mathematics, vol. 141, Springer-Verlag, New York, 1993, A computational approach to commutative algebra, In cooperation with Heinz Kredel.

[Cla69] James R. Clay, *The punctured plane is isomorphic to the unit circle*, J. Number Theory **1** (1969), 500–501.

[Con97] John H. Conway, *The sensual (quadratic) form*, Carus Mathematical Monographs, vol. 26, Mathematical Association of America, Washington, DC, 1997, With the assistance of Francis Y. C. Fung.

[Cox11] David A. Cox, *Why Eisenstein proved the Eisenstein criterion and why Schönemann discovered it first*, Amer. Math. Monthly **118** (2011), no. 1, 3–21.

[Cox12] _____, *Galois theory*, second ed., Pure and Applied Mathematics (Hoboken), John Wiley & Sons, Inc., Hoboken, NJ, 2012.

[Cuo00] Al Cuoco, *Meta-problems in mathematics*, The College Mathematics Journal **31** (2000), 373–378.

[CW03] Daniel Cass and Gerald Wildenberg, *Math bite: A novel proof of the infinitude of primes, revisited*, Math. Mag. **76** (2003), 203.

[DD05] Greg Dresden and Wayne M. Dymàček, *Finding factors of factor rings over the Gaussian integers*, Amer. Math. Monthly **112** (2005), no. 7, 602–611.

[DF04] David S. Dummit and Richard M. Foote, *Abstract algebra*, third ed., John Wiley & Sons, Inc., Hoboken, NJ, 2004.

[Elk88] Noam D. Elkies, *On $A^4 + B^4 + C^4 = D^4$*, Math. Comp. **51** (1988), no. 184, 825–835.

[Ens99] Douglas E. Ensley, *Invariants under group actions to amaze your friends*, Math. Mag. **72** (1999), no. 5, 383–387.

[Fed51] Yu. G. Fedorov, *On infinite groups of which all nontrivial subgroups have a finite index*, Uspehi Matem. Nauk (N.S.) **6** (1951), no. 1(41), 187–189.

[FT63] Walter Feit and John G. Thompson, *Solvability of groups of odd order*, Pacific J. Math. **13** (1963), 775–1029.

[Gal93] Joseph Gallian, *On the converse of the Lagrange's theorem*, Math. Mag. **66** (1993), no. 1, 23.

[Gal01] _____, *The classification of groups of order $2p$*, Math. Mag. **74** (2001), no. 1, 60–61.

[Gol96] Solomon W. Golomb, *A symmetry criterion for conjugacy in finite groups*, Math. Mag. **69** (1996), 373–375.

[Goo98] Frederick M. Goodman, *Algebra, abstract and concrete*, Prentice Hall, Upper Saddle River, New Jersey, 1998.

[Goo99] Geoffrey R. Goodson, *Inverse conjugacies and reversing symmetry groups*, Amer. Math. Monthly **106** (1999), no. 1, 19–26.

[Gre] Ben Green, *Approximate groups and their applications: work of Bourgain, Gamburd, Helfgott and Sarnak*, Current Events Bulletin of the AMS, 2010, pp. 1–25, available at `arXiv:0911.3354v2 [math.NT]`.

[Gre00] Bo Green, *A project for discovery, extension, and generalization in abstract algebra*, College Math. J. **31** (2000), no. 4, 329–332.

[Had78] C. R. Hadlock, *Field theory and its classical problems*, The Mathematical Association of America, 1978.

[Har04] Malcolm Harper, $\mathbb{Z}[\sqrt{14}]$ *is Euclidean*, Canad. J. Math. **56** (2014), no. 1, 55–70.

[Her75] I. N. Herstein, *Topics in algebra*, second ed., John Wiley & Sons, New York, 1975.

[HH76] B. Hartley and T. O. Hawkes, *Rings, modules and linear algebra*, Chapman and Hall, London, 1976.

[Hil98]   David Hilbert, *The theory of algebraic number fields*, Springer-Verlag, Berlin,
          1998, Translated from the German and with a preface by Iain T. Adamson,
          With an introduction by Franz Lemmermeyer and Norbert Schappacher.

[HN52]   Graham Higman and B. H. Neumann, *Groups as groupoids with one law*, Publ.
          Math. Debrecen **2** (1952), 215–221.

[Hog96]   Guy T. Hogan, *More on the converse of Lagrange's theorem*, Math. Mag. **69**
          (1996), 375–376.

[HSY90]   Dean Hickerson, Sherman Stein, and Kenya Yamaoka, *When quasinormal implies
          normal*, Am. Math. Monthly **97** (1990), no. 6, 514–518.

[Inf78]   Leopold Infeld, *Whom the gods love: The story of Evariste Galois*, The National
          Council of Teachers of Mathematics, 1978.

[Isa94]   I. Martin Isaacs, *Algebra, a graduate course*, Brooks/Cole, Pacific Grove, 1994.

[IM98]   I. Martin Isaacs and David Petrie Moulton, *Real fields and repeated radical ex-
          tensions*, J. Algebra **201** (1998), no. 2, 429–455.

[Iwa41]   Kenkiti Iwasawa, *Über die endlichen Gruppen und die Verbände ihrer Unter-
          gruppen*, J. Fac. Sci. Imp. Univ. Tokyo. Sect. I. **4** (1941), 171–199.

[Joh00]   Colonel Johnson, Jr., *Group operation tables and normalizers*, College Math. J.
          31 (2000), no. 1, 50–51.

[Koh68]   J. Kohler, *A note on solvable groups*, J. London Math. Soc. **43** (1968), 235–236.

[Kun92]   Kenneth Kunen, *Single axioms for groups*, J. Automat. Reason. **9** (1992), no. 3,
          291–308.

[Lam04]   Tsit Yuen Lam, *On subgroups of prime index*, Amer. Math. Monthly **111** (2004),
          no. 3, 256–258.

[Lan01]   Charles Lanski, *A characterization of infinite cyclic groups*, Math. Mag. **74**
          (2001), no. 1, 61–65.

[LP66]   L. J. Lander and T. R. Parkin, *Counterexample to Euler's conjecture on sums
          of like powers*, Bull. Amer. Math. Soc. **72** (1966), 1079.

[LP67]   Leon J. Lander and Thomas R. Parkin, *A counterexample to Euler's sum of
          powers conjecture*, Math. Comp. **21** (1967), 101–103.

[Mac95]   George Mackiw, *Permutations as products of transpositions*, Amer. Math.
          Monthly **102** (1995), no. 5, 438–440.

[Mac96]   George Mackiw, *Finite groups of* 2×2 *integer matrices*, Math. Mag. **69** (1996),
          no. 5, 356–361.

[McL12]   Colin McLarty, *Theology and its discontents, the origin myth of modern mathe-
          matics*, Circles Disturbed : The Interplay of Mathematics and Narrative (Apos-
          tolos Doxiadis and Barry Mazur, eds.), Princeton University Press, Princeton,
          2012, pp. 105–129.

[MO00]   Peter Malcolmson and Frank Okoh, *Rings without maximal ideals*, Amer. Math.
          Monthly **107** (2000), no. 1, 60–61.

[MS96]   W. McCune and A. D. Sands, *Computer and human reasoning: Single implicative
          axioms for groups and for abelian groups*, Amer. Math. Monthly **103** (1996),
          no. 10, 888–892.

[Neu79]   Peter M. Neumann, *A lemma that is not Burnside's*, Math. Sci. **4** (1979), no. 2,
          133–141.

[Neu11]   _____, *The mathematical writings of Évariste Galois*, Heritage of European
          Mathematics, European Mathematical Society (EMS), Zürich, 2011.

[Oso99]   Barbara L. Osofsky, *Nice polynomials for introductory Galois theory*, Math. Mag. **72** (1999), no. 3, 218–222.

[Pec02]   Paul Peck, *Problem 691*, College Math. J. **33** (2002), no. 1, 55.

[Poo95]   David G. Poole, *The stochastic group*, Amer. Math. Monthly **102** (1995), no. 9, 798–801.

[Rei96]   Constance Reid, *Hilbert*, Copernicus, New York, 1996, Reprint of the 1970 original.

[Rog71]   Kenneth Rogers, *The axioms for Euclidean Domains*, The American Mathematical Monthly **78** (1971), no. 10, pp. 1127–1128.

[Ron06]   Mark Ronan, *Symmetry and the monster*, Oxford University Press, Oxford, 2006, One of the greatest quests of mathematics.

[Ser03]   Jean-Pierre Serre, *On a theorem of Jordan*, Bull. Amer. Math. Soc. (N.S.) **40** (2003), no. 4, 429–440.

[Sha70]   Parviz Shahriari, *Raveshhaye jabr (the methods of algebra)*, second Persian ed., Amir Kabir, Tehran, 1970 (Persian).

[Sha87]   Shahriar Shahriari, *On normal subgroups of capable groups*, Arch. Math. **48** (1987), 193–198.

[Sol01]   Ronald Solomon, *A brief history of the classification of the finite simple groups*, Bull. Amer. Math. Soc. (N.S.) **38** (2001), no. 3, 315–352.

[ST79]   I. N. Stewart and D. O. Tall, *Algebraic number theory*, Chapman and Hall, London, 1979.

[Ste15]   Ian Stewart, *Galois theory*, fourth ed., CRC Press, Boca Raton, FL, 2015.

[SW12]   John Shareshian and Russ Woodroofe, *A new subgroup lattice characterization of finite solvable groups*, J. Algebra **351** (2012), 448–458.

[Tan96]   Lin Tan, *The group of rational points on the unit circle*, Math. Mag. **69** (1996), no. 3, 163–171.

[TR96]   Laura Toti Rigatelli, *Evariste Galois*, Vita Mathematica, vol. 11, Birkhäuser Verlag, Basel, 1996, 1811–1832, Translated from the Italian by John Denton.

[Wag93]   Stan Wagon, *The Banach-Tarski paradox*, Cambridge University Press, Cambridge, 1993, With a foreword by Jan Mycielski, Corrected reprint of the 1985 original.

[Wil07]   Gerald Wildenberg, *An integral domain lacking unique factorization into irreducibles*, Mathematics Magazine **80** (February 2007), 75–76(2).

[Win06]   Peter Winkler, *Names in boxes puzzle*, College Math. J. **37** (2006), no. 4, 260, 285, & 289.

[Win07]   ———, *Mathematical mind-benders*, A K Peters Ltd, Wellesley, Massachusetts, 2007.

[Wri81]   E. M. Wright, *Burnside's lemma: a historical note*, J. Combin. Theory Ser. B **30** (1981), no. 1, 89–90.

Index

Published Titles in This Series